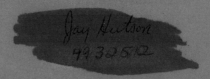
To convert from	to	Multiply by
Moment dyne-centimeter	newton-meter (N-m)	$1.000\ 000\ (10^{-7})$*
pound-foot	newton-meter (N-m)	$1.355\ 818$
Power Btu/hour	watt (W)	$2.930\ 711\ (10^{-1})$
erg/second	watt (W)	$1.000\ 000\ (10^{-7})$*
foot-pound/second	watt (W)	$1.355\ 818$
horsepower (550 ft–lbs)	watt (W)	$7.456\ 999\ (10^{2})$
Pressure and Stress atmosphere	pascal (Pa)	$1.013\ 25\ (10^{5})$
bar	pascal (Pa)	$1.000\ 000\ (10^{5})$*
centimeter of mercury (0^{8}C)	pascal (Pa)	$1.333\ 22\ (10^{3})$
dyne/centimeter^2	pascal (Pa)	$1.000\ 000\ (10^{-1})$*
pound/foot^2	pascal (Pa)	$4.788\ 026\ (10^{1})$
pound/inch^2 (psi)	pascal (Pa)	$6.894\ 757\ (10^{3})$
Speed foot/second	meter/second (m/s)	$3.048\ 000\ (10^{-1})$*
inch/second	meter/second (m/s)	$2.540\ 000\ (10^{-2})$*
kilometer/hour	meter/second (m/s)	$2.777\ 778\ (10^{-1})$
knot	meter/second (m/s)	$5.144\ 444\ (10^{-1})$
mile/hour	meter/second (m/s)	$4.470\ 400\ (10^{-1})$*
Volume foot^3	meter^3 (m^3)	$2.831\ 685\ (10^{-2})$
gallon (U.S. liquid)	meter^3 (m^3)	$3.785\ 412\ (10^{-3})$
inch^3	meter^3 (m^3)	$1.638\ 706\ (10^{-5})$
liter	meter^3 (m^3)	$1.000\ 000\ (10^{-3})$*
quart (U.S. liquid)	meter^3 (m^3)	$9.463\ 529\ (10^{-4})$

*Denotes an exact quantity.

1 lbm = .4535 kg   1 kg = 2.205 lbm = 0.06854 slug

slug = 14.59 kg   1 slug = 32.17 lbm

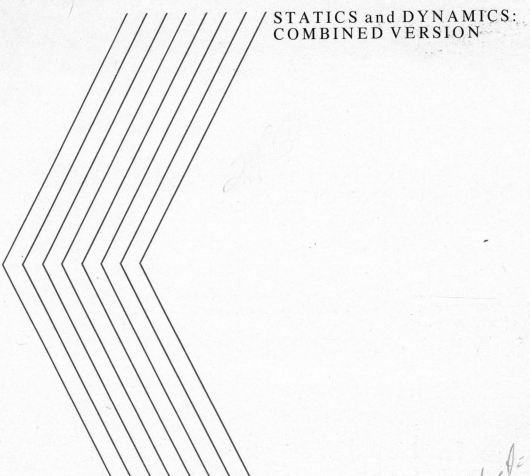

STATICS and DYNAMICS:
COMBINED VERSION

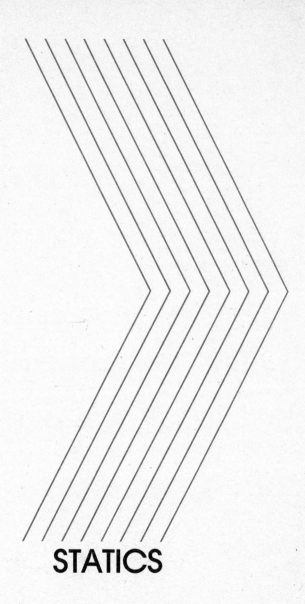

STATICS

JERRY H. GINSBERG
JOSEPH GENIN

School of Mechanical Engineering
Purdue University

JOHN WILEY & SONS
New York   Santa Barbara   London
Sydney   Toronto

DYNAMICS: COMBINED VERSION

This book was printed and bound by Kingsport Press.
It was set in Times Roman by Graphic Arts Composition, Inc.
The copyeditor was Lynne Lackenbach.
Kenneth R. Ekkens supervised production.

Cover and text design by Edward A. Butler.

Library of Congress Cataloging in Publication Data

Ginsberg, Jerry H.   1944-
    Statics and Dynamics.

    Includes indexes.
    1. Mechanics, Applied.   I. Genin, Joseph,
1932-   joint author.   II. Ginsberg, Jerry H.,
1944-   Statics. 1977.   III. Title: Statics.
TA350.G56   620.1   76-30664
ISBN 0-471-01795-7

Printed in the United States of America

10 9 8 7 6 5 4 3 2 1

PREFACE There are two fundamental goals to be attained in the basic engineering mechanics courses of statics and dynamics. The first is to develop an understanding of the physical laws governing the response of engineering systems to forces. The second is to enhance the reasoning powers required in engineering; that is, to develop the ability to solve problems logically, using the concept of mathematical models for physical systems.

To meet these objectives this book is written in modular form. The modules are broader in extent than the conventional grouping by chapters, being largely self-contained in order to minimize the amount of cross-referencing necessary to develop the material. This organization was also chosen because it is our desire to let the book communicate directly to the student, with a minimum of amplification and clarification required of the instructor.

To achieve this goal it was necessary to develop a consistent approach to problem solving that could be applied for a broad class of problems. This approach is founded on the recognition that there are two distinct aspects to the learning experience in mechanics. Clearly, one must first develop an understanding of the fundamental principles. Only then can a group of these principles be applied selectively to the solution of a broad class of problems.

In this book the comprehension phase is addressed in a conventional manner. After each principle or technique is derived, it is keynoted by remarks regarding common systems that illustrate its implications, as well as critical comments regarding its applicability. This is then followed by one or more solved examples. The examples directly illustrate how to employ the derived principle, and, equally important, serve to enhance the understanding of the physical meaning of the derived principles and of the system responses that result from these principles.

A primary goal in several of the modules is the synthesis of the basic principles and procedures into a unified approach for solving problems. Toward this end, at key locations in the modules, a set of sequential steps detailing the multiple operations necessary for the solution of general problems is presented. These steps are merely a logical sequence to follow. (They are certainly not the only possible sequence.) With these steps, where appropriate, we indicate places where common errors are made.

By presenting a systematized approach to problem solving we hope to enhance the student's senses of logic and deductive reasoning. Thus, the steps are not intended for memorization. Instead, as a student gains proficiency in problem solving, the steps will be performed intuitively.

The steps are then employed in a series of three or more solved problems, called illustrative problems, whose solution requires the synthesis of the concepts previously developed. The illustrative problems are cross-referenced directly to the steps for problem solving. This allows the student to isolate a particular aspect of the solution that may prove troublesome. Numerous homework problems are presented after both examples and illustrative problems; in general, those following the latter are broader in scope.

The sole difference between the approach outlined above and the approach for the applications modules is that steps for solving problems are not presented in the latter. Emphasis in the discussions and solutions of examples in the applications modules is placed on the development of a logical approach using the basic methodology, and each group of solved examples is followed by a broad range of homework problems.

The development of physical understanding in mechanics is addressed in the solved problems, as well as in the formal text material. Where appropriate in these problems, care is taken to discuss the qualitative aspects of the solutions. Also, the mathematics presented does not overwhelm the physical aspects of the problem, for each solution is implemented with the aid on only that level of mathematics appropriate to solve the problem at hand. Hopefully, this overall philosophy will give the student better insight as to how an engineer thinks.

Note from the Contents that the conventional ordering of the subject matter has been retained. Nevertheless, the viewpoint and treatment of many of the topics are not contained in other texts. In particular, the systematized treatments of virtual work and energy methods, and of the kinematics and kinetics of rigid bodies in three-dimensional motion, are entirely new. Our methodology makes these topics accessible to students with a broad range of backgrounds.

Another feature of the book, beyond the systematization it provides, is the flexibility it affords the instructor. It was written to provide the greatest opportunity for adjustments and accommodations in instruction, in order to communicate clearly with students having a broad range of individual experience. The text is equally suitable for courses using innovative instructional approaches, such as self-paced and self-taught classes, as well as for those using the more conventional methods.

Note also that this book does not attempt to address the organizational problems involved in self-paced and self-taught courses. However, it is possible for instructors wishing to utilize such an approach to employ the book as a framework that is to be supplemented by material addressing the problem of the interaction between the student and the instructor. The self-contained aspect of the modules makes the text especialy valuable for such courses.

A final technical matter for consideration here is the question of physical units. We strongly believe that the SI system of metric units is the best for engineering. Hopefully, by the time this is read, the SI system will have been adopted universally. However, recognizing the transitory nature of this problem, and noting the large number of physical systems which have already been built according to the U.S.-British system of units, the numerical problems have been divided approximately into a two to one proportion between SI and British units. This proportion allows for a sufficient number of problems for courses using only one system of units.

We gratefully acknowledge the encouragement of our colleagues at Purdue University, whose helpful comments proved invaluable. We also thank the faculty of the École Nationale Supérieure d'Électricité et de Mécanique at Nancy, France, particularly Professor Michel Lucius. By providing an academic environment for Professor Ginsberg during his sabbatical leave, they greatly expedited the completion of this work. We are indebted to Rona Ginsberg for her excellent editorial comments, as well as for typing the manuscript, and to Dr. Joseph W. Klahs for his technical review of the *Dynamics* manuscript. Finally, we note the knowledgeable assistance of the staff of John Wiley, especially our editor, Thurman R. Poston, in all phases of the production process.

Jerry H. Ginsberg

Joseph Genin

CONTENTS
STATICS

CONTENTS
DYNAMICS

618185403

MODULE I
FUNDAMENTAL
CONCEPTS

A. MECHANICS

Many times when touching an object we wonder how it has been affected physically. Thoughts of this kind are basic to the field of *mechanics,* where the general concern is the study of the effect of *forces* acting on a body.

To a certain extent the concept of a force is intuitive, because forces are not actual objects. In spite of this, *force* does have a basis in our fundamental sensing system, especially in the sense of touch. For instance, the sensations our brain processes when our fingertips contact another body are a measure of how hard and in which direction we push or pull on the object. In mechanics this is referred to as the external force exerted on the object.

The study of mechanics is divided into two broad categories, *statics* and *dynamics*. *Statics* is the study of physical systems that remain at rest under the action of a set of forces. *Dynamics* is the study of physical systems in motion. This text is devoted to the study of *statics*.

A key phrase used above is "physical system." By it we mean a collection of basic atomic elements that combine to form bodies. In the field of mechanics, for the convenience of problem solving, we tend to categorize these collections of elements into three broad groupings: particles, rigid bodies, and deformable bodies.

PARTICLE   A body whose dimensions are negligible is said to be a particle. It follows, then, that a particle occupies only a single point in space.

RIGID BODY   A body occupying more than one point in space is said to be rigid if all of the constituent elements of matter within the body are always at fixed distances from each other.

DEFORMABLE BODY   A body is said to be deformable if its constituent elements of matter experience changes in their distances from each other that are significant to the problem being investigated.

Clearly there is a certain amount of ambiguity in these definitions. For instance, is the group of molecules that compose a gas a system of particles or is it a deformable body? Another basic question we could ask is whether any body can correctly be considered to be rigid; for all real materials deform when forces are exerted upon them.

There are no absolute answers to these questions, for the *model* of the system we form depends on what knowledge we wish to gain about the response of the system. This leads to the next topic, which is the modeling process.

B. THE MODELING PROCESS

The general approach in the static analysis of a physical system is to consider first the nature of the forces acting on the system and the type of information we seek from the subsequent analysis. On the basis of those considerations we construct a conceptual model of the system. That is, we consciously consider its components to be either particles, rigid bodies, or deformable bodies. Then, by applying the laws of mechanics we determine the interrelationships among the forces acting on the system. An analysis of these forces should provide us with the information we desire. This general approach to the subject of statics can be summarized in the conceptual diagram shown in Figure 1.

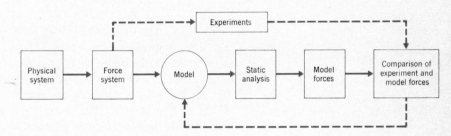

Figure 1

The dashed paths in Figure 1 illustrate a key feature of the modeling process. That is, to be sure of the validity of the model initially chosen to represent the physical system, it is necessary that the model display any relevant phenomena that are experimentally observed. If not, the model must be improved. Hence, modeling is an educated trial-and-error procedure. That is, the modeling process is, in part, an art based on prior experience and physical intuition. On the other hand, the modeling process is also a science based on a knowledge of the function that the system is designed to perform, and of the analytical methods available for studying the phenomena exhibited by the system.

As an example, consider an object in the shape of a sphere. If this sphere is a small ball bearing resting on a flat level surface, as a first attempt one would tend to model it as a particle occupying a single point in space. We shall see in later modules that this model would result in a successful analysis.

Suppose that we now step on this ball bearing, placing the ball bearing at the edge of the sole of our shoe. Our later studies will show that considering the ball bearing as a particle leaves us unable to explain why it tends to slide out from under foot when we press on it harder and harder. In this case the nature of the forces acting on the ball bearing changes drastically from the case where it was merely resting on the ground. Modeling the ball bearing as a rigid body, as opposed to a single particle, would allow us to resolve this problem.

Continuing with this example, suppose that instead of stepping on a steel ball bearing, we were stepping on a tennis ball. Needless to say, the shape of the ball would be altered considerably as we stepped on it.

Clearly, the model of a rigid body should then be discarded in favor of the model of a deformable one.

From the foregoing examples we see that the model chosen to represent the physical system is a crucial element in the solution process. In this text, a first course on the statics of physical systems, we focus on the particle and rigid body models. The relationship between the forces acting on a body and its deformation are considered in the subject of mechanics of materials, which builds on the techniques we will develop here.

C. VECTOR QUANTITIES—FORCES

A quantity that can be completely characterized by its numerical value is called a *scalar*. Length, area, volume, weight, mass, temperature, time, and energy are scalars. Scalar quantities are merely numbers, and as such obey all the laws of ordinary algebra.

Several quantities that occur in mechanics require a description in terms of their direction, as well as the numerical value of their magnitude. Such quantities behave as *vectors*. Typical vectors are force, velocity, position, displacement, and acceleration. Because vectors have both *magnitude* and *direction*, they have a unique algebra.

Pictorially, a vector may be represented by a directed line segment, that is, by a straight arrow. The length of the arrow, to any convenient scale, represents the magnitude of the vector, and the direction of the arrow is the direction of the vector. Thus, in Figure 2 we show a vector quantity $\bar{A}$.

Figure 2

In this book we will denote vector quantities by placing a bar over the symbol adopted for the vector. Thus, the vector in Figure 2 is $\bar{A}$. In contrast, when we wish to refer to the magnitude of a vector, such as the magnitude of $\bar{A}$, we will do so by writing either $|\bar{A}|$ or simply A. Furthermore, because of the significant role of forces in mechanics, an arrow representing a force will be represented more boldly in diagrams than arrows representing other types of vector quantities.

For some vector quantities, a specification of their magnitude and direction describes them completely; that is, their action is not confined to, or associated with, a unique line in space. Such quantities are said to be *free vectors*. An example of a free vector is the statement that a person walked two kilometers (2 km) in a southeasterly direction, which is a description of the displacement resulting from a motion. Velocity and acceleration are other free vectors.

In contrast, when describing a force we must provide some specification of where the force is applied. To see this, consider the horizontal force $\bar{F}$ applied to corner A of the table shown in Figure 3. If the force $\bar{F}$ is sufficient to move the table, in addition to pushing the table forward it will cause the table to rotate counterclockwise as viewed looking downward.

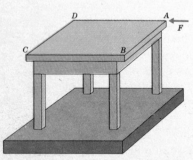

Figure 3

(Try it.) On the other hand, if a force having the same magnitude and direction as $\bar{F}$ is applied at corner B (as opposed to applying $\bar{F}$ at corner A), the result is a clockwise rotation. Hence, we can conclude that in this case we also need to specify the point of application of the force. The vectors described in the foregoing example are *fixed vectors,* for each had a unique point of application.

At times we may regard a force as a *sliding vector,* one that acts anywhere along a unique line in space. In such cases we describe the force by saying that it is collinear with a specific straight line, called its *line of action,* and then giving its magnitude and sense. The actual point of application of the force along its line of action is then considered to be immaterial to the problem. For example, in Figure 3 if we choose to regard $\bar{F}$ as a sliding vector, we can imagine it being applied at point A *or* D and we would get the same resulting motion for the table.

To gain some insight into the seemingly ambiguous cases of fixed vectors and sliding vectors, consider a stalled car that is to be moved with a tow truck. As shown in Figure 4a, the truck may pull on the front bumper with a horizontal force $\bar{F}_A$. Alternatively, as shown in Figure 4b, it may push on the rear bumper with a horizontal force $\bar{F}_B$.

In this case the forces $\bar{F}_A$ and $\bar{F}_B$ are fixed vectors, having specific points of application, although they share the same line of action and have the same sense. If these forces also have the same magnitude, they will produce the same motion of the automobile. Hence, if the resulting motion was what we wanted to know, we could also treat the force applied to either bumper as a sliding vector.

Suppose now that we consider what happens if the tow truck that is attempting to move the stalled vehicle exerts too large a force. If $\bar{F}_A$ is too large, the front bumper will be torn off, whereas if $\bar{F}_B$ is too large, the rear bumper will be crushed. To account for what happened, we would have to consider the specific point of application of the force. In other words, the force could then be regarded only as a fixed vector.

These observations are a result of the *principle of transmissibility,* which states that

> The effect on a rigid body, in its entirety, of a force external to the body depends only on the magnitude, direction, and line of action of the force.

When we consider questions such as the relationship between the external forces acting on a rigid body, or the motion of a rigid body resulting from a set of external forces, we may apply the principle of transmissibility. In such cases the forces are regarded as sliding vectors. However, if we are concerned with the effect of forces on one portion of an assembly of rigid bodies, or on one region within a single rigid body, we may *not* invoke the

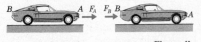

Figure 4a   **Figure 4b**

principle of transmissibility. The forces should then be regarded as fixed vectors.

The principle of transmissibility is a corollary of a theorem developed in the study of the dynamics of rigid bodies. Because of its importance, we will regard it as an axiom in statics.

Thus far we have discussed forces without really considering how they are generated. There are two broad categories of forces: those resulting from the *contact* of two bodies, and those associated with a *force field* created by the interaction of two bodies. Typical of the latter are force fields associated with gravity and electromagnetism. We will have more to say about gravity later in this module.

An important type of contact force is the one exerted by a lightweight cable fastened to a body, as shown in Figure 5. Pictorially, in Figure 5 we see a cable (or string) fastened to an arbitrary body at a point A with a force $\bar{F}$ applied at its opposite end B. The line of action of $\bar{F}$ is line AB connecting the ends of the cable, with $\bar{F}$ pulling on the end of the cable. The magnitude $|\bar{F}|$ of this force is called the *tension* in the cable because it is transmitted uniformly along the length of the cable. That is, if we ask what force is applied on the region to the left of the dashed line in Figure 5 by the portion of the cable to its right, we would find the same force $\bar{F}$.

Figure 5

A more subtle feature of $\bar{F}$ is that it can pull on end B with any magnitude up to the breaking force of the cable, but it cannot push on it. When we try to push on end B, the cable becomes slack.

The foregoing discussion contains a *modeling assumption*. The model we considered was for a cable that is inextensible. Cables made of any material will elongate due to a tension force. A clear example of this is a rubber band. We will refer to cables that undergo a significant amount of stretching as *elastic cables*. Defining the elongation, Δ, as the increase in the length of the cable due to an applied force, we may plot the tension as a function of elongation. Two examples are shown in Figure 6.

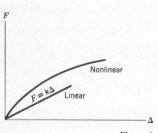

Figure 6

If the plot is a curved line, we have a nonlinear elastic cable. The straight-line plot represents a linear elastic cable, for which we can write

$$\boxed{F = k\Delta} \qquad (1)$$

The constant k is called the *stiffness* of the cable, because it tells us how hard the cable pulls for a given elongation.

The case of a coil spring, such as that in Figure 7, is identical to that of an elastic cable, with one important exception; the force $\bar{F}$ may be reversed from the direction shown, resulting in the spring pushing on the body with a force whose line of action is AB. The magnitude of $\bar{F}$ is then called the *compressive force*. Equation (1) is also valid for a linear spring, except that now negative Δ corresponds to the distance the spring is shortened, and negative F denotes compression.

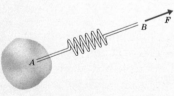

Figure 7

D. VECTOR ALGEBRA

We have thus far established that force, being a vector quantity, is specified by its magnitude and direction. For the moment we shall leave aside the question of the point of application of a force. We now wish to define some basic algebraic operations: (1) multiplication of a vector by a scalar, (2) addition of vectors, and (3) subtraction of vectors. The more subtle operations of vector algebra will be developed in later modules in situations where they are naturally motivated.

1 Multiplication by a Scalar

The expression "double the force" means that the magnitude of the force is increased by a factor of 2, but the direction of the force is unchanged. Mathematically, we state this by letting q be a scalar number and $\bar{F}$ be an arbitrary force. The resulting vector $q\bar{F}$ is defined to be parallel to $\bar{F}$. If q is positive, then $q\bar{F}$ is in the same sense as $\bar{F}$, whereas when q is negative $q\bar{F}$ is in the opposite sense from $\bar{F}$. The (positive) magnitude of the resulting vector is defined to be

$$|q\bar{F}| \equiv |q|\,|\bar{F}|$$

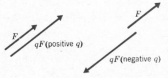

Figure 8

This is illustrated in Figure 8.

2 Addition of Vectors

The basic rule for adding any two vectors $\bar{A}$ and $\bar{B}$ is the parallelogram law. The vector sum $\bar{C} \equiv \bar{A} + \bar{B}$, which is called the *resultant*, is obtained by drawing the vectors $\bar{A}$ and $\bar{B}$ with their tails coincident. This forms two sides of a parallelogram. The resultant $\bar{C}$ is then the diagonal vector of this parallelgram whose tail is coincident with those of $\bar{A}$ and $\bar{B}$, as depicted in Figure 9.

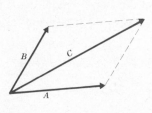

Figure 9

Notice that the order of addition is not important, so $\bar{A} + \bar{B} \equiv \bar{B} + \bar{A}$. This equality suggests an alternative rule for adding vectors. To do so we form a triangle by placing the tail of one vector at the head of the other. The resultant $\bar{C} = \bar{A} + \bar{B}$ is then the third side of the triangle, extending from the tail of the first vector to the head of the second. This is depicted in Figure 10 for the same vectors as were used in Figure 9.

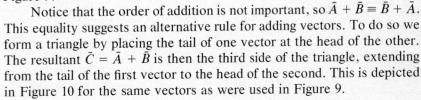

Figure 10

3 Subtraction of Vectors

The rule for subtraction of vectors is a corollary of the addition rule. This

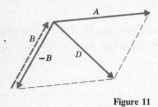

Figure 11

Figure 12

follows because we may write $\bar{D} = \bar{A} - \bar{B} \equiv \bar{A} + (-\bar{B})$. Then, according to the rule for multiplication by a scalar, the vector $-\bar{B} \equiv (-1)\bar{B}$ is a vector having the same magnitude as $\bar{B}$, but opposite direction. For the vectors $\bar{A}$ and $\bar{B}$ of Figure 9, the construction of the difference (subtraction) of vectors, according to the parallelogram law, is shown in Figure 11.

Noting that the difference $\bar{D}$ forms the minor diagonal in the parallelogram of Figure 9, we see that an alternative rule for constructing $\bar{D}$ is to place the two vectors $\bar{A}$ and $\bar{B}$ so that their tails coincide. The difference then goes from the head of the negative vector to the head of the positive one, as shown in Figure 12. This figure also demonstrates that $\bar{B} + \bar{D} \equiv \bar{B} + (\bar{A} - \bar{B}) = \bar{A}$, as it should.

The foregoing vector operations are illustrated in the following example.

EXAMPLE 1

Given two vectors $\bar{A}$ and $\bar{B}$, where $|\bar{A}| = 27$ directed east and $|\bar{B}|$ is directed 40° north of east, determine (a) $\bar{A} + \bar{B}$, (b) $\bar{B} - \bar{A}$, (c) $\bar{A} - 5\bar{B}$.

Solution

For the purpose of drawing the vectors let us consider a horizontal line to the right to define the easterly direction and an upward line to be north. The given vectors are then as shown in the sketch.

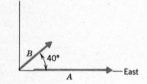

Part a

Using the triangle construction, we form the sum by moving the tail of $\bar{B}$ to the head of $\bar{A}$, giving the resultant $\bar{C}$ shown in the second sketch.

From the law of cosines we have

$$|\bar{C}|^2 = |\bar{A}|^2 + |\bar{B}|^2 - 2|\bar{A}||\bar{B}| \cos \theta$$

$$= (27)^2 + (15)^2 - 2(27)(15) \cos 140°$$

$$|\bar{C}| = 39.68$$

Then, from the law of sines we have

$$\frac{\sin \alpha}{|\bar{B}|} = \frac{\sin \theta}{|\bar{C}|}$$

$$\sin \alpha = \frac{15}{39.68} \sin 140°$$

$$\alpha = 14.06°$$

Hence

$$\bar{C} = \bar{A} + \bar{B} = 39.68 \text{ directed } 14.06° \text{ north of east}$$

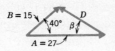

Part b

The triangle for evaluating $\bar{D} = \bar{B} - \bar{A}$ is formed by placing the vectors tail to tail. Then $\bar{D}$ goes from the head of $\bar{A}$ (the negative vector) to the head of $\bar{B}$, as shown in the sketch.

The law of cosines now gives

$$|\bar{D}|^2 = |\bar{A}|^2 + |\bar{B}|^2 - 2|\bar{A}||\bar{B}| \cos 40°$$

$$= (27)^2 + (15)^2 - 2(27)(15) \cos 40°$$

$$|\bar{D}| = 18.26$$

From the law of sines we then find

$$\frac{\sin \beta}{|\bar{B}|} = \frac{\sin 40°}{|\bar{D}|}$$

$$\sin \beta = \frac{15}{18.26} \sin 40°$$

$$\beta = 31.87°$$

Hence

$$\bar{D} = \bar{B} - \bar{A} = 18.26 \text{ directed } 31.87° \text{ north of west}$$

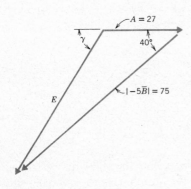

Part c

We may regard $\bar{A} - 5\bar{B}$ to be either $\bar{A} + (-5\bar{B})$ or $\bar{A} - (5\bar{B})$. Let us consider it as the former. Multiplying $\bar{B}$ by -5, we obtain a vector of magnitude 75 directed opposite of $\bar{B}$, that is, directed 40° south of west. Thus, the triangle construction for $\bar{E} = \bar{A} + (-5\bar{B})$ is as shown in the sketch.

The law of cosines gives

$$|\bar{E}|^2 = |\bar{A}|^2 + |-5\bar{B}|^2 - 2|\bar{A}||-5\bar{B}| \cos 40°$$

$$= (27)^2 + (75)^2 - 2(27)(75) \cos 40°$$

$$|\bar{E}| = 57.02$$

whereas the law of sines gives

$$\frac{\sin(180° - \gamma)}{|-5\bar{B}|} = \frac{\sin 40°}{|\bar{E}|}$$

$$\sin(180° - \gamma) = \frac{75}{57.02} \sin 40°$$

$$180° - \gamma = 57.72°, 122.28°$$

From the computation of $|\bar{E}|$ we know that $|-5\bar{B}|$ is the longest side in the triangle. It then follows that the corresponding vertex angle, $180° - \gamma$, must be the largest interior angle. Hence, choosing $180° - \gamma = 122.28$, we find

$$\gamma = 57.72°$$

so

$$\bar{E} = \bar{A} - 5\bar{B} = 57.02 \text{ directed } 57.72° \text{ south of west}$$

HOMEWORK PROBLEMS

I.1 and I.2 Determine the magnitude and direction of the resultant of the two forces shown (N = newton).

In Problems I.3 through I.7: $\bar{A} = 10$ due north, $\bar{B} = 13$ at 45° north of east, $\bar{C} = 27$ at 30° south of east, $\bar{D} = 18$ at 15° west of north.

I.3 Find (a) $\bar{A} + \bar{B}$, (b) $\bar{A} - \bar{B}$, and (c) $3\bar{A} - 2\bar{B}$.

I.4 Find $\bar{A} + \bar{B} - \bar{C}$.

I.5 Find $\bar{A} + \bar{B} + \bar{C} + \bar{D}$.

I.6 Find $\bar{A} + \bar{B} - \bar{C} - \bar{D}$.

I.7 Find the vector $\bar{E}$ for which $\bar{A} + \bar{B} - \bar{C} + \bar{E} = \bar{0}$.

I.8 The vectors $\bar{A}$, $\bar{B}$, and $\bar{C}$ shown represent the medians of triangle PQR, each being drawn from a vertex to the midpoint of the opposite side. Prove that $\bar{A} + \bar{B} + \bar{C} = \bar{0}$. *Hint:* Represent each side of the triangle as a vector and describe $\bar{A}$, $\bar{B}$, and $\bar{C}$ is terms of these vectors.

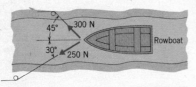

Prob. I.1

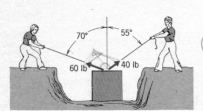

Prob. I.2

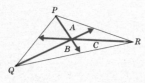

Prob. I.8

E. NEWTONIAN MECHANICS

The laws of motion stated by Sir Isaac Newton (1666–1727) in his historic work *Principia* (1687) are the foundation for the study of engineering mechanics. Here we state his laws in modern language.

FIRST LAW   A particle will remain at rest or move with constant speed along a straight line, unless it is acted upon by a resultant force.

SECOND LAW   When a resultant force is exerted upon a particle, the acceleration of that particle is parallel to the direction of the force and the magnitude of the acceleration is proportional to the magnitude of the force.

THIRD LAW   Each force exerted upon a body is the result of an interaction with another body, the forces of action and reaction being equal in magnitude, opposite in direction, and collinear.

Our interest in statics lies primarily in the first and third laws; we leave the study of accelerating systems for dynamics. However, a system that is at rest or moving with constant velocity is a special case of an accelerating system, so the first law is really included in the second law. It is stated separately in order to emphasize that the state of motion will be unchanged unless there is a resultant force acting on a system. The third law helps us to define what forces are acting on a body.

In this text on statics, we will use the second law later in this module to discuss the force of gravity and to establish a consistent system of units for describing physical quantities. The second law may be reworded to state that the resultant force $\bar{F}$ acting on a particle is proportional to the acceleration $\bar{a}$ of the particle. As you probably know, the constant of proportionality in this relation is the mass m of the particle. Thus, we have the familiar equation

$$\bar{F} = m\bar{a} \tag{2}$$

A remarkable aspect of Newton's laws is that, as stated, they apply only to a particle. Part of the study of mechanics is devoted to extending these laws to deal with rigid and deformable bodies.

F.   SYSTEMS OF UNITS

The quantities appearing in Newton's second law, $\bar{F} = m\bar{a}$, involve measurements of length, time, force, and mass, which we denote for brevity as $L, T, F,$ and M, respectively. The units chosen for the measurement of these four dimensional quantities cannot be defined independently. They must obey the *law of dimensional homogeneity,* which requires that the physical units of all terms in an equation be identical. Stated in colloquial language, "apples cannot be equated to oranges."

In terms of the four basic measurements, the dimensions of acceleration are L/T^2. Dimensional homogeneity of equation (2) requires that

$$F = M\frac{L}{T^2}$$

In other words, we are free to choose the units of three of the four quantities, $L, T, F,$ and M. The fourth must be derived. Historically, the units by which length and time are measured are well defined, and, through the efforts of international standards committees, universally accepted.

This leaves us with the choice of defining either a standard unit of mass or of force, and then deriving the units for the undefined quantity. When the unit of mass is defined according to some standard, the system is called *absolute,* whereas a system in which the unit of force is defined is called *gravitational.* This terminology originates from the relationship between the weight W and the acceleration g of a particle that is falling freely in a vacuum at the surface of the earth. According to Newton's second law, this relationship is

$$\boxed{W = mg} \tag{3}$$

In using equation (3) we must be aware that the values of W and g depend on the location of the particle with respect to the surface of the earth, and equally important on the choice of measurement system. Presently, the two systems that are in wide usage are the metric absolute system and the British gravitational system. In terms of the units of these systems, for most problems we will use the value $g = 9.806$ meters/second^2 (m/s^2) (metric) or $g = 32.17$ feet/second^2 (ft/s^2) (British) as average values. These values are looked at in considerable detail in the dynamics text.

In Table 1 we summarize the commonly used systems of units and indicate how the derived units are obtained. The basic units in each system are defined with respect to standard bodies or with respect to physical phenomena. For example, a kilogram is the mass of a particular bar of platinum alloy. On the other hand, a meter is defined as 1,650,763.73 wavelengths in a vacuum of the orange-red line of krypton 86. The derived quantities in the table are the starred values.

Table 1   System of units

UNIT/SYSTEM	LENGTH	TIME	MASS	FORCE
Metric absolute (SI)	Meter (m)	Second (s)	Kilogram (kg)	Newton (N)* = kg-m/s^2
Metric gravitational	Meter (m)	Second (s)	Metric slug* = kg-s^2/m	Kilogram (kg)
British absolute	Foot (ft)	Second (s)	Pound (lb)	Poundal (pdl)* = lb-ft/s^2
British gravitational	Foot (ft)	Second (s)	Slug* = lb-s^2/ft	Pound (lb)

The choice for systems of units presented in Table 1 is further complicated by the fact that certain multiples and submultiples of the basic units in each system are given their own names, such as the mile, which is 5280 feet, or the dyne, which is one hundred-thousandth of a newton. To simplify this situation the SI (Standard International) system is now being adopted worldwide.

Essentially, the SI system is the metric absolute system with only one unit being used to describe each type of quantity. The SI units for length, time, mass, and force are meters, seconds, kilograms, and newtons, respectively. Then, if one desires, decimal multiples and submultiples of the basic units may be indicated by the appropriate prefixes. Thus, in the SI system length dimensions of 0.2 meters (m) and 200 millimeters (mm) are identical and correct. On the other hand, $2(10^{-3})$ newtons (N) or equivalently 2 millinewtons (mN) is the correct description for a force whose magnitude is 200 dynes. Lists of the SI units and of the preferred prefixes may be found in Appendix A.

Many of the examples and problems in this text will be presented in SI units. Unfortunately, this system is not yet universally used in the United States, where many engineers, as well as nontechnical people, still use the British gravitational system. We will therefore present examples and problems that are posed in either system of units. In general, we will solve problems in the same system of units as that used for the given information, and we shall not convert between different systems of units. Obviously, the units of any physical quantity can be changed with the aid of conversion factors. These conversion factors are numbers that are physically unit values, but not numerically, for example, 25.40 mm = 1 in. A brief set of factors are given in Appendix A.

Conversion factors can be used consecutively. For instance, to change from a speed in miles per hour to a speed in meters per second, we can compute

$$1 \frac{\text{mile}}{\text{hr}} \left(\frac{1 \text{ hr}}{3600 \text{ s}} \right) \left(\frac{5280 \text{ ft}}{1 \text{ mile}} \right) \left(\frac{12 \text{ in.}}{1 \text{ ft}} \right) \left(\frac{25.40 \text{ mm}}{1 \text{ in.}} \right) \left(\frac{1 \text{ m}}{1000 \text{ mm}} \right) = 0.4470 \text{ m/s}$$

Remember, the units of all quantities in an equation must be consistent in order to satisfy the law of dimensional homogeneity.

As practice for obtaining a consistent set of units, and also to gain an intuitive understanding of the magnitude of British units relative to SI units, we present the following example.

EXAMPLE 2
Using the facts that 1 in. equals 25.40 mm and that a body weighing 1 lb on the surface of the earth has a mass of 0.4536 kg, determine the factors for

converting the British gravitational units of pounds force, slugs, mass, and density (slugs per cubic foot) into their SI equivalents.

Solution

The conversion factors between pounds and newtons is determined by using $W = mg$. Let us compute the weight in newtons of a 0.4536-kg body. Not knowing where the body was weighed, we use the average value $g = 9.806$ m/s² to obtain

$$W = (0.4536 \text{ kg})(9.806 \text{ m/s}^2) = 4.448 \text{ N}$$

In other words,

$$1 \text{ lb} = 4.448 \text{ N}$$

Then, since 1 lb-s²/ft is 1 slug, we have

$$1 \text{ slug} = \left(1 \frac{\text{lb-s}^2}{\text{ft}}\right)\left(\frac{4.448 \text{ N}}{1 \text{ lb}}\right)\left(\frac{1 \text{ ft}}{12 \text{ in.}}\right)\left(\frac{1 \text{ in.}}{25.4 \text{ mm}}\right)\left(\frac{1000 \text{ mm}}{1 \text{ m}}\right)$$

$$= 14.593 \frac{\text{N-s}^2}{\text{m}} = 14.593 \text{ kg}$$

To convert the units of density, we use

$$1 \frac{\text{slug}}{\text{ft}^3} = \left(1 \frac{\text{slug}}{\text{ft}^3}\right)\left(\frac{14.593 \text{ kg}}{\text{slug}}\right)\left(\frac{1 \text{ ft}}{12 \text{ in.}}\right)^3\left[\left(\frac{1 \text{ in.}}{25.40 \text{ mm}}\right)\left(\frac{1000 \text{ mm}}{1 \text{ m}}\right)\right]^3$$

$$= 515.3 \text{ kg/m}^3$$

The differences between the values obtained here and those in the table in Appendix A are attributable to the fact that we were only using four significant figures for our calculations. The limitation of four significant figures follows from our choice for the value of g.

HOMEWORK PROBLEMS

In the following problems, t is time, x is distance (length), v is speed (length/time), a is the magnitude of acceleration (length/time²), and m is mass.

I.9   Derive conversion factors for changing the following British gravitational units to their SI equivalents. (a) area: inch², (b) volume: foot³, (c) force: ounce, (d) pressure: pounds/inch², (e) pressure: pounds/foot², (f) speed: foot/second, (g) speed: in./hr, (h) acceleration: miles/hour².

I.10   In each of the following formulas, c_1, c_2, and so on, denote constants and θ is an angle in radians. Determine the units of these constants if the formula is to be dimensionally correct. (a) $a = c_1 v^2 / x$, (b) $\frac{1}{2} m v^2 = \frac{1}{2} c_1 x^2$, (c) $x = c_1 x_0 + c_2 v_0 + c_3 t^2$, (d) θ (degrees) $= c_1 \theta$ (radians), (e) $d\theta/dt = c_1 + c_2 t$, (f) $t = c_1 \sqrt{x}$, (g) $mc_1^2 \, d^2\theta/dt^2 = Fx$.

I.11 At a certain large university known for its athletic prowess, the following new set of physical units are in use. The basic length unit is the *touchdown* which is 100 yards, the basic time unit is the *dash* which is 9 s, and the basic force unit is the *golf ball*, which is 1.620 oz. Determine the conversion factors between these units and their SI equivalents.

I.12 At the university in Problem I.11 the unit of mass is the *basket*, 1 basket being 1 golf ball-dash²/touchdown. Determine the number of kilograms in one basket.

I.13 When a linear coil spring in the suspension system of an automobile is compressed by 2.5 in., it exerts a compressive force of 375 lb. Determine the stiffness of the spring in British gravitational and in SI units.

I.14 When a certain linear spring has a length of 180 mm the tension within it is 150 N. For a length of 160 mm the compressive force within the spring is 100 N. Determine the unstretched length and stiffness of the spring (a) in SI units, (b) in British gravitational units.

I.15 A widely employed fluid mechanics equation for the pressure drop along a pipe of length l and diameter D is

$$P_1 - P_2 = f \frac{l}{D} \rho \frac{v^2}{2}$$

where P denotes pressure, ρ is the density of the fluid, v is its velocity, and f is the resistance coefficient, which is given by the expression

$$\sqrt{f} = 2.30 \log_{10} \left(\frac{Dv}{\nu} \sqrt{f} \right) - 0.91$$

Determine the dimensions of ν, the kinematic viscosity.

I.16 A restaurant buys 1000 lb of coffee at a standard locality and also a spring scale calibrated at that locality. Determine what the coffee weighs on the spring scale in a locality where (a) $g = 32.11$ ft/s², (b) $g = 9.791$ m/s².

I.17 The work done by a force is Fx and the power developed by this force is Fv. In the SI system the basic unit of work and energy is joules (J),

1 J being the work done by a 1-N force in moving a body a distance of 1 m. The basic SI unit for power is a watt (W), 1 W being the power developed by a 1-N force when it moves a body at a speed of 1 m/s. Determine what (a) 1 J is in terms of meters, kilograms, and seconds, (b) 1 W is in terms of meters, kilograms, and seconds, and (c) 1 W is in terms of joules and seconds.

I.18 The power developed by a force is Fv. In the SI system the basic unit of power is a watt, where a watt is as defined in Problem I.17. In the British gravitational system the basic unit of power is a horsepower (hp), 1 hp being the power developed by a 550-lb force when it moves a body at a speed of 1 ft/s. Determine (a) how many kilowatts there are in 1 hp, and (b) how many horsepower there are in 1 kW.

MODULE II
CONCURRENT FORCE
SYSTEMS AND
EQUILIBRIUM OF
A PARTICLE

In this and the following module we shall establish the basic laws governing the forces acting on bodies that are in a state of static equilibrium. The concepts, techniques, and methods in this development are central to all engineering studies concerning the way in which physical systems respond to the forces acting on them. It is natural that in the initial phase, presented in this module, we consider the case of a particle, for it is the simplest model that can be used to represent a physical body.

A. BASIC CONCEPTS

By definition, a particle is a body that is considered to occupy a single point in space. Hence, the forces acting on a particle must be *concurrent* (intersect) at the particle. Consider the two forces $\bar{F}_1$ and $\bar{F}_2$ acting on the particle P shown in Figure 1. Forces $\bar{F}_1$ and $\bar{F}_2$, being vectors, add according to the parallelogram law. Their total effect on the particle is equivalent to the force $\bar{R}$ in Figure 1, where

$$\bar{R} = \bar{F}_1 + \bar{F}_2$$

The vector $\bar{R}$ is called the *resultant* of $\bar{F}_1$ and $\bar{F}_2$. Note that we use a double slash on the arrow representing the resultant to distinguish it from an actual force acting on the particle.

If there were more than two forces acting on the particle the resultant of such a *system of forces* would be

$$\boxed{\bar{R} = \sum_{i=1}^{N} \bar{F}_i} \tag{1}$$

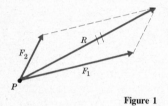

Figure 1

In writing equation (1) we made use of the fact that the sum of vectors is independent of the order in which they are added. Further, it is apparent that the point of application of the resultant force is the point occupied by the particle.

Let us consider a number of ways in which the resultant force vector can be obtained. Clearly, as was done in Figure 1, we can use the parallelogram law to add the vectors consecutively. For example, in the case of three forces, we could first obtain the vector $\bar{R}_1 = \bar{F}_1 + \bar{F}_2$ and then find the total resultant according to

$$\bar{R} = (\bar{F}_1 + \bar{F}_2) + \bar{F}_3 = \bar{R}_1 + \bar{F}_3$$

This is depicted in Figure 2a on the following page.

Another view of the process for determining the resultant is obtained from the alternative rule for the addition of vector quantities. In this method the individual vectors in the sum are moved in space, with their orientations constant, until the tail of each vector in the sum coincides

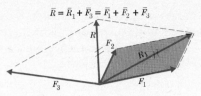

$$\bar{R} = \bar{R}_1 + \bar{F}_3 = \bar{F}_1 + \bar{F}_2 + \bar{F}_3$$

Figure 2a

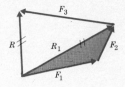

Figure 2b

with the tip of the previous vector, as illustrated in Figure 2b. The result-ant $\bar{R}$ then extends from the tail of the first vector ($\bar{F}_1$) to the tip of the last vector ($\bar{F}_3$). Note that in this approach it is not necessary to first determine the intermediate resultant force $\bar{R}_1$. In general, then, the resultant of a system of forces acting on a particle is the closing vector of a polygon whose other sides are composed of all the forces in the system.

Two major difficulties arise when using the approaches of Figure 2 to determine resultant vectors. Clearly, it is difficult to construct the appro-priate diagrams for a system of three-dimensional vectors. Equally impor-tant, even in the case of planar force systems, the determination of the resultant usually involves the geometry and trigonometry of scalene (gen-eral) triangles.

A graphical solution in which the vector polygon is constructed to scale circumvents the need to use trigonometry. However, this approach places unnecessary limitations on the accuracy of the solution, and still does not directly address the question of three-dimensional force systems.

Another procedure, which we shall use throughout the text, is to represent vectors by their components with respect to a set of mutu-ally orthogonal (i.e., perpendicular) axes. These axes form an *XYZ* coor-dinate system which we refer to as the *frame of reference*. This compo-nent approach for treating vectors has the significant advantage of simplifying the geometry and trigonometry to that of right triangles, while simultaneously providing a consistent approach for both planar and three-dimensional problems.

B.   COMPONENTS OF A VECTOR

Before we can use the component approach for vector operations, such as addition, it is necessary for us to develop the ability to represent vectors in terms of their components with respect to an *arbitrarily chosen* rectan-gular Cartesian coordinate system *XYZ*. A typical situation is depicted in Figure 3 where we wish to determine the components of a vector $\bar{A}$. The origin O of the coordinate system is chosen, for convenience, to coincide with the tail of the vector.

The dotted lines in Figure 3 are constructed to intersect the tip of the vector $\bar{A}$ and to intersect each coordinate axis at a right angle. From the intersections of these perpendiculars with the axes, we can form a rectan-

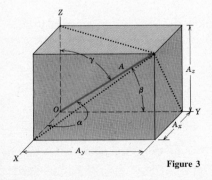

Figure 3

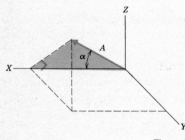

Figure 4

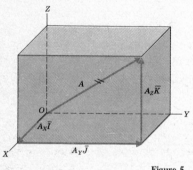

Figure 5

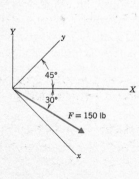

gular parallelepiped, as depicted in the figure. The lengths A_X, A_Y, and A_Z are defined as the components of the vector $\bar{A}$ with respect to the X, Y, and Z coordinate axes, respectively. Because $\bar{A}$ is the diagonal of the parallelepiped, we may write

$$\left|\bar{A}\right| = \sqrt{A_X{}^2 + A_Y{}^2 + A_Z{}^2} \qquad (2)$$

If we passed a plane through the vector $\bar{A}$ and the X axis, the angle that would appear between the two lines is α. This is depicted by the shaded plane in Figure 4. Clearly, $A_X = \left|\bar{A}\right| \cos \alpha$. More generally, in terms of the angles between the vector $\bar{A}$ and the three coordinate axes, we may write

$$A_X = \left|\bar{A}\right| \cos \alpha \qquad A_Y = \left|\bar{A}\right| \cos \beta \qquad A_Z = \left|\bar{A}\right| \cos \gamma \qquad (3)$$

The cosines of the *direction angles* α, β, and γ are called the *direction cosines*. Placing equations (3) into equation (2) yields

$$\cos^2 \alpha + \cos^2 \beta + \cos^2 \gamma \equiv 1$$

A way of interpreting this identity is to note that only two of the three direction angles can be independent quantities.

From the foregoing we see that the properties of a free vector are completely described by its components with respect to a specific coordinate system. The coordinate system itself is specified by introducing a set of unit vectors (vectors of magnitude one) $\bar{I}$, $\bar{J}$, and $\bar{K}$ which are oriented in the directions of the positive X, Y, and Z axes, respectively. It then follows, as shown in Figure 5, that $A_X\bar{I}$ is a vector parallel to the X axis whose magnitude is the component A_X. Figure 5 also shows that the sum of the three vectors $A_X\bar{I}$, $A_Y\bar{J}$, and $A_Z\bar{K}$ is the vector $\bar{A}$. Hence

$$\boxed{\bar{A} = A_X\bar{I} + A_Y\bar{J} + A_Z\bar{K}} \qquad (4)$$

From equation (4) we observe that the unit vectors have no physical units. The physical units of the vector are the same as those of its components.

EXAMPLE 1
Represent the force $\bar{F}$ shown in the sketch in terms of its components with respect to the two coordinate systems XYZ and xyz.

Solution
The force lies in the plane of the paper, so the direction angle between $\bar{F}$ and the Z axis is 90°. Because $\cos 90° \equiv 0$, there is no component perpen-

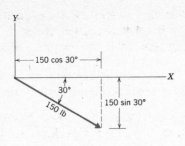

dicular to the paper. A sketch of the components of $\bar{F}$ with respect to the X and Y axes shows that

$$F_X = 150 \cos 30° = 129.90 \text{ lb}$$
$$F_Y = -150 \sin 30° = -75.00 \text{ lb}$$

Note that F_Y is negative because it is along the negative Y axis.

Equation (4) then gives

$$\bar{F} = 129.90\bar{I} - 75.00\bar{J} \text{ lb}$$

We now follow the same approach for the components with respect to xyz, beginning with a sketch.

From the sketch,

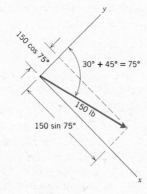

$$F_x = 150 \sin 75° = 144.89 \text{ lb}$$
$$F_y = 150 \cos 75° = 38.82 \text{ lb}$$

Thus

$$\bar{F} = 144.89\bar{i} + 38.82\bar{j} \text{ lb}$$

where $\bar{i}, \bar{j}, \bar{k}$ are the unit vectors for the xyz axes.

This result illustrates the obvious fact that the components of a vector have meaning only when the coordinate system is specified. Using equation (2) we can easily verify that both expressions represent the same vector; the differences in the component values result from the fact that the $\bar{I}$ and $\bar{J}$ unit vectors refer to the XYZ system, whereas the $\bar{i}$ and $\bar{j}$ unit vectors refer to the xyz system.

EXAMPLE 2

Determine the vector that locates a point P whose rectangular Cartesian coordinates are $(400, 400, -200)$ mm with respect to the origin O. What are the direction angles of this vector?

Solution

A sketch shows that the desired vector is $\bar{r}_{P/O}$, which is the position of point P with respect to point O (the notation should be read as r of P with respect to O).

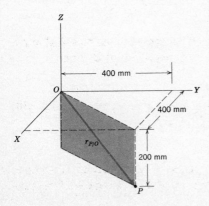

We see from the sketch that the components of the vector $\bar{r}_{P/O}$ in the X and Y directions are both 400 mm, whereas the Z component is -200 mm; the negative sign arises because $\bar{r}_{P/O}$ projects onto the negative Z

axis. Hence we write the vector as

$$\bar{r}_{P/O} = 400\bar{I} + 400\bar{J} - 200\bar{K} \text{ mm}$$

This expression illustrates the general result that the components of $\bar{r}_{P/O}$ are the coordinates of point P with respect to a coordinate system whose origin is point O.

We now find the direction cosines by calculating $\left|\bar{r}_{P/O}\right|$ and then applying equations (3), as follows.

$$\left|\bar{r}_{P/O}\right| = \sqrt{(400)^2 + (400)^2 + (200)^2} = 600 \text{ mm}$$

$$\cos \alpha = \frac{(r_{P/O})_X}{\left|\bar{r}_{P/O}\right|} = \frac{400}{600} = 0.6667$$

$$\cos \beta = \frac{(r_{P/O})_Y}{\left|\bar{r}_{P/O}\right|} = \frac{400}{600} = 0.6667$$

$$\cos \gamma = \frac{(r_{P/O})_Z}{\left|\bar{r}_{P/O}\right|} = \frac{-200}{600} = -0.3333$$

In order to eliminate any ambiguity, it is common practice to consider the direction angles to have values between 0° and 180°. Thus

$$\alpha = \beta = \cos^{-1}(0.6667) = 48.19°$$

$$\gamma = \cos^{-1}(-0.3333) = 109.47°$$

Recalling the definition of the direction angles, can you explain the significance of γ being larger than 90°?

The preceding examples show that the description of a vector quantity according to equation (4) fully describes the vector, provided we have a sketch of the coordinate system. Hence, when solving problems it is acceptable to leave a solution for a vector quantity in the form of equation (4), unless the magnitude and direction angles of the vector are specifically requested.

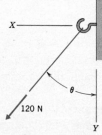

Prob. II.1

HOMEWORK PROBLEMS

II.1 The cable pulls on the eyebolt with a tensile force of 120 N. If $\theta = 25°$, determine the components of the force with respect to the XYZ coordinate system.

II.2 In Problem II.1 the X component of the 120-N force is -90 N. Determine the Y component of the force and the corresponding angle θ.

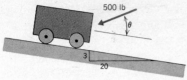

Prob. II.4

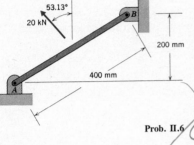

Prob. II.6

II.3 In Problem II.1 the X component of the 120-N force is twice as large as the Y component. Determine these components and the corresponding angle θ.

II.4 The cart is pushed up the 15% grade by the 500-lb force. If $\theta = 30°$, determine the components of this force (a) parallel and perpendicular to the hill, (b) horizontally and vertically.

II.5 In Problem II.4 the horizontal component of the 500-lb force is 400 lb. Determine (a) the angle θ, (b) the vertical component of the force, (c) the components of the force parallel and perpendicular to the hill.

II.6 The 20-kN force is applied to the bar AB as shown. Determine the components of this force (a) horizontally and vertically, (b) parallel and perpendicular to the longitudinal axis of the bar.

II.7 A spring whose unstretched length is 2 ft and whose stiffness is 200 lb/ft is tied between points A and B. Determine the horizontal and vertical components of the force the spring exerts (a) on point A, (b) on point B.

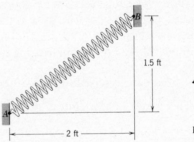

Prob. II.7

Prob. II.8

II.8 The x and y axes shown form nonorthogonal Cartesian coordinates in the plane. It is desired to write the force $\bar{F}$ in the form $\bar{F} = F_x\bar{i} + F_y\bar{j}$. Determine the values of F_x and F_y. Are these components the perpendicular projections onto the corresponding axes?

II.9 An observer on the ground at point A is looking at the top of a 200-m tall building at point B. (a) Determine the components of the position of point B with respect to point A and write the result as a vector. (b) Determine the direction cosines of this position vector with respect to the southerly and easterly directions.

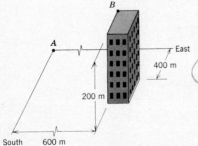

Prob. II.9

II.10 A force $\bar{F}$ has a magnitude of 4 tons and the direction angles be-

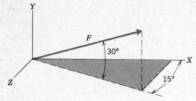

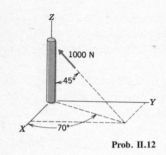

Prob. II.11

tween this force and the Y and Z axes are $\beta = 45°$, $\gamma = 120°$. Determine the components of this force and write the result as a vector. Also, determine the direction angle between the force and the X axis.

II.11 Write the force $\bar{F}$ shown in terms of its components and determine the direction angles for this force with respect to the XYZ reference frame.

II.12 A 1000-N force acts on the vertical pole shown. (a) Determine the components of the force with respect to the XYZ coordinate system and write the force as a vector. (b) Determine the direction angles between the force and the XYZ coordinate axes.

II.13 The X, Y, and Z components of a 140-N force are in the proportion of 3:-2:6. Determine these components and also the direction angles between the force and the coordinate axes.

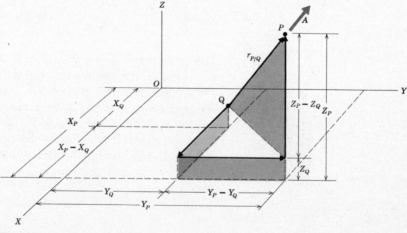

Prob. II.12

C. GENERAL UNIT VECTORS

An asset of the component representation of vectors is that the vector operations take on the form of algebraic equations. For instance, if we wish to multiply a vector $\bar{A}$ by a scalar number c, by definition the result of this operation is a vector parallel to $\bar{A}$ whose magnitude is c times the magnitude of $\bar{A}$. From equation (4) it then follows that

$$c\bar{A} = cA_X\bar{I} + cA_Y\bar{J} + cA_Z\bar{K} \tag{5}$$

This result is particularly useful in the situation where we must de-

Figure 6

scribe a vector that is parallel to a specific line. This situation is depicted in Figure 6 on the preceding page, where a vector $\bar{A}$, having known magnitude $|\bar{A}|$, is formed such that it is parallel to the line from point Q to point P, and thus parallel to $\bar{r}_{P/Q}$.

Denoting the coordinates of the end points of $\bar{r}_{P/Q}$ as (X_P, Y_P, Z_P) and (X_Q, Y_Q, Z_Q), we see from the figure that the components of $\bar{r}_{P/Q}$ can be written as the differences in the coordinates; that is

$$\bar{r}_{P/Q} = (X_P - X_Q)\bar{I} + (Y_P - Y_Q)\bar{J} + (Z_P - Z_Q)\bar{K} \tag{6}$$

Once $\bar{r}_{P/Q}$ is known, we may obtain a unit vector $\bar{e}_{P/Q}$ in the direction of $\bar{r}_{P/Q}$ (and hence $\bar{A}$) by writing

$$|\bar{r}_{P/Q}| \ \bar{e}_{P/Q} = \bar{r}_{P/Q}$$

so

$$\bar{e}_{P/Q} = \frac{\bar{r}_{P/Q}}{|\bar{r}_{P/Q}|}$$

Then, to obtain $\bar{A}$ we need only multiply the unit vector by the magnitude of $\bar{A}$, which gives

$$\boxed{\bar{A} = |\bar{A}|\bar{e}_{P/Q} = |\bar{A}|\frac{\bar{r}_{P/Q}}{|\bar{r}_{P/Q}|}} \tag{7}$$

EXAMPLE 3
The magnitudes of the forces $\bar{F}_1$ and $\bar{F}_2$ shown in the sketch are 1.20 and 0.80 kN, respectively. Express these forces in terms of their components.

Solution
The forces $\bar{F}_1$ and $\bar{F}_2$ are directed along the lines from point B to point A, and point A to point C, respectively. A method for constructing the necessary position vectors $\bar{r}_{A/B}$ and $\bar{r}_{C/A}$ is to picture ourselves moving from the tail to the tip of each vector. For instance, to move from point B to point A we must go 400 mm in the negative Z direction, 240 mm in the positive X direction, and 320 mm in the positive Y direction, so that

$$\bar{r}_{A/B} = 240\bar{I} + 320\bar{J} - 400\bar{K} \text{ mm}$$

Similarly,

$$\bar{r}_{C/A} = 240\bar{I} - 320\bar{J} \text{ mm}$$

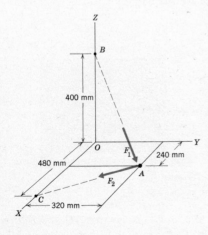

From equation (7) we then form

$$\bar{F}_1 = |\bar{F}_1|\,\bar{e}_{A/B}$$

$$= (1.2)\,\frac{\bar{r}_{A/B}}{|\bar{r}_{A/B}|} = (1.2)\frac{240\bar{I} + 320\bar{J} - 400\bar{K}}{\sqrt{(240)^2 + (320)^2 + (400)^2}}$$

$$= 0.3091\bar{I} + 0.6788\bar{J} - 0.8485\bar{K} \text{ kN}$$

$$\bar{F}_2 = (0.8)\,\frac{\bar{r}_{C/A}}{|\bar{r}_{C/A}|} = (0.8)\frac{240\bar{I} - 320\bar{J}}{\sqrt{(240)^2 + (320)^2}}$$

$$= 0.480\bar{I} - 0.640\bar{J} \text{ kN}$$

As an aside, note that $\bar{F}_2$, being a force in the XY plane, could have been obtained just as easily by determining the angle between the line AC and one of the coordinate axes and then using the method of Example 1. However, this method is awkward to employ in the case of the three-dimensional force $\bar{F}_1$.

HOMEWORK PROBLEMS

II.14 The piston is pushed by the force $\bar{F}$. Write an expression for the unit vector in the direction of $\bar{F}$ in terms of the X and Y axes shown.

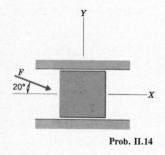

Prob. II.14

II.15 The structure shown is pulled by cable CD. If the tensile force within the cable is 8000 N, determine the components with respect to the XYZ axes of the force exerted by the cable on the structure at point D. Also, determine the unit vector parallel to this force.

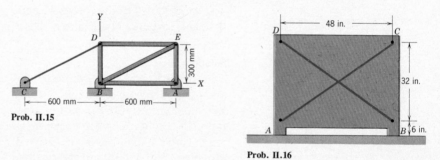

Prob. II.15

Prob. II.16

II.16 The cabinet shown is braced by tensioned crosswires AC and BD. The tensile forces in both cables are 40 lb. Determine the X and Y components of the unit vectors associated with the forces exerted by the cables on the cabinet at the four corners.

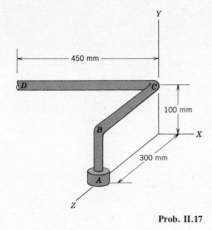

Prob. II.17

II.17 Determine the unit vectors extending from point A to points B, C, and D of the bent bar shown.

II.18 The rectangular plate is supported in part by cables AB and CD. If the tensile force in each cable is 150 lb, determine the components of the forces each cable exerts on the plate.

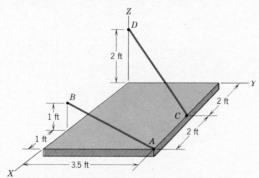

Prob. II.18

II.19 The package is being pulled at point A by cable AB. For what tensile force in the cable will the X component of the force exerted by the cable on the package be 200 N?

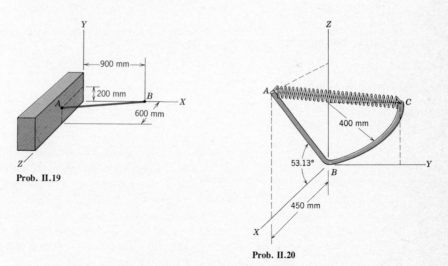

Prob. II.19

Prob. II.20

II.20 A spring whose stiffness is 400 N/m and whose unstretched length is 400 mm is stretched between ends A and C of bent rod ABC. Determine the force exerted by the spring on end C of the rod.

D. DETERMINATION OF THE RESULTANT
OF CONCURRENT FORCES

It was shown in equation (1) that the resultant of a set of concurrent forces acting on a particle is the sum of the forces. Let us consider Figure 7, where two forces $\bar{F}_1$ and $\bar{F}_2$ are concurrent at the origin O, with the aim of evaluating the resultant of these forces. Toward this end we move $\bar{F}_2$ so as to align the vectors "head to tail." This allows us to form the resultant according to the polygon rule for addition.

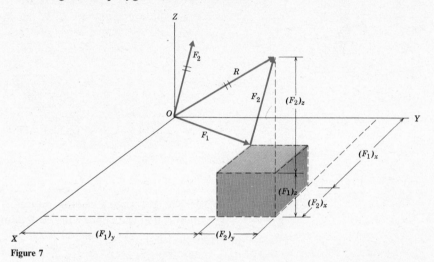

Figure 7

The figure shows that the components of $\bar{R}$ are the sum of the corresponding components of $\bar{F}_1$ and $\bar{F}_2$. This conclusion can be extended (by induction) to the case of the resultant of several forces, with the result that

$$\bar{R} = \Sigma\, F_X\, \bar{I} + \Sigma\, F_Y\, \bar{J} + \Sigma\, F_Z\, \bar{K} \qquad (8)$$

where $\Sigma\, F_X$, $\Sigma\, F_Y$, and $\Sigma\, F_Z$ denote the sum of the force components in the X, Y, and Z directions, respectively. Also, remember that because $\bar{R}$ is the resultant of the system of forces acting on the particle at point O, the vector $\bar{R}$ must be applied at *the point of concurrency* of the force system.

Clearly, although equation (8) was derived by considering the resultant of a set of forces, it is valid for any set of vectors. Thus, each component of the sum of a set of vectors is the sum of the corresponding components of the individual vectors.

The rule for the difference of two vectors $\bar{A}$ and $\bar{B}$ is a corollary to the foregoing development. For example, given the vector $\bar{B}$ we may write

$$-\bar{B} \equiv (-1)\bar{B} = -\, B_X\bar{I} - B_Y\bar{J} - B_Z\bar{K}$$

Thus, the difference between $\bar{A}$ and $\bar{B}$ is

$$\bar{A} - \bar{B} = \bar{A} + (-\bar{B}) = (A_X - B_X)\bar{I} + (A_Y - B_Y)\bar{J} + (A_Z - B_Z)\bar{K} \qquad (9)$$

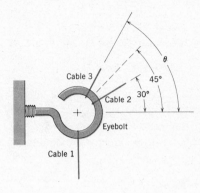

EXAMPLE 4

Three cables exert a concurrent set of forces on the eyebolt to which they are attached. Cables 1 and 2 pull on the eyebolt with forces of 6 and 8 kN, respectively. It is desired that the force exerted by cable 3 be the minimum value that causes the resultant of the three forces to be along the indicated dashed line. Determine the magnitude of this force and the corresponding value of the angle θ.

Solution

Recalling the discussion in Module I, we know that cables can exert only pulling forces on the objects to which they are attached; that is, cables can act only in tension. From the given information we know the tensile forces of cables 1 and 2. Further, we let $\bar{F}_3$ denote the tensile force in cable 3. Each of these forces is along the corresponding cable, as illustrated in the sketch.

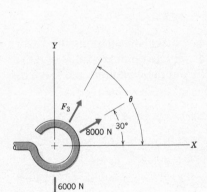

Also shown in the sketch is the coordinate system to be used for expressing components. Note that we chose one axis, the X axis, horizontal because all angles are referred to this line. The forces are all coplanar, so there are no components in the Z direction.

To determine the components of the forces we write each vector as the product of its magnitude times its unit vector. Cable 1 pulls downward, so

$$\bar{F}_1 = 6000(-\bar{J}) = -6000\bar{J} \text{ N}$$

whereas cable 2 is 30° above the X axis, so

$$\bar{F}_2 = 8000(\cos 30° \, \bar{I} + \sin 30° \, \bar{J})$$

$$= 6928\bar{I} + 4000\bar{J} \text{ N}$$

We do not as yet know the magnitude of $\bar{F}_3$ or the angle θ, hence for $\bar{F}_3$ we write

$$\bar{F}_3 = F_3(\cos \theta \, \bar{I} + \sin \theta \, \bar{J})$$

Before adding these forces to equate the sum to the resultant, let us use the given information to express vectorially the requirement that $\bar{R}$ must be parallel to the indicated 45° line. This means that

$$\bar{R} = R(\cos 45° \, \bar{I} + \sin 45° \, \bar{J})$$

$$= 0.7071R(\bar{I} + \bar{J})$$

Note that in writing this expression we have considered $\bar{R}$ to be directed outward from the origin along the dashed line. The sense of the resultant will be indicated by the sign of R.

In terms of the force vectors of the system, equation (1) gives

$$\bar{R} = \bar{F}_1 + \bar{F}_2 + \bar{F}_3$$

$$0.7071R(\bar{I} + \bar{J}) = (-6000\bar{J}) + (6928\bar{I} + 4000\bar{J}) + F_3(\cos\theta\,\bar{I} + \sin\theta\,\bar{J})$$

$$= (6928 + F_3\cos\theta)\bar{I} + (-2000 + F_3\sin\theta)\bar{J}$$

This equation expresses the fact that the vector on the left side of the equality sign must be the same as the vector on the right side. Because two vectors are identical only when they have the same components, we must equate the corresponding components on both sides of the equation. Thus

$$\bar{I} \text{ components:} \qquad 0.7071R = 6928 + F_3\cos\theta$$

$$\bar{J} \text{ components:} \qquad 0.7071R = -2000 + F_3\sin\theta$$

There are three unknowns in these two scalar equations; however, we want the smallest value of F_3. Hence, we must find the value of θ that minimizes F_3. To do this we eliminate R from the equations and solve for F_3 in terms of θ, as follows.

$$6928 + F_3\cos\theta = -2000 + F_3\sin\theta$$

$$F_3 = \frac{8928}{(\sin\theta - \cos\theta)}$$

For F_3 to be a minimum, $dF_3/d\theta = 0$, hence

$$\frac{dF_3}{d\theta} = 0 = -\frac{8928}{(\sin\theta - \cos\theta)^2}(\cos\theta + \sin\theta)$$

from which we get

$$\cos\theta + \sin\theta = 0 \quad \text{so} \quad \tan\theta = -1 \qquad \theta = 135° \qquad \triangleleft$$

$$F_3 = \frac{8928}{\sin(135°) - \cos(135°)} = 6313 \text{ N} \qquad \triangleleft$$

How do we know that this value of θ, which gives $dF_3/d\theta = 0$, represents the minimum value of F_3, and not the maximum? To answer this question notice that the vector $\bar{F}_3$ that we obtained is perpendicular to the desired direction of $\bar{R}$. That this should be so becomes apparent when we consider the addition of the forces as shown in the diagram. Clearly the shortest length of $\bar{F}_3$ is obtained when $\bar{F}_3$ is perpendicular to the resultant force.

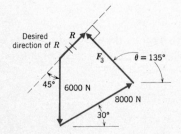

EXAMPLE 5

A radio antenna is supported by three guy wires. The tensile force in cables *AB*, *AC*, and *AD* are 6 kilopounds (kips), 5 kips, and 8 kips, respectively. Determine the resultant force exerted on the antenna by these cables.

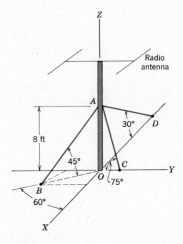

Solution

The first step is to draw a sketch showing the forces applied by the cables upon the mast; recall that cables have a pulling effect.

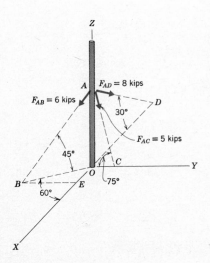

Let us first write the vector expressions for $\bar{F}_{AC}$ and $\bar{F}_{AD}$, because they have the simplifying feature of coinciding with the coordinate planes. From the sketch we see that the components of $\bar{F}_{AD}$ are oriented along the negative X and negative Z axes. Then, because the angle between the line AD and the X axis is 30°, it follows that

$$\bar{F}_{AD} = 8(-\cos 30° \bar{I} - \sin 30° \bar{K})$$
$$= -6.928\bar{I} - 4.0\bar{K} \text{ kips}$$

In a similar manner we find that

$$\bar{F}_{AC} = 5(\cos 75° \bar{J} - \sin 75° \bar{K})$$
$$= 1.292\bar{J} - 4.830\bar{K} \text{ kips}$$

We express $\bar{F}_{AB}$ using the unit vector oriented from point A to point B, $\bar{e}_{B/A}$. From equation 7 we have

$$\bar{F}_{AB} = 6\bar{e}_{B/A} = 6\frac{\bar{r}_{B/A}}{|\bar{r}_{B/A}|} \text{ kips}$$

The XYZ components of $\bar{r}_{B/A}$ are the lengths of the lines OE, EB, and AO, respectively (with the appropriate signs). To determine these lengths we note that points B, O, and A form a 45° right triangle, so that $BO = AO = 8$ ft. Now, because triangle OEB is also a right triangle, it follows that

$$EB = BO \sin 60° = 6.928 \text{ ft}$$

$$OE = BO \cos 60° = 4.00 \text{ ft}$$

Noting that a displacement from point A to point B requires displacements in the directions of the positive X and negative Y and Z axes, we have

$$\bar{r}_{B/A} = OE\bar{I} - EB\bar{J} - AO\bar{K} = 4\bar{I} - 6.928\bar{J} - 8\bar{K}$$

Thus

$$\bar{F}_{AB} = 6 \frac{4\bar{I} - 6.928\bar{J} - 8\bar{K}}{\sqrt{4^2 + (6.928)^2 + 8^2}}$$

$$= 2.121\bar{I} - 3.674\bar{J} - 4.234\bar{K} \text{ kips}$$

Finally, we determine the resultant by adding the corresponding components of the forces; that is,

$$\bar{R} = \bar{F}_{AB} + \bar{F}_{AC} + \bar{F}_{AD} = (2.121 + 0 - 6.928)\bar{I}$$

$$+ (-3.674 + 1.294 + 0)\bar{J} + (-4.234 - 4.830 - 4.0)\bar{K}$$

$$= -4.807\bar{I} - 2.380\bar{J} - 13.604\bar{K} \text{ kips}$$

HOMEWORK PROBLEMS

II.21-II.23 Determine the magnitude and direction of the resultant of the two forces shown.

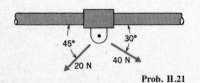

Prob. II.21

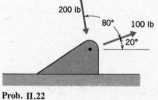

Prob. II.22

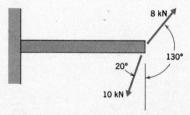

Prob. II.23

II.24 Two forces act at point C on the beam. Knowing that $F = 10$ tons and $\alpha = 40°$, determine the magnitude and angle relative to the beam of the resultant of these forces.

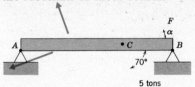

5 tons

Probs. II.24 and II.25

II.25 Two forces act at point C on the beam. Determine the smallest magnitude of the force $\bar{F}$ and the corresponding angle α for which the resultant force is (a) horizontal, (b) vertical.

II.26 Three people pull horizontally on cables attached to an automobile. If each person exerts a 200-N pull, determine the magnitude and the angle θ for the resultant force exerted on the vehicle.

II.27 Cables 2 and 3 are pulled by 200-N forces. It is desired that the resultant of the forces exerted by the three cables be oriented along the dashed line indicated by the angle θ. Determine the required pull of cable 1 to have (a) $\theta = 0$, (b) $\theta = 20°$.

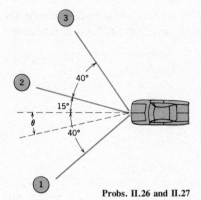

Probs. II.26 and II.27

II.28 Three forces are applied to the collar on the inclined bar. The angle between the two 50-N forces is constant at $20°$, but the angle α is variable. Determine the value of α for which the resultant of the three forces is horizontal.

II.29 Determine the required value of α for the resultant of the three forces to be parallel to the inclined bar.

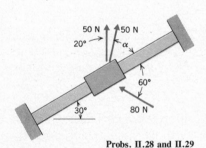

Probs. II.28 and II.29

II.30 Joint E of the bridge structure is shown isolated at the right. Knowing that the magnitudes of forces $\bar{F}_1$, $\bar{F}_2$, and $\bar{F}_3$ are 4, 5, and 6 kN, respectively, determine the magnitude of $\bar{F}_4$ such that the resultant of these forces at joint E is vertical. Note that the members are bars, not cables, so they can sustain compressive forces.

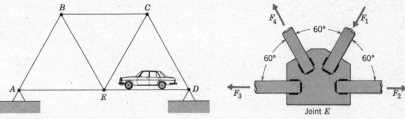

Joint E

Prob. II.30

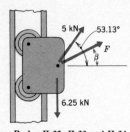

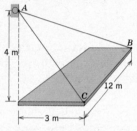

Probs. II.32, II.33 and II.34

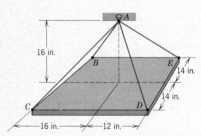

Probs. II.35 and II.36

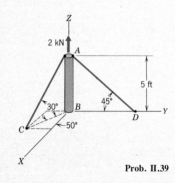

Prob. II.37

Prob. II.39

II.31 In Problem II.30 the magnitudes of $\bar{F}_3$ and $\bar{F}_1$ are 2 kN and 3 kN, respectively. Determine the magnitudes of $\bar{F}_2$ and $\bar{F}_4$ for which the resultant of these forces at joint E is zero.

II.32 It is desired that the resultant of the three forces acting on the roller guide be horizontal. If $F = 8$ kN, determine the corresponding angle β.

II.33 Find the smallest value of F and the corresponding angle β for which the resultant of the three forces shown is (a) horizontal, (b) vertical.

II.34 Knowing that $\beta = 30°$, determine the magnitude of $\bar{F}$ for which the resultant of the force system shown has a magnitude of 5 kN.

II.35 The tensile force in cable AB is 2 kN, and that in cable AC is 3 kN. Determine the resultant force exerted by the cables on the support at point A.

II.36 The tensile force in cable AB is 2 kN. For what value of the tension force in cable AC will the resultant force exerted by the cables on the support at point A have a magnitude of 4 kN?

II.37 The square plate is suspended by four cables as shown. The tensile forces in cables AB and AC are 90 lb, and those in cables AD and AE are 60 lb. Determine the resultant force exerted by the cables on the support at point A.

II.38 In Example 5 the tensile force in cable AD is 8 kips. Determine the tensile forces in the other two cables for which the resultant force exerted by the cables on the antenna is vertical. Determine the magnitude of the resultant force in this case.

II.39 Two cables are attached to the vertical pole AB to steady it. An upward force of 2 kips is applied to the pole with the effect that the resultant of the three forces acting at point A on the pole is parallel to the X axis. Determine the tensions in the two cables and the magnitude of the resultant.

II.40 A 200-kg block is held in position on the 36.87° inclined ramp by two cables. Knowing that the tensile force in cable AB is 2500 N and that in cable AC is 3000 N, determine the resultant force exerted by the two cables and the force of gravity on the block.

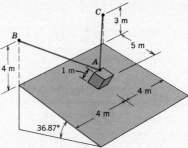

Prob. II.40

II.41 In Problem II.40 the resultant force exerted by the two cables and the gravitational force on the 200-kg block is normal to the 36.87° plane. Determine the tensile force in each cable.

II.42 In order to hold the television tower erect, the resultant loading exerted on end D must be 20 kN in the negative Z direction. Determine the tensile force in each cable required to accomplish this.

II.43 Solve Problem II.42 for the case where an additional force of 1 kN is applied to the tower at point D in the positive Y direction.

Prob. II.42

E. DOT (SCALAR) PRODUCTS

In the preceding section we confined our efforts to determining the components of a force parallel to each of the three axes of a system of rectangular Cartesian coordinates. Essentially, this was done by multiplying the magnitude of the vector by the cosine of the angle it made with the axis. This operation can be put in a more general form in order to facilitate the determination of the components of a vector in a direction that is not a coordinate axis. The operation we will derive to accomplish this is called the *dot product* of two vectors.

Let us consider the two intersecting vectors $\bar{A}$ and $\bar{B}$ shown in Figure 8. (They have been aligned to make their tails coincide.) The length $|\bar{A}|$ cos θ is the component of $\bar{A}$ parallel to $\bar{B}$. and $|\bar{B}|$ cos θ is the component of $\bar{B}$ parallel to $\bar{A}$, where θ is the angle between the two vectors. Obviously, the scalar number $|A|(|B|$ cos $\theta)$ is the same as the scalar number $|\bar{B}|(|\bar{A}|$ cos $\theta)$. We write this product symbolically as $\bar{A} \cdot \bar{B}$, which

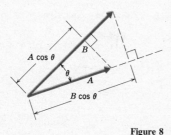

Figure 8

should be read as $\bar{A}$ dot $\bar{B}$. Thus, the definition of a dot (or scalar) product is

$$\boxed{\bar{A} \cdot \bar{B} = \bar{B} \cdot \bar{A} \equiv |\bar{A}| |\bar{B}| \cos \theta} \tag{10}$$

In words, the dot product is the magnitude of one vector times the component of the other vector parallel to the first one. The result is a *scalar number*. Note that when $\theta > 90°$, the dot product is negative. Physically, this means that the component of $\bar{A}$ parallel to $\bar{B}$ is opposite the sense of $\bar{B}$ (or vice versa).

When the two vectors are perpendicular to each other, equation (10) gives

$$\bar{A} \cdot \bar{B} = |\bar{A}| |\bar{B}| \cos 90° = 0$$

If we calculate the dot product of a vector with itself, equation (10) gives

$$\bar{A} \cdot \bar{A} = |\bar{A}| |\bar{A}| \cos 0 = |\bar{A}|^2$$

Two useful properties of the dot product are

$$p(\bar{A} \cdot \bar{B}) = (p\bar{A}) \cdot \bar{B} = \bar{A} \cdot (p\bar{B})$$

$$\bar{A} \cdot (\bar{B} + \bar{C}) = \bar{A} \cdot \bar{B} + \bar{A} \cdot \bar{C} \tag{11}$$

Both identities in equations (11) are easily proven by use of the definition of the dot product, equation (10).

Let us now turn to the subject of the determination of a dot product in terms of the components of the vectors. This is accomplished by first calculating the dot products of the unit vectors. For example, using equation (10),

$$\bar{I} \cdot \bar{I} = (1)(1) \cos 0° = 1$$

Hence, for the unit vectors of a rectangular Cartesian coordinate system,

$$\bar{I} \cdot \bar{I} = \bar{J} \cdot \bar{J} = \bar{K} \cdot \bar{K} = 1 \tag{12a}$$

Any other combination of unit vectors leads to zero, for instance,

$$\bar{I} \cdot \bar{J} = (1)(1) \cos 90° = 0$$

Therefore,

$$\bar{I} \cdot \bar{J} = \bar{J} \cdot \bar{I} = \bar{J} \cdot \bar{K} = \bar{K} \cdot \bar{J} = \bar{K} \cdot \bar{I} = \bar{I} \cdot \bar{K} = 0 \tag{12b}$$

We may now compute a dot product of two arbitrary vectors in terms of their components. To do this we use equations (11) to write

$$\bar{A} \cdot \bar{B} = (A_X \bar{I} + A_Y \bar{J} + A_Z \bar{K}) \cdot \bar{B}$$

$$= A_X \bar{I} \cdot (B_X \bar{I} + B_y \bar{J} + B_Z \bar{K})$$

$$+ A_Y \bar{I} \cdot (B_X \bar{I} + B_Y \bar{J} + B_Z \bar{K})$$

$$+ A_Z \bar{K} \cdot (B_X \bar{I} + B_Y \bar{J} + B_Z \bar{K})$$

Then, in view of equations (12), this becomes

$$\bar{A} \cdot \bar{B} = A_X B_X + A_Y B_Y + A_Z B_Z \tag{13}$$

Let us now return to our original reason for introducing the concept of a dot product: the determination of the component of a force parallel to a specific line that is not a coordinate axis. Let $\bar{e}$ be a unit vector parallel to the specified line, as is shown in Figure 9. We desire the component of the vector $\bar{A}$ in the specified direction. Using equation (10) we have

$$\bar{A} \cdot \bar{e} = |\bar{A}||\bar{e}| \cos \theta = |\bar{A}| \cos \theta \tag{14}$$

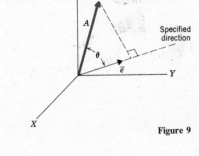

Figure 9

Clearly, this is the desired component because θ is, by definition, the angle between $\bar{A}$ and the unit vector $\bar{e}$.

EXAMPLE 6
It is common practice in describing the resultant force acting on the surface of a body to refer to the component normal to the surface as a tension force if it pulls on the surface or as a compression force if it pushes on the surface. Also, the component parallel to the surface is called the shearing force. Determine the magnitudes of the tension (or compression) and shear forces acting on the shaded rectangular area shown in the sketch.

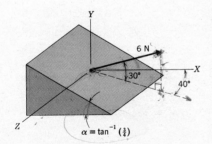

Solution
We first express the 6-kN force in terms of its components. Note in the given sketch that the projection of the force onto the XZ plane is $6(\cos 30°)$ and that the angle between this projection and the X axis is 40°. Thus

$$\bar{F} = 6 \cos 30°(\cos 40° \, \bar{I} + \sin 40° \, \bar{K}) + 6 \sin 30 \, \bar{J}$$

$$= 3.980\bar{I} + 3.00\bar{J} + 3.340\bar{K} \text{ kN}$$

To determine the component of the force normal to the rectangular area we will need the unit vector $\bar{e}_n$ normal to the surface. In this particular case we see that $\bar{e}_n$ does not have a $\bar{K}$ component; that is, the normal vector is parallel to the XY plane, as shown in the sketch.

Because $\bar{e}_n$ is a unit vector, we have

$$\bar{e}_n = |\bar{e}_n|(\sin \alpha \, \bar{I} + \cos \alpha \, \bar{J}) = 0.60\bar{I} + 0.80\bar{J}$$

The component of the 6-kN force in the direction of $\bar{e}_n$ is found from the dot product.

$$F_n = \bar{F} \cdot \bar{e}_n = (3.980\bar{I} + 3.00\bar{J} + 3.340\bar{K}) \cdot (0.60\bar{I} + 0.80\bar{J})$$
$$= 3.980(0.60) + 3.00(0.80) + 0 = 4.788 \text{ kN}$$

Because this component is positive it means that it has the same sense as the $\bar{e}_n$ vector. Thus the component of $\bar{F}$ normal to the surface is directed outward from the surface. We conclude then that F_n is a tensile force.

Finally, the shear component F_s is the component parallel to the inclined surface, and thus perpendicular to the direction of F_n. Therefore, we can find F_s from the Pythagorean theorem.

$$F_s = \sqrt{|\bar{F}|^2 - F_n^2} = \sqrt{(6.0)^2 - (4.788)^2} = 3.616 \text{ kN}$$

HOMEWORK PROBLEMS

II.44 Prove the identities of equations (11) starting with the definition of a dot product, equation (10).

II.45 Determine the dot product of the two forces shown by (a) using the formal definition of this product, (b) calculating the dot product of the component expressions for each force.

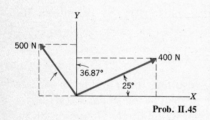

Prob. II.45

II.46 Use the dot product of $\bar{F}_1$ and $\bar{F}_2$ to derive the trigonometric identity for $\cos(\phi_2 - \phi_1)$.

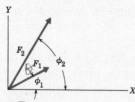

Prob. II.46

II.47 Three cables brace a newly planted tree. Determine the angles between (a) cables AB and AC, (b) cables AB and AD, (c) cables AC and AD.

II.48 Each of the cables has a tension of 250 N. Determine the component of the resultant force exerted by these cables on the tree (a) parallel to line AB, (b) parallel to line AC, (c) parallel to line AD.

Probs. II.47 and II.48

II.49 Two cables are attached to the building at point D as shown. Determine the angle between these two cables.

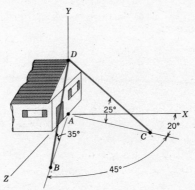

Probs. II.49 and II.50

II.50 Cable BD has a tension of 200 lb, whereas the tension in cable CD is 100 lb. Determine the component parallel to line CD of the resultant force exerted by the cables on point D.

II.51 Rectangle $ABCD$ is situated in space as shown. Determine the unit vector that is oriented from point A toward point B.

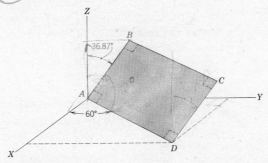

Prob. II.51

II.52 The 6-kN force in Example 6 was expressed in terms of its tension and shear components relative to the inclined surface. Express these two components as vectors, thus replacing the given force by a tension force and a shear force.

II.53-II.55 A 5000-N vertical force is applied to the surface shown at point B. (a) Determine the magnitude of the tension and shear forces produced by this force with respect to the tangent to the surface at point

B. (b) Write these tension and shear forces as vectors with respect to the *XYZ* reference frame shown.

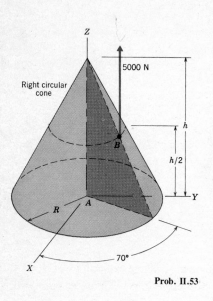

Prob. II.53

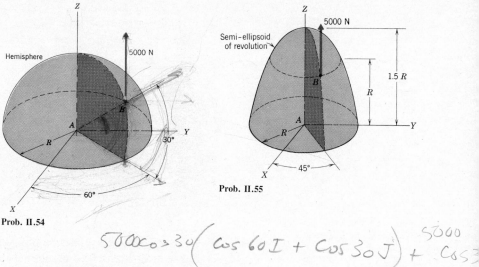

Prob. II.54

Prob. II.55

$5000 \cos 30 \left(\cos 60 \, I + \cos 30 \, J \right) + \frac{5000}{\cos 3}$

F.   EQUILIBRIUM OF A PARTICLE

1   Basic Equations

The conditions for the static equilibrium of a particle are addressed explicitly in Newton's first law. It tells us that a particle will remain at rest only if the resultant of all forces acting on the particle is zero. Equation (8) relates the resultant to the components of a system of forces. Thus, for *static equilibrium*

$$\bar{R} = \Sigma \, F_X \, \bar{I} + \Sigma \, F_Y \, \bar{J} + \Sigma \, F_Z \, \bar{K} \equiv \bar{0}$$

As a zero vector can have only zero components, it follows that we have the following three scalar equations for static equilibrium of a particle.

$$\boxed{\Sigma \, F_X = 0 \qquad \Sigma \, F_Y = 0 \qquad \Sigma \, F_Z = 0} \tag{15}$$

In words, static equilibrium of a particle requires that the sum of all force components in each coordinate direction be zero. Equations (15) are three scalar equations that can be used to determine the *XYZ* components of a force necessary to maintain the static equilibrium of a particle.

In the special case of a planar system of forces, by definition there are no forces perpendicular to the plane. Equations (15) then reduce to the two scalar equations obtained by summing force components in the plane.

2 Free Body Diagrams

In general engineering practice, before writing the equations governing the behavior of a system, one first models the essential features of the system. This model identifies the important characteristics of the system, and equally important, serves to remove extraneous information from the problem. In mechanics, we create such models by drawing *free body diagrams*. For particle systems these diagrams isolate each particle to which forces are applied and show *every* force exerted on the particle. With the free body diagram we also sketch an *XYZ* coordinate reference system in order to minimize the chance for error in describing the force components, and also in order to simplify the task of checking our work.

In any problem it will be apparent that certain forces must be included in the free body diagram. Typical forces in this category are those exerted by cables, springs, and gravity (the weight force), and those that are given in the statement of the problem.

Forces that are less obvious, but equally important, are those that arise from the contact between the particle and its fixed surroundings. These contact forces are called *constraint forces*. A constraint restricts the possible movement of the particle. Although the topic of constraint forces will be considered in detail in the next module, where we study the equilibrium of rigid bodies, let us here look at a simple example of a physical system with a constraint force.

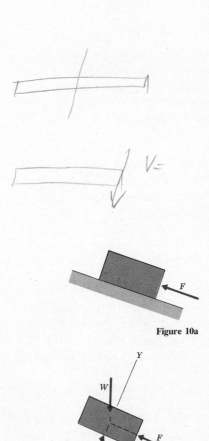

Figure 10a

Figure 10b

Consider a box resting on a smooth (frictionless) inclined plane, as shown in Figure 10a. Clearly, the box cannot penetrate the solid surface of the incline. This is a constraint condition. The constraint condition is obtained by the constraint force, $\bar{N}$, normal to the plane, which is simply the interaction force between the box and the incline. In essence, in the free body diagram of Figure 10b we *remove* the plane and *account* for its presence by the interaction force $\bar{N}$. In addition, to complete the free body diagram, we show the given force $\bar{F}$ and the weight force $\bar{W}$. Only X and Y axes are shown in the sketch, because we have a planar system of forces. The orientation of the axes may be chosen arbitrarily. In this case they were picked to facilitate the description of the components of the forces.

In the foregoing study the box was modeled as a particle. This means that we assumed that a system of concurrent forces exists. The shortcoming of such a model is that it cannot be used to study the effect of the point of application of $\bar{F}$ to the box. However, the two planar equilibrium equations that can be written for the box ($\Sigma F_X = 0$, $\Sigma F_Y = 0$) will yield the values of F and N (in terms of W) required to have a state of static equilibrium.

It may be confusing to you that at times we can take the liberty of regarding physical systems occupying a region in space as particles. Throughout the text we will frequently make assumptions when modeling a system. These assumptions will not be loosely decided upon. Rather, they will be engineering decisions that result from our study of the various models that can be employed and the physical phenomena that these models can exhibit. Ultimately however, the verification of an analytical study, and thus of all the assumptions it contains, comes from agreement with experimental results.

3 Problem Solving

At this juncture we can deduce a series of formal steps that will enable us to attack logically many problems relating to the equilibrium of bodies acted on by a system of concurrent forces. They are *not* items to be followed by rote; memorization does not lead to understanding. Rather, these steps should be used as a guideline to assist you in developing your own analytical abilities.

1   Draw a complete free body diagram. It should exhibit all constraint forces, all gravity forces, and all external forces acting on the system. When isolating one body from a larger system, be sure to account for the forces exerted by all other parts of the system that are in contact with the body of interest.
2   Choose an *XYZ* reference frame that best fits the way in which the dimensions for the system are given. Show the axes in the free body diagram.
3   Evaluate the geometrical parameters necessary to describe all forces by their components. In planar problems quantities, such as angles, that are not obvious from the given information should be shown in the free body diagram. For three-dimensional problems the forces should be written out in component form. Here, it may be necessary to use the concept of a general unit vector.
4   Write the equations of static equilibrium corresponding to the free body diagram drawn in step 1.
5   Count the number of scalar equations that result from step 4 and the number of unknowns contained in these equations. If there are more unknowns than equations, look for another free body diagram in the original system that will give information about one or more of the unknowns. Now repeat steps 1, 3, and 4. Eventually, a solvable set of equations will be obtained.
6   Solve the equations.

ILLUSTRATIVE PROBLEM 1

The 3000-lb automobile rests on an ice-covered hill. A cable-pulley system is connected between the tow truck and the automobile. Because the friction within the pulleys is negligible, the tension within the cable is unmodified as the cable passes over the pulleys. Also, the pulleys have negligible mass. Considering the hill to be smooth and frictionless, determine the force that the tow truck must apply to the cable to hold the automobile in position.

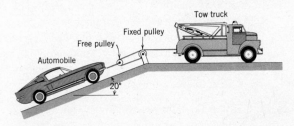

Solution

Step 1   The fact that the cable tension is constant along the length of a cable that passes over a frictionless pulley will be proven in the next module. The terminology used for referring to a pulley that is modeled as massless and frictionless is to say that it is *ideal*. Clearly, an ideal pulley is an approximation.

The automobile seems to be the focal point of the problem. Hence let us begin with a free body diagram of it. Note that the free body diagram must account for the constraint force $\bar{N}$ caused by the interaction with the hill, the weight of the car, and the force $\bar{F}$ representing the influence of the cable we cut to isolate the automobile.

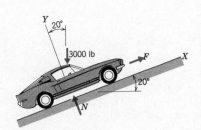

As an aside, note that if the automobile were to be modeled as a rigid body, as opposed to a particle, we would be concerned with the points of application of these forces. For example, there would be a contact force between each tire and the hill, rather than the single force $\bar{N}$, the weight force would be applied at the mass center of the vehicle, and $\bar{F}$ would have a distinct point of application.

Steps 2 and 3   The orientation of the XY axes was chosen to facilitate the writing of the equations of static equilibrium, and the necessary angles are shown in the free body diagram.

Step 4   Summing force components with respect to the axes shown in the free body diagram, the equilibrium equations are

$$\Sigma F_X = F - 3000 \sin 20° = 0$$

$$\Sigma F_Y = N - 3000 \cos 20° = 0$$

Note that the force components that act opposite to the direction of the axes were summed as negative values. Also, because this is a planar problem, force components could be summed without following the intermediate step of writing the weight force in terms of its components before summing forces.

Steps 5 and 6   The two foregoing equilibrium equations have two unknowns, F

and N, and therefore may be solved. In this problem we are interested in the cable force transmitted to the tow truck, but not the normal force $\bar{N}$. Hence, solving the first equation we obtain

$$F = 3000 \sin 20° = 1026 \text{ lb}$$

Is this the cable tension we seek? The answer is no, for $\bar{F}$ is merely the force exerted between the automobile and the free pulley. To determine the desired cable tension we note that the tension force of the cable and the force $\bar{F}$ are both applied to the free pulley, thus suggesting that we examine the equilibrium of that body. We will now repeat the procedures for studying equilibrium, using them now to study this pulley.

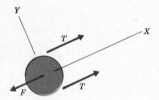

Step 1a The easiest way in which we may isolate the pulley is to imagine "cutting" the cables. We then account for the effect of the cable by applying tension forces $\bar{T}$ to the cut ends of the cables, as shown. We also must remember to include the tension force $\bar{F}$ resulting from the link between the pulley and the automobile. The weight of the pulley is negligible. The given information tells us that the tension forces in the cables on either side of the pulley are both $\bar{T}$.

Steps 2a and 3a Not applicable, for these steps were performed earlier.

Steps 4a-6a The only nontrivial equilibrium equation for the pulley comes from summing forces in the X direction. Thus

$$\Sigma F_X = 2T - F \equiv 2T - 1026 = 0$$

$$T = 513 \text{ lb}$$

This is the force in the cable that is attached to the tow truck. Hence we have the required solution.

Note that the magnitude of the force $|\bar{F}|$ exerted on the automobile is twice that exerted on the tow truck. In technical language we sometimes refer to the ratio of the output force of a system to the input force as a *mechanical advantage*. The mechanical advantage of the pulley system in this problem is 2.

ILLUSTRATIVE PROBLEM 2
A 500-kg crate is suspended by cables that are joined at point A. Determine the tension in each of these cables.

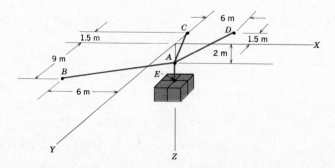

Solution

Step 1 There are two free body diagrams to be drawn here. One isolates the crate and the other isolates the focal point A at which the cables meet. The order in which they are drawn is arbitrary. Let us draw the diagram for point A first, because it will contain all the forces we seek.

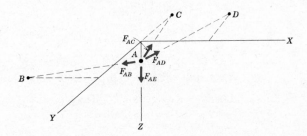

Step 2 The given XYZ coordinate system is suitable for computations.

Step 3 Because the cables are oriented in three dimensions, it is best to express the forces in component form. The forces are oriented along lines from point A to points B, C, D, and E. Hence, from equation (7) we have

$$\bar{F}_{AB} = F_{AB}\,\bar{e}_{B/A} = F_{AB}\,\frac{\bar{r}_{B/A}}{|\bar{r}_{B/A}|} = F_{AB}\,\frac{-6\bar{I} + 9\bar{J} - 2\bar{K}}{\sqrt{6^2 + 9^2 + 2^2}}$$

$$= F_{AB}(\tfrac{-6}{11}\bar{I} + \tfrac{9}{11}\bar{J} - \tfrac{2}{11}\bar{K})$$

$$\bar{F}_{AC} = F_{AC}\,\bar{e}_{C/A} = F_{AC}\,\frac{\bar{r}_{C/A}}{|\bar{r}_{C/A}|} = F_{AC}\,\frac{-1.5\bar{J} - 2\bar{K}}{\sqrt{(1.5)^2 + 2^2}}$$

$$= F_{AC}\left(-\tfrac{3}{5}\bar{J} - \tfrac{4}{5}\bar{K}\right)$$

$$\bar{F}_{AD} = F_{AC}\,\frac{6\bar{I} - 1.5\bar{J} - 2\bar{K}}{\sqrt{6^2 + (1.5)^2 + 2^2}}$$

$$= F_{AD}\left(\tfrac{12}{13}\bar{I} - \tfrac{3}{13}\bar{J} - \tfrac{4}{13}\bar{K}\right)$$

$$\bar{F}_{AE} = F_{AE}\,\bar{K}$$

Step 4 For static equilibrium the sum of the force components at point A equals zero, therefore

$$\Sigma\,F_X = -\tfrac{6}{11}\,F_{AB} + \tfrac{12}{13}F_{AD} = 0$$

$$\Sigma\,F_Y = \tfrac{9}{11}F_{AB} - \tfrac{3}{5}F_{AC} - \tfrac{3}{13}F_{AD} = 0$$

$$\Sigma\,F_Z = -\tfrac{2}{11}F_{AB} - \tfrac{4}{5}F_{AC} - \tfrac{4}{13}F_{AD} + F_{AE} = 0$$

Step 5 There are *three* equations for *four* unknowns. As mentioned earlier, the additional information comes from considering the equilibrium of the crate.

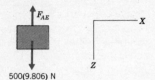

500(9.806) N

Step 1a   The crate is acted upon only by cable AE and gravity, hence the free body diagram is as shown. Note that the tensile force exerted by cable AE on the crate is directed upward.

Steps 2a and 3a   Not applicable.

Step 4a   The only nontrivial equilibrium equation comes from summing forces in the Z direction. Thus

$$\Sigma F_Z = 4903 - F_{AE} = 0$$

Step 5a   Referring to the equilibrium equations obtained in step 5, we now have the fourth equation required to obtain a solution.

Step 6   The solution of the equation just obtained is obviously

$$F_{AE} = 4903 \text{ N}$$

Placing this result into the previous set of equilibrium equations reduces the system to three simultaneous equations in the three unknown cable tensions. From the first two equations we can find F_{AD} and F_{AC} in terms of F_{AB}.

$$F_{AD} = \tfrac{13}{12}(\tfrac{6}{11})F_{AB} = 0.5909F_{AB}$$

$$F_{AC} = \tfrac{5}{3}(\tfrac{9}{11}F_{AB} - \tfrac{3}{13}F_{AD})$$

$$= 1.6667[0.8182F_{AB} - 0.2308(0.5909)F_{AB}]$$

$$= 1.1364F_{AB}$$

The last equation for ΣF_Z then gives

$$\tfrac{2}{11}F_{AB} + \tfrac{4}{5}(1.1364)F_{AB} + \tfrac{4}{13}(0.5909)F_{AB} = 4903$$

$$F_{AB} = 3852 \text{ N}$$

The other forces are then found to be

$$F_{AD} = 0.5909(3852) = 2276 \text{ N}$$

$$F_{AC} = 1.1364(3852) = 4377 \text{ N}$$

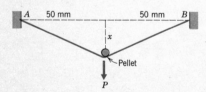

A   50 mm   50 mm   B

x

Pellet

P

ILLUSTRATIVE PROBLEM 3

The system shown in the sketch represents a sling shot formed by tying a rubber band between points A and B. When the rubber band is deformed from its rest position (dashed line) it develops a tensile force that is proportional to its elongation; that is, $F = k\Delta$, where Δ is the elongation of the rubber band and k is the spring constant. Assuming that the rubber band has no tension when in the unde-formed position, determine the force P required to hold the pellet in static equilib-rium as a function of the distance x.

Solution

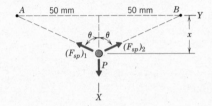

Step 1 In view of the fact that the spring forces and applied force $\bar{P}$ meet at the pellet, we draw a free body diagram of this body. (The mass of the pellet is not given, so the weight force is ignored.)

Steps 2 and 3 The X and Y axes shown in the free body diagram were selected to fit the manner in which the dimensions are given. The angle θ for the spring forces depends on the value of x, in meters, according to

$$\theta = \sin^{-1}\left(\frac{0.05}{\sqrt{(0.05)^2 + x^2}}\right) = \cos^{-1}\left(\frac{x}{\sqrt{(0.05)^2 + x^2}}\right)$$

Step 4 For equilibrium, we have

$$\Sigma F_X = P - [(F_{sp})_1 + (F_{sp})_2]\cos\theta = 0$$

$$\Sigma F_Y = [(F_{sp})_2 - (F_{sp})_1]\sin\theta = 0$$

Step 5 Clearly the second of the foregoing equilibrium equations yields $(F_{sp})_2 = (F_{sp})_1$, as one would expect from symmetry considerations. To determine the value of P in terms of x we now eliminate $(F_{sp})_1$ from the first equation. This is accomplished by using the given information that $(F_{sp})_1 = k\Delta$, where Δ is the increase in the length of the rubber band from the unstretched position. The unstretched length of the rubber band is given as 0.10 m, and the stretched length of the band is $2((0.05)^2 + x^2)^{1/2}$. Hence

$$(F_{sp})_1 = k\Delta = k(2\sqrt{(0.05)^2 + x^2} - 0.10)$$

Step 6 Using the foregoing expression for $(F_{sp})_1$ and the equation for the angle θ obtained earlier, we have

$$P = 2(F_{sp})_1\cos\theta = 2k(2\sqrt{(0.05)^2 + x^2} - 0.10)\frac{x}{\sqrt{(0.05)^2 + x^2}}$$

$$= 4kx\left(1 - \frac{0.05}{\sqrt{0.0025 + x^2}}\right)$$

Particular attention is drawn to the fact that the force P is not proportional to the distance x, even though the springs are linear. Can you explain why this is so?

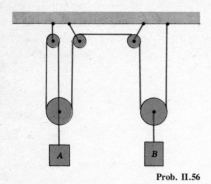

HOMEWORK PROBLEMS

II.56 Block A has a mass of 15 kg. Determine the mass of block B for static equilibrium. The pulleys are ideal.

II.57 If $W_1 = 10$ lb, determine the values of W_2 and W_3 for static equilibrium. All pulleys have ideal properties.

Prob. II.56

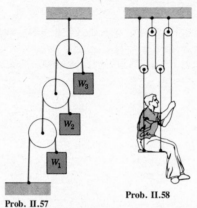

Prob. II.57

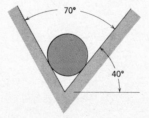

Prob. II.58

II.58 A 70-kg sign painter sitting on a 10-kg scaffold holds the free end of the cable. The pulleys have ideal properties. Determine the tension in the cable and the reaction force exerted between the painter and the scaffold.

II.59 If $\alpha = 75°$, determine the magnitude of the force $\bar{F}$ and the angle β corresponding to static equilibrium. The pulleys are ideal.

II.60 Determine the smallest angle α for which the system will be in static equilibrium. Also find the corresponding values of the magnitude of the force F and the angle β. The pulleys are ideal.

II.61 A 50-g ball bearing rests in a smooth groove, as shown in the vertical cross section. Determine the forces acting on the bearing.

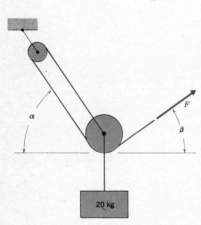

Probs. II.59 and II.60

Prob. II.61

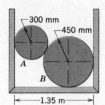

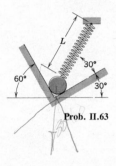

Prob. II.62

Prob. II.63

II.62 The 50-kg cylinder A and the 100-kg cylinder B are held inside the crate shown in the vertical cross-sectional view. All surfaces of contact are smooth. Determine the contact forces exerted between the cylinders, and between each cylinder and the sides of the crate.

II.63 A 150-g ball bearing is attached to a spring and then brought into contact with the smooth groove whose vertical cross section is shown. The stiffness of the spring is 20 N/m and its unstretched length is 100 mm. Knowing that $L = 120$ mm, determine the reaction forces exerted by the walls of the groove on the ball bearing.

II.64 In Problem II.63, determine the maximum value of L for which the ball bearing will maintain contact with both walls of the groove.

II.65 Each end of a small-radius cylinder of mass m is connected to a spring to hold the cylinder in position against the smooth horizontal semicylinder. (Only one spring is shown in the side view.) Both springs have a stiffness k and their unstretched length is R. Relate the angle θ for static equilibrium to the other parameters.

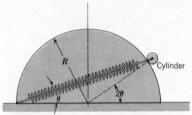

Prob. II.65

II.66-II.68 A collar on a smooth vertical rod is held in position by two springs of identical unstretched length L, but different stiffnesses k_1 and k_2. (a) Derive an expression for the distance Δ representing the downward displacement of the collar from the position where the springs are unstretched. (b) Determine the stiffness of a single equivalent spring (k_{eq}). The criterion for a spring to be equivalent is that it give the same value of Δ as obtained in part (a).

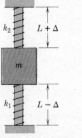

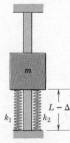

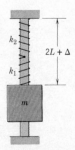

Prob. II.66   Prob. II.67   Prob. II.68

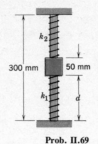

Prob. II.69

II.69   The friction between the 80-g collar and its vertical guide is negligible. The lower spring has a stiffness $k_1 = 4$ N/m and an unstretched length $L_1 = 200$ mm, whereas the corresponding properties for the upper spring are $k_2 = 3$ N/m, $L_2 = 150$ mm. Determine the distance d defining the position of static equilibrium.

II.70   Block A, whose weight is 90 lb, is supported in the position shown. Determine the weight of block B.

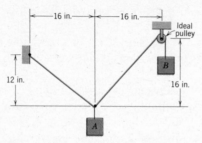

Prob. II.70

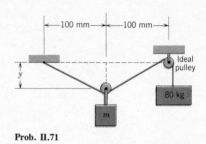

Prob. II.71

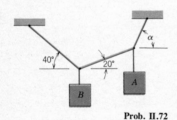

Prob. II.72

II.71   The block of unknown mass m is suspended by means of a cable attached to an 80-kg counterweight, as shown. Determine the relationship between the value of m and the distance y.

II.72   The masses of blocks A and B are 20 and 40 kg, respectively. Determine the value of α and the tension forces in the cables for static equilibrium.

II.73   Determine the tensile force in chain D. All chains are of negligible weight.

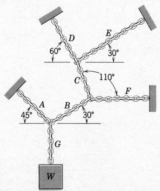

Prob. II.73

II.74 The spring has an unstretched length of 14 in. and a stiffness of 80 lb/ft. Determine the height x corresponding to static equilibrium in the position shown.

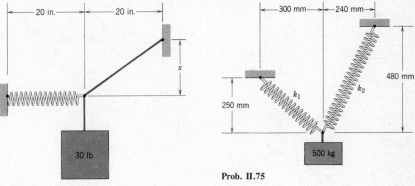

Prob. II.74

Prob. II.75

II.75 The stiffnesses of the springs are $k_1 = 20$ kN/m and $k_2 = 10$ kN/m. If the 500-kg block is suspended in static equilibrium in the position shown, determine the unstretched lengths of the springs.

II.76 A 5-lb block is held in the position shown on the smooth semicylinder by a cable that is attached to a counterweight. Determine the weight W of the counterweight.

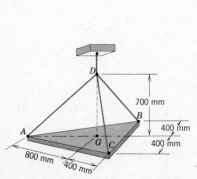

Prob. II.76

II.77 A 3000-lb automobile is suspended on a platform by four cables of equal length. The platform weighs 200 lb. Determine the tension in each of the cables.

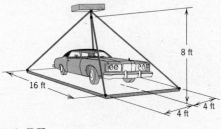

Prob. II.77

II.78 A 200-kg triangular plate is supported by three cables as shown. Determine the tension in each cable. (*Hint:* The weight of the plate may be considered to act at the centroid G of the plate.)

Prob. II.78

II.79 A 50-kg traffic light is suspended 6 m above the ground by cables from three 8-m-tall poles located as shown. Determine the tension in each cable.

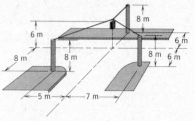

Prob. II.79

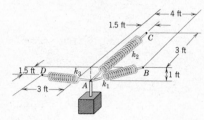

Prob. II.80

II.80 A 10-lb block is supported by three springs as shown. The stiffnesses of the springs are $k_1 = 5$ lb/in., $k_2 = 3$ lb/in., $k_3 = 4$ lb/in. Determine the unstretched length of each spring.

II.81 The 20-kg cylinder is supported by cables from three 15-kg counterweights that are equally spaced at 600 mm. Knowing that the pulleys have ideal properties, determine the dimension s representing the sag.

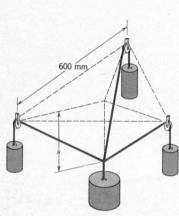

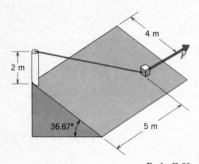

Prob. II.81

II.82 A 100-kg crate is supported on the smooth inclined plane by the cable and the horizontal force $\bar{F}$ tangent to the incline, as shown. Determine the magnitude of $\bar{F}$ and the tension in the cable.

II.83 The 3000-lb automobile on the ice-covered 20° hill is being held by cables from two tow trucks, as shown. Considering the hill to be smooth and frictionless, determine the tensile force in each cable.

Prob. II.82

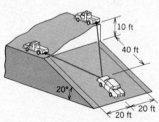

Prob. II.83

II.84 A 1-kg ball bearing is pushed into the smooth right-angle corner by the 20-N force shown, which passes through the center of the sphere.

Determine the reaction forces exerted between the ball bearing and the vertical walls and horizontal floor.

II.85 The diagram shows the side view and cross section of a V-block that is holding a 200-g steel sphere. The reaction between the sphere and the fixed vertical wall is 1000 N. Determine the reaction forces exerted between the sphere and the walls of the groove, and between the sphere and the inclined movable block.

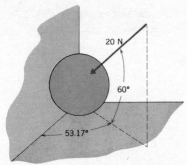

Prob. II.84

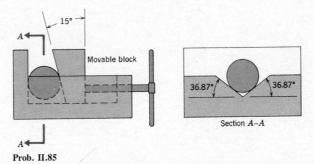

Prob. II.85

II.86 A grooved block, whose cross section is as shown, is tilted up at a 20° angle against a vertical wall, and a 1.5-kg sphere is placed in the corner. Determine the reaction forces exerted between the walls of the groove and the ball bearing. *Hint:* Use the *XYZ* reference frame shown whose *Z* axis is tangent to the groove.

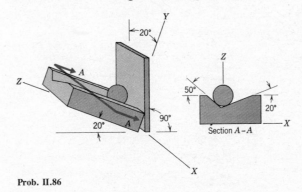

Prob. II.86

MODULE III
GENERAL FORCE
SYSTEMS AND
EQUILIBRIUM
OF RIGID BODIES

A rigid body, as compared to a particle, occupies more than a single point in space. As a result, the body will generally be subjected to a system of nonconcurrent loads having a tendency to cause the body to rotate. To describe this rotational tendency we will develop the concept of the *moment of a force*. We shall see that the moment is dependent on the line of action of a force, as well as its magnitude and direction.

The methods for determining moments, in conjunction with the methods for describing forces that were developed in the previous module, are the tools that will enable us to establish and study the equations of static equilibrium for rigid bodies. The computation of the moment of a force is addressed in the first set of topics in this module. The general topic of static equilibrium is presented in the last portion of this module. Module III, therefore, is the focal point of the study of statics.

Simply stated, a rigid body is a collection of particles that are always at a fixed distance from each other. Consider the two-particle systems shown in Figure 1.

Figure 1a shows two particles connected by an elastic spring. If the spring is extended or compressed, the particles will no longer be at their original distance from each other; hence the system is not a rigid body. In Figure 1b, if we neglect the secondary (small) effects of the deformation of the steel bar, the particles are always at a fixed distance apart and therefore the system is termed a rigid body.

In general, the concept of a rigid body is a mathematical model, in that no real material can remain undeformed under the influence of forces. What we are saying is that for the class of problems we will study, the microscopic deformation effects will not influence the results.

Incidentally, if we did not neglect the deformation of the steel bar we would be concerned with the elastic properties of flexible bodies, which is an important topic (an integral part of the design process) that you will encounter in the study of mechanics of materials.

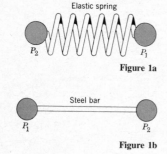

Elastic spring

P_2　　　P_1

Figure 1a

Steel bar

P_1　　　P_2

Figure 1b

A. MOMENTS OF A FORCE ABOUT A LINE AND ABOUT A POINT

Let us begin this study by considering the system shown in Figure 2, where a force $\bar{F}$ is applied at point D on the perimeter of a disk mounted on a shaft. The Y axis of the coordinate system shown is chosen parallel to the shaft AB. Because $\bar{F}$ is parallel to the XY plane it may be replaced by the two vector components $\bar{F}_X$ and $\bar{F}_Y$ shown. The distance d is the perpendicular distance from the shaft to the line of action of the vector $\bar{F}_X$. In the terminology of statics, d is the *lever arm* of the force $\bar{F}_X$ about the axis AB.

For the situation of Figure 2, the component $\bar{F}_X$ will cause the disk to rotate about the axis of the shaft in the sense of the curved arrow. The

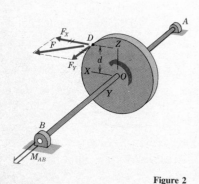

Figure 2

other component $\bar{F}_Y$ causes no rotation about axis AB. Rather, it merely pulls on the disk parallel to that axis.

To quantify this rotational effect we define the concept of the *moment of a force about an axis*. Specifically, we define the moment $\bar{M}_{AB}$ of the force $\bar{F}$ about line AB such that its magnitude is the product of the component of the force $\bar{F}$ perpendicular to the line and the lever arm of the force (perpendicular distance) to the line. Thus

$$|\bar{M}_{AB}| = F_X d \qquad (1)$$

Dimensionally, moment carries the physical units of force times distance. Throughout this text we shall consistently give the force unit of the moment first and then the length unit. This is done in order to avoid confusion with the units of energy, which also involve the product of force and length.

In addition to having a magnitude, the moment is a vector quantity having direction. This direction is associated with the sense of the rotational effect. We could use the curved line in Figure 2 to describe this direction by saying that the moment of $\bar{F}$ about line AB is counterclockwise as viewed from point B toward point A. Such descriptions prove to be inadequate in many situations.

The accepted way to describe the moment about a line is to employ the *right-hand rule,* as depicted in Figure 3. The fingers of the right hand are curled in the sense of the rotational effect of the moment about the axis. The direction of the extended thumb is then defined as the direction of the moment. Accordingly, we now can describe the $\bar{M}_{AB}$ of Figure 2 as $|\bar{M}_{AB}|\bar{J}$, so

$$\bar{M}_{AB} = F_X d \, \bar{J}$$

In an effort to avoid confusion between force and moment vectors, we shall use a double line throughout the text to depict moment vectors, as is shown in Figures 2 and 3.

The concept of the moment of a force about a line will prove to be valuable in two-dimensional situations. For example, by confining our attention to the moment $\bar{M}_{AB}$ in Figure 2, in essence we were looking at the effect of $\bar{F}$ on equilibrium in the XZ plane; the force component $\bar{F}_Y$ did not have any effect in this plane. To consider the effect of $\bar{F}$ on the entire system, we form the *moment of the force about a point.*

Consider an arbitrary point, C, on the shaft AB, as shown in Figure 4. To indicate that the desired moment is about point C, we write this quantity as $\bar{M}_C$, using a single subscript. This contrasts with a moment with double subscripts, which denotes that the moment is about an axis.

The distance r is the perpendicular distance from the line of action of the force to point C. This distance is called the *lever arm of the force*

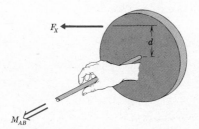

Figure 3

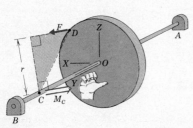

Figure 4

about point C. As is the case of the moment about an axis, we define the magnitude of $\bar{M}_C$ to be the product of this lever arm and the magnitude of $\bar{F}$; that is,

$$|\bar{M}_C| = |\bar{F}|r \tag{2}$$

The right-hand rule is again used to define the direction of the moment. To do this we form an analogy to the way in which we regarded the plane of the disk when defining the moment of the force about the axis of the shaft. The fingers of the right hand are curled in the sense of the rotation that would be imparted to the shaded plane formed by point C and the line of action of $\bar{F}$. The direction of $\bar{M}_C$ is then perpendicular to the shaded plane, in the sense of the extended thumb of the right hand.

The similarities in the definitions of the moment of a force about a line and the moment of a force about a point suggest that the two quantities are directly related, and they are. Later we will study this relationship. Before doing so, we must increase our ability to describe the moment of a force. For example, in Figure 4 the determination of the lever arm r and the direction of $\bar{M}_C$ can be quite complicated. To avoid cumbersome computational procedures, we will now study the remaining vectorial tool we need, the cross (vector) product.

B. THE CROSS (VECTOR) PRODUCT

The dot product of two vectors, defined in the previous module, gives a scalar value involving the component of one vector parallel to the other vector. Here, we will see that the *cross product* gives a vector result in terms of the component of one vector perpendicular to the other vector. Because of the difference in the type of result of these two multiplication operations, the dot product is sometimes called the *scalar product*, and the cross product is sometimes called the *vector product*.

Consider the two vectors $\bar{A}$ and $\bar{B}$ shown in Figure 5. Without loss of generality, let the plane of the vectors coincide with the plane of the paper. Denoting the angle between the two vectors as θ, the component of $\bar{A}$ perpendicular to $\bar{B}$ is $|\bar{A}| \sin \theta$, whereas the component of $\bar{B}$ perpendicular to $\bar{A}$ is $|\bar{B}| \sin \theta$, as shown. The magnitude of the cross product $\bar{A} \times \bar{B}$ is defined to be the component of $\bar{B}$ perpendicular to $\bar{A}$ multiplied by the magnitude of $\bar{A}$:

$$|\bar{A} \times \bar{B}| = |\bar{A}||\bar{B}| \sin \theta \tag{3}$$

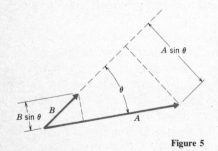

Figure 5

Notice that $|\bar{A} \times \bar{B}| \equiv |\bar{B} \times \bar{A}|$ and that $\bar{A} \times \bar{B}$ is zero whenever $\bar{A}$ and $\bar{B}$ are parallel.

The direction of the cross product is defined to be normal to the plane formed by the two vectors. To define the sense of this normal, we again employ the right-hand rule, curling the fingers of the right hand about the common tail of the two vectors, from the first vector in the product to the second. The extended thumb is then oriented in the sense of the product.

According to this definition for the direction, $\bar{A} \times \bar{B}$ in Figure 5 is outward from the plane of the diagram, whereas $\bar{B} \times \bar{A}$ is inward. Because the order of multiplication does not affect the magnitude of the product given by equation (3), we can conclude that the two products have equal magnitude and opposite direction, so

$$\bar{B} \times \bar{A} = -\bar{A} \times \bar{B} \tag{4}$$

Primarily, we will employ the cross product operation in conjunction with a description of the vectors in terms of their components with respect to a coordinate system XYZ. To do this we must first consider the algebraic rules for this operation. In view of the fact that the order in which the vectors appear in the cross product affects the result, as shown by equation (4), we may wonder whether the other algebraic rules are different in form from those for the dot product. The answer is no. Specifically, we can prove that

$$p(\bar{A} \times \bar{B}) = (p\bar{A}) \times \bar{B} = \bar{A} \times (p\bar{B})$$
$$\bar{A} \times (\bar{B} + \bar{C}) = (\bar{A} \times \bar{B}) + (\bar{A} \times \bar{C}) \tag{5}$$
$$(\bar{A} + \bar{B}) \times \bar{C} = (\bar{A} \times \bar{C}) + (\bar{B} \times \bar{C})$$

We begin the discussion of vector components by first considering the cross products of the unit vectors. In doing this it is essential that we use a right-handed coordinate system in order to be consistent with the use of the right-hand rule in defining the vector product. Figure 6 shows a typical coordinate system.

As each of the unit vectors is obviously parallel to itself, it follows that

$$\bar{I} \times \bar{I} = \bar{J} \times \bar{J} = \bar{K} \times \bar{K} = \bar{0} \tag{6}$$

In contrast, the angle between any two different unit vectors is 90°, so

$$|\bar{I} \times \bar{J}| = |\bar{I}| \, |\bar{J}| \, \sin 90° = 1$$
$$= |\bar{J} \times \bar{K}| = |\bar{K} \times \bar{I}|$$

Also, the third unit vector is perpendicular to the coordinate plane of the other two unit vectors. The right-hand rule gives the appropriate sign. For example, when we curl our fingers from $\bar{J}$ to $\bar{I}$ in order to evaluate $\bar{J} \times \bar{I}$,

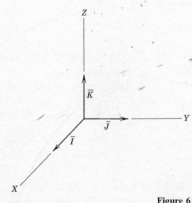

Figure 6

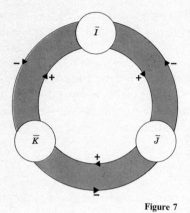

Figure 7

our thumb is oriented opposite the direction of positive $\bar{K}$. The full set of relations are

$$\bar{I} \times \bar{J} = -\bar{J} \times \bar{I} = \bar{K}$$

$$\bar{J} \times \bar{K} = -\bar{K} \times \bar{J} = \bar{I} \tag{7}$$

$$\bar{K} \times \bar{I} = -\bar{I} \times \bar{K} = \bar{J}$$

We will employ equations (7) frequently. As an aid in remembering them, we can consider the cross product of two unit vectors to be positive for positive alphabetical order and negative for negative alphabetical order, as depicted in Figure 7.

Finally, we find the cross product of two general vectors by expressing them in terms of their components and then employing equations (5) through (7). Thus

$$
\begin{aligned}
\bar{A} \times \bar{B} &= (A_X \bar{I} + A_Y \bar{J} + A_Z \bar{K}) \times \bar{B} \\
&= A_X [\bar{I} \times (B_X \bar{I} + B_Y \bar{J} + B_Z \bar{K})] \\
&\quad + A_Y [\bar{J} \times (B_X \bar{I} + B_Y \bar{J} + B_Z \bar{K})] \\
&\quad + A_Z [\bar{K} \times (B_X \bar{I} + B_Y \bar{J} + B_Z \bar{K})] \\
&= A_X (B_Y \bar{K} - B_Z \bar{J}) + A_Y (-B_X \bar{K} + B_Z \bar{I}) \\
&\quad + A_Z(B_X \bar{J} - B_Y \bar{I}) \\
\bar{A} \times \bar{B} &= (A_Y B_Z - A_Z B_Y)\bar{I} + (A_Z B_X - A_X B_Z)\bar{J} \\
&\quad + (A_X B_Y - A_Y B_X)\bar{K} \tag{8}
\end{aligned}
$$

It is noteworthy that in most problems one or both of the vectors will have some zero components, thereby simplifying the above steps. Alternatively, the computation of a cross product may be pursued by expanding a determinant, for example

$$\bar{A} \times \bar{B} = \begin{vmatrix} \bar{I} & \bar{J} & \bar{K} \\ A_X & A_Y & A_Z \\ B_X & B_Y & B_Z \end{vmatrix} \tag{9}$$

If you are familiar with determinants, you may use equation (9) to perform your calculations. However, the method of equation (8) will be used throughout this book.

Before we apply the vector product to the evaluation of moments, let us consider the following example, which demonstrates another use of this operation.

EXAMPLE 1
Three points A, B, and C are located in the sketch on the following page.

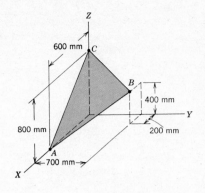

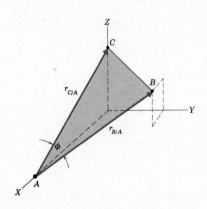

Determine the angle ϕ between the lines AB and AC, the area of triangle ABC, and a unit vector normal to the plane of this triangle.

Solution

Considering the two vectors $\bar{A}$ and $\bar{B}$ in Figure 5, we can see that the component of one of these vectors perpendicular to the other is the altitude of the triangle formed by $\bar{A}$, $\bar{B}$, and the line connecting their heads. Thus, for the vectors in Figure 5 the area is

$$\text{Area} = \tfrac{1}{2}|\bar{A} \times \bar{B}|$$

For the problem at hand, we can obtain the angle ϕ from the same cross product as that used for the area if we employ the position vectors $\bar{r}_{B/A}$ and $\bar{r}_{C/A}$ as shown in the sketch. We then form

$$\text{Area} = \tfrac{1}{2}|\bar{r}_{B/A} \times \bar{r}_{C/A}| \equiv \tfrac{1}{2}|\bar{r}_{B/A}||\bar{r}_{C/A}| \sin \phi$$

Referring to the given sketch to determine the position of the various points, we see that a displacement from point A to point B is equivalent to displacements of 700 mm in the $+Y$ direction, 400 mm in the $-X$ direction, and 400 mm in the $+Z$ direction, so that

$$\bar{r}_{B/A} = 0.70\bar{J} - 0.40\bar{I} + 0.40\bar{K} \text{ m}$$

For $\bar{r}_{C/A}$ we follow the coordinate axes from point A to point C to find

$$\bar{r}_{C/A} = -0.60\bar{I} + 0.80\bar{K} \text{ m}$$

The process of evaluating the vector product may be thought of as being similar to that for multiplying two scalar polynomials. Thus

$$
\begin{aligned}
\bar{r}_{B/A} \times \bar{r}_{C/A} &= (0.7\bar{J} - 0.4\bar{I} + 0.4\bar{K}) \times (-0.6\bar{I} + 0.8\bar{K}) \\
&= -0.7(0.6)\bar{J} \times \bar{I} + 0.7(0.8)\bar{J} \times \bar{K} \\
&\quad +0.4(0.6)\bar{I} \times \bar{I} - 0.4(0.8)\bar{I} \times \bar{K} \\
&\quad -0.4(0.6)\bar{K} \times \bar{I} + 0.4(0.8)\bar{K} \times \bar{K} \\
&= 0.42\bar{K} + 0.56\bar{I} + 0.32\bar{J} - 0.24\bar{J} \\
&= 0.56\bar{I} + 0.08\bar{J} + 0.42\bar{K}
\end{aligned}
$$

Incidentally, as you become more familiar with the cross product, you will find that you can skip the second of the above computational steps. Using equation (3) we have

$$\sin \phi = \frac{|\bar{r}_{B/A} \times \bar{r}_{C/A}|}{|\bar{r}_{B/A}||\bar{r}_{C/A}|}$$

The magnitudes of these vectors are

$$\left| \bar{r}_{B/A} \right| = \sqrt{(0.7)^2 + (0.4)^2 + (0.4)^2} = 0.90 \text{ m}$$

$$\left| \bar{r}_{C/A} \right| = \sqrt{(0.6)^2 + (0.8)^2} = 1.00 \text{ m}$$

$$\left| \bar{r}_{B/A} \times \bar{r}_{C/A} \right| = \sqrt{(0.56)^2 + (0.08)^2 + (0.42)^2} = 0.7046 \text{ m}^2$$

Thus, because $\left| \bar{r}_{B/A} \times \bar{r}_{C/A} \right|$ is twice the area of triangle ABC, we have

$$\text{Area} = \tfrac{1}{2}(0.7046) = 0.3523 \text{ m}^2 \qquad \triangle$$

Also, from the expression for $\sin \phi$, we find that

$$\sin \phi = \frac{0.7046}{0.90 \, (1.00)} \qquad \phi = 51.52° \qquad \triangle$$

Finally, we divide $\bar{r}_{B/A} \times \bar{r}_{C/A}$ by its magnitude to obtain a *unit* vector normal to the plane of $\bar{r}_{B/A}$ and $\bar{r}_{C/A}$. This gives

$$\bar{e}_n = \frac{0.56\bar{I} + 0.08\bar{J} + 0.42\bar{K}}{0.7046} = 0.7948\bar{I} + 0.1135\bar{J} + 0.5961\bar{K} \qquad \triangle$$

As all of the components of $\bar{e}_n$ are positive, we can conclude that $\bar{e}_n$ is directed outward from the origin.

HOMEWORK PROBLEMS

III.1 Compute $\bar{A} \times \bar{B}$ for the following vectors: (a) $\bar{A} = 5\bar{I} - \bar{J} - \bar{K}$, $\bar{B} = -4\bar{I} + 2\bar{J} + 4\bar{K}$, (b) $\bar{A} = 2\bar{I} + \bar{J}$, $\bar{B} = 3\bar{I} - 4\bar{K}$, (c) $\bar{A} = \bar{I} + 2\bar{J} + \bar{K}$, $\bar{B} = -3\bar{I} + 2\bar{J} - \bar{K}$.

III.2 Given that $\bar{A} = A_X\bar{I} - 5\bar{J} + 2\bar{K}$, $\bar{B} = -3\bar{I} + 2\bar{J} - B_Z\bar{K}$, determine the values of A_X and B_Z for which $A \times B$ is parallel (a) to the Y axis, (b) to the X axis.

III.3 Given that $\bar{A} = \bar{I} + \bar{J} - \bar{K}$, $\bar{B} = \bar{I} - 2\bar{J} + 3\bar{K}$, and $\bar{C} = -\bar{I} + 2\bar{J} - 4\bar{K}$, compute (a) $(\bar{A} \times \bar{B}) \times \bar{C}$, (b) $\bar{A} \times (\bar{B} \times \bar{C})$.

III.4 For the vectors given in Problem III.3, compute (a) $\bar{A} \cdot \bar{B} \times \bar{C}$, (b) $\bar{C} \cdot \bar{A} \times \bar{B}$.

III.5 The product $\bar{A} \cdot \bar{B} \times \bar{C}$ of three arbitrary vectors $\bar{A}$, $\bar{B}$, and $\bar{C}$ is called the scalar triple product. Prove the following identities: (a) $\bar{A} \cdot \bar{B} \times \bar{C} \equiv \bar{A} \times \bar{B} \cdot \bar{C}$, (b) $\bar{A} \cdot \bar{B} \times \bar{C} \equiv \bar{B} \cdot \bar{C} \times \bar{A}$.

III.6 It is desired to pass a plane through point A whose unit normal vector is $0.4286\bar{I} - 0.8571\bar{J} + 0.2857\bar{K}$. Determine the values of Y_B and Z_C, thus locating where the plane intersects the corresponding coordinate axes.

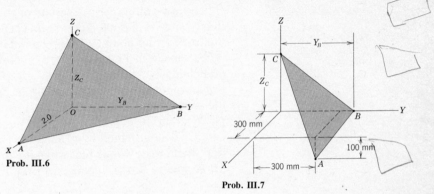

Prob. III.6

Prob. III.7

III.7 The plane formed by points A, B, and C has a unit normal vector given by $-0.4444\bar{I} + 0.1111\bar{J} + 0.8889\bar{K}$. Determine the values of Y_B and Z_C locating points B and C.

III.8 Forces $\bar{F}_1$ and $\bar{F}_2$ are both applied at point A. The force $\bar{F}_1 = 500\bar{I} - 400\bar{J} + Z\bar{K}$ N is to form a 60° angle with force $\bar{F}_2 = -200\bar{I} + 100\bar{J} + 300\bar{K}$ N. Determine (a) the value of Z, (b) the unit vector normal to the plane formed by $\bar{F}_1$ and $\bar{F}_2$.

III.9 Line AB is perpendicular to line AC. Determine the relationship between the distances X_B and Z_C, and describe the unit vector normal to the plane ABC.

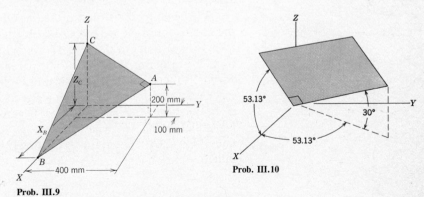

Prob. III.9

Prob. III.10

III.10 A rectangle is situated in space as shown. Determine a unit vector normal to its surface.

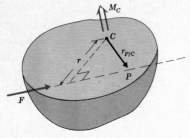

Figure 8

C. MOMENTS OF A FORCE USING CROSS PRODUCTS

Figure 8 depicts a general sitation where a force $\bar{F}$ is applied along the designated line of action. We wish to determine the moment of this force about point C, which has been chosen arbitrarily.

By definition, the magnitude of the moment $\bar{M}_C$ is $|\bar{F}|r$ in the direction of an axis that intersects point C and is normal to the plane containing the force $\bar{F}$ and point C. Letting point P be any point on the line of action of $\bar{F}$, we see in Figure 8 that the component of the position vector $\bar{r}_{P/C}$ perpendicular to $\bar{F}$ is the lever arm r. From the definition of a cross product, equation (3), it follows that the cross product of $\bar{F}$ and $\bar{r}_{P/C}$ has the magnitude of $\bar{M}_C$. Also, this cross product is perpendicular to the plane, and thus is parallel to $\bar{M}_C$. The proper order of multiplication of $\bar{F}$ and $\bar{r}_{P/C}$ is determined by placing the tail of $\bar{F}$ at point C. Then, using the right-hand rule, we see that when the thumb of the right hand is parallel to $\bar{M}_C$ in Figure 8, the fingers curl from $\bar{r}_{P/C}$ to $\bar{F}$. Thus

$$\bar{M}_C = \bar{r}_{P/C} \times \bar{F} \tag{10}$$

Equation (10) says that

> The moment of a force about a point C is obtained by crossing a position vector from point C to any point on the line of action of the force into the force vector.

With equation (10) as an aid for the determination of the moment about a point, we can now establish the relationship between this type of moment and the moment about a line. Consider Figure 9, where we wish to determine the moment of the force $\bar{F}$ about the line AB.

We begin by constructing the line OP, which is perpendicular to both the line of action of the force and the line AB. Thus, the length OP is the lever arm of $\bar{F}$ about the axis AB. We next define a convenient reference frame by orienting the Y axis to be parallel to line AB and the Z axis to be parallel to the line OP, as shown in Figure 9. Now, using the fact that a force may be applied anywhere along its line of action, we can replace $\bar{F}$ by its components $\bar{F}_X$ and $\bar{F}_Y$ ($F_Z \equiv 0$) acting at point P. The component $\bar{F}_Y$ is parallel to line AB, whereas the component $\bar{F}_X$ is perpendicular to both lines AB and OP. Then, from equation (2) for the moment about a line, we obtain

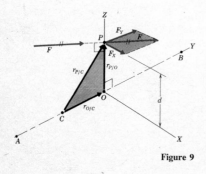

Figure 9

$$\bar{M}_{AB} = F_X \, d \, \bar{J}$$

Let us now consider the moment of the force $\bar{F}$ about any point C

loaded on line AB. From equation (10) we have

$$\bar{M}_C = \bar{r}_{P/C} \times \bar{F}$$

From Figure 9 we note that

$$\bar{r}_{P/C} = \bar{r}_{O/C} + \bar{r}_{P/O} = |\bar{r}_{O/C}|\bar{J} + (d)\bar{K}$$

so upon replacing $\bar{F}$ by its components we have

$$\bar{M}_C = [|\bar{r}_{O/C}|\bar{J} + (d)\bar{K}] \times (F_X\bar{I} + F_Y\bar{J})$$
$$= -|\bar{r}_{O/C}|F_X\bar{K} + (d)F_X\bar{J} - (d)F_Y\bar{I}$$

The second term in this expression is the component of $\bar{M}_C$ parallel to the Y axis (the line AB). This component is identical in magnitude and direction to the moment $\bar{M}_{AB}$. Hence, we can conclude that

$$\bar{M}_C \cdot \bar{J} = F_X d = |\bar{M}_{AB}|$$

In other words:

> The magnitude of the moment about a line is the component parallel to that line of the moment about any point on the line.

Finally, we express the direction of $\bar{M}_{AB}$ by again recalling equation (2), to get

$$\bar{M}_{AB} = F_X d \bar{J} = (\bar{M}_C \cdot \bar{J})\bar{J}$$

In Figure 9, the reference frame was intentionally chosen such that the Y axis was parallel to the line of interest, AB. To express the foregoing relationship in a situation where the Y axis of the coordinate system is *not* parallel to the axis AB, it is only necessary to replace $\bar{J}$ by a unit vector $\bar{e}_{B/A}$ parallel to line AB. Thus

$$\boxed{\bar{M}_{AB} = (\bar{M}_C \cdot \bar{e}_{B/A})\,\bar{e}_{B/A}} \qquad (11)$$

The significance of equations (10) and (11) is that they enable us to determine the moment of a force by expressing the force and position vectors in terms of the XYZ components. Before solving some examples, let us consider the general situation in Figure 10. By studying the resulting moment of the force $\bar{F}$ about point C, we can gain increased understanding of the significance of the vector operations.

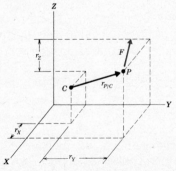

Figure 10

We compute $\bar{M}_C$ by expressing the position vector $\bar{r}_{P/C}$ and the general force $\bar{F}$ in terms of their components with respect to the XYZ coordinate system. Thus

$$\bar{r}_{P/C} = r_X \bar{I} + r_Y \bar{J} + r_Z \bar{K}$$

$$\bar{F} = F_X \bar{I} + F_Y \bar{J} + F_Z \bar{K}$$

Equation (10) then gives

$$\bar{M}_C = \bar{r}_{P/C} \times \bar{F} = (r_Y F_Z - r_Z F_Y)\bar{I} + (r_Z F_X - r_X F_Z)\bar{J} + (r_X F_Y - r_Y F_X)\bar{K} \quad (12)$$

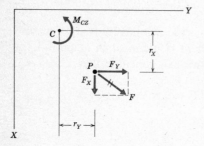

Figure 11

The meaning of the component values in equation (12) becomes apparent when we consider any one component, for example, by looking down the Z axis in Figure 10. This view is shown in Figure 11, where M_{CZ} denotes the Z component of $\bar{M}_C$. In view of the relationship between the moment about a line and the moment about a point along that line, $\bar{M}_{CZ}$ is *also the moment about the line parallel to the Z axis intersecting point C.* The moment $\bar{M}_{CZ}$ is shown as a counterclockwise curling arrow in this figure, corresponding to the way the fingers of our right hand curl according to the right-hand rule when M_{CZ} is positive.

Suppose that we consider the moment of each force component in Figure 11 and then sum the individual effects. The lever arm for $\bar{F}_Y$ is r_X. This component causes a counterclockwise rotation about point C, so its moment is positive for the axes shown. Similarly, the lever arm of $\bar{F}_X$ is r_Y. The corresponding moment is negative because $\bar{F}_X$ would cause a clockwise rotation about point C. Combining these individual effects, we find

$$M_{CZ} = r_X F_Y - r_Y F_X$$

which, of course, is the same result for M_{CZ} found in equation (12) by using the cross product.

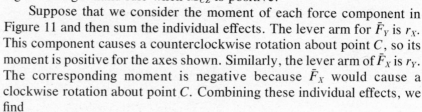

Similar arguments for the other components of $\bar{M}_C$ lead to the following conclusion. Equation (12) states that we can determine the moment about a point C by vectorially adding the moments of *each component* of a force about the lines parallel to the coordinate axes intersecting point C. This fact is known as *Varignon's theorem,* after the mathematician who formulated the result before the advent of vector algebra. This theorem provides a method for computing moments without evaluating a cross product.

When dealing with three-dimensional forces, the use of Varignon's theorem to compute the components of $\bar{M}_C$ requires that we draw three diagrams such as Figure 11, representing the views looking down each coordinate axis. This is a cumbersome procedure. For such situations we will rely on equations (10) and (11) to evaluate moments. In contrast, for problems *where the plane formed by the force and point C is parallel to a*

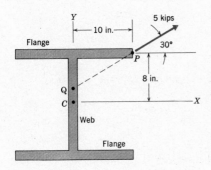

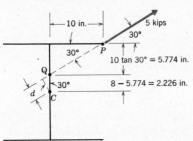

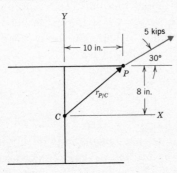

coordinate plane, the moment about C is identical to the moment about the line normal to that plane intersecting point C. For such problems, considering the force components individually offers a definite computational advantage, as is demonstrated in the next example.

EXAMPLE 2

A 5-kip force is applied to the flange of the H beam as shown. Determine the moment of this force about the center C of the cross section by

a   evaluating the lever arm of the force,
b   evaluating a vector product,
c   considering force components acting at the flange point P,
d   considering force components acting at the web point Q.

Solution

Method A
We first draw a sketch to evaluate the lever arm d. From this diagram

$$d = (2.226) \cos 30° = 1.928 \text{ in.}$$

so

$$M_C = |\bar{F}|d = 5(1.928) = 9.640 \text{ kip-in. clockwise}$$

Method B
For the cross product formulation we need a position vector from point C to a point on the line of action of $\bar{F}$. Referring to the given diagram, we see that the dimensions for point P are given in terms of horizontal and vertical distances. We therefore express $\bar{r}_{P/C}$ in terms of the coordinate system shown in the sketch. From this sketch we have

$$\bar{r}_{P/C} = 10\bar{I} + 8\bar{J} \text{ in.}$$

$$\bar{F} = 5(\cos 30° \, \bar{I} + \sin 30° \, \bar{J})$$

$$= 4.330\bar{I} + 2.5\bar{J} \text{ kips}$$

$$\bar{M}_C = \bar{r}_{P/C} \times \bar{F} = (10\bar{I} + 8\bar{J}) \times (4.330\bar{I} + 2.5\bar{J})$$

$$= [10(2.5) - 8(4.330)]\bar{K}$$

$$= -9.640\bar{K} \text{ kip-in.}$$

Because $\bar{M}_C$ is in the negative Z direction, it is clockwise as viewed in the plane.

Method C
Point P is located with respect to point C in terms of horizontal and

vertical distances, so we draw a sketch showing the horizontal and vertical components of the force.

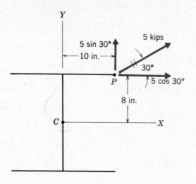

Because the Z axis is outward, counterclockwise moments are positive according to the right-hand rule. Multiplying each component by its lever arm, we then have

$$M_{CZ} = 10(5 \sin 30°) - 8(5 \cos 30°)$$

$$= -9.640 \text{ kip-in.}$$

Method D

This approach is based on the principle of transmissibility, according to which a force may be applied anywhere along its line of action. The force components are the same as in the two previous methods. The location of point Q is given in Method A. From the sketch we see that the vertical component of the force passes through point C, and that the horizontal component exerts a clockwise (negative) moment. Thus

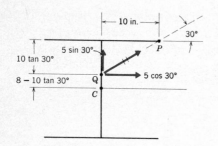

$$M_{CZ} = -(8 - 10 \tan 30°)(5 \cos 30°)$$

$$= -9.640 \text{ kip-in.}$$

Assessing the four methods, the geometrical calculations in the first were fairly simple. However, if we were interested in the moment about some other point, such as the right tip of the lower flange, the determination of the lever arm would have been more complicated. The second and third methods make maximum use of the given geometrical information, but the third method yields the result more directly, because we do not have to write explicit expressions for the vectors. The last method is somewhat more cumbersome because the force was applied at a point whose location we had to determine. In general, for planar problems, we will employ either the first or third method for the evaluation of moments, depending on the given geometrical information.

EXAMPLE 3

A 10-kN force is applied as shown to the free end, point D, of the bent rod. Compute the moment of this force about point A and about line BC.

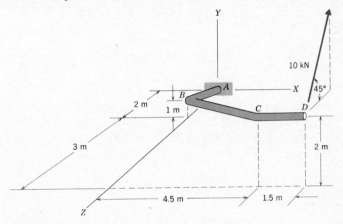

Solution

This is a three-dimensional problem. Therefore, we compute the moment about point A by constructing the position vector from this point to point D on the line of action of the force and then applying equation (10). The diagram indicates that the force is parallel to the YZ plane at an angle of $45°$ above the negative Z direction. Thus

$$\bar{r}_{D/A} = 5\bar{K} + 6\bar{I} + 2\bar{J} \text{ m}$$

$$\bar{F} = 10(-\cos 45° \ \bar{K} + \sin 45° \ \bar{J})$$

$$= 7.071(\bar{J} - \bar{K}) \text{ kN}$$

$$\bar{M}_A = \bar{r}_{D/A} \times \bar{F} = (5\bar{K} + 6\bar{I} + 2\bar{J}) \times 7.071(\bar{J} \times \bar{K})$$

$$= 7.071[5(\bar{K} \times \bar{J}) - 5(\bar{K} \times \bar{K}) + 6(\bar{I} \times \bar{J})$$

$$- 6(\bar{I} \times \bar{K}) + 2(\bar{J} \times \bar{J}) - 2(\bar{J} \times \bar{K})]$$

$$= 7.071(-5\bar{I} + 6\bar{K} + 6\bar{J} - 2\bar{I})$$

$$= -49.50\bar{I} + 42.43\bar{J} + 42.43\bar{K} \text{ kN-m}$$

For the second part of the problem, where the moment about line BC is required, we will first determine the moment about some point along this line and then find the moment about the line using equation (11). Let us use point C, because the position vector from point C to the force is easiest to construct. Thus

$$\bar{M}_C = \bar{r}_{D/C} \times \bar{F} = 1.5\bar{I} \times 7.071(\bar{J} - \bar{K})$$

$$= 10.607(\bar{K} + \bar{J}) \text{ kN-m}$$

The unit vector parallel to line BC is

$$\bar{e}_{C/B} = \frac{\bar{r}_{C/B}}{|\bar{r}_{C/B}|} = \frac{3\bar{K} + 4.5\bar{I} + \bar{J}}{\sqrt{(3)^2 + (4.5)^2 + (1)^2}}$$

$$= 0.8182\bar{I} + 0.1818\bar{J} + 0.5455\bar{K}$$

Hence, the component of $\bar{M}_C$ parallel to line BC is

$$\bar{M}_C \cdot \bar{e}_{C/B} = 10.607(\bar{J} + \bar{K}) \cdot (0.8182\bar{I} + 0.1818\bar{J} + 0.5455\bar{K})$$

$$= 10.607(0.1818 + 0.5455) = 7.714 \text{ kN-m}$$

This is the magnitude of $\bar{M}_{BC}$. To express its direction we write

$$\bar{M}_{BC} = (\bar{M}_C \cdot \bar{e}_{C/B})\bar{e}_{C/B} = 7.714(0.8182\bar{I}$$

$$+ 0.1818\bar{J} + 0.5455\bar{K})$$

$$= 6.313\bar{I} + 1.402\bar{J} + 4.208\bar{K} \text{ kN-m}$$

This vector is oriented from point B to point C. If $\bar{M}_C \cdot \bar{e}_{C/B}$ had been negative, the moment vector would then be oriented from point C to point B.

HOMEWORK PROBLEMS

III.11 A 300-lb force is applied to the rectangular plate at corner C. Determine the moment of this force about pin E (a) by computing its lever arm, (b) by replacing the force by its components acting at point C, (c) by replacing the force by its components acting at some other convenient point on its line of action.

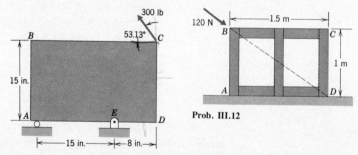

Prob. III.11

Prob. III.12

III.12 A crate resting on the ground is acted upon by the 120-N force shown. Determine the moment of this force about corner A (a) by determining the lever arm of the force, (b) by replacing the force by its components acting at point B, (c) by replacing the force by its components acting at some other convenient point on its line of action.

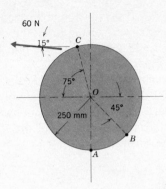

Probs. III.13, 14, and 15

III.13 The 60-N force is applied to the disk at point C. Determine the moment of the force about the center O (a) by determining the lever arm of the force, (b) by replacing the force by its X and Y components.

III.14 Using the two methods given in Problem III.13, determine the moment of the 60-N force about point A.

III.15 Using the two methods given in Problem III.13, determine the moment of the 60-N force about point B.

III.16 A 5-kN force is applied to the bent arm at point A. Determine the moment of the force about point C (a) for $\theta = 0°$, (b) for $\theta = 30°$.

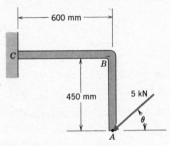

Prob. III.16

III.17 In Problem III.16, determine the angle θ such that the 5-kN force applied at point A exerts the maximum moment about point C. What is the corresponding moment?

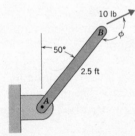

Prob. III.18

III.18 Control bar AB is inclined at 50° from the vertical. A 10-lb force is applied to the bar at end B. Determine the moment of this force about pin A (a) if the force is horizontal, (b) if $\phi = 120°$.

III.19 In Problem III.18, determine the value of the angle ϕ for which the 10-lb force exerts the maximum moment about pin A. Also determine the corresponding moment.

III.20 An 80-N force is applied to the special purpose wrench as shown. Determine the moment of this force about the center of the bolt when $\alpha = 75°$.

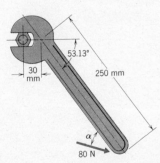

Prob. III.20

III.21 In Problem III.20, determine the angle α for which the magnitude of the moment of the 80-N force about the center of the bolt is (a) a minimum, (b) a maximum.

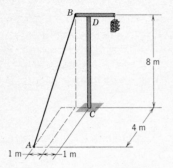

Prob. III.22

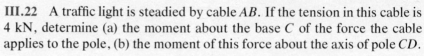

III.22 A traffic light is steadied by cable AB. If the tension in this cable is 4 kN, determine (a) the moment about the base C of the force the cable applies to the pole, (b) the moment of this force about the axis of pole CD.

III.23 A socket wrench is used to tighten a screw. A 120-N horizontal force is applied to the handle as shown. (a) Determine the moment of this force about the head of the screw. (b) What portion of this moment has the effect of tightening the screw? (c) What are the lever arms of the force about the axis of the screw and about the screw head?

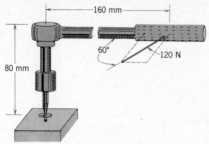

Prob. III.23

III.24 Cables AE and BE are used to support the trapezoidal plate. The tension in cable AE is 3 kips and that in cable BE is 2 kips. Determine the moment of the force that each cable exerts on the plate (a) about corner D, (b) about corner C.

III.25 In Problem III.24, determine the moment about edge BC of the force exerted by each of the cables on the plate.

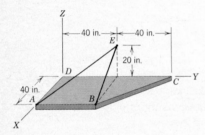

Prob. III.24

III.26 The rectangular sheet of plywood is held in position by the cable. The tension in the cable is 150 N. Determine the moment of the force exerted by the cable on the plywood (a) about corner A, (b) about edge AB.

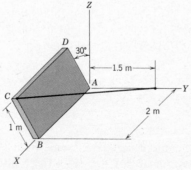

Prob. III.26

III.27 In Problem III.26, determine the moment of the force exerted by the cable on the plywood sheet (a) about corner *D*, (b) about edge *AD*.

III.28 A 1.5-kN force is applied to the pipe assembly as shown. Determine the moment of this force about each of the joints (points *A*, *B*, and *C*).

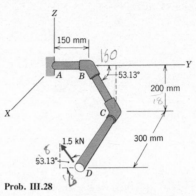

Prob. III.28

III.29 In Problem III.28, determine the moment of the 1.5-kN force about the axis of pipe segment *AB* and about the axis of pipe segment *BC*. Why is the latter value zero?

III.30 A 2-ton force is applied as shown to one end of the curved rod. Section *AB* coincides with the *XY* plane, whereas section *BC* coincides with the *YZ* plane. Determine the moment of this force (a) about the welded end *A* of the bar, (b) about the midpoint *B*.

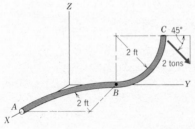

Prob. III.30

III.31 In Problem III.30, determine the moment of the 2-ton force (a) about line *AB*, (b) about line *BC*. Why is the latter value zero?

D. FORCE COUPLES

We have now established how the magnitude, direction, and line of action of a force are characterized mathematically. With this knowledge as a foundation, the next sequence of topics we will study is how to describe the total effect of a system of forces acting on a rigid body. Our concern here is solely with the way in which a body in influenced by its surroundings. In later modules we will study the force interaction of individual elements of a rigid body.

The most fundamental system of forces is formed by two forces that are equal in magnitude and opposite in direction. Such a system is said to form a *couple*. Figure 12 depicts a couple formed by a force $\bar{F}$ applied at point *A* and a force $-\bar{F}$ applied at point *B*. For convenience the diagram is drawn in the plane of the forces.

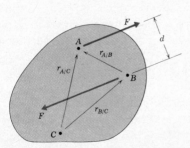

Figure 12

If the forces in this figure were collinear, we know from our study of concurrent force systems that $\bar{F}$ and $-\bar{F}$ would have a zero resultant, thereby cancelling each other. In the illustrated situation the forces still do not push or pull on the body; they only exert a twisting effect. Let us compute the total moment these two forces exert about an arbitrary point C, which need not necessarily be coplanar with the forces. Using the position vectors shown in Figure 12, we have

$$\bar{M}_C = \bar{r}_{A/C} \times \bar{F} + \bar{r}_{B/C} \times (-\bar{F}) = (\bar{r}_{A/C} - \bar{r}_{B/C}) \times \bar{F}$$

However, the diagram shows that

$$\bar{r}_{A/C} = \bar{r}_{B/C} + \bar{r}_{A/B} \quad \text{so} \quad \bar{r}_{A/B} = \bar{r}_{A/C} - \bar{r}_{B/C}$$

The foregoing expression for $\bar{M}_C$ then becomes

$$\boxed{\bar{M}_C = \bar{r}_{A/B} \times \bar{F}} \tag{13}$$

This result has several noteworthy features. First, we see that only the relative position of the points of application of the couple, as expressed by $\bar{r}_{A/B}$, affects the resultant moment. Thus, regardless of which point is selected for summing moments, a couple exerts the same moment. We call this resultant moment the torque of the couple, and denote it by $\bar{M}$. Furthermore, because the component of $\bar{r}_{A/B}$ perpendicular to $\bar{F}$ is shown in Figure 12 to be the perpendicular distance d between the lines of action of the forces forming the couple, it follows that the magnitude of the torque is given by

$$\boxed{|\bar{M}| = |\bar{F}|d} \tag{14}$$

Finally, the direction of the torque is perpendicular to the plane formed by the couple forces, in the sense obtained from the right-hand rule when the fingers are curled in the direction of the forces.

From these observations we can conclude that different couples can have equivalent effects on a rigid body provided that the forces forming the couples are in parallel planes. Equivalence then requires that the products $|\bar{F}|d$ for the couples are the same, and that their torques have the same sense. Typical examples of equivalent couples are shown on the following page in Figure 13a for a planar system and Figure 13b for a three-dimensional system.

It is important to realize that the torques of individual couples are additive as vectors. This is a consequence of the fact that the torque is simply the total moment of the forces forming the couple, and a moment is a vector quantity. One result of this statement is that we may describe the

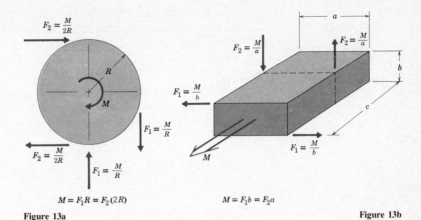

$$M = F_1 R = F_2(2R)$$

Figure 13a

$$M = F_1 b = F_2 a$$

Figure 13b

torque of a couple as the sum of its components with respect to a Cartesian coordinate system; these components represent the rotational tendency about each coordinate axis.

The foregoing discussion on components of couples can be generalized as follows:

> A set of couples acting on a rigid body is equivalent to a single couple whose torque is the sum of the moments, about an arbitrary point, of all forces forming the couples.

EXAMPLE 4

Determine the smallest forces applied at ends A and B of the beam shown that are equivalent to the given loading.

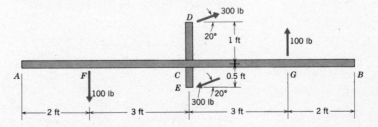

Solution

The planar force system consists of two couples formed by the 100-lb forces and the 300-lb forces. In order to find an equivalent couple we must determine the total torque of the given couples. This torque may be evaluated by summing the moments of each force about any point; the geometry of the system suggests that we use point C.

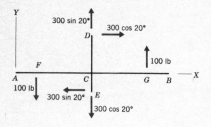

The points of application of the forces are located by horizontal and vertical dimensions. Hence we choose a coordinate system having horizontal and vertical axes. The coordinate system and force components are shown in the sketch.

The Z axis is outward for a right-hand coordinate system, so counterclockwise moments are positive according to the right-hand rule. The vertical force components at points D and E pass through point C, making their lever arms about this point zero. Hence, the resultant torque is

$$M = -300 \cos 20°(1.0) - 300 \cos 20°(0.5)$$

$$+ 100(3.0) + 100(3.0)$$

$$= 177.14 \text{ lb-ft}$$

The problem now is to form a 177.14 lb-ft couple from the smallest forces $\bar{F}$ at point A and $-\bar{F}$ at point B we can find. The magnitude of the torque of this couple is $|\bar{F}|d$, so we must align $\bar{F}$ and $-\bar{F}$ to obtain the largest value for d. As shown in the sketch, this situation is obtained when $\bar{F}$ is perpendicular to the beam. Notice that the sense of the forces is determined from the requirement that they form a counterclockwise couple. From the sketch we then have

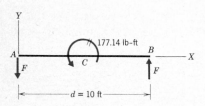

$$M = 177.14 = |\bar{F}|d = 10|\bar{F}|$$

$$|\bar{F}| = 17.714 \text{ lb}$$

$$\bar{F} = -17.714\bar{J} \text{ lb}$$

◁

EXAMPLE 5

A bus driver applies two 60-N forces to the steering wheel as shown. What is the torque of this couple, and what portion of this torque has the effect of turning the steering wheel?

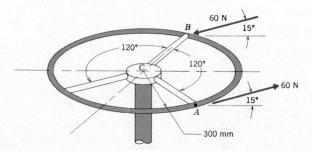

Solution

We will evaluate the torque of the couple by summing moments about the center point C, for the position vectors from this point to the points of

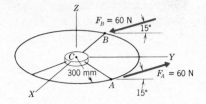

application of the forces are easily described. The XYZ reference frame shown in the sketch is chosen because points A, B, and C then all coincide with a coordinate plane, and also for the convenience it affords in describing force components.

Denoting the forces at points A and B as $\bar{F}_A$ and $\bar{F}_B$, we have

$$\bar{F}_A = -\bar{F}_B = 60(\cos 15° \, \bar{J} + \sin 15° \, \bar{K})$$

$$= 57.96\bar{J} + 15.53\bar{K} \text{ N}$$

The position vectors from point C to the forces are

$$\bar{r}_{B/C} = -0.30\bar{I} \text{ m}$$

$$\bar{r}_{A/C} = 0.30(\sin 30° \, \bar{I} + \cos 30° \, \bar{J})$$

$$= 0.15\bar{I} + 0.2598\bar{J} \text{ m}$$

Hence, the torque of the couple is

$$\bar{M} = \bar{r}_{A/C} \times \bar{F}_A + \bar{r}_{B/C} \times \bar{F}_B$$

$$= (0.15\bar{I} + 0.2598\bar{J}) \times (57.96\bar{J} + 15.53\bar{K})$$

$$+ (-0.30\bar{I}) \times (-57.96\bar{J} - 15.53\bar{K})$$

$$= 0.15(57.96)\bar{K} + 0.15(15.53)(-\bar{J})$$

$$+ 0.2598(15.53)(\bar{I}) + 0.30(57.96)\bar{K}$$

$$+ 0.30(15.53)(-\bar{J})$$

$$= 4.03\bar{I} - 6.99\bar{J} + 26.08\bar{K} \text{ N-m}$$

Only the component of the torque that is parallel to the shaft of the steering wheel has the effect of turning the wheel. This is the Z component of $\bar{M}$.

$$M_Z = 26.08 \text{ N-m}$$

Furthermore, because M_Z is positive, the driver is trying to turn the wheel counterclockwise looking down the Z axis.

HOMEWORK PROBLEMS

III.32 The drive shaft from the engine applies a counterclockwise torque, as viewed from the front, of 1.8 kN-m to the propeller of the aircraft shown. Determine vertical reactions between the wheels and the ground at points A and B that are equivalent to this couple.

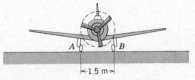

Prob. III.32

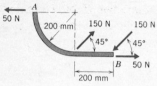

Prob. III.33

III.33 Determine the smallest forces at ends A and B of the curved bar that are equivalent to the two couples shown. Specify the magnitude and direction for these forces.

III.34 The three couples shown exert a resultant moment of 60 ft-lb on the wheel. Determine the magnitude of $\bar{F}$.

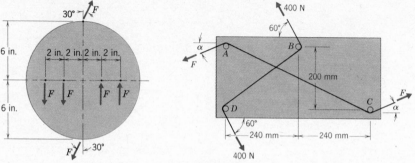

Prob. III.34   **Prob. III.35**

III.35 Cables are passed over smooth pegs in the block as shown. The tension in the cable passing over pegs B and D is 400 N. Knowing that the tension in the cable passing over pegs A and C is 300 N and that $\alpha = 30°$, determine the couple $\bar{M}$ that is equivalent to the force system exerted by the cables on the block.

III.36 In Problem III.35, determine the minimum value of F and the corresponding angle α for which the effect of the four cables on the block is a counterclockwise couple whose torque is 250 N-m.

III.37 The drive shaft applies a torque $M_1 = 5$ kN-m to the gear box and the driven shaft applies a torque $M_2 = 15$ kN-m, as shown. Determine the resultant couple acting on the gear box, specifying its magnitude and its direction angles with respect to the XYZ coordinate system.

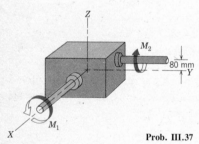

Prob. III.37

III.38 The wide flange beam shown is loaded by four forces. Determine the couple $\bar{M}$ that is equivalent to this loading.

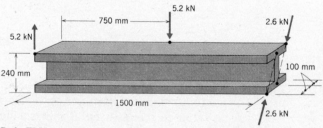

Prob. III.38

III.39 The four forces acting on the rectangular parallelepiped are equivalent to a couple $\bar{M}$. Determine the magnitude of $\bar{F}$, the angle α, and $\bar{M}$.

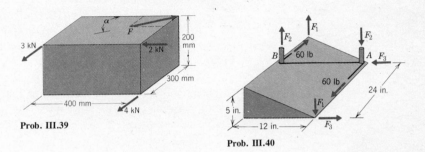

Prob. III.39

Prob. III.40

III.40 A cord is passed over the smooth pegs A and B on the triangular block. Tensile forces of 60 lb are applied at its ends. Determine the magnitude of the forces $\bar{F}_1$, $\bar{F}_2$, and $\bar{F}_3$ for which the resultant couple acting on the block is zero.

E. RESULTANT OF A GENERAL SYSTEM OF FORCES

In the preceding section we showed that the effect of a couple acting on a rigid body is a torque representing the twisting tendency of the body. Intuitively, we know that a general system of forces can also have a rotational effect, as well as having the effect of pushing or pulling a body.

Because we have already seen how to obtain the resultant of a concurrent system of forces, let us begin by investigating how noncurrent forces may be transferred to a common point of application. Consider a single force $\bar{F}$ applied to point P of a rigid body, as shown in Figure 14a. It is our desire to transfer $\bar{F}$ to the arbitrary point C.

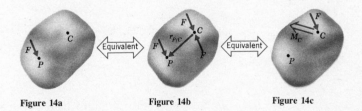

Figure 14a   Figure 14b   Figure 14c

To achieve the transfer we apply two forces $\bar{F}$ and $-\bar{F}$ at point C. These two forces are concurrent. Their resultant is zero, and therefore the force system in Figure 14b is equivalent to the force $\bar{F}$ in Figure 14a. However, the force $\bar{F}$ at point P and the force $-\bar{F}$ at point C form a couple

$\bar{M}_C$ which is given by

$$\bar{M}_C = \bar{r}_{P/C} \times \bar{F} \qquad (15)$$

This couple is simply the moment of the force $\bar{F}$ about point C. This means that *the original force $\bar{F}$ is equivalent to the force $\bar{F}$ at point C and the couple $\bar{M}_C$, which is the moment of $\bar{F}$ about C.* This is called a *force-couple system*. The resulting equivalent force-couple system is shown in Figure 14c.

Now consider the problem of a general force system, such as the one shown in Figure 15a. Each force in the system may be transferred to a common point C, as is shown in Figure 15b, where $(\bar{M}_C)_i$ denotes the couple whose torque is the moment of force $\bar{F}_i$ about point C.

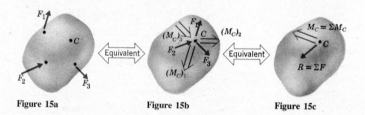

Figure 15a   Figure 15b   Figure 15c

The forces are now concurrent, so they may be summed to obtain their resultant $\bar{R}$. Further, because couples are additive, the individual couples $(\bar{M}_C)_i$ may be summed vectorially to obtain the resultant couple $\bar{M}_C$. The value of $\bar{M}_C$ is the sum of the moments about point C of the forces in the original system. It then follows that the original system of forces is equivalent to the force-couple system in Figure 15c. In other words:

A general system of forces is equivalent to a force $\bar{R}$ applied at any point C, and a couple $\bar{M}_C$ about point C, provided that

$$\bar{R} = \Sigma \bar{F} \qquad \bar{M}_C = \Sigma \bar{M}_C \qquad (16)$$

where $\Sigma \bar{F}$ denotes the sum of the forces in the original system and $\Sigma \bar{M}_C$ denotes the sum of the moments of the forces in the original system about point C.

Clearly, the resultant force $\bar{R}$ in the equivalent force-couple system is independent of the choice for point C; but the resultant couple $\bar{M}_C$ depends on which point we choose. This observation leads to a corollary of equation (16). *Two different force systems have equivalent effects on a rigid body if the forces in each system exert the same total moment about any point, as well as having identical sums.*

EXAMPLE 6

Replace the force and couple by an equivalent force-couple system at end B of the curved beam.

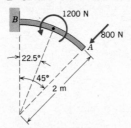

Solution

There is only one force in the given system, so the resultant $\bar{R}$ applied at point B is simply the 800-N force. The resultant couple $\bar{M}_B$ is the moment of the given force system about point B. When summing moments, we use the fact that a couple exerts the same moment about any point. Thus, the 22.5° angle is not needed in the solution. As an aid, we draw a sketch of the force system, choosing the Y axis vertical because angles are referred to this line.

The geometry is circular. Hence the lever arm of the 800-N force about point B is easily determined, as shown in the sketch. Noting that the Z axis is outward, clockwise moments are negative by the right-hand rule. Therefore,

$$M_{Bz} = -800(2 \sin 45°) + 1200 = 68.63 \text{ N-m}$$

In terms of the chosen coordinate system, the resultant force $\bar{R}$ is

$$\bar{R} = \Sigma \bar{F} = 800(-\sin 45° \, \bar{I} - \cos 45° \, \bar{J})$$

$$= -565.6(\bar{I} + \bar{J}) \text{ N}$$

The equivalent system is illustrated in the margin.

EXAMPLE 7

Determine the force-couple system at point A that is equivalent to the forces acting on the socket wrench shown.

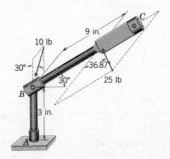

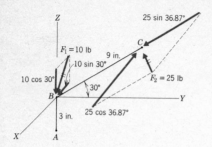

Solution

We begin with a sketch showing the forces and the XYZ reference frame, whose orientation is chosen to fit the manner in which the dimensions are given. The force components shown in the sketch are the same as those in the given diagram.

We must represent all forces by their XYZ components. The appropriate expression for $\bar{F}_1$ is easily obtained from the sketch.

$$\bar{F}_1 = (10 \sin 30° \, \bar{I} - 10 \cos 30° \, \bar{K}) = 5.0\bar{I} - 8.66\bar{K} \text{ lb}$$

The illustrated components of $\bar{F}_2$ are described with respect to bar BC. The component parallel to this bar is in the YZ plane, at an angle of 30° from the negative Y direction, whereas the component perpendicular to the bar is in the negative X direction. Hence

$$\bar{F}_2 = (25 \sin 36.87°)(-\cos 30° \, \bar{J} - \sin 30° \, \bar{K}) - (25 \cos 36.87°)\bar{I}$$

$$= -12.99\bar{J} - 7.5\bar{K} - 20.0\bar{I} \text{ lb}$$

The equivalent force system we seek consists of the resultant force $\bar{R}$ applied at point A and the resultant couple $\bar{M}_A$. To find $\bar{R}$ we sum forces.

$$\bar{R} = \bar{F}_1 + \bar{F}_2 = (5.0 - 20.0)\bar{I} - 12.99\bar{J} + (-8.66 - 7.5)\bar{K}$$

$$= -15.00\bar{I} - 12.99\bar{J} - 16.16\bar{K} \text{ lb}$$

To find $\bar{M}_A$ we sum moments about point A.

$$\bar{M}_A = \bar{r}_{B/A} \times \bar{F}_1 + \bar{r}_{C/A} \times \bar{F}_2$$

A displacement from point A to point B involves a movement of 3 in. in the positive Z direction, so

$$\bar{r}_{B/A} = 3.0\bar{K} \text{ in.}$$

Continuing to point C from point B requires an additional movement of 9 in. parallel to bar BC at an angle of 30° above the Y axis. Thus

$$\bar{r}_{C/A} = (3 + 9 \sin 30°)\bar{K} + 9 \cos 30° \, \bar{J} = 7.5\bar{K} + 7.794\bar{J} \text{ in.}$$

The moment is now found to be

$$\bar{M}_A = (3.0\bar{K}) \times (5.0\bar{I} - 8.66\bar{K}) + (7.5\bar{K}$$

$$+ 7.794\bar{J}) \times (-12.99\bar{J} - 7.5\bar{K} - 20.0\bar{I})$$

$$= 3.0(5.0)(\bar{J}) - 7.5(12.99)(-\bar{I})$$

$$- 7.5(20.0)(\bar{J}) - 7.794(7.5)(\bar{I}) - 7.794(20.0)(-\bar{K})$$

$$= -39.0\bar{I} - 135.0\bar{J} + 155.9\bar{K} \text{ lb-in.}$$

The equivalent force-couple system is sketched in the left margin.

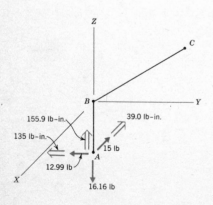

HOMEWORK PROBLEMS

III.41 A force of 100 N is applied as shown to the tire wrench. Determine the equivalent force-couple system acting (a) at point A, (b) at the center of the wheel.

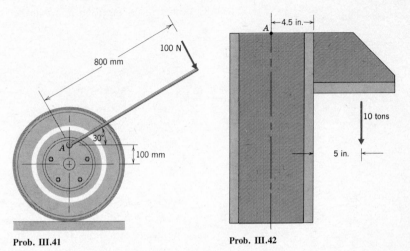

Prob. III.41

Prob. III.42

III.42 A 10-ton load is applied to the bracket, which is welded onto the I-beam. Determine the equivalent force-couple system acting at point A on the centerline of the I-beam.

III.43 A 4-kN force is applied to the bent bar as shown. Determine an equivalent force-couple system (a) at point A, (b) at point B, (c) at point C.

III.44 A cable is passed over two ideal pulleys that are pinned to a block. The tension within the cable is 300 N. Determine an equivalent force-couple system acting (a) at pin A, (b) at pin B.

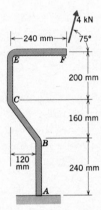

Prob. III.43

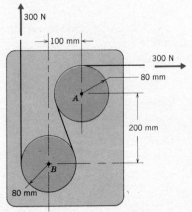

Prob. III.44

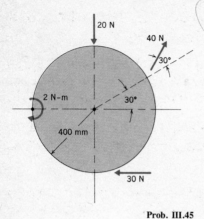

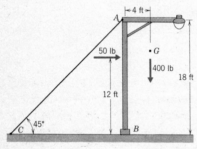

Prob. III.45

III.45 Replace the loading system shown by an equivalent force-couple system acting at the center of the wheel.

III.46 The stepped gear is subjected to two forces as shown. Knowing that $F = 200$ N, determine the force-couple system acting at the center C of the gear that is equivalent to these forces.

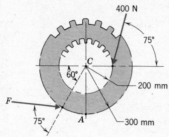

Prob. III.46

III.47 In Problem III.46, the value of F is such that the two given forces are equivalent to a single force acting at the center C. Determine this value of F and the corresponding equivalent force at C.

III.48 The street light is supported by cable AC. The total weight of the structure is 400 lb acting at point G, and the tension in the cable is 900 lb. The 50-lb horizontal force represents the effect of the wind. Determine the force-couple system at the base B that is equivalent to the combined effect on the structure of gravity, wind, and the cable tension.

Prob. III.48

III.49 In Problem III.48, the combined effect on the structure of gravity, wind, and the cable tension is equivalent to a single force acting at the base B. Determine the tension in the cable and the equivalent single force.

III.50 It is desired to replace the 400-kN force shown by two parallel forces $\bar{F}_A$ and $\bar{F}_B$ acting at points A and B on the shelf bracket. Determine the required forces $\bar{F}_A$ and $\bar{F}_B$.

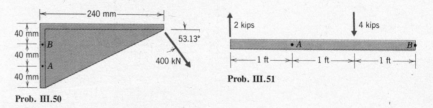

Prob. III.50

Prob. III.51

III.51 Determine parallel forces acting at points A and B that are equivalent to the vertical loads shown.

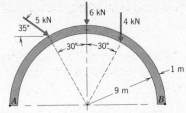

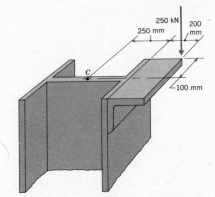

Prob. III.52

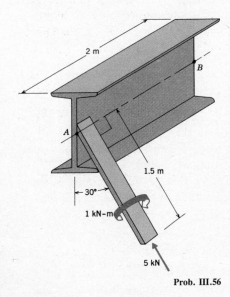

Prob. III.56

III.52 Determine the force-couple system at corner A of the circular arch that is equivalent to the system of forces shown.

III.53 In Problem III.52, it is desired to replace the system of forces shown acting on the semicircular arch by a vertical force $\bar{F}_A$ at point A and a force $\bar{F}_B$ at point B having an equivalent effect. Determine $\bar{F}_A$ and $\bar{F}_B$.

III.54 A 250-kN force is applied to the bracket, which is welded to the vertical column. Determine the equivalent force-couple system acting at point C on the centerline of the column.

III.55 A 40-lb horizontal force is applied to the wrench. Determine the equivalent force-couple system acting (a) at joint A, (b) at joint B. The pipe assembly is parallel to the vertical plane.

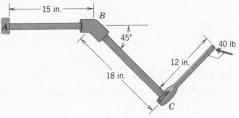

Prob. III.55

III.56 A bar is welded to the end of the I-beam. A force and a couple are applied to the end of the bar. Determine the equivalent force-couple system acting (a) at point A on the centerline of the beam, (b) at point B on the centerline of the beam.

III.57 A valve is tightened with the T-bar assembly by applying 75-N forces at points B and C as shown. (a) Replace these forces by an equivalent force-couple system acting at point A. (b) What portion of this equivalent system has the effect of turning the valve about axis AD?

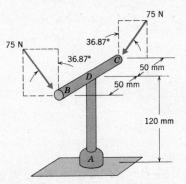

Prob. III.57

Prob. III.54

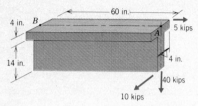

Prob. III.58

III.58 The concrete T-beam is loaded as shown. Determine the equivalent force-couple system acting (a) at point A, (b) at point B.

III.59 Replace the forces acting on the rectangular parallelepiped by an equivalent force-couple system (a) at corner A, (b) at corner B.

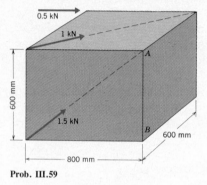

Prob. III.59

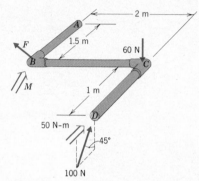

Prob. III.60

III.60 The pipe assembly is loaded as shown. Knowing that the force $\bar{F}$ and couple $\bar{M}$ are zero, determine the equivalent force-couple system acting at end A.

III.61 In Problem III.60, the force-couple system at end A equivalent to the force system shown is zero. Determine the force $\bar{F}$ and couple $\bar{M}$ at joint B.

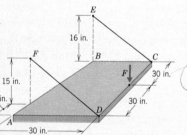

Prob. III.62

III.62 The tension in cable CE is 3400 lb and the tension in cable DF is 4200 lb. The force $\bar{F}$ is a downward load of 1000 lb. Determine the force-couple system acting at corner A that is equivalent to the effect on the plate of the cables and $\bar{F}$.

III.63 In Problem III.62, the vertical force $\bar{F}$ and the given cable tensions are equivalent to a force $\bar{R}$ and couple $\bar{M}_A$ acting at corner A. The magnitude of $\bar{F}$ is unknown, but it is known that $\bar{M}_A$ has no component parallel to edge AB. Determine (a) the magnitude of $\bar{F}$, (b) the corresponding force $\bar{R}$ and couple $\bar{M}_A$.

F.   SINGLE FORCE EQUIVALENTS

The results of the preceding section demonstrate that when we represent a general system of forces by a force-couple system, the torque of the couple depends on the point to which we transfer the forces. This gives rise to the question of whether there is some point O that eliminates the couple, thereby enabling us to replace a general system of forces by a single force.

Suppose that we have determined the force $\bar{R}$ at point C and the couple $\bar{M}_C$ which are the equivalent of a system of forces. We may transfer this resultant to another point O, such as that shown in Figure 16. Both

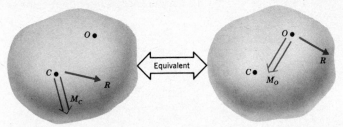

Figure 16

forces depicted in the figure are identical, so the two force-couple systems will be equivalent if they exert equal moments about any point. Choosing point C for the moment computation, we have

$$\bar{M}_C = \bar{M}_O + \bar{r}_{O/C} \times \bar{R} \qquad (17)$$

Although equation (17) is the relationship between the couples for any two equivalent force-couple systems, our desire here is to determine the conditions for which $\bar{M}_O = \bar{0}$. Setting $\bar{M}_O$ equal to zero in equation (17) gives

$$\bar{M}_C = \bar{r}_{O/C} \times \bar{R} \qquad (18)$$

In view of the properties of a cross product, equation (18) tells us that $\bar{M}_C$ must be perpendicular to $\bar{R}$ in order for $\bar{M}_O$ to be zero. Recalling the meaning of $\bar{M}_C$, it follows that a general system of forces may be replaced by a single force $\bar{R}$ acting at point O only when the moment of the forces in the original system about any point C is perpendicular to the resultant force $\bar{R}$. This condition is satisfied in three special cases.

1 Concurrent Forces

This case was studied in Module II. The point O of zero moment is any point along a line parallel to $\bar{R}$ which passes through the point of concurrency P, as illustrated in Figure 17. This follows because the vector $\bar{r}_{P/O}$ is parallel to $\bar{R}$, so

$$\bar{r}_{P/O} \times \bar{R} \equiv \bar{0} \equiv \bar{r}_{P/O} \times \Sigma\bar{F} \equiv \Sigma\bar{M}_O$$

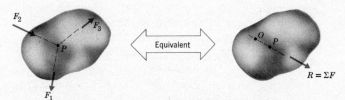

Figure 17

2 Planar Force Systems

When all forces coincide with a plane, their moment about any point in that plane is perpendicular to the plane. Consider the planar force system and its single force equivalent depicted in Figure 18.

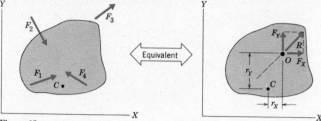

Figure 18

The components of the resultant $\bar{R}$ are the sum of the corresponding components of the original force system. To locate the point O of zero moment we sum moments about any convenient point C. Equating the moments of the forces in the two equivalent systems yields

$$\Sigma\, M_C = (\Sigma\, F_Y)r_X - (\Sigma\, F_X)r_Y$$

where counterclockwise moments are positive for the chosen coordinate system. This equation may be solved for the ratio of r_Y to r_X, corresponding to *any point* O along the line of action of $\bar{R}$, as indicated by the dashed line in Figure 18.

3 Parallel Forces

When all the forces are parallel to a given line, we can align one of the coordinate axes, such as the Z axis, with this line. In this case the force sum $\bar{R}$ will have only a Z component, as shown in Figure 19. A force in the

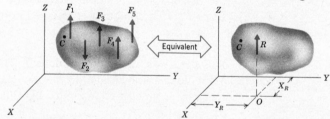

Figure 19

Z direction does not exert a moment about any line parallel to the Z axis, hence the sum of the moments about any point C will have only X and Y components. It then follows that the moment $\bar{M}_C$ is perpendicular to the resultant force $\bar{R}$. If $\bar{R}$ is nonzero, that is, if the given force system does not combine to form a couple, we may match the X and Y components of $\bar{M}_C$ for the actual force system to the corresponding components of the moment of the resultant $\bar{R}$ about point C. This will yield the values of X_R and Y_R illustrated in Figure 19. They locate where the line of action of the force resultant intersects the XY plane. Any point O along the indicated line of action of $\bar{R}$ yields $\Sigma\, \bar{M}_O = \bar{0}$.

4 General System of Forces

In general, there is no point C for which the torque $\bar{M}_C$ of the equivalent force-couple system is perpendicular to the resultant force $\bar{R}$. A slight simplification may be obtained by representing $\bar{M}_C$ by its components parallel and perpendicular to $\bar{R}$. We can then find a point O for which the torque $\bar{M}_O$ of the equivalent force-couple system is parallel to $\bar{R}$. Such a force system is called a *wrench*. The action of a wrench is typified by the manner in which we use a screwdriver, simultaneously pushing and twisting on the shaft on the screwdriver. The actual determination of a wrench is academic. It does little to increase understanding of the effect of the force system. Therefore, we shall not pursue the required procedure.

EXAMPLE 8

Three people exert horizontal forces on a rectangular table as shown. What single force has the same effect as these people, and where along edge BC should this force be applied?

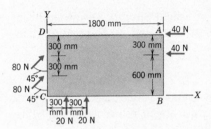

Solution

The force system is planar, so it has a single equivalent force. This force is the sum of the given forces. We therefore sum force components using the coordinate system illustrated in the sketch. Note that we could first replace each pair of parallel forces by their single equivalent, but there is no advantage in doing that.

The force sum gives

$$\bar{R} = \Sigma \bar{F} = \Sigma F_X \bar{I} + \Sigma F_Y \bar{J}$$

$$= [2(80 \cos 45°) - 2(40)]\bar{I} + [+2(80 \sin 45°)$$

$$+ 2(20)]\bar{J} = 33.14\bar{I} + 153.14\bar{J} \text{ N}$$

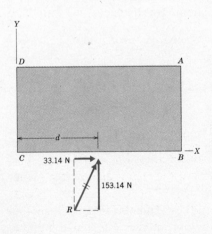

We now determine the point of application along edge BC with the aid of a sketch in which $\bar{R}$ is applied at an undetermined point along the edge. Then the moment of $\bar{R}$ about any point must equal the moment of the actual force system about that point.

The choice of the point for the moment sum is arbitrary; let us use point C. The Z axis is outward, so counterclockwise moments are positive. The lever arms of the force components are shown in the given sketch. The Y components of the 80-N forces pass through point C, so we compute the moment of the given forces to be

$$\Sigma M_C = -(80 \cos 45°)(0.3) - (80 \cos 45°)(0.6)$$

$$+ 20(0.3) + 20(0.6) + 40(0.6) + 40(0.9)$$

$$= 27.09 \text{ N-m}$$

The moment of the force resultant $\bar{R}$ about point C is obtained with the aid of the second sketch. Noting that the component R_X passes through point C and that the lever arm of component R_Y is the unknown distance d, we find

$$\Sigma\, M_C = (153.14)d$$

Equating the two expressions for $\Sigma\, M_C$ yields

$$d = \frac{27.09}{153.14} = 0.1769 \text{ m}$$

EXAMPLE 9
Two people are standing on the wing of a light aircraft, resulting in the loading shown. Determine the equivalent force and its point of application.

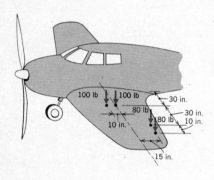

Solution
In order to facilitate the usage of the given dimensions and to simplify the representation of the parallel forces, we use the reference frame illustrated in the second sketch. The resultant force $\bar{R}$ is the sum of the given forces. Because these forces are all in the negative Z direction, we have

$$\bar{R} = -(80 + 80 + 100 + 100)\bar{K} = -360\bar{K} \text{ lb}$$

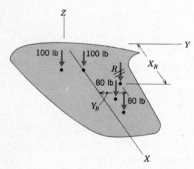

To determine the dimensions X_R and Y_R of the point of application of $\bar{R}$, we equate the moments about any point of the given forces and the resultant force $\bar{R}$. Using the origin of the coordinate system for the moment sum and referring to the given diagram for distances, we formulate

$$\Sigma\, M_C = (30\bar{I}) \times (-100\bar{K}) + (30\bar{I} - 10\bar{J}) \times (-100\bar{K})$$

$$+ (60\bar{I} + 15\bar{J}) \times (-80\bar{K}) + (70\bar{I} + 15\bar{J}) \times (-80\bar{K})$$

$$= (X_R\,\bar{I} + Y_R\,\bar{J}) \times \bar{R} = (X_R\,\bar{I} + Y_R\,\bar{J}) \times (-360\bar{K})$$

$$-3000(-\bar{J}) - 3000(-\bar{J})$$

$$+ 1000\bar{I} - 4800(-\bar{J}) - 1200\bar{I} - 5600(-\bar{J}) - 1200\bar{I}$$

$$= -360 Y_R\,\bar{I} - 360 X_R(-\bar{J}) - 1400\bar{I} + 16{,}400\bar{J}$$

$$= -360 Y_R\,\bar{I} + 360 X_R\,\bar{J} \text{ lb-in.}$$

Because two vectors are equal only if the components are equal, equating corresponding components for the vectors on the left- and right-hand sides of the equation yields

$$\bar{I} \text{ component:} \qquad -1400 = -360 Y_R$$

$$\bar{J} \text{ component:} \qquad 16,400 = 360 X_R$$

Solving these equations, we obtain

$$X_R = 45.55 \text{ in.} \qquad Y_R = 3.89 \text{ in.}$$

HOMEWORK PROBLEMS

III.64 Four vertical loads are applied to the cantilever shown. Determine the resultant of this system of forces and locate where it is applied along the beam (a) when $F = 0$, (b) when $F = 6$ kips, (c) when $F = 12$ kips.

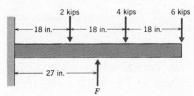

Prob. III.64

III.65 Knowing that the distance $d = 1.5$ m, determine the resultant of the loads acting on the beam shown. Specify the point of application of this resultant on the beam.

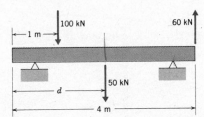

Prob. III.65

III.66 Determine the distance d for which the resultant of the three forces shown passes through the midpoint of the beam.

III.67 In a certain drive mechanism a cable is passed over the idler pulley assembly shown. The pulleys are ideal, and the tension within the pulleys is 300 N. Determine the magnitude and direction of the resultant force exerted by the cable on the assembly, and also determine the point of application of this force along edge AB.

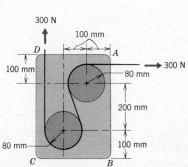

Prob. III.67

III.68 Determine the resultant of the force system shown. Specify its magnitude, direction, and point of application.

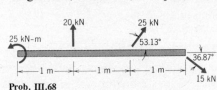

Prob. III.68

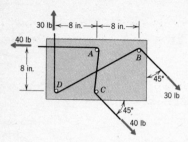

Prob. III.69

III.69 Two cords are pulled over smooth pegs embedded in the rectangular board. The tension in the cord passing over pegs A and C is 40 lb and the tension in the cord passing over pegs B and D is 30 lb. (a) Determine the magnitude and direction of the resultant force exerted on the board by the cords. (b) Determine the point of application of this resultant force along line AB. (c) Determine the point of application of the resultant force along line CD.

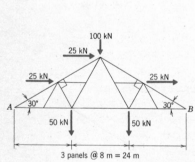

Prob. III.70

III.70 A crate supported by two cables is pushed to one side by a person who exerts the 50-lb horizontal force. The weight of the crate is 400 lb, acting at point G. The tension in cable AC is 280 lb and the tension in cable BD is 240 lb. Determine the resultant of the forces acting on the crate, and specify its point of application along edge AB.

III.71 A street lamp supported by two cables is subjected to the loading shown. Determine the resultant of this system of forces, and specify its point of application along line AB.

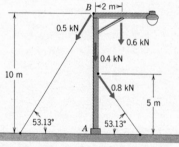

Prob. III.71

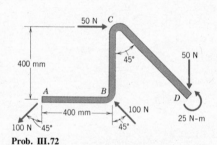

Prob. III.72

III.72 The bent bar is subjected to the loading shown. (a) Determine the magnitude and direction of the resultant of this loading. (b) Determine the point of application of the resultant force along line AB. (c) Determine the point of application of the resultant force along line BC.

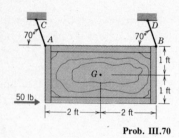

Prob. III.73

III.73 The truss structure is loaded as shown. Determine the resultant of this loading, specifying its magnitude, direction, and point of application along line AB.

III.74 The circular arch is subjected to the loading shown. Determine the magnitude and direction of the resultant of these forces, and determine where its line of action intersects the arch.

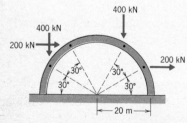

Prob. III.74

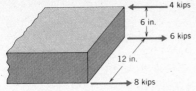

Prob. III.75

III.75 Three axial forces are exerted on the concrete beam as shown. Determine the point of application of the resultant force on the cross section of the beam.

III.76 A T-beam section is loaded as shown. Determine the resultant of these forces and its point of application on the top of the beam.

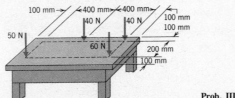

Prob. III.76

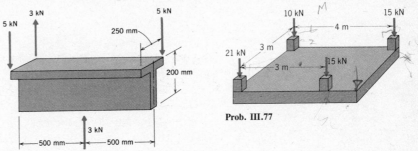

Prob. III.77

III.77 Four columns rest on the rectangular foundation slab. For the column loadings shown determine the resultant force and its point of application on the slab.

III.78 Three parallel columns equally spaced around the perimeter of the circular foundation slab carry the loads shown. Determine the resultant of this loading, and specify its point of application on the slab.

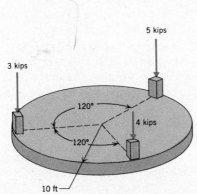

Prob. III.78

III.79 A semicircular rod embedded in a wall is loaded as shown. Determine the resultant of these forces and its point of application in the XY plane.

Four packages resting on the table exert the forces shown. Determine the resultant of these forces, and specify its point of application on the top of the table.

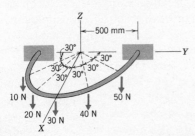

Prob. III.79

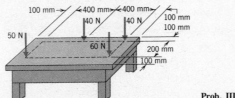

Prob. III.80

III.81 In Problem III.80, it is desired to apply to the perimeter of the table an upward force that will cause the resultant of the forces acting on the table top to be at the center of the table. Determine this force and its point of application.

G. EQUILIBRIUM OF RIGID BODIES

A fundamental application of the techniques for describing forces is the determination of the forces required to prevent a body from moving. This is the topic we will treat here. However, before we can investigate the relationship between the forces acting on a body in static equilibrium, we must learn to properly account for all forces.

1 Reactions and Free Body Diagrams

One aid in describing forces is the free body diagram. As we saw in Module II when we studied concurrent force systems, the free body diagram isolates the body of interest from its surroundings. In the diagram we show *all* forces being *exerted on the body,* bearing in mind that Newton's third law tells us that forces are the result of the interaction between bodies. The importance of a clear free body diagram cannot be overemphasized. The process of constructing the diagram aids us in understanding which parameters of the system are important. Equally important, the resulting diagram is a vital visual aid in making certain that we consider all forces during the solution process.

We will have no difficulty in depicting certain classes of forces in the diagram. Typical forces in this category are given forces, tensile forces exerted by cables, and the weight force. The latter, in particular, acts vertically downward, passing through the center of mass of the body. (The methods for determining this point are discussed in Module V.)

Free body diagrams for rigid bodies are somewhat more difficult to construct than those we encountered in treating particles. The primary reason for this difficulty lies in the fact that the diagram accompanying the statement of the problem will usually depict the real system. That is, we will be shown the body (or bodies) of interest, as well as the means by which the body is supported and thus prevented from moving.

It is part of the task of constructing a free body diagram to examine the supports in order to deduce what types of force they apply to the body. These forces are called *constraint forces* because they constrain (restrict) the motion of the body. They are also called *reactions* because they represent the way that the objects to which the body is attached react.

The characteristics of the constraint forces can be established by considering the manner in which the body is supported. In determining how to depict the constraint forces we need only two facts. First, because a force represents a pushing or pulling effect, a point in a body is prevented from moving in a specific direction by the action of a reaction force in the opposite direction. Second, because a couple represents a twisting effect, a body is prevented from rotating about a specific axis by the action of a reaction couple whose torque is parallel to the axis of rotation, twisting opposite the sense in which rotation would occur.

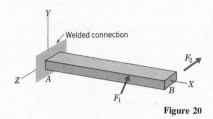

Figure 20

To see how these criteria are applied when examining the forces transmitted to a body by its supports, let us consider the welded connection between the rigid bar and the wall shown in Figure 20. A weld connection between two bodies fully joins their material. Thus, end A of the bar is not free to move in the direction of any of the coordinate axes, so there must be reaction force components in each coordinte direction. We do not know in advance the value of these reaction forces, nor even whether they are oriented in the positive or negative coordinate direction; this information depends on how the bar would tend to move under the action of the applied forces. We will denote a reaction force at a specific point by the letter corresponding to this point. Therefore, a portion of the constraint at end A consists of a reaction force $\bar{A}$ having components $\bar{A}_X$, $\bar{A}_Y$, and $\bar{A}_Z$.

To complete the determination of the nature of the constraint imposed by a welded connection, we must consider the type of rotations the end may undergo. Because the material of the bar is continuous with the material of the wall, and the wall itself is assumed to be rigid and fixed, it follows that there can be no rotation about any of the coordinate axes. This means that there must be a reaction couple $\bar{M}_A$, whose components $\bar{M}_{AX}$, $\bar{M}_{AY}$, and $\bar{M}_{AZ}$ represent the constraints necessary to prevent the rotations. We usually show the components of general reaction forces and couples in the free body diagram in order to be certain that we remember to account for each unknown component. Figure 21 depicts the free body diagram of the rigid bar.

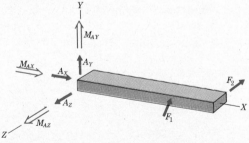

Figure 21

A welded joint represents the greatest degree of fixity (lack of freedom to move) that can be imposed by a connection. In fact, a welded support is usually described as a *fixed end*.

Whenever the type of support allows the body to move in a specific direction, or rotate, about a specific axis, there will be no reaction force, or couple, in that direction. For example, suppose that a bar is supported at one end by a pin connection, such as that shown in Figure 22a. The pin acts like a miniature shaft. The end of the bar is fully constrained from moving in any direction by the pin and the brackets. The only possible way in which the bar may move is to rotate about the axis of the pin,

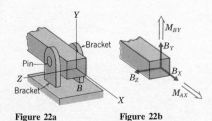

Figure 22a   **Figure 22b**

which is the Z axis in the figure. Thus, the reaction force and couple corresponding to this support are as shown in Figure 22b. Note that there is no couple in the Z direction, because rotation about this axis is not restricted.

The foregoing discussion implicitly assumed that the pin fit closely in the hole of the bar. Further, it was also assumed that the pin is rigid. This, of course, may not be the situation. For instance, a very common method of constructing a pin connection is to use a hinge, such as those found on doors. Anyone who has ever tried to support a door with just one hinge knows that it affords very little constraint against rotation in any direction. Thus, depending on specific knowledge of the type of construction, we may choose to retain or neglect the couples depicted in Figure 22b. We shall differentiate between the two alternatives by referring to the former condition as an ideal *pin,* and the latter one as a *hinge,* although this is not standard terminology. In a planar problem where the axis of the pin is perpendicular to the plane of interest, as shown in Figure 23a, we will be concerned only with reaction forces that are parallel to the plane and couples whose torque is perpendicular to the plane. In this case there is no difference between an ideal pin and a hinge, as depicted in Figure 23b.

As an aid for our future studies, Table 1 depicts other common methods of connecting and supporting bodies, accompanied by a description of the nature of the corresponding restriction on the motion of the body and a sketch of the reaction force and couple components. This tabulation is presented in order to supplement, but not replace, our ability to examine the means of supporting a body. This is especially so because on some occasions we will encounter a support that is not depicted in Table 1.

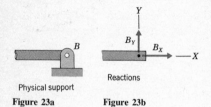

Physical support

Figure 23a   **Figure 23b**

Reactions

Table 1   Common Supports and Reactions

TYPE OF SUPPORT	MOTION RESTRICTION	REACTIONS
Cable	Attached point A cannot move outward, parallel to the cable. Free rotation.	$\bar{T}$ is a tensile force parallel to the cable.

Table 1 *(continued)*

TYPE OF SUPPORT	MOTION RESTRICTION	REACTIONS
Short link 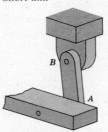	Attached point A cannot move parallel to the line AB of the link.	$\bar{F}_{AB}$ can be either a tensile or compressive force parallel to the link. 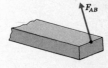
Smooth surface 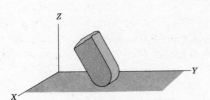	No movement that penetrates the surface. Free rotation.	$\bar{N}$ can only be normal upward
Rough surface or sharp edge 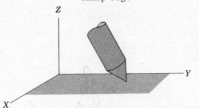	No movement that either penetrates the surface or is parallel to it. Free rotation. (For the effects of friction, see Module VII.)	A_X A_Y N
Small roller on a rough surface 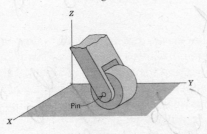	No movement that penetrates the surface or is perpendicular to the plane of the roller. Free rotation. The case of a small roller in a smooth groove is identical.	In the absence of a groove or friction, $A_x = 0$.  A_X N

(continued)

Table 1 *(continued)*

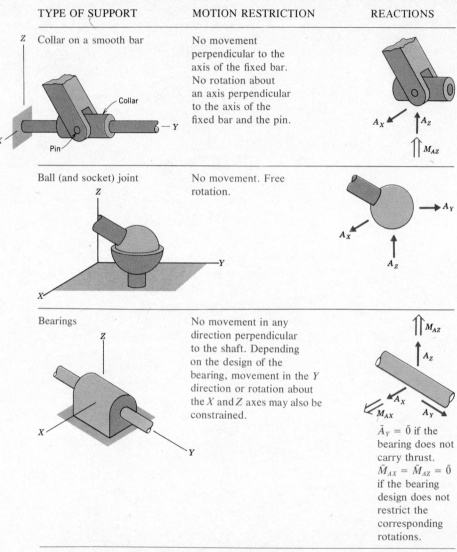

TYPE OF SUPPORT	MOTION RESTRICTION	REACTIONS
Collar on a smooth bar	No movement perpendicular to the axis of the fixed bar. No rotation about an axis perpendicular to the axis of the fixed bar and the pin.	
Ball (and socket) joint	No movement. Free rotation.	
Bearings	No movement in any direction perpendicular to the shaft. Depending on the design of the bearing, movement in the Y direction or rotation about the X and Z axes may also be constrained.	$\bar{A}_Y = \bar{0}$ if the bearing does not carry thrust. $\bar{M}_{AX} = \bar{M}_{AZ} = \bar{0}$ if the bearing design does not restrict the corresponding rotations.

A question that arises in dealing with bodies that are supported by bearings, as depicted in Table 1, is whether or not to include the couples $\bar{M}_{AX}$ and $\bar{M}_{AZ}$ in drawing the free body diagram. In that the presence of these couples complicate the problem, frequently making it insoluble, our approach in this text will be to assume that there are no couples, unless stated otherwise. In other words, we shall regard the action of a bearing and of a pin to be identical.

The following example will demonstrate how the concepts of constraints and constraint forces are implemented in drawing free body diagrams.

EXAMPLE 10
Draw free body diagrams for the systems shown. The centers of mass are denoted as points G.

a   A ladder

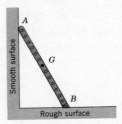

b   A disk attached to a block by a cable

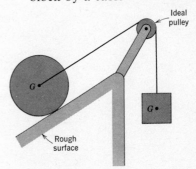

c   Each of the two bars separately, and joined together

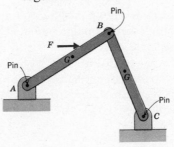

d   A door suspended by two hinges

e   The entire truss structure (neglect gravity)

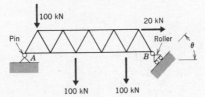

f   An antenna tower (neglect gravity)

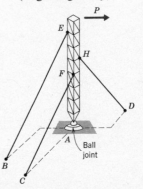

Solution

As an aid to including all forces, it is axiomatic that anything touching a body exerts a force on it.

Part a

There is a force normal to the smooth wall at point A, whereas the rough surface exerts normal and tangential forces at point B. We show the weight force acting downward, passing through point G.

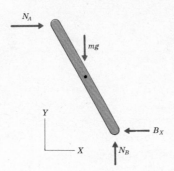

The force $\bar{B}_X$ is shown pushing to the left because it opposes the way in which point B would tend to move. However, it is also correct to show $\bar{B}_X$ in the opposite direction, acting to the right, for we would then find a negative value for B_X. In contrast, it would make no sense to show $\bar{N}_A$ and $\bar{N}_B$ in opposite directions, because the surfaces can only push on the ladder. Nevertheless, an *incorrect* choice will merely result in a negative value for the reactive force.

Part B

There are three connected bodies: the disk, the block, and the pulley. We shall draw free body diagrams of each. A useful technique to facilitate using Newton's third law is to *lay out the diagrams to resemble an assembly drawing*.

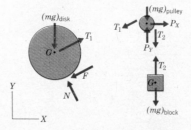

The force F is the tangential component of the reaction between the disk and the rough surface. The forces $\bar{P}_X$ and $\bar{P}_Y$ represent the reactions of the shaft on the pulley. The tensile forces T_1 and T_2 pull on the bodies they contact. We have already seen that for an ideal pulley, $T_1 = T_2$.

Part c

Each bar is pinned at both of its ends. Further, the reactions exerted between the two bars at pin B must satisfy Newton's third law of action and reaction. Thus,

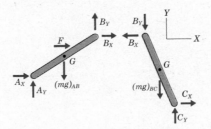

When we draw a free body of the two bars joined together, the reactions at pin B are internal to the system, so they do not appear in such a diagram. In effect, the diagram of the system is an overlay of the individual diagrams sketched above.

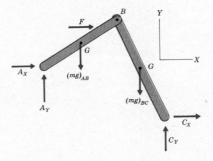

Part d

We agreed to neglect the couple reactions of hinges, hence

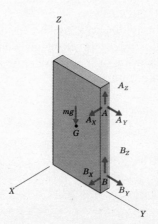

Part e

Note that the reaction of the roller is perpendicular to the inclined surface.

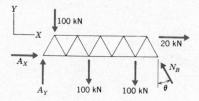

Part f

The cables exert tensile forces and the ball joint prevents movement of the base.

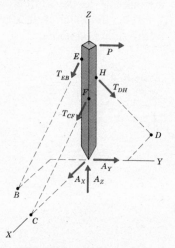

HOMEWORK PROBLEMS

Sketch free body diagrams for the systems shown. In each case the given forces are illustrated. In problems where the weight force is to be included, the location of the mass center, point G, is given. For simplicity, dimensions are omitted.

III.82   The wrecking bar

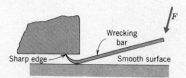

Prob. III.82

III.83   The stepped disk

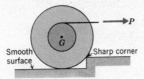

Prob. III.83

III.84   The bent bar

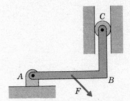

Prob. III.84

III.85   The bar *AB*

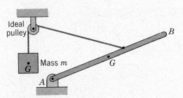

Prob. III.85

III.86   The bar *AB*

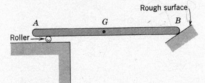

Prob. III.86

III.87   The tape guide assembly

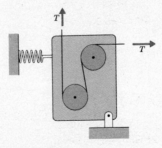

Prob. III.87

III.88   The automobile with only the brakes at the rear wheels applied

Prob. III.88

III.89   The balance arm

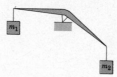

Prob. III.89

III.90   The two-beam assembly, and beams *AB* and *BC* individually

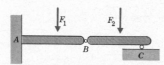

Prob. III.90

III.91   The two-bar assembly, and bars *AB* and *CD* individually

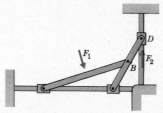

Prob. III.91

III.92   The bent bar

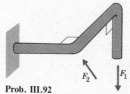

Prob. III.92

III.93 The flywheel and bar *AB*, each individually

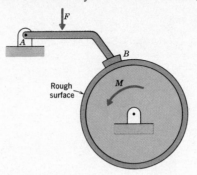

Prob. III.93

III.94 The hand drill

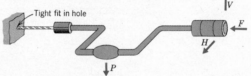

Prob. III.94

III.95 The T-bar

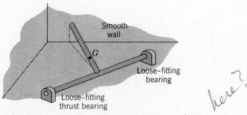

Prob. III.95

III.96 The hatch cover

moments here?

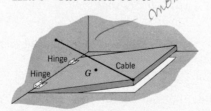

Prob. III.96

III.97 The truss structure — all connections are ball joints

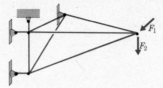

Prob. III.97

III.98 The bar assembly

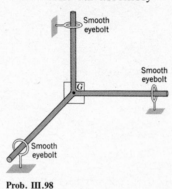

Prob. III.98

2 Conditions for Static Equilibrium

We have already seen that a general system of forces is equivalent to a force $\bar{R}$ passing through an arbitrary point C and a couple $\bar{M}_C$. The force $\bar{R}$, which is the sum of the forces in the given system, $\bar{R} = \Sigma \bar{F}_i$, describes the equivalent pushing or pulling effect on point C. The couple $\bar{M}_C$ is the sum of the moments of the forces about point C, $\bar{M}_C = \Sigma \bar{r}_{i/C} \times \bar{F}_i$. This resultant couple $\bar{M}_C$ describes the twisting effect of the force system about point C. It follows that

> The resultant force and couple should both be zero if a rigid body is to be in static equilibrium.

Notice that the equilibrium condition for concurrent force systems is a special case of the general statement.

The foregoing is an intuitive approach for determining the conditions for static equilibrium. A rigorous derivation is presented in the companion dynamics text, where mathematical proofs accounting for the effects of motion are required. The conditions of static equilibrium are obtained, as a special case, by arresting the motion of the body.

The most direct way of setting the resultant force and couple equal to zero is to actually compute the sum of the forces and the sum of the moments about any arbitrary point. We then equate the two resultant vectors to zero, thus obtaining

$$\Sigma \bar{F} = \bar{0}$$
$$\Sigma \bar{M}_C = \bar{0}$$

(19a)

This is the approach we shall usually apply. In the case of a three-dimensional force system the force and moment sums have three components each, so equations (19a) are equivalent to the following six scalar *equations of static equilibrium*:

$$\Sigma F_X = 0 \qquad \Sigma F_Y = 0 \qquad \Sigma F_Z = 0$$
$$\Sigma M_{CX} = 0 \qquad \Sigma M_{CY} = 0 \qquad \Sigma M_{CZ} = 0$$

(19b)

Because the point we choose for summing moments is arbitrary, in practice we shall choose a point to be situated along the line of action of an unknown reaction force, thereby eliminating that force from the moment equation.

The fact that we can eliminate an unknown force from the moment equation suggests that we should explore the possibility of summing moments about more than one point. Because of the frequency with which planar force systems occur, we shall first discuss such a case.

We begin by considering Figure 24, which illustrates the resultant force and couple with respect to point C for an arbitrary set of planar forces. To investigate an alternative condition for static equilibrium, let us choose another point D and resolve $\bar{R}$ into its components parallel and perpendicular to the line connecting points C and D, as shown in the figure.

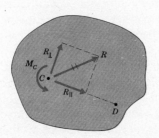

Figure 24

Consider the case where $\bar{M}_C$ has already been set to zero, leaving the nonzero force $\bar{R}$. If we also equate the moment about D of the force system to zero, this will cause $\bar{R}$ to be zero, by the definition of the moment of a force. The remaining force component $\bar{R}_{\parallel}$ will then vanish by equating to zero the sum of the force components in any direction other than the one that is perpendicular to line CD. Alternatively, in this case we can make $\bar{R}_{\parallel}$ vanish by equating the moment about a third point to zero, provided that this third point is not coincident with line CD.

Summarizing, *for planar forces the scalar equations for static equilibrium yield only three nontrivial equations.* The scalar equations may be obtained from three alternative formulations, as follows:

1   Equate to zero the moment about one point and the force sums in two directions. This is the approach of equations (19).
2   Equate to zero the moment sums about two points and the force sum in any direction that is not parallel to the line connecting the two chosen points.
3   Equate to zero the moment sums about three points that are not collinear.

We could present similar alternatives for the case of a three-dimensional system of forces. However, as we have already seen, the determination of moments in the three-dimensional case is somewhat lengthier than that for planar forces. Thus, the simplification in the resulting equilibrium equations would be offset by the additional computations required to obtain these equations. In order to have a unified approach for planar and three-dimensional forces, in both cases we will generally employ the formulation of equation (19b), where the force and moment results are equated to zero. Hence, for a planar set of forces lying in the XY plane, the equilibrium equations we shall employ are

$$\Sigma F_X = 0, \qquad 2\,F_Y = 0, \qquad \Sigma M_{CZ} = 0 \tag{19c}$$

3 Static Indeterminacy and Partial Constraints

The static equilibrium conditions we have presented describe the relationships that must be satisfied if a body is to remain at rest. Based on these conditions, we will soon consider problems where the reactions and any other forces required to maintain the static equilibrium condition are to be determined. However, we have not yet considered whether it is possible to determine all unknown forces from the equations of statics, or indeed whether it is possible for a body to remain at rest.

In the case of a general force system there are six scalar equations of equilibrium, obtained by setting the resultant force and moment vectors to zero. In the case of a planar force system only three of these equations are nontrivial. Clearly, if there are more unknown components of the reaction forces and couples than the number of equilibrium equations available, we will not be able to evaluate the reactions. The reactions of such a body are then said to be *statically indeterminate.*

A second possibility is that there are fewer unknown components to the reactions than the number of nontrivial equations for static equilibrium. In that case there can be no set of reactions that satisfy all of the equilibrium equations. We then say that the body is *partially constrained,* because for each equilibrium equation that is unsatisfied there is a a corresponding type of motion: translational motion for a force equation and rotational motion for a moment equation. (In the text on dynamics, precise definitions are given for these motion terms.)

Clearly then, in order to be able to solve the equations of static equilibrium, and have *statically determinate* reactions, the number of unknown reaction components must be the same as the number of nontrivial equilibrium equations. However, this is merely a necessary condition, not a sufficient one. There are special situations where one of the equilibrium equations still cannot be satisfied. These situations arise when the supports of the body have not been properly arranged, with the result that motion of the body may still occur. Such bodies are termed *improperly constrained*.

To gain insight into these possibilities, let us consider the beam shown in Figure 25 along with its free body diagram. It will be shown that

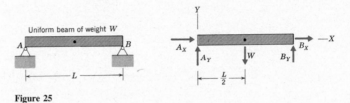

Figure 25

the beam is statically indeterminate. The pins at both ends of the bar exert unknown horizontal and vertical forces. Choosing point A for summing moments, in order to eliminate the reactions at this point from the moment equation, the equilibrium equations are

$$\Sigma F_X = A_X + B_X = 0$$

$$\Sigma F_Y = A_Y + B_Y - W = 0$$

$$\Sigma M_{AZ} = B_Y L - W \frac{L}{2} = 0$$

The solution of these equations shows that $A_Y = B_Y = W/2$. However, we can only determine that $A_X = -B_X$. It is very tempting to say that A_X and B_X are both zero, because no loading force is acting in that direction. However, we cannot prove this contention on the basis of the available equations, and indeed it is not true. Furthermore, if a horizontal force was also applied to the bar, we would have no idea of the separate values of A_X and B_X.

The condition of static indeterminancy is one situation where it is inadequate to model a system as a rigid body. The value of the indeterminate reactions can be obtained only when we consider the manner in which the body deforms, a subject that is treated in the study of mechanics of materials.

When a system is indeterminate, we can solve the basic equilibrium equations for some of the reactions in terms of the remaining ones, as was done for the system of Figure 25. The remaining reactions, which in this case was only the component B_X, are said to be *redundant* because they are not required to support the system. By modifying the supports to remove the redundant reactions, we can obtain a statically determinate system. For example, we can replace the pin at B by a roller as shown in Figure 26.

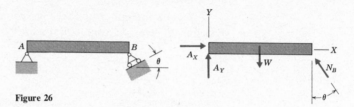

Figure 26

The system in Figure 26 is statically determinate, because all reactions may be obtained from the corresponding equilibrium equations. The solutions are $A_Y = W/2$, $A_X = W/(2 \cos \theta)$ and $N_B = (W/2) \tan \theta$. These answers are acceptable, except for the case where $\theta = 90°$, for which A_X and N_B become infinite.

Notice that when $\theta = 90°$, the reaction $\bar{N}_B$ in Figure 26 is the same as the reaction $\bar{B}_X$ in Figure 25, so the reaction that has been removed in that case is not the redundant one. The system is then capable of rigid body motion, as illustrated in Figure 27. This occurs because the roller permits point B to move in the vertical direction, which is the same direction as that in which the end would move if the bar was attached only to pin A. This particular modification of the supports causes the beam to be partially constrained.

Figure 27

The condition of partial constraint describes the situation under a general system of loads. Clearly, if the intention is to always hold the body stationary, as we would desire for structures such as a building, the supports must provide total constraint. However, many systems, such as vehicles and machines, are designed to permit motion. In such cases we have partially constrained objects. It is possible to hold these types of systems at rest by requiring that some of the loading forces be sufficiently general to allow the equations of static equilibrium to be satisfied. In such cases the problem will be statically determinate or indeterminate, depending on whether we can solve the equilibrium equations for the reactions and the additional required forces.

To illustrate this matter, consider the beams mounted on rollers in Figures 28a and 28b. Clearly, for a general set of loads, both beams will move horizontally. Therefore, they are partially constrained. However, if the loading system is such that the sum of their components in the X direction is zero, as is the case for the loads shown, this motion will not occur. For the beam in Figure 28a we may determine the normal reactions of the two rollers by summing moments about a roller and summing forces in the Y direction. Hence, this beam is statically determinate. On the other hand, the static equilibrium equations are inadequate for the determination of the three normal reactions of the roller supports in Figure 28b. Hence, this is a condition of static indeterminancy.

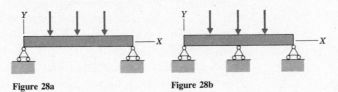

Figure 28a   Figure 28b

Our main concern in this test is with statically determinate systems that are *properly constrained,* either because of their support conditions or because of the loads to which they are subjected. Nevertheless, you should be able to recognize situations of static indeterminancy and partial constraint on the basis of the solvability of the equilibrium equations.

4   Solving Problems in Static Equilibrium

The development of the ability to determine the forces acting on a body by applying the equations of static equilibrium is a primary goal of a course in statics. This capability is a prerequisite to the study of most of the topics in this text.

Problem solving is not an art, but rather a professional decision made by recognizing the proper keys. In what follows we present a logical way of thinking to identify the keys. It is not meant for memorization, but rather as a guide while working problems. Work enough problems *correctly* and the keys will become part of your thinking process.

1   Select the body (or bodies) to be isolated in a free body diagram (FBD) by examining where the forces of interest are applied. Draw a simple sketch of this body. Label any significant points if they are not already labeled.

2   Choose an XYZ coordinate system for the description of the components of the force and position vectors. Usually, the coordinate axes should coincide with the directions in which the dimensions of the system are given. Show this coordinate system in the sketch of step 1.

3   Complete the FBD begun in step 1 by showing all forces acting on the body. In doing this, examine each support of the body to determine what types of reactions exist. The weight force, acting at the center of mass,

should be included unless it is small compared to the other forces, or unless no information about it is given. If the dimensions describing a force do not appear in the given diagram, show the necessary dimensions in the FBD.

4 Choose a convenient point along the line of action of some unknown constraint forces and compute the sum of the moments of the forces about this point. Equate the components of the moment sum to zero.

5 Equate to zero each component of the sum of the forces appearing in the FBD. (At times, this step may be omitted if it can be seen that the force sums involve unknown forces whose values are not desired, provided that these unknown forces have not appeared in the moment equation obtained in step 4.)

6 Solve the simultaneous equations that result from steps 4 and 5. Interpret the results, for example, by checking any initial assumptions.

Note that all of the simplifications of planar force systems, such as being able to sum force components directly and being able to compute moments without actually using the cross product, may be employed.

ILLUSTRATIVE PROBLEM 1

A 5-kN force and a 4-kN-m couple are applied to the forked bar. Determine the reaction forces at supports A and B when $\theta = 36.87°$.

Solution

Steps 1 and 2 We choose the forked bar for the FBD, because we wish to determine the forces required to support this body against the given set of loads. The X and Y axes for this planar problem are selected as horizontal and vertical because the dimensions are given with respect to these directions.

Step 3 The reaction at pin B is normal to the surface of the smooth groove, so we have the following FBD.

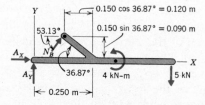

The dimensions shown in the FBD are needed to define the orientation and point of application of the normal force $\bar{N}_B$.

Step 4   We choose pin A for the moment sum in order to eliminate the unknown reactions at that pin. Thus, we have

$$\Sigma \, M_{AZ} = -N_B \cos 53.13°(0.090) + N_B \sin 53.13° \, (0.250$$

$$- \, 0.120) + 4 - 5(0.600) = 0 \text{ kN-m}$$

Step 5   The equilibrium equations obtained by summing force components in the directions of the coordinate axes are

$$\Sigma \, F_X = N_B \cos 53.13° + A_X = 0 \text{ kN}$$

$$\Sigma \, F_Y = N_B \sin 53.13° + A_Y - 5 = 0 \text{ kN}$$

Step 6   We may now solve the moment equation for N_B and substitute this result into the force equations. Therefore

$$N_B[-0.6(0.9) + 0.8(0.13)] + 1.0 = 0 \quad \text{so} \quad N_B = -20 \text{ kN} \qquad \triangleleft$$

$$-20(0.60) + A_X = 0 \quad \text{so} \quad A_X = 12 \, kN \qquad \triangleleft$$

$$-20(0.80) + A_Y - 5 = 0 \quad \text{so} \quad A_Y = 21 \text{ kN} \qquad \triangleleft$$

The negative value of N_B means that this force acts in the opposite direction from the one shown (guessed at) in the FBD. What does this mean in regard to which surface of the groove pin B is bearing against?

ILLUSTRATIVE PROBLEM 2
A lightweight rod is wedged between smooth pins A and B, and rests on the smooth floor. Determine all reaction forces as functions of the magnitude of the force $\bar{F}$ and the angle θ at which it is applied. Then determine the range of values of θ for which static equilibrium is possible.

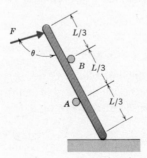

Solution

Steps 1 and 2   We isolate the bar for the FBD because all of the forces are applied

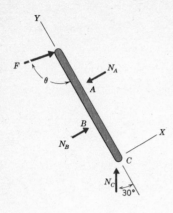

to this body. For this planar problem, we incline the Y axis parallel to the bar because the dimensions are referred to the axis of the bar.

Step 3 Because the floor and the pins are smooth, the reaction forces are normal to the surface of contact. Hence, the FBD is as shown.

Step 4 The unknown force $\bar{N}_C$ passes through point C, so the corresponding moment equilibrium equation about point C is

$$\Sigma M_{CZ} = -N_A\left(\frac{L}{3}\right) + N_B\left(\frac{2L}{3}\right) - F \sin \theta(L) = 0$$

Step 5 Equating the force sums in the X and Y direction to zero yields

$$\Sigma F_X = N_C \sin 30° + N_A - N_B + F \sin \theta = 0$$

$$\Sigma F_Y = N_C \cos 30° + F \cos \theta = 0$$

Step 6 We have three equilibrium equations for the three unknowns N_A, N_B, and N_C, which may be solved determining N_C from the second equation and using that result in the other two equations. Thus

$$N_C = -\frac{F \cos \theta}{\cos 30°} = -1.1547 \, F \cos \theta$$

$$\left(-F\frac{\cos \theta}{\cos 30°}\right) \sin 30° + N_A - N_B + F \sin \theta = 0$$

$$-N_A + 2N_B - 3F \sin \theta = 0$$

The solution of the last two of these equations is

$$N_B = F(2 \sin \theta + \cos \theta \tan 30°)$$

$$= F(2 \sin \theta + 0.5774 \cos \theta)$$

$$N_A = F(\sin \theta + 2 \cos \theta \tan 30°)$$

$$= F(\sin \theta + 1.1547 \cos \theta)$$

For static equilibrium the three normal reactions we have determined may not be negative because the surfaces of contact can only press against each other. Thus, for N_C non-negative we have

$$-\cos \theta \le 0$$

which is satisfied when $90° \le \theta \le 270°$. For N_B positive or zero we have

$$2 \sin \theta + 0.5774 \cos \theta \ge 0$$

The equality sign applies when $\theta = 164.4°$ of $344.4°$. It is easily verified that the inequality is satisfied when $0 \leq \theta \leq 164.4°$ or $344.4° \leq \theta \leq 360°$. Finally, for N_A non-negative we have

$$\sin \theta + 1.1547 \cos \theta \geq 0$$

The equality sign applies when $\theta = 130.9°$ or $310.9°$ and the inequality is satisfied when $0 \leq \theta \leq 130.9°$ or $310.9° \leq \theta > 360°$. Thus, the region where all reactions are positive is

$$90° \leq \theta < 130.9°$$

What happens if θ is not within this range?

ILLUSTRATIVE PROBLEM 3

The 2000-kg rectangular plate is supported by a hinge at point A, a short horizontal link at corner E, and two cables, as shown. The weight force acts at the geometric center G. Determine the tensions in the cables and the reaction force at support A.

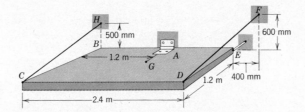

Solution

Steps 1 and 2 To determine the reaction forces required to support the plate against its own weight, we choose this body for the FBD. We select the Z axis vertical and the X and Y axis parallel to the edges of the plate.

Step 3 A hinge is considered to prevent movement of the supported point, but not prevent rotation of the body. The short link exerts a force parallel to its axis, thereby preventing movement of the body in that direction. These considerations lead to the following free body diagram.

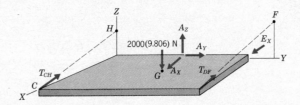

Step 4 Before considering the equilibrium equations, we shall describe the cable tensions in terms of their components. Writing each of these forces as the product of its (unknown) magnitude times its unit vector, we have

$$\bar{T}_{CH} = T_{CH}\frac{\bar{r}_{H/C}}{|\bar{r}_{H/C}|} = T_{CH}\frac{-1.2\bar{I} + 0.5\bar{K}}{\sqrt{(1.2)^2 + (0.5)^2}}$$

$$= T_{CH}(-0.9231\bar{I} + 0.3846\bar{K})$$

$$\bar{T}_{DF} = T_{DF}\frac{\bar{r}_{F/D}}{|\bar{r}_{F/D}|} = T_{DF}\frac{-1.2\bar{I} + 0.4\bar{J} + 0.6\bar{K}}{\sqrt{(1.2)^2 + (0.4)^2 + (0.6)^2}}$$

$$= T_{DF}(-0.8571\bar{I} + 0.2857\bar{J} + 0.4286\bar{K})$$

The moment equilibrium equations may now be formulated. In order to eliminate the reaction forces at hinge A, we sum moments about this point. Thus, for moment equilibrium we write

$$\Sigma\,\bar{M}_A = \bar{r}_{C/A} \times \bar{T}_{CH} + \bar{r}_{D/A} \times \bar{T}_{DF} + \bar{r}_{E/A} \times E_x\bar{I}$$

$$+ \bar{r}_{G/A} \times [-2(9.806)\bar{K}] = \bar{0}\ \text{kN-m}$$

Referring to the original diagram of the system to determine the position vectors, and using the foregoing expressions for $\bar{T}_{CH}$ and $\bar{T}_{DF}$, yields

$$\Sigma\,\bar{M}_A = (1.2\bar{I} - 1.2\bar{J}) \times T_{CH}(-0.9231\bar{I} + 0.3846\bar{K})$$

$$+ (1.2\bar{I} + 1.2\bar{J}) \times T_{DF}(-0.8571\bar{I} + 0.2857\bar{J}$$

$$+ 0.4286\bar{K}) + 1.2\bar{J} \times E_x\bar{I} + 0.6\bar{I} \times [-2(9.806)\bar{K}]$$

$$= \bar{0}\ \text{kN-m}$$

Carrying out the cross products gives

$$\Sigma\,\bar{M}_A = T_{CH}\ (-0.4615\bar{J} - 1.1077\bar{K} - 0.4615\bar{I})$$

$$+ T_{DF}(+0.3428\bar{K} - 0.5143\bar{J} + 1.0285\bar{K}$$

$$+ 0.5143\bar{I}) - 1.2E_X\bar{K} + 11.767\bar{J} = \bar{0}$$

The scalar equations are now obtained by equating the sums of like components to zero, in effect thereby summing moments about the three lines passing through point A that are parallel to the coordinate axes. Hence,

$$\Sigma\,M_{AX} = -0.4615T_{CH} + 0.5143T_{DF} = 0\ \text{kN-m}$$

$$\Sigma\,M_{AY} = -0.4615T_{CH} - 0.5143T_{DF} + 11.767 = 0\ \text{kN-m}$$

$$\Sigma\,M_{AZ} = -1.1077T_{CH} + (0.3428 + 1.0285)T_{DF} - 1.2E_X = 0\ \text{kN-m}$$

Step 5 The FBD shows that the reaction at hinge A has components in three directions and that the weight force is in the negative Z direction. Using the vector expressions for the cable forces derived in step 4, the sums of the force components parallel to each of the coordinate axes are

$$\Sigma\, F_X = -0.9231 T_{CH} - 0.8571 T_{DF} + A_X + E_X = 0 \text{ kN}$$

$$\Sigma\, F_Y = 0.2857 T_{DF} + A_Y = 0$$

$$\Sigma\, F_Z = 0.3846 T_{CH} + 0.4286 T_{DF} + A_Z - 2(9.806) = 0 \text{ kN}$$

Step 6 The equilibrium equations in steps 4 and 5 may be solved by using the first two moment equations to determine T_{CH} and T_{DF}, then substituting these results into the last moment equation to find E_X. These values are then substituted into the force equations to find the forces at hinge A. This procedure yields

$$T_{CH} = 12.75 \qquad T_{DF} = 11.44 \qquad E_X = 1.30 \text{ kN}$$

$$A_X = 20.27 \qquad A_Y = 3.27 \qquad A_Z = 9.81 \text{ kN}$$

Note that the values of the cable tensions are positive, as they must be because cables can only pull on other bodies.

ILLUSTRATIVE PROBLEM 4

An ideal pulley is pinned to end A of the bent bar ABC and a cable is wrapped around the pulley as shown. A tensile force of 200 lb is applied to the free end of the cable. (a) Prove that the tensile force within the cable is constant. (b) Determine the reactions at pin B and roller C. Is the roller bearing against the upper or lower surface?

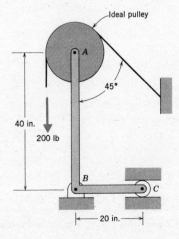

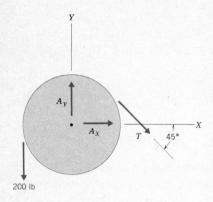

200 lb

Solution

Up to now, we have accepted the fact that the tension in a cable that passes over an ideal pulley is constant. We shall now prove it.

Steps 1 a and 2a   In part (a) we are concerned with the interaction between the cable and the pulley, so we isolate the pulley in a FBD. Horizontal and vertical coordinate axes in this planar problem suit the directions in which dimensions are given.

Step 3a   To the left of the pulley the tension within the cable is the 200-lb force applied to its end, whereas the tension on the right is considered to be some unknown value T. The pulley is also attached to a pin at its center, so there are two reaction force components at that point, as shown.

Step 4a   Choosing pin A for the moment sum in order to eliminate the reactions at this location, the moment equilibrium equation is

$$\Sigma M_{AZ} = -T(5) + 200(5) = 0 \text{ lb-in.}$$

Step 5a   The force sums for the planar system of forces are

$$\Sigma F_X = T \cos 45° + A_X = 0 \text{ lb}$$

$$\Sigma F_Y = A_Y - T \sin 45° - 200 = 0 \text{ lb}$$

Step 6a   The moment equation gives the desired result:

$$T = 200 \text{ lb}$$

If the tension were not constant, it would be necessary to have a couple acting on the pulley to resist the difference in the moments of the two tension forces.

The remaining force equilibrium equations give the values of A_X and A_Y, which were not requested. Nevertheless, they are

$$A_X = -141.4 \text{ lb} \qquad A_Y = 341.4 \text{ lb}$$

We will have use for these values in the following.

Steps 1b and 2b   In order to determine the reactions at points B and C, we draw a FBD of the bent bar. The coordinate system utilized in part (a) will also suffice for this portion of the problem.

Step 3b   Forces are exerted on the bent bar by the pin at point B, the roller at point C, and pin A which holds the pulley. Hence, we have the FBD shown in the left margin.

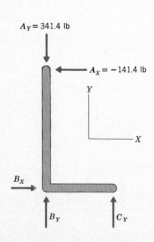

$A_Y = 341.4$ lb

$A_X = -141.4$ lb

B_X

B_Y C_Y

Particular attention is drawn to the way in which the reaction forces at point A are entered into the FBD. We show $\bar{A}_X$ and $\bar{A}_Y$ opposite from the directions in which they were assumed when we drew the FBD of the pulley, in accordance with Newton's third law. Then, after depicting the algebraic forces, we enter their numerical values with the same sign as they were calculated to have. Thus, we see that the pulley really pulls the bar to the right while pressing it down. This procedure is employed in order to prevent serious, and also common, sign errors. We will have more to say about this matter in the next module.

Step 4b   Summing moments about pin B, we have

$$\Sigma M_{BZ} = C_Y(20) + (-141.4)(40) = 0 \text{ lb-in.}$$

Step 5b   The force sums are

$$\Sigma F_X = B_X - (-141.4) = 0 \text{ lb}$$

$$\Sigma F_Y = B_Y + C_Y - 341.4 = 0 \text{ lb}$$

Step 6b   The solutions of the equilibrium equations are readily found to be

$$C_Y = 282.8 \qquad C_Y = 58.6 \qquad B_X = 141.4 \text{ lb} \qquad \triangleleft$$

Because C_Y is positive, it acts in the same sense as that shown in the FBD. Therefore, roller C is being pushed upward, which means that it is bearing on the lower surface.

As a final comment, let us consider how we would have approached the solution of part (b) if we were allowed to consider as an established fact that the tension within a cable is unchanged when the cable passes over an ideal pulley. If we then were to isolate the assembly of bar ABC and the pulley in a FBD, we would only have the reaction forces at points B and C and the two cable tensions, as shown.

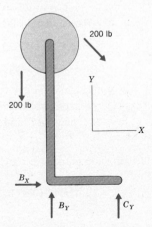

This would enable us to determine the desired reactions, without going through the intermediate step of studying the interaction between the pulley and the bent bar.

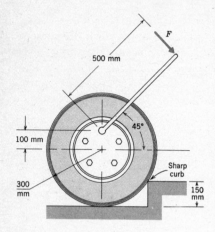

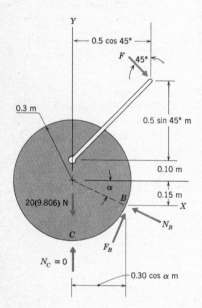

ILLUSTRATIVE PROBLEM 5

A lightweight wrench is attached to the 20-kg automobile tire as shown. Determine the magnitude of the force $\bar{F}$ that will cause the tire to begin to roll over the sharp curb whose height is 150 mm.

Solution

Steps 1 and 2 We wish to study the effect on the tire of the force $\bar{F}$ which is applied to the wrench. In order to avoid studying the interaction forces exerted between these two bodies, we draw a FBD of their assembly. Noting the way in which the dimensions of the planar system are given, we choose the X axis horizontal and the Y axis vertical.

Step 3 At the point of contact with the curb there are normal and tangential reactions because the corner is sharp. In general, there would also be a normal force exerted between the tire and the horizontal surfaces at point C, but in this problem we want to determine the condition where rolling is impending (i.e., on the verge of occurring). We therefore set this force to zero. The resulting FBD is as shown. Do not forget the weight force.

The dimensions shown locate the point of application of the given force $\bar{F}$ and the point of contact B. It is readily seen that

$$\alpha = \sin^{-1}\left(\frac{0.15}{0.30}\right) = 30°$$

Step 4 We choose point B for the moment sum in order to eliminate the unknown reaction forces $\bar{N}_B$ and $\bar{F}_B$ which are applied at this point. This yields

$$\Sigma M_{BZ} = 20(9.806)(0.30 \cos 30°) - F \cos 45°(0.15$$
$$+ 0.10 + 0.50 \sin 45°) - F \sin 45°(0.50 \cos 45°$$
$$- 0.30 \cos 30°) = 0 \text{ N-m}$$

Step 5 We omit consideration of force equilibrium, because the moment equation may be solved to obtain the value of F and we are not interested in the reaction components N_B and F_B.

Step 6 The solution of the moment equation is

$$F = 103.3 \text{ newtons}$$

If the magnitude of the force $\bar{F}$ exceeds this value, it will not be possible to maintain static equilibrium in the position where the wheel is almost in contact with the ground.

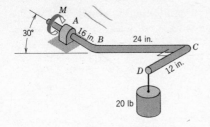

ILLUSTRATIVE PROBLEM 6

A 20-lb block is suspended from the bent bar, which is supported by a thrust bearing at end A. In the position shown, section ABC of the bar coincides with the vertical plane, and section CD is horizontal. Considering the thrust bearing to be tight fitting, it will permit rotation of the rod only about axis AB. Determine the couple $\bar{M}$ that must be exerted about this axis in order to maintain static equilibrium.

Solution

Steps 1 and 2 The problem is to determine the couple that should be applied to the bent bar to resist the loading of the suspended block, so we isolate the bar in the FBD, aligning one coordinate axis with the horizontal portion BC and choosing one of the other axes vertical.

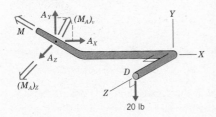

Step 3 The reactions at bearing A consist of a force-couple system, $\bar{A}$ and $\bar{M}_A$. These are general vectors, except that $\bar{M}_A$ can have no component parallel to axis AB because the bearing does not prevent rotations about that axis. Other forces acting on the bar are the couple $\bar{M}$ parallel to axis AB, and a 20-lb force representing the tension in the cable holding the block. As can be seen, we represent the bearing reactions by their components.

Step 4 Although summing moments about point A in the conventional manner will give the value of M eventually, a shortcut is obtained by noting that we can eliminate all unknown forces and couples except for $\bar{M}$ by considering only the component of $\Sigma \bar{M}_A$ parallel to line AB. By definition, this is the sum of the moments about line AB. Clearly, if $\Sigma \bar{M}_A = 0$, it must also be true that $\Sigma M_{AB} = 0$.

To compute the moment of the 20-lb force about line AB, we first compute the moment of this force about any convenient point along line AB. Clearly, point B is best, because $\bar{r}_{D/B}$ is easily determined to be

$$\bar{r}_{D/B} = 24\bar{I} + 12\bar{K} \text{ in.}$$

Thus

$$\bar{M}_B = (24\bar{I} + 12\bar{K}) \times (-20\bar{J}) = 240(\bar{I} - 2\bar{K}) \text{ lb-in.}$$

We next evaluate the component of $\bar{M}_B$ parallel to line AB by calculating the dot product of $\bar{M}_B$ with a unit vector parallel to that line, directed from point A to point B. This gives

$$\bar{M}_B \cdot \bar{e}_{B/A} = 240(\bar{I} - 2\bar{K}) \cdot (\cos 30°\, \bar{I} - \sin 30°\, \bar{J})$$

$$= 207.8 \text{ lb-in.}$$

Finally, because the couple $\bar{M}$ is directed along the line from B to A, we have

$$\Sigma M_{AB} = -M + 207.8 = 0 \text{ lb-in.}$$

Step 5 We omit the computation of the force equilibrium equations, because the moment equation may be solved for M and we are not interested in any of the reactions exerted by the bearing.

Step 6 The solution of the moment equation is

$$M = 207.8 \text{ lb-in.}$$

As an aid in seeing why this method of solution was employed, you should try the alternative of solving $\Sigma \bar{M}_A = 0$.

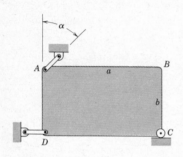

ILLUSTRATIVE PROBLEM 7

Determine the value of the angle α for the short link at point A that results in improper constraint of the rectangular plate.

Solution

Recall that the condition of partial constraint concerns equilibrium under a general set of loading forces. Hence, we will assume an arbitrary set of loads.

Steps 1 and 2 We want to study the nature of the constraints of the plate, so we draw a FBD of the plate. The given dimensions suit choosing coordinate axes parallel to the edges of the plate.

Step 3 Let two arbitrary forces $\bar{F}_1$ and $\bar{F}_2$ represent the general set of loads. The short links exert forces only along their axes, so the FBD is as shown.

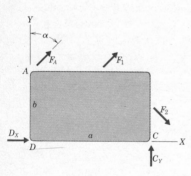

Step 4 We select point C for the moment sum because the reaction forces $\bar{C}_Y$ and $\bar{D}_X$ both intersect this point. The moment equilibrium equation is then

$$\Sigma M_{CZ} = -(F_A \sin \alpha)b - (F_A \cos \alpha)a + M_1 + M_2 = 0$$

where M_1 and M_2 are the moments of $\bar{F}_1$ and F_2 respectively, about point C.

Step 5 The force equilibrium equations are

$$\Sigma F_X = D_X + F_A \sin \alpha + (F_1)_X + (F_2)_X = 0$$

$$\Sigma F_Y = C_Y + F_A \cos \alpha + (F_1)_Y + (F_2)_Y = 0$$

Step 6 The three equilibrium equations are now solved for the values of F_A, C_Y, and D_X in terms of the angle α, yielding

$$F_A = \frac{M_1 + M_2}{b \sin \alpha + a \cos \alpha}$$

$$D_X = -(F_1)_X - (F_2)_X - \frac{(M_1 + M_2) \sin \alpha}{b \sin \alpha + a \cos \alpha}$$

$$C_Y = -(F_1)_Y - (F_2)_Y - \frac{(M_1 + M_2)\cos\alpha}{b\sin\alpha + a\cos\alpha}$$

We now look for the condition of partial constraint, which is evidenced by an infinite value for at least one of the reactions. For this system, all three reactions are infinite when the denominator $b\sin\alpha + a\cos\alpha = 0$. Thus

$$\alpha = -\tan^{-1}\left(\frac{a}{b}\right)$$

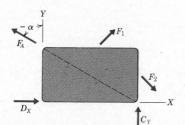

The reason that this value of α leads to partial constraint becomes obvious when we draw the FBD corresponding to this value of α, as depicted to the left. In this situation all of the reactions intersect point C, so they cannot provide a moment to counteract the moment of the loading forces about point C.

In general, there are two ways of having improper constraint of a body for the case of a planar set of forces. As happened in this problem, all of the reaction forces may be concurrent. Alternatively, as occurred for the beams in Figure 26, all of the reaction forces may be parallel.

HOMEWORK PROBLEMS

III.99 A wrecking bar wedged into a crevice is loaded at its end by a 200-N force, as shown. Considering the surface of contact at point B to be smooth, determine the forces exerted on the bar at end A.

III.100 The 10-lb, 4-ft-long bar is suspended between two smooth walls as shown. The center of mass of the bar is at the midpoint G. Determine the reactions exerted by the walls on the bar.

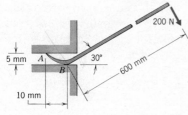

Prob. III.99

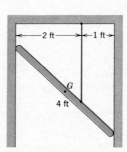

Prob. III.100

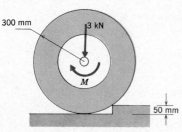

Prob. III.101

III.101 The rear axle of an automobile exerts the vertical 3-kN force and couple $\bar{M}$ on the 20-kg tire, which is at rest. Determine the magnitude of $\bar{M}$ required to cause the tire to roll over the curb.

III.102 The 1300-kg automobile is at rest on the 20° hill. The weight force acts at point G. Determine the normal force exerted between the rough ground and each tire. Compare these forces to those obtained when the vehicle is on level ground. Can the reaction forces tangent to the hill be determined? The brakes on all four wheels are applied.

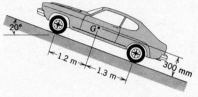

Prob. III.102

III.103 Solve Problem III.102 if the automobile is facing downward on the hill.

III.104 The bar AD is supported by collars B and C, which ride on smooth horizontal rods. End A of the bar rests against the smooth vertical wall. Determine the reactions resulting from the 20-lb force shown.

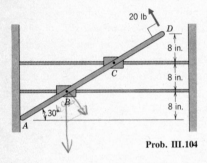

Prob. III.104

III.105 The symmetrical balance beam scale shown is used to compare the unknown mass m_B to the standard mass m_A. The support C consists of a frictionless knife edge. Derive an expression for m_B/m_A in terms of the angle θ. The center of mass of arm ACB coincides with point C.

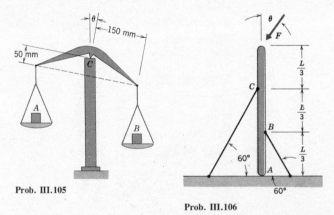

Prob. III.105

Prob. III.106

III.106 A lightweight pole, which rests on a smooth horizontal surface at end A, is supported by two cables, as shown. (a) Determine the tension in each cable as a function of the angle θ at which the force $\bar{F}$ is applied. (b) From the result of part (a), determine the allowable range of values of θ for which the pole will be in static equilibrium.

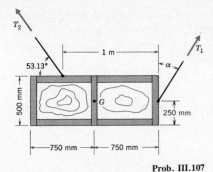

Prob. III.107

III.107 Determine the tension in the cables and the angle α required to hold the 1000-kg crate in the position shown. Point G is the center of mass.

III.108 The stiffness of the spring is 2 kN/m and its unstretched length is 400 mm. Bar AB is in equilibrium in the position shown. Determine the mass of the bar, knowing that its center of mass is at the midpoint G.

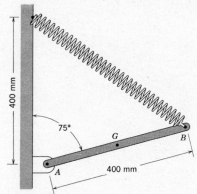

Prob. III.108

III.109 An 80-lb force and a 40 ft-lb couple are applied to the bent bar as shown. Determine the reaction at roller B. Which surface of the groove does this roller bear against?

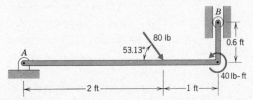

Prob. III.109

III.110 The bent bar is supported by the fixed pin A and pin B which fits into the smooth groove. Knowing that $d = 750$ mm, determine the reactions for the loading shown.

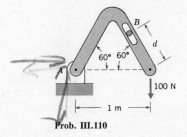

Prob. III.110

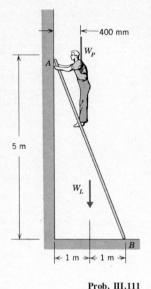

Prob. III.111

III.111   A 65-kg painter is standing on a 10-kg ladder which rests against the smooth wall and the rough ground. The weights of the painter and the ladder act as shown. Determine the reactions acting on the ladder.

III.112   Determine the horizontal force $\bar{F}$ that must be applied to corner C of the 100-kg crate to hold it in position on the hill as shown. Both surfaces of contact are smooth. The center of mass of the crate is at the geometric center G.

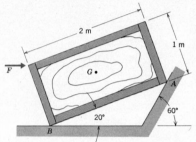

Prob. III.112

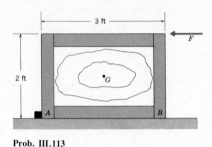

Prob. III.113

III.113   The 2-ton crate resting on the smooth horizontal surface is restrained by cleat A. Determine the maximum force $\bar{F}$ that may be exerted on the crate without causing it to begin to tip over. The center of mass of the crate is at the geometric center G.

III.114   The tape is wrapped over ideal pulleys A and B in the tape guide. The system is horizontal. Determine the force exerted by the spring at point C.

III.115   The crane boom shown is supported by a pin at end A and cable CD. Determine the tension in cable CD required to hold the 10-ton crate in the position shown.

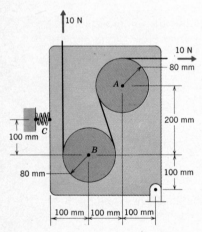

Prob. III.114

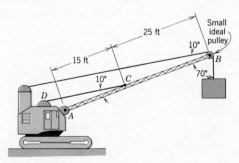

Prob. III.115

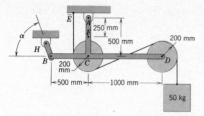

Prob. III.116

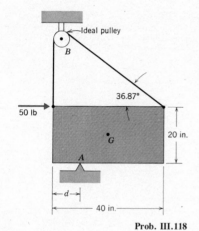

Prob. III.118

III.116 The forked bar is supported by fixed pin *A* and short link *BH*. A 50-kg block is suspended from a cable that passes over ideal pulleys *C* and *D* which are pinned to the bar, as shown. Knowing that $\alpha = 45°$, determine the reaction force exerted by the short link.

III.117 Solve Problem III.116 if, instead of being fastened at point *E*, the cable is fastened to the bar at point *F*.

III.118 The 100-lb box is supported by a knife edge at point *A* and by a cable that passes over the ideal pulley *B*. The center of mass of the box is at the geometric center *G*. For the loading shown, determine the tension in the cable (a) when $d = 10$ in., (b) when $d = 28$ in.

III.119 A gear box mounted on a beam is subjected to two torques, as shown. Determine the reactions at the supports of the beam.

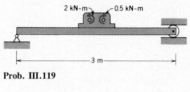

Prob. III.119

III.120–III.123 Determine the reactions required to support the beam shown.

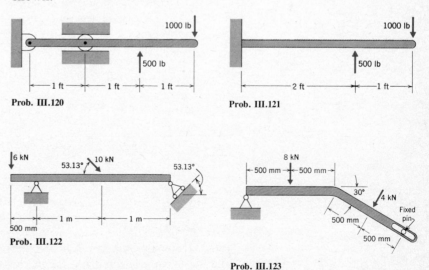

Prob. III.120

Prob. III.121

Prob. III.122

Prob. III.123

III.124–III.126 Determine the reactions required to support the structure shown.

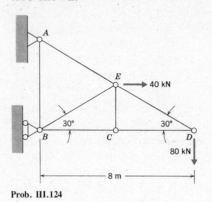

Prob. III.124

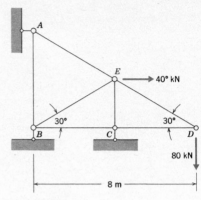

Prob. III.125

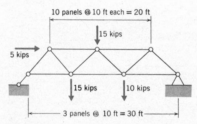

Prob. III.126

III.127 The structure shown is supported by three short links at points *A, B,* and *C*. Determine the reactions of these links when the angle for the link at point *C* is (a) $\alpha = 90°$, (b) $\alpha = 36.87°$.

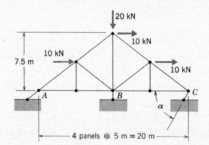

Prob. III.127

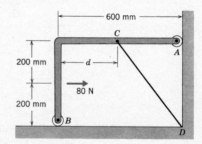

Prob. III.128

III.128 The bent bar is supported by rollers at ends *A* and *B* and by the cable. Knowing that $d = 300$ mm, determine the reactions at the rollers and the tension in the cable for the loading shown.

III.129   A tow truck exerts the horizontal force $\bar{F}$ to hold the 1500-kg automobile at rest on the hill in the position shown. The brakes are not applied. The center of mass of the automobile is point G. Determine the magnitude of the force $\bar{F}$.

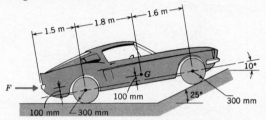

Prob. III.129

III.130   The winch supports a mass m in the position shown. Derive an expression for the required vertical force $\bar{V}$ in terms of the other parameters of the system.

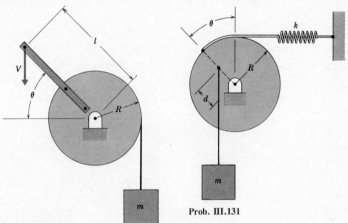

Prob. III.130

Prob. III.131

III.131   A block of mass m is suspended from the cylindrical drum and a spring is attached to a cable that may wrap around the drum, as shown. When $\theta = 0$ the spring is unstretched. Determine an expression for the value of θ in terms of the other parameters.

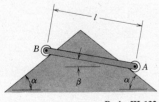

Prob. III.132

III.132   Rollers A and B of masses m_A and m_B, respectively, are interconnected by a lightweight rod of length l and placed on the smooth surface shown. Knowing that in the equilibrium position shown $\alpha = 30°$ and $\beta = 10°$, determine the ratio m_A/m_B.

III.133   For the system in Problem III.132, derive a general expression for the ratio m_A/m_B in terms of the angles α and β.

III.134–III.136 Derive an expression for the angle θ corresponding to equilibrium for the systems shown. In each case the bar has a mass m and its center of mass is at the midpoint G. All surfaces are smooth.

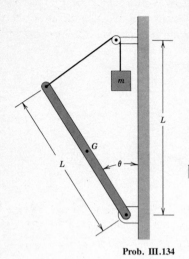

Prob. III.134

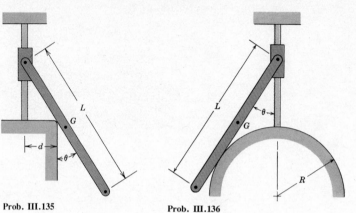

Prob. III.135

Prob. III.136

III.137 In each case, determine whether (a) the plate is completely, partially, or improperly constrained, and (b) whether the reactions are statically determinate or indeterminate for cases of complete constraint. (*Hint:* Refer to the last paragraph of Illustrative Problem 7.)

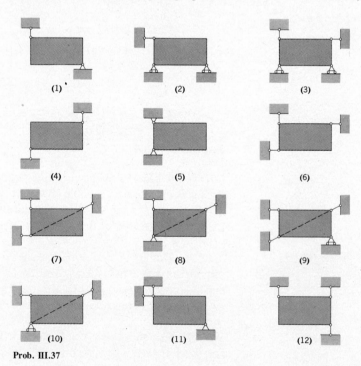

(1)   (2)   (3)

(4)   (5)   (6)

(7)   (8)   (9)

(10)   (11)   (12)

Prob. III.37

III.138 (a) For the forked bar in Illustrative Problem 1, determine the reactions at pins A and B as a function of the angle θ. (b) From the results in part (a), determine whether there are any values of θ for which the bar is improperly constrained.

For the systems in the following problems, determine the value of the dimension d that results in improper constraint of the system.

III.139   Problem III.110

III.140   Problem III.118

III.141   Problem III.128

For the systems in the following problems, determine the angle α that results in improper constraint of the system.

III.142   Problem III.116

III.143   Problem III.127

III.144   Point C on the top flange of the wide flange beam is subjected to the loads shown. Determine the reactions acting at point A on the axis of symmetry.

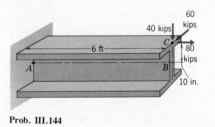

Prob. III.144

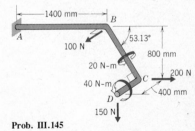

Prob. III.145

III.145   The bent bar, which is welded to the wall at end A, carries the loads shown. Determine the reactions.

III.146   Two identical rods of 2 kg mass each are welded at right angles and suspended by vertical wires. Section AB is horizontal and section BC is at 20° below the horizontal. Determine the tension in each wire. The center of gravity of each rod is at its midpoint.

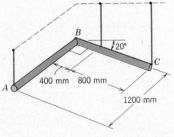

Prob. III.146

III.147 A package is placed on the three-legged circular table, whose mass is 30 kg. The reactions between the legs of the table and the ground are known to be $N_A = 100$, $N_B = 120$, $N_C = 130$ newtons. Determine the mass of the package and the values of R and θ describing its location.

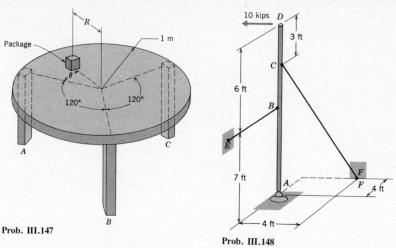

Prob. III.147

Prob. III.148

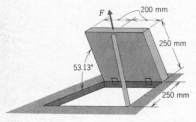

Prob. III.149

III.148 The antenna mast supported by a ball-and-socket joint at end A and two guy wires is subjected to the horizontal 10 kip load. Determine the tensions in the guy wires and the reactions at the ball-and-socket joint.

III.149 The cabinet door is supported by a rod that exerts an axial force $\bar{F}$ as shown. Knowing that the door has a mass of 10 kg and that its center of mass coincides with its geometric center, determine the magnitude of $\bar{F}$ for the open position shown.

III.150 The 10-kg horizontal flagpole is supported by a ball-and-socket joint at point A and two cables. Knowing that $d = 2$ m, determine the tension in the cables. The center of mass of the boom is at the midpoint.

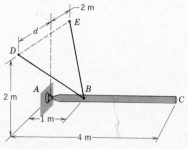

Prob. III.150

III.151 Solve Problem III.150 for $d = 1$ m.

III.152 The horizontal boom is held in position at an angle of 60° with respect to the vertical wall by the ball-and-socket joint and the two cables. Determine the reactions at the ball-and-socket joint and the tension in the cables resulting from the vertical 50-kN loads.

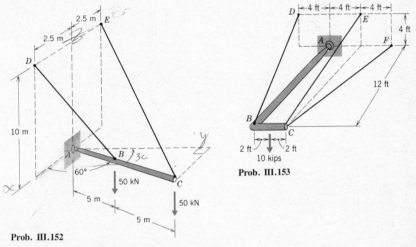

Prob. III.152

Prob. III.153

III.153 The bent bar *ABD* is supported by three cables and a ball-and-socket joint at end *A*. Determine the tension in the cables for the loading shown.

III.154 The rectangular plate is oriented vertically and supported by four short links at points *A* and *B* and by two cables. Determine the reaction forces of the links for the loading shown. The connections for the short links are ball-and-socket joints.

III.155 The bent bar is supported by three smooth eyebolts as shown. Is the system statically determinate and properly constrained?

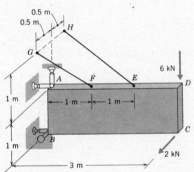

Prob. III.154

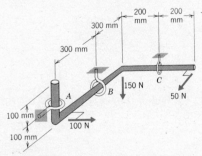

Prob. III.155

III.156 A rigid bent tube guides a flexible shaft that is transmitting a 20 lb-ft torque, as shown. The tube is supported by the smooth eyebolts at A, B, and C. Is the system properly constrained?

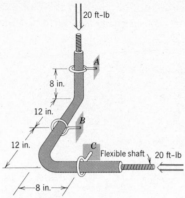

Prob. III.156

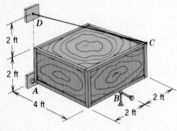

Prob. III.157

III.157 A 2-ton crate is supported by a ball joint at point A, two short links at point B, and cable CD. The connections for the short links are also ball joints. The center of mass of the crate is at its geometric center. Determine the tensile force in the cable and the reaction forces of the links at point B.

III.158 The winch is holding a 100-kg package. Determine the vertical force $\bar{F}$ and the reactions at ball bearings A and B for the position shown. The drum of the winch has a diameter of 400 mm.

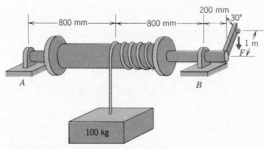

Prob. III.158

III.159 The T-bar is supported by thrust bearing A, bearing B, and the smooth wall at end D. The bar fits loosely in both bearings. Knowing that sections *AB* and *CD* each have a mass of 2 kg, and that their centers of mass are at their respective midpoints, determine the reactions exerted by each bearing and by the wall.

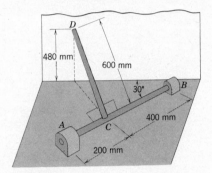

Prob. III.159

III.160 Ends *A* and *B* of the 5-kg rigid bar are attached by ball joints to lightweight collars that may slide over the smooth fixed rods shown. The center of mass of the bar is at its midpoint. Determine the horizontal force $\bar{F}$ applied to collar *A* that will result in static equilibrium in the position where $d = 1$ m. The length of the bar is 1.5 m.

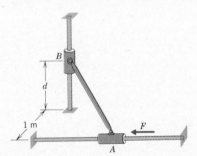

Prob. III.160

III.161 Solve Problem III.160 for the position where $d = 0.5$ m.

MODULE IV
STRUCTURES

The methods for evaluating the forces acting on a rigid body were developed in Module III. A few systems encountered there contained interconnected bodies. In those situations it was sometimes necessary to investigate each body in the system, using the law of action and reaction.

The systems we will consider in this module all share the feature of being composed of many interconnected bodies. These systems are *engineering structures,* which are designed for the purpose of safely resisting, transmitting, and, at times, modifying the various forces applied to them. The individual bodies forming a structure are called its *members.* In this module we will treat two categories of structures. They differ in the manner in which their members transfer forces.

A.   TRUSSES

1   Two Force Members

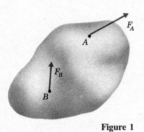

A primary characteristic of a truss is that is is composed *solely* of two force members, that is, members having forces applied at only two locations. To see why this feature is important we have illustrated an arbitrary two force member in Figure 1.

In this figure $\bar{F}_A$ and $\bar{F}_B$ are the forces acting at points A and B, respectively. Let us investigate the relationship that must exist between these two forces for static equilibrium of the body. Choosing point A for the moment sum, the equilibrium equations are

Figure 1

$$\Sigma \bar{M}_A = \bar{r}_{B/A} \times \bar{F}_A = \bar{0}$$

$$\Sigma \bar{F} = \bar{F}_A + \bar{F}_B = \bar{0} \tag{1}$$

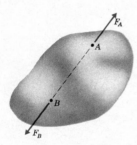

The force sum requires that $\bar{F}_A = -\bar{F}_B$; in other words, $\bar{F}_A$ and $\bar{F}_B$ must have equal magnitude and be oriented in opposite directions. Recalling that the cross product of two vectors is zero only when they are parallel, we see that the moment sum requires that $\bar{F}_B$ be parallel to the line connecting points A and B. Hence, for static equilibrium the forces must be directed as shown in Figure 2.

In summary, the equilibrium equations require that the forces at each end must be equal in magnitude, opposite in direction, and collinear. Clearly, it is possible for the forces to be reversed from the orientations shown.

Figure 2

A further restriction on the members of a truss is that they must be straight bars, such as the one illustrated in Figure 3a on the following page. The force system shown there is said to place the bar in a state of *tension,* because it tends to pull the bar apart, as if it were a cable. In fact, in this case the bar acts exactly like a cable, carrying a constant tension F_{AB} on every cross section, as shown in Figure 3a.

The difference between the bar in Figure 3a and a cable is that the forces in the bar can also have the effect of pushing the ends together.

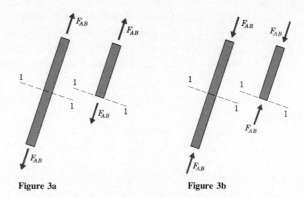

Figure 3a   **Figure 3b**

This is the state of *compression,* illustrated in Figure 3b.

2   Characteristics of a Truss

A *truss* is a rigid framework of straight two-force members that are joined together at their ends. In order that the reactions at the ends of the members consist of forces, without couples, we model the connections of the members as pins in the case of planar trusses, or ball-and-socket joints in the case of three-dimensional trusses. The latter are usually referred to as *space trusses*. To gain some appreciation for the type of structure in which we are interested, Figure 4 shows some typical planar trusses; those with a peak are generally used to support roofs, whereas the flat ones are generally used to support bridges. The dots in each diagram indicate the pin connections.

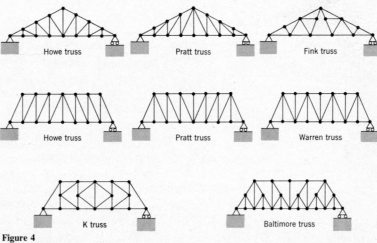

Figure 4

In an actual structure the members may be I-beams, channels, angle bars, or bars with other types of cross sections. The most common way of fastening members together is by means of welding, riveting, or bolting to gusset plates, as shown in Figure 5a. However, the model of a pin connection shown in Figure 5b is sufficiently accurate, provided that the center-lines of all members meeting at that connection are concurrent. (The proof

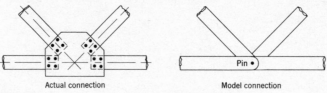

Actual connection   Model connection

Figure 5a   **Figure 5b**

of the validity of this approximation is beyond the scope of this book.) The terminology used to describe the model connection is to refer to it as a *joint*.

It will be noted that the members in each of the trusses illustrated in Figure 4 are interconnected to form traingles. Comparing the basic truss component in Figure 6a to the quadrilateral in Figure 6b, we see that the triangular shape is essential to the formation of a rigid truss. If we try to

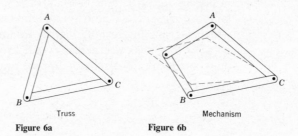

Truss   Mechanism

Figure 6a   **Figure 6b**

form a polygon having more than three sides, the pin joints allow the bars to rotate relative to one, thus forming a nonrigid structure called a *mechanism*.

Using the truss of Figure 6a as a basis, we can increase the size of the structure by attaching more members. If this addition results in the formation of additional basic triangular components, we have a *simple truss,* such as the planar one shown in Figure 7a and the space truss shown in Figure 7b on the following page.

As we will see, simple trusses are statically determinate, allowing for the evaluation by the basic equilibrium equations of the forces being transmitted by each member. Simple trusses will be the focus of our attention. However, not all trusses are simple. For example, the planar

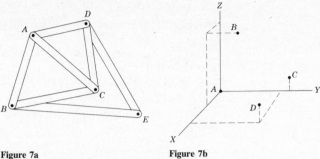

Figure 7a   **Figure 7b**

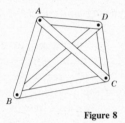

Figure 8

truss in Figure 8 was formed by adding three members to the basic truss component in Figure 6a. Not all of the members in this truss are needed to form the polygon *ABCD;* either of the diagonal members *AD* or *BC* could be removed to form a simple truss. The extra member is *redundant.* It can be shown that this structure is statically indeterminate.

For statically indeterminate trusses, in addition to the equations of static equilibrium, the evaluation of the force in each member requires that the small deformations of the bars resulting from the loading be considered. In other words, the model of a rigid body is not adequate for treating statically indeterminate structures. The subject of indeterminate trusses is treated in courses on strength of materials and structural analysis.

There is a simple relationship between the number of joints and the number of members necessary to form a simple truss. Starting from the basic triangle in Figure 6a, where there are three members and three joints, we see that for a planar truss, such as that in Figure 7a, additional triangles are formed by adding two bars for every joint added. Thus, letting m be the number of members and j be the number of joints, we find that

$$m = 2j - 3 \qquad \text{for a simple planar truss} \qquad (2a)$$

In case of a simple space truss, we see that the requirement of rigid triangles leads to the formation of tetrahedrons, such as that in Figure 7b. Each additional joint now requires three additional members, so

$$m = 3j - 6 \qquad \text{for a simple space truss} \qquad (2b)$$

Another characteristic of a truss pertains to the way in which it is loaded. Because the ends of the members are interconnected at the joints, it follows that, in order for there to be two force members, the external forces acting on the structure must be applied at the joints only. Any loading that is not applied at the joints requires that the member to which it is applied transfer more than just a tensile or compressive force. Hence, we have the following criteria for determining when a structure acts like a truss:

1   All members are straight.
2   The centerlines of all members that are connected at a joint must be concurrent at that joint.
3   The external forces acting on the structure must be applied at the joints.

A problem arises in considering the question of the weight of each member, because this force acts at the midpoint of the member, and thus violates the second of the above criteria. In most cases the weight force can be neglected, because it will be small compared to the other forces acting on the structure.

There are two methods that may be employed to evaluate the member forces. We shall make it a standard practice when applying either method *to evaluate first the reaction force being exerted by the supports,* so that these forces may be regarded as known when we determine the forces in the members. (There are truss problems where it is not necessary to first determine the reactions. We will discuss these cases after both methods have been fully developed.)

3   Method of Joints

To describe this method, let us consider a single triangular truss element, such as that shown in Figure 9a.

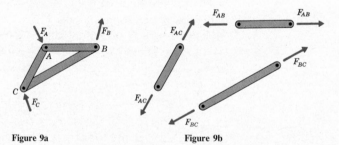

Figure 9a   Figure 9b

The external forces $\bar{F}_A$, $\bar{F}_B$, and $\bar{F}_C$ are known. To ensure equilibrium of the structure, we now consider each member of the structure individually in a free body diagram, as shown in Figure 9b. In view of the earlier discussion, equilibrium for each member requires that its end forces be equal in magnitude, opposite in direction, and collinear. The notation we use is to denote the force in each member by subscripts corresponding to the joints at each end. Also note that in Figure 9b we begin by assuming, for the sake of convenience, that each member is in tension.

Each member is automatically in equilibrium by considering it to be in tension. To determine these axial forces we recognize that the structure is formed from pin joints, as well as straight members. The pins forming the joints must also be in equilibrium, so we draw free body diagrams of each joint; hence the name *method of joints*. Employing Newton's third law and recalling that we have assumed each member to be in tension, the force exerted by each member on a joint must pull away from that joint, toward the other end of the member. Thus, the free body diagrams of the members and joints of the truss in Figure 9a are as shown in Figure 10. Note that the external force acting at each joint must also be shown to complete the free body diagram.

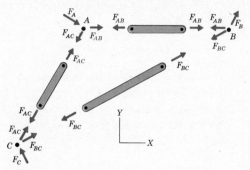

Figure 10

The unknown forces in the free body diagrams are those of the members of the truss. It is only necessary for us to consider equilibrium of the joints, for the members shown in Figure 10 are identically in equilibrium, as we have already noted. Because the forces at each joint are concurrent, each joint will be in equilibrium if the resultant force acting on it is zero. Thus, choosing a convenient coordinate system, such as that shown in Figure 10 for this planar system we have $\Sigma F_X = 0$ and $\Sigma F_Y = 0$ at each joint. In all, we have six scalar equations for the system in Figure 10.

At this juncture it might seem that we have more equations than we need, in that the only unknowns are the three member forces. This is a subtle point. Recall that earlier we decided to begin by obtaining the reactions of the truss. This procedure enables us to say that the structure as a whole is in equilibrium. However, the entire structure will also be in equilibrium if the forces acting on each joint are equilibrated. In other words, we have three excess equations, corresponding to equilibrium of the entire structure. The extra equations may be employed to check the solution, as will be seen in the examples that follow.

We have assumed that all the members are in tension. This assumption is confirmed when the magnitude of the member force is found to be positive, whereas a negative value indicates a state of compression. Clearly, it is not necessary for us to regard tension as positive, but this convention allows for an orderly interpretation of the results. It is paramount for the design of a member to know whether the force being transmitted is tensile or compressive, for the design procedure used depends entirely on this knowledge.

The foregoing discussion can be extended to treat general planar trusses, as well as space trusses. Briefly stated, we see that

---

The method of joints consists of drawing free body diagrams for all joints and then solving the equilibrium equations for the concurrent force systems acting at each joint.

In the case of a planar truss, we have $\Sigma F_X = 0$ and $\Sigma F_Y = 0$ for each joint, so we obtain a total of $2j$ equilibrium equations for the j joints of a planar truss. However, the equilibrium of the entire structure is ensured by the determination of the reactions; only $2j - 3$ of the joint equations are independent. The other three equations can be used to check the results.

Hence, in a planar truss we see that in order for the member forces to be statically determinate, there must be $2j - 3$ members. Notice that this is the number of members required for a simple planar truss, as given by equation (2a). A planar truss having more than $2j - 3$ members is *internally indeterminate*, even though its reactions may be determinate. On the other hand, if there are fewer than $2j - 3$ members, the equilibrium equations for the joints cannot be satisfied. The truss is then not a rigid structure.

The foregoing are merely necessary conditions for having an internally determinate, rigid planar truss. It is possible to construct planar trusses that have j joints and $2j - 3$ members that are not rigid and statically determinate. As always, the ultimate verification of these conditions rests in the solvability of the equilibrium equations for the joints.

In the case of a space truss we have $\Sigma F_X = 0$, $\Sigma F_Y = 0$, and $\Sigma F_Z = 0$ for each joint. Thus, a space truss having j joints will have $3j$ equilibrium equations for these joints. Deducting the six checking equations corresponding to the static equilibrium of the entire structure, we see that the conditions of internal determinancy and rigidity require $3j - 6$ members. This is the same number of members as that given by equation (2b) for a simple space truss. A space truss with more than $3j - 6$ members is internally indeterminate, whereas a space truss having fewer than this number of members is not rigid.

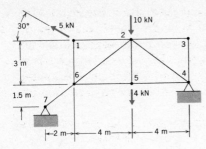

EXAMPLE 1

Determine the force in each member of the truss shown using the method of joints.

Solution

We begin by determining the reactions. Member 6-7 is a two-force member, so it can apply only an axial force to the rectangular portion of the truss. The other point of support of the truss is at pin joint 4, so we have the following free body diagram for the entire truss.

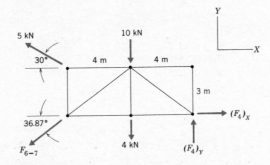

For consistency we have assumed that $\bar{F}_{6-7}$ is a tensile force. The dimensions of the truss are given in terms of horizontal and vertical distances, so the coordinate axes are aligned in these directions.

The Z axis is outward for the coordinate system shown, making counterclockwise moments positive. In order to eliminate two reaction components, we choose joint 4 for the moment sum. The equilibrium equations are then

$$\Sigma M_{4Z} = 10(4) + 4(4) + (F_{6-7} \sin 36.87°)8$$

$$- (5 \sin 30°)8 + (5 \cos 30°) \, 3 = 0 \text{ kN-m}$$

$$\Sigma F_X = -F_{6-7} \cos 36.87° + (F_4)_X - 5 \cos 30° = 0$$

$$\Sigma F_Y = -F_{6-7} \sin 36.87° + (F_4)_Y - 4 - 10 + 5 \sin 30° = 0$$

Solving the moment equation for F_{6-7} and using the force sums, we determine $\bar{F}_4$. Thus

$$F_{6-7} = -10.206 \text{ kN} \quad (F_4)_Y = 5.376 \quad (F_4)_X = -3.835 \text{ kN}$$

The equilibrium of each joint of the truss may now be considered. To assist in the calculations we sketch the truss showing the angles between the members and the coordinate directions, and all loadings, including the reactions, in their correct orientation.

Note that the force $(\bar{F}_4)_X$ is pushing to the left, because its value was negative. Also note that the principle of transmissibility does not apply

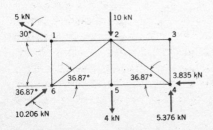

when finding the forces in the internal members of the structure, so each external force must be shown at the joint at which it acts.

This sketch can now be used to draw free body diagrams of each joint, using the convention of considering the members to be in tension. Hence the members pull on the joint, as shown. Note that each joint is labeled in the free body diagrams.

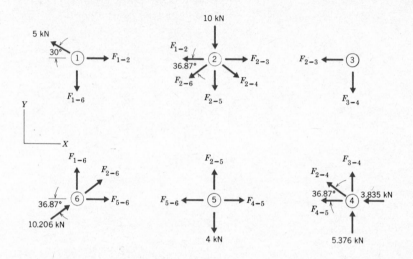

For computational purposes *it is best to consider joints connecting the fewest members first.* This allows us to solve successively for the member forces rather than dealing with a large number of simultaneous equations. In other words, in planar problems we solve the equilibrium equations at joints that contain two or less unknowns first. After referring to the free body diagrams, the solution order we choose, by joints, is 1-3-4-5-6-2.

Joint 1:

$$\Sigma F_X = F_{1-2} - 5 \cos 30° = 0$$

$$\Sigma F_Y = -F_{1-6} + 5 \sin 30° = 0$$

$$F_{1-2} = 4.330 \qquad F_{1-6} = 2.50 \text{ kN}$$

Joint 3:

$$\Sigma F_X = -F_{2-3} = 0 \qquad \Sigma F_Y = -F_{3-4} = 0$$

$$F_{2-3} = F_{3-4} = 0$$

Joint 4:

$$\Sigma F_X = -F_{4-5} - F_{2-4} \cos 36.87° - 3.835$$

$$\equiv -F_{4-5} - 0.8F_{2-4} - 3.835 = 0$$

$$\Sigma F_Y = F_{3-4} + F_{2-4} \sin 36.87° + 5.376$$

$$\equiv 0 + 0.6F_{2-4} + 5.376 = 0$$

$$F_{2-4} = -8.960 \qquad F_{4-5} = 3.333 \text{ kN}$$

Joint 5:

$$\Sigma F_X = F_{4-5} - F_{5-6} \equiv 3.333 - F_{5-6} = 0$$

$$\Sigma F_Y = F_{2-5} - 4 = 0$$

$$F_{5-6} = 3.333 \qquad F_{2-5} = 4 \text{ kN}$$

Joint 6:

$$\Sigma F_X = F_{5-6} + F_{2-6} \cos 36.87° + 10.206 \cos 36.87°$$

$$\equiv 3.333 + 0.8F_{2-6} + 8.165 = 0$$

$$\Sigma F_Y = F_{1-6} + F_{2-6} \sin 36.87° + 10.206 \sin 36.87°$$

$$\equiv 2.50 + 0.6F_{2-6} + 6.124 = 0$$

$$F_{2-6} = -14.373 \text{ kN}$$

Both equilibrium equations for joint 6 give the same value of F_{2-6}. This is the first check of the computations. The equilibrium equations for the remaining joint, number 2, give two more computational checks because the forces in all members are already known.

Joint 2:

$$\Sigma F_X = F_{2-3} + F_{2-4} \cos 36.87° - F_{2-6} \cos 36.87° - F_{1-2}$$

$$\equiv 0 + (-8.960)0.8 - (-14.373)0.8 - 4.330 \equiv 0$$

$$\Sigma F_Y = -(F_{2-6} + F_{2-4}) \sin 36.87° - F_{2-5} - 10$$

$$\equiv -(-14.373 - 8.960)0.6 - 4 - 10 \equiv 0$$

Because all of the check equations have been verified, we can be fairly certain that our calculations of the member forces, as well as of the reactions, are correct. The member forces that are negative indicate that the member is in compression.

When solving a problem by the method of joints, it may happen that you have chosen to consider the joints in the wrong order, so that the member forces for a certain joint cannot be found immediately (e.g., more than two unknowns at a joint in a planar problem). If that should happen, proceed to the other joints until a sufficient number of forces have been calculated to allow the joint to be reconsidered.

As a concluding remark, recall that we began the solution by considering member 6–7 as a support of the rectangular truss. This allowed us to express the reaction on the left side of the truss as the axial force in that member. Suppose we had instead considered member 6–7 to be part of the truss, and ignored the fact that this member is a two-force member. The corresponding reaction forces would have been those of pin 7, so we would have had four reaction components (two for pin 7 and two for pin 4). The equilibrium equations for the entire structure would have then been insoluble by themselves. In that case it would have been necessary to consider the equilibrium of the truss joints first, and then to use the results for the joints to reexamine the reactions. Clearly, it is better to be able to recognize two-force members that support a structure.

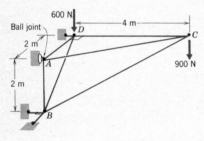

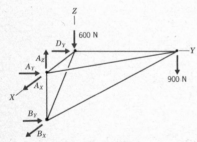

EXAMPLE 2

Using the method of joints, determine the forces in the members of the space truss shown.

Solution

We first determine the reactions by drawing a free body diagram of the truss, noting that the short links at joints B and D can only exert axial forces. Assuming that these forces are tensile, we have the free body diagram of the truss on the left. As always, the XYZ coordinate system is chosen to fit the directions in which dimensions are given.

In order to eliminate the three reaction components from the ball joint, point A is chosen for the moment sum. The equilibrium equations are then

$$\Sigma M_A = \bar{r}_{D/A} \times (D_Y \bar{J} - 600\bar{K}) + \bar{r}_{B/A} \times (B_X \bar{I} + B_Y \bar{J})$$

$$+ \bar{r}_{C/A} \times (-900\bar{K})$$

$$= (-2\bar{I}) \times (D_Y \bar{J} - 600\bar{K}) + (-2\bar{K}) \times (B_X \bar{I} + B_Y \bar{J})$$

$$+ (-2\bar{I} + 4\bar{J}) \times (-900\bar{K})$$

$$= 2D_Y \bar{K} - 1200\bar{J} - 2B_X \bar{J} + 2B_Y \bar{I} - 1800\bar{J} - 3600\bar{I} = \bar{0} \text{ N-m}$$

$$\Sigma \bar{F} = (A_X \bar{I} + A_Y \bar{J} + A_Z \bar{K}) + (D_Y \bar{J}) + (B_X \bar{I} + B_Y \bar{J})$$

$$- 900\bar{K} - 600\bar{K} = \bar{0} \text{ N}$$

Equating the components of the moment sum to zero yields

$\bar{I}$ component: $\quad\quad 2B_Y - 3600 = 0 \quad$ so $\quad B_Y = 1800$ N

$\bar{J}$ component: $\quad\quad -2B_X - 3000 = 0 \quad$ so $\quad B_X = -1500$ N

$\bar{K}$ component: $\quad\quad -2D_Y = 0 \quad$ so $\quad D_Y = 0$

Using these values of B_X, B_Y, and D_Y in equating the components of the force sum to zero yields

$\bar{I}$ component: $\quad\quad A_X + B_X = 0 \quad$ so $\quad A_X = 1500$ N

$\bar{J}$ component: $\quad\quad A_Y + D_Y + B_Y = 0 \quad$ so $\quad A_Y = -1800$ N

$\bar{K}$ component: $\quad\quad A_Z - 1500 = 0 \quad$ so $\quad A_Z = 1500$ N

We now consider the equilibrium of the joints. The member forces are collinear with the corresponding members. To describe their directions we employ unit vectors parallel to each bar. Recalling that in our notation a unit vector extends *from* the point denoted by the second subscript *to* the point denoted by the first subscript, these unit vectors are

$$\bar{e}_{B/A} = -\bar{e}_{A/B} = \frac{\bar{r}_{B/A}}{|\bar{r}_{B/A}|} = -\bar{K}$$

$$\bar{e}_{C/A} = -\bar{e}_{A/C} = \frac{\bar{r}_{C/A}}{|\bar{r}_{C/A}|} = \frac{-2\bar{I} + 4\bar{J}}{\sqrt{2^2 + 4^2}}$$

$$= -0.4472\bar{I} + 0.8944\bar{J}$$

$$\bar{e}_{D/A} = -\bar{e}_{A/D} = \frac{\bar{r}_{D/A}}{|\bar{r}_{D/A}|} = -\bar{I}$$

$$\bar{e}_{C/B} = -\bar{e}_{B/C} = \frac{\bar{r}_{C/B}}{|\bar{r}_{C/B}|} = \frac{-2\bar{I} + 4\bar{J} + 2\bar{K}}{\sqrt{2^2 + 4^2 + 2^2}}$$

$$= -0.4082\bar{I} + 0.8165\bar{J} + 0.4082\bar{K}$$

$$\bar{e}_{D/B} = -\bar{e}_{B/D} = \frac{-2\bar{I} + 2\bar{K}}{\sqrt{2^2 + 2^2}} = -0.7071\bar{I} + 0.7071\bar{K}$$

$$\bar{e}_{D/C} = -\bar{e}_{C/D} = -\bar{J}$$

When drawing the free body diagrams for each joint, we assume that all members are in tension. Therefore, the axial force of each member at a joint is written as the product of the magnitude of the force and the unit vector from that joint to the other end of that member. For instance, the force member BC exerts on joint B is $F_{BC}\,\bar{e}_{C/B}$, whereas the force that this member exerts on joint C is $F_{BC}\,\bar{e}_{B/C}$. Thus, we have the free body diagrams on the left.

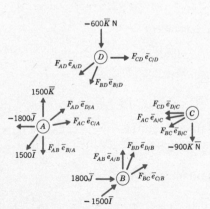

Particular attention is called to the fact that the reaction forces are entered into the free body diagram in the direction in which they were calculated to act.

In the space truss we are analyzing each joint connects only three members, so the joints may be considered in any order we desire. Starting with joint D, the equilibrium equations are

Joint D:

$$\Sigma F = F_{AD}\,\bar{e}_{A/D} + F_{BD}\,\bar{e}_{B/D} + F_{CD}\,\bar{e}_{C/D} - 600\bar{K}$$

$$\equiv F_{AD}(+\bar{I}) + F_{BD}(+0.7071\bar{I} - 0.7071\bar{K})$$

$$+ F_{CD}(+\bar{J}) - 600\bar{K} = \bar{0}$$

The following three scalar equations are found by equating the sum of corresponding components to zero.

$\bar{I}$ component: $F_{AD} + 0.7071F_{BD} = 0$

$\bar{J}$ component: $F_{CD} = 0$

$\bar{K}$ component: $-0.7071F_{BD} - 600 = 0$

$$F_{BD} = -849 \qquad F_{AD} = 600 \text{ N}$$

Next we write the equilibrium equations for joint C, which are

Joint C:

$$\Sigma \bar{F} = F_{AC}\,\bar{e}_{A/C} + F_{BC}\,\bar{e}_{B/C} + F_{CD}\,\bar{e}_{D/C} - 900\bar{K}$$

$$\equiv F_{AC}(0.4472\bar{I} - 0.8944\bar{J}) + F_{BC}(0.4082\bar{I}$$

$$- 0.8165\bar{J} - 0.4082\bar{K}) + (0)(-\bar{J}) - 900\bar{K} = \bar{0}$$

Equating like components to zero yields

$\bar{I}$ component: $0.4472F_{AC} + 0.4082F_{BC} = 0$

$\bar{J}$ component: $-0.8944F_{AC} - 0.8165F_{BC} = 0$

$\bar{K}$ component: $-0.4082F_{BC} - 900 = 0$

$$F_{BC} = -2205 \qquad F_{AC} = 2013 \text{ N}$$

These values for F_{BC} and F_{AC} satisfy all three equations (the $\bar{J}$ equation is -2 times the $\bar{I}$ equation), so we have validation of the calculations from the first excess equation.

The last unknown member force is found from the equilibrium equations for joint B, which is

Joint B:

$$\Sigma \bar{F} = F_{AB}\,\bar{e}_{A/B} + F_{BC}\,\bar{e}_{C/B} + F_{BD}\,\bar{e}_{D/B} - 1500\bar{I} + 1800\bar{J}$$

$$\equiv F_{AB}\bar{K} - 2205(-0.4082\bar{I} + 0.8165\bar{J} + 0.4082\bar{K})$$

$$- 849(-0.7071\bar{I} + 0.7071\bar{K}) - 1500\bar{I} + 1800\bar{J} = \bar{0}$$

In component form this yields

$\bar{I}$ component: $900 + 600 - 1500 \equiv 0$

$\bar{J}$ component: $-1800 + 1800 \equiv 0$

$\bar{K}$ component: $F_{AB} - 900 - 600 = 0$ $F_{AB} = 1500$ N

The first two equations are checks on the calculations. With the forces in all members known, the component equations resulting from equilibrium of joint A will be three more checks. We leave it to you to perform the necessary computations for their verification. Once again, remember that member forces that are negative indicate compression.

HOMEWORK PROBLEMS

IV.1–IV.14   Determine the forces in the members of each truss shown. State whether each member is in tension or compression.

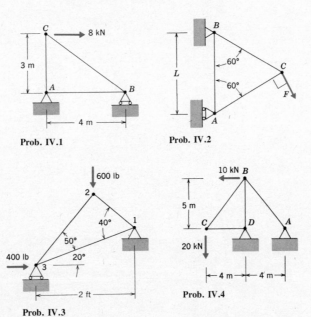

Prob. IV.1

Prob. IV.2

Prob. IV.3

Prob. IV.4

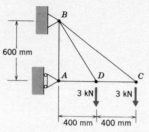

Prob. IV.5

B

600 mm

A   D   C

3 kN   3 kN

400 mm   400 mm

3 @ 6 ft = 18 ft

8 ft

4   3   2   1

5   6   7   8

10 kips   10 kips   10 kips

Prob. IV.6

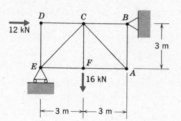

D   C   B

12 kN

3 m

E   F   A

16 kN

3 m   3 m

Prob. IV.7

2.5 m

A

G   F

2.5 m

B   C   D   E

8 kN   4 kN   8 kN

3 @ 4 m = 12 m

Prob. IV.8

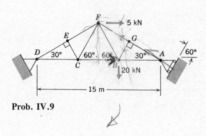

F   5 kN

E   G

D   30°   60°   60°   30°   A   60°

C   B   20 kN

15 m

Prob. IV.9

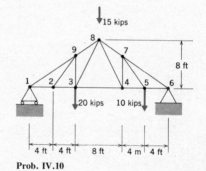

15 kips

8

9   7

8 ft

1   2   3   4   5   6

20 kips   10 kips

4 ft   4 ft   8 ft   4 m   4 ft

Prob. IV.10

4 m

4   5   5 kN

3   5 kN

6

3 @ 3 m = 9 m

2   5 kN

7

1   8

36.87°

Prob. IV.11

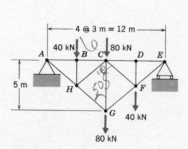

4 @ 3 m = 12 m

40 kN   80 kN

A   B   C   D   E

5 m

H   F

G   40 kN

80 kN

Prob. IV.12

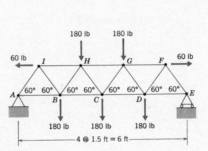

180 lb   180 lb

60 lb   I   H   G   F   60 lb

A 60° 60° 60° 60° 60° 60° 60° 60° E
      B   C   D

180 lb   180 lb   180 lb

—— 4 @ 1.5 ft = 6 ft ——

Prob. IV.13

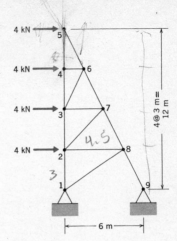

4 kN ← 5

4 kN → 6
        4

4 kN → 7
        3

4 kN → 8
        2

        1   9

|← 6 m →|

4 @ 3 m = 12 m

Prob. IV.14

IV.15 Determine the forces in all members of the symmetrical space truss when it is subjected to the self-equilibrating set of loads shown. State whether the members are in tension or compression.

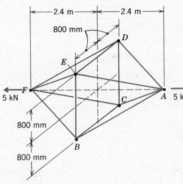

|← 2.4 m →|← 2.4 m →|

800 mm
                    D
        E

5 kN ← F   A 5 kN
                    C

800 mm

        B

800 mm

Prob. IV.15

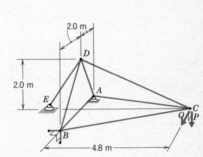

2.0 m
            D

2.0 m
        E   A
                            C
                          Q P
        B

—— 4.8 m ——

Prob. IV.16

IV.16 Determine the forces in the members of the space truss if $P = 30$ kN and $Q = 0$. Indicate whether the members are in tension or compression.

IV.17 Solve Problem IV.16 if $P = 30$ kN and $Q = 20$ kN.

IV.18 Determine the forces in all members of the space truss if $P = 4$ kips and $Q = 0$. Indicate whether the members are in tension or compression.

IV.19 Solve Problem IV.18 for $P = 4$ kips and $Q = 3$ kips.

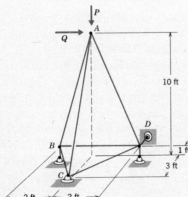

P

Q →   A

10 ft

        D

B   1 ft
                      3 ft
    C

|← 2 ft →|← 2 ft →|

Prob. IV.18

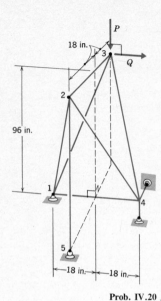

Prob. IV.20

IV.20 Determine the forces in all members of the space truss if $P = 400$ lb and $Q = 0$.

IV.21 Solve Problem IV.20 if $P = 400$ lb and $Q = 500$ lb.

IV.22 Determine the forces in all members of the space truss if $P = 12$ kN and $Q = 0$.

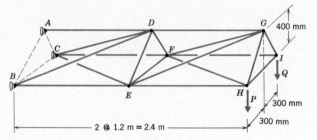

Prob. IV.22

IV.23 Solve Problem IV.22 if $P = 12$ kN and $Q = 18$ kN.

4 Method of Sections

The primary objective of the method of joints is the determination of the forces in all members of a truss. To this end it is necessary to solve a large number of algebraic equations. Suppose that we are interested only in the forces in a few members of the truss. If we use the method of joints, we will have to determine the forces in many members that are of no interest to us. (There are instances, of course, where the member force desired can still be found from the equilibrium equations for one or two joints. However, such situations are not typical.)

The method of sections is intended specifically to be used for the determination of the forces in a few members of a truss. Equally important, in learning to use the method of sections we will strengthen our ability to disassemble a structure to create useful free body diagrams.

Recall that we derived the method of joints by breaking the truss up into its individual components of bars and pins. In the method of sections we leave the structure assembled, decomposing the truss into two distinct portions. This decomposition is achieved by passing an imaginery *cutting plane* through the structure. For example, suppose that we wanted to know the force in member *CH* of the truss in Figure 11 on the following page. In order to determine $\bar{F}_{CH}$ it must appear as an external force in a free body diagram. Hence, *the cutting plane must cut through the member of interest.*

In the process of passing the cutting plane *a-a* through the structure,

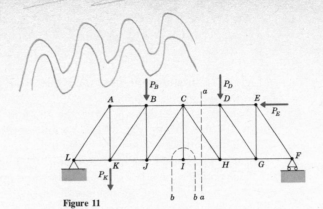

Figure 11

other members are also cut. In this manner the structure is decomposed into two parts, each of which must be in static equilibrium. The forces in the cut members are now external forces for these two truss segments. Once again, we assume the member forces to be tensile, so they pull on the isolated portions of the structure. The free body diagrams for the cut portions to the left and right of plane *a-a* are as shown in Figure 12.

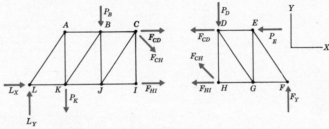

Figure 12

Recalling that our standard procedure is to determine the reaction forces first, the only unknown forces in these free body diagrams are those corresponding to the cut members: F_{CD}, F_{CH}, F_{HI}. Each isolated portion of the structure forms a rigid body carrying a planar force system, so we have three equations of static equilibrium for each body. The equations for *either* part will yield the values of the member forces being sought. Because the equilibrium of the entire structure was assured by the determination of the reactions exerted by the supports, the equilibrium equations for the remaining portion of the truss merely repeat the other equilibrium equations. The decision as to which portion of the cut truss should be equilibrated is arbitrary, although the portion having the smaller number of external forces and simpler geometry is usually the more convenient one to use for calculations.

The cutting plane need not be flat; it can be a curved plane. Also, it need not cut through an entire truss. In fact, we could even use a curved plane to isolate a joint. For example, let us determine the force in member *CI* of the truss in Figure 11. Passing plane *b−b* through the truss (actually around the joint), we can isolate the joint in the free body diagram shown

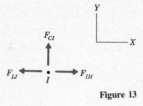

in Figure 13. Summing forces in the Y direction shows that member IC is a zero force member.

The major decision in using the method of sections involves the choice for the cutting plane. Let us first consider the case of a planar truss. If the cutting plane cuts through no more than three members, as was done in Figure 11, we may determine the member forces directly from the three equilibrium equations for an isolated portion of the structure. On the other hand, if a cutting plane cuts through four or more members, it will not be possible to determine all of the cut member forces, although it might be possible to obtain values for a few of the forces.

A situation where it is not possible to determine the member force of interest with one cutting plane is typified by the truss in Figure 14.

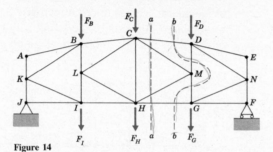

Figure 14

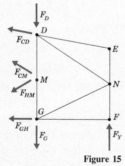

Figure 15

Suppose that we wish to determine the force in member CM. Any cutting plane, such as plane a-a, that cuts this member also cuts at least three other members. The free body diagram resulting from plane a-a is shown in Figure 15, where $\bar{F}_Y$ is the known reactive force. For the isolated section there are only three equations of static equilibrium, so the four unknowns F_{CD}, F_{CM}, F_{HM}, and F_{GH} cannot be found solely from these equations.

Let us now consider another section, formed from the curved cutting plane b−b appearing in Figure 14; the corresponding free body diagram is shown in Figure 16. Once again, we cannot solve for all the unknown forces, for the free body diagram leads to three equations of static equilibrium in the four unknowns, F_{CD}, F_{DM}, F_{GM}, F_{GH}. However, by summing moments about joint G, we can determine the force in member CD. This value may then be used in the static equilibrium equations obtained from the free body diagram shown in Figure 15, which then enables us to solve for the force in member CM. From this example we see that, at times, more than one cutting plane may be necessary to determine the force in a specific member.

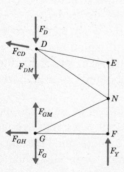

Figure 16

The method of sections may also be applied to space trusses, in which case equilibrium of the isolated portion of the structure results in six scalar equations. If the cutting plane cuts through no more than six

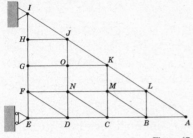

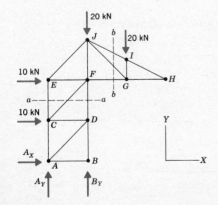

members, we will be able to determine all of the forces in the cut members. A cutting plane intersecting seven or more members will leave us unable to determine at least some of the member forces. In the examples that follow we shall concentrate on planar trusses, because the only true difference between these trusses and three-dimensional ones involves the techniques for forming the equilibrium equations.

Finally, the discussion of the method of sections would be incomplete if we did not note that it is not universally applicable. In Figure 17 we have a simple truss that requires an excessive number of cutting planes to evaluate the force in an interior member, such as member NM. In such a case we resort to the method of joints, as we would if we wanted to know the forces in many or all of the members of any truss.

Figure 17

EXAMPLE 3
Determine the forces in members CF and FG of the truss shown.

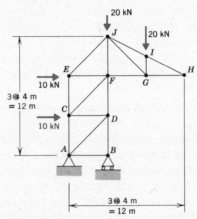

Solution
We begin with a free body diagram of the entire structure, which is used for calculating the reactions. The XYZ coordinate system is chosen horizontal and vertical, because the structure's dimensions are given in that manner.

Summing forces and summing moments about point A, we find

$$\Sigma M_{AZ} \equiv 4B_Y - 10(4) - 10(8) - 20(4) - 20(8)$$

$$\equiv 4B_Y - 360 = 0$$

$$\Sigma F_X = A_X + 10 + 10 = 0$$

$$\Sigma F_Y = A_Y + B_Y - 20 - 20 = 0$$

$$B_Y = 90 \qquad A_Y = -50 \qquad A_X = -20 \text{ kN}$$

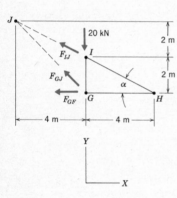

We next choose the cutting planes to be employed. If possible, the cutting planes should cut through no more than three members in the process of cutting the member of interest. Such a plane is plane $a-a$ for member CF and plane $b-b$ for member FG.

Considering the section $a-a$, we will investigate the equilibrium of the upper portion of the cut structure because there are fewer external forces acting there than on the lower portion. Equally important, we do not have to use any computed results, that is, the reactions. In so doing, we minimize the chance of carrying an error forward.

The free body diagram of the portion of the truss above plane $a-a$ is shown in the sketch. Remember that we assume all members to be in tension. The value of F_{CF} can be determined from a single equation by summing forces in the X direction. Thus

$$\Sigma F_X = -F_{CF} \cos 45° + 10 = 0 \qquad F_{CF} = -14.14 \text{ kN}$$

For plane $b-b$ let us consider the portion to the right, because only one force acts there. The free body diagram for the right portion is as shown. The value of F_{FG} is obtained in a single equilibrium equation by summing moments about joint J. (Note that all three equilibrium equations would be required for this determination if we chose joint I or H for the moment sum.) The moment equilibrium equation is

$$\Sigma M_{JZ} = -4F_{GF} - 20(4) = 0 \qquad F_{GF} = -20 \text{ kN}$$

Finally, note that in this particular problem it was not necessary to know the reaction forces, because the cutting planes utilized enabled us to isolate portions of the structure that did not contain the supports. Nevertheless, until you develop insight for cutting and isolating appropriate sections of trusses, it is a good idea to get into the habit of determining the reactions first, for their values will generally be needed.

EXAMPLE 4
Determine the tension in members 4−8 and 7−8 of the truss shown using the method of sections.

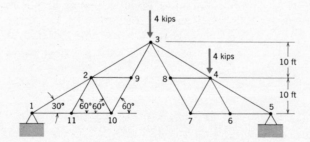

Solution

Perhaps the most notable feature of this truss is that it is supported by two pins, as shown by the reactions in the free body diagram.

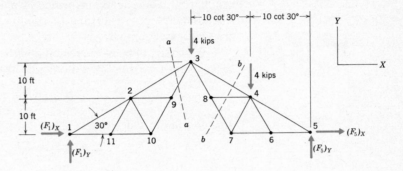

For equilibrium of the truss we have

$$\Sigma F_X = (F_1)_X + (F_5)_X = 0$$

$$\Sigma F_Y = (F_1)_Y + (F_5)_Y - 4 - 4 = 0$$

$$\Sigma M_{1Z} = (F_5)_Y [4(10 \cot 30°)] - 4[2(10 \cot 30°)]$$

$$- 4[3(10 \cot 30°)] = 0$$

$$(F_5)_Y = 5 \qquad (F_1)_Y = 3 \qquad (F_5)_X = -(F_1)_X \text{ kips}$$

We cannot determine $(F_5)_X$ and $(F_1)_X$ from the foregoing equations. At first glance it might seem that these reactions are statically indeterminate. This is not so. Recall that an indeterminate structure has redundant supports that are not necessary for holding the system in static equilibrium. In the truss shown, changing one of the pin supports to a roller will result in a nonrigid structure, for the roller support will move, allowing joint 3 to fall down. On this basis we can intuitively conclude that the truss shown is statically determinate.

The reason for the apparent ambiguity is that the truss is really composed of two simple trusses, each of which is rigid by itself. The left truss is supported at pin 1 and joint 3, whereas the right truss is supported at pin 5 and joint 3. In general, a truss formed from several simple trusses is called a *compound truss*. The Fink truss that was shown in Figure 4 is another compound truss. It resembles the truss in this problem, but it is rigid and statically determinate with a pin and roller support because of the presence of the additional horizontal member connecting the two portions.

From these observations, we conclude that the determination of the values of $(F_1)_X$ and $(F_5)_X$ requires that we consider the equilibrium of the

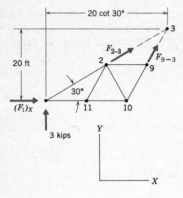

simple trusses from which the compound truss is formed, as well as of the entire structure. For this consideration note that cutting plane $a-a$ shown in the free body diagram only cuts through two members, although equilibrium of the sections formed by plane $a-a$ will yield three additional equations. Hence, choosing to equilibrate the left section, we have the free body diagram shown.

We have not been requested to determine F_{2-3} and F_{9-3}, so we obtain $(F_1)_X$ directly by summing moments about joint 3. Thus

$$\Sigma M_{3Z} = (F_1)_X(20) - 3(20 \cot 30°) = 0$$

It then follows that

$$(F_1)_X = 3 \cot 30° = 5.196 \text{ kips}$$

$$(F_5)_X = -(F_1)_X = -5.196 \text{ kips}$$

Now that we know the reactions, we can choose a cutting plane for determining the forces in members $4-8$ and $7-8$. We will use the cutting plane $b-b$ shown in the free body diagram of the entire truss because it cuts through both members of interest while cutting through a total of only three members. Consider the portion of the truss to the right of the cutting plane for equilibrium. The corresponding free body diagram is as shown.

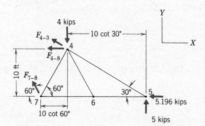

We choose joint 4 for the moment sum in order to eliminate F_{4-3} and F_{4-8}. The equilibrium equations are

$$\Sigma M_{4Z} = -F_{7-8} \sin 60° (10 \cot 60°) - F_{7-8} \cos 60°(10)$$

$$+ 5(10 \cot 30°) + (-5.196)(10) = 0$$

$$\Sigma F_X = -F_{7-8} \cos 60° - F_{4-8} - F_{4-3} \cos 30° - 5.196 = 0$$

$$\Sigma F_Y = F_{3-4} \sin 30° + F_{7-8} \sin 60° + 5 - 4 = 0$$

$$F_{7-8} = 3.464 \qquad F_{4-3} = -8.00 \qquad F_{4-8} = 0 \text{ kips}$$

The value of F_{4-3} was obtained in the process of solving the equations, even though it was not requested.

Finally, note that member $4-8$ could have been readily shown to be a zero force member by the method of joints; a free body diagram shows that it is the only force acting on joint 8 in the direction perpendicular to the line from joint 3 to joint 7. However, determining the force in member 7−8 by the method of joints requires that other joints be solved first.

HOMEWORK PROBLEMS

Using the method of sections, determine the axial force in the members indicated.

IV.24   Member *BD* in the truss of Problem IV.5

IV.25   Members 3−4 and 3−6 in the truss of Problem IV.6

IV.26   Member *CE* in the truss of Problem IV.7

IV.27   Members *CD* and *DG* in the truss of Problem IV.8

IV.28   Members *BG* and *BF* in the truss of Problem IV.9

IV.29   Members 2−9 and 8−9 in the truss of Problem IV.10

IV.30   Members 2−3 and 2−7 in the truss of Problem IV.11

IV.31   Member *CF* in the truss of Problem IV.12

IV.32   Members *BC* and *GH* in the truss of Problem IV.13

IV.33   Members 2−7, 3−7, and 6−7 in the truss of Problem IV.14

IV.34   Determine the axial force in members *DE* and *DF* of the truss shown.

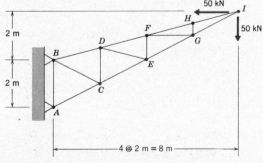

Prob. IV.34

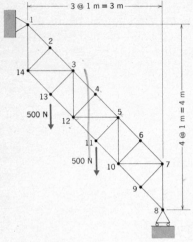

Prob. IV.35

IV.35 Determine the axial forces in members 4–12 and 5–12 of the truss shown.

IV.36 Determine the axial force in members *BH* and *BJ* of the truss shown. Note that the acute angles formed by the members are either 30°, or 60°.

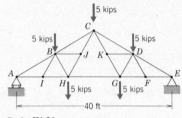

Prob. IV.36

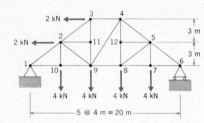

Prob. IV.37

IV.37 Determine the axial force in members 4–9 and 9–11 of the truss shown.

IV.38 Determine the axial forces in members *AB* and *AC* of the truss shown.

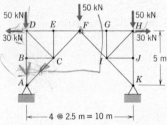

Prob. IV.38

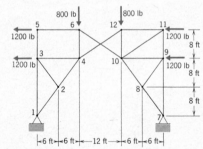

Prob. IV.39

IV.39 Members 4–12 and 6–10 cross, but are not connected. Determine the forces in these members, as well as in members 1–2 and 1–3.

IV.40 Can the axial forces in members *DE* and *GH* of the truss be determined? Explain your answer.

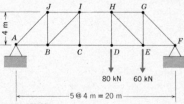

Prob. IV.40

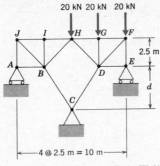

20 kN 20 kN 20 kN

2.5 m

d

4 @ 2.5 m = 10 m

Prob. IV.41

IV.41 Determine the axial forces in members *DC* and *FG* of the truss shown when $d = 5$ m. Can these forces be found when $d = 2.5$ m? Explain your response.

IV.42 Determine the axial forces in members 5–8 and 6–8 of the truss shown.

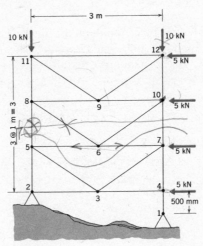

Prob. IV.42

IV.43 Determine the axial forces in members *IJ* and *IO* of the truss shown.

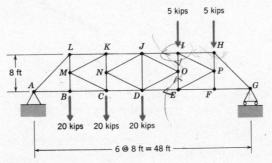

Prob. IV.43

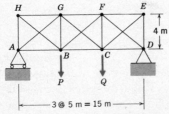

3 @ 5 m = 15 m

Prob. IV.44

IV.44 The diagonal members in the truss shown exert negligible forces when required to be in compression because of a phenomenon called "buckling." Hence, they may be considered to sustain tensile axial forces only. Such members are called *counters*. These counters cross, but are not connected to each other. Determine the axial forces in counters *BF* and *CG* when $P = 40$ kN and $Q = 0$.

IV.45 Solve Problem IV.44 for $P = 40$ kN and $Q = 80$ kN.

IV.46 Determine the forces in members DE, DF, and DG in the truss of Problem IV.22.

IV.47 Determine the force in members DE, DF, and DG in the truss of Problem IV.23.

IV.48 The tower of an oil drilling rig is formed in the shape of a pyramid. The supports A, B, and C are equally spaced on a circle of 1 m radius, as shown in the plan view, and the vertical line passing through the center of this circle intersects the top joint D. Determine the forces in members 1 and 2.

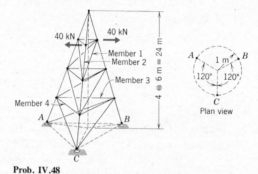

Prob. IV.48

IV.49 For the oil drilling rig of Problem IV.48, determine the forces in members 3 and 4.

B. FRAMES AND MACHINES

A truss is a special structure. It has only straight members which are connected concurrently at its joints, and the loads it transmits are applied only at these joints. We will now explore how to determine the forces transmitted by the members of more general structures for more general loading conditions.

In this section we shall examine two types of general structures. They differ only in their application. *Frames* are rigid structures intended to support a given loading, whereas *machines* are nonrigid structures that transmit and modify a given set of input loading forces into another set of output loadings. Our interest in machines in this text on statics is with the relationship of input and output forces necessary for static equilibrium. We leave the question of the possible movement of a machine for the companion text on dynamics. We shall not distinguish between the methods of solution for frames and machines. They both require that we

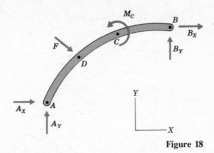

Figure 18

study the equilibrium of structures composed of many *multiforce members*, that is, members that carry forces that can be applied anywhere along its boundary.

In that a frame or machine is composed of multiforce members, we can no longer formulate the solution on the basis of all members carrying axial tension or compression forces. For example, consider the curved member AB in Figure 18, which is pinned at its ends to a larger structure. The nature of the reaction components at pins A and B cannot be determined in advance. Rather, it is necessary for us to write the equations of static equilibrium in order to relate these reactions to the given loading system, which in this case consists of a force at point D and a couple at point C. In Module VI, where we study the internal forces in structural members, we will learn how a multiforce member can sustain general internal forces and couples, as contrasted to just the axial tension or compression of a two force member.

It appears that the member in Figure 18 is statically indeterminate because there are only three equilibrium equations for the member, whereas there are four reaction components. However, recall that it was stipulated that this is a member in a structure. Hence, we will obtain additional equations when we consider the static equilibrium of the other members of the structure and of the structure as a whole. It is only after counting all such equations and the corresponding unknowns that we can judge the determinacy of the structure.

The foregoing is really a statement of the basic method of studying frames and machines. It is a process of simultaneously considering the static equilibrium (a) of the entire structure, and (b) of each of its members individually.

As always when dealing with bodies that are connected to each other, special care must be taken to satisfy Newton's third law of action and reaction when drawing free body diagrams. That is, when drawing free body diagrams, the forces exerted by one member on a second member must be equal, opposite, and collinear to the forces exerted by the second member on the first.

As an example, consider the frame in Figure 19, whose weight is negligible compared to the force $\bar{F}_D$. If we draw the free body diagram for the structure and the two members ACD and BC, we arrive at the results shown in Figure 20 on the following page, where the directions of the unknown forces are arbitrarily assumed using Newton's third law.

In these free body diagrams there are six unknown reaction components, for which we have nine equations of static equilibrium (three for each isolated body). As was true in our study of trusses, equilibrium of the entire truss is assured if the individual members are in equilibrium, so the excess equations may be regarded as checks for our calculations.

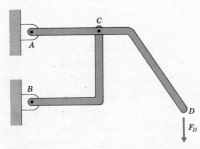

Figure 19

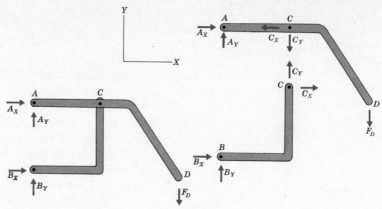

Figure 20

In drawing the free body diagrams in Figure 20 we guessed at the directions of the components of the force exerted by bar ACD on bar BC at joint C. We then drew the reaction forces exerted by bar BC on bar ACD *consistently* with this initial guess. As usual, we will know that a direction has been improperly assumed if the force component is found to be negative.

> A useful check on whether the free body diagrams for the individual members have been drawn consistent with Newton's third law is to superimpose the individual diagrams in order to form the original structure. If the reaction forces between the members cancel each other, with the result that the superposition exactly produces the free body diagram of the entire structure, then it can be concluded that the internal reaction forces are consistent.

We drew the free body diagrams in Figure 20 on the premise that we had not first carefully considered the nature of the members. However, with a little observation before we start sketching, we can save a good deal of computational effort. It will be noted that member BC is loaded only at its ends. Therefore, even though it is not a straight bar, it is a two-force member. The discussion of two-force members at the beginning of this module showed that the two forces applied to curved, as well as straight members must be equal in magnitude, opposite in direction, and collinear with the line connecting their points of application. Thus, the observation that bar BC is a two-force member results in the alternative set of free body diagrams shown in Figure 21 on the next page.

The computational savings previously alluded to result from the fact that member BC is a self-equilibrant member, as well as from the fact that we now have only one unknown, F_{BC}, in place of the four unknowns, B_X,

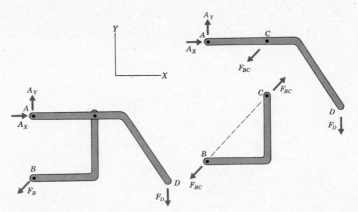

Figure 21

B_Y, C_X, and C_Y in Figure 19. Nevertheless, it should be emphasized that the observation that certain members are two-force members is useful, but not crucial, to the solution process.

The foregoing discussion may be generalized in the following series of steps for determining the forces exerted between the members of a frame or a machine.

1 Inspect the structure to determine if there are any two-force members.

2 Draw free body diagrams of the entire structure and of each member of the structure. In doing so, be certain to depict the forces acting on two force members consistently, that is, equal in magnitude, opposite in direction, and collinear. In drawing the free body diagrams, give special attention to satisfying Newton's third law for the forces exerted between the members.

3 Successively write and solve the equilibrium equations corresponding to the free body diagrams. The two-force members are automatically in equilibrium if their forces have been properly accounted for, so such members need not be considered. In following this step it is only necessary to formulate a sufficient number of equations to obtain the forces of interest. However, if all of the possible equilibrium equations are written, the excess equations may be used to check earlier computations.

These steps are equally valid for investigating planar and three-dimensional structures. For a given number of members, three-dimensional structures result in twice as many equilibrium equations as planar structures. The force and moment computations for three-dimensional structures are considerably more involved than those for planar structures. For these reasons, we shall restrict our attention to planar problems. In actuality, the solution of the equations for most space structures, as well as for many complicated planar structures, requires the aid of an electronic computer.

Finally, we note that it is possible for frames to be statically indeterminate. As always, this is manifested by an excess of unknown forces in comparison with the number of available equilibrium equations.

EXAMPLE 5

A 200-N force is applied to the crusher at point C, as shown. Determine the force the piston D exerts on the object within the cylinder and also determine the reaction at pin A.

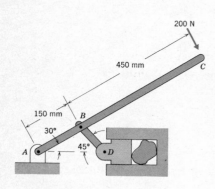

Solution

There are three bodies of interest in this problem — the two bars and the piston. Of these, bar BD is a two-force member. In accounting for the reaction forces, we show a transverse force and a couple exerted by the walls of the cylinder on the piston, because the cylinder does not allow the piston to move transversely or to rotate. Thus, we have the following free body diagrams.

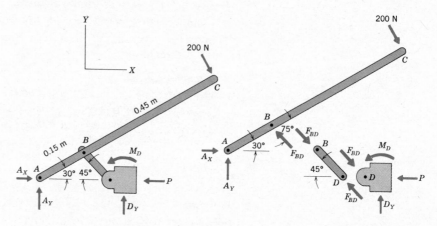

We have depicted $\bar{F}_{BD}$ as a compressive force, because it can be seen that the ends of member BD are being squeezed together. This is contrary to our general procedure when treating two-force members, in which we assume a state of tension. We ignored the practice here because there is no chance that we will obtain, and thereby misinterpret, a negative sign for the axial force.

The values of A_X, A_Y, and F_{BD} can be found from the equilibrium equations for bar ABC, which are

$$\Sigma M_{AZ} = F_{BD} \sin 75° \, (0.15) - 200(0.60) = 0$$

$$\Sigma F_X = A_X - F_{BD} \cos 45° + 200 \cos 60° = 0$$

$$\Sigma F_Y = A_Y + F_{BD} \sin 45° - 200 \sin 60° = 0$$

$$F_{BD} = 828.2 \qquad A_X = 485.6 \qquad A_Y = -412.4 \text{ N}$$

We may now relate F_{BD} to P by summing forces in the X direction for the piston. Thus

$$\Sigma F_X = -P + F_{BD} \cos 45° \equiv -P + 828.2(0.7071) = 0$$

$$P = 585.6 \text{ N}$$

Notice that the value of P could also have been obtained from a single equilibrium equation by summing forces in the X direction for the entire structure after determining the reaction $\bar{A}$. We leave it to you to write and use the checking equations for this structure.

EXAMPLE 6

In the compensating trailer hitch shown, the spring is stretched to a tension force $\bar{T}$ in order to better distribute the loading of the 1000-kg trailer on the automobile. Compare the reactions between all tires and the ground when $T = 3$ kN to that obtained when $T = 0$. The centers of mass of the automobile and the trailer are at points G_1 and G_2, respectively.

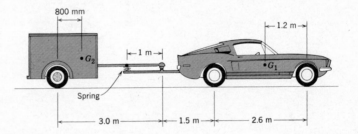

Solution

The connection between the trailer and the hitch is a ball joint, so we have the following free body diagrams. Notice that there are no horizontal forces acting on the tires, which is consistent with the entire system being at rest on a level surface.

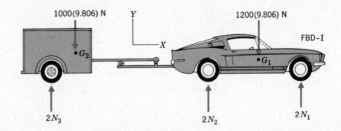

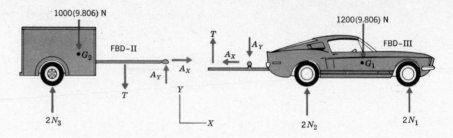

In this problem we are not interested in the reaction force components $\bar{A}_X$ and $\bar{A}_Y$ at the ball joint, so we write only the equilibrium equations that do not involve these forces. Thus, letting point C be the center of the tire of the trailer, we have

Car and trailer (FBD I):

$$\Sigma M_{CZ} = 2N_2(4.5) + 2N_1(7.1) - 1000(9.806)(0.8) - 1200(9.806)(5.9) = 0$$

$$\Sigma F_X \equiv 0$$

$$\Sigma F_Y = 2N_1 + 2N_2 + 2N_3 - 1200(9.806) - 1000(9.806) = 0$$

Car (FBD III):

$$\Sigma M_{AZ} = -T(1.0) + 2N_2(1.5) + 2N_1(4.1) - 1200(9.806)(2.9) = 0$$

Trailer (FBD II):

$$\Sigma M_{AZ} = T(1.0) + 1000(9.806)(2.2) - 2N_3(3.0) = 0$$

In the four algebraic equations above there are three unknowns: N_1, N_2, and N_3, because T is a known parameter. Thus, there is one check equation. Solution of the equation for the trailer yields

$$N_3 = \frac{T}{6} + 3596 \text{ newtons}$$

whereas the simultaneous solution of the two remaining moment equations is

$$N_1 = 0.2885T + 2414 \text{ newtons} \qquad N_2 = -0.4552T + 4777 \text{ newtons}$$

We now check these computations with the force sum in the Y direction, which gives

$$\Sigma F_Y = 2(0.2885T + 2414) + 2(-0.4552T + 4777) + 2\left(\frac{T}{6} + 3596\right) - 21573 \equiv 0$$

Hence, we conclude that the solutions are correct. Therefore, for $T = 3000$ newtons we find

$$N_1 = 3280 \qquad N_2 = 3411 \qquad N_3 = 4096 \text{ newtons}$$

whereas for $T = 0$ we find

$$N_1 = 2414 \qquad N_2 = 4777 \qquad N_3 = 3596 \text{ newtons}$$

These values show that an appreciable equalization of the forces on the front and rear wheels of the automobile is achieved through the action of the spring. Notice that it also increases the load on the wheels of the trailer.

EXAMPLE 7

The frame shown supports a 100-lb load at point E. Determine the reactions at supports A and B.

Solution

None of the members are two-force members, so we show arbitrary force components at each pin joint. The pin at D rides in a groove, which means that the reaction is normal to the groove. This results in the following free body diagrams.

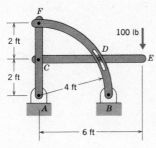

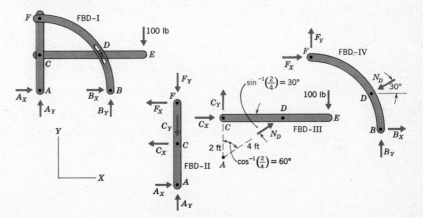

There are nine unknown force components in the free body diagrams. We begin by noting that the free body diagram for bar CD can be solved for all of the forces acting on that body, so we write:

Bar CD (FBD III):

$$\Sigma M_{CZ} = N_D \sin 30°(4 \sin 60°) - 100(6) = 0$$

$$\Sigma F_X = C_X + N_D \cos 30° = 0$$

$$\Sigma F_Y = C_Y + N_D \sin 30° - 100 = 0$$

The solutions are

$$N_D = 346.4 \qquad C_X = -300.0 \qquad C_Y = -73.2 \text{ lb}$$

The values of F_X and F_Y were not requested, so we next sum moments about joint F on bar AF and on bar BF. This yields:

Bar AF (FBD II):

$$\Sigma M_{FZ} = A_X(4) - C_X(2) \equiv 4A_X - (-300.0)(2) = 0 \qquad A_X = -150 \text{ lb}$$

Bar BF (FBD IV):

$$\Sigma M_{FZ} = -N_D \cos 30°(2) - N_D \sin 30°(4 \sin 60°) + B_Y(4) + B_X(4)$$
$$\equiv -346.4(0.8660)(2) - 346.4(0.50)4(0.8660) + 4B_Y + 4B_X = 0$$

$$B_X + B_Y = 300.0 \text{ lb}$$

We now use the free body diagram of the entire structure to determine the values of A_Y, B_X, and B_Y. Thus:

Entire structure (FBD I):

$$\Sigma M_{AZ} = B_Y(4) - 100(6) = 0$$
$$\Sigma F_X = A_X + B_X = 0$$
$$\Sigma F_Y = A_Y + B_Y - 100 = 0$$

These equations yield

$$B_Y = 150 \qquad B_X = 150 \qquad A_X = -150 \qquad A_Y = -50 \text{ lb}$$

In these results the value of A_X is the same as that obtained by considering bar AF, thus validating (in part) the calculations.

HOMEWORK PROBLEMS

IV.50–IV.52   Determine the reactions at pins A and C in the frames shown.

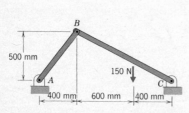

Prob. IV.50

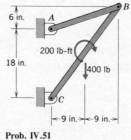

Prob. IV.51

Prob. IV.52

IV.53–IV.54 Determine the clamping force on the bolt and the force on pin *A* for the pliers shown.

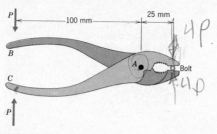

Prob. IV.53

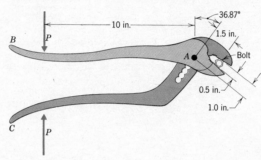

Prob. IV.54

IV.55 A 200-N force is applied to the piston. For the position where $\theta = 90°$, determine the couple $\bar{M}$ that should be applied to the crankshaft to hold the system in static equilibrium.

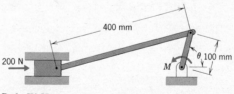

Prob. IV.55

IV.56 Solve Problem IV.55 for $\theta = 60°$.

IV.57 A 500 lb-ft couple is applied to bar *AB*, as shown. This bar is pinned to collar *B*, which slides on the smooth bar *CD*. Determine the couple $\bar{M}_{CD}$ that must be applied to bar *CD* to maintain static equilibrium at $\theta = 90°$.

IV.58 Solve Problem IV.57 for $\theta = 45°$.

IV.59 The mechanism shown is in equilibrium in the vertical plane. Collar *C* rides on the smooth vertical guide and the unstretched length of the

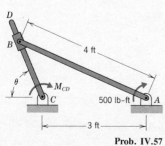

Prob. IV.57

spring is 1 m. Knowing that the bars have a mass of 2 kg/m, determine the stiffness of the spring and the reaction at pin A.

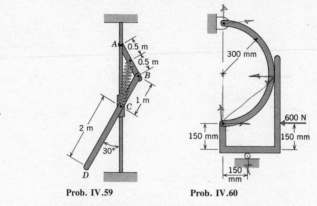

Prob. IV.59   **Prob. IV.60**

IV.60   Determine the forces acting on the smooth semicircular bar in the frame.

IV.61–IV.63   Determine the reactions at ends A and B for each system of beams shown.

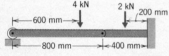

Prob. IV.61

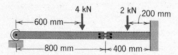

Prob. IV.62

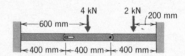

Prob. IV.63

IV.64 and IV.65   Determine the forces acting on members AC and CE of the frame shown.

IV.66   Determine the forces acting on the members of the frame (a) if $P = 200$ lb and $Q = 0$, (b) if $P = 0$ and $Q = 200$ lb, (c) if $P = Q = 200$ lb.

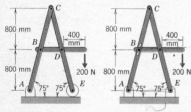

Prob. IV.64   **Prob. IV.65**

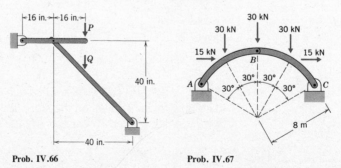

Prob. IV.66   **Prob. IV.67**

IV.67   The three-pin arch is subjected to the set of loads shown. Determine the reactions at pins A, B, and C.

IV.68 The spring has a stiffness of 50 lb/in. and an unstretched length of 20 in. Determine the reactions at pins A, B, and D.

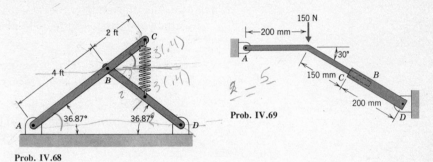

Prob. IV.68

Prob. IV.69

IV.69 Bar AB, having a circular cross section, fits closely in the smooth pipe CD. Determine the reactions at pins A and D.

IV.70 Determine the reactions at pins A, B, and C when the frame supports a 100-kg package.

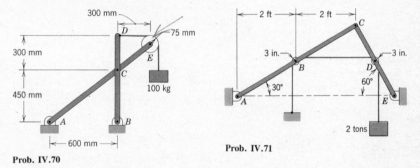

Prob. IV.70

Prob. IV.71

IV.71 The frame shown is used to support the 2-ton crate. Determine the reactions at pins A and E.

IV.72 Curved bar AB is pinned at end B to the cantilevered bar BC. End A rests on the smooth incline. Determine the reactions at end C.

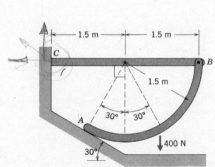

Prob. IV.72

IV.73 Each of the bars in the structure shown has weight W. Their centers of mass are at their midpoints. Consider the situation where $P = 0$. (a) Determine the reactions at the ends of each member. (b) Compare the results of part (a) to those obtained when the structure is considered to be a truss, where the weight *must* act through the joints. Thus, a fair approximation is that one-half the weight of each member is applied to the joints at the ends of that member. *Hint:* Use the method of joints for part (b).

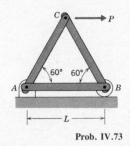

Prob. IV.73

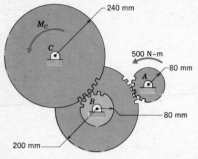

Prob. IV.75

IV.74 Solve Problem IV.73 for $P = 10W$.

IV.75 A 500 N-m couple is applied to gear A, as shown. Stepped gear B, which transmits this couple to gear C, rides freely on its shaft. Determine the magnitude and sense of the couple $\bar{M}_C$ acting on gear C necessary for equilibrium.

IV.76 Solve Problem IV.75 if a counterclockwise couple of 100 N-m is acting on the stepped gear B in addition to the 500 N-m couple on gears A.

IV.77 A counterclockwise couple $\bar{M}_1$ is applied to the planetary gear system shown. The radius of the central gear is r_1 and that of the planetary gears is r_2. Determine the couples $\bar{M}_2$ and $\bar{M}_3$ acting on the spider and the outer gear, respectively, for the system to be in static equilibrium.

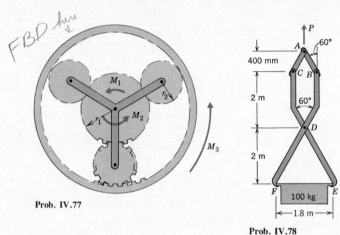

Prob. IV.77

Prob. IV.78

IV.78 A 100-kg radioactive container is held by applying the force $\bar{P}$ to the pair of tongs. Determine the forces acting on member BF.

IV.79 Determine the reactions at pins A and F for the loading shown.

IV.80 Determine all forces acting on member AC when $P = 8$ kN and $Q = 0$.

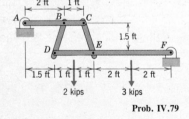

Prob. IV.79

Prob. IV.80

IV.81 Solve Problem IV.80 for $P = 0$ and $Q = 8$ kN.

IV.82 Determine the force exerted on the bolt when the handle of the bolt cutter is gripped by the pair of 30-lb forces, as shown.

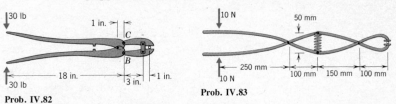

Prob. IV.82

Prob. IV.83

IV.83 The mechanism shown is a device for clamping onto inaccessible objects. The spring has a stiffness of 2 kN/m and an unstretched length of 40 mm. Determine the clamping force on the object if the handles are squeezed with a pair of 10-N forces.

IV.84 Determine the forces exerted on the bolt and on member CD when the locking pliers is gripped by the pair of 400-N forces, as shown.

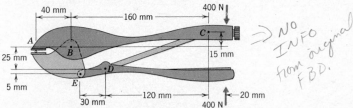

Prob. IV.84

IV.85 A 50-lb force is applied to the press punch, as shown. Determine the force exerted on the smooth plate F and the reaction forces at pins A and E.

IV.86 A 500-kg package rests on the platform, which is supported by the mechanism shown. Determine the force in the hydraulic cylinder and the reactions at pins A and F necessary for static equilibrium as a function of θ.

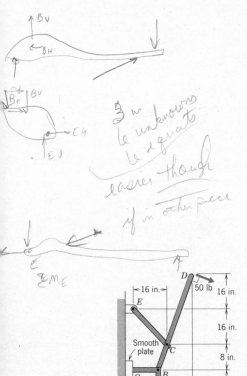

Prob. IV.85

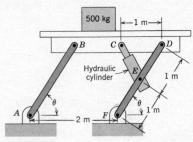

Prob. IV.86

IV.87 The position of the scoop on the tractor is controlled by two parallel mechanisms, only one of which appears in the side view. Determine the forces acting on the main arm ADF when the scoop is carrying an 8-kN load in the position shown.

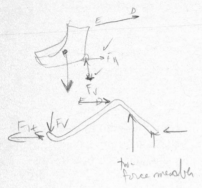

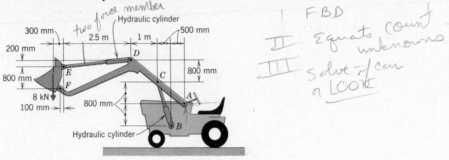

Prob. IV.87

IV.88 A mining trencher is subjected to the forces shown. The position of the bucket is controlled by the hydraulic cylinders BC, DE, and GH. Determine the forces acting on arms AF and EI.

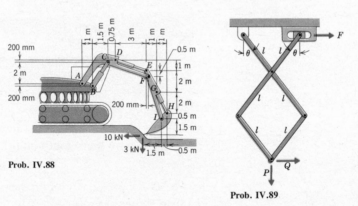

Prob. IV.88

Prob. IV.89

IV.89 Determine the horizontal force $\bar{F}$ required to hold the mechanism in position as a function of the angle θ and the magnitude of the force $\bar{P}$ if $Q = 0$.

IV.90 Solve Problem IV.89 if $Q = 1.50P$.

IV.91 The suspension system shown transmits the 4-kN reaction between the tires and the ground to the frame of an automobile. Determine the force in the spring and the forces exerted by members AB and CD on the frame.

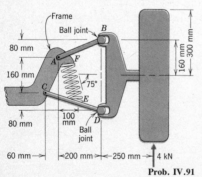

Prob. IV.91

MODULE V
DISTRIBUTED FORCES

The focus of attention in this module will be situations where forces are spread out over a region of surface area or volume of a body. Such forces are called *distributed forces*. In the previous modules our concern was with the effects of forces acting at points, that is, *concentrated forces*. Such forces do not exist in reality, for any material will deform when a force is applied to it, resulting in the distribution of force over a small amount of contact area. Thus, when we consider a force to be concentrated, we are assuming that the contact area is sufficiently small to represent it as a point.

Our objective here will be to determine how to represent distributed forces by equivalent concentrated forces in situations where the region of application of the distributed force system is not negligible.

A. GRAVITATIONAL FORCE

A common distributed force system is the gravitational attraction exerted on a body by the planet earth. The *weight* of a body is the magnitude of the resultant of the gravity force on each particle of the body. The point of application of this resultant force is the *center of mass,* or equivalently, the *center of gravity*. Previously, we regarded the location of the center of mass as a given quantity. Now the task is to learn how to determine its location.

1 Centers of Mass and Geometric Centroids

We begin the investigation by recalling that the gravitational force acting on a particle of mass m near the surface of the earth is simply mg directed toward the center of mass of the earth. In the model of the earth that we employ in statics, the earth is a sphere and its center of mass is the center of the sphere.

We may consider a rigid body on the surface of the earth to be composed of many particles i. We will denote the mass of each particle as Δm_i. Because the center of the earth is at a large distance from these particles, the gravitational forces on all particles are essentially parallel, in the vertical direction. Without loss of generality, and also for consistency with the notion used in most references, let us denote a convenient coordinate system by the lower case letters xyz. Corresponding to this coordinate system are the unit vectors $\bar{i}, \bar{j}$, and $\bar{k}$. Thus, choosing the z axis to be vertically outward from the center of the earth, the gravitational forces acting on the body are as shown in Figure 1a on the following page.

The force system in Figure 1a consists of a set of parallel forces. From our studies in Module III we know that their effect can be represented by an equivalent resultant force $\bar{R}$, as shown in Figure 1b. The magnitude of this resultant force is the weight mg. To verify this we

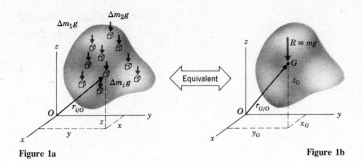

Figure 1a   **Figure 1b**

equate the sum of the forces in Figure 1a to the resultant $\bar{R}$.

$$\Sigma F_z = - \sum_{i=1}^{N} \Delta m_i g = -R$$

Because the mass m of the body is the sum of the mass of all of its particles, the force sum yields

$$R = \left(\sum_{i=1}^{N} \Delta m_i \right) g \equiv mg$$

Clearly, summing all of the mass elements Δm_i by summing the atoms of matter is not feasible. Instead, we consider the body to be a continuum of an infinite number of infinitesimal masses $\Delta m_i = dm$. The process of summation then becomes a process of integration over the space $\mathcal{V}$ occupied by the body. Thus

$$\boxed{\begin{array}{c} R = mg \\ m = \int_{\mathcal{V}} dm \end{array}} \tag{1}$$

The relationship between the space $\mathcal{V}$ and the elements dm will be addressed below.

The foregoing discussion defines the magnitude and direction of the resultant weight force. By definition, this force acts at the center of mass G shown in Figure 1b. To determine the coordinates of this point we use the requirement that the equivalent weight force in Figure 1b must exert the same moment about an arbitrary point as the distributed force system in Figure 1a. For this point let us choose the origin O of the coordinate system. The position of the center of mass is denoted $\bar{r}_{G/O}$. By equating the moments of the two-force systems we find

$$\Sigma \bar{M}_O = \bar{r}_{G/O} \times \bar{R} \equiv \bar{r}_{G/O} \times (-mg\bar{k})$$

$$= \sum_{i=1}^{N} \bar{r}_{i/O} \times (-\Delta m_i g \bar{k})$$

The rectangular Cartesian coordinates of mass element Δm_i are (x, y, z). Thus, as shown in Figure 1a, the components of $\bar{r}_{i/O}$ are these coordinates, that is,

$$\bar{r}_{i/O} = x\bar{\imath} + y\bar{\jmath} + z\bar{k}$$

Similarly, the components of $\bar{r}_{G/O}$ are the coordinates of the center of mass G.

$$\bar{r}_{G/O} = x_G\,\bar{\imath} + y_G\,\bar{\jmath} + z_G\,\bar{K}$$

Further, to account for the entire mass, we replace Δm_i by the infinitesimal mass element dm, thus transforming the summation in the moment equation to an integral. This gives

$$(x_G\,\bar{\imath} + y_G\,\bar{\jmath} + z_G\,\bar{k}) \times (-mg\bar{k}) = \int_{\mathscr{V}} (x\bar{\imath} + y\bar{\jmath} + z\bar{k})(-dm\,g\bar{k})$$

$$mgx_G\,\bar{\jmath} \times mgy_G\bar{\imath} = g \int_{\mathscr{V}} (x\,dm\,\bar{\jmath} - y\,dm\,\bar{\imath})$$

We can now equate the corresponding components in this equation, cancelling the factor g, with the following result:

$$mx_G = \int_{\mathscr{V}} x\,dm \tag{2a}$$

$$my_G = \int_{\mathscr{V}} y\,dm \tag{2b}$$

These integrals are the *first moments of mass* with respect to the x and y coordinates. They yield the coordinates of the mass center, x_G and y_G.

The value of z_G cannot be determined from the foregoing equations because the gravity forces in Figures 1 do not exert a moment about the z axis. This coordinate value may be determined by reorienting the gravity field of Figures 1, for instance, by orienting the force of gravity to be parallel to the x axis. Summing moments in the same manner as was done above then yields

$$mz_G = \int_{\mathscr{V}} z\,dm \tag{2c}$$

Equations (2) are the basic ones required to locate the center of mass of a body. Because we usually know the mass density ρ (mass per unit volume) of the material composing the body, we may perform the required integrations by writing

$$dm = \rho \, d\mathcal{V}$$

(3)

where $d\mathcal{V}$ is an infinitesimal element of volume.

For a body with arbitrary properties, equations (2) require the evaluation of three first moments of mass. However, a computational simplification can be achieved for bodies having a plane of symmetry. To investigate this matter, consider the body in Figure 2, where for convenience, we selected the x and y axes to coincide with the plane of symmetry.

The property of symmetry with respect to the xy plane has the geometric significance that the portion of the body to the right of the plane is the mirror image of the portion to the left. Mathematically, this means that at every cross section perpendicular to the y axis at some value y_P, such as the one illustrated in Figure 2, each mass element dm_R having coordinates (x, y_P, z_R) to the right of the symmetry plane has a companion element dm_L to the left of this plane whose coordinates are $(x, y_P, -z_R)$. By virtue of the symmetry, we also know that both elements have the same density, $\rho_L = \rho_R$. These elements also have identical volumes, so it follows that $dm_L = dm_R$. Thus, in the integral for the first moment of mass with respect to the z coordinate, corresponding to each term $z_R \, dm_R$, there is a term $z_L \, dm_L \equiv - z_R \, dm_R$. Obviously, these terms cancel when added, so that the first moment of mass is zero. Equation (2c) then yields $z_G = 0$, which enables us to conclude that:

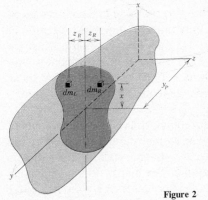

Figure 2

> The center of mass of a body having a plane of symmetry is situated somewhere on that plane.

Note that the foregoing statement applies only to those situations where the geometric shape *and* the mass distribution, as given by the density ρ, are symmetric with respect to a plane. In the common situation where a body is *homogeneous,* that is, where its material properties do not depend on the location within the body, the density ρ is constant. It is then only necessary to consider geometrical symmetry.

A further simplification resulting for a homogeneous body is that the density ρ, being a constant, may be removed from the integrals over the space $\mathcal{V}$. In this case equation (1) gives

$$m = \int_{\mathcal{V}} dm = \rho \int_{\mathcal{V}} d\mathcal{V} = \rho \mathcal{V}$$

(4)

Substituting equations (3) and (4) into equations (2), we find that the density ρ cancels, yielding the following results:

$$
\begin{aligned}
\mathcal{V} x_G &= \int_{\mathcal{V}} x \, d\mathcal{V} \\[2mm]
\mathcal{V} y_G &= \int_{\mathcal{V}} y \, d\mathcal{V} \\[2mm]
\mathcal{V} z_G &= \int_{\mathcal{V}} z \, d\mathcal{V}
\end{aligned}
\tag{5}
$$

The integrals in equations (5) are the *first moments of volume*. In all situations, homogeneous or otherwise, equations (5) define the location of the *geometric centroid of the volume* $\mathcal{V}$ occupied by the body. Thus we see that when the body is homogeneous, the center of mass coincides with the geometric centroid of the volume. It is not necessarily true that these points will coincide if the body is not homogeneous, *even* if it is a symmetric body.

It is important to understand the difference between the meanings of the center of mass and of the centroid. The former point is affected by the way the mass of the body is distributed in space, whereas the latter is affected only by the geometry of the volume. The distinction is analogous to the difference between the population center and the geographic center of a region, such as a state or nation.

Equations (2) and (5) are the basic ones we will need to locate the mass center. The most obvious way to employ them is to evaluate the integrals. To do this it will be necessary for us to describe the position of points in a body, as well as the body's element of volume $d\mathcal{V}$. The required techniques are the subject of the next section.

2 Integration Techniques

Let us recall two methods of analytic geometry for locating a point. One method is to write the rectangular Cartesian coordinates (x, y, z) of the point. The other method is to specify the location of the point using *cylindrical coordinates*. In the latter, first we choose a coordinate plane and locate the point with respect to the plane. For instance, arbitrarily choosing the xy plane, we have the situation depicted in Figure 3, where the distance z locates the point P above the xy plane. Then, we locate point P parallel to the xy plane by using the polar coordinates (R, ϕ). The resulting set of coordinates (R, ϕ, z) are the cylindrical coordinates. For the polar coordinates defined in Figure 3, we have the following coordinate transformation from the Cartesian to the polar variables:

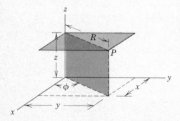

Figure 3

$$ x = R \cos \phi \qquad y = R \sin \phi $$

Transformations such as these enable us to use either Cartesian or cylindrical coordinates in formulating the integrals in equations (2) and (5). However, the decision to use either of these sets of coordinates is not arbitrary. In general, the coordinates we shall select are the ones that most easily permit us to describe the boundaries of the body.

For an arbitrary body, we may form an element of mass dm from either Cartesian or cylindrical coordinates, as shown in Figure 4. The

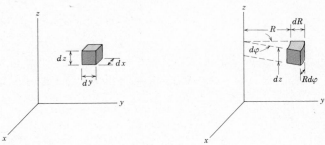

Figure 4

angle $d\phi$ is infinitesimal, so both elements dm in Figure 4 are parallelepipeds. Hence, using equation (3) we have

$$dm = \rho \, dx \, dy \, dz \quad \text{or} \quad dm = \rho R \, d\phi \, dR \, dz \qquad (6)$$

We should note that larger elements of mass can be formed, but they seldom represent a significant advantage over the ones shown in Figure 4.

Another item that may require some review at this time is the formulation of the limits of an integral. This matter is treated in the following examples.

EXAMPLE 1
Determine by integration the (x, y, z) coordinates of the center of mass of the homogeneous orthogonal tetrahedron shown.

Solution
Because the body is homogeneous, the center of mass coincides with the geometric centroid, so we will use equations (5). The solution will be formulated in terms of the Cartesian coordinates (x, y, z) because three of the faces of the tetrahedron coincide with the coordinate planes and the fourth is a slanted plane. All lines lying in a flat plane are straight, so we know that the values of the (x, y, z) coordinates of points on this slanted plane are linearly related. Thus, the general equation of the plane is

$$z = k_1 x + k_2 y + k_3$$

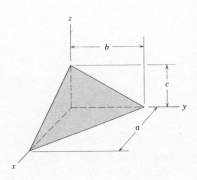

where k_1, k_2, and k_3 are constants.

To determine these constants we note that three points on the plane are $(a, 0, 0)$, $(0, b, 0)$, and $(0, 0, c)$. Substituting these values into the general equation of the plane yields

$$0 = k_1 a \quad + k_2(0) + k_3$$

$$0 = k_1(0) + k_2 b \quad + k_3$$

$$c = k_1(0) + k_2(0) + k_3$$

The solution of these equations of the plane is

$$k_3 = c \qquad k_2 = -\frac{c}{b} \qquad k_1 = -\frac{c}{a}$$

so the equation defining the slanted face is

$$z = c\left(1 - \frac{x}{a} - \frac{y}{b}\right)$$

In Cartesian coordinates an element of volume is

$$d\mathcal{V} = dx\, dy\, dz$$

To integrate all of these elements we first consider a cross section corresponding to a fixed value of one of the coordinates. Choosing a cross section parallel to the yz plane at an arbitrary position x, we have the situation depicted in the following sketch.

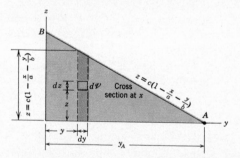

strip for any value of y on the cross section. We will account for all elements $d\mathcal{V}$ on the vertical strip by integrating first over z in the range 0 strip for any value of y on the cross section. We will acount for all elements $d\mathcal{V}$ on the vertical strip by integrating first over z in the range $0 \leq z \leq c(1 - x/a - y/b)$. Then we will account for all such vertical strips by integrating over all values of y on the cross section, which is the range $0 \leq y \leq y_A$. To determine the value of y_A, we note that at point A,

$$z = c\left(1 - \frac{x}{a} - \frac{y_A}{b}\right) = 0 \quad \text{so} \quad y_A = b\left(1 - \frac{x}{a}\right)$$

Finally, after having included all elements dm on the cross section, we account for all cross sections x by integrating over the range of values of x $(0 \leq x \leq a)$ for the tetrahedron.

To use equations (5) we will need the value for the volume of the tetrahedron. To this end we integrate each of the elements of volume (writing the integrals in the order in which they will be integrated) and find

$$
\begin{aligned}
\mathcal{V} &= \int_0^a \int_0^{b(1-x/a)} \int_0^{c(1-x/a \ - \ y/b)} dz \ dy \ dx \\
&= \int_0^a \int_0^{b(1-x/a)} c\left(1 - \frac{x}{a} - \frac{y}{b}\right) dy \ dx \\
&= c \int_0^a \left[b\left(1 - \frac{x}{a}\right) - \frac{x}{a}(b)\left(1 - \frac{x}{a}\right) - \frac{1}{2b}(b)^2\left(1 - \frac{x}{a}\right)^2 \right] dx \\
&= \frac{1}{6} abc
\end{aligned}
$$

Now we may calculate the first moments of volume. Beginning with the first of equations (5), we have

$$
\begin{aligned}
\left(\frac{1}{6}abc\right)x_G &= \int_0^a \int_0^{b(1-x/a)} \int_0^{c(1-x/a \ - \ y/b)} x \ dz \ dy \ dx \\
&= \int_0^a \int_0^{b(1-x/a)} cx\left(1 - \frac{x}{a} - \frac{y}{b}\right) dy \ dx \\
&= c \int_0^a \left[bx\left(1 - \frac{x}{a}\right) - \frac{x^2}{a}(b)\left(1 - \frac{x}{a}\right) - \frac{x}{2b}(b)^2\left(1 - \frac{x}{a}\right)^2 \right] dx \\
&= \frac{1}{24} a^2 bc
\end{aligned}
$$

The solution for x_G is

$$
x_G = \frac{a}{4}
$$

In other words, the centroid is at one-quarter of the distance from the base in the yz plane to the apex on the x axis.

At this point we could continue to evaluate the other first moments arising in equations (5). However, this is not necessary here, because the geometry with respect to all of the coordinate axes is identical. It follows that the results for the values of y_G and z_G can be obtained by permuting the symbols. This yields

$$
y_G = \frac{b}{4} \qquad z_G = \frac{c}{4}
$$

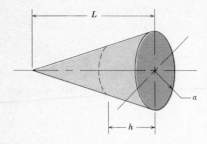

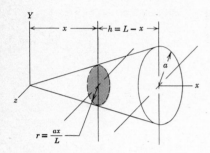

EXAMPLE 2

Determine the mass and the location of the center of mass of the right circular cone shown. Its density varies with distance h from the base according to $\rho = \rho_0(1 - h/2L)$.

Solution

Recall that when determining the center of mass, the concept of symmetry of the body about a plane requires that the density at matching points on either side of the geometric symmetry plane be identical. For the given cone the density at all points on a cross section perpendicular to the axis of the cone at any value of h is constant. Furthermore, this cross section is circular. Therefore, any plane containing the axis of the cone is a plane of symmetry, and it follows that the center of mass is situated on this axis.

In order to exploit the symmetry fully, we choose the xyz coordinate system such that one coordinate axis, say the x axis, coincides with the axis of the cone. For convenience we place the origin at the apex of the cone, as shown.

As was noted earlier, each cross section parallel to the yz plane, at any value x, is circular. The radius of this circle is shown in the sketch at the left to be $r = ax/L$. This suggests that we should use polar coordinates to locate the mass elements on a cross section. The polar coordinates to be employed are shown in the sketch below.

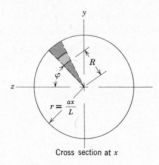

Cross section at x

For the polar coordinates selected, we see that

$$y = R \sin \phi \qquad z = R \cos \phi$$

Adapting equation (6) to the case where x measures the distance perpendicular to the plane of the polar coordinates, we have

$$dm = \rho R \, d\phi \, dR \, dx$$

The given expression for ρ is in terms of h. Noting that $h = L - x$, we have

$$\rho = \rho_0\left(1 - \frac{L - x}{2L}\right) = \frac{1}{2}\rho_0\left(1 + \frac{x}{L}\right)$$

To account for the elements of mass on the cross section we shall first integrate over the range of all possible radial distances, which is $0 \leq R \leq ax/L$. (This forms the shaded circular sector shown in the preceding sketch.) We then include all of these sectors by integrating over $0 \leq \phi \leq 2\pi$. (An alternative approach would be to form a circular ring by first integrating over $0 \leq \phi \leq 2\pi$ and then to include all such rings by integrating over $0 \leq R \leq ax/L$.) Finally, all cross sections are accounted for by integrating over $0 \leq x \leq L$.

The mass can be determined by using equation (1). Writing the integrals in the order in which they are to be evaluated, this yields

$$m = \int_0^L \int_0^{2\pi} \int_0^{ax/L} \left[\frac{1}{2}\rho_0\left(1 + \frac{x}{L}\right)\right] R \, dR \, d\phi \, dx$$

$$= \frac{1}{2}\rho_0 \int_0^L \int_0^{2\pi} \left(1 + \frac{x}{L}\right)\frac{1}{2}\left(\frac{ax}{L}\right)^2 d\phi \, dx$$

$$= \frac{1}{4}\rho_0 \int_0^L 2\pi\left(1 + \frac{x}{L}\right)\left(\frac{ax}{L}\right)^2 dx = \frac{7\pi}{24}\rho_0 a^2 L$$

Because the body is not homogeneous, its center of mass does not coincide with its geometric centroid. Hence, we must employ equations (2) to evaluate the first moments of mass. We know that the center of mass is situated on the x axis, so we need only the first moment of mass with respect to the x coordinate; specifically

$$\left(\frac{7\pi}{24}\rho_0 a^2 L\right)x_G = \int_0^L \int_0^{2\pi} \int_0^{ax/L} x\left[\frac{1}{2}\rho_0\left(1 + \frac{x}{L}\right)\right] R \, dR \, d\phi \, dx$$

$$= \frac{1}{4}\rho_0 \int_0^L 2\pi x\left(1 + \frac{x}{L}\right)\left(\frac{ax}{L}\right)^2 dx = \frac{9\pi}{40}\rho_0 a^2 L^2$$

Solving for x_G we then have

$$x_G = 0.771L$$

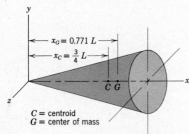

C = centroid
G = center of mass

The location of point G is illustrated in the sketch. For contrast, the location of the centroid C of the volume of the cone is also shown. The center of mass is closer to the base of the cone because the density was prescribed to be greater near the base, resulting in a greater concentration of mass there.

HOMEWORK PROBLEMS

V.1–V.7   Set up and evaluate the integrals required for the determination of the location of the center of mass of the homogeneous body shown.

V.1   Triangular prism

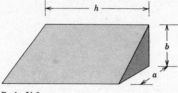

Prob. V.1

V.2   Hemisphere

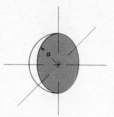

Prob. V.2

V.3   Paraboloid of revolution

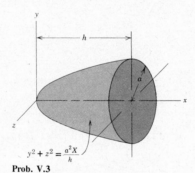

$$y^2 + z^2 = \frac{a^2 X}{h}$$

Prob. V.3

V.4   Spherical segment

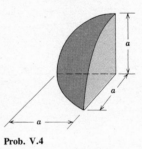

Prob. V.4

V.5   Semi-ellipsoid

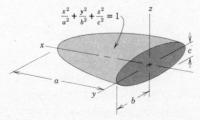

$$\frac{x^2}{a^2} + \frac{y^2}{b^2} + \frac{z^2}{c^2} = 1$$

Prob. V.5

V.6   Circular cone segment

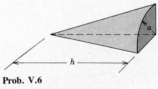

Prob. V.6

V.7 Sliced circular cylinder

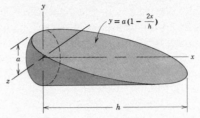

$$y = a\left(1 - \frac{2x}{h}\right)$$

Prob. V.7

V.8 The density of the rectangular parallelepiped varies with the distance h from one face according to $\rho = \rho_0 \cos(\pi h/3a)$. Determine the mass and location of the center of mass of this body.

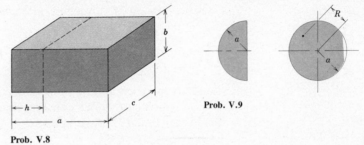

Prob. V.8

Prob. V.9

V.9 The density of a hemisphere of radius a varies with the radial distance R from the axis of symmetry, according to $\rho = \rho_0(1 + R/a)$. Determine the mass and the location of the center of mass of this body. Is the center of mass closer or farther from the flat base than the geometric centroid? Explain your answer.

V.10 The upper surface of the body shown is a ruled surface known as a hyperbolic paraboloid, which is defined by the equation $z = cxy/ab$. Knowing that the body is homogeneous, determine the mass and the location of the center of mass of this body.

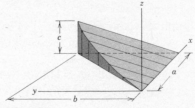

Prob. V.10

V.11 Solve Problem V.10 for the case where the mass density of the body varies according to $\rho = \rho_0(1 + z^2/c^2)$.

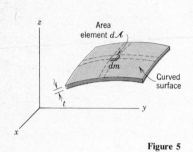

Figure 5

3 Centroids of Surfaces and Lines

The developments in the preceding sections enable us to determine the location of the center of mass for general three-dimensional bodies. Certain simplifications in the equations are possible in two cases, pertaining to thin bodies. The first type of thin body we shall consider is a thin *shell*. An egg shell is a familiar example of a thin shell. An arbitrary shell is shown in Figure 5, where t is the thickness and $\mathcal{A}$ is the surface area of the shell.

To determine the center of mass for a shell we consider the thickness to be so small that each element of mass dm is located at the curved surface. This means that the (x, y, z) coordinates locating the mass element are the same as those of the corresponding point on the surface. Furthermore, the infinitesimal volume of the element is then $t\,d\mathcal{A}$, where $d\mathcal{A}$ is an element of surface area. Equation (3) then gives

$$dm = \rho t\,d\mathcal{A}$$

The product ρt describes the mass per unit surface area. In most situations the shell will be homogeneous and it will have constant thickness, in which case ρt is a constant that may be factored out of integrals. Equations (1) and (2) then become

$$
\begin{aligned}
m &= \rho t \int_{\mathcal{A}} d\mathcal{A} = \rho t \mathcal{A} \\
\mathcal{A} x_G &= \int_{\mathcal{A}} x\,d\mathcal{A} \\
\mathcal{A} y_G &= \int_{\mathcal{A}} y\,d\mathcal{A} \\
\mathcal{A} z_G &= \int_{\mathcal{A}} z\,d\mathcal{A}
\end{aligned}
\tag{7}
$$

Recall that the coordinates (x, y, z) in equations (7) locate points on the surface $\mathcal{A}$. The integrals in these equations are first moments of the surface area.

In general, equations (7) locate the *geometric centroid of the surface area $\mathcal{A}$*. Hence, the center of mass of a homogeneous shell with constant thickness coincides with the centroid of its surface area. In the case where the surface area is flat, the centroid of the area will be situated on the planar surface. It is then only necessary to utilize two of the first moments in equations (7). Conversely, if the surface is curved, the centroid need not, and generally will not, be situated on the surface.

As an aside, it is interesting to note that in addition to their application for thin shells, equations (7) may be employed to locate the center of mass of another type of homogeneous body. Suppose that a body has a constant thickness t (not necessarily small) in a direction perpendicular to two planar sides. Such bodies are called *cylinders* in analytic geometry. Figure 6 shows an arbitrary cylinder, for which the xy plane was chosen to be parallel to the planar sides whose area is $\mathcal{A}$. In view of the symmetry of this body, the center of mass G must be situated somewhere along the midplane that cuts the thickness in half. This is indicated by a long dashed curve in Figure 6. Hence, we need only determine the value of x_G and y_G to locate the center of mass. Because $t\,d\mathcal{A}$ is an element of volume for this body, just as it was for an arbitrary thin shell, we have $m = \rho t \mathcal{A}$. It then follows that the values of x_G and y_G are given by the corresponding equations (7). Thus, as indicated in Figure 6, we can locate the center of mass of a cylindrical body by merely locating the centroid of the area of one of its planar faces.

Let us now turn our attention to a second class of thin bodies. We have seen that a shell has one small dimension, its thickness, so its mass can be regarded to be concentrated at its surface. There are also bodies, such as the one depicted in Figure 7, whose mass is concentrated along a curve. This is the case of a *slender curved bar*. Such bodies have the characteristic that the dimensions defining the cross sectional area $\mathcal{A}_c$ perpendicular to its axis are very small compared to the arc length along the axis.

The element of mass for this body may be formed from a section of the bar of arc length ds, as indicated in Figure 7. The volume of this element is $\mathcal{A}_c\,ds$, so the corresponding expression for dm is

$$dm = \rho\,\mathcal{A}_c\,ds$$

The product $\rho\,\mathcal{A}_c$ is the mass per unit length of the bar. If the bar is homogeneous and has a constant cross-sectional area, we can factor $\rho\,\mathcal{A}_c$ out of the integrals in equations (1) and (2). Letting $\mathcal{S}$ denote the total arc length of the bar, these equations become

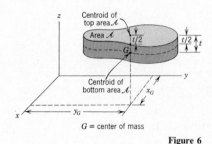

Centroid of top area $\mathcal{A}$

Area $\mathcal{A}$

Centroid of bottom area $\mathcal{A}$

x_G

y_G

G = center of mass

Figure 6

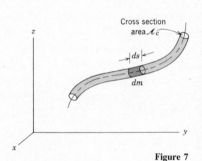

Cross section area $\mathcal{A}_c$

ds

dm

Figure 7

$$
\begin{aligned}
m &= \rho\,\mathcal{A}_c \int_{\mathcal{S}} ds = \rho\,\mathcal{A}_c\,\mathcal{S} \\[1mm]
\mathcal{S}\,x_G &= \int_{\mathcal{S}} x\,ds \\[1mm]
\mathcal{S}\,y_G &= \int_{\mathcal{S}} y\,ds \\[1mm]
\mathcal{S}\,z_G &= \int_{\mathcal{S}} z\,ds
\end{aligned}
$$

(8)

Recall that we are considering the mass of the bar to be concentrated along the axis of the bar, so the coordinates (x, y, z) in equations (8) are those of points along this axis.

Equations (8) define the *geometric centroid of the curve S*. It follows that the center of mass of a homogeneous bar of constant cross-sectional area coincides with the centroid of its axial curve. In general, this centroid probably will not be situated on the curve.

The primary difficulty in locating the centroids of surfaces and curves lies in relating the area $d\mathcal{A}$ for a surface, or arc length ds for a curve, to the x, y, and z coordinates. In the examples and homework problems we will restrict ourselves to geometries, particularly planar surfaces, surfaces of revolution, and planar curves, which can be readily formulated for solution.

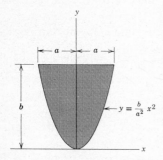

EXAMPLE 3
Locate the center of mass of the parabolic steel plate of constant thickness t shown.

Solution
A flat plate is simply a planar shell. The given plate is composed of one material, steel, and it has a constant thickness. Therefore, the center of mass is situated midway in the thickness, having the same x and y coordinates as the centroid of the parabolic area. Further, noting that the z axis is outward from the plane of the diagram, we see that the yz plane is a plane of symmetry. This indicates that the center of mass must lie on the y axis. As a result, the problem reduces to the evaluation of the first moment of area with respect to the y coordinate.

The perimeter of the area is easily described in terms of equations relating the x and y coordinates. We shall formulate the solution in terms of Cartesian coordinates, for which an element of area is $d\mathcal{A} = dx\,dy$, as shown in the sketch.

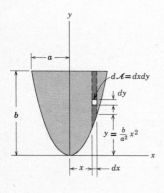

To account for all such elements, we could first integrate over all possible values of z for a particular value of y, and thus form a horizontal strip. However, as shown in the sketch, we choose to form a vertical strip by integrating over all possible values of y for a particular value of x. For this strip $bx^2/a^2 \leq y \leq b$. All vertical strips are then accounted for by integrating over $-a \leq x \leq a$.

The first step in applying equations (7) is to find the area. This is obtained by writing

$$\mathcal{A} = \int_{-a}^{a} \int_{bx^2/2}^{b} dy\,dx = \int_{-a}^{a} \left(b - \frac{bx^2}{a^2} \right) dx = 2ba - \frac{2ba^3}{3a^2} = \frac{4}{3}ba$$

Now, forming the first moment of area with respect to the y coordinate, equation (7) yields

$$\left(\frac{4}{3}ba\right)y_G = \int_{-a}^{a}\int_{bx^2/a^2}^{b} y\, dy\, dx = \frac{1}{2}\int_{-a}^{a}\left[b^2 - \left(\frac{bx^2}{a^2}\right)^2\right] dx = \frac{4}{5}b^2a$$

Solving for y_G, we find

$$y_G = \frac{3}{5}b$$

Thus the center of mass is midway in the plate thickness t with coordinates $x_G = 0$, $y_G = \frac{3}{5}b$.

EXAMPLE 4
An aluminum bar of constant mass per unit length σ is bent into a quarter-circle of radius r. Determine the (x, y) coordinates of the center of mass.

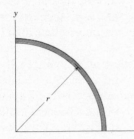

Solution
We begin by noting that the bar is symmetric about the plane formed by the z axis and the 45° line shown in the second sketch. This means that the centroid must lie along the 45° line, so $y_G = x_G$. Also, because the mass per unit length is constant, the center of mass coincides with the centroid of the circular arc.

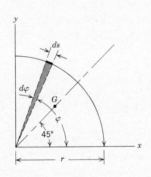

We employ polar coordinates because the mass elements are all situated at a constant radial distance r from the origin. The coordinate transformations for the chosen coordinate systems are

$$x = r\cos\phi \qquad y = r\sin\phi$$

where the values of ϕ range from zero to $\pi/2$ radians. Noting that the element ds is the arc of the circle subtended by the angle $d\phi$, we have

$$ds = r\, d\phi$$

Similarly, we know that the total length of the circular arc is

$$\mathcal{S} = \frac{1}{4}(2\pi r) = \frac{\pi}{2}r$$

Thus, in this case it is not necessary to evaluate an integral to determine $\mathcal{S}$; we need only determine the value of x_G. From the given information we know that the mass per unit length is constant, so we can use the first moment with respect to the x coordinate in equations (8).

$$\left(\frac{\pi}{2}r\right)x_G = \int_0^{\pi/2} x\, ds = \int_0^{\pi/2} (r\cos\phi)(r\, d\phi) = r^2\sin\frac{\pi}{2} = r^2$$

$$x_G = \frac{2r}{\pi} \equiv y_G$$

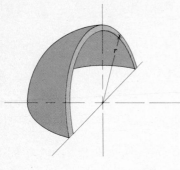

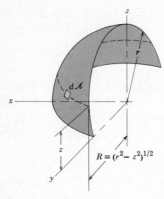

EXAMPLE 5

Locate the centroid of the aluminum shell of constant thickness formed by cutting a spherical shell of radius r into quarter-sections.

Solution

The shell is homogeneous and it has constant thickness. Therefore, its center of mass is coincident with the centroid of its spherical surface. We begin the solution by selecting an xyz coordinate system. In view of the shape of the body, we choose an origin at the center of the sphere and align the coordinate axes with the centerlines, as illustrated. Because the xz plane is a plane of symmetry, we know that $y_G = 0$. Furthermore, a plane containing the y axis at a 45° angle above the xy plane also cuts the surface in half, which means that $x_G = z_G$.

The spherical shape is conducive to the use of cylindrical coordinates. The polar coordinates R and ϕ may be defined to locate points parallel to any of the coordinate planes; our choice is the xy plane. As shown in the sketch to the left, cross sections of the surface parallel to the xy plane are semicircles of radius $R = (r^2 - z^2)^{1/2}$. The polar coordinates are defined with the aid of the sketch shown below of this cross section. Also, in order to depict the area element $d\mathcal{A}$, we draw the cross section formed by a cutting plane that contains the z axis and the element $d\mathcal{A}$.

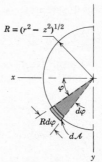

Cross section parallel to xy plane

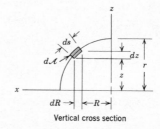

Vertical cross section

For the polar coordinates shown in the xy cross section, we may write

$$x = R \cos \phi = (r^2 - z^2)^{1/2} \cos \phi$$

$$y = R \sin \phi = (r^2 - z^2)^{1/2} \sin \phi$$

Deriving the expression for the area element $d\mathcal{A}$ is the only new feature of the formulation. The edge views of the element $d\mathcal{A}$ presented in the foregoing sketches of the cross sections show that

$$d\mathcal{A} = (R d\phi)\, ds = (r^2 - z^2)^{1/2}\, d\phi\, ds$$

The Pythagorean theorem is now used to relate the arc length ds to the increments of dz and dR illustrated in the sketch of the vertical cross section. This yields

$$ds = [(dz)^2 + (dR)^2]^{1/2}$$

Then, because $R = (r^2 - z^2)^{1/2}$, we find

$$dR \equiv \frac{dR}{dz}\ dz = \frac{1}{2}(r^2 - z^2)^{-1/2}(-2z)\ dz = \frac{-z\ dz}{(r^2 - z^2)^{1/2}}$$

$$ds = \left[(dz)^2 + \frac{(z\ dz)^2}{(r^2 - z^2)}\right]^{1/2} = \frac{r\ dx}{(r^2 - z^2)^{1/2}}$$

Hence

$$d\mathscr{A} = (r^2 - z^2)^{1/2}(d\phi)\left[\frac{r\ dz}{(r^2 - z^2)^{1/2}}\right] = r\ d\phi\ dz$$

As an aside, we note that this simple expression for $d\mathscr{A}$ is a result of the spherical geometry. For other surfaces of revolution, where the radial distance R is a different function of the axial distance z, this expression will not result.

Now that an expression for $d\mathscr{A}$ has been obtained, we may proceed to apply equations (7). All elements $d\mathscr{A}$ at an xy cross section are accounted for by integrating over $-\pi/2 < \phi < \pi/2$ radians. All cross sections are then accounted for by integrating over $0 \le z \le r$. Thus

$$\mathscr{A} = \int_0^r \int_{-\pi/2}^{\pi/2} d\mathscr{A} = \int_0^r \int_{-\pi/2}^{\pi/2} r\ d\phi\ dz = \pi r^2$$

As noted earlier, the symmetry of the shell means that we only have to calculate x_G (or z_G). Therefore, we form the first moment of surface area with respect to the x coordinate.

$$(\pi r^2)x_G = \int_0^r \int_{-\pi/2}^{\pi/2} x\ d\mathscr{A} = \int_0^r \int_{-\pi/2}^{\pi/2} [(r^2 - z^2)^{1/2} \cos \phi]r\ d\phi\ dz$$

$$= r\int_0^r (r^2 - z^2)^{1/2}\left(2 \sin \frac{\pi}{2}\right)\ dz = r\left[z(r^2 - z^2)^{1/2} + r^2 \sin^{-1}\frac{z}{r}\right]_0^r$$

$$= r^3 \sin^{-1}(1) = \frac{\pi}{2}r^3$$

A table of integrals was employed to evaluate the integral over the z variable. Solving for x_G, we find

$$x_G = \frac{r}{2} = z_G \qquad y_G = 0$$

HOMEWORK PROBLEMS

V.12–V.17 Determine the location of the center of mass of the homogeneous flat plate of constant thickness shown.

V.12 Right triangle

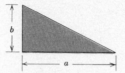

Prob. V.12

V.13 Elliptical quadrant

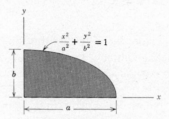

Prob. V.13

V.14 Circular sector

Prob. V.14

V.15 Cubic spandrel

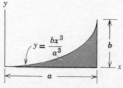

Prob. V.15

V.16   Spiral area

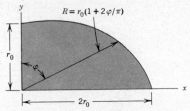

$R = r_0(1 + 2\varphi/\pi)$

Prob. V.16

V.17   Hyperbolic quadrant

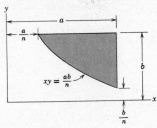

$xy = \dfrac{ab}{n}$

Prob. V.17

V.18–V.20   Determine the location of the center of mass of the bent wire shown. It has constant mass per unit length.

V.18   Circular arc

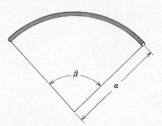

Prob. V.18

V.19   Parabolic arc

$y = \dfrac{x^2}{a}$

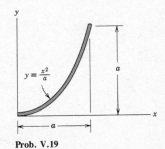

Prob. V.19

V.20   Spiral segment — *Hint:* In polar coordinates $ds = [(dR)^2 + (R\,d\phi)^2]^{1/2}$.

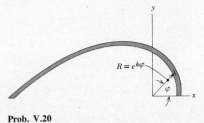

$R = e^{k\varphi}$

Prob. V.20

V.21–V.24   Determine the location of the center of mass of the thin shell shown having constant mass per unit length.

V.21   Conical shell

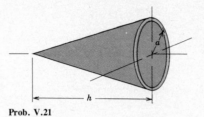

Prob. V.21

V.22   Conical shell segment

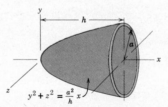

Prob. V.22

V.23   Parabolic shell

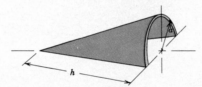

Prob. V.23

V.24   Circular cylindrical shell segment — *Hint:* Formulate the solution by locating points parallel to the xy plane in terms of polar coordinates and evaluate the integral over the z variable first.

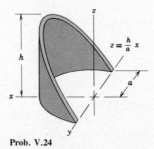

Prob. V.24

4   Composite Bodies

It would not make sense to evaluate integrals every time the location of the center of mass of a recognizable geometric shape is needed. To avoid doing so, it is common practice to tabulate the geometric and mass properties of a large number of basic (that is, recognizable) shapes. Typical tables are given at the end of this text; Appendix B contains information for planar bodies, and Appendix C contains information for three-dimensional shapes.

When a homogeneous body can be recognized as a tabulated shape, its center of mass may be located by substituting its dimensions into the appropriate tabulated formula. In other cases it may be possible to form the body of interest from two or more of the basic shapes. Such a body is called a *composite body*. An arbitrary composite body, formed from the basic bodies 1 and 2, is illustrated in Figure 8.

The space $\mathcal{V}$ occupied by the composite body is the combination of the spaces $\mathcal{V}_1$ and $\mathcal{V}_2$ of the basic recognizable shapes. From equation (1) we find that the total mass is

$$m = \int_{\mathcal{V}} dm = \int_{\mathcal{V}_1} dm + \int_{\mathcal{V}_2} dm$$

$$\boxed{m = m_1 + m_2} \tag{9}$$

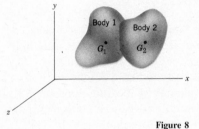

Figure 8

This result says that the total mass of the body is the sum of the mass of its individual components.

Similarly, first moments of mass may be decomposed into the contributions of the individual components. For instance, equation (2a) becomes

$$mx_G = \int_{\mathcal{V}} x\ dm = \int_{\mathcal{V}_1} x\ dm + \int_{\mathcal{V}_2} x\ dm$$

Here, the integrals are the first moments for components 1 and 2, individually. Applying equation (2a) again yields

$$\boxed{mx_G = m_1(x_G)_1 + m_2(x_G)_2} \tag{10a}$$

Following similar steps for the other first moments, we get

$$\boxed{my_G = m_1(y_G)_1 + m_2(y_G)_2} \tag{10b}$$

$$\boxed{mz_G = m_1(z_G)_1 + m_2(z_G)_2} \tag{10c}$$

In equations (10), $((x_G)_i, (y_G)_i, (z_G)_i)$ are the coordinates of the center of mass of the ith component body. Knowing the mass and location of the center of mass of each component, it is a simple matter to solve equations (10) for the coordinates of the center of mass G of the entire composite body.

Equations (9) and (10) are easily extended to the case of composite bodies formed from more than two basic shapes by summing the contributions of all components. Also, in the case where the body contains a hole, equations (9) and (10) may be applied by considering the hole to contain a negative mass. Then for example, equation (10a) takes the form

$$mx_G = m_1(x_G)_1 - m_2(x_G)_2$$

where body 1 corresponds to the entire body considering the hole to be filled and body 2 corresponds to the material missing from the entire body because of the hole.

As was true in the case of a general body, when a composite body is homogeneous, we may locate its center of mass by locating the centroid of the volume occupied by the body. For such cases the equations for the first moments of volume are identical in form to equations (10). That is, noting that $m = \rho\mathcal{V}$, where ρ is a constant, the symbol m becomes the symbol $\mathcal{V}$ throughout the equations. Similarly, the equations for the center of mass of a shell of constant mass per unit surface area or a slender bent bar of constant mass per unit length are identical to equations (10), except that the symbol m becomes the symbol $\mathcal{A}$ or $\mathcal{S}$, whichever is applicable.

EXAMPLE 6

The trapezoidal plate shown is made of aluminum. It has a thickness of 40 mm. Determine the location of its center of mass.

Solution

The plate is homogeneous and it has a constant thickness. Hence, its center of mass is midway in its thickness, having the same location in the plane as the centroid of the trapezoid. Appendix B does not give the location of the centroid of a trapezoid, so we treat it as a composite shape. There are several possible decompositions that result in components having areas whose properties are tabulated. We will consider the trapezoid to be formed from a rectangle and triangle, as shown in the sketch.

The dimensions locating the center of mass of each component body are taken from Appendix B. The last step required before employing equations (10) is to select a coordinate system. We choose x and y axes

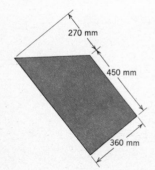

270 mm

450 mm

360 mm

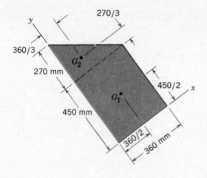

parallel to the known dimensions, as illustrated in the sketch.

The areas are found to be

$$\mathcal{A}_1 = 0.45(0.36) = 0.1620 \text{ m}^2$$

$$\mathcal{A}_2 = \tfrac{1}{2}(0.27)(0.36) = 0.0486 \text{ m}^2$$

$$\mathcal{A} = \mathcal{A}_1 + \mathcal{A}_2 = 0.2106 \text{ m}^2$$

Using the dimensions given in the sketch and the computed areas, the first moments of area are

$$0.2106 x_G = \mathcal{A}_1(x_G)_1 + \mathcal{A}_2(x_G)_2$$

$$= 0.1620[\tfrac{1}{2}(0.36)] + 0.0486[\tfrac{1}{3}(0.36)]$$

$$x_G = 0.1662 \text{ m}$$

$$0.2106 y_G = \mathcal{A}_1(y_G)_1 + \mathcal{A}_2(y_G)_2$$

$$= 0.1620[\tfrac{1}{2}(0.45)] + 0.0486[0.45 + \tfrac{1}{3}(0.27)]$$

$$y_G = 0.2977 \text{ m}$$

EXAMPLE 7

A 4-in.-diameter, 12-in.-long steel rod fits tightly into a hole in the rectangular aluminum bar, as shown. Determine the mass and the location of the center of mass of this assembly.

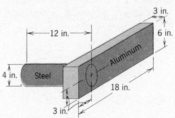

Solution

The assembly is composed of two different materials, so we must determine the center of mass using first moments of mass, not volume. We can break this composite body into three tabulated components. First, we have a solid rectangular parallelepiped, representing the aluminum bar. From this body we shall remove an aluminum cylinder 2 having the dimensions of the hole in the aluminum bar. The third body is, of course, the cylindrical steel rod.

The origin of the xyz coordinate system may be located at any convenient point. Our choice for xyz is shown in the sketch to the left. In this sketch we also show the dimensions locating the center of mass G_i of each

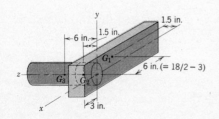

of the components. This sketch shows that the coordinates of these points are

point G_1: $(-6, 0, 1.5)$;   point G_2: $(0, 0, 1.5)$;   point G_3: $(0, 0, 6)$

To avoid errors in writing these coordinates, particular care is required to place the points in the correct octants of the coordinate system.

We begin the calculation of the mass of the basic components by converting the mass densities of aluminum and steel given in Appendix D to units of inches.

$$\rho_{al} = \frac{168 \text{ lb/ft}^3}{g} = \frac{168 \text{ lb/ft}^3}{32.17 \text{ ft/s}^2}\left(\frac{1 \text{ ft}}{12 \text{ in.}}\right)^4 = 2.518(10^{-4}) \text{ lb-s}^2/\text{in.}^4$$

$$\rho_{st} = \frac{490 \text{ lb/ft}^3}{g} = 7.345(10^{-4}) \text{ lb-s}^2/\text{in.}^4$$

The mass of the hole in the aluminum bar is considered to be negative, because it is removed from the system. Thus, writing $m_i = \rho_i \mathcal{V}_i$, we have

$$m_1 = \rho_{al}(3)(6)(18) = 0.08158 \text{ lb-s}^2/\text{in.}$$

$$m_2 = -\rho_{al}\pi(2)^2(3) = -0.00949$$

$$m_3 = \rho_{st}\pi(2)^2(12) = 0.11076$$

$$m = m_1 + m_2 + m_3 = 0.18285$$

We may now employ equations (10), using the coordinates of the centers of mass determined earlier. This gives

$$0.18285x_G = m_1(x_G)_1 + m_2(x_G)_2 + m_3(x_G)_3$$

$$= 0.08158(-6) + (-0.00949)(0) + 0.11076(0)$$

$$x_G = -2.68 \text{ in.}$$

$$0.18285y_G = m_1(y_G)_1 + m_2(Y_G)_2 + m_3(Y_G)_3 = 0$$

$$y_G = 0$$

$$0.18285z_G = 0.08158(1.5) + (-0.00949)(1.5) + 0.11076(6)$$

$$z_G = 4.23 \text{ in.}$$

Note that we could have foreseen that $y_G = 0$, because the xz plane is a plane of symmetry.

EXAMPLE 8

A copper wire having a circular cross section of 5 mm diameter is bent into the closed curve shown. Determine the mass and center of mass of the wire.

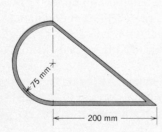

Solution
The density of copper is $8.91(10^3)$ kg/m³, so the mass per unit length has the value

$$\rho \mathcal{A}_c = 8.91(10^3)\pi \left(\frac{0.005}{2}\right)^2 = 0.17495 \text{ kg/m}$$

Because this parameter is constant we may locate the center of mass by computing first moments of the arc length. The bent wire is a composite of a circular section 1 and two straight sections 2 and 3, as shown in the sketch. The origin of the chosen xyz coordinate system is placed at a convenient point. The sketch also contains the locations of each center of mass of the components of the composite. It is only necessary to refer to Appendix C for the location of point G_1, in that the locations of the other points are apparent from symmetry. This composite is a planar shape, so we need only the (x, y) coordinates of each component. The sketch shows that these are

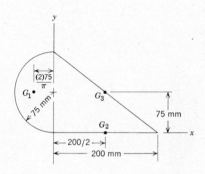

point G_1: $\left(-\dfrac{0.150}{\pi}, 0.075\right)$;   point G_2: (0.10, 0);   point G_3: (0.10, 0.075)

The arc lengths of the basic components are readily calculated to be

$$\mathcal{S}_1 = \pi(0.075) = 0.2356 \text{ meters}$$

$$\mathcal{S}_2 = 0.20 \text{ meters}$$

$$\mathcal{S}_3 = \sqrt{(0.20)^2 + (0.15)^2} = 0.250 \text{ meters}$$

$$\mathcal{S} = \mathcal{S}_1 + \mathcal{S}_2 + \mathcal{S}_3 = 0.6856 \text{ meters}$$

The mass of the bent wire is its total length times its mass per unit length. Thus

$$m = \rho \mathcal{A}_c \mathcal{S} = (0.17495)(0.6856) = 0.1199 \text{ kg} \quad \triangleleft$$

Adjusting the form of equations (10) to the calculation of first moments of arc length, we have

$$\mathcal{S} x_G = \mathcal{S}_1(x_G)_1 + \mathcal{S}_2(x_G)_2 + \mathcal{S}_3(x_G)_3$$

$$(0.6856)x_G = 0.2356\left(-\frac{0.150}{\pi}\right) + 0.20(0.10) + 0.25(0.10)$$

$$x_G = 0.0493 \text{ meters} = 49.3 \text{ mm} \quad \triangleleft$$

$$\mathcal{S} y_G = \mathcal{S}_1(y_G)_1 + \mathcal{S}_2(y_G)_2 + \mathcal{S}_3(y_G)_3$$

$$(0.6856)y_G = 0.2356(0.075) + 0.20(0) + 0.25(0.075)$$

$$y_G = 0.0531 \text{ meters} = 53.1 \text{ mm} \quad \triangleleft$$

HOMEWORK PROBLEMS
V.25–V.34   Determine the location of the center of mass of the homogeneous plates of constant thickness shown.

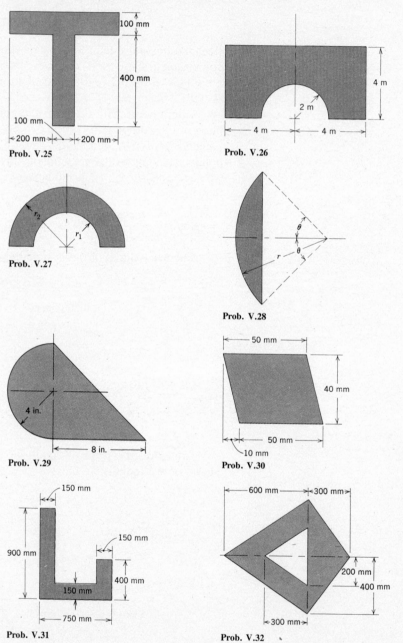

Prob. V.25

Prob. V.26

Prob. V.27

Prob. V.28

Prob. V.29

Prob. V.30

Prob. V.31

Prob. V.32

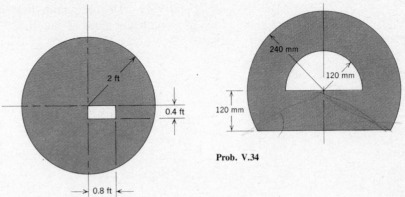

Prob. V.34

Prob. V.33

V.35–V.40 Determine the location of the center of mass of the bent wire shown having constant mass per unit length.

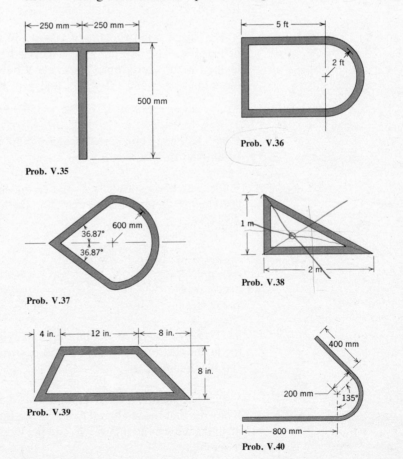

Prob. V.35

Prob. V.36

Prob. V.37

Prob. V.38

Prob. V.39

Prob. V.40

V.41 and V.42 The aluminum disk shown has a steel insert whose faces are flush with the faces of the disk. The thickness of the disk is as indicated. Determine the location of the center of mass of this body.

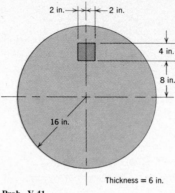

Prob. V.41   Prob. V.42

V.43 Three small weights of mass $m/50$ are attached to the circular disk of mass m. The assembly is suspended in the vertical plane from a pin located at A. Determine the angle θ between line OA and the vertical direction if $\alpha = 180°$, $\beta = 45°$, and $\gamma = 90°$.

V.44 Solve Problem V.43 if $\alpha = 150°$, $\beta = 30°$, and $\gamma = 90°$.

V.45 Knowing that $\alpha = 130°$ and $\gamma = 30°$ in Problem V.43, determine the value of β that causes θ to be zero.

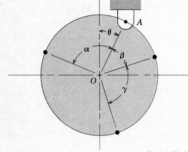

Prob. V.43

V.46 The body shown is a solid composed of plastic having a density of 2000 kg/m³. Determine the mass and the location of the center of mass of this body.

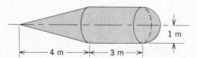

Prob. V.46

V.47 Solve Problem V.46 if only the cylindrical portion is composed of the given plastic, while the conical portion is steel and the hemispherical cap is aluminum.

V.48 The identical shelf brackets are not spaced equally on the shelf. Knowing that the shelf and brackets are both composed of wood weighing 48 lb/ft³, determine the weight and the location of the center of mass of the assembly.

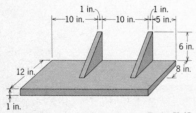

Prob. V.48

V.49 The container shown is formed from 3-mm-thick aluminum. Determine the mass and the location of the center of mass of the container.

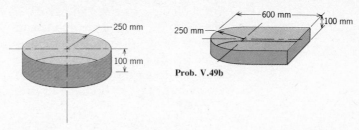

Prob. V.49b

Prob. V.49a

V.50 and V.51 The body shown consists of a solid pine interior that has been completely coated with lead to a uniform thickness of 10 mm. Determine the mass and the location of the center of mass of this body.

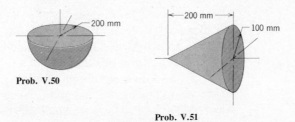

Prob. V.50

Prob. V.51

V.52 The crankpin shown is composed of steel. Determine the mass and the location of the center of mass of this body.

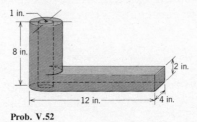

Prob. V.52

B. SURFACE LOADINGS

The gravitational forces we considered in Section A are examples of distributed force systems that are applied over the space occupied by a body; they are termed *body forces*. Another important type of force system is one that acts over the surface area of a body; this is the case of a *surface force*. Typically, distributed surface forces can result from the interaction of two bodies along an edge or contacting surfaces, or from pressure exerted by fluids, such as air and water.

Here we will study two types of surface forces. When a force is distributed over a line, we have a *line load*. The magnitude of a line load has the units of force per unit length. When a force is distributed over an area we have an *area load*. The magnitude of an area load has the units of force per unit area. In the following we will learn how to determine the resultant of each type of loading. For simplicity, we shall restrict our attention to cases where the loaded surface is flat.

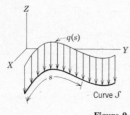

Figure 9

1 Line Loads

Frequently a structure having a flat surface, such as a plate or a beam, is required to support a normal force that is distributed along a line. We will denote this line load as the vector $\bar{q}$, and its magnitude as q, having units of force per unit length. A typical line load acting on a curve lying in the XY plane is shown in Figure 9.

The magnitude of q may vary with its location along the curve. To define this position dependency, we denote the arc length from an end of the curve to a point on the curve by the symbol s, as shown in the figure, and then give the function $q(s)$ defining the magnitude of q at each value of s. In order to depict this line load at a series of points along the curve, we draw arrows in the directions of the load whose heights are proportional to the magnitude of q at that point. As can be seen in Figure 9, the result is a curved surface, which we shall call the *loading surface*.

The forcing effect of the line load at some point on the surface may be described by multiplying the value of $\bar{q}$ at that point by the element of arc length ds situated at that point. The result is an infinite number of infinitesimal concentrated forces $\bar{q}\,ds$ acting along the curve. For clarity, only one of these concentrated forces is illustrated in Figure 10a.

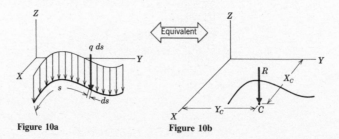

Figure 10a   Figure 10b

The system of concentrated forces formed in Figure 10a are all parallel to the Z axis. From our previous studies we know that such a system of forces can be represented by a single equivalent force $\bar{R}$, parallel to the Z axis. This force is depicted in Figure 10b. For equivalence, the magnitude of $\bar{R}$ must be identical to the sum of the magnitudes of the infinitesimal concentrated forces $\bar{q}\, ds$. The process of summing infinitesimal quantities requires integration. Specifically, we have

$$R = \int_{\mathscr{S}} q\, ds \tag{11}$$

where $\mathscr{S}$ denotes the curve carrying the load.

In view of the fact that the resultant is parallel to the Z axis, its line of action may be specified by determining the point of application C of this force in the XY plane. The coordinates of this point are determined by equating the moment of $\bar{R}$ to the sum of moments of all forces $\bar{q}\, ds$ about any convenient point. Computing moments about the origin of the coordinate system, we have

$$(X_C\bar{I} + Y_C\bar{J}) \times (-R\bar{K}) = \int_{\mathscr{S}}(X\bar{I} + Y\bar{J}) \times (-q\, ds\, \bar{K})$$

$$RX_C\bar{J} - RY_C\bar{I} = \int_{\mathscr{S}}(Xq\bar{J} - Yq\bar{I})\, ds$$

Equating corresponding components then yields

$$RX_C = \int_{\mathscr{S}} Xq\, ds$$
$$RY_C = \int_{\mathscr{S}} Yq\, ds \tag{12}$$

Equations (12) apply to the situation where a line load perpendicular to the XY plane acts on a curve $\mathscr{S}$ in that plane. Formulas for line loads acting on the other coordinate planes may be obtained by permuting the symbols in equations (12).

The pictorial representation of a line load that we introduced in Figure 9 allows for a useful interpretation of equations (11) and (12). Recall how the loading surface was formed; at each point along the curve the height represents the magnitude of $\bar{q}$ at that point. Thus, a differential element of area for this surface can be formed from a rectangular strip whose height is q and whose base is ds. Several such strips are depicted in Figure 10a. It follows that an element of area for the loading surface is $q\, ds$. Hence, the integral in equation (11) gives the area of the loading surface and the integrals in equation (12) are first moments of area for this surface. In this interpretation, equations (12) have the same meaning as equations (7) for centroids of a surface. Hence, we have the following analogy.

> The resultant of a line load $\bar{q}$ perpendicular to a flat plane is the area of the loading surface, perpendicular to the flat plane in the sense of positive q. This resultant intersects the centroid of the loading surface.

The importance of this analogy is that it enables us to use the tabulated values in Appendices B and C to determine the location of the point of application of the resultant of a line load whenever the loading surface is a basic shape or a composite of basic shapes.

EXAMPLE 9

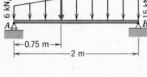

A beam supports the trapezoidal line load and the concentrated force shown. Determine (a) the resultant of the line load, (b) the reactions at supports A and B.

Solution

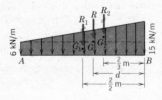

The given force diagram shows that the surface of the loading function is a planar trapezoid. To solve part (a) we sketch this loading surface and its resultant. Because the line load is downward, the resultant is also downward. Note that in order to distinguish between concentrated and distributed forces in a sketch, we draw the arrows for the former more boldly.

To employ the analogy between the loading surface and the resultant, we need the area of the trapezoid and the horizontal distance d to its centroid. As this information does not appear in Appendix B, we consider the trapezoid to be a composite consisting of a rectangle and a triangle, as shown. For each of these basic shapes, there is a resultant force $\bar{R}_i$ $(i = 1, 2)$ that intersects the centroid of the corresponding area. These individual resultant forces are depicted in the sketch with hatchmarks in order to indicate that their combined effect is equivalent to the total resultant $\bar{R}$. The sketch also shows the location of the centroids of the triangular and rectangular areas.

The magnitude of each resultant is the area of the corresponding loading surface. Thus

$$R_1 = (6 \text{ kN/m}) (2 \text{ m}) = 12 \text{ kN}$$

$$R_2 = \tfrac{1}{2}[(15 - 6)\text{kN/m}](2 \text{ m}) = 9 \text{ kN}$$

It then follows that

$$R = R_1 + R_2 = 21 \text{ kN}$$

To determine the value of d, we equate the moment of $\bar{R}$ to that of the resultants of the basic areas. By summing moments about end B, we find

$$R(X_G) = R_1(1 \text{ m}) + R_2(\tfrac{2}{3} \text{ m})$$

$$21(X_G) = 12(1) + 9(\tfrac{2}{3})$$

$$X_G = 0.8571 \text{ m}$$

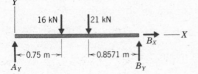

To solve part (b) we draw a free body diagram of the system, replacing the trapezoidal line load by its resultant. The XYZ coordinate system shown matches the directions in which dimensions are given. The equilibrium equations are

$$\Sigma M_{BZ} = -A_Y(2) + 21(0.8571) + 16(2-0.75) = 0$$

$$\Sigma F_X = B_X = 0$$

$$\Sigma F_Y = A_Y + B_Y - 16 - 21 = 0$$

Thus, we have

$$A_Y = 19 \text{ kN} \qquad B_Y = 18 \text{ kN} \qquad B_X = 0$$

Before we leave this problem, let us consider the situation where the problem statement requests that we solve part (b) only. In such a case, by combining the sketch for evaluating the line load with the free body diagram as illustrated in the sketch, we can eliminate the intermediate step of determining the total resultant $\bar{R}$. All that is necessary then is to determine the values of R_1 and R_2, as we did in the preceding solution, and apply the equilibrium equations to the system of concentrated forces appearing in the free body diagram.

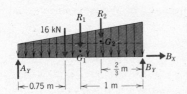

EXAMPLE 10
A bar, bent into a quarter-circular curve 18 in. in radius, is welded to a wall at end A. A line load whose intensity increases in proportion to the arc length along the bar from end A is applied transversely to the bar. The magnitude of the line load at end B is 4 lb/ft. Determine the reactions at end A.

Solution
From the given information we deduce that

$$q(s) = \frac{4}{\mathcal{S}} s \text{ lb/ft}$$

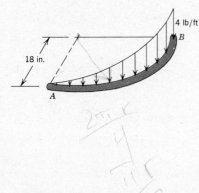

where $\mathcal{S} = (\pi/2)(18 \text{ in.})$ is the total length of the bar and s is the arc length from end A to a point on the bar. The resulting loading surface depicted in

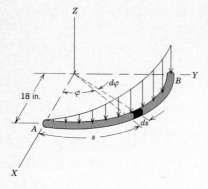

the given sketch does not appear in Appendix C, so the analogy between the resultant and the properties of the loading surface is of no use to us. We must therefore employ equations (11) and (12) to determine the resultant. For this we require a sketch describing the geometry of the system and a suitable XYZ coordinate system.

It will be necessary for us to relate the values of X and Y at some point on the bar to the value of s at that point. This is most conveniently done with the aid of the polar angle ϕ shown in the sketch, for then we have

$$X = 18 \cos \phi \qquad Y = 18 \sin \phi \text{ inches}$$

$$s = 18 \phi \qquad ds = 18 \, d\phi \text{ inches}$$

We see that points on the bent bar are contained in the range $0 \leq \phi \leq \pi/2$. For consistency of units, we write q in units of pounds per inch.

$$q = \frac{4 \text{ lb/ft}}{\mathscr{s}} s = 4 \frac{\text{lb}}{\text{ft}} \left(\frac{1 \text{ ft}}{12 \text{ in.}} \right) \frac{s}{(\pi/2)18} = 1.1789(10^{-2})s \text{ lb/in.}$$

where s is measured in inches. Equation (11) then gives

$$R = \int q(s) \, ds = \int_0^{\pi/2} [1.1789(10^{-2})(18\phi)](18 \, d\phi) = 4.712 \text{ lb}$$

This force is in the negative Z direction, as is $\bar{q}$. We next employ equations (12), which gives

$$4.712 X_C = \int_{\mathscr{s}} Xq(s) \, ds$$

$$= \int_0^{\pi/2} (18 \cos \phi)[1.1789(10^{-2})(18\phi)](18 \, d\phi)$$

$$= 1.1789(10^{-2})(18)^3 [\cos \frac{\pi}{2} + \frac{\pi}{2} \sin \frac{\pi}{2} - \cos (0) - (0) \sin (0)] = 39.24$$

$$X_C = 8.328 \text{ in.}$$

$$4.712 Y_C = \int_{\mathscr{s}} Yq(s) \, ds = \int_0^{\pi/2} (18 \sin \phi)[1.1789(10^{-2})(18 \phi)](18 \, d\phi)$$

$$= 1.1789(10^{-2})(18)^3 [\sin \frac{\pi}{2} - \frac{\pi}{2} \cos \frac{\pi}{2} - \sin (0) - (0) \cos (0)] = 68.75$$

$$Y_C = 14.590 \text{ in.}$$

Now that the resultant of the line load has been determined, we may proceed to the determination of the reactions at the support of the bar by drawing a free body diagram. The reactions consist of a force $\bar{F}_A$ and

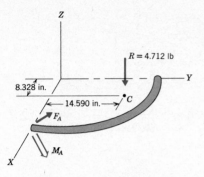

couple $\bar{M}_A$ at the welded end. The only other force is the resultant $\bar{R}$, so the free body diagram is as shown.

Summing the moments about point A, the equilibrium equations are

$$\Sigma \bar{M}_A = \bar{M}_A + [-(18 - 8.328)\bar{I} + 14.590\bar{J}] \times (-4.712\bar{K}) = 0 \text{ lb-in.}$$

$$\Sigma \bar{F} = \bar{F}_A - 4.712\bar{K} = 0 \text{ lb}$$

Thus

$$\bar{F}_A = 4.71\bar{K} \text{ lb}$$

$$\bar{M}_A = 68.7\bar{I} + 46.6\bar{J} \text{ lb-in.}$$

HOMEWORK PROBLEMS

V.53–V.57 Determine the resultant of the distributed loading acting on the beam shown. Also determine the support reactions.

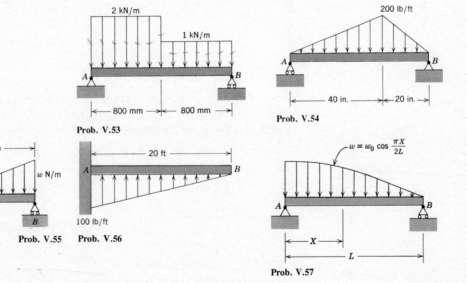

Prob. V.53

Prob. V.54

Prob. V.55   Prob. V.56

Prob. V.57

V.58–V.62 Determine the reactions at the supports of the beams shown.

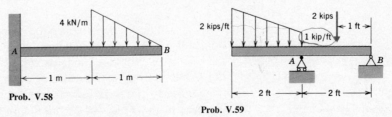

Prob. V.58

Prob. V.59

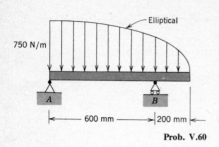

750 N/m

Elliptical

A

B

600 mm | 200 mm

Prob. V.60

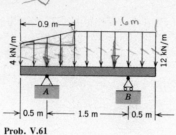

4 kN/m

0.9 m

1.6 m

12 kN/m

A

B

0.5 m | 1.5 m | 0.5 m

Prob. V.61

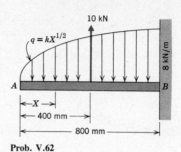

10 kN

$q = kX^{1/2}$

8 kN/m

A

B

X

400 mm

800 mm

Prob. V.62

V.63 and V.64   For the type of loading shown, determine the values of q_1 and q_2 that result in the reactions at the fixed support A being (a) a force with no couple, (b) a couple with no force. In each case give the corresponding reaction.

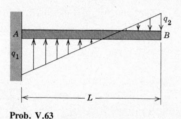

A q_2 B

q_1

L

Prob. V.63

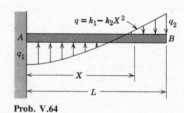

$q = k_1 - k_2 X^2$ q_2

A   B

q_1

X

L

Prob. V.64

V.65   To support the concentrated force $\bar{R}$ acting on the foundation, the ground develops the trapezoidal loading shown. Determine the values of q_1 and q_2 in terms of the magnitude R of the force and the ratio of distances d/L. Answer: $q_1 = 2(2 - 3d/L)(R/L)$, $q_2 = 2(3d/L - 1)(R/L)$.

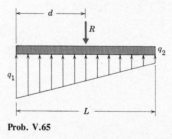

d

R

q_2

q_1

L

Prob. V.65

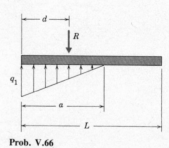

d

R

q_1

a

L

Prob. V.66

V.66   It is known that the ground can only push upward on the foundation in Problem V.65. (a) Determine the range of values of d/L for which the results of Problem V.65 are valid. (b) When the force $\bar{R}$ is applied to the left of the region of validity determined in part (a), the foundation loading is as shown. Determine the values of q_1 and the distance ratio a/L in terms of the values of R and d/L in this case.

V.67–V.70 Determine the reactions at the fixed end A of the curved beam shown.

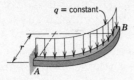

$q = \text{constant}$

Prob. V.67

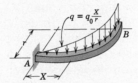

$q = q_0(1 - \frac{s}{\mathcal{S}})$

Prob. V.68

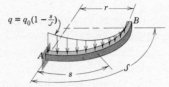

$q = q_0 \frac{X}{r}$

Prob. V.69

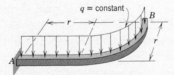

$q = \text{constant}$

Prob. V.70

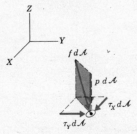

Figure 11

2 Surface Loads — Pressure Distributions

As is the case for all force systems, a surface load is a vector having magnitude and direction, which we shall denote as $\bar{f}$. In treating such a load, we can discuss its effect on each point of the surface by multiplying the vector $\bar{f}$ by the infinitesimal area $d\mathcal{A}$ on the surface. This is depicted in Figure 11, where the XY plane was chosen such that it coincides with the planar area $\mathcal{A}$ being loaded.

The surface load $\bar{f}$ may be replaced by three components, as depicted in Figure 11. The component $\bar{p}$, called the *pressure*, is normal to the surface, pushing on it. When $\bar{p}$ has a pulling (suction) effect, we will consider its magnitude to be negative.

In mechanics, quantities that have the units of force per unit area are termed *stresses*. Thus, the pressure can alternatively be termed a *normal stress*. The other two components of $\bar{f}$, $\bar{\tau}_X$ and $\bar{\tau}_Y$ tangent to the plane of the area $d\mathcal{A}$, are called *shear stresses*. In general, the magnitudes of the pressure and shear stresses can vary from point to point along the surface $\mathcal{A}$, and thus are functions of position. The function $p(X, Y)$ is the *pressure distribution* on the area, and $\tau_X(X, Y)$ and $\tau_Y(X, Y)$ are the *shear stress distributions* on the area.

Because the area load $\bar{f}$ is a vector quantity, we may consider the effect of each of these distributions separately, and then add the results vectorially. In the remainder of this section we will study the effects of pressure distributions on bodies and not consider the effects of shear stress distributions, although the results for the latter are straightforward extensions of those for pressure.

A typical pressure distribution $p(X, Y)$ acting on an area $\mathcal{A}$ in the XY plane is shown in Figure 12. In depicting $p(X, Y)$ we have used arrows whose length is proportional to the pressure p at the point where the arrow intersects the area $\mathcal{A}$. Thus the pressure distribution depicted in Figure 12 forms an imaginary body, which we shall call the *pressure space,* whose height above each point in the plane is the magnitude of the pressure function $p(X, Y)$ for that point.

Recall our earlier observation that the normal force acting on an infinitesimal area $d\mathcal{A}$ is $p\,d\mathcal{A}$. Therefore, there are an infinite number of parallel infinitesimal normal forces acting throughout the area $\mathcal{A}$. A set of parallel forces may be replaced by a single equivalent force. The pressure distribution of Figure 12 and its equivalent single resultant force $\bar{R}$ acting at point C are shown in Figure 13.

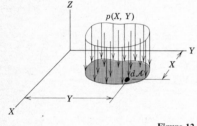

Figure 12

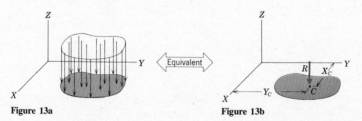

Figure 13a   Figure 13b

The force $\bar{R}$ is the sum of all of the parallel forces. To account for each of the infinitesimal forces, the summation becomes an integral. Thus

$$R = \int_{\mathcal{A}} p\, d\mathcal{A} \tag{13}$$

In view of the fact that $p(X, Y)$ is the height above the area $\mathscr{A}$ of the pressure space formed by the pictorial representation of the pressure distribution, equation (13) tells us that

> The resultant force $\bar{R}$ is the volume of the pressure space. It is normal to the planar surface being loaded, in the sense of the pushing effect for positive pressure.

A negative pressure should be regarded as a negative contribution to the volume of the pressure space.

The coordinates X_C and Y_C required to specify the line of action of the resultant are obtained by equating the moments of the two equivalent force systems in Figure 13. Computing the moment about the origin O, the position vector from point O to the point of application of the infinitesimal force $p\,d\mathscr{A}$ can be seen from the figure to be $X\bar{I} + Y\bar{J}$. Similarly, the corresponding position vector for the resultant $\bar{R}$ is $X_C\bar{I} + Y_C\bar{J}$. As before, accounting for all of the infinitesimal forces requires an integral formulation. Equating the moment sums yields

$$\Sigma\,\bar{M}_O = (X_C\bar{I} + Y_C\bar{J}) \times (-R\bar{K}) = \int_{\mathscr{A}}(X\bar{I} + Y\bar{J}) \times (-p\,d\mathscr{A}\,\bar{K}\,)$$

$$R(X_C\bar{J} - Y_C\bar{I}) = \int_{\mathscr{A}}(X\bar{J} - Y\bar{I})p\,d\mathscr{A}$$

and the result of matching corresponding components is

$$
\boxed{\begin{aligned}
RX_C &= \int_{\mathscr{A}} X\,p\,d\mathscr{A} \\
RY_C &= \int_{\mathscr{A}} Y\,p\,d\mathscr{A}
\end{aligned}}
\tag{14}
$$

The position on the area $\mathscr{A}$ whose coordinates are (X_C, Y_C) is frequently called the *center of pressure*.

Equations (14) also have significance with respect to the properties of the pressure space. For this body $p\,d\mathscr{A}$ is an infinitesimal element of volume (height times area). Thus, the integrals in equation (14) are equivalent to the first moments of volume with respect to the X and Y coordinates. Recalling that R is the volume of the fictional body, by comparing equations (14) to equations (5), we deduce that X_C and Y_C are the X and Y coordinates of the centroid of this volume. This allows us to state that

> The line of action of the resultant force $\bar{R}$ intersects the geometric centroid of the pressure space.

The importance of the analogy between the resultant $\bar{R}$ and the pressure space is that it will frequently enable us to make use of tabulated geometric properties when determining the resultant force. This occurs whenever the pressure space can be represented in terms of the fundamental shapes given in Appendix C.

The results for a line load, which we obtained in the previous section, can be seen to be a special case of those for a pressure distribution, wherein the pressure space has the shape of a thin shell. Further, in the commonly occurring situation where we have a constant pressure distribution with respect to the width of a rectangular area, we can easily reduce the pressure space to an equivalent line load. The intensity of this line load is $q = pb$, where b is the width of the rectangle in the direction of constant pressure. This line load acts along the centerline of the rectangular area, as illustrated in Figure 14 for a pressure distribution which only depends on the coordinate X.

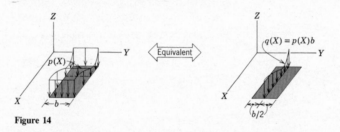

Figure 14

EXAMPLE 11

The pressure distribution on a rectangular plate is as shown, with a maximum value of $2(10^5)$ pascals. Determine the resultant of this pressure distribution and the corresponding center of pressure.

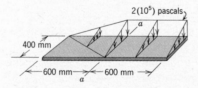

Solution

Although the pressure is applied to a rectangular area, it is not constant in the direction of either edge. Therefore, we do not seek an equivalent line load. Noting that the given diagram depicts the pressure space to be an orthogonal tetrahedron to the left of line a-a and a prism to the right of line a-a, we may employ the analogy between the resultant and the geometric properties of the pressure space.

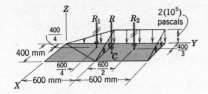

We begin with a sketch that shows the total resultant $\bar{R}$ and the resultants $\bar{R}_i$ ($i = 1, 2$) for the basic shapes forming the pressure space. The forces $\bar{R}_i$ intersect the centroids of their spaces, so the sketch also shows the location of the centroid of each basic shape in the XY plane, as found in Appendix C.

The values of R_1 and R_2 are calculated as the volume of the corresponding pressure spaces. By definition 1.0 pascal $\equiv$ 1.0 N/m², so we have

$$R_1 = \tfrac{1}{6}(0.4)(0.6)2(10^5) \text{ N} = 8.00 \text{ kN}$$

$$R_2 = \tfrac{1}{2}(0.4)(0.6)2(10^5) \text{ N} = 24.00 \text{ kN}$$

$$R = R_1 + R_2 = 32.0 \text{ kN} \qquad \triangleleft$$

The center of pressure, being the point where the resultant $\bar{R}$ is applied to the plate, is determined by equating the moment of $\bar{R}$ to the total moment of $\bar{R}_1$ and $\bar{R}_2$. Referring to the sketch for the position vectors, we then have

$$(X_C\bar{I} + Y_C\bar{J}) \times \bar{R} = [0.10\bar{I} + (0.60 - 0.15)\bar{J}] \times \bar{R}_1$$
$$+ [0.133\bar{I} + (0.60 + 0.30)\bar{J}] \times \bar{R}_2 \text{ N-m}$$

$$(X_C\bar{I} + Y_C\bar{J}) \times (-32.0\bar{K}) = (0.10\bar{I} + 0.45\bar{J}) \times (-8.0\bar{K})$$
$$+ (0.133\bar{I} + 0.90\bar{J}) \times (-24.0\bar{K})$$

$$-32.0Y_C\bar{I} + 32.0X_C\bar{J} = -25.2\bar{I} + 3.992\bar{J}$$

Equating corresponding components then yields

$$X_C = 0.125 \text{ m} \qquad Y_C = 0.788 \text{ m} \qquad \triangleleft$$

HOMEWORK PROBLEMS

V.71–V.75 The traffic sign shown is subjected to a $5(10^3)$ N/m² uniform pressure caused by the wind. Determine the reactions at the fixed base of the pole resulting from this loading.

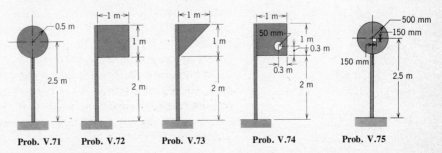

Prob. V.71   Prob. V.72   Prob. V.73   Prob. V.74   Prob. V.75

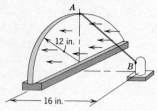

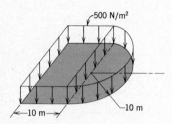

Prob. V.76

V.76 The semicircular plate shown is supported in a wind tunnel by a groove along its lower edge and by cable *AB*. The lateral pressure on the plate is 40 psi. Determine (a) the resultant wind force on the plate, (b) the tension in cable *AB*. The wind is blowing from the right.

V.77–V.79 The snow load on a flat roof is as shown. Determine the resultant force exerted by the snow on the roof. Specify the point of application of this force.

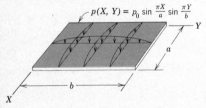

Prob. V.77

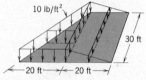

Prob. V.78

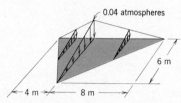

Prob. V.79

V.80–V.82 Determine the resultant of the pressure distribution acting on the flat plate shown. Also locate the corresponding center of pressure.

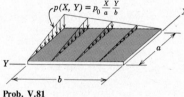

Prob. V.80

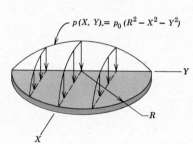

Prob. V.81

Prob. V.82

C. STATICS OF LIQUIDS

A common source of distributed loads on bodies is a fluid, such as oil, water, or air. In this course on statics we are concerned with fluids that are at rest. The matter of fluids in motion is treated in courses on fluid mechanics, which are traditionally offered after the elements of dynamics have been studied.

To identify the special properties of a fluid, recall our observation in the previous section that when two arbitrary bodies are in contact over a region of surface area, they may exert pressure and shear stresses on one another. The distinguishing characteristic of a *fluid* is that it is capable of applying shear stresses to a body only when it is in motion relative to the body. In other words, fluids at rest exert only pressure loads on stationary bodies.

The density of a *gas* changes radically with large changes in pressure, whereas the density of a *liquid* changes little with pressure, so it may be modeled as a fluid of constant density. In other words, a liquid is essentially *incompressible*. Here we shall only study liquids, for the study of gases requires the consideration of the temperature of the material.

When a liquid is placed in an open container it forms a horizontal surface, called the *free surface*. The pressure in a liquid is related directly to the vertical distance below the free surface. To develop this relationship, consider the rectangular parallelepiped element of liquid shown in Figure 15, which extends to a depth h below the free surface. Note that the XY plane is chosen to coincide with the free surface.

Only the vertical forces are depicted in Figure 15. The force $p_a \, dX \, dY$ is the result of the atmosphere pushing on the free surface (for standard conditions, $p_a = 1$ atm $= 1.0133(10^5)$ N/m² $= 14.7$ psi), whereas the force $p_{ab} \, dX \, dY$ is the result of the fluid below the element pushing on the lower horizontal face. Note that there are no vertical force components on the vertical faces, for these would require the existence of shear stresses. Summing forces in the vertical direction yields

$$\Sigma F_Z = p_{ab} \, dX \, dY - p_a \, dX \, dY - mg = 0$$

The mass m of the element of fluid may be calculated by multiplying the density ρ by the fluid volume $h \, dX \, dY$. Doing this and solving for p_{ab} gives

$$p_{ab} = p_a + \rho g h$$

The pressure p_{ab} defined by this equation is the *absolute pressure*. In most cases we will be concerned only with the increase of pressure above atmospheric, because for design purposes we usually regard bodies at atmospheric pressure to be unloaded. In order to describe this pressure increase we define a new pressure p, called the *gage pressure*, which is the difference between the absolute pressure and atmospheric pressure. Hence, the gage pressure is given by

$$\boxed{p = \rho g h} \tag{15}$$

Equation (15) defines a *hydrostatic pressure distribution*.

As it was derived, the pressure given by equation (15) is that acting on a horizontal surface. We will now prove that this pressure is exerted

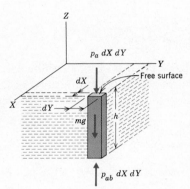

Figure 15

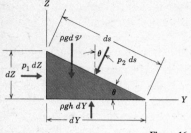

Figure 16

equally on surfaces having any orientation at the depth h below the free surface. To do this, consider the small prismatic element of fluid dm in Figure 16. Its width in the horizontal direction perpendicular to the plane of the diagram is unity.

We denote the pressures on the inclined and vertical faces of the element by the unknown values p_1 and p_2. The force on each face depicted in the figure is obtained by multiplying the pressure on that face by the area of the face (remembering that the width is unity).

The equations of equilibrium for the element of fluid are

$$\Sigma\, F_Y = p_1\, dZ - p_2\, ds \sin\theta = 0$$

$$\Sigma\, F_Z = \rho gh\, dY - p_2\, ds \cos\theta - \rho g\, d\mathcal{V} = 0$$

where ρg is the *specific weight* (units of force per unit volume) of the fluid. The weight of the element $\rho g\, d\mathcal{V} = \rho g\, dy\, dz/2$, being a second-order differential, is negligible when compared to the other terms in the Z force sum. The $\Sigma\, F_Z$ equation then reduces to

$$\Sigma\, F_Z = \rho gh\, dY - p_2\, ds \cos\theta = 0$$

From Figure 16 we note that $ds \sin\theta = dZ$ and $ds \cos\theta = dY$. Thus the equilibrium equations yield

$$p_1 = p_2 = \rho gh$$

Noting that this result was obtained for an arbitrary value of θ, we may conclude that

> The gage pressure at a point in a fluid is proportional to the depth h from the free surface to the point ($p = \rho gh$). This pressure is exerted equally on all surfaces at that depth, regardless of their orientation.

The second sentence in this statement is known as Pascal's law. It should be noted that in the U.S.-British system of units it is common to prescribe the specific weight ρg of a material, whereas in the SI system the density ρ of a material is the preferred prescribed parameter.

Now that the relationship governing the pressure in a liquid has been established, we proceed to the study of the resultant force exerted by a fluid on a surface. Here we will consider three specific types of surfaces. They are typical of cases occurring in actual practice. The case of an arbitrary surface is left for texts on fluid mechanics.

1 Rectangular Surfaces

Consider the situation where one edge of a submerged rectangular surface is horizontal. Then any line in the rectangle parallel to this edge is situated

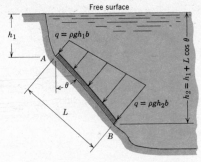

Figure 17

at a constant depth h. Therefore, the hydrostatic pressure is constant along this line. This permits us to convert the pressure distribution into a line load by multiplying the pressure by the horizontal width b of the surface. A typical case of the pressure on a rectangular plate of length L that is exposed to a liquid on only one side is shown in side view in Figure 17, where b is the width of the surface perpendicular to the plane of the figure.

At the upper edge A the depth is h_1. Here, the gage pressure is $\rho g h_1$, and the line load is $q = pb \equiv \rho g h_1 b$. At the lower edge B the depth is h_2, so the line load at this location is $q = \rho g h_2 b$. Noting that the distance along the line AB is linearly related to the depth, the line load is trapezoidal, as illustrated. Obvious exceptions occur when edge A is at the free surface, in which case $h_1 = 0$ and the loading is a triangular area; and when the plate is horizontal ($\theta = 90°$), for then $h_1 = h_2$ and the surface of the loading function is a rectangle.

EXAMPLE 12

The automatic valve shown in cross section consists of a 2- by 3-m rectangular plate AB that can rotate about a horizontal shaft passing through point C. Neglecting the weight of this plate, determine the depth of water d in the reservoir for which the valve will open.

Solution

We are interested in the effect of the water on the valve plate. Hence, we wish to draw a free body diagram of the plate. In the condition where the valve is about to open, there will be no contact at edges A and B, so there will be no reactive force at either point. At edge B the depth is the unknown value d, whereas at edge A the depth is $d - 3 \sin 53.13°$. Recalling that the pressure is normal to the surface it acts upon, we obtain the free body diagram shown. In the values for the line load shown, the factor 2 is the width of the plate.

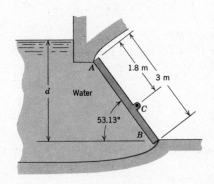

As indicated in the free body diagram, we consider the trapezoidal load to be a composite of two triangular loads. Computationally, this is somewhat more efficient than considering a composite of a triangular and rectangular load. Equating each force to the area of the corresponding triangle yields

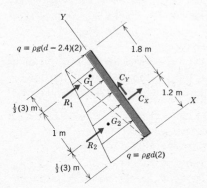

$$R_1 = \tfrac{1}{2}[\rho g(d - 2.4)(2)](3) = \rho g(3d - 7.2) \text{ newtons}$$

$$R_2 = \tfrac{1}{2}[\rho g d(2)](3) = \rho g(3d) \text{ newtons}$$

We are not interested in the reactions at the shaft C, so the only equilibrium equation we need is the moment sum at that point.

$$\Sigma M_{CZ} = -R_1(1.8 - 1.0) + R_2(1.2 - 1.0) = 0 \text{ N-m}$$

Substituting the expressions for R_1 and R_2 found above, we find

$$-\rho g(3d - 7.2)(0.8) + \rho g(3d)(0.2) = 0 \qquad d = 3.20 \text{ m} \qquad \triangleleft$$

If d is less than this value, the portion of the trapezoidal loading below shaft C exerts a greater moment about this shaft than the portion of the loading above the shaft, thus keeping the valve shut. When d exceeds 3.20 m, the opposite is true and the valve swings open. As an aside, we note that if we had needed it, the density ρ would have been for water. However, because this parameter does not affect the value of d, the valve would open at a depth of 3.20 m for any liquid.

HOMEWORK PROBLEMS

V.83 The fish tank shown holds water to a depth of 18 in. For each face, including the bottom, determine the resultant force exerted by the water. Also determine the point of application of each resultant force (the center of pressure).

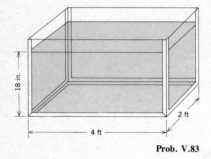

Prob. V.83

V.84 and V.85 Determine the hydrostatic force acting on the 10-m section of seawall shown when the water is at flood stage, and locate the corresponding center of pressure. The density of sea water is 2.5% greater than that of fresh water.

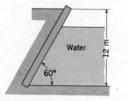

Prob. V.84

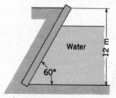

Prob. V.85

V.86 and V.87 The 3-m-wide sliding gate shown is at the bottom of a retaining wall. Determine the force exerted by the groove at the lower edge of the gate to resist the water pressure (a) when $d = 3$ m, (b) when $d = 6$ m.

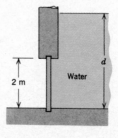

Prob. V.86

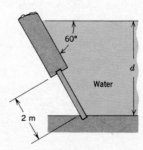

Prob. V.87

V.88 and V.89 The 1.5-m-wide gate shown, which may pivot about the horizontal shaft A, is held in the closed position by the preloaded spring. Determine the initial compressive force in the spring for which the gate will open when the depth of the water is (a) $d = 2$ m, (b) $d = 4$ m.

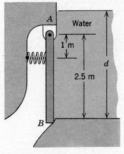

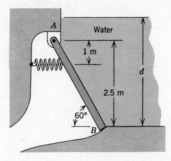

Prob. V.88   Prob. V.89

V.90 and V.91 The cross section of a 60-ft-long section of concrete formwork is shown. The left upright panel is attached to the anchored panel on the right by 40 equally spaced tie rods. Determine the tension in the tie rod when concrete, having a density of 150 lb/ft³, is in its liquid state. Assume that the lower edge of the upright panel, edge A, is simply supported.

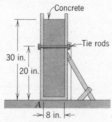

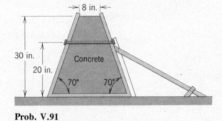

Prob. V.91

Prob. V.90

V.92 A 2-m-long tank, whose cross section is shown, is used to separate a pool of water from a pool of mercury. Four sets of opposing cables (only one set is shown in the cross-sectional view) support the metal panel separating the two fluids. The cables are slack when the tank is empty. Determine the force in the cables, specifying which side is tensioned, when the depth of mercury is (a) $d = 250$ mm, (b) $d = 500$ mm.

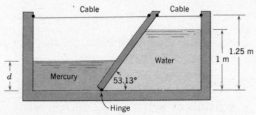

Prob. V.92

V.93 Water has accumulated in the trap of the oil storage tank, as shown. Determine the minimum horizontal force the hydraulic cylinder must exert to keep the 800-mm-wide gate shut.

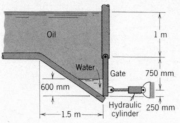

Prob. V.93

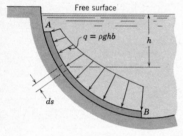

Figure 18

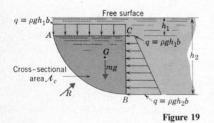

Figure 19

2 Cylindrical Surfaces

In this section we will consider a surface that can be described as a sector of a horizontal (arbitrary) cylinder. Such a surface, formed from the curve AB, is shown in the side view of Figure 18. Once again let the horizontal width of the surface be b.

For this case a line load may be formed by multiplying the pressure by the width b, as shown in Figure 18. The corresponding differential elements of force $\bar{q}\, ds$ change direction because the pressure at each point is parallel to the unit vector normal to the surface at each point. One way to account for this distributed force is to consider the horizontal and vertical components of $\bar{q}\, ds$ separately, expressing the slope of the curved surface at each location. However, a more convenient approach is available.

Consider the cylindrical body of liquid having cross-sectional area $\mathcal{A}_c$ depicted in Figure 19. For this body of liquid the line load is constant along the horizontal face AC, because the depth is constant for that face. The line load on the vertical face is trapezoidal. Another force acting on this body of liquid is its weight mg, which acts at center of mass G of the body. As viewed in Figure 19, the center of mass G coincides with the centroid of the cross-sectional area $\mathcal{A}_c$. Noting that this body has a constant width, we have

$$m = \rho \mathcal{V} = \rho \mathcal{A}_c b$$

Finally, to complete the free body diagram, we have the force $\bar{R}$, which is the reaction corresponding to the force exerted by the water on the curved surface AB.

The force $\bar{R}$ is what we seek when we wish to know the effect of the liquid on surface AB. To determine it we write the equations of static equilibrium. The force sums in the horizontal and vertical directions define the magnitude and direction of $\bar{R}$, whereas the moment sum defines a point of application of $\bar{R}$.

EXAMPLE 13

The cylindrical surface shown is formed from the circular arc AB. In this case, because the pressure at each point is normal to the local tangent plane, the distributed forces acting on the surface are concentric at the center of the arc, point C. It follows that the resultant must also pass through point C. Considering the equilibrium of the shaded element of fluid, prove this hypothesis and also determine the resultant force $\bar{R}$ acting on the cylindrical surface.

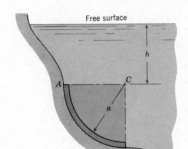

Solution

The free body diagram for the given body of liquid is shown to the left. For this diagram the location of the center of mass G is taken from Appendix B. Regarding the trapezoidal loading as the composite of the two triangular loadings shown, we have

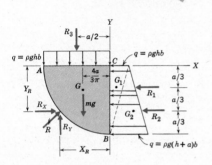

$$R_1 = \tfrac{1}{2}(\rho g h b)a = \tfrac{1}{2}\rho g h a b$$

$$R_2 = \tfrac{1}{2}[\rho g(h + a)b]a = \tfrac{1}{2}\rho g(ha + a^2)b$$

$$R_3 = (\rho g h b)a = \rho g h a b$$

The cross-sectional area of the body of water is $\pi a^2/4$, so its mass is

$$m = \rho \mathcal{A}_c b = \frac{\pi}{4}\rho a^2 b$$

Choosing point C for a moment sum, the equilibrium equations are

$$\Sigma M_{CZ} = R_X(Y_R) - R_Y(X_R) - R_1\left(\frac{a}{3}\right) - R_2\left(\frac{2a}{3}\right)$$

$$+ R_3\left(\frac{a}{2}\right) + mg\left(\frac{4a}{3\pi}\right)$$

$$\equiv Y_R R_X - R_Y X_R - \tfrac{1}{6}\rho g h a^2 b - \tfrac{1}{3}\rho g(ha + a^2)ab$$

$$+ \tfrac{1}{2}\rho g h a^2 b + \left(\frac{\pi}{4}\rho g a^2 b\right)\left(\frac{4a}{3\pi}\right)$$

$$\equiv R_X Y_R - R_Y X_R = 0$$

$$\Sigma F_X = R_X - R_1 - R_2 \equiv R_X - \tfrac{1}{2}\rho g(2ha + a^2)b = 0$$

$$\Sigma F_Y = R_Y - mg - R_3 \equiv R_Y - \rho g a b\left(\frac{\pi}{4}a + h\right) = 0$$

Because the moments of the other forces cancel each other, the moment equation states that the moment of $\bar{R}$ about point C is zero. This can be true only if the line of action of $\bar{R}$ intersects point C, thus proving the hypothesis.

The expression for R_X and R_Y are obtained by solving the force equilibrium equations. These give

$$R_X = \rho g a b \left(h + \frac{1}{2} a \right)$$

$$R_Y = \rho g a b \left(h + \frac{\pi}{4} a \right)$$

The force exerted by the water on the surface is the reaction corresponding to these values.

The problem did not request that we determine the value of X_R (or Y_R) defining the line of action of the resultant $\bar{R}$. If this information were desired, there are two ways in which we could proceed. The approach that is valid for all types of cylindrical surfaces, not just circular ones, is first to write the relationship between X_R and Y_R, which is this case is $Y_R = (a^2 - X_R^2)^{1/2}$. Substituting this relationship and the results for R_X and R_Y into the moment equation yields an equation for X_R.

The alternative method is valid for circular cylindrical surfaces only. In this approach we perform a geometrical calculation using the fact that the resultant $\bar{R}$ intersects the center of the cylinder. This is illustrated in the sketch to the left.

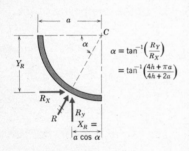

$$\alpha = \tan^{-1}\left(\frac{R_Y}{R_X} \right)$$
$$= \tan^{-1}\left(\frac{4h + \pi a}{4h + 2a} \right)$$

EXAMPLE 14

The 3-ft-wide bent plate is supported by a horizontal shaft at edge B and bears against the bottom of the wall at edge A. Determine the reaction at edge A for a depth of water $d = 8$ ft.

Solution

We could solve this problem in two parts. First, we could consider an isolated body of water to find the resultant, and then apply the resultant to the bent plate in order to evaluate the desired reaction. However, note that the problem is to determine the reaction force produced by the pressure of the water; it is not required that the resultant of the pressure force be found. This suggests that we can solve the problem in one step, by considering the equilibrium of the body of water and the plate *jointly*. In such an approach the resultant force $\bar{R}$, which is due to the interaction of the bent plate and the isolated body of water, becomes an internal force.

The isolated body of water we form is one having horizontal and vertical faces, with the remaining face being the curved surface of interest. The combination of this body of water and the bent plate is shown in the free body diagram on the following page.

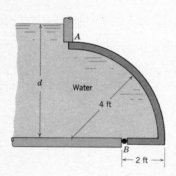

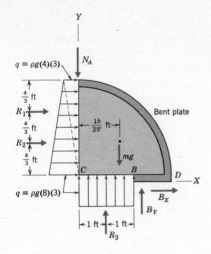

Note that on the horizontal face, the pressure acts only on the segment BC. For the segment BD the pressure is an internal force for the combination of bodies we have isolated.

Decomposing the trapezoidal loading into two triangular loadings, as indicated, we have

$$R_1 = \tfrac{1}{2}(12\rho g)(4) = 24\rho g \qquad R_2 = \tfrac{1}{2}(24\rho g)(4) = 48\rho g$$

$$R_3 = (24\rho g)(2) = 48\rho g$$

The weight of the body of water is

$$mg = \rho\left(\frac{\pi}{4}\right)(4^2)(3)g = 37.70\rho g$$

We only want the value of N_A, so we sum moments about shaft B. Thus

$$\Sigma M_{BZ} = -R_1\left(\frac{8}{3}\right) - R_2\left(\frac{4}{3}\right) - R_3(1) + mg\left(2 - \frac{16}{3\pi}\right) + N_A(2) = 0 \text{ lb-ft}$$

$$2N_A = 24\rho g\left(\frac{8}{3}\right) + 48\rho g\left(\frac{4}{3}\right) + 48\rho g(1) - 37.70\rho g(0.3023)$$

$$N_A = 82.30\rho g = 5.136 \text{ kips}$$

where the value of ρg in U.S.-British units is 62.4 lb/ft³. The value of this reaction demonstrates the enormous forces required to contain large bodies of water.

HOMEWORK PROBLEMS

V.94 Cross sections of two alternative designs for the face of a dam are shown. Considering a 1-m-wide section, resolve the pressure of the water into an equivalent force-couple system at the base B when the water is at the level of the top of the dam.

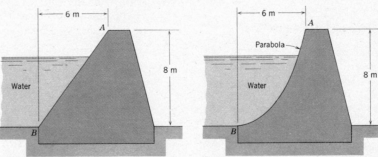

Prob. V.94

V.95 and V.96 A viewing window in an aquarium is in the form of a 3-m-long semicylinder, as shown in the cross-sectional view. Determine the magnitude, direction, and point of application along line AB of the resultant force exerted by the water on the window.

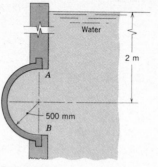

Prob. V.95

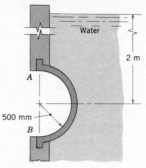

Prob. V.96

V.97 A window in the hull of a boat, through which fish can be observed, has the parabolic cross section shown. The window is 4 ft wide. Knowing that the depth $h = 2.5$ ft, determine the resultant force exerted on the window by the water. Specify the point of application on the window of this resultant.

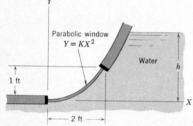

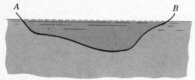

Prob. V.97

V.98 According to Archimedes' principle of buoyancy, a liquid exerts an upward force on a body equal to the weight of the liquid displaced by the body. Demonstrate that this principle applies to the arbitrary cylindrical body shown, which is formed from the curve AB. Also determine the line of action of this buoyant force. The colored area defines the body of liquid that has been displaced.

Prob. V. 98

V.99–V.101 A water conduit has the cross section shown. The conduit is supported by hinges along the lower edge A and by cables spaced at 400-mm intervals along edge B. Determine the tension in the cables when the conduit is filled to the top.

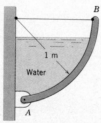

Prob. V.99

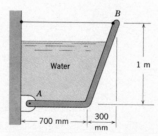

Prob. V.100

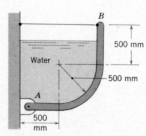

Prob. V.101

V.102 and V.103 A quarter-circular plate 6 ft long is used as a gate for a dam. Determine the minimum horizontal force exerted by the hydraulic cylinder needed to open the gate when the depth of the water is (a) $d = 3$ ft, (b) $d = 9$ ft.

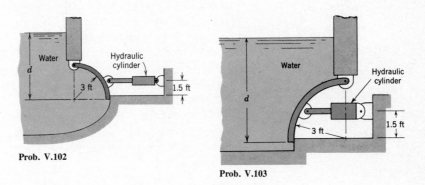

Prob. V.102

Prob. V.103

V.104 A 20-m-long section of a breakwater has the parabolic cross section shown. Determine the force-couple system at edge C representing the reaction of the *ground on the breakwater* when flooding is about to occur. The density of seawater is 1025 kg/m³ and that of concrete is 2400 kg/m³.

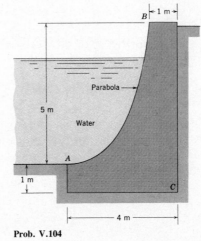

Prob. V.104

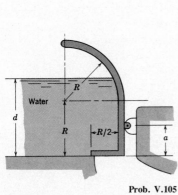

Prob. V.105

V.105 The cross section of a gate in a channel is shown. Determine the ratio of dimensions a/R for which the gate will open when the depth of water in the channel is $d = 2R$.

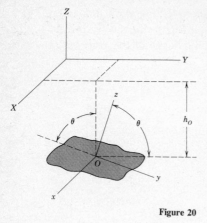

Figure 20

3 Arbitrary Planar Surfaces

We were able to treat the pressure distributions appearing the last two sections by converting them into line loads. In contrast, when the surface in contact with the liquid has an arbitrary width, we must deal with the actual pressure distribution. Consider the situation shown in Figure 20, where the XY plane locates the free surface and the colored area is a flat surface inclined at an angle θ from the vertical.

For convenience in locating points on this inclined surface, we define an auxiliary coordinate system xyz, for which the xy plane coincides with the plane of the inclined surface, with the x axis horizontal. It then follows that the yz plane is vertical and that the angle between the negative y axis and a vertical line is θ, as shown.

Because the x axis is horizontal, all points in the surface at a constant value of y are at the same depth. Letting h_0 denote the depth of the origin O, whose location will be specified later, the depth of other points is given by

$$h = h_0 + y \cos \theta$$

It follows that the pressure distribution is

$$p = \rho g h = \rho g (h_0 + y \cos \theta) \tag{16}$$

To describe the resultant of this pressure distribution we could resort to the analogy with the geometric properties of the pressure space, as developed in Section B.2 of this module. However, except in rare situations, the pressure space resulting from equation (16) will not be easily represented in terms of the basic shapes in Appendix C. We therefore shall employ equations (13) and (14) found in the Section B.2. Letting $d\mathcal{A}$ be an element of area for the inclined surface, equation (13) gives

$$R = \int_\mathcal{A} p \, d\mathcal{A} = \rho g h_0 \int_\mathcal{A} d\mathcal{A} + \rho g \cos \theta \int_\mathcal{A} y \, d\mathcal{A}$$

Clearly, the first integral on the right side gives the area $\mathcal{A}$, whereas the second integral is the first moment of area with respect to the y coordinate. In order to eliminate the latter term, we now *locate the origin O at the centroid of the surface area $\mathcal{A}$*. The result is

$$\boxed{R = \rho g h_0 \mathcal{A}} \tag{17}$$

This resultant force is perpendicular to the plane in the negative z direction. An interpretation of equation (17) is to say that $\rho g h_0$, where h_0 is the depth of the centroid of the area, is the *average pressure* acting on the area.

Now that the resultant force has been determined, we may proceed to locate the center of pressure. Note that $\bar{R}$ does *not* act at the centroid. Equations (14) give

$$Rx_R \equiv \rho g h_o \mathcal{A} x_R = \int_{\mathcal{A}} xp \, d\mathcal{A}$$

$$= \rho g h_o \int_{\mathcal{A}} x \, d\mathcal{A} + \rho g \cos \theta \int_{\mathcal{A}} xy \, d\mathcal{A}$$

$$Ry_R \equiv \rho g h_o \mathcal{A} y_R = \int_{\mathcal{A}} yp \, d\mathcal{A}$$

$$= \rho g h_o \int_{\mathcal{A}} y \, d\mathcal{A} + \rho g \cos \theta \int_{\mathcal{A}} y^2 \, d\mathcal{A}$$

As a result of the fact that the origin is located at the centroid, the first moments of area in the foregoing equations vanish. Let us define the following symbols:

$$\boxed{\begin{aligned} I_x &= \int_{\mathcal{A}} y^2 \, d\mathcal{A} \\ I_{xy} &= \int_{\mathcal{A}} xy \, d\mathcal{A} \end{aligned}} \tag{18}$$

The expressions for x_R and y_R may then be written as

$$\boxed{x_R = \frac{I_{xy} \cos \theta}{h_o \mathcal{A}} \qquad y_R = \frac{I_x \cos \theta}{h_o \mathcal{A}}} \tag{19}$$

The quantity I_x is called the *second moment of area about the x axis*, or alternatively, the *area moment of inertia about the x axis*. The reason for associating it with the x axis is that y is the distance of an element of area from the x axis. The value of y^2 can never be negative, so the integrand for I_x is never negative, and the value of I_x will always be a positive quantity. Hence, the center of pressure is never higher then the centroid of the surface area.

The parameter I_{xy} is the *area product of inertia*. This quantity is studied in detail in Module IX. Briefly, I_{xy} is a measure of the asymmetry of the area $\mathcal{A}$. If the area is symmetrical with respect to either the x or y axes, then $I_{xy} = 0$. In that case equations (19) give $x_R = 0$, which means that the center of pressure is situated on the y axis. In all of the problems we shall consider in this section the area will have this type of symmetry.

Values of the area moments of inertia for some basic geometric shapes about centroidal axes are given in Appendix B. Methods for calculating this property for geometric shapes not appearing in Appendix B are described in Module IX.

Before we apply equations (17) and (19), let us make some fundamental observations regarding them. The fact that the pressure at the centroid of the area is the average for the entire area is logical. What is surprising is that the angle of inclination θ does not affect the magnitude of the resultant. The expression for y_R shows that, for a given area $\mathcal{A}$, the center of pressure moves upward (y_R decreases) if the depth h_O or the angle of inclination θ is increased. These tendencies are a result of the fact that an increase of either of these parameters increases the ratio of the pressure of the highest point on the area to the pressure at the lowest point. It is also interesting to note that similar expressions arise in the study of mechanics of materials when the state of stress in beams carrying axial and transverse loads is studied.

EXAMPLE 15

The 1-m diameter circular plate allows access to the interior of the spherical water tank when the tank is empty. Determine the resultant water force exerted on the plate and the location of the center of pressure when the water tank is full, but a free surface is still present.

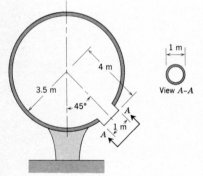

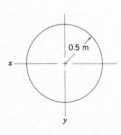

Solution

We need a coordinate system xyz for the circular plate. In order to employ equations (17) and (19), we let x and y coincide with the plane of the plate with the x axis horizontal. Locating the origin at the centroid of the plate, which of course is the center of the circle, we obtain the coordinate system depicted in the next sketch.

From the diagram accompanying the statement of the problem, we calculate that when the free surface is at the top of the tank, the depth of the center of the plate is

$$h_O = 3.5 + 4.0 \cos 45° = 6.328 \text{ m}$$

The angle of inclination of the plate is seen to be $\theta = 45°$. From Appendix B we have

$$\mathcal{A} = \pi(0.5)^2 = 0.7854 \text{ m}^2$$

$$I_x = \frac{\pi}{4}(0.5)^4 = 0.04909 \text{ m}^4$$

Also, the plate is symmetric about both the x and y axes, so we know that the center of pressure is situated on the y axis. (Symmetry about one axis is sufficient for this.)

Setting $\rho = 1000$ kg/m³, equation (17) gives

$$R = 1000(9.806)(6.328)(0.7854) = 49.7 \text{ kN}$$

For the location of the center of pressure, equation (19) gives

$$y_R = \frac{0.04909(\cos 45°)}{6.328(0.7854)} = 6.98(10^{-3}) \text{ m}$$

Hence, the center of pressure is 6.98 mm below the center of the 500-mm-radius plate.

HOMEWORK PROBLEMS

V.106 A circular window in an aquarium is situated on a vertical wall. Determine the resultant force of the water on the window and the center of pressure if the depth is (a) $d = 1$ ft, (b) $d = 10$ ft.

V.107 Three possibilities for a throttle valve for the 1-m inside diameter circular pipe leading from the water tank are shown. In each design the plate forming the valve is pivoted about the horizontl axis BC. For a depth $d = 1$ m, determine the force exerted by the water on each plate and the corresponding center of pressure.

Prob. V.106

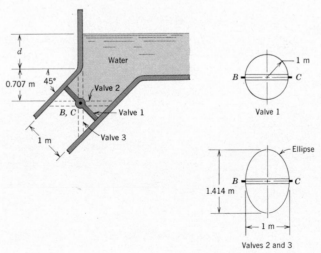

Prob. V.107

V.108 Solve Problem V.107 when $d = 0$.

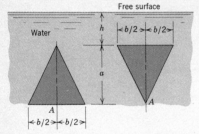

Prob. V.109

V.109 Two identical triangular plates are given opposite orientations on a vertical wall in the tank of water, as shown. (a) Determine the resultant force of the water on each plate and the corresponding center of pressure. (b) Replace the resultant forces determined in part (a) by equivalent force-couple systems acting at point A. Give a physical explanation for the difference in the results for the two plates.

V.110–V.112 The flat plate shown fits into a matching opening in the vertical wall of a tank of oil. The density of the oil is 800 kg/m³. The plate is hinged about the horizontal axis AB. Determine the force $\bar{P}$ applied at point C required to keep the plate in place.

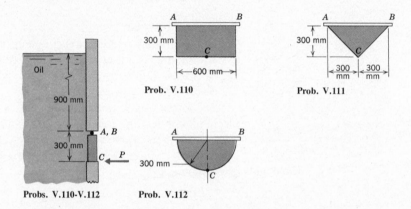

Prob. V.110   Prob. V.111

Probs. V.110-V.112   Prob. V.112

V.113 and V.114 The end of a water trough is inclined at 36.87° from the vertical. The end plate shown is attached to the walls of the trough by bolts at points A, B, and C. Determine the tension in the bolts when the trough is filled to the top.

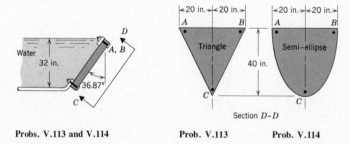

Probs. V.113 and V.114   Prob. V.113   Prob. V.114

MODULE VI
INTERNAL
FORCE ANALYSIS

The focus of Module IV was structures and machines whose intercon-nected members support and modify the loads that are applied to them. In the process of studying the static equilibrium of such systems, we did not consider how the forces applied at one location on a member were trans-mitted internally to another location on the member. The determination of the internal forces is the subject of this module.

In earlier modules we implicitly considered the internal forces in some structural members. For example, a lightweight cable that is loaded only at its ends carries a constant tension force. We have also seen that a straight two-force member in a truss carries a constant axial tensile or compressive force. In fact, one of the methods we formulated for analyz-ing the forces in the members of a truss, the method of sections, is typical of the type of analysis we will perform here.

Recall that in the method of sections we passed an imaginary cutting plane through a truss in order to have the internal axial forces in certain members appear as external forces acting on the isolated section. The equilibrium equations for the isolated section then provided the required relations for determining the internal forces.

The methods we shall develop in this module will extend this type of analysis to treat more general types of structural members. We will utilize cutting planes to isolate a portion of the member of interest, in order to make the internal forces appear explicitly in a free body diagram. It is apparent that this approach is necessary, for the equations of static equilibrium only relate the external forces acting on a system.

Knowledge of the distribution and flow of internal forces in members of a structure is essential for their design. In the developments that follow we shall consider two of the basic elements of a structural system: (1) *beams* and (2) *cables* that serve as multiple force members.

A.   BEAMS

1   Basic Definitions

Certain bodies occurring in previous modules were referred to as beams, without really considering the precise meaning of this term. A primary physical characteristic of a beam is that it is a slender bar. That is, the arc length of some line in the body is considerably larger than any dimension measured perpendicular to that line.

The line whose arc length we measure is the locus of the centroids of the cross sections. The bar is said to be uniform or nonuniform according to whether or not the cross sections have a constant shape. Also, the bar is said to be straight or curved in accordance with the shape of its centroi-dal axis, as depicted in Figure 1.

The property of slenderness is not sufficient to identify that a bar acts

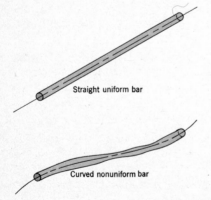

Straight uniform bar

Curved nonuniform bar

Figure 1

like a beam, although all beams must have that property. The second item of importance is the type of internal forces the bar transmits. To investigate this matter let us pass an imaginary cutting plane through an arbitrary bar and consider the force system exerted by one portion of the cut bar on the other.

In the most general situation the internal forces will be equivalent to a force-couple system. The point of reference that we choose for this force-couple system is the centroid of the cross-sectional area. The reason for choosing the centroid is that many formulas derived in mechanics of materials to describe the load-carrying ability of bars require such a resolution for the internal forces. Thus, considering the cut bar in Figure 2, the internal forces exerted on the cross-sectional surface to the right of the cutting plane by the portion of the bar to the left of the plane are equivalent to a resultant force $\bar{R}$ acting at the centroid C and a couple $\bar{M}_C$.

Although they are not depicted in Figure 2, it follows from Newton's third law that the internal forces exerted by the portion of the bar to the right of the cutting plane on the exposed cross section to the left of the cutting plane consist of a resultant force $-\bar{R}$, applied at the centroid, and a couple $-\bar{M}_C$.

To describe the internal forces further, let us choose an xyz coordinate system having its origin at the centroid C. The x axis is chosen to be outward from the exposed face, tangent to the centroidal axis of the bar. In other words, the x axis is the outward normal to the cross section. It follows that the yz plane is coincident with the cross section.

Lowercase letters are used to denote this coordinate system in order to emphasize that it may be different from the XYZ coordinate system used to evaluate the reactions acting on the beam. Also, aside from the case of a straight bar, where the cross sections all lie in parallel planes, the orientation of the xyz system will depend on which cross section of the bar is being considered.

We shall use the xyz system to express the internal force $\bar{R}$ and couple $\bar{M}_C$ in terms of their components. Figure 3 depicts a blow-up of the cross section featured in Figure 2.

Special terms are used to describe the components of the internal force and moment. The force components $\bar{R}_y$ and $\bar{R}_z$, which are parallel to the cross section, are called the *shear forces*. As was true when we treated straight two-force members, the component $\bar{R}_x$, normal to the cross section, is called a *tensile* or *compressive* force, according to whether $\bar{R}_x$ is in the same or opposite sense of the x axis. The components of the internal moment parallel to the cross section, that is, $(\bar{M}_C)_y$ and $(\bar{M}_C)_z$, are called *bending moments* because they have the effect of tending to bend the bar. Finally, the component $(\bar{M}_C)_x$ normal to the cross section is called the *torsional moment* because it has the effect of tending to twist

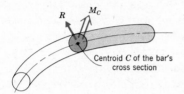

Centroid C of the bar's cross section

Figure 2

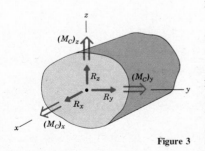

Figure 3

the bar about its centroidal axis.

The primary factor in deciding whether a bar acts like a beam is that

> In a beam the shear forces and bending moments are significant in comparison to the axial (tensile or compressive) force and the torsional moment.

Clearly, the foregoing definition excludes cables and the straight members of a truss, for both transmit only axial forces. It also excludes a straight bar, frequently called a torsion bar, whose only loads are torsional moments.

The majority of multiforce members in the frames and machines we analyzed in Module IV are beams. Beams of all descriptions may be found in the structural framework of buildings, automobiles, trucks and airplanes. They are crucial to many industrial machines, and parts of many hand tools, such as the handles of pliers and wrenches, which function like beams. A beam is one of the fundamental structural elements for a designer.

The basic method we will develop for evaluating the internal forces in a beam will be equally valid for curved and straight beams. We shall focus attention on straight beams, because of the simplicity of their analysis and because of the frequency with which they occur in practice. The added difficulty in analyzing curved beams is merely that of describing the geometry of its curvature.

It frequently proves convenient to classify straight beams according to the manner in which they are supported. Some typical types are displayed in Figure 4. They have been depicted in the horizontal position solely for convenience; in practice, beams may have any orientation in space.

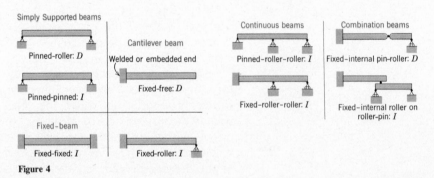

Figure 4

All except one of the beams exhibited in Figure 4 have two names. The more descriptive name explicitly states the type of supports, whereas the other name (heading each entry) is the one more commonly employed by engineers.

Another feature of the beams in Figure 4 is that in several cases, the supports exert more reactions than the number of available static equilibrium equations. These are statically indeterminate systems whose analysis requires consideration of the deformation of the beam caused by the beam loading. Note that each system in Figure 4 is labeled D or I to indicate that it is statically determinate or indeterminate, respectively. Statically indeterminate systems are treated in courses on mechanics of-materials and structures. We treat only statically determinate systems in this text.

2   Determination of the Internal Forces

A beam may be subjected to concentrated or distributed loads only, or to a combination of both. In view of the fact that we are restricting ourselves to the consideration of statically determinate beams, it follows that the reaction forces exerted by the supports of a beam to prevent the beam from moving can be fully determined by using the methods of Modules III, IV, and V to formulate the equations of equilibrium. After such a determination, we may treat the reactions as additional known loads.

At this juncture, we can discuss straight and curved beams with equal ease, so let us consider an arbitrary curved beam, for which the full set of loads, including the reactions, are known. Suppose that we wish to know the internal forces acting on a cross section at some specific location C along the centroidal axis of the beam. This determination requires that the beam be (fictitiously) separated into two portions by a cutting plane at location C, as depicted in Figure 5.

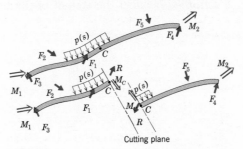

Figure 5

A significant feature of Figure 5 is that all forces acting on the portions of the beam to either side of the cutting plane are depicted. Thus, with the inclusion of the internal force $\bar{R}$ and the couple $\bar{M}_C$, Figure 5 shows a free body diagram for each portion of the cut beam.

From Figure 5, it is apparent that the internal force $\bar{R}$ and couple $\bar{M}_C$ are merely the force-couple system acting at the centroid C of the cross section required to hold each portion of the cut beam in equilibrium. We have stipulated that the entire beam is in static equilibrium under the system of external forces. It follows that the equilibrium equations for either cut portion of the beam give the same force $\bar{R}$ and couple $\bar{M}_C$, because the equilibrium of the entire beam and either portion guarantees the equilibrium of the other portion.

When we write the moment equilibrium equation, it is advantageous to sum moments about point C (the centroid of the cut cross section) in order to eliminate the unknown force $\bar{R}$ from the equation. Thus, the process of determining the internal force-couple system acting on a specific cross section of a beam can be summarized as follows.

> First, isolate in a free body diagram the portion of the beam and of the loading system to *either* side of the cutting plane for that cross section. Then, formulate the equilibrium equations for the chosen portion of the beam by summing moments about the centroid of the cross section and summing forces for the isolated portion.

It should be observed that two coordinate systems are utilized in our formulation: the XYZ that is used for determining the reactions and the xyz that is used for describing the internal forces. For simplicity, we shall make it standard practice to also employ the XYZ coordinate system when formulating the static equilibrium equations for the portion of the beam isolated by the cutting plane.

EXAMPLE 1
Determine the internal forces acting on the cross section of bar ABC at location E. Neglect the weight of the bar.

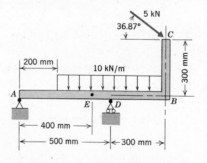

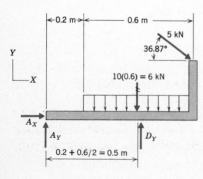

Solution
We first determine the reaction forces. For this, we draw a free body diagram replacing the 10-kN/m distributed load by its resultant acting at the centroid of the area of the loading function. The equilibrium equations for the entire beam are formulated in terms of any convenient coordinate system, such as the one shown. Choosing pin A for the moment sum, we have

$$\Sigma M_{AZ} = D_Y(0.5) - 6.0(0.5) - (5 \cos 36.87°)(0.3) - (5 \sin 36.87°)(0.8) = 0$$

$$\Sigma F_X = A_X + 5 \cos 36.87° = 0$$

$$\Sigma F_Y = A_Y + D_Y - 6.0 - 5 \sin 36.87° = 0$$

These equations give

$$D_Y = 13.2 \qquad A_X = -4 \qquad A_Y = -4.2 \text{ kN}$$

Now that the reactions have been found, we pass a cutting plane through point E, perpendicular to the centroidal axis of the bar. Either portion of the cut bar may be chosen for consideration; we will choose the left end because there are fewer forces on that side. In the free body diagram for the isolated portion chosen, the reaction forces at pin A are depicted in the direction in which they were calculated to act.

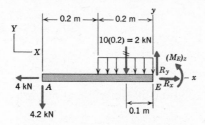

It should be noted that the xyz coordinate system illustrated in the sketch is chosen such that the x axis is the outward normal for the exposed face. We use this system solely for describing the internal force components. Because this is a planar problem, it can be seen that the only non-zero internal reactions are the axial force R_x, the shear force R_y, and the bending moment $(M_E)_z$. The XYZ coordinate system, which is also shown in the sketch, shall be employed for the formulation of the static equilibrium equations.

The free body diagram contains only that portion of the distributed load acting on the isolated section. The nontrivial static equilibrium equations are

$$\Sigma M_{EZ} = 2(0.1) + 4.2(0.4) + (M_E)_z = 0$$

$$\Sigma F_X = R_x - 4.0 = 0 \qquad \Sigma F_Y = -2.0 - 4.2 + R_y = 0$$

which give

$$(M_E)_z = -1.88 \text{ kN-m} \qquad R_x = 4.0 \text{ kN} \qquad R_y = 6.6 \text{ kN}$$

These results may be interpreted to mean that at point E the beam carries a tensile force (positive R_x) of 4.0 kN, a vertical shear force of 6.6 kN, and a bending moment of 1.88 kN-m about the horizontal axis parallel to the cross section.

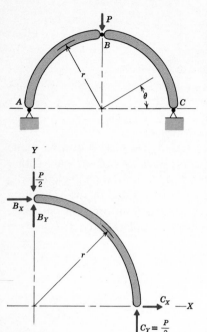

EXAMPLE 2

The three-pinned arch shown carries a vertical force $\bar{P}$ at pin B. Determine the internal forces acting on the cross section at $\theta = 30°$.

Solution

The arch consists of two curved bars AB and BC. The analysis of internal forces requires that we first find the reactions. By symmetry (or equivalently, from the equilibrium equations for the entire structure), we find that the vertical reactions at pins A and C are each $P/2$ upward. We also find that the horizontal reactions at pins A and C are equal in magnitude and opposite sense. To determine these reactions, as well as the force exerted between the two bars at pin B, we draw a free body diagram for one bar and write the corresponding equations for static equilibrium.

$$\Sigma M_{BZ} = C_X(r) + \frac{P}{2}(r) = 0$$

$$\Sigma F_X = C_X + B_X = 0$$

$$\Sigma F_Y = B_Y + \frac{P}{2} - \frac{P}{2} = 0$$

The solutions of these equations are

$$B_Y = 0 \qquad B_X = -C_X = \frac{P}{2}$$

Now that the reactions are known, we pass a cutting plane through the bar at the location where we wish to determine the internal forces. We have chosen to isolate the lower portion of the beam formed by the cutting plane, as shown in the free body diagram to the left. An xyz coordinate system with the x axis being the outward normal for the cross section is selected, as illustrated in the free body diagram. The internal forces are depicted as an axial force F in the sense of the x axis, a shear force V in the sense of the y axis, and a bending moment M about the z axis. There are no other components of internal force resulting from the planar system of forces.

These internal forces are determined from the equilibrium equations corresponding to the free body diagram. As always, we use the XYZ coordinate system for formulating these equations. Hence, we have

$$\Sigma M_{QZ} = M + \frac{P}{2}(r - r\cos 30°) - \frac{P}{2}(r\sin 30°) = 0$$

$$\Sigma F_X = -V\cos 30° - F\sin 30° - \frac{P}{2} = 0$$

$$\Sigma F_Y = -V\sin 30° + F\cos 30° + \frac{P}{2} = 0$$

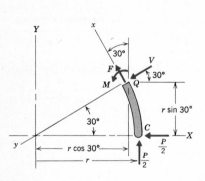

These equations yield

$$V = -0.1830P \qquad F = -0.6830P \qquad M = 0.1830Pr$$

The axial force is negative, so it is compressive. It should be noted that members AB and BC are two-force members, because loads are applied to them only at their ends. However, because the bars are curved there is internal shear and a bending moment. Such internal forces are not present in straight two-force members.

HOMEWORK PROBLEMS

VI.1–VI.14 Determine the axial and shear forces and the bending moment acting on the cross section at (a) location A, (b) location B for the bar shown.

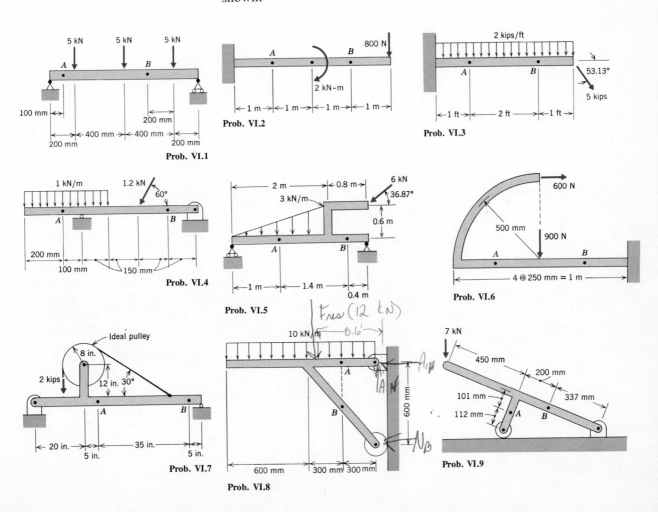

Prob. VI.1

Prob. VI.2

Prob. VI.3

Prob. VI.4

Prob. VI.5

Prob. VI.6

Prob. VI.7

Prob. VI.8

Prob. VI.9

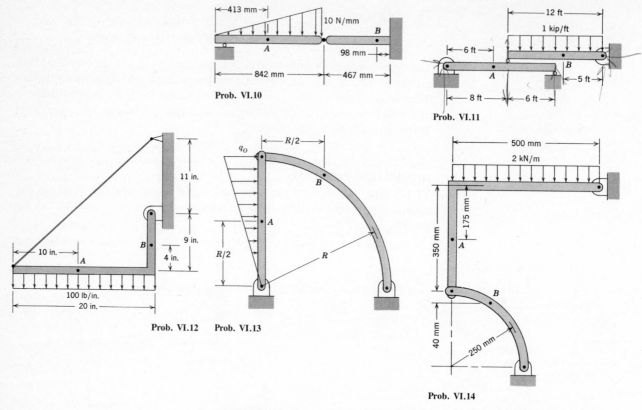

Prob. VI.10

Prob. VI.11

Prob. VI.12   Prob. VI.13

Prob. VI.14

VI.15–VI.17 Determine the axial force, the resultant shear force, the resultant bending moment, and the torsional moment at location A of the bar shown.

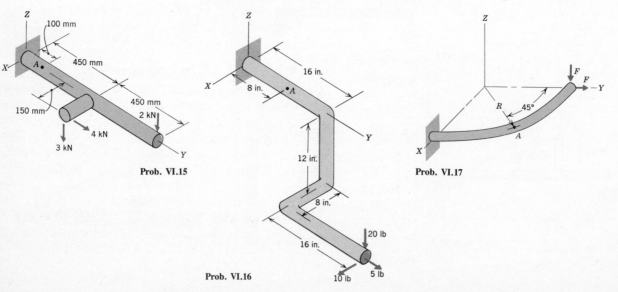

Prob. VI.15

Prob. VI.16

Prob. VI.17

3 Shear and Bending Moment Diagrams

The design of structural members requires knowledge of the internal forces at all locations within the member, not just at selected cross sections. This information can be obtained by systematically employing the method developed in the previous section for finding internal forces. The remainder of the material on beams in this module is devoted to such a determination.

We shall now abandon consideration of curved beams, and confine the development to the more common situation that arises in practice: straight beams that are loaded by a planar system of transverse forces and couples. In so doing we eliminate the need for an xyz coordinate system to describe the internal force component. This follows from the observation that the straight geometry means that the xyz and XYZ would always be mutually parallel.

A typical loading on a straight beam is depicted in Figure 6a. Now, the variable x is utilized to locate the cutting plane, consistent with standard usage.

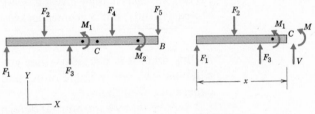

Figure 6a   Figure 6b

When we consider equilibrium of any section of the beam, such as the one of length x in Figure 6b, we see that the internal forces consist of a shear force V and a bending moment M. Therefore, an investigation of the internal forces on all cross sections of the beam reduces to the determination of the manner in which the shear force and bending moment depend on the distance x locating the cross section.

The functional dependence of the shear force and bending moments are usually exhibited by drawing *shear and bending moment diagrams*. Such diagrams require that we have nonambiguous definitions for positive shear and positive bending moment, both in order to avoid using inconsistent signs in the computations and in order to enable someone else to interpret correctly the diagrams we draw. These definitions are called *sign conventions*. (For example, we have already been using a sign convention by saying that an internal force normal to a cross section is positive if it is a tensile force.)

The sign convention we will use here is obtained by considering the type of deformation a section of the beam of length Δx would undergo if it carried only a shear force or only a bending moment. These situations are depicted on the following page in Figures 7a and b, respectively.

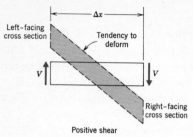

Positive shear

Figure 7a

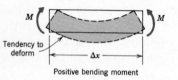

Positive bending moment

Figure 7b

For convenience, we will refer to a cross section whose outward normal faces left as a *left-facing cross section,* and a cross section whose outward normal faces right as a *right-facing cross section.* Figure 7a shows that in our sign convention,

> A shear force is positive if it pushes the left-facing cross section upward, or equivalently if it pushes the right-facing cross section downward.

Alternately, we can look upon this definition as: *Positive shear causes a clockwise rotation of the beam element being analyzed.*

Figure 7b shows that we will consider *positive bending moment* to be one that has the effect of bending the bar upward, that is, one that *causes the centroidal axis to assume a shape that is concave upward.* This means that in our sign convention,

> A bending moment is positive if it is clockwise on the left-facing cross section, or equivalently, if it is counterclockwise on the right-facing cross section.

At this time, it is necessary that you memorize these sign conventions, although they will become quite natural to use after you have had some experience with them.

Notice that we introduced deformation effects in Figure 7. Until now we considered a beam to be rigid. There is no contradiction here because, as is true for all systems that we model as rigid bodies, we are assuming that the beam undergoes a very small amount of deformation, so that the equations of static equilibrium may be formulated without considering deformations. (The validity of this assumption is shown in mechanics of materials to be quite good.)

Now that we have established the sign conventions for shear and bending moment, let us turn our attention to how we may go about constructing shear and bending moment diagrams. The most direct procedure is to construct a free body diagram of a section of beam of length x, for which the equations of static equilibrium yield the shear V and bending moment M in terms of x. Then the desired diagrams are plotted from the functions $V(x)$ and $M(x)$.

To develop this procedure, consider a lightweight simply supported beam loaded by a downward force $\bar{F}$ at its midpoint. After evaluating the reactions, the free body diagram of the entire beam is as shown in Figure 8. To determine V and M we must isolate a portion of the beam formed by a cutting plane. We may isolate the portion lying to either side of the cutting plane; let us use the left portion.

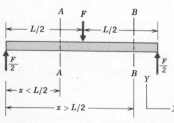

Figure 8

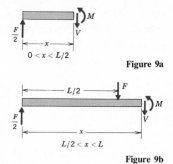

Figure 9a

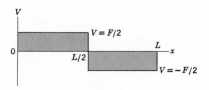

Figure 9b

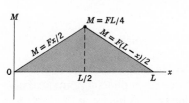

Figure 10

As depicted in Figure 8, we shall need to consider two cutting planes. This is because the concentrated force $\bar{F}$ represents a *singularity*. (The word "singularity" is used in the mathematical sense of an undefined situation.) The free body diagrams resulting from each cutting plane are shown in Figures 9a and b. It is important to note that in each diagram V and M are depicted in the positive direction as defined by our sign convention.

The static equilibrium equations for each free body diagram give

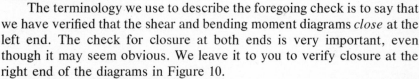

$$0 < x < L/2: \qquad V = \frac{F}{2} \qquad M = \frac{F}{2}x$$

$$\frac{L}{2} < x < L: \qquad V = -\frac{F}{2} \qquad M = \frac{F}{2}x - F\left(x - \frac{L}{2}\right) = \frac{F}{2}(L - x)$$

These functions are now plotted, to yield the shear and bending moment diagrams shown in Figure 10.

The correctness of these diagrams may be partially verified by checking that the values of the shear and bending moments at the ends match the force-couple systems acting at these locations. For instance, from Figure 8 we see that there is an upward force $F/2$ and no couple acting at the left end of the beam. According to our sign convention, a positive shear is upward on a left-facing cross section, which is exactly the situation for the left end. Therefore, the external force-couple system acting at the left end is equivalent to a positive shear force $F/2$ and no bending moment. These values are confirmed by the shear and bending moment diagrams in Figure 10.

The terminology we use to describe the foregoing check is to say that we have verified that the shear and bending moment diagrams *close* at the left end. The check for closure at both ends is very important, even though it may seem obvious. We leave it to you to verify closure at the right end of the diagrams in Figure 10.

A noteworthy feature of the diagrams in Figure 10 is that at the midpoint, where the force is applied, the value of the shear force and the derivative (slope) of the bending moment function are discontinuous. This is a result of the singularity introduced by the concentrated force $\bar{F}$.

Concentrated forces are one type of singularity resulting from the loads on the beam. Recall that this singularity resulted from a change in the nature of the force system acting on the isolated portion of the beam. Other types of singularities arise when a couple is applied to the beam at some location, and in the case of distributed loading, whenever the loading function changes.

For example, consider the beam shown in Figure 11. Singularities occur at point B, where the concentrated force $\bar{F}_B$ is applied; at point C, where the distributed load function changes from q_1 to q_2; at point D,

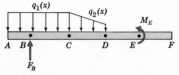

Figure 11

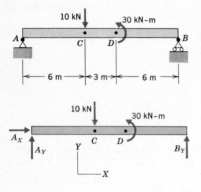

where the distributed load changes from q_2 to zero; and at point E, where the couple M_E is applied. To obtain the shear and bending moment diagrams in this case we would need to use cutting planes that pass between each pair of singular points, forming a total of five different cases to be considered.

EXAMPLE 3

Sketch the shear and bending moment diagrams for the simply supported beam shown.

Solution

The first step in any analysis of internal forces is the determination of the reactions. Thus, we draw a free body diagram of the beam and write the equations of equilibrium.

$$\Sigma \, M_{AZ} = 30 - 10(6) + B_Y(15) = 0 \text{ kN-m}$$

$$\Sigma \, F_X = A_X = 0$$

$$\Sigma \, F_Y = A_Y + B_Y - 10 = 0 \text{ kN}$$

This gives

$$B_Y = 2 \qquad A_X = 0 \qquad A_Y = 8 \text{ kN}$$

We next ascertain where singularities occur in the loading. There is a concentrated force applied at point C and a couple applied at point D. This means that we will require three cutting planes, one between the left end A and point C, the second between points C and D, and the last between point D and the right end B.

It is best to perform the computation in an orderly fashion. Experience has shown that students are less prone to errors if they progress across the beam in one direction, usually the direction of increasing values of x. Measuring x from end A, we will isolate the section of the beam to the left of each of the three aforementioned cutting planes.

When we draw the free body diagrams, we depict the shear and bending moment in the positive direction according to the appropriate sign convention. On a right-facing cross section we consider downward shear and counterclockwise bending moment positive, so the free body diagrams and corresponding computations of V and M are as shown below.

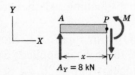

Region: $0 < x < 6$ m

$$\Sigma \, M_{PZ} = M - 8(x) = 0 \qquad \Sigma \, F_Y = -V + 8 = 0$$

$$V = 8 \text{ kN} \qquad M = 8x \text{ kN-m}$$

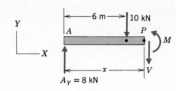

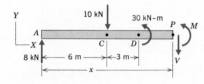

Region: $6 < x < 9$ m

$$\Sigma M_{PZ} = M + 10(x - 6) - 8(x) = 0 \qquad \Sigma F_Y = -V - 10 + 8 = 0$$

$$V = -2 \text{ kN} \qquad M = 30 - 2x \text{ kN-m}$$

Region: $9 < x < 15$ m

$$\Sigma M_{PZ} = M + 30 + 10(x - 6) - 8(x) = 0$$

$$\Sigma F_Y = -V - 10 + 8 = 0$$

$$V = -2 \text{ kN} \qquad M = 30 - 2x \text{ kN-m}$$

We now have sufficient information to plot the shear and bending moment diagrams. An accurate plot may be obtained by a point-by-point evaluation of the expressions. However, this is not really necessary, because we can evaluate the expressions corresponding to each region at their two limits of validity and then connect these points by the curve corresponding to the equation derived for the region. For example, in the range $0 < x < 6$ m, we calculate and plot the values of V and M at $x = 0$ and $x = 6$. The value of V is constant, so it is represented by a horizontal line. The value of M is linearly dependent on the value of x, so it is represented by a straight sloping line. The result of this procedure is the shear and bending moment diagrams shown. Notice that the singularity introduced by a couple loading is such that the value of the bending moment is discontinuous, but there is no discontinuity in the shear at that point. Also, notice that the singularity introduced by a concentrated force is such that the value of the shear is discontinuous, but there is no discontinuity in the bending moment at that point.

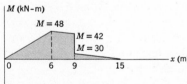

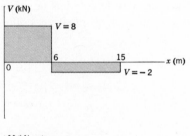

To verify that the shear and moment diagrams close, we refer back to the free body diagram of the entire beam. At end A we see that because $A_Y = 8$ kN upward and there is no couple, V should be positive 8 kN (end A is a left-facing cross section) and M should be zero. At end B we see that because $B_Y = 2$ kN upward and there is no couple, V should be negative 2 kN (right-facing cross section) and M should be zero. These values are all confirmed by, and consistent with, the diagrams we have drawn.

EXAMPLE 4

Sketch the shear and bending moment diagrams for the horizontal over-hanging beam shown. The beam has constant weight per unit length w.

Solution

The first step in the solution is the determination of the reactions. The

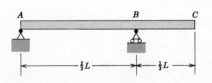

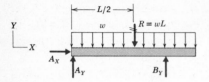

beam carries a constant distributed loading w, as shown in the free body diagram. The equilibrium equations for the beam are

$$\Sigma \, M_{AZ} = B_Y\left(\frac{2L}{3}\right) - wL\left(\frac{L}{2}\right) = 0$$

$$\Sigma \, F_X = A_x = 0 \qquad \Sigma \, F_Y = A_Y + B_Y - wL = 0$$

and the reactions are found to be

$$A_X = 0 \qquad A_Y = \tfrac{1}{4}wL \qquad B_Y = \tfrac{3}{4}wL$$

We next locate the singularities. The distributed loading is the same along the whole length of the beam, so the only change in the loading occurs when the cutting plane passes point B, where a reaction force is applied. Thus, we must analyze two regions, one to the left and one to the right of point B. Following the procedure outlined in the previous example, we measure x from the left end A and determine the internal forces by isolating the section of the beam to the left of the cutting plane. The required free body diagrams and equilibrium equations are given below. Notice that because we are concerned with the shear and bending moment acting on the right-facing cross section, the shear V is depicted as positive downward and the bending moment M is depicted as positive counterclockwise.

Region: $0 < x < 2L/3$

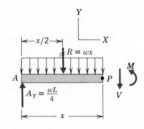

$$\Sigma \, M_{PZ} = M + wx\left(\frac{x}{2}\right) - \frac{wL}{4}(x) = 0$$

$$\Sigma \, F_Y = -V - wx + \frac{wL}{4} = 0$$

$$V = w\left(\frac{L}{4} - x\right) \qquad M = -\frac{w}{4}(2x^2 - Lx)$$

Region: $2L/3 < x < L$

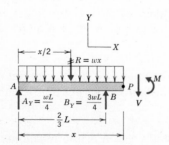

$$\Sigma \, M_{PZ} = M - \frac{3wL}{4}\left(x - \frac{2L}{3}\right) + wx\left(\frac{x}{2}\right) - \frac{wL}{4}(x) = 0$$

$$\Sigma \, F_Y = -V + \frac{3wL}{4} + \frac{wL}{4} - wx = 0$$

$$V = w(L - x) \qquad M = -w\left(\frac{x^2}{2} - Lx + \frac{1}{2}L^2\right)$$

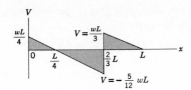

To sketch the shear diagram, we note that both expressions for V are linear in x, so the curves for V in both ranges are straight lines. Evaluating each function for V at the limits of its region of validity and plotting the corresponding points, we obtain the shear diagram shown.

To sketch the bending moment, we note that both expressions for M are quadratic polynomials in which the coefficient of the x^2 term is negative. This means that the curves representing M in either range are parabolas that are concave downward (in the mathematical sense of curvature). We may determine the location of the horizontal tangent of each parabola by setting $dM/dx = 0$. This yields

Region: $0 < x < 2L/3$

$$\frac{dM}{dx} = -\frac{w}{4}(4x - L) = 0 \quad \text{so} \quad x = L/4$$

Region: $2L/3 < x < L$

$$\frac{dM}{dx} = -w(x - L) = 0 \quad \text{so} \quad x = L$$

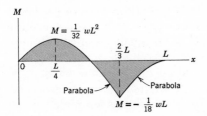

The foregoing knowledge of the shape of the bending moment curve, together with the plotted points representing the values of each bending moment function at the limits of its range of validity, allow us to draw the bending moment diagram. Can you verify that the shear and bending moment diagrams close?

It is interesting to note that the bending moment has its extreme values ($dM/dx = 0$) at the locations $x = L/4$ and $x = L$ where the shear is also zero. This is not a coincidence, as will be proven in the next section.

EXAMPLE 5
Sketch the shear and bending moment diagrams for the cantilever beam carrying the triangular distributed load shown.

Solution
We begin with a free body diagram of the entire beam, which we use to determine the reactions at the fixed end A. In the free body diagram the force $\bar{R}$ is the resultant of the distributed load.

$$\Sigma M_A = M_A - 1260(21 + 7) = 0$$

$$\Sigma F_X = A_X = 0 \qquad \Sigma F_Y = A_Y - 1260 = 0$$

$$A_X = 0 \qquad A_Y = 1260 \text{ lb} \qquad M_A = 35{,}280 \text{ lb-in.}$$

We next identify the singularities. For the loading shown there is only one singular point, specifically the midpoint where the distributed load

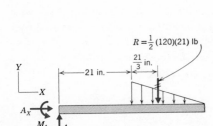

begins. Measuring x from the left end, we first consider a cutting plane to the left of the distributed load. In accordance with the procedure we have developed, we draw a free body diagram and write the equilibrium equations for the section of beam of length x formed by the plane. This is described below.

Region: $0 < x < 21$ in.

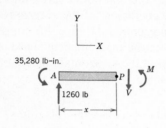

$$\Sigma M_{PZ} = M + 35{,}280 - 1{,}260(x) = 0 \qquad \Sigma F_Y = -V - 1{,}260 = 0$$

$$V = 1{,}260 \text{ lb} \qquad M = 1{,}260x - 35{,}280 \text{ lb-in.}$$

We now consider the situation arising from a cutting plane to the right of the midpoint of the beam. If we were to follow the approach developed in the two previous examples, we would isolate the portion of the beam to the left of the cutting plane. It is not difficult to see that this section carries a trapezoidal distributed loading. (Recall that we consider only the portion of the loads that are actually applied to the isolated section of the beam.) Solution of the equilibrium equations would then require that we determine the resultant of a trapezoidal loading.

Alternatively, should we choose to isolate the portion of the beam to the right of the cutting plane, we find that the resulting section carries a triangular loading, whose resultant is considerably easier to determine. Thus, the free body diagram and corresponding equations of equilibrium are as shown below. Note that in our sign convention, for a left-facing cross section positive shear is upward and positive bending moment is clockwise, as depicted. Also note that the position of the cross section of the beam is still measured from the left end. By doing this we minimize the chance of error when plotting the shear and moment equations obtained for each region of the beam.

Region: $21 < x < 42$ in.

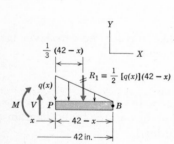

$$\Sigma M_{PZ} = -M - R_1[\tfrac{1}{3}(42 - x)] = 0 \qquad \Sigma F_Y = V - R_1 = 0$$

$$V = R_1 = \tfrac{1}{2}q(x)(42 - x)$$

$$M = -\tfrac{1}{3}R_1(42 - x) = -\tfrac{1}{6}q(x)(42 - x)^2$$

The determination of V and M in terms of x now requires that we determine the expression for the loading function $q(x)$. This is obtained by observing that the load decreases linearly with increasing values of x, so $q(x) = k_1 - k_2 x$. The values of k_1 and k_2 are found by equating $q(x)$ to 120 lb/in. at $x = 21$ in. and to zero at $x = 42$ in. Thus

$$q(21) = k_1 - 21k_2 = 120 \qquad q(42) = k_1 - 42k_2 = 0$$

so

$$k_1 = 240 \qquad k_2 = \frac{120}{21} \quad \text{and} \quad q(x) = 120\left(2 - \frac{x}{21}\right)$$

The equations for V and M then become

Region: $21 < x < 42$ in.

$$V = \frac{1}{2}\left[120\left(2 - \frac{x}{21}\right)\right](42 - x) = 1260\left(2 - \frac{x}{21}\right)^2 \text{ lb}$$

$$M = -\frac{1}{6}\left[120\left(2 - \frac{x}{21}\right)\right](42 - x)^2 = -8820\left(2 - \frac{x}{21}\right)^3 \text{ lb-in.}$$

We may now sketch the shear and bending moment diagrams. To do this we note that in the region $0 < x < 21$ in., V is constant and M increases linearly with x, so both functions plot as straight lines. In the region $21 < x < 42$ in., V is a quadratic polynomial and M is a cubic polynomial. The curve for V is readily identified as an upward-curving parabola (in the mathematical sense of curvature) having its lowest point at $x = 42$ in. where $V = 0$. Similarly, we see that the curve for M is a downward-curving cubic whose maximum value is zero at $x = 42$ in. Knowing this, we evaluate the expressions for M and V in the region $0 < x < 21$ at $x = 0$ and $x = 21$, and the expressions for M and V in the region $21 < x < 42$ at $x = 21$. Then, by plotting these points and connecting them by the appropriate curves, we obtain the accompanying diagrams.

These diagrams close at both ends, because there is no force-couple system acting at the right end B, and the reactions at end A, which is a left-facing cross section, are equivalent to a positive (upward) shear and a negative (counterclockwise) bending moment.

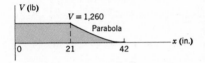

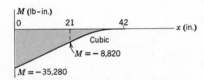

HOMEWORK PROBLEMS

VI.18–VI.35 Draw shear and bending moment diagrams for the beam shown. Give all critical values.

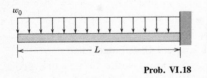

Prob. VI.18

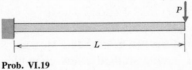

Prob. VI.19

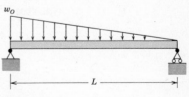

Prob. VI.20

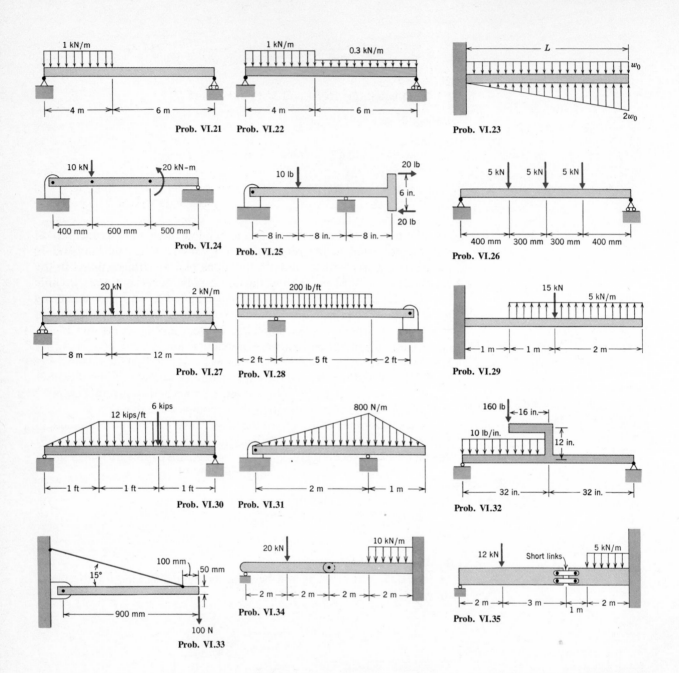

Prob. VI.21

Prob. VI.22

Prob. VI.23

Prob. VI.24

Prob. VI.25

Prob. VI.26

Prob. VI.27

Prob. VI.28

Prob. VI.29

Prob. VI.30

Prob. VI.31

Prob. VI.32

Prob. VI.33

Prob. VI.34

Prob. VI.35

4 Relations Between Bending Moment, Shear, and Distributed Load

After having studied a few examples and solved some homework problems, it should be apparent to you that although the theory for the determination of the shear and bending moment diagrams is conceptually

straightforward, in practice the computations can get quite involved. To reduce the amount of calculations, in this section we develop a method for constructing shear and bending moment diagrams without explicit consideration of the equilibrium of isolated sections of a beam.

Let us begin by determining the general relations between the internal forces and the loading on a beam. In Figure 12 we isolate a small segment of a beam, of length Δx, carrying a distributed load $\bar{q}$, but no concentrated loads or couples.

Acting on the isolated segment are the internal forces on the left-facing cross section at position x and on the right-facing cross section at position $x + \Delta x$. The shear and bending moment are shown in Figure 12 in their positive sense on each cross section. As indicated there, these quantities need not be the same on both faces. The free body diagram for the element is completed by replacing the distributed load by its equivalent concentrated force. As shown, the magnitude of this force is obtained by multiplying the mean value of q, corresponding to some point x_m within the region $x \leq x_m \leq x + \Delta x$, by the length Δx. Further, the force is applied at a distance $\epsilon\,\Delta x$ from the left end, where ϵ is a finite number.

Let us write the equilibrium equations for this beam element. Summing moments about point P, we have

$$\Sigma\,M_{PZ} = M(x + \Delta x) - M(x) - [V(x + \Delta x)]\Delta x - q(x_m)\,\Delta x(\epsilon\,\Delta x) = 0$$

$$\Sigma\,F_X \equiv 0 \qquad \Sigma\,F_Y = V(x) - V(x + \Delta x) - q(x_m)\,\Delta x = 0$$

After we rearrange terms, these relations become

$$\frac{V(x + \Delta x) - V(x)}{\Delta x} = -q(x_m) \tag{1}$$

$$\frac{M(x + \Delta x) - M(x)}{\Delta x} = V(x + \Delta x) - \epsilon q(x_m)\,\Delta x \tag{2}$$

Consider the situation where the length of the element becomes smaller and smaller, that is, $\Delta x \to 0$. In the limit the left sides of equations (1) and (2) are the definitions of dV/dx and dM/dx, respectively. Further, as $\Delta x \to 0$, $x_m \to x$, because the value of x_m is intermediate between x and $x + \Delta x$. Therefore, in the limit, equations (1) and (2) yield

$$\frac{dV}{dx} = -q(x) \tag{3a}$$

$$\frac{dM}{dx} = V(x) \tag{3b}$$

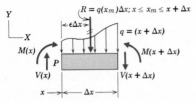

Figure 12

Equations (3) are called *differential equations* because they equate the derivative of one variable to another function. We will find these equations to be useful when plotting the shear and bending moment diagrams. For example, equation (3b) tells us that the bending moment has a horizontal tangent at the location where the shear is zero. (We noted this fact in Example 4.)

Another use for equations (3) is associated with the determination of the shear and bending moment resulting from a given loading. This calls for an integral representation of the relations. Suppose that we know the shear and bending moment at some value x_A and wish to determine the value of these forces at some other value x_B. Equations (3) may be used for this determination by multiplying each equation by dx and integrating between the limits corresponding to the two points. This gives

$$\int_{V_A}^{V_B} dV \equiv V_B - V_A = -\int_{x_A}^{x_B} q(x)\, dx$$

$$\int_{M_A}^{M_B} dM \equiv M_B - M_A = \int_{x_A}^{x_B} V(x)\, dx$$

Rearranging terms, we obtain the desired result

$$V_B = V_A - \int_{x_A}^{x_B} q(x)\, dx \tag{4a}$$

$$M_B = M_A + \int_{x_A}^{x_B} V(x)\, dx \tag{4b}$$

From the fact that an integral of a function is the area under the curve representing that function, equation (4a) can be seen to be equivalent to the statement that

The shear force at x_B is the shear force at x_A *minus* the area under the distributed load curve in the region between x_A and x_B.

Similarly, equation (4b) states that

The bending moment at x_B is the bending moment at x_A *plus* the area under the shear curve in the region between x_A and x_B.

Our interest in this text is the quantitative determination of the internal forces at a few key locations along a beam, for example, where the bending moment and shear assume extreme values, as well as the determi-

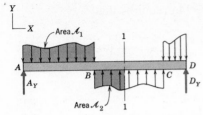

Figure 13

nation of a pictorial representation of the way in which the internal forces vary with location. To see how we can use the foregoing statements toward this goal, consider the situation depicted in Figure 13, where a beam is carrying a distributed load and concentrated forces (reactions) at its ends.

According to our sign conventions, the vertical force $\bar{A}_Y$ at end A is equivalent to a positive shear force. Using equation (4a), we see that the shear force at the location of the cutting plane 1-1 appearing in the figure is given by

$$V_1 = V_A - (\mathcal{A}_1 - \mathcal{A}_2) = A_Y - \mathcal{A}_1 + \mathcal{A}_2$$

where $\mathcal{A}_1$ and $\mathcal{A}_2$ are areas formed by the loading function, as indicated in Figure 13. In regard to the area $\mathcal{A}_2$, it is important to realize that between points B and C, the load, and thus the area, is negative because we considered downward loads to be positive in deriving equations (3).

Let us determine the shear diagram for the beam in Figure 13. We begin by recalling that discontinuous changes in the value of shear occur only at locations where there are concentrated transverse forces. Thus, the curves representing the shear distribution in the regions bounded by the singular points B and C are continuous. We may use equation (4a) to obtain the value of shear at any critical point, particularly the ends of the beam and the singular points B and C. Then, the character of the shear curves connecting the plotted points may be explored by using equation (3a), $dV/dx = -q$, to ascertain how dV/dx (the slope of the shear curve) and V depend on x.

The foregoing procedure for constructing the shear diagram, and a similar procedure for bending moment diagrams, will be demonstrated in the next two examples. Before proceeding, however, we will consider the cases of loadings by concentrated forces and couples in order to gain further insight into the discontinuities they cause in the shear and bending moment diagrams.

To investigate concentrated forces and couples, consider an isolated beam element of length Δx. We apply either a concentrated downward force $\bar{F}_C$ or a couple $\bar{M}_C$ at the midpoint C of the element, having position x_C, as illustrated in Figures 14a and b, respectively.

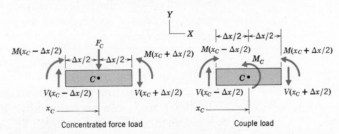

Figure 14a

Concentrated force load

Figure 14b

Couple load

Equilibrium of the element in Figure 14a requires that

$$\Sigma\,M_{CZ} = M\left(x_C + \frac{\Delta x}{2}\right) - M\left(x_C - \frac{\Delta x}{2}\right) - \left[V\left(x_C - \frac{\Delta x}{2}\right) + V\left(x_C + \frac{\Delta x}{2}\right)\right]\left(\frac{\Delta x}{2}\right)$$
$$= 0 \tag{5a}$$

$$\Sigma\,F_Y = V\left(x_C - \frac{\Delta x}{2}\right) - V\left(x_C + \frac{\Delta x}{2}\right) - F_C = 0 \tag{5b}$$

Because of the presence of the finite term F_C in the force equation, we cannot simply divide each term in equations (5) by Δx and take the limits as $\Delta x \to 0$, as we did in the case of a distributed load. This would give rise to the improper division of a finite number by zero.

Instead consider what happens to equations (5) as Δx becomes infinitesimal, but nonzero. As Δx decreases, the left and right cross sections both approach point C. Denoting the position of the cross section to the left of point C by x_C^-, and the position of the cross section to the right of point C as x_C^+, and letting Δx become the infinitesimal quantity dx, equations (5) become

$$\Sigma\,M_{CZ} = M(x_C^+) - M(x_C^-) - [V(x_C^-) + V(x_C^+)]\left(\frac{dx}{2}\right) = 0$$

$$\Sigma\,F_Y = V(x_C^-) - V(x_C^+) - F_C = 0$$

Neglecting terms of order dx, these equations reduce to

$$V(x_C^+) = V(x_C^-) - F_C \tag{6a}$$
$$M(x_C^+) = M(x_C^-) \tag{6b}$$

In words, equations (6) state that

The value of the bending moment is continuous at the singular point of a downward concentrated force, but the shear is discontinuous, being less to the right of the point of application of the force by an amount equal to the magnitude of the force.

Note that in the case where the concentrated force is upward, it is only necessary to regard F_C as a negative quantity in equation (6a).

Similar steps can be followed to investigate the equilibrium of the beam element in Figure 14b. Such an investigation shows that

$$V(x_C{}^+) = V(x_C{}^-) \tag{7a}$$

$$M(x_C{}^+) = M(x_C{}^-) - M_C \tag{7b}$$

In words, equations (7) state that

> The value of the shear is continuous at the singular point of a clockwise couple, but the bending moment is discontinuous, being less to the right of the point of application of the couple by an amount equal to the magnitude of the couple.

Note that a clockwise couple is treated by regarding M_C as a negative quantity.

Referring back to any of the shear and bending moment diagrams previously determined, particularly in Examples 3 and 4, you will observe that equations (6) and (7) validate the results given there.

Placing equations (6) and (7) in the context of our general goal of determining shear and moment diagrams, we see that equations (4) may be used to relate shears or bending moments within the regions lying between points where concentrated forces and couples are applied, whereas equations (6) and (7) may be used to go from one side of these singular points to the other. In this manner, the values of shear and bending moment at the singular points and other critical points may be determined. After these values are plotted, the characteristics of the curves connecting the plotted values in each region bounded by the singular points can be investigated with the aid of equations (3).

The remarkable feature of this method for constructing the shear and bending moment diagrams is that once the reactions have been obtained, it is no longer necessary to write the equations of static equilibrium, as is demonstrated in the following examples.

EXAMPLE 6
The simply supported beam carries the loading shown. Use the derived relations between load, shear, and bending moment to sketch the shear and bending moment diagrams. Give all critical values.

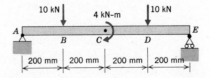

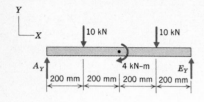

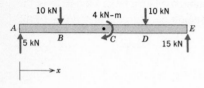

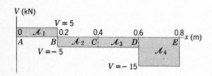

Solution

We begin by determining the reactions. The necessary free body diagram and computations are given below.

$$\Sigma M_{AZ} = -10(0.20) - 4 - 10(0.60) + E_Y(0.80) = 0$$

$$\Sigma F_X \equiv 0 \qquad \Sigma F_Y = A_Y + E_Y - 20 = 0$$

$$E_Y = 15 \text{ kN} \qquad A_Y = 5 \text{ kN}$$

We now draw the loading diagram by simply repeating the free body diagram of the beam, with the unknown reactions replaced by their calculated values. Singularities are readily identified to exist at points B, C, and D.

From the load diagram we may construct the shear diagram. At point A, the force is equivalent to a positive shear (upward on a left-facing cross section), $V = 5$ kN. From equation (6a) we observe that the concentrated *downward* forces result in discontinuous *decreases* of the shear by 10 kN each as points B and D are passed from left to right when sketching the shear diagram. Further, with respect to the couple, we observe from equation (7a) that the shear is continuous at point C. Finally, noting that there is no loading between the singular points, we may conclude that the shear is constant in these regions. These observations lead to the shear diagram shown. As a check, we note that the shear diagram closes at end E, because the upward 15-kN force at this end is equivalent to a negative shear.

We now use the shear diagram to construct the moment diagram. From the load diagram, we see that there is no moment at end A. From equation (6b) we note that the bending moment is continuous at points B and D, where there are concentrated forces, whereas from equation (7b) we observe that the *clockwise* 4 kN-m couple at point C results in a discontinuous *increase* in bending moment in that amount as point C is passed in the diagram from left to right.

The values of bending moment at the singularities may then be calculated by using equation (4b) to proceed from one singularity to the next. In terms of the areas indicated in the shear diagram, this equation gives

$$M(x_B = 0.20) = M(x_A = 0) + \mathcal{A}_1 = 0 + 5(0.20) = 1.0 \text{ kN-m}$$

$$M(x_C = 0.40^-) = M(x_B = 0.20) - \mathcal{A}_2 = 1.0 - 5(0.20) = 0$$

$$M(x_C = 0.40^+) = M(x_C = 0.40^-) + 4 = 4.0 \text{ kN-m}$$

$$M(x_D = 0.60) = M(x_C = 0.40^+) - \mathcal{A}_3 = 4.0 - 5(0.20) = 3.0 \text{ kN-m}$$

$$M(x_E = 0.80) = M(x_D = 0.60) - \mathcal{A}_4 = 3.0 - 15(0.2) = 0$$

Note that the bending moment closes at end E.

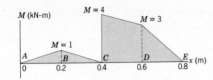

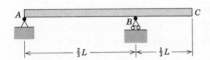

We next plot these values. To determine the type of curves connecting the plotted points, we recall equation (3b), $dM/dx = V$. Because V is constant between each pair of singular points, we see that each curve has constant slope. In other words, the curves are straight lines. Thus, we obtain the bending moment diagram shown to the left.

EXAMPLE 7

The overhanging beam of weight per unit length w studied in Example 4 is depicted again in the sketch. Use the relations between distributed load, shear, and bending moment to sketch the shear and bending moment diagrams. Give all critical values.

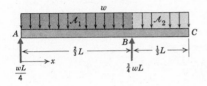

Solution

The necessary preliminary step of determining the reactions was performed in Example 4, so it is not repeated here. After the reactions are determined the loading diagram is drawn by showing all loads, including the reactions, in terms of their known magnitude and sense. This diagram indicates that the only singularity occurs at point B, where there is a concentrated upward force.

The shear diagram is obtained from the loading diagram. The force $wL/4$ at end A is equivalent to a positive shear. From equation (6a) we find that the concentrated *upward* force at point B results in a discontinuous *increase* in shear by $3wL/4$ as this point is passed in the diagram from left to right. Then, using equation (4a) and the areas indicated in the loading diagram, we calculate the shear at the singularities. This yields

$$V\left(x_B = \frac{2L^-}{3}\right) = V(x_A = 0) - \mathcal{A}_1 = \frac{1}{4}wL - w\left(\frac{2L}{3}\right) = -\frac{5}{12}wL$$

$$V\left(x_B = \frac{2L^+}{3}\right) = V\left(x_B = \frac{2L^-}{3}\right) - F_B = -\frac{5}{12}wL - \left(-\frac{3}{4}wL\right) = \frac{1}{3}wL$$

$$V(x_C = L) = V\left(x_B = \frac{2L^+}{3}\right) - \mathcal{A}_2 = \frac{1}{3}wL - w\left(\frac{L}{3}\right) = 0$$

As a check, we note that the shear closes at end C, because no force is applied at this end.

These points are now plotted and equation (3a), $dV/dx = -q$, is used to ascertain the types of curves connecting the plotted points. Because q is constant, it follows that V is linearly dependent on x (integration of a constant). In other words, the shear curves are straight lines. Further, because q is positive, these straight lines are sloping downward. The shear diagram resulting from these considerations is as shown.

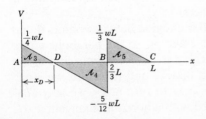

The determination of the bending moment proceeds in a similar manner, using the areas indicated in the shear diagram to evaluate the integrals arising from equation (4b). The determination of these areas requires that we first obtain the distance x_D locating point D where $V = 0$. Using similar triangles, we have

$$\frac{x_D}{(\frac{1}{4}wL)} = \frac{(2L/3)}{(\frac{1}{4}wL + \frac{5}{12}wL)} \quad \text{so} \quad x_D = \frac{L}{4}$$

The area $\mathcal{A}_4$ is negative because the shear is negative. Also, the beam does not carry any couple loads, so the moment diagram is continuous. Thus, the computations of the bending moment at the singular points proceed as follows, starting from $x_A = 0$, where we know that $M = 0$.

$$M\left(x_D = \frac{L}{4}\right) = M(x_A = 0) + \mathcal{A}_3 = 0 + \frac{1}{2}\left(\frac{1}{4}wL\right)\left(\frac{L}{4}\right) = \frac{1}{32}wL^2$$

$$M\left(x_B = \frac{2L}{3}\right) = M\left(x_D = \frac{L}{4}\right) - \mathcal{A}_4$$

$$= \frac{1}{32}wL^2 - \frac{1}{2}\left(\frac{5}{12}wL\right)\left(\frac{2L}{3} - \frac{L}{4}\right) = -\frac{1}{18}wL^2$$

$$M(x_C = L) = M\left(x_B = \frac{2L}{3}\right) + \mathcal{A}_5$$

$$= -\frac{1}{18}wL^2 + \frac{1}{2}\left(\frac{1}{3}wL\right)\left(\frac{L}{3}\right) = 0$$

Notice that the bending moment closes at end C because there is no couple at that location.

The bending moment diagram is obtained by first plotting the values of bending moment that were determined. To draw the curves connecting the plotted points, we recall equation (3b), $dM/dx = V$. The shear curves are straight lines between the singularities, so V is linearly dependent on x. It then follows that M is dependent on x^2 (integrating x), so the moment curves are parabolas.

The nature of these parabolas is investigated by using $dM/dx = V$ to study their slope. Proceeding first from point A to point D, we see that the slope of the parabola decreases from a positive value at $x_A = 0$ to zero at $x_D = L/4$ and then becomes large negatively for increasing values of x beyond x_D. This can be true only if the parabola between points A and B curves downward with a horizontal tangent at x_D.

Similar reasoning may be applied to the region between points B and C. There is a discontinuity in the shear diagram, so the parabolic curve depicting the bending moment in this region is different from that in the region between points A and B. Noting that V, and therefore the slope dM/dx, decreases from a positive value at x_B to zero at x_C, we may conclude that the parabola curves downward with a horizontal tangent at x_C. Thus, we obtain the following bending moment diagram.

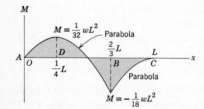

Clearly, the diagrams we obtained here are identical to those obtained in Example 4.

The advantages of the procedure we have just presented become evident when we note that the majority of the effort in the solution of the examples was devoted to explanations, not calculations. By evaluating the areas under a few curves, we can quickly and accurately obtain the required diagrams. It is no longer necessary to study the free body diagrams corresponding to different cutting planes. Needless to say, the formulas for the areas of some elementary shapes given in Appendix C may be valuable for computations by this method.

HOMEWORK PROBLEMS

VI.36–VI.47 Using the relationships between the transverse load, the shear, and the bending moment, sketch shear and bending moment diagrams for the beams indicated below. Give all critical values and show all computations.

Homework Problem	VI.36	VI.37	VI.38	VI.39	VI.40
Beam in Problem	VI.19	VI.24	VI.26	VI.25	VI.18
Homework Problem	VI.41	VI.42	VI.43	VI.44	VI.45
Beam in Problem	VI.20	VI.22	VI.27	VI.29	VI.30
Homework Problem	VI.46	VI.47			
Beam in Problem	VI.31	VI.34			

VI.48 and VI.49 The shear diagram for a simply supported beam is partially known, as depicted in the accompanying figure. It is also known that no couples are applied to the beam. (a) Determine the distributed and concentrated forces acting on the beam, (b) the length of the beam and all distances required to specify where the loads are applied, (c) the magnitude and location of the maximum bending moment in the beam.

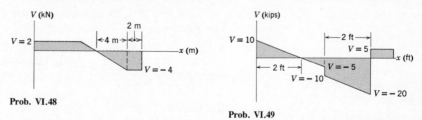

Prob. VI.48

Prob. VI.49

VI.50 The shear diagram for a simply supported beam carrying no couples is known to the extent shown. The maximum bending moment in

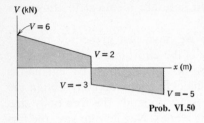

Prob. VI.50

this beam is $M = 20$ kN-m. Make a sketch of the beam, showing all loads, reaction forces, and lengths along the beam.

VI.51 The overhanging beam is loaded by a sinusoidal distributed force, given in units of newtons per meter by $q(x) = 400 \sin \pi x/20$, where x is the distance in meters from the left end. Sketch the shear and moment diagrams for the beam, specifying all critical values.

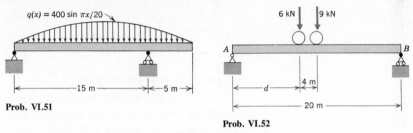

Prob. VI.51

Prob. VI.52

VI.52 The load exerted by the chassis of a movable crane on one set of wheels is shown. Determine the distance d, specifying the location of the chassis that results in the maximum possible bending moment in the supporting beam AB. What is the corresponding value of this moment?

B. FLEXIBLE CABLES

Like a beam, a *flexible cable* is a slender body. The primary difference between these two types of structural components is that the ratio of the length to the largest cross-sectional dimension for a cable (of the order of 1000 or more) is much larger than that for a beam. Because of this, a cable offers negligible resistance to bending and it bows out if one tries to apply a compressive force to it. Thus, the only internal force of any significance that it may withstand is an axial tension force.

Flexible cables are used in a variety of ways in structures and machines. They are a main element in suspension bridges, conveyor belts, and alpine cable cars. Their behavior is important to the design of electric and telephone transmission lines.

One use of a cable that we have already treated is to tie it between two points having known positions as a means of interconnecting these points. In this usage the tensile force is usually so large that we may consider the cable to be taut, meaning that the tensile force is collinear with the line connecting the end points.

Our concern in this module is with situations where cables carry transverse loads that are sufficiently large to make it unacceptable to consider the cable to be stretched in a straight line between its ends. We then say that the cable *sags*.

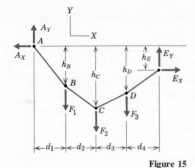

Figure 15

F Discrete Transverse Loads

Here we will consider the case where a cable supports a set of transverse loads at discrete points along its length. Such a situation is exemplified by Figure 15, which shows the free body diagram of a cable that is anchored to pins at ends A and E as it supports a series of transverse loads $\bar{F}_i$. The lines of action of the vertical loads are specified by the dimensions d_1, d_2, d_3, and d_4, which we will regard as known quantities.

If the loads are much larger than the weight of the cable, we may regard the segments of the cable between the points of application of the loads to be taut. Each segment then carries a constant tensile force, although the tension may vary from segment to segment. Thus, the analysis of the forces within the cable in Figure 15 reduces to the determination of the tensile forces T_{AB}, T_{BC}, T_{CD}, and T_{DE}.

Let us assume that the location of ends A and E of the cable are known, so that h_E is a known dimension. Then, additional information we seek is the vertical distances h_B, h_C, and h_D. These distances represent the *sag* of the cable. Including the four unknown reaction components, the analysis of this cable requires that we determine 11 unknowns (4 tensions, 4 reactions, and 3 sag distances).

Considering that the cable is functioning as a series of straight two-force members, it might seem that the problem can be solved by the same methods as those used for trusses. This is true, but there is one new feature here that complicates the solution. Specifically, in a truss the location of all joints is known, whereas the locations of joints B, C, and D for the cable are part of the information we seek.

Let us investigate the number of available equations. There are five joints (points A through E). Because the forces at each joint are collinear, there are two equilibrium equations for each joint, making a total of 10 available equilibrium equations. In practice, we usually consider three of the equilibrium equations for the joints as checks. These equations are replaced by the three equilibrium equations corresponding to the free body diagram of the entire structures, Figure 15. Nevertheless, there are only 10 independent equations of static equilibrium, whereas there are 11 unknowns; we need one more equation.

The cable is *not* statically indeterminate. Rather, the difficulty lies in the fact that we have not fully specified the geometry of the system.

One class of problems we can pursue pertains to situations where we know the length of the cable. In that case we would use the Pythagorean theorem to express the length of each segment of the cable in terms of the sags h_B, h_C, and h_D. The required additional equation is then obtained by equating the sum of the individual lengths to the total length.

A complication arises in a problem of this type, because the Pythagorean theorem gives rise to terms in the equation for the length that are the square roots of the unknown sags. Thus, although the formulation of the required equations is fairly straightforward, the solution of these equations is laborious.

A simpler situation occurs when one of the sags is specified. This, of course, reduces by one the number of unknowns to be determined, so that the static equilibrium equations are sufficient. There is a simple procedure available for formulating the solution for this class of problems.

For example, suppose that we know h_C in Figure 15. We employ the method of sections for trusses, using a cutting plane that cuts the cable to one side of pin C. Choosing a cutting plane to the left of this point, the corresponding free body diagram for the portion of the cable to the left of the plane is shown in Figure 16.

When we sum moments about point C, we obtain an equation relating A_Y and A_X. In conjunction with the three equilibrium equations for the entire system, we can determine all of the reactions. Further, the force equilibrium equations for the section in Figure 16 give the tension T_{BC} and the angle of elevation θ_{BC} of cable BC.

Once the reactions are known, the method of joints may be used to determine the tension and angle of elevation of each cable segment. Finally, the sag of each point can be determined from the laws of trigonometry.

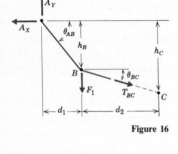

Figure 16

EXAMPLE 8
The cable carries three transverse loads, as shown. Determine (a) the elevations of joints B and D, (b) the maximum tension, (c) the maximum angle of elevation.

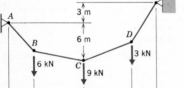

Solution
The free body diagram and equilibrium equations for the entire system are as given below.

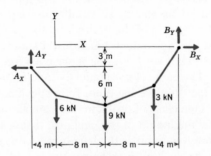

$$\Sigma M_{AZ} = 24B_Y - 3B_X - 6(4) - 9(12) - 3(20) = 0 \text{ kN-m}$$

$$\Sigma F_X = -A_X + B_X = 0$$

$$\Sigma F_Y = A_Y + B_Y - 18 = 0 \text{ kN}$$

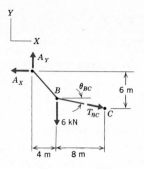

These equations are not sufficient to solve for the reactions. Additional equations are obtained by passing a cutting plane through the segment of cable on either side of point C, whose elevation is known, and isolating one of the cable sections formed by the plane. Our choice is shown in the free body diagram. The corresponding equilibrium equations are

$$\Sigma M_{CZ} = A_X(6) - A_Y(12) + 6(8) = 0$$

$$\Sigma F_X = T_{BC} \cos \theta_{BC} - A_X = 0$$

$$\Sigma F_Y = A_Y - T_{BC} \sin \theta_{BC} - 6 = 0$$

The simultaneous solutions of the moment equation for the section and the equations for the entire system are

$$A_X = B_X = 9.60 \qquad A_Y = 8.80 \qquad B_Y = 9.20 \text{ kN}$$

The force equations for the section of cable then become

$$T_{BC} \cos \theta_{BC} = A_X = 9.60$$

$$T_{BC} \sin \theta_{BC} = A_Y - 6 = 2.80$$

The solutions of these equations are

$$\theta_{BC} = \tan^{-1}\left(\frac{2.80}{9.60}\right) = 16.26°$$

$$T_{BC} = \frac{9.60}{\cos 16.26°} = 10.00 \text{ kN}$$

We now proceed to investigate the joints, starting with either end. Considering pin A first, the free body diagram and equilibrium equations are as follows.

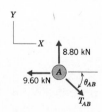

$$\Sigma F_X = T_{AB} \cos \theta_{AB} - 9.60 = 0$$

$$\Sigma F_Y = 8.80 - T_{AB} \sin \theta_{AB} = 0$$

The foregoing equations yield

$$\theta_{AB} = 42.51° \qquad T_{AB} = 13.023 \text{ kN}$$

Because we now know the tension and angle of elevation of segments AB and BC, the equilibrium equations for joint B only give checks on the computations. Skipping this step, we proceed to joint C. The free body diagram and equilibrium equations for this joint are described on the next page, assuming that point D is higher than point C.

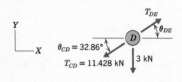

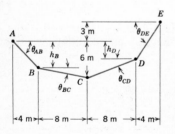

$$\Sigma F_X = T_{CD} \cos \theta_{CD} - 10.00 \cos 16.26° = 0$$

$$\Sigma F_Y = T_{CD} \sin \theta_{CD} + 10.00 \sin 16.26° - 9 = 0$$

$$\theta_{CD} = 32.86° \qquad T_{CD} = 11.428 \text{ kN}$$

Finally, for joint D we have

$$\Sigma F_X = T_{DE} \cos \theta_{DE} - 11.428 \cos 32.86° = 0$$

$$\Sigma F_Y = T_{DE} \sin \theta_{DE} - 11.428 \sin 32.86° - 3 = 0$$

Thus

$$\theta_{DE} = 43.78° \qquad T_{DE} = 13.297 \text{ kN}$$

We have found all of the tensions and all of the angles of elevation, so we omit consideration of the equilibrium of joint E, which would only serve as a check on the calculations. From these results we may answer parts (b) and (c), for we see that the maximum tension and angle of elevation are those for segment DE, being

$$\theta_{max} = \theta_{DE} = 43.8° \qquad T_{max} = T_{DE} = 13.30 \text{ kN} \qquad \triangleleft$$

We should not be surprised that the tension is maximum in the cable having the steepest slope, for equilibrium requires that the horizontal component T_i and θ_i of the tension in the ith segment be a constant equal to $A_X (= -E_X)$.

To determine the sags from the angle of elevation we draw a sketch of the geometry. From this sketch we see that

$$h_B = 4 \tan \theta_{AB} = 3.667 \text{ m} \qquad \triangleleft$$

and

$$h_D + 3 = 4 \tan \theta_{DE} \quad \text{so} \quad h_D = 0.833 \text{ m} \qquad \triangleleft$$

The sketch also suggests another check, in addition to the excess equilibrium equations, on the computations. Specifically, we may ascertain whether, starting from either pin A or E, the sag h_C of point C actually is found to be 6 m. Starting from pin A, we have

$$h_C = h_B + 8 \tan \theta_{BC} = 6.000 \text{ m}$$

whereas starting from pin E, we find

$$h_C = h_D + 8 \tan \theta_{CD} = 6.001 \text{ m}$$

Because both values of h_C are in agreement with the given value, the check validates the results.

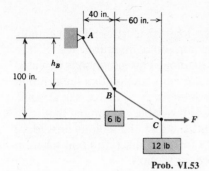

Prob. VI.53

HOMEWORK PROBLEMS

VI.53 Two blocks are supported by the lightweight cable *ABC*, as shown. Determine (a) the magnitude of the horizontal force $\bar{F}$, (b) the corresponding sag h_B.

VI.54 A lightweight cable carries the two concentrated forces shown. Knowing that $h_B = 8$ m, determine (a) the sag h_C, (b) the maximum tension, (c) the maximum angle of elevation.

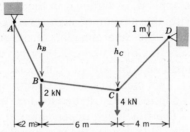

Prob. VI.54 and VI.55

VI.55 A lightweight cable supports the two concentrated forces shown. Knowing that $h_C = 8$ m, determine (a) the sag h_B, (b) the maximum tension, (c) the maximum angle of elevation.

VI.56 A cable supports three forces, as shown. Determine (a) the sags h_B and h_D, (b) the maximum tension, (c) the maximum angle of elevation.

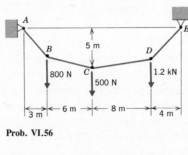

Prob. VI.56

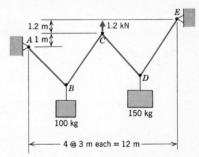

Prob. VI.57

VI.57 A 1.2-kN vertical force is applied to the cable at point *C* in order to hold the blocks in the position shown. Determine (a) the maximum tension in the cable, (b) the length of the cable.

VI.58 A set of horizontal and vertical forces are applied to cable *ABCDE*. Determine (a) the distances d_B and d_D, (b) the maximum tension, (c) the maximum angle of elevation.

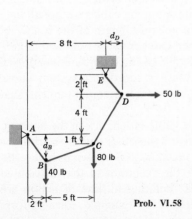

Prob. VI.58

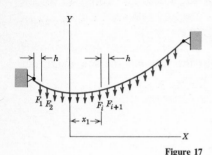

Figure 17

2 Distributed Loads — Parabolic Sag

Consider a cable supporting many discrete loads that are applied at equally spaced intervals a distance h apart in the horizontal direction. We wish to determine how the cable sags under the loading. This situation is depicted in Figure 17. It was demonstrated in the previous section that the solution of this situation requires additional information, such as the sag at some point along the cable or the total length of the cable. Assuming that such information is given, we may determine the sag by the tedious procedure of considering the equilibrium equations for each discrete point where a load is applied, as outlined in the preceding section.

With the advent of high-speed electronic computers, such an approach for the determination of the sag is not unreasonable. However, engineers in earlier times did not have such computational aids, so they developed an approximate, but more convenient, method for determining the sag. This method was based on replacing the discrete loads in Figure 17 by a continous distributed load $q(x)$, where q has units of force per unit length. The functional dependence of $q(x)$ is defined such that at point i along the cable, having the horizontal coordinate x_i as shown in the figure, $q(x_i) = F_i/h$. In other words, $q(x)$ is the average line load.

The case where $q(x)$ is a constant value, q_0, is the one that is most commonly encountered in practice. This is the situation that arises in the design of a suspension bridge, such as the one depicted in Figure 18, where a number of vertical cables are suspended from the spanning cables

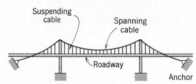

Figure 18

to support the roadway. If we neglect the weight of the cables and the stiffness of the roadway, we may treat the spanning cables as though they were loaded only by the constant weight per unit length of the roadway.

Equations pertaining to the sag of the cable may be obtained in general terms, independent of the precise details of the distributed load $q(x)$. For the convenience it affords, we select an xyz coordinate system whose origin is placed at the point where the cable has a horizontal tangent, that is, at the lowest point of the cable.

Let us consider a finite section of the cable of length x in the horizontal direction, having one end at the origin and the other end at an arbitrary point P. The free body diagram for such an element is shown in Figure 19.

The force $\bar{R}$ shown in the figure is the resultant of the distributed load. Its magnitude and the distance αx defining its line of action are known if the loading function $q(x)$ is known. For example, for a uniform load q_0, $R = q_0 x$ and $\alpha = \frac{1}{2}$. The other forces acting on the element are the

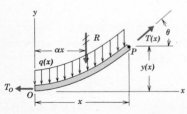

Figure 19

tensile forces tangent to the cable at each end. The term T_O denotes the tension at the horizontal tangent, point O.

Figure 19 shows that the sag of the cable is defined by the function $y(x)$. To determine this function, as well as the tension within the cable, we formulate the equations of equilibrium for the element. Choosing point P on the right for the moment sum, we obtain

$$\Sigma M_{PZ} = R(x - \alpha x) - T_0(y) = 0$$

$$\Sigma F_x = T \cos \theta - T_0 = 0 \qquad \Sigma F_y = T \sin \theta - R = 0$$

The solution of these equations is

$$T = (T_0{}^2 + R^2)^{1/2} \tag{8a}$$

$$\tan \theta = \frac{R}{T_0} \tag{8b}$$

$$y = \frac{(1 - \alpha)xR}{T_0} \tag{8c}$$

Two features of these results are of particular importance. First, from the equation for ΣF_X we see that the cable sags such that the tension within it always has a constant horizontal component T_0. This quantity will be a key parameter for the solution of problems. Second, because θ defines the tangent to the cable, it follows that

$$\tan \theta = \frac{dy}{dx} \tag{9}$$

In other words, equations (8b) and (8c) are not independent. After substitution of the expression for R into equation (8c), the resulting expression for y as a function of x must give the same result for $\tan \theta$ in equation (9) as that obtained from equation (8b). This serves to verify our assumption that the cable carries only a tensile force.

To proceed further, we must define the load function. As noted earlier, our primary interest is with the case of a uniform load. Setting $q(x) = q_0$, $R = q_0 x$, and $\alpha = \frac{1}{2}$ in equations (8a and c), we obtain

$$y = \frac{q_0}{2T_0} x^2 \tag{10a}$$

$$T = (T_0{}^2 + q_0{}^2 x^2)^{1/2} \tag{10b}$$

The equation for y shows that a cable carrying a *uniform horizontal load* sags into a *parabolic* shape. It is readily verified that the equation for $\tan \theta$ obtained from equations (9) and (10a) is identical to that given by equation (8b), confirming the correctness of the analysis.

As was true in the case of cables carrying concentrated loads, there are two standard types of problems we can analyze, assuming that the locations of the ends of the cables are known. In one, we know the sag of some point on the cable. Equations (10) will prove sufficient for such cases.

The second type of problem is the one where we know the length of the cable. To solve problems of this type we need an expression relating the length of the cable to the other parameters for the system. This expression is obtained by evaluating the arc length of the parabola formed by the cable.

Consider the situation depicted in Figure 20. The total length of the cable is the sum of the arc lengths s_A and s_B measured from the origin to each end. Both quantities are arc lengths along the parabola measured from the origin, so we need only consider how an expression for one arc length is obtained. In courses on calculus, it is shown that the arc length s_B may be obtained by evaluating the following integral.

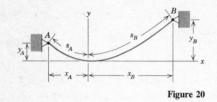

Figure 20

$$s_B = \int_0^{x_B} \left[1 + \left(\frac{dy}{dx} \right)^2 \right]^{1/2} dx$$

Upon substitution for y from equation (10), the integral becomes

$$s_B = \int_0^{x_B} \left[1 + \frac{q_0^2}{T_0^2} x^2 \right]^{1/2} dx \tag{11}$$

An exact solution for this integral may be found using a table of integrals; however, the resulting expression is cumbersome for calculations. A more convenient expression may be obtained by expanding the integrand in equation (11) in an infinite series using the binomial theorem, which is

$$(1 + Z)^n = 1 + nZ + \frac{n(n-1)}{2!} Z^2 + \frac{n(n-1)(n-2)}{3!} Z^3 + \ldots$$

With this expression, equation (11) becomes

$$s_B = \int_0^{x_B} \left(1 + \frac{q_0^2 x^2}{2T_0^2} - \frac{q_0^4 x^4}{8T_0^4} + \ldots \right) dx$$

$$= x_B \left(1 + \frac{q_0^2 x_B^2}{6T_0^2} - \frac{q_0^4 x_B^4}{40T_0^4} + \ldots \right)$$

Finally, using equation (10) to eliminate q_0/T_0, we find

$$s_B = x_B \left[1 + \frac{2}{3} \left(\frac{y_B}{x_B} \right)^2 - \frac{2}{5} \left(\frac{y_B}{x_B} \right)^4 + \ldots \right] \tag{12}$$

The infinite series in equation (12) is convergent when $y_B/x_B < 0.5$. If y_B/x_B is substantially lower than this value, as it is in many engineering applications, then the convergence is very rapid and the listed terms provide sufficient accuracy. On the other hand, if y_B/x_B exceeds 0.5, the exact integral of equation (11) is required.

EXAMPLE 9

A cable supporting a pipeline weighing 1 kip/ft is to be suspended between pins A and B, whose positions are as shown. The design calls for the pipeline to be 8 ft below the elevation of pin A and for the low point on the cable to be at the pipeline. Determine (a) the maximum tension in the cable, (b) the required length of the cable.

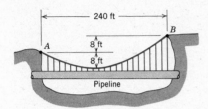

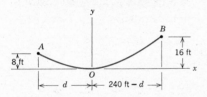

Solution

In this problem the location of the horizontal tangent (the low point) is not completely given. This point is the origin from which the position of all points on the cable are measured, so its location must be found. This evaluation is achieved with the aid of a sketch of the geometry and the coordinate system.

Letting d be the horizontal distance of the origin O from pin A, we see from the sketch that the coordinates of point A are ($x_A = -d$, $y_A = 8$), whereas those for point B are ($x_B = 240 - d$, $y_B = 16$). Both of these points lie on the parabola defined by equation (10). Therefore, we have

$$y_A = 8 = \frac{q_0}{2T_O}(x_A)^2 = \frac{q_0}{2T_O}d^2$$

$$y_B = 16 = \frac{q_0}{2T_O}(x_B)^2 = \frac{q_0}{2T_O}(240 - d)^2$$

The distributed load is the weight of the pipeline, $q_0 = 1000$ lb/ft. Therefore, the foregoing represents two equations whose simultaneous solution gives the values of d and T_O. We first solve for q_0/T_O from the first equation.

$$\frac{q_0}{T_O} = \frac{8}{d^2}$$

The second equation then becomes a quadratic in d. Specifically, it gives

$$16 = \left(\frac{8}{d^2}\right)(240 - d)^2$$

$$d^2 + 480d - 5.76(10^4) = 0$$

$$d = 99.41, -579.4 \text{ ft}$$

The negative value is an extraneous root for this problem, so $d = 99.41$ ft. Substituting the value of q_0 and d, we then find T_O.

$$T_O = q_0\left(\frac{d^2}{8}\right) = 1000\frac{(99.41)^2}{8} = 1.2353(10^6) \text{ lb}$$

We now have the parameters required to obtain the information requested. From equation (10b) we see that the tension is a maximum at the point that is farthest from the horizontal tangency. This is at pin B because the distance d is less than half the horizontal distance between pins A and B. Thus, from equation (10b) we have, with $x_B = 240 - d = 140.59$ ft,

$$T_{\max} = (T_O^2 + q_0^2 x_B^2) = [(1.2353)^2(10^{12}) + (10)^6(140.59)^2]$$

$$= 1.2433(10^6) \text{ lb}$$

To determine the length of the cable we note that y_A/x_A and y_B/x_B are both extremely small. We therefore may employ equation (12). This yields

$$s_A = x_A\left[1 + \frac{2}{3}\left(\frac{y_A}{x_A}\right)^2 - \frac{2}{5}\left(\frac{y_A}{x_A}\right)^4\right]$$

$$= 99.41\left[1 + \frac{2}{3}\left(\frac{8}{99.41}\right)^2 - \frac{2}{5}\left(\frac{8}{99.41}\right)^4\right]$$

$$= 99.41 + 0.4275$$

$$s_B = 140.59\left[1 + \frac{2}{3}\left(\frac{16}{140.59}\right)^2 - \frac{2}{5}\left(\frac{16}{140.59}\right)^4\right]$$

$$= 140.59 + 1.2045$$

$$s = s_A + s_B = 240 + 1.632 = 241.63 \text{ ft}$$

These results show that because of the relatively small amount by which the cable sags, the tension within the cable is much larger than the total weight of the pipeline, (1000 lb/ft)(240 ft), and the required length of the cable is only 1.63 ft longer than the horizontal distance spanned.

HOMEWORK PROBLEMS

VI.59 The span of a suspension bridge between two towers of equal height is 1000 m and the sag of its cables at midspan is 100 m. The roadway, having a mass of $20(10^3)$ kg/m, is supported equally by two suspension cables. Determine (a) the maximum and minimum tensions in the cable, (b) the length of the cable between the towers.

VI.60 Cable AB supports a 200-kN total load which is distributed uniformly along the horizontal. The slope of the cable is zero at the lower

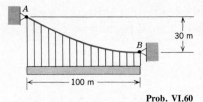

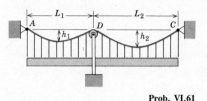

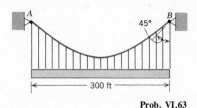

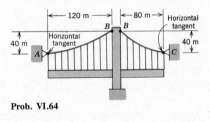

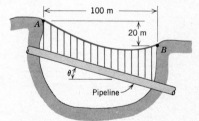

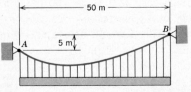

Prob. VI.60

Prob. VI.61

Prob. VI.63

Prob. VI.64

Prob. VI.65

Prob. VI.67

end. Determine (a) the maximum and minimum tensions in the cable, (b) the location of the point on the cable where the tension is the average of the values determined in part (a).

VI.61   A cable carrying a constant load q_0 along the horizontal direction is anchored at ends A and C. The cable passes over a roller at the intermediate support D, as shown. Derive an expression for L_2/h_z in terms of L_1, h_1, and L_2.

VI.62   A cable spanning a distance of 300 ft between pins located at the same elevation supports a uniform load of 3 lb/ft. The maximum tension in the cable is 2000 lb. Determine (a) the maximum sag, (b) the slope of the cable at its supports, (c) the length of the cable.

VI.63   Cable AB, carrying a uniform load of 500 lb/ft, has a span of 300 ft. Knowing that the tangent to the cable at an end is at a 45° angle from the vertical, determine the maximum sag and the maximum tension in the cable.

VI.64   Cables AB and BC supporting a uniform horizontal load of 4 kN/m are both pinned to tower B, as shown. Determine the resultant force exerted by both cables on the tower.

VI.65   The pipeline, whose mass per unit length is 200 kg/m, is oriented horizontally ($\theta = 0$) as it passes over the gorge. The ends of the cable are located as shown. It is known that at its lowest point the cable is 5 m below pin B. Determine (a) the maximum tension within the cable, (b) the length of the cable.

VI.66   Solve Problem VI.65 if the pipeline is inclined at $\theta = 20°$.

VI.67   A cable spanning the distance between pins A and B supports a uniform load of 200 N/m, distributed along the horizontal. Knowing that the lowest point on the cable is 10 m horizontally from pin A, determine the lowest elevation of the cable below pin A and the length of the cable.

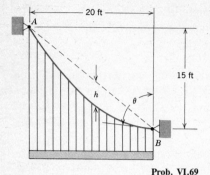

Prob. VI.69

VI.68 Solve Problem VI.67 if, instead of knowing the horizontal location of the low point of the cable, it is known that the minimum tension in the cable is 20 kN.

VI.69 A cable carrying a uniform horizontal load of 4 kips/ft is anchored at pins A and B. It is known that at end B the cable slopes upward at $\theta = 75°$ from the vertical. Determine (a) the maximum tension, (b) the length of the cable, (c) the maximum value, h_{max}, by which the cable sags below line AB.

VI.70 and VI.71 The homogeneous body shown has a weight W. It is known that the cable has a horizontal tangency at end A. Considering the span L and the constant horizontal component T_O of the tension force to be known parameters, derive expressions for (a) the equation of the curve formed by the sagging cable, (b) the distance h by which pin B is higher than pin A, and (c) the maximum tension in the cable.

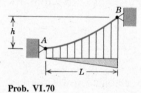

Prob. VI.70

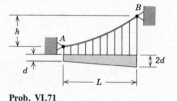

Prob. VI.71

3 Catenary

Frequently, we are interested in the sag of a cable under its own weight. Restricting ourselves to the case of a uniform cable having a weight per unit length w_0, it follows that the gravitational force acting on an element of cable of infinitesimal arc length ds is $w_0\,ds$, as depicted in Figure 21.

The situation in Figure 21 is a special case of a cable carrying an arbitrary distributed load, described by Figure 19 and equations (8). Let us therefore consider the resultant of the distributed gravity force acting on the section of cable. Because w_0 is constant, it follows that $w_0 s$ is the resultant gravitational force, or equivalently, the weight of the section of cable. Hence, $R = w_0 s$. Further, by definition, the weight force is applied at the center of mass. This creates a problem, because the location of the center of mass depends on the shape of the curve formed by the cable in Figure 21, and this shape is what we wish to determine. Therefore, we cannot locate the line of action of the resultant force $\bar{R}$, which means that the parameter α in equation (8c) is unknown. It follows that we cannot determine the function $y(x)$ describing the sag of the cable solely by solving the equilibrium equations for the element.

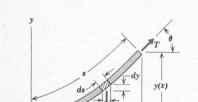

Figure 21

To resolve this difficulty, we recall that equations (8b) and (8c) are not independent because equation (9), $\tan \theta = dy/dx$, must be satisfied. Equation (8b) does not depend on the parameter α, so we will use that equation in place of equation (8c). Substituting $R = w_0 s$ into equations (8a) and (8b), we get

$$T = [T_0{}^2 + w_0{}^2 s^2]^{1/2} \tag{13a}$$

$$\tan \theta = \frac{dy}{dx} = \frac{w_0}{T_0} s \tag{13b}$$

Equation (13b) is a *differential equation*. It relates the differentials dy and dx of the infinitesimal element shown in Figure 21. This figure also shows that, by the Pythagorean theorem,

$$(ds)^2 = (dx)^2 + (dy)^2$$

Noting that s appears in the right side of equation (13b), we use the foregoing relationship to eliminate dy by writing

$$\frac{dy}{dx} = \frac{[(ds)^2 - (dx)^2]^{1/2}}{dx} = \frac{w_0}{T_0} s$$

$$\left[\left(\frac{ds}{dx} \right)^2 - 1 \right]^{1/2} = \frac{w_0}{T_0} s$$

$$\frac{ds}{dx} = 1 + \left(\frac{w_0}{T_0} \right)^2 s^2$$

Solving for dx, we find

$$dx = \frac{ds}{1 + (w_0/T_0)^2 s^2} \tag{14}$$

This last equation may be integrated with the aid of a table of integrals. As shown in Figure 21, s is measured from the origin, so $s = 0$ when $x = 0$. Thus, a definite integral of equation (14) gives

$$\int_0^x dx = \int_0^s \frac{ds}{[1 + (w_0/T_0)^2 s^2]^{1/2}}$$

$$x = \frac{T_0}{w_0} \left[\sinh^{-1} \left(\frac{w_0}{T_0} s \right) - \sinh^{-1}(0) \right] \tag{15}$$

The function $v = \sinh u$ is the hyperbolic sine of u, and the function $u = \sinh^{-1} v$ is its inverse, called the arc hyperbolic sine. There is some similarity between the hyperbolic and trigonometric functions, resulting from their definitions in terms of complex numbers. For our purposes here, we will need the following properties of the hyperbolic sine and cosine.

$$\sinh u \equiv \tfrac{1}{2}(e^u - e^{-u}) \qquad \cos u \equiv \tfrac{1}{2}(e^u + c^{-u}) \tag{16a}$$

$$\frac{d}{du} \sinh u = \cosh u \qquad \frac{d}{du} \cosh u = \sinh u \tag{16b}$$

$$\cosh^2 u - \sinh^2 u = 1 \tag{16c}$$

From the definitions, equations (16a), we see that

$$\sinh(-u) = -\sinh u \qquad \cosh(-u) = \cosh u \tag{16d}$$

$$\sinh (0) = 0 \qquad \cosh (0) = 1 \tag{16e}$$

Values of the hyperbolic functions may be found in standard numerical tables, as well as by evaluating them directly. In addition, they are keyboard functions on some electronic calculators.

Returning to equation (15), we apply equations (16e) and solve for s, with the result that

$$s = \frac{T_o}{w_0} \sinh\left(\frac{w_0 x}{T_o}\right) \tag{17}$$

This result is important because it provides a relationship between the horizontal distance and the length of the cable, analogous to equation (12) for a cable supporting a uniform horizontal load. Equation (17) also enables us to determine the sag $y(x)$.

Substitution of the expression for s into equation (13) yields

$$\frac{dy}{dx} = \sinh\left(\frac{w_0 x}{T_o}\right)$$

It then follows from the second of equations (16b) that

$$y = \frac{T_o}{w_0} \cosh \frac{w_0 x}{T_o} + C$$

where C is a constant of integration. Because the cable passes through the origin, it must be that $y = 0$ when $x = 0$, which means that

$$0 = \frac{T_o}{w_0} \cosh(0) + C \quad \text{so} \quad C = -\frac{T_o}{w_0}$$

Thus, the expression for y becomes

$$y = \frac{T_O}{w_0}\left[\cosh\left(\frac{w_0 x}{T_O}\right) - 1\right]$$ (18)

The curve formed by plotting $y(x)$ is called a *catenary*.

In addition to the curve formed by the sagging cable, we also need to know the tension within the cable. The required expression is found by substituting equation (17) for s into equation (13a) and then using the identity given by equation (16c). This yields

$$T = \left[T_O{}^2 + w_0{}^2\left(\frac{T_O}{w_0}\right)^2 \sinh^2\left(\frac{w_0 x}{T_O}\right)\right]$$

$$T = T_O \cosh\left(\frac{w_0 x}{T_O}\right)$$ (19)

Equations (17), (18), and (19) are sufficient for solving problems. Assuming that we know the location of the pins holding the ends of the cable, a key parameter to be determined is the constant T_O, which is the constant horizontal component of the tension. Let us consider what is involved in such a determination for a cable whose pins are at the same elevation, as illustrated in Figure 22.

By symmetry, the origin O, which is located at the horizontal tangent, must be located midway in the span, as shown. Letting h be the maximum sag, we see that the end points, having coordinates $(x_A = -L/2,$ $y_A = h)$ and $(x_B = L/2, y_B = h)$, are points on the catenary. Substituting these coordinates into equation (18), we find that

$$h = \frac{T_O}{w_0}\left[\cosh\left(\frac{w_0 L}{2T_O}\right) - 1\right]$$

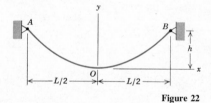

Figure 22

Two *nondimensional* combinations of the parameters may be identified in the foregoing, specifically, h/L and $w_0 L/T_O$. In terms of these quantities, we then find

$$\frac{h}{L}\left(\frac{w_0 L}{T_O}\right) = \cosh\left(\frac{w_0 L}{2T_O}\right) - 1$$ (20)

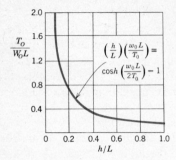

Figure 23

If we know the sag h and span L, the solution of equation (20) will give the value of T_0. Unfortunately, a closed-form solution is not possible. Nevertheless, it is possible to solve equation (20) by a trial-and-error procedure. The graph in Figure 23 shows the relationship between the nondimensional parameters (h/L) and (wL/T_0) resulting from such a solution. This graph may be used by itself to solve problems, or alternatively, as the starting point of your own trial-and-error computation if more accurate results are required.

In closing, it should be noted that there is no resemblance between the formulas obtained here and those obtained in the previous section for a cable carrying a uniform horizontal load. This is attributable to the fact that the weight of a cable is distributed uniformly along its length, not along the horizontal. However, in the case where the tension T_0 is sufficiently large to make the sag ratio very small ($h/L < 0.10$), the cable is almost oriented horizontally. In such a case we may obtain an approximate solution by considering the weight to be a uniform horizontal load, and thus set $q_0 = w_0$ in the formulas for a uniform load in the previous section.

EXAMPLE 10

A uniform cable whose mass per unit length is 10 kg/m hangs as shown. Determine the maximum tension in the cable and its length.

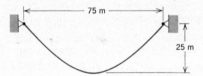

Solution

From the given information we calculate the weight per unit length and the ratio of maximum sag to span

$$w_0 = (10 \text{ kg/m})(9.806 \text{ m/sec}^2) = 98.06 \text{ N/m}$$

$$\frac{h}{L} = \frac{25}{75} = \frac{1}{3}$$

The other important parameter for an anlysis is $T_0/w_0 L$. The value of $T_0/w_0 L$ that we read from the graph given in Figure 23 corresponding to $h/L = 1/3$ is

$$\frac{T_0}{w_0 L} = 0.42$$

It may be that this is all the accuracy that is desired, in which case substitution of the values of w_0 and L will yield T_0. However, let us assume that we want more accuracy. We may follow a trial-and-error procedure by using the estimate of T_0/w_0L to evaluate the difference between the left- and right-hand sides of equation (20). The estimate is then improved by modifying the value of T_0/w_0L to reduce this difference. This computation may be performed in tabular form, as indicated below, where

$$\text{LHS} = \left(\frac{h}{L}\right)\left(\frac{w_0L}{T_0}\right) = \left(\frac{1}{3}\right)\left(\frac{w_0L}{T_0}\right)$$

$$\text{RHS} = \cosh\left[\frac{1}{2}\left(\frac{w_0L}{T_0}\right)\right] - 1$$

T_0/w_0L	LHS	RHS	LHS – RHS
0.42	0.7937	0.7964	−0.0027
0.419	0.7955	0.8006	−0.0051
0.421	0.7918	0.7921	−0.0003
0.4211	0.7916	0.7917	−0.0001
0.4212	0.7914	0.7913	0.0001

The second trial value of T_0/w_0L in the tabulation was based on a guess that the initial value was too high. This guess was shown to be wrong by the fact that there was a larger difference between the values of LHS and RHS corresponding to the second estimate than there was for the initial estimate. The succeeding values of T_0/w_0L were chosen in accord with the trend indicated by the first two trial values. The last two trials indicate that the value T_0/w_0L is somewhere between 0.4211 and 0.4212. Using the former value, we calculate that

$$T_\rightarrow = 0.4211 w_0L = 0.4211(98.06)(75) = 3097 \text{ N}$$

The tension in the cable is given by equation (19). This equation shows that T has its maximum value at the location where x is a maximum, which is one of the ends. Setting $x = L/2 = 37.5$ m, we find

$$T_{\text{max}} = T\left(x = \frac{L}{2}\right) = 3097 \cosh\left[\frac{(98.06)(37.5)}{3097}\right] = 5549 \text{ N} \quad \triangle$$

The total length of the cable is twice the arc length from the lowest point on the cable, which is where the origin is located. Therefore, from equation (17) we find

$$s_{\text{total}} = 2s\big|_{x=37.5} = 2\left(\frac{3097}{98.06}\right)\sinh\left[\frac{(98.06)(37.5)}{3097}\right] = 93.9 \text{ m} \quad \triangle$$

HOMEWORK PROBLEMS

VI.72   A uniform 1-kg/m cable is suspended as shown. (a) Determine the maximum and minimum values of the tension in the cable, and the cable length. (b) Compare the values found in part (a) to the approximate values obtained by considering the weight of the cable to be 1 kg/m along the horizontal.

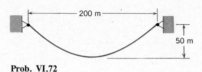

Prob. VI.72

VI.73   The maximum tension that an electrical transmission cable weighing 10 lb/ft can withstand without failure is 4 kips. The cable is suspended from two towers at equal elevations 200 ft apart. Determine the minimum length of cable that can be used and the corresponding sag. *Hint:* There are two values of T_O that give a maximum tension of 4 kips.

VI.74   A 0.5-kg/m cable is suspended between two buildings with its ends at equal elevations. The length of the cable is 24 m and the maximum sag is 10 m. Determine (a) the horizontal distance spanned by the cable, (b) the minimum and maximum tensions. *Hint:* Formulate expressions for the length and the maximum sag of the cable and then use identity (16c).

VI.75 and VI.76   The cable shown, whose weight per unit length is w_0, is pinned at end B and attached to collar A which may slide along the smooth bar. Knowing that $h = d$, determine the magnitude of the force $\bar{P}$ required to hold the collar in position in terms of d and w_0.

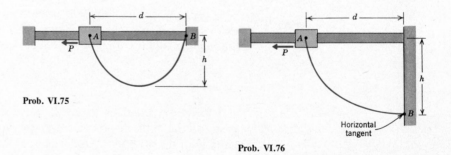

Prob. VI.75

Prob. VI.76

VI.77   Solve Problem VI.75 in the case where (a) $h = 2d$, (b) $h = 3d$.

VI.78 An electrical transmission cable having a mass per unit length of 10 kg/m is suspended by a series of towers, as shown. Determine the resultant force exerted by the cables on tower B.

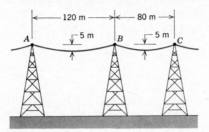

Prob. VI.78

VI.79 A 50-ft-long cable weighing 2 lb/ft spans a distance L between two pins at equal elevation. Knowing that the maximum sag $h = 0.2L$, determine (a) the distance L, (b) the horizontal and vertical components of the reaction force at one of the pin supports.

VI.80 Determine the sag to span ratio h/L of a cable whose total weight equals the minimum tension in the cable.

VI.81 Determine the sag to span ratio h/L of a cable whose total weight equals the maximum tension in the cable.

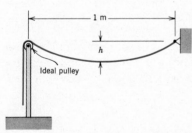

Probs. VI.80 and VI.81

VI.82 A uniform 2-m-long cable having a total mass of 1 kg is suspended as shown. Determine the *two* values of h for which the cable is in equilibrium.

VI.83 The mass per unit length of the cable is 0.4 kg/m. Determine the *two* values of the sag h for which the maximum tension in the cable is 300 N.

Ideal pulley

Prob. VI.82

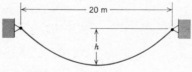

Prob. VI.83

MODULE VII
FRICTION

Reactive forces that frequently appeared in earlier modules were the normal forces exerted between two bodies having surfaces in contact. It was usually stipulated in those situations that the surfaces were smooth. As a result, the normal forces (which have the effect of preventing the contact surfaces from penetrating each other) were the only possible reactions.

In this module we shall consider the situation where the contact surfaces are rough. This results in the possibility of forces tangent to the plane of contact between the two bodies, whose effect is to oppose the movement of one surface relative to the other. Such forces are called *friction forces*.

Obviously, there is no surface that is absolutely smooth, but the concept of a system without friction is frequently a useful model to study. We refer to the condition of a smooth surface as an *ideal* model. The friction that we account for in a *real* model may originate from several sources. In general, friction forces can be classified on the basis of whether the contacting surfaces are dry, or whether a fluid medium is involved.

Fluid friction forces arise partially, but not entirely, from the viscosity of fluids. Such forces are important phenomena in the design of conduits for conveying liquids, in the design of transportation systems such as airplanes, automobiles, and ships, and in the study of lubricants. The study of fluid friction requires knowledge of the mechanics of fluids, and will not be treated here.

Dry friction is itself a complicated phenomenon that is still not fully understood. Current thinking is that it results from the microscopic irregularities present in all surfaces (hence the phrase: rough surface) and also from molecular attraction. Fortunately, elementary laws, first reported by Coulomb in 1781, are available for forming a simple model to describe dry friction. Coulomb's laws, which are the subject of the next section, should not be regarded as exact, because they possess certain anomalies. Nevertheless, they do provide accurate information in a broad range of applications that we will study in the following sections.

A negative connotation is usually associated with the word friction, because it gives rise to additional energy expenditures, and because it results in wear of surfaces in contact. This represents a limited view, because the existence of friction is very often useful, if not essential. For instance, how could we walk without friction?

A.   DRY FRICTION

1   Coulomb's Laws

A primary feature of the friction forces exerted between surfaces in contact is that they oppose the direction in which the surfaces move, or would *tend* to move, relative to each other. To illustrate this, consider the block

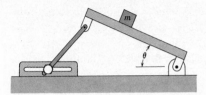

Figure 1a

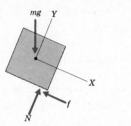

Figure 1b

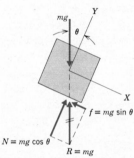

Figure 2

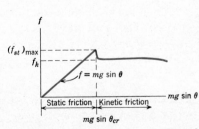

Figure 3

on the inclined plane AB in Figure 1, whose angle of elevation θ can be set at any desired value.

In the absence of friction, the block would certainly slide down the plane. Thus, the free body diagram, Figure 1b, shows the friction force $\bar{f}$ acting on the block to be in the negative X direction, opposite to the direction in which the block is tending to move relative to the incline.

Note that we have not considered a free body diagram of the incline because, in effect, it is the ground. However, if we did draw a free body diagram of the incline, the friction force $-\bar{f}$, which is the reaction exerted on the incline by the block, would be in the positive X direction. This is consistent with the foregoing general characteristic of a friction force, because if the block were to slide down the plane the inclined surface would seem to move in the negative X direction when viewed from the block. Remember that the friction force is opposite to the *relative* movement.

Let us now consider a simple experiment with the apparatus in Figure 1a, in which the value of θ is set at successively larger values, starting with $\theta = 0$, after which the block is placed at rest on the incline. There is a range of values of the angle θ for which the block will not slide. In this situation, the total reaction force exerted on the block $\bar{R} = -f\bar{I} + N\bar{J}$ is known from the equations of equilibrium to be as shown in Figure 2.

In the figure, $mg \sin \theta$ is the component of the external (gravity) force acting on the block in the direction in which the block would tend to slide, whereas f is the component of the reaction force resisting this tendency to move. For static equilibrium,

$$f = mg \sin \theta$$
$$N = mg \cos \theta \tag{1}$$

Now consider the situation for larger values of the angle θ. At some angle, call it the *critical angle* θ_{cr}, the block will slide, rather than remaining at rest. The friction force during this sliding motion may be determined by means of Newton's second law.

Let us plot a graph showing along the ordinate the magnitude of the friction force f in this series of trials for different values of θ. Along the abscissa of this graph we plot the component of the gravitational load parallel to the direction of movement, $mg \sin \theta$. The result is shown in Figure 3.

The graph illustrates that when there is static equilibrium, $f = mg \sin \theta$. The maximum friction force, $(f_{st})_{max}$ acting on the block occurs at an angle θ infinitesimally smaller than θ_{cr} for which sliding occurs, so

$$(f_{st})_{max} = mg \sin \theta_{cr}$$

The graph also shows that once sliding begins, the friction force decreases to the value f_k and varies only slightly as the force component in the

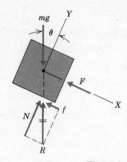

Figure 4

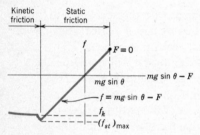

Figure 5

direction of motion increases. In the region where we have static equilibrium, we say that $\bar{f}$ is a *static friction force*, $\bar{f}_{st}$, whereas $\bar{f}$ is said to be a *kinetic friction force*, $\bar{f}_k$, when movement occurs.

Let us now consider a second experiment involving the apparatus of Figure 1, where we apply a force $\bar{F}$ tangent to the incline, in the uphill sense. Let the angle of elevation be set at a constant value θ smaller than θ_{cr}. The resulting free body diagram of the block is shown in Figure 4. The component of the external load acting on the block in the X direction is now $mg \sin \theta - F$.

Let us plot f versus $mg \sin \theta - F$ for this experiment, as F is increased from zero. From Figure 4 it is apparent that $f = mg \sin \theta - F$ when there is static equilibrium. The resulting graph appears in Figure 5.

From this graph we see that there is again a maximum magnitude for the static friction force, followed by a decrease in magnitude to the kinetic friction force. Note that the plotted values of $(f_{st})_{max}$ and f_k are not the same as those plotted in Figure 3. Also, the negative values for f represent a reversal in the direction of the friction force, corresponding to $F \geq mg \sin \theta$. This is because the friction force opposes the relative movement that would occur if the surfaces were smooth.

Suppose that we perform a series of these experiments for many values of θ. We now that $N = mg \cos \theta$ in all cases, because even when the block slides, the sliding motion is parallel to the X axis so there is always a force balance in the Y direction. When we form the ratios $|(\bar{f}_{st})_{max}|/N$ and $|\bar{f}_k|/N$ for each value of θ, it is found that these ratios are constants. The first is called the *coefficient of static friction*, μ_s, and the second is called the *coefficient of kinetic friction*, μ_k. These observations may be generalized as follows.

a *Case of no relative movement.* For any pair of contact surfaces that are not moving relative to each other, the magnitudes of the friction and normal forces satisfy

$$\boxed{|\bar{f}| \leq \mu_s |\bar{N}|} \tag{2}$$

where the equality sign corresponds to the situation where sliding is about to occur, that is, the case of *impending motion*. It should be emphasized that the question of whether or not movement will occur depends on the magnitude, but not the sense, of the friction force. The sense in which the friction force actually acts on one of the surfaces is opposite to the sense in which that surface would move relative to the other surface if there were no friction.

b *Case of relative movement.* For any pair of surfaces that are sliding over each other, the magnitudes of the friction and normal forces satisfy

$$\boxed{|\bar{f}| = \mu_k |\bar{N}|}$$ (3)

where μ_k is a lower value than μ_s. In this situation the direction of $\bar{f}$ is opposite the direction of the relative movement.

Equations (2) and (3) are Coulomb's laws of friction. As originally stated, the values of μ_s and μ_k are regarded as dependent only on the types of materials in contact (for example, steel in contact with aluminum) and the condition of each surface (for example, highly polished or corroded). This is an approximation, for it gives rise to certain subtle differences with observed phenomena, such as the different tractive capabilities of two automobile tires of the same construction, but different dimensions. Further, as is exhibited by Figures 3 and 5, the value of μ_k is not truly a constant. A more refined experiment would show that, in reality, μ_k is slightly dependent on the relative velocity of the sliding surfaces.

Equations (2) and (3) are quite useful, even though they have the limitations discussed above. Typical values of μ_s and μ_k are presented in Table 1. Note that there is no reason why the coefficients of friction should be less than unity, as exemplified by the value of μ_s given for contact between two cast iron surfaces. In actuality a broad range of values exists for each case given. The values shown in Table 1 merely give an idea of the general phenomena. For specific cases these values may be incorrect by more than 100%. Thus the tabulation should be employed only if values resulting from experiments on the actual system are unavailable.

Table 1

MATERIALS	μ_s	μ_k
Mild steel on mild steel	0.74	0.57
Aluminum on mild steel	0.61	0.47
Copper on mild steel	0.53	0.36
Cast iron on cast iron	1.10	0.15
Brake material on cast iron	0.40	0.30
Oak on cast iron	0.60	0.32
Leather on cast iron	0.60	0.56
Stone on cast iron	0.45	0.22
Rubber on metal	0.40	0.30
Rubber on wood	0.40	0.30
Rubber on pavement	0.90	0.80
Leather on wood	0.40	0.30
Glass on nickel	0.78	0.56

The coefficient of static friction is sometimes represented in terms of the angle θ_{cr} at which the block in Figure 1 begins to slide down the incline. Corresponding to θ_{cr} (impending motion), we have $f = \mu_s N$. Equations (1) then give

$$N = mg \cos \theta_{cr}$$

$$f = mg \sin \theta_{cr} = \mu_s N = \mu_s \, mg \cos \theta_{cr}$$

from which we find

$$\mu_s = \tan \theta_{cr} \tag{4}$$

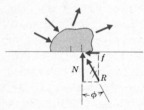

Figure 6

Equation (4) is useful for a pictorial representation of the static friction force. For example, let us consider a body in contact with a rough plane and subjected to an arbitrary set of loads, as shown in Figure 6. The friction force $\bar{f}$ and normal force $\bar{N}$ are equivalent to the resultant reaction $\bar{R}$ at an angle ϕ from the normal to the plane of contact. Clearly,

$$\tan \phi = \frac{f}{N}$$

Substituting for f from equations (2) and (4) results in

$$\tan \phi \leq \left(\frac{\mu_s N}{N}\right) = \tan \theta_{cr}$$

$$\phi \leq \theta_{cr}$$

In words, if there is no sliding, the angle ϕ for the total reaction $\bar{R}$ at the contact surface will be less than θ_{cr}, whereas the case of impending motion corresponds to

$$\phi_s = \theta_{cr} = \tan^{-1} \mu_s \tag{5}$$

The angle ϕ_s is called the *angle of static friction*.

For the situation where slipping occurs, equation (3) tells us that $f = \mu_k N$, so we may replace μ_s with μ_k in equation (5). Thus

$$\phi_k = \tan^{-1} \mu_k \tag{6}$$

where ϕ_k is the *angle of kinetic friction*. From the fact that $\mu_k < \mu_s$, it follows that $\phi_k < \phi_s$. It can be shown that ϕ_k is the angle of elevation of the apparatus in Figure 1 for which the block will continue to slide down the incline at a constant speed once it is set in motion.

Equations (5) and (6) represent alternative ways of specifying the coefficient of friction. However, our approach to statics involves treating reaction forces in terms of their components rather than their magnitude and direction, so we shall have little use for the concept of the angle of friction when we solve problems.

EXAMPLE 1

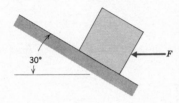

The coefficients of friction between the inclined plane and the 2.5-kg block are $\mu_s = 0.25$ and $\mu_k = 0.20$. The block is initially at rest when the horizontal force $\bar{F}$ is applied to it. Determine (a) the friction force when $F = 20$ N, (b) the friction force when $F = 5$ N, (c) the minimum magnitude of $\bar{F}$ for which the block will slide uphill.

Solution

In parts (a) and (b) we cannot tell by inspection whether there is sliding or not, for this depends on the comparison of the friction and normal reactions. We therefore will assume that there is no sliding, and compare the values of the friction force $\bar{f}$ and normal force $\bar{N}$ obtained from the equations of static equilibrium. On the other hand, in part (c) we are interested in the condition where uphill motion is impending, so for that case, we set $|\bar{f}| = \mu_s |\bar{N}|$. Note that in parts (a) and (b) we can assume a direction for f, because a negative sign in the solution for the magnitude of $\bar{f}$ will indicate that the direction was assumed incorrectly. However, in part (c) the friction force must be opposite the direction of the impending movement of the block relative to the hill, so $\bar{f}$ *must* be depicted as being downhill.

For parts (a) and (b) let us assume that the friction force is down the incline, so the free body diagram for these cases is as shown.

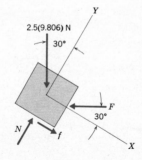

Because this problem does not involve consideration of the lines of action of the forces, we need only consider the equations for force equilibrium. These are

$$\Sigma F_X = -F \cos 30° + f + 2.5(9.806) \sin 30° = 0 \text{ newtons}$$

$$\Sigma F_Y = N - F \sin 30° - 2.5(9.806) \cos 30° = 0 \text{ newtons}$$

Thus

$$f = 0.866F - 12.26$$

$$N = 0.50F + 21.23$$

Substituting the given values of the applied force F in parts (a) and (b) yields

part (a): $f_a = 5.06$ $N_a = 31.73$ newtons

part (b): $f_b = -11.83$ $N_b = 23.73$ newtons

The negative value of f_b tells us that the friction force is up the incline, which means that the tendency of the block is to move downward.

We cannot accept the preceding values of f until they have been checked against the maximum static friction force possible in each case, $|\bar{f}| = \mu_s N$. Thus

part (a): $\mu_s N_a = 0.25(31.73) = 7.93$ newtons $> |\bar{f}_a|$

part (b): $\mu_s N_b = 0.25(23.73) = 5.93$ newtons $< |\bar{f}_b|$

We observe that equation (2) is satisfied in part (a), but not in part (b). Thus, in part (a) there is no sliding, so

$$\bar{f} = f_a \bar{I} = 5.06 \bar{I} \text{ newtons}$$

In part (b) the contact surfaces cannot sustain a friction force sufficient to prevent sliding. The direction in which the sliding will occur can be deduced to be down the plane, because that is the tendency indicated by the result for f_b. Therefore

$$\bar{f} = -\mu_k N_b \bar{I} = -4.75 \bar{I} \text{ newtons}$$

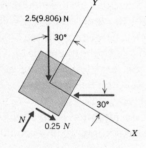

2.5(9.806) N

To solve part (c) we draw a new free body diagram, because in this situation of impending motion, $\bar{f}$ is $\mu_s N$ down the plane. The equilibrium equations are now

$$\Sigma F_X = -F \cos 30° + 0.25N + 2.5(9.806) \sin 30° = 0 \text{ newtons}$$

$$\Sigma F_Y = N - F \sin 30° - 2.5(9.806) \cos 30° = 0 \text{ newtons}$$

Solving these equations simultaneously yields

$$F = 23.7 \text{ newtons}$$

HOMEWORK PROBLEMS

Prob. VII.1

VII.1   Determine the horizontal force $\bar{P}$ that must be exerted on the 50-kg block to cause the block to move. The coefficient of static friction is 0.30.

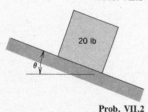

Prob. VII.2

VII.2   The coefficients of friction between the 20-lb block and the inclined plane are $\mu_s = 0.20$ and $\mu_k = 0.15$. Determine the friction force acting on the block if (a) $\theta = 5°$, (b) $\theta = 20°$.

VII.3   For the system in Example 1, determine the minimum value of the force $\bar{F}$ for which the block will not slide down the incline.

VII.4   The coefficients of friction between the 5-kg block and the inclined

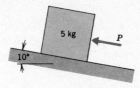

5 kg

P

10°

Prob. VII.4

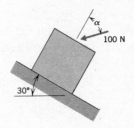

α

100 N

30°

Probs. VII.5 and VII.6

plane are $\mu_s = 0.40$, $\mu_k = 0.32$. Draw a graph showing how the magnitude of the friction force acting on the block depends on the magnitude of the force $\bar{P}$, which is always oriented up the plane.

VII.5   The coefficient of static friction between the block and the inclined plane is 0.60. A 100-N force is applied to the block at an angle α with respect to the normal to the incline, as shown. Knowing that this force moves the block up the plane if $\alpha > 45°$, determine the mass of the block.

VII.6   The mass of the block is 6 kg, and the 100-N force is applied at an angle $\alpha = 15°$. Knowing that the block is held in position by the force, determine (a) the friction force acting on the block, (b) the corresponding minimum allowable value of the coefficient of static friction.

2   Systems

The example and homework problems in the last section illustrate the fact that there are basically two classes of static friction problems. The first class of problems involves situations where the external loads are given and we must evaluate the friction force. To do this, we start by considering the case where sliding does not occur. The friction force is depicted in the free body diagram as being tangent to the contact plane in any sense we choose. Then, the equations of equilibrium yield the unknown friction force, including its actual sense. If it is found that the magnitude f of the friction force required for static equilibrium exceeds the maximum possible static friction force $\mu_s N$, it can be concluded that sliding will occur. This means that the friction force is $\mu_k N$. Furthermore, the kinetic friction force will be in the same direction as the friction force required for static equilibrium, because the static force is opposite the tendency of the motion.

The second class of static friction problems involves situations where motion is impending and we must evaluate some characteristic of the system or of the set of loads acting on the system. The friction force is then $\mu_s N$ (the maximum value for which sliding will not occur), acting opposite the direction of the impending relative movement. The equations of static equilibrium then yield the desired information.

The preceding discussion focuses on only one aspect of a system, specifically, the reaction force generated between two rough surfaces in contact. The system itself may be a single particle, a system of particles, a single rigid body, or a system of rigid bodies. The equations of static equilibrium to be employed are those that are appropriate to the system. Thus, the basic methods for studying static equilibrium developed in earlier modules will be required for a successful solution of a static friction problem.

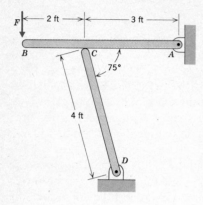

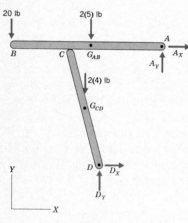

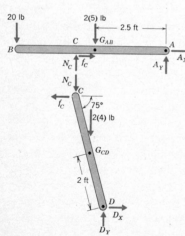

EXAMPLE 2

The system of bars supports a vertical force $\bar{F}$ at end B. Each bar weighs 2 lb/ft. Determine (a) the friction and normal forces acting on bar CD at end C, (b) the minimum value of the coefficient of static friction between the bars required for the system to be in static equilibrium.

Solution

The system consists of two multiforce members. We wish to examine the internal interaction force at point C. Following the method developed in Module IV for frames, we draw free body diagrams for the entire system and for each member, as shown in the left margin. The contact forces exerted at point C are depicted by noting that the contact plane is parallel to bar AB, so the normal component of the reaction is vertical and the friction force is horizontal. The weight forces are known from the given information.

In drawing the diagrams, we assumed a sense for the friction force acting on bar AB, and then depicted the friction force acting on bar CD consistent with Newton's third law. There is no need to concern ourselves with depicting the friction forces in their proper sense. If we are wrong in our guess, the equilibrium equations will indicate it by giving a negative value for f_C. (In this system it is possible to determine in advance the sense of the friction forces by considering the direction in which the bars would move relative to each other if there were no friction between them. We have not pursued this consideration here because it is difficult to apply for systems where the tendency to move is not obvious.)

Returning to the solution of the problem, we write only the equations for moment equilibrium of each bar about its fixed pin, because we are not interested in the reactions at these pins. Therefore, we have

$$\text{bar } AB: \quad \Sigma M_{AZ} = -N_C(3) + 10(2.5) + 20(5) = 0 \text{ lb-ft}$$

$$\text{bar } CD: \quad \Sigma M_{DZ} = 8(2 \cos 75°) + N_C(4 \cos 75°)$$

$$+ f_C(4 \sin 75°) = 0 \text{ lb-ft}$$

Solving the first equation for N_C and substituting the result into the second equation enables us to solve for f_C, with the result that

$$N_C = 41.67 \qquad f_C = -12.24 \text{ lb}$$

This is the solution to part (a). To solve part (b) we use the fact that the minimum allowable coefficient of static friction corresponds to the condition of impending motion: $|\bar{f}_C| = \mu_s N_C$. Thus, we have

$$(\mu_s)_{min} = \frac{|\bar{f}_C|}{N_C} = \frac{12.24}{41.67} = 0.293$$

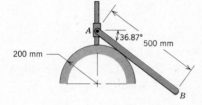

EXAMPLE 3

The coefficients of static friction between collar A and the vertical rod, and between bar AB and the semicylindrical surface, are 0.25 and 0.20, respectively. Determine if the bar is in static equilibrium in the position shown.

Sol

One approach to this problem is to try to determine the friction forces acting on the bar, assuming that the system is in static equilibrium, and to compare the ratio of the friction and normal force at each point to the corresponding given coefficient of static friction. The difficulty with this approach is that there are four unknown forces acting on the bar (the friction and normal forces between the bar and the semicylindrical surface, and between the vertical rod and collar A). Because there are only three equilibrium equations for the bar, the problem is statically indeterminate when formulated in this manner.

We require a different approach. Note that in the condition of impending sliding, the friction forces are known in terms of the normal forces. Let us therefore apply a force $\bar{P}$ of unknown magnitude that can bring the system to a condition of impending motion. We may determine the range of values of the magnitude of $\bar{P}$ for which the system remains in static equilibrium. Then, if $|\bar{P}| = 0$ is within this range, we know that the given system (without $\bar{P}$) is in equilibrium.

The only requirement in choosing the type of force $\bar{P}$ to apply is that it must have the possibility of causing the system to move. Thus, it cannot be normal to a contact plane, acting at the point of contact for that plane. The force $\bar{P}$ we will apply is a vertical force acting at collar A. If such a force is sufficiently large, collar A will certainly slide.

We can draw a general free body diagram by denoting the friction forces as f_A and f_C, and then explore the two types of impending motion by substituting either $f = \mu_s N$ or $f = -\mu_s N$, according to the direction in which the contact surfaces tend to move. This diagram is shown below, along with a separate diagram for geometric calculations.

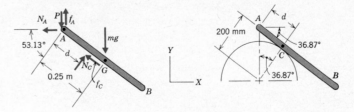

The distance d is found from the geometric sketch to be

$$d = 0.20 \tan 36.87° = 0.15 \text{ m}$$

Choosing point A for the moment sum, the equilibrium equations are

$$\Sigma M_{AZ} = N_C(0.15) - mg(0.25 \cos 36.87°) = 0$$

$$\Sigma F_X = N_C \sin 36.87° - f_C \cos 36.87° - N_A = 0$$

$$\Sigma F_Y = f_A - P - mg + N_C \cos 36.87° + f_C \sin 36.87° = 0$$

We shall first consider impending downward motion of collar A. In this case, point C on bar AB has a tendency to slide downward, tangent to the semicylinder. Considering the friction forces to oppose these tendencies, and using the given values of the coefficients of friction, we have

$$f_A = 0.25N_A \qquad f_C = 0.20N_C$$

Thus, the equilibrium equations become

$$\Sigma M_{AZ} = 0.15N_C - 0.20mg = 0$$

$$\Sigma F_X = 0.60N_C - 0.80(0.20N_C) - N_A = 0$$

$$\Sigma F_Y = (0.25N_A) - P - mg + 0.80N_C + 0.60(0.20N_C) = 0$$

The solutions of these equations are

$$N_C = 1.3333mg \qquad N_A = 0.5867mg \qquad P = 0.3733mg$$

Next we consider impending upward motion of collar A. In this case the friction forces are reversed from their sense in the preceding case of impending motion. Therefore,

$$f_A = -0.25N_A \qquad f_C = -0.20N_C$$

The general equilibrium equations then become

$$\Sigma M_{AZ} = 0.15N_C - 0.20mg = 0$$

$$\Sigma F_X = 0.60N_C - 0.80(-0.20N_C) - N_A = 0$$

$$\Sigma F_Y = (-0.25N_A) - P - mg + 0.80N_C + 0.60(-0.20N_C) = 0$$

The solutions are

$$N_C = 1.3333mg \qquad N_A = 1.0133mg \qquad P = -0.3467mg$$

From these results it can be deduced that sliding will not occur in the range $-0.3467mg < P < 0.3733mg$. The system of interest corresponds to $P = 0$, so we can conclude that the system is in static equilibrium.

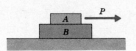

Prob. VII.7

HOMEWORK PROBLEMS

VII.7 Determine the force $\bar{P}$ required to cause motion in the system shown. At which surfaces will sliding occur? The coefficient of static friction between the 2-kg block A and the 8-kg block B is $\mu_s = 0.2$ and the coefficient of static friction between block B and the horizontal plane is $\mu_s = 0.1$.

VII.8 The coefficient of static friction between the person and the floor is 0.50, whereas the coefficient of static friction between the crate and the floor is 0.25. Determine the largest mass for the crate that the person can move by pulling on the cable, if the person's mass is 70 kg.

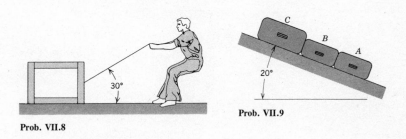

Prob. VII.9

Prob. VII.8

VII.9 Three suitcases A, B, and C are placed on a chute. Their weights are $w_A = w_B = 30$ lb, $w_C = 45$ lb. The coefficients of static friction between the suitcases and the chute are $\mu_A = \mu_B = 0.40$, $\mu_C = 0.20$. Determine which, if any, of the suitcases will slide down the chute. *Hint:* Successively consider the possibility of impending motion of suitcase A alone, then of both suitcases A and B, and finally of all three suitcases.

VII.10 Solve Problem VII.9 if the suitcases are arranged such that A is the lowest and B is the highest.

VII.11 The masses of blocks A and B are 5 and 10 kg, respectively. The coefficient of static friction between block A and its incline is 0.20. Determine the minimum coefficient of static friction between block B and its incline required to maintain equilibrium. If this coefficient is not sufficiently large, in which direction will the blocks slide?

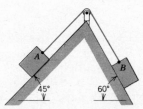

Prob. VII.11

VII.12 Solve Problem VII.11 if block A has a mass of 10 kg and block B has a mass of 5 kg.

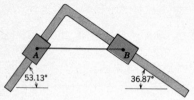

Prob. VII.13

VII.13 Two 3-kg collars are connected by a cord and are in equilibrium in the position shown. The coefficient of static friction between collar A and its inclined guide is 0.30. Determine the minimum allowable value of the coefficient of static friction between collar B and its guide.

VII.14 The masses of blocks A and C are 10 and 20 kg, respectively. These blocks are interconnected by cord ABC. Knowing that the system is in equilibrium, determine (a) the friction force acting on each block, (b) the minimum allowable coefficient of static friction between each block and its surface for static equilibrium.

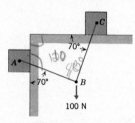

Prob. VII.14

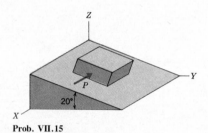

Prob. VII.15

VII.15 A 10-lb block resting on a plane inclined at 20° above the horizontal is subjected to a horizontal force $\bar{P}$ parallel to the plane, as shown. It is observed that the block will start to move if the magnitude of $\bar{P}$ exceeds 3 lb. Determine (a) the coefficient of static friction between the block and the inclined plane, (b) the direction in which the block starts to slide if the magnitude of $\bar{P}$ exceeds 3 lb.

Prob. VII.16

VII.16 Bar AB, whose mass is 100 kg, is supported by a roller resting on the horizontal surface at end A and by the rough inclined plane at end B. Knowing that the angle $\theta = 60°$ and that the bar is in static equilibrium, determine (a) the friction force acting on the bar at end B, (b) the minimum value of the coefficient of static friction between the bar and the incline for which static equilibrium is possible.

VII.17 For the bar in Problem VII.16, the coefficient of static friction between the inclined plane and end B is 0.15. Determine the maximum value of the angle θ for which equilibrium is possible.

VII.18 The uniform 60-kg beam rests against the smooth vertical wall and the rough horizontal wall ($\mu_s = 0.25$, $\mu_k = 0.21$). Knowing that the

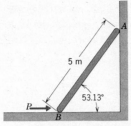

5 m

53.13°

P→

B

Probs. VII.18 and VII.19

horizontal force $\bar{P}$ is 100 N, acting to the right, determine the friction force acting on the lower end of the beam.

VII.19 The uniform 60-kg beam AB rests against the smooth vertical wall and the rough horizontal floor ($\mu_s = 0.25$). A horizontal force $\bar{P}$ is applied to end B. Determine the range of values for the magnitude of $\bar{P}$ such that the beam will not move from the position shown.

VII.20 Solve Problem VII.19 if the wall is also rough ($\mu_s = 0.15$).

VII.21 The 3-ft-diameter cylinder weighs 500 lb. The coefficient of friction between the cylinder and both surfaces is $\mu_s = 0.32$. Determine the maximum magnitude of the counterclockwise couple $\bar{M}$ for which sliding will not occur.

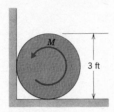

Prob. VII.21

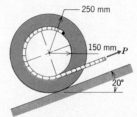

250 mm

150 mm

P

20°

Prob. VII.22

VII.22 A cable is wrapped around the 50-kg stepped drum and a tensile force $\bar{P}$ is applied to the free end of the cable, as shown. The drum is in static equilibrium. Determine (a) the magnitude of $\bar{P}$ and of the friction force acting on the drum, (b) the direction in which the point on the drum in contact with the inclined surface is tending to move, (c) the minimum value of the coefficient of static friction for which static equilibrium is possible.

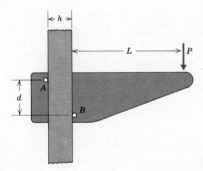

←h→

←—— L ——→ | P

d

A

B

Prob. VII.23

VII.23 The movable bracket transfers the downward load $\bar{P}$ to the vertical support arm. The coefficient of static friction between pins A and B and the vertical arm are both μ. Derive an expression for the minimum distance L for which the bracket will not slip regardless of the magnitude of $\bar{P}$.

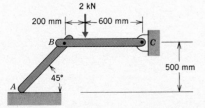

2 kN

200 mm ←|→ 600 mm →

B   C

500 mm

45°

A

Prob. VII.24

VII.24 A 2-kN load is applied to the linkage, as shown. Knowing that the system is in equilibrium, determine (a) the friction force acting on bar AB at end A, (b) the minimum value of the coefficient of static friction for which static equilibrium is possible. The mass of the bars is negligible.

VII.25 Solve Problem VII.24 for the case where the uniform bars have masses $AB = 100$ kg and $BC = 150$ kg.

VII.26 Determine the smallest horizontal distance d between the center of mass G of the 150-lb worker and the wall for which the 30-lb ladder will not slip. The coefficients of static friction between the ladder and the house, and the ladder and the ground are 0.25 and 0.50, respectively.

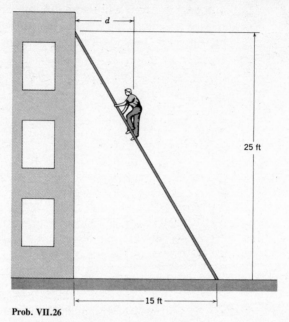

25 ft

15 ft

Prob. VII.26

75°   90°   15°

Prob. VII.27

VII.27 The semicylinder of mass m and radius R is in a condition of impending motion. The coefficients of static friction between the semi-cylinder and the inclined planes are the identical values μ. Determine μ.

VII.28 The coefficient of static friction between the 20-kg block C and the incline is 0.40. Knowing that the mass of bars AB and BC is negligible, determine the range of values of the magnitude of the horizontal force $\bar{P}$ for which the block will not move.

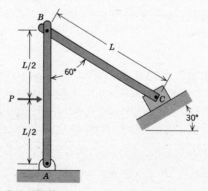

Prob. VII.28

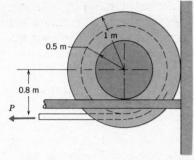

0.5 m

1 m

0.8 m

P

Prob. VII.29

VII.29 A spool of wire having a mass of 400 kg is supported by two horizontal rails (only one is shown in the side view) and rests against the vertical wall. The coefficient of static friction between all surfaces of contact is 0.40. Determine the smallest horizontal force $\bar{P}$ applied to the end of the wire which will cause the spool to rotate.

VII.30 The coefficient of static friction between the oil drum and all surfaces it contacts is 0.50. The drum has a mass of 500 kg. Determine the minimum force the hydraulic cylinder must exert to move the oil drum up the incline. At which surface is slip impending for this force? *Hint:* There is no contact between the drum and the horizontal surface when the drum begins to move up the hill. Assume slipping at one contact surface and check if the solution satisfies this assumption.

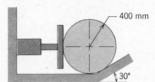

400 mm

30°

Prob. VII.30

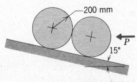

200 mm

P

15°

Prob. VII.31

VII.31 Two identical 30-kg cylinders are held on the ramp by the horizontal force $\bar{P}$. The coefficient of friction between all contacting surfaces is 0.25. Determine the maximum force $\bar{P}$ that can be applied without causing the cylinders to move. *Hint:* Assume that the cylinders tend to roll over the inclined plane without slipping, and then verify this assumption.

VII.32 The coefficient of friction between the uniform 30-kg bar and both surfaces is $\mu_s = 0.20$. Is the system in static equilibrium?

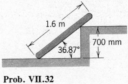

1.6 m

700 mm

36.87°

Prob. VII.32

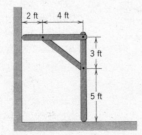

2 ft   4 ft

3 ft

5 ft

Prob. VII.33

VII.33 The coefficients of static friction between the frame and the wall, and the frame and the floor are 0.20 and 0.05, respectively. Each member of the frame weighs 10 lb/ft. Is the frame in static equilibrium?

VII.34 A hydraulic cylinder applies the vertical force $\bar{P}$ to the lifting frame that is supporting a 1000-kg homogeneous crate. The crate is

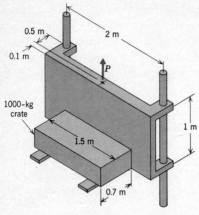

Prob. VII.34

situated symmetrically on the frame and the weight of the frame is negligible. Knowing that the coefficient of static friction at each contact surface between the frame and the vertical guide rails is 0.10, determine the minimum force $\bar{P}$ required to move the frame upward from rest.

VII.35 In Problem VII.34, determine the magnitude and sense of the minimum force $\bar{P}$ required to move the frame downward from rest.

VII.36 The 200 lb-in. couple is applied to rod AB in order to resist the 70-lb force applied to piston C. Knowing that the linkage is in static equilibrium at the position where $\theta = 90°$, determine (a) the friction force acting on the piston, (b) the minimum coefficient of static friction between the piston and its cylinder for which this situation is possible.

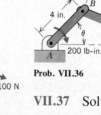

Prob. VII.36

VII.37 Solve Problem VII.36 for the position where $\theta = 135°$.

VII.38 Collar B may slide over rod CD. The coefficient of static friction between these bodies is 0.16. Determine the range of values of the couple $\bar{M}_A$ for which the linkage is in static equilibrium under the 100-N load at end D.

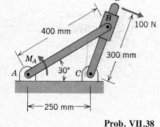

Prob. VII.38

VII.39 A couple $\bar{M}_C$ is applied to the brake drum C. Determine the smallest force exerted by the hydraulic cylinder on brake arm AB for which the brake drum will rotate if $\bar{M}_C$ is 300 N-m (a) clockwise, (b) counterclockwise. The coefficient of static friction between the drum and the arm is 0.75.

VII.40 The coefficients of static and kinetic friction between the brake drum C and the brake arm AB are 0.50 and 0.40, respectively. Knowing that the hydraulic cylinder pushes end B to the left with a force of 1.0 kN, determine the friction force acting on the drum if the couple $\bar{M}_C$ is 300 N-m (a) clockwise, (b) counterclockwise.

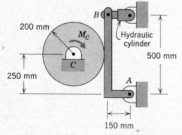

Probs. VII.39 and VII.40

VII.41 Rod CD, which weighs 4 lb, is used to prop up board AB, which weighs 16 lb. Determine the minimum values of the coefficients of static friction between each pair of surfaces in contact for which equilibrium can be maintained.

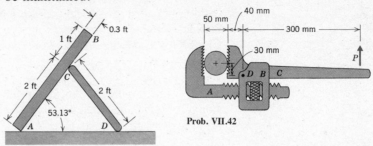

Prob. VII.41

Prob. VII.42

VII.42   A 50-mm-diameter rod is being gripped by the Stilson wrench shown. Members *A* and *B* may be regarded as a single rigid body connected to member *C* only by pin *D*. It is observed that, regardless of the magnitude of the force $\bar{P}$, the wrench does not slip over the rod (the wrench is said to be *self-locking*). Determine the minimum coefficients of static friction at both points of contact between the wrench and the pipe.

VII.43   The thin hemispherical shell of mass *m* and radius *R* is acted upon by a horizontal force $\bar{P}$ applied at its rim. Determine the angle ϕ for impending motion if the coefficient of static friction between the shell and the surface is 0.18.

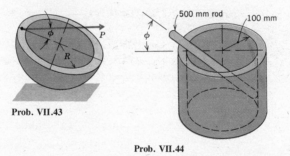

Prob. VII.43

Prob. VII.44

VII.44   A uniform rod 500 mm long is placed against the rim of a cylinder as shown. Determine the range of values of the angle of elevation ϕ of the rod for which the rod is in static equilibrium. The coefficient of static friction between the cylinder and the rod is 0.25.

VII.45   A 5-ft-long bar is connected to the floor by a ball-and-socket joint that is 4 ft from the vertical wall supporting the other end of the bar. The bar weighs 20 lb and the coefficient of static friction between the bar and the wall is 0.35. Determine the maximum value of the angle θ for which static equilibrium is possible. *Hint:* Because of the support at end *A*, end *B* tends to move tangent to the dashed circle.

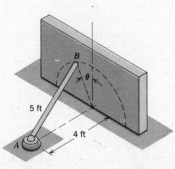

Prob. VII.45

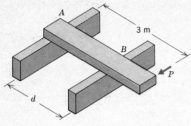

Prob. VII.46

VII.46 A 30-kg plank is resting on two horizontal joists, perpendicular to the axis of the joists in the horizontal plane. The coefficient of static friction between the plank and the joists is 0.40. Determine the force $\bar{P}$, parallel to the joists, required to move the plank when $d = 2$ m. *Hint:* The friction forces acting on the plank may be considered to be parallel to the joists.

VII.47 In Problem VII.46, determine the range of values of the dimension d for which a sufficiently large force $\bar{P}$ would cause the plank to slide (a) at joist A, (b) at joist B.

3 Sliding or Tipping

The accidental overturning of a glass of water that was resting on a table is a classic illustration of the fact that trying to overcome the force of friction can have surprising consequences. A comparable problem arises when a filing cabinet is pushed across a floor, as depicted in Figure 7a.

Figure 7a   Figure 7b   Figure 7c

In Figure 7b we see the cabinet slide along the floor, as we would expect it to, whereas in Figure 7c we see the unexpected result in which the cabinet tips over. To determine which situation will occur, let us draw a free body diagram of the filing cabinet, as shown in Figure 8, and evaluate the forces. Note that $\bar{P}$ is the force being applied to move the cabinet.

The force $\bar{R}$ is the resultant of the normal and frictional forces exerted by the floor, and the distance q locates the point of application of this force. Physically, because the cabinet is standing on edge in the tipping condition of Figure 7c, it must be that the reaction $\bar{R}$ acts at $q = b$ for tipping.

Note that because we are limited to the equations of static equilibrium, we cannot determine whether or not the cabinet will actually fall over on its side. This is so because, as soon as the bottom of the cabinet rotates away from the floor, the cabinet is in motion, requiring the application of the laws of dynamics. We can, however, determine if the force $\bar{P}$ will cause tipping to begin.

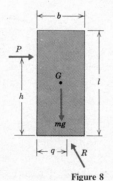

Figure 8

To determine if the cabinet will tip, we use the equations of static equilibrium to find the distance q at which the resultant force must act for static equilibrium. If $q < b$, we can conclude that the bottom of the cabinet remains in contact with the floor. In contrast, if $q > b$, we have a geometrically impossible situation, for the resultant force required for equilibrium cannot be applied to the cabinet. We then would conclude that tipping will occur. Clearly, $q = b$ is the transitional condition for *impending tipping*.

It is best to be consistent in the approach to problems involving the possibility of both slipping and tipping. In situations where we are uncertain about which, if either, of these conditions will occur, we shall begin by assuming that slipping will occur. Then, if the normal force $\bar{N}$ required for equilibrium is physically possible (for example, if $q < b$ for the filing cabinet in Figure 8) we know that the impending motion is one in which there is no tipping. On the other hand, if we determine that the normal force $\bar{N}$ corresponds to an impossible situation, then we can conclude that tipping will occur. This is the case when $q > b$ for the filing cabinet. The tipping condition may then be investigated by placing the normal force at the point of contact when tipping occurs, for instance, the lower right-hand corner of the crate in Figure 8. Note that in the tipping condition, the point of contact will generally not be in a condition where slip is also impending. Therefore, the friction force will be an unknown; that is, in general, $|\bar{f}| < \mu_s|\bar{N}|$.

These observations are illustrated in the following example.

EXAMPLE 4

The 200-kg crate, whose center of mass is point G, is at rest on the hill. The crate is supported by small skids at points A and B. The coefficient of static friction between the incline and the skids is 0.25. Determine the minimum horizontal force $\bar{P}$ that will move the crate.

Solution

We will start by assuming that if the magnitude of $\bar{P}$ is sufficiently large, the crate will slide uphill without tipping. Thus, the friction force at each skid will be considered to be μ_s times the normal force at that skid, acting downhill, as shown in the free body diagram below.

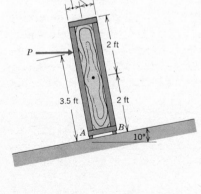

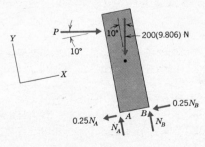

Because we are concerned with the points of application of forces, we write the equations for force and moment equilibrium of the crate. Choosing point A for the moment sum, we have

$$\Sigma M_{AZ} = N_B(1.0) - (P \cos 10°)(3.5) - 200(9.806) \cos 10° (0.5)$$

$$+ 200(9.806) \sin 10° (2.0) = 0 \text{ N-m}$$

$$\Sigma F_X = P \cos 10° - 200(9.806) \sin 10° - 0.25N_A$$

$$- 0.25N_B = 0 \text{ newtons}$$

$$\Sigma F_Y = N_A + N_B - P \sin 10° - 200(9.806) \cos 10° = 0 \text{ newtons}$$

We now have three equations for the three unknowns N_A, N_B, and P. Their solutions are

$$P = 875 \qquad N_B = 3299 \qquad N_A = -1216 \text{ newtons}$$

These are the magnitudes that the unknown forces must have if the crate is to be in a condition of impending slipping without tipping. Notice that the value of N_A is negative. Clearly, this is physically impossible, for it requires that the incline pull downward on corner A. We may therefore conclude that applying the force $\bar{P}$ in the indicated manner will cause the crate to tip before it slides.

To determine the force $\bar{P}$ required to produce the condition of impending tipping, we set N_A to zero and draw a free body diagram of the crate in its given equilibrium position. Recall that in this condition, the friction force at the point of contact is unknown. Thus, the free body diagram for this portion of the investigation is as shown.

Choosing point B for the moment sum, the equilibrium equations for this case are

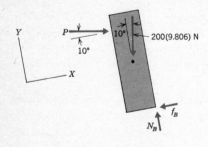

$$\Sigma M_{BZ} = -(P \cos 10°)(3.5) + (P \sin 10°)(1.0) + 200(9.806) \cos 10° (0.5)$$

$$+ 200(9.806) \sin 10° (2.0) = 0 \text{ N-m}$$

$$\Sigma F_X = P \cos 10° - 200(9.806) \sin 10° - f_B = 0 \text{ newtons}$$

$$\Sigma F_Y = -P \sin 10° - 200(9.806) \cos 10° + N_B = 0 \text{ newtons}$$

These equations yield

$$P = 503 \text{ newtons}$$

$$f_B = 155 \qquad N_B = 2019 \text{ newtons}$$

Notice that the value of P we have obtained is smaller than that found earlier by considering the slipping condition. This confirms our earlier conclusion that the crate would tip before it slides. Also, note that the friction force f_B given above should be smaller than $\mu_s N_B$, for if it were not, we would have a contradiction with the results of the study of the condition of impending slip. It is readily verified that this requirement is fulfilled.

An interesting aside to this problem is its similarity to the problem posed in part (c) of Example 1. It can be seen that concern about the possibility of impending tipping, which always requires that we consider moment equilibrium, complicates the solution of problems involving impending motion.

HOMEWORK PROBLEMS

VII.48 The coefficient of static friction between the 1-ton granite block and the floor is 0.35. Knowing that $h = 2.5$ ft, determine the smallest horizontal force $\bar{P}$ that will cause the block to move.

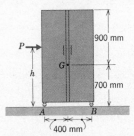

2.5 ft

P

4 ft

h

Prob. VII.48

VII.49 Determine the range of values of h in Problem VII.48 for which slipping will occur before tipping.

VII.50 In Example 4, let h denote the distance above the bottom of the crate at which the horizontal force $\bar{P}$ is applied. Determine the maximum value of h for which the crate can be pushed up the hill without tipping.

VII.51 A 60-kg refrigerator, having center of mass G, is mounted on four casters. The coefficient of static friction between a locked caster and the floor is 0.50, and the frictional resistance of a rolling caster is negligible. Determine the magnitude of the horizontal force $\bar{P}$ required to move the refrigerator when $h = 1.0$ m, (a) if all casters are locked, (b) if only casters B are locked, (c) only casters A are locked.

P

G

h

900 mm

700 mm

A B

400 mm

Prob. VII.51

VII.52 In Problem VII.51, determine the range of values of h for which the refrigerator will not tip (a) if all casters are locked, (b) if only casters B are locked. (c) Explain why it is impossible to tip the refrigerator over with the force $\bar{P}$ when only casters A are locked.

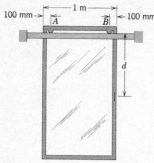

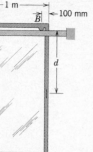

Prob. VII.53

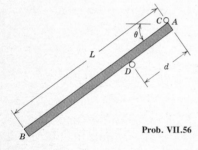

Prob. VII.56

VII.53   The 25-kg door is mounted on the horizontal rail by means of runners A and B. The coefficients of friction for these runners are $\mu_A = \mu_B = 0.15$. The door handle is pulled to the right to open the door. Determine (a) the maximum distance d for the door handle for which the door will not tip when it is opened, (b) the force required to open the door if d equals the value found in part (a), (c) the force required to open the door if d equals one-half the value found in part (a).

VII.54   Solve Problem VII.53 if $\mu_A = 0.20$, $\mu_B = 0.10$.

VII.55   Solve Problem VII.53 for the case where the door is being closed (moved to the left) if $\mu_A = 0.20$, $\mu_B = 0.10$.

VII.56   Bar AB of mass m and length L is supported in the vertical plane by pins C and D, which are separated by the distance d. The coefficient of static friction between the pins and the bar is μ_s. Derive an expression for the range of values of L/d for which static equilibrium is possible.

VII.57   The angle β of the incline is very gradually increased until the block of mass m moves. Develop formulas that indicate whether the block slides or tips in terms of the coefficient of friction μ, the ratio b/h, and the angle β at which movement occurs.

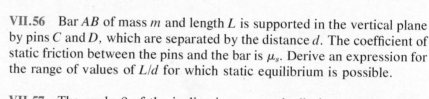

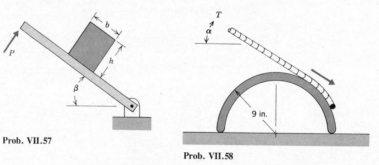

Prob. VII.57

Prob. VII.58

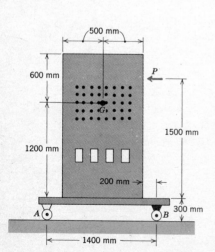

Prob. VII.59

VII.58   The 8-lb thin semicylindrical shell is to be towed to the left. The coefficient of static friction between the shell and surface is 0.35. Determine the largest angle α at which the shell may be towed without causing it to tip. What is the corresponding cable tension?

VII.59   A 500-kg electronic computer, having center of mass G, is placed on the 20-kg dolly. The casters A of the dolly are locked. The coefficients of friction between the computer and the dolly, and between the casters and the floor, are 0.50 and 0.30, respectively. Determine the maximum force $\bar{P}$ that can be applied without causing the computer to move.

B.   BASIC MACHINES HAVING FRICTION

There are a multitude of simple mechanical devices, which we refer to as basic machines, that can be used to either move or prevent the motion of physical objects. These devices can magnify or convert the input forces applied to them into a different set of output forces that are applied to other systems. For example, a wheel fixed to an axle has the capability of transforming a torque exerted by its axle into a forward force that can propel a vehicle.

In earlier modules, particularly Module IV, we developed the basic techniques for investigating the equilibrium of machines. The common feature of the machines we shall consider here is the important, and sometimes useful, effect of friction.

1   Wedges

A wedge is merely a block having two flat surfaces that form a small angle relative to each other. Consider the triangular wedge in Figure 9a, which is being pushed into a crack in a large body by the force $\bar{P}$.

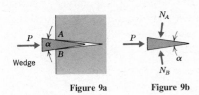

Figure 9a   Figure 9b

In the situaton where the surfaces of contact between the wedge and the large body are smooth, the free body diagram of the wedge is as shown in Figure 9b. It is not difficult to see that, if the angle α is sufficiently small, the normal forces $\bar{N}_A$ and $\bar{N}_B$ will be much larger than the applied force $\bar{P}$. Thus, this wedge could be used to enlarge the crack. An example of this type of action is the head of an axe.

However, Figure 9b is not a correct representation of the action of the wedge, because friction cannot be neglected. To study the effect of friction we must distinguish between the situation where the tendency is for the wedge to be forced further into the crack, as shown in the free body diagram in Figure 10a, and the case where the tendency is for the wedge to be squeezed out of the crack, as shown in the free body diagram in Figure 10b.

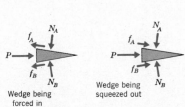

Wedge being   Wedge being
forced in   squeezed out

Figure 10a   Figure 10b

Considering $\bar{P}$ to be known, we see that there are four unknown forces acting on the wedge in each case illustrated in Figure 10. Because we are not concerned here with the points of application of the forces acting on the wedge, there are only two scalar equations of force equilibrium. We can conclude then that the free body diagrams in Figure 10 represent statically indeterminate systems having two redundant forces.

Fortunately, our interest in wedges lies in the study of impending motion. One problem we wish to investigate is how the wedge can be forced further into the crack. This is the case of impending inward motion. It may be studied by using the free body diagram of Figure 10a with $f = \mu_s N$ for each friction force. For such a problem the friction forces are no longer independent unknowns, so the system becomes a statically determinate one.

Another problem we wish to investigate is how the wedge can be pulled out of the crack. In this case we have impending outward motion, so we use the free body diagram of Figure 10b, setting $f = \mu_s N$ for each friction force. Once again, the friction forces are no longer independent unknowns, so the system is a statically determinate one.

The preceding discussion enables us to observe that the existence of friction has both good and bad effects in wedges. The bad effect arises because the friction force for impending inward motion, Figure 10a, opposes the input force $\bar{P}$. Therefore, the existence of friction means that a larger force $\bar{P}$ is needed to drive the wedge in. The beneficial effect of friction is associated with removing the wedge from the crack. If the coefficients of friction are very small, impending outward motion of the wedge, Figure 10b, still corresponds to an inward force $\bar{P}$. In other words, it will be necessary to hold the wedge in position. In contrast, if the coefficients of static friction are sufficiently large, for impending outward motion a force $\bar{P}$ in the opposite sense of the one depicted in the free body diagram (that is, a pulling force) will be required to remove the wedge. In such cases the wedge is said to be *self-locking*.

EXAMPLE 5

Wedges A and B are used to raise block C against the 5-kN load that is acting on it. The coefficient of friction between wedge B and all surfaces in contact with it is 0.10. Cleat D holds wedge A in position. (a) Derive an expression for the minimum force $\bar{P}$ required to raise block C. (b) Determine whether the wedge is self-locking.

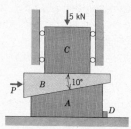

Solution

The first consideration here is which free body diagrams to draw. In view of the presence of cleat D, the wedge A is in effect part of the ground, so there is no reason to consider this wedge. On the other hand, wedge B and block C are individual bodies whose equilibrium must be assured. Thus, we will draw free body diagrams of these bodies. In part (a) we are interested in impending motion of wedge B to the right, whereas, to investigate self-locking we want to consider impending motion of wedge B to the left. In order to have one set of free body diagrams for both cases, we denote

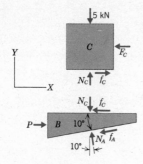

the friction forces on this wedge as $\bar{f}_C$ and $\bar{f}_A$ and then set $f = \pm\mu_s N$, choosing the sign to fit the type of impending motion. For the purposes of drawing free body diagrams, $\bar{f}_C$ and $\bar{f}_A$ have been depicted in the diagrams as though block B where tending to move to the right. Note that the forces exerted by the wedge on block C must satisfy Newton's third law.

The equilibrium equations for these bodies are

$$\text{wedge } B: \quad \Sigma F_X = P - N_A \sin 10° - f_C - f_A \cos 10° = 0$$

$$\Sigma F_Y = N_A \cos 10° - f_A \sin 10° - N_C = 0$$

$$\text{block } C: \quad \Sigma F_Y = N_C - 5 = 0 \text{ kN}$$

We are not interested in $\bar{F}_C$, so we have disregarded ΣF_X for block C. First, to solve part (a) we set

$$f_C = + 0.10 N_C \qquad f_A = + 0.10 N_A$$

Substituting these expressions, as well as $N_C = 5$ kN found from the last equilibrium equation, into the first two equilibrium equations, we obtain

$$P - 0.1736 N_A - (0.50) - 0.9848(0.10 N_A) = 0$$

$$0.9848 N_A - 0.1736(0.10 N_A) - 5 = 0$$

Solution of these equations yields

$$N_A = 5.168 \qquad P = 1.961 \text{ kN} \qquad \qquad \triangleleft$$

We now turn our attention to part (b). In order to evaluate whether the wedge is self-locking, we will determine if the force $\bar{P}$ should be to the right or left when motion is impending to the left. In this case, we have

$$f_A = -\mu_s N_A \qquad f_C = -\mu_s N_C$$

The equilibrium equation for block C still gives $N_C = 5$ kN, so the equilibrium equations for the wedge now become

$$P - 0.1736 N_A - (-0.50) - 0.9848(-0.10 N_A) = 0$$

$$0.9848 N_A - 0.1736(-0.10 N_A) - 5 = 0$$

The corresponding solutions are

$$N_A = 4.989 \qquad P = -0.125 \text{ kN}$$

The fact that P is negative means that the force $\bar{P}$ must pull on the wedge to the left to move the wedge in that direction. Thus, the wedge is self-locking. $\qquad \qquad \triangleleft$

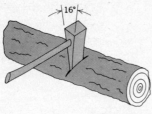

16°

Prob. VII.60

HOMEWORK PROBLEMS

VII.60   The faces of the blade of the axe form a 16° angle. Determine the minimum coefficient of static friction between the blade and the log for which the axe is self-locking.

VII.61 and VII.62   Determine the downward force $\bar{P}$, applied to the 20-kg block, required to move the 500-kg block. The coefficient of static frction between all surfaces is 0.15.

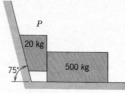

Prob. VII.61

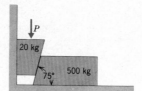

Prob. VII.62

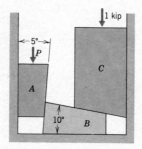

Prob. VII.63

VII.63   Determine the downward force $\bar{P}$, applied to wedge A, required to raise block C against the 1-kip load. The coefficient of static friction between all surfaces is 0.45.

VII.64   Two 100-lb flat plates resting on horizontal surfaces support a lightweight wedge, as shown. The coefficients of static friction between the plates and the wedge, and between the plates and the horizontal surfaces, are 0.40 and 0.30, respectively. Determine (a) the maximum overhang d for which the wedge will cause the plates to slide without tipping, (b) the vertical force $\bar{F}$ that should be applied to the wedge in order to move the plates when d is smaller than the value obtained in part (a).

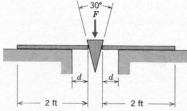

Prob. VII.64

VII.65   A 2000-kg lathe having center of mass G is to be leveled by means of a wedge at corner B. The coefficient of static friction between the wedge and the surfaces it contacts is 0.10. Determine (a) the minimum

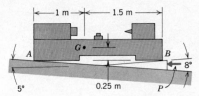

Prob. VII.65

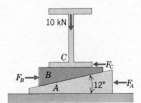

Prob. VII.66

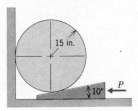

Prob. VII.68

horizontal force $\bar{P}$ required to raise corner B of the lathe, (b) the minimum coefficient of static friction between corner A of the lathe and the floor for which the lathe will not slide when the force $\bar{P}$ in part (a) is applied to the wedge. (c) Is the system self-locking? The lathe may be considered to be in the horizontal position.

VII.66 The H-beam is to be adjusted into its final position by applying the force $\bar{F}_B$ to wedge B. The reaction on the end of the beam is 10 kN. The coefficient of static friction between all surfaces is 0.20. (a) What is the minimum force $\bar{F}_B$ required to raise the beam? (b) What are the corresponding forces $\bar{F}_C$ and $\bar{F}_A$ required to prevent the beam and lower wedge from shifting?

VII.67 Solve Problem VII.66 in the case where the sense of $\bar{F}_B$ is reversed in order to lower the beam.

VII.68 Determine the minimum force $\bar{P}$ required to lift the 500-lb drum shown. The mass of the wedge is negligible and the coefficient of friction between all surfaces is 0.25. Which surfaces are on the verge of sliding for this force $\bar{P}$?

VII.69 A vertical force $\bar{P}$ is to be applied to wedge E in order to push end C of bar CD to the right. The coefficient of static friction between all surfaces is 0.15. Determine the minimum force $\bar{P}$ that will achieve this goal.

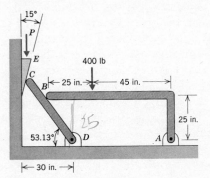

Prob. VII.69

VII.70 Determine whether the system in Problem VII.69 is self-locking. That is, will the wedge remain in place if the force $\bar{P}$ is no longer applied to the wedge?

2   Square Threaded Screws

A screw is a basic machine for transforming a couple applied about an axis into an axial force that can be used to move objects or to fasten objects together. Familiar examples of screw-type machines are an automobile bumper jack (for moving objects) and a bolt and nut set (for fastening objects).

There are several common shapes for the threads of a screw. We will limit our efforts here to the analysis of square threads, because the analysis of such threads is quite similar to that of wedges. A blow-up of a square-threaded screw carrying a couple $\bar{M}$ that would advance the screw in opposition to a load $\bar{F}$ is shown in Figure 11.

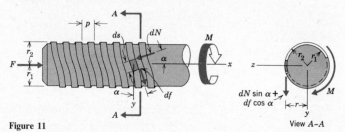

Figure 11   View A–A

In order for the screw to sustain the axial force $\bar{F}$ and the couple $\bar{M}$, there are reaction forces distributed along the screw threads resulting from their contact with the body to which the screw is fastened. Let us consider an infinitesimal portion of a screw thread of length ds, such as the one shown in Figure 11.

A screw thread is helical, so it can be thought of as being developed by wrapping an inclined plane around a cylinder. The inclined plane analogy indicates that we can consider the forces acting on the infinitesimal element of thread ds to be a normal force $d\bar{N}$ and a frictional force $d\bar{f}$. Note that if the screw is tending to advance (move in the negative x direction) against the force $\bar{F}$, then $d\bar{f}$ is in the sense shown in Figure 11, whereas if it is tending to retract, then $d\bar{f}$ has the opposite sense from that shown. The advancing movement corresponds to tightening a screw, and the retracting movement corresponds to loosening it.

The angle α describing the orientation of the normal and frictional forces is the *lead angle* of the helix, and is a constant. As we will soon see, this angle is related to the *pitch p* of the screw thread, which is shown in Figure 11 to be the axial distance between matching points on two adjacent screw threads.

Considering the screw threads to be rigid and to mesh perfectly with the threads of the body to which the screw is fastened, it follows that dN and df are the same for all elements ds.

Let us now write the equilibrium equations for the screw. We require the equations for the force sum in the x direction and the moment sum about the x axis; the polar symmetry of the many elements ds about the x axis automatically satisfies the other equilibrium equations. From the side view in Figure 11 we see that the forces acting on the element ds have a resultant axial component dF_x, given by

$$dF_x = df \sin \alpha - dN \cos \alpha \tag{7}$$

Further, the sectional view in Figure 11 shows that the y component of these forces, which is $df \cos \alpha + dN \sin \alpha$, exerts a positive moment dM_z about the z axis, given by

$$dM_z = (df \cos \alpha + dN \sin \alpha)r \tag{8}$$

where r is the average of the inner and outer radii for the screw thread, that is,

$$r = \tfrac{1}{2}(r_1 + r_2) \tag{9}$$

The sum of an infinite number of infinitesimal quantities is an integral, so the basic equilibrium equations for the screw are

$$\Sigma F_x = F + \int dF_x = 0 \tag{10}$$

$$\Sigma M_z = -M + \int dM_z = 0 \tag{11}$$

Substituting the expressions for dF_x and dM_z from equations (7) and (8) into equations (10) and (11) and noting that the angle α is constant, the equilibrium equations become

$$\Sigma F_X = F + \int(df \sin \alpha - dN \cos \alpha)$$

$$\equiv F + f \sin \alpha - N \cos \alpha = 0 \tag{12}$$

$$\Sigma M_Z = -M + r \int(df \cos \alpha + dN \sin \alpha)$$

$$\equiv -M + r(f \cos \alpha + N \sin \alpha) = 0 \tag{13}$$

When using a screw we are usually interested in the relationship between the force $\bar{F}$ and the couple $\bar{M}$. Considering F as a known quantity, equations (12) and (13) are only two equations for the three unknowns, N, f, and M. Thus, as it stands, the system is statically indeterminate.

Our interest, however, lies solely in cases of impending motion. Basically, we would like to know the couple required to advance (tighten) a screw and the couple required to retract (loosen) a screw. Noting the sense in which the friction force was assumed to act in Figure 11, we have

$$f = +\mu_s N: \quad \text{advancing} \tag{14a}$$

$$f = -\mu_s N: \quad \text{retracting} \tag{14b}$$

Either of equations (14) provides the additional relationship that will enable us to determine M as a function of F.

Before we present the solution, let us see how we can determine the lead angle α for a particular type of screw. Clearly, this is a key parameter in the equations. To do this, recall that a helix is formed by wrapping a triangle around a cylinder. If we unwrap a portion of the helix equivalent to one circumference ($2\pi r$), we obtain the triangle shown in Figure 12.

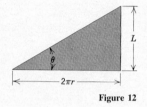

L

θ

2πr

Figure 12

The distance L is called the *lead*. It is the distance the screw will advance (or retract) when given one full turn. In most common screws there is only one thread, which is continuously wrapped around the axis of the screw. Such a screw is said to be *single threaded,* and the lead L equals the pitch p. Other screws have two or three parallel sets of threads. Such cases are termed *double-* or *triple-threaded screws,* respectively. By induction we can state the for an *n-threaded screw,* we have

$$L = np$$

It is then apparent from Figure 12 that

$$\sin \alpha = \frac{np}{[(np)^2 + (2nr)^2]^{1/2}} \qquad \cos \alpha = \frac{2\pi r}{[(np)^2 + (2\pi r)^2]^{1/2}} \qquad (15)$$

When we substitute the preceding expressions and either of the friction relations of equations (14) into the equilibrium equations, we can eliminate N to find

$$\text{advance:} \qquad M = Fr\,\frac{np + 2\pi\mu_s r}{2\pi r - \mu_s np} \qquad (16a)$$

$$\text{retract:} \qquad M = Fr\,\frac{np - 2\pi\mu_s r}{2\pi r + \mu_s np} \qquad (16b)$$

Equations (16) describe the situation where the screw is at rest and we wish to advance or retract it further. In the situation where the screw is rotating slowly, the threads are sliding over each other. The force analysis in this case is essentially unchanged (assuming that inertial effects from the rotation are negligible). Thus, equations (16) may be modified to describe a rotating screw by replacing μ_s by μ_k, the coefficient of kinetic friction.

This analysis demonstrates that a square screw thread functions like a wedge wrapped around a cylinder. Characteristic of the similarities between a screw and a wedge is the fact that both systems are statically indeterminate when motion is not impending. A further similarity is the possibility that a screw, like a wedge, may be self-locking. A self-locking screw will retract only if the couple $\bar{M}$ is in the sense of the rotation required to retract the screw. This property is exhibited by the value of M in equation (16b) being negative. Thus,

$$\mu_s > \frac{np}{2\pi r} \equiv \tan \alpha \text{ for self-locking} \tag{17}$$

This result has a simple interpretation. Consider once again the element of thread ds in Figure 11. Now we wish to consider the case where the screw is about to retract. This situation is depicted in Figure 13. The

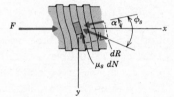

Figure 13

friction and normal forces on the thread element are equivalent to the resultant $d\bar{R}$, which for the condition depicted, is a force at an angle ϕ_s measured from the normal to the thread. Angle ϕ_s is the angle of static friction; that is, $\phi_s = \tan^{-1} \mu_s$. If ϕ_s is larger than α, we see that the y component of $d\bar{R}$ is negative. This means that $d\bar{R}$ will oppose a rotation that would tend to retract the screw. Now, noting the relationship between ϕ_s and μ_s, it can be deduced that equation (17) is equivalent to the requirement $\phi_s > \alpha$.

EXAMPLE 6

A square-threaded single-pitch bolt whose thread has a mean diameter of 12 mm and a pitch of 4 mm is screwed into a fixed metal block. The coefficients of static and kinetic friction between the block and the bolt are 0.35 and 0.27, respectively. Friction between the head of the bolt and the block is negligible. Determine the axial force developed in the bolt if it is tightened with a socket wrench that applies a 10 N-m-couple in order to (a) turn the bolt slowly, (b) start the bolt rotating from rest.

Solution

The primary difference between the two given cases is that in case (a) the bolt is actually turning as it advances, so there is kinetic friction. This requires that we replace μ_s by $\mu_k = 0.27$ in equation (16a). In contrast, the advancing movement is impending in case (b), so we will apply equation (16a) directly, with $\mu_s = 0.35$. For both cases, the given information tells us that

$$r = \tfrac{1}{2}(12 \text{ mm}) = 0.006 \text{ m} \qquad p = 0.004 \text{ m} \qquad n = 1$$

Solving part (a) first, for a 10 N-m couple equation (16a) yields

$$10 = F(0.006)\frac{(1)(0.004) + 2\pi(0.27)(0.006)}{2\pi(0.006) - (0.27)(1)(0.004)}$$

$$F = 4304 \text{ N}$$

Similarly, for part (b) we have

$$10 = F(0.006)\frac{(1)(0.004) + 2\pi(0.35)(0.006)}{2\pi(0.006) - (0.35)(1)(0.004)}$$

$$F = 3518 \text{ N}$$

The difference between these results has an important, subtle consequence in the general problem of tightening screws to a predetermined clamping force by means of a torque wrench. (A torque wrench has a dial that indicates the moment being applied to the screw.) Clearly, it is easy to see that a screw is rotating, and almost impossible to sense by hand that the screw is on the verge of rotating. Therefore, torque specifications corresponding to a desired axial force in a screw usually require that the screw be turned with the wrench until the wrench indicates the specified torque. After achieving this adjustment it would require a considerably higher torque to start the screw rotating again to tighten it further.

HOMEWORK PROBLEMS

VII.71 The C clamp holds two pieces of wood together. The clamp has a single square thread with a pitch of 3 mm and a mean diameter of 18 mm. The coefficient of static friction is 0.20. Determine the torque required to tighten the clamp further when it is at rest, applying a clamping force of 800 N.

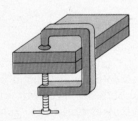
Prob. VII.71

VII.72 In Problem VII.71, determine the torque required to tighten the clamp further if it is triple threaded.

VII.73 In Problem VII.71, determine the torque required to loosen the clamp.

VII.74 The metal plate A shown in the diagram is fastened to block B by the double-threaded screw, whose square threads have a pitch of $\frac{1}{16}$ in. The mean radius of the threads is $\frac{1}{4}$ in. The coefficients of static and kinetic friction for the threads are 0.30 and 0.25, respectively, and friction between plate A and the head of the screw is negligible. The screw is tightened by means of a torque wrench. Determine the clamping force on plate A if the torque wrench reads 40 ft-lb (a) when the screw is about to turn, (b) when the screw is turning.

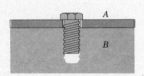
Prob. VII.74

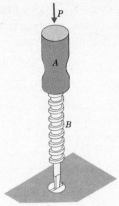

Prob. VII.75

VII.75   The automatic screwdriver shown operates by pushing the handle A downward with a force $\bar{P}$ to turn the threaded shaft B, which holds the bit. The square threads have a pitch of 90 mm, and the mean radius of the thread is 15 mm. The coefficients of static and kinetic friction for the threads are 0.15 and 0.12, respectively. If shaft B is single threaded, determine the force $\bar{P}$ required to tighten the wood screw to 2 N-m in one slow continuous motion from the position where the screw is first started in the hole.

VII.76   Solve Problem VII.75 if the shaft of the screwdriver is double threaded and all other properties are unchanged.

VII.77   The turnbuckle is used to keep the cable taut. Its eyebolts have square single threads whose mean radius and pitch are 0.25 in. and 0.10 in., respectively. The coefficient of static friction between the threaded sleeve and the eyebolts is 0.40. Determine the torque that must be applied to the sleeve to tighten the turnbuckle further if it has already been adjusted to produce a tensile force of 1000 lb in the cable. Eyebolt A has a right-hand thread and eyebolt B has a left-hand thread.

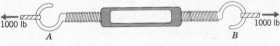

Prob. VII.77

VII.78   Solve Problem VII.77 if the eyebolts are double threaded and all other information is as given.

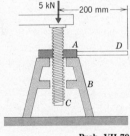

Prob. VII.79

VII.79   The jack shown in cross section consists of the threaded collar A, which bears on the frame B as it rides on screw C. The collar may be rotated by turning bar AD. The screw itself is prevented from rotating by the object being supported. The single square thread has inner and outer radii of 20 mm and 25 mm, and its pitch is 10 mm. The coefficient of friction for the screw threads is 0.40. The friction between collar A and the frame is negligible. Determine the smallest force applied perpendicular to the diagram at end D of bar AD that will raise the object carrying a 5-kN reaction.

VII.80   For the jack in Problem VII.79, determine the smallest force applied perpendicular to the diagram at end D of bar AD that will lower the object carrying a 5-kN reaction.

VII.81   Worm gear CD having a single square thread of 50 mm mean radius and 10 mm pitch meshes with gear A. Gear A is rigidly attached to

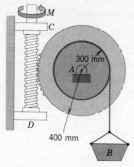

Prob. VII.81

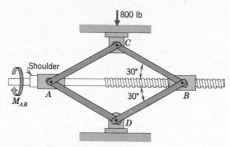

the drum that holds the cable suspending the 100-kg bucket B. The coefficient of static friction between the threads of the worm gear and the teeth of gear A is 0.20. Determine the torque $\bar{M}$ that must be applied to the shaft CD to lift the bucket slowly.

VII.82 The scissors jack shown supports an automobile exerting the 800-lb load. The screw has a double square thread whose pitch is $\frac{1}{8}$ in. The mean radius of the thread is $\frac{3}{4}$ in. Knowing that the coefficient of static friction is 0.33, determine the couple $\bar{M}_{AB}$ that should be applied to the screw (a) to raise the automobile further, (b) to lower the automobile. The screw fits loosely in collar A. Friction between the shoulder in the screw and this collar is negligible.

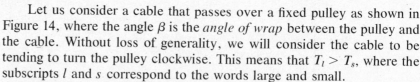

Prob. VII.82

3 Belt Friction

Cables that are guided over pulleys are effective devices for transferring forces between bodies. If it were not for friction, a cable would slide freely over a pulley and the tension in the cable on either side of the pulley would be identical. In some, but not all, cases this would be a desirable result. At times, friction may be required to provide a propelling force. For example, friction is needed for the functioning of the belt drive used in gasoline engines to drive electric generators and other auxiliary equipment.

Let us consider a cable that passes over a fixed pulley as shown in Figure 14, where the angle β is the *angle of wrap* between the pulley and the cable. Without loss of generality, we will consider the cable to be tending to turn the pulley clockwise. This means that $T_l > T_s$, where the subscripts l and s correspond to the words large and small.

It then follows that the cable exerts a moment $\bar{M}_O$ about the center of the pulley whose magnitude is given by

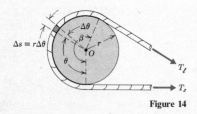

Figure 14

$$M_O = (T_l - T_s)r \tag{18}$$

and whose sense is the same as that of the moment of $\bar{T}_l$ about center O.

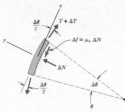

Figure 15

We wish to explore the relationship between T_l and T_s in the case where sliding between the cable and the pulley is impending. To do this, let us isolate a small element of cable Δs that is in contact with the pulley in the situation where the tendency of the cable is to rotate the pulley clockwise. Because of the contact between the cable and the pulley, a frictional force Δf and normal force ΔN are developed. The element Δs is tending to move tangentially to the pulley, in the positive x direction shown in Figure 15. Therefore, when sliding is impending, the friction force acting on the element is $\Delta f = \mu_s \Delta N$ in the negative x direction. To complete the free body diagram we have the tension force T at the position θ on one side of the element and the tension force $T + \Delta T$ at the position $\theta + \Delta\theta$ on the other side of the element. The x axis is chosen to be tangent to the pulley at the midpoint of the element, so it follows that the angle between either tensile force and the x axis is $\Delta\theta/2$.

The equations for force equilibrium of the element are then found to be

$$\Sigma F_x = -T \cos \frac{\Delta\theta}{2} + (T + \Delta T) \cos \frac{\Delta\theta}{2} - \mu_s \Delta N = 0$$

$$\Sigma F_y = -T \sin \frac{\Delta\theta}{2} - (T + \Delta T) \sin \frac{\Delta\theta}{2} \Delta N = 0$$

These reduce to

$$\Delta T \cos \frac{\Delta\theta}{2} = \mu_s \Delta N \tag{19a}$$

$$(2T + \Delta T) \sin \frac{\Delta\theta}{2} = \Delta N \tag{19b}$$

Let us now consider what happens when we isolate smaller and smaller elements, which means that the value of $\Delta\theta$ is reduced. In the limit of an infinitesimal element, $\Delta\theta$ approaches $d\theta$. The trigonometric terms in equations (19) then become

$$\cos \frac{d\theta}{2} = 1 \qquad \sin \frac{d\theta}{2} = \frac{d\theta}{2}$$

Therefore, retaining only first-order differentials, equations (19) reduce to

$$dT = \mu_s \, dN \quad \text{and} \quad T \, d\theta = dN \tag{20}$$

Our interest is with the tension force, not the normal force, so we eliminate dN and obtain

$$\frac{dT}{T} = \mu_s \, d\theta \tag{21}$$

Equation (21) may be integrated. Referring to Figure 14, the tension at $\theta = 0$ is T_s, while that at $\theta = \beta$ is T_l. Thus we have

$$\int_{T_s}^{T_l} \frac{dT}{T} = \int_0^{\beta} \mu_s \, d\theta$$

$$\ln T_l - \ln T_s \equiv \ln \frac{T_l}{T_s} = \mu_s \beta$$

A more useful form for this result is

$$\boxed{\frac{T_l}{T_s} = e^{\mu_s \beta}}$$

(22)

Let us review the meaning of each term in equation (22). The angle β, measured in radians, is the angle of wrap of the cable around the pulley. It is possible for β to be larger than 2π. For instance, if there are n full turns of the cable around the pulley, then $\beta = 2\pi n$. The symbols T_l and T_s represent the larger and the smaller of the tension forces on either side of the pulley; the tendency of the cable is to slide toward the side where the tension is T_l.

Note that equation (22) describes the situaton where a cable is about to slide over *any fixed cylindrical object,* such as a binding post. It may also be employed to describe a cable that is wrapped around a rotating cylindrical object when there is no slipping of the cable, provided that inertial effects are negligible. Further, equation (22) may be modified to describe the situation where the cable is actually sliding over the cylindrical object. In this case, the symbol μ_s should be replaced by the symbol μ_k, the coefficient of kinetic friction.

Finally, note that equation (22) was derived for the case of sliding or impending motion. It is not valid for the general, statically indeterminate case. A further restriction on equation (22) is that it is valid only if contact between the cable and the fixed object is along a cylindrical surface. It must be modified if we wish to analyze a V-belt and its pulley. This modification is the subject of Homework Problem VII.95.

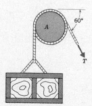

EXAMPLE 7

The tensile force $\bar{T}$ is applied to the free end of a cable that is wrapped around the horizontal binding post A for one complete turn and then attached to the 60-lb crate. The coefficient of static friction between the post and the cable is 0.30. Determine (a) the minimum tension T for which the crate will not descend, (b) the maximum tension T for which the crate will not rise.

Solution

The tensile force in the vertical portion of the cable is obviously the weight of the crate, 60 lb. When the crate is to be lowered, the cable is tending to move in the direction that the crate is pulling it, so we have $T_l = 60$ lb. The value T we seek then corresponds to T_s when the cable is on the verge of slipping. Conversely, when the crate is to be raised, the cable is about to move in the direction in which its free end is being pulled. In that case T corresponds to T_l and $T_s = 60$ lb.

The angle of wrap is determined from the diagram and the given information. Adding the 150° of contact depicted in the diagram to the one additional turn specified in the problem, we have

$$\beta = 150°\left(\frac{\pi}{180°}\right) + 2\pi = 8.901 \text{ rad}$$

Equation (22) for part (a) then yields

$$\frac{60}{T} = e^{(0.30)(8.901)} = 14.444$$

$$T = 4.15 \text{ lb}$$

Similarly, for part (b) we have

$$\frac{T}{60} = e^{(0.30)(8.901)}$$

$$T = 866 \text{ lb}$$

We can see that it is easy for a person to lower the crate and quite difficult for a person to raise it. This is a consequence of the fact that the exponential function increases rapidly with an increase in its argument.

EXAMPLE 8

A motor applies a torque $M_A = 20$ N-m to the small pulley A. The coefficient of static friction between the flat belt and the pulley is 0.40. Knowing that the given torque is the maximum for which the belt will not slip over pulley A, determine (a) the tension in the belt on either side of pulley A, and (b) the torque M_B on pulley B for equilibrium. (c) Prove that the belt is not slipping over pulley B.

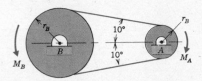

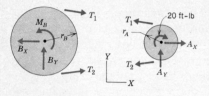

Solution

There are two rigid bodies, specifically, the pulleys, whose equilibrium must be assured. Therefore, let us cut the cable and draw the free body diagrams of each pulley. In doing this, we must not forget to include the forces exerted by the shaft of each pulley to withstand the pulling effect of the cables.

We are not interested in the reaction forces exerted by the shaft, so we write only the equations for moment equilibrium of each pulley about its center. It is given that M_A = 20 N-m in the sense shown, so we have

$$\text{pulley } A: \quad \Sigma M_{A\dot{z}} = -20 + (T_1 - T_2)r_A$$

$$\text{pulley } B: \quad \Sigma M_{BZ} = M_B + (T_2 - T_1)r_B$$

These equations allow us to solve part (b) immediately, for they show that

$$T_1 - T_2 = \frac{20}{r_A} = \frac{M_B}{r_B}$$

$$M_B = 20\frac{r_B}{r_A} \text{ N-m} \qquad \triangle$$

The ratio of the radii of the pulleys is the *mechanical advantage* of the system.

To solve part (a) we require another relationship for the belt tensions, specifically, equation (22). The information given in the problem states that slip is impending at pulley A when M_A = 20 N-m. Therefore, this is the pulley we shall describe. From the given diagram of the system, we determine that the belt is wrapped around pulley A by

$$\beta = 160°\left(\frac{\pi \text{ rad}}{180°}\right) = 2.793 \text{ rad}$$

Further, the equation for moment equilibrium of pulley A tells us that T_1 is larger than T_2, so T_1 corresponds to T_l. Then, for μ_s = 0.40, equation (22) gives

$$\frac{T_1}{T_2} = e^{(0.40)(2.793)} \quad \text{so} \quad T_2 = 0.3272T_1$$

Substituting this expression into the moment equilibrium equation results in

$$T_1 - T_2 \equiv T_1 - 0.3272T_1 = \frac{20}{r_A}$$

$$T_1 = \frac{29.73}{r_A} \qquad T_2 = \frac{9.73}{r_A} \text{ N-m} \qquad \triangle$$

We now must show that the belt is not slipping over pulley B. To do this, we note that the angle of wrap for pulley B is larger than that for pulley A. Therefore, because the value of μ_s is the same for both pulleys, the ratio $T_1/T_2 \equiv T_l/T_s$ required for impending slip over pulley B is larger than that required for pulley A. It follows that the value of T_1/T_2 in the system, which is just sufficient to cause impending slip over pulley A, is insufficient to cause slip over pulley B.

HOMEWORK PROBLEMS

VII.83 A 160-lb person stands on a 20-lb scaffold attached to one end of a rope. The rope passes over a fixed cylinder and the free end of the cable is grasped by the person. The coefficient of static friction between the cable and the cylinder is 0.50. Determine the force the person must apply to (a) lower the scaffold, (b) raise the scaffold.

Prob. VII.83

VII.84 A ship is secured by wrapping a rope around the capstan. A dock worker can apply a 200-N force to counteract a 7-kN force by the ship. Determine the number of complete turns of the rope about the capstan required to keep the rope from slipping if the coefficient of static friction is 0.25.

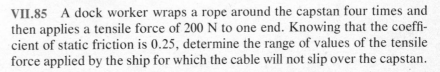

Probs. VII.84 and VII.85

VII.85 A dock worker wraps a rope around the capstan four times and then applies a tensile force of 200 N to one end. Knowing that the coefficient of static friction is 0.25, determine the range of values of the tensile force applied by the ship for which the cable will not slip over the capstan.

VII.86 A cable is wrapped around a fixed cylinder through an angle β. If the tension T on one end of the cable is constant, show that $T = (T_1 T_2)^{1/2}$, where T_1 and T_2 are the maximum and minimum values of the tension on the other end of the rope for which the cable will not slip.

VII.87 Three rough pegs protrude from the 50-kg block A, which slides in smooth vertical guides. Determine the tension T at the lower end of the cable for which block A is in static equilibrium in a condition of impending motion. The coefficient of static friction between the cable and the pegs is 0.60.

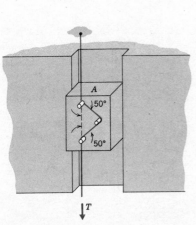

Prob. VII.87

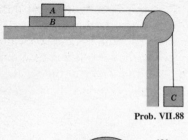

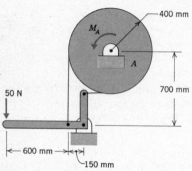

Prob. VII.88

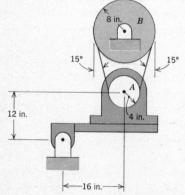

Prob. VII.89

VII.88   Block B weighs 10 lb and block C weighs 12 lb. The coefficients of static friction are 0.30 between the cable and the fixed circular guide, 0.50 between blocks A and B, and 0.40 between block B and the horizontal surface. Determine the minimum weight of block A for which the system is in static equilibrium.

VII.89   A 50-N force is applied to a band brake which restrains flywheel A against the counterclockwise couple $\bar{M}_A$. The coefficients of static and kinetic friction between the belt of the brake and the flywheel are 0.40 and 0.30, respectively. Determine the couple $\bar{M}_A$ required to (a) hold the flywheel at rest in a condition of impending slip, (b) rotate the flywheel in the direction of the couple at a constant rate of rotation.

VII.90   Solve Problem VII.89 for the case where the couple $\bar{M}_A$ is clockwise.

VII.91 and VII.92   Determine the maximum couple $\bar{M}_A$ that the motor may apply to pulley A without exceeding the maximum allowable belt tension of 2500 N. Also determine the corresponding couple $\bar{M}_B$ exerted on pulley B by its drive shaft. Where will slipping first occur? The coefficient of static friction between the belt and the pulleys is 0.35 and the pulleys are rotating at a constant rate.

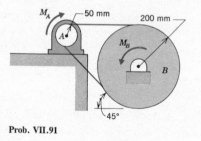

Prob. VII.91

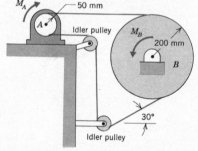

Prob. VII.92

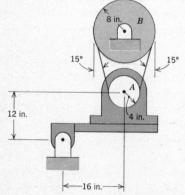

Prob. VII.93

VII.93   The belt drive shown is called a Rockwood mount. In it the 40-lb motor, whose center of mass coincides with its shaft, is mounted on a pivoted platform. Determine the maximum clockwise couple $\bar{M}_A$ the motor may apply to the drive pulley A, and also determine the corresponding couple $\bar{M}_B$ applied to pulley B by its shaft. The coefficient of friction between the belt and the pulley is 0.50. Neglect the mass of the platform.

VII.94   Solve Problem VII.93 in the case where the motor applies a counterclockwise couple to pulley A.

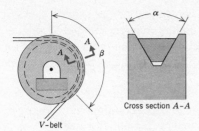

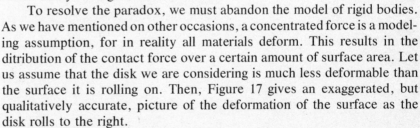

Cross section A-A

V-belt

Prob. VII.95

VII.95   A V-belt and its pulley are in contact along the nonparallel sides of the belt, as shown in the cross sectional view A-A. Contact is not made along a cylindrical surface, so equation (22) is not valid. Isolate an element ds of the V-belt and follow similar steps to those employed in deriving equation (22) to show that, for impending slip,

$$\frac{T_l}{T_s} = e^{\mu_s \beta / \sin(\alpha/2)}$$

Hint: Determine the radial component of the resultant of the normal forces exerted by the pulley on the sides of an element ds of the V-belt.

4   Rolling Friction

Perhaps the simplest and most important of the basic machines is the wheel. However, it is not without reason that we have made consideration of the resistance to rolling of a wheel the last item in this module, for as you will see it is a very complicated phenomenon to analyze.

Consider a homogeneous disk of mass m that is rolling to the right along a flat horizontal surface at a constant speed v, being propelled by a horizontal force $\bar{P}$ applied to the center O. Assuming that the disk and the surface are rigid, the only reaction forces possible are friction and normal forces exerted by the surface on the disk at the point of contact C. These are shown in the free body diagram, Figure 16. The disk is rolling along at a constant speed, so we may apply the equations of static equilibrium. By summing moments about center O, we find that $\bar{f} = 0$, which means that $\bar{P}$ must also equal zero. This leaves us with a paradox, for its says that there can be no resistance, so that the wheel will roll at a constant velocity without being pushed. However, we know that there must be resistance, for all freely rolling bodies slow down.

To resolve the paradox, we must abandon the model of rigid bodies. As we have mentioned on other occasions, a concentrated force is a modeling assumption, for in reality all materials deform. This results in the ditribution of the contact force over a certain amount of surface area. Let us assume that the disk we are considering is much less deformable than the surface it is rolling on. Then, Figure 17 gives an exaggerated, but qualitatively accurate, picture of the deformation of the surface as the disk rolls to the right.

The deformation pattern of the surface moves ahead with the disk, in a fashion similar to the bow wave preceding a boat in water. Thus, the phenomenon of rolling resistance bears little resemblance to the concept of dry friction described by Coulomb's laws, for which the model of rigid bodies are sufficient.

As shown in Figure 17, the creation of an area of contact between the disk and the surface means that the resultant contact force $\bar{R}$ exerted

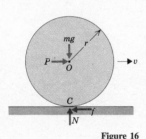

Figure 16

Figure 17

between these bodies need no longer be applied at the lowest point C. Instead, it may be applied at point D, which is an angle θ ahead of point C. Now, the equilibrium equation obtained by summing moments about the center of the disk requires that the line of action of the resultant force $\bar{R}$ be the line OD.

As shown in Figure 17, the horizontal component of $\bar{R}$ may be interpreted as the friction force $\bar{f}$, whereas the vertical component of $\bar{R}$ may be interpreted as the reaction force $\bar{N}$ normal to the undeformed surface of the floor. Thus, we have

$$
\begin{aligned}
f &= R \sin \theta \\
N &= R \cos \theta
\end{aligned}
\tag{23}
$$

It is apparent from the figure that the horizontal force $\bar{P}$ required to move the wheel forward at a constant velocity is no longer zero. We have resolved the paradox caused by assuming that the floor is rigid.

We noted earlier that Figure 17 gives an exaggerated picture of the deformation pattern. In most instances, the area of contact, and therefore the angle θ, is very small. It proves more convenient in such cases to replace the angle θ by the horizontal distance a locating the point D where the resultant force acts. It can be seen from Figure 17 that

$$
\sin \theta = \frac{a}{r} \qquad \cos \theta = \left(1 - \frac{a^2}{r^2} \right)^{1/2}
$$

Substituting these expressions into equations (23) and eliminating the magnitude R of the resultant force yields

$$
f = \frac{\sin \theta}{\cos \theta} N = \frac{a/r}{(1 - a^2/r^2)^{1/2}} N
$$

Then, because a/r is much smaller than unity, this reduces to

$$
\boxed{f = \frac{a}{r} N}
\tag{24}
$$

The distance a is called the *coefficient of rolling resistance,* and it has units of length. Early investigaors, such as Coulomb, regarded it as a parameter that is independent of the magnitude of the normal force $\bar{N}$, and also independent of the radius r. If this were true, a/r would be a factor of proportionality between the friction and normal forces, similar to the coefficients of dry friction. A further implication of equation (24) in that case is that the rolling resistance on a wheel is inversely proportional to its radius. Typical experimental values for the magnitude of the coefficient of rolling resistance a that have been reported are: mild steel on mild steel, $a = 0.18$ to 0.38 mm; hardened steel on hardened steel, $a = 0.005$ to 0.012 mm.

The foregoing analysis is contradicted by recent experimental and theoretical work. Although differing among each other, the investigators found that a is slightly dependent on the magnitude of $\bar{N}$ and strongly dependent on the radius r. In view of this disagreement, equation (24) should be regarded as useful for discussion and preliminary analysis purposes only. We will therefore not pursue the application of this formulation to solve problems.

MODULE VIII
VIRTUAL WORK AND
ENERGY METHODS

The methods developed thus far for studying the static equilibrium of systems require that we isolate each of the rigid bodies forming the system. The equations for force and moment equilibrium then provide the basis for determining all unknown forces, provided that the system is statically determinate. In such analyses the interaction forces exerted between bodies in a system must be accounted for, even when knowledge of these forces is not of interest.

Unlike these methods, the *principle of virtual work* and its corollaries considers a system of rigid bodies in its entirety. This *systems viewpoint* frequently allows us to ascertain unknown forces without considering the internal interaction forces within the system. The principle of virtual work has many specialized applications, particularly in the field of structural mechanics.

We will see that the usefulness of the principle of virtual work decreases as the complexity of the geometry of the system increases, whereas in treatments employing the equilibrium equations the geometry of the system has little effect on the procedures followed. Thus, the principle of virtual work is merely a supplement to, but not a replacement for, the basic equations of static equilibrium.

The starting point for a treatment of this topic is the concept of the *work done* by a force. For simplicity, we shall limit ourselves to planar problems, although the resulting principles are equally applicable to three-dimensional problems.

A. WORK DONE BY A FORCE

When we think of doing work to perform a physical task, we intuitively think of the force required to move an object by a specific distance. Loosely speaking, that is the basis for our analyses.

Let us consider the work done by a force $\bar{F}$ applied to a point P when that point follows a fixed path S in space. A typical situation is shown in Figure 1.

When point P moves an infinitesimal arc length ds, the position vector $\bar{r}_{P/O}$ of that point changes by a small amount $d\bar{r}_{P/O}$. Because the arc is infinitesimal, the magnitude of $d\bar{r}_{P/O}$ is ds and the direction of $d\bar{r}_{P/O}$ is tangent to the path, as illustrated in the figure.

The result of this differential displacement is that the force does a differential amount of work dU. Mathematically, this can be expressed as

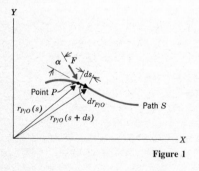

Figure 1

$$dU \equiv \bar{F} \cdot d\bar{r}_{P/O} \equiv |\bar{F}| \, ds \cos \alpha \tag{1}$$

where, as shown in Figure 1, α is the angle between $\bar{F}$ and the local tangent.

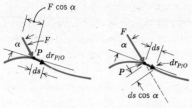

Figure 2a   **Figure 2b**

Recalling the definition of the dot product of two vectors, the work dU given by equation (1) may be pictured in either of two ways. We can say that dU is the component of force tangent to the path S multiplied by the distance ds traveled along the path, as depicted in Figure 2a. Alternatively, we can say that dU is the component of the displacement vector $d\bar{r}_{P/O}$ parallel to the force $\bar{F}$ multiplied by the magnitude of this force, as shown in Figure 2b.

Some special features of the concept of work can be found by studying equation (1). The most obvious is that work is a scalar quantity having units of force times length. The value of the work dU may be positive or negative, depending on whether α is less than, or greater than 90°, respectively. The case where dU is negative means that the force opposes the movement, because in that case the component of force parallel to the path is opposite the direction of movement along the path. Also, note that when the force is perpendicular to the displacement, that is, when $\alpha = 90°$, the work done by the force is zero. Finally, note that we can state that the work done is the force times the distance traveled only when $\alpha = 0$, that is, only when the force is parallel to the displacement.

So far, we have only considered the work done by a single force. Because work is a scalar quantity, the total work done by a system of forces acting on a body may be obtained by algebraically adding the work done by each force. Hence, letting dU_i denote the work done by the ith force in a set of N forces, we have

$$dU = \sum_{i=1}^{N} dU_i \qquad (2)$$

Equation (2) expresses the means by which we may determine the infinitesimal work done by a system of forces. In the special case where two forces form a couple, it may be specialized even further. A couple $\bar{F}$ and $-\bar{F}$ acting on an arbitrary rigid body is depicted in Figure 3. The moment of this couple is $M = F l$ counterclockwise, or vectorially, $\bar{M} = F l \bar{K}$. We now wish to determine the work done by this couple.

The most general infinitesimal change in the position of a rigid body can be shown to consist of a superposition of an infinitesimal displacement of an arbitrarily chosen point O', such as the point shown in Figure 3, and an infinitesimal rotation about point O'. (Proofs of this statement are contained in Modules IV and VI of the companion text on dynamics.) With this in mind, to simplify the development that follows we use the principle of transmissibility to transfer the forces $\bar{F}$ and $-\bar{F}$ to points P and Q on their respective lines of action, as illustrated in Figure 3. Note that line PQ intersects point O' and is perpendicular to the forces.

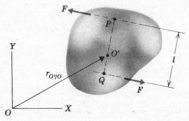

Figure 3

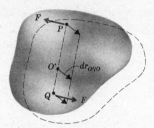

Figure 4a

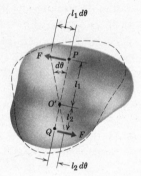

Figure 4b

Let us consider the infinitesimal displacement $d\bar{r}_{O'/O}$ of point O separately from the infinitesimal rotation $d\theta$ about point O'. The first type of movement is illustrated in Figure 4a and the second is shown in Figure 4b. The total work done by the couple will then be the sum of the work done in each type of movement.

For the situation in Figure 4a, which is called a translational displacement, points O', P, and Q all undergo the same displacement $d\bar{r}_{O'/O}$. Hence, the work done by the couple is

$$dU_{\text{trans}} = \bar{F} \cdot d\bar{r}_{O'/O} + (-\bar{F}) \cdot d\bar{r}_{O'/O} \equiv 0$$

thus demonstrating that a couple does no work when a body translates. This makes sense, because we have seen that a couple does not exert a pushing effect on a rigid body.

Considering the rotational movement in Figure 4b, we see that point P moves in the direction of $\bar{F}$ and point Q moves in the direction of $-\bar{F}$, so each force does positive work. Also, because the angle of rotation is infinitesimal, each displacement is parallel to the corresponding force, and its magnitude equals the arclength. Adding the work done by each force, we have

$$dU_{\text{rot}} = F(l_1\, d\theta) + F(l_2\, d\theta) = F(l_1 + l_2)\, d\theta$$

where l_1 and l_2 are the distances from point O' to points P and Q, respectively.

It can be seen from Figure 4b that the lever arm l for the couple is $l = l_1 + l_2$. It then follows that the total work done by the couple in the infinitesimal translation and rotation is

$$dU = dU_{\text{rot}} + dU_{\text{trans}} = Fl\, d\theta$$

$$\boxed{dU = |\bar{M}|\, d\theta} \tag{3}$$

Note that the rotation $d\theta$ was considered to be in the same sense as the moment of the couple, that is, counterclockwise, so the work done is positive. In situations where the rotation is in the opposite sense from the moment, the work done is negative, indicating that the couple is opposing the rotation. In general, the best way to avoid a sign error is to choose the signs of moments and rotations according to the right-hand rule with respect to the Z axis normal to the plane.

Equations (1) and (3) bring to mind the important matter of physical units. Dimensionally, equation (1) has units of force times length. This is consistent with equation (3), because the angle of rotation must be measured in radians. In order to avoid confusing moment and work, we generally describe a quantity of work by giving its length unit first and then its force unit, which is the opposite order from that used to describe a moment. Also, in the SI system work has its own derived unit, the joule (J), 1 J being defined as 1 m-N.

B.   VIRTUAL MOVEMENT

Because static systems are at rest, and we now know that a force does no work unless its point of application moves, the question that arises is how the concept of work can be used to solve statics problems. To answer this question we resort to the concept of a *virtual movement*. The word *virtual* is a classic term; what is really meant is *fictitious* or *imaginary*. In a virtual movement we imagine that the system moves away from its static equilibrium position. In the virtual movement all forces acting on the system are considered to maintain their magnitude and direction. Also, the points at which the forces are applied to the bodies in the system are unchanged. For example, in the case of a fully constrained structure, in which a virtual movement can be obtained only by imagining that a physical support is removed, the force corresponding to this support is still considered to act on the system.

In order to indicate that we are not considering a real displacement of the system, we use the Greek symbol δ instead of d to denote a differential virtual quantity. Thus, the *virtual displacement* of a point P resulting from the virtual movement we give the system is denoted as $\delta \bar{r}_{P/O}$, and the *virtual work* δU of a force $\bar{F}$ applied at this point is

$$\delta U = \bar{F} \cdot \delta \bar{r}_{P/O} \qquad (4)$$

A simple system that is useful for illustrating a virtual movement is the inclined bar shown in Figure 5, which is carrying a load $\bar{F}$ perpendicu-

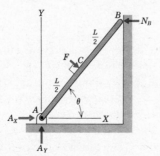

Figure 5

lar to the bar. A virtual movement of this bar may be produced by imagining what would happen if the wall at end B were removed, thus allowing

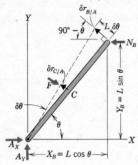

Figure 6

the angle θ to change. Because the virtual displacement is entirely a figment of our imagination, we can consider the rotation of the bar in this virtual movement to be either clockwise or counterclockwise. In order to minimize the opportunity for making an error is sign, the virtual movement depicted in Figure 6 shows a counterclockwise rotation corresponding to an increase of the angle θ by $\delta\theta$.

As a result of the virtual rotation, the forces acting on the bar do work. The computation of this virtual work requires that the virtual displacement of the point of application of each force be related to the value of $\delta\theta$. There are basically two ways of achieving this. One, called the geometric approach, involves applying the laws of geometry and trigonometry to the diagram of the virtual displacement, which is Figure 6 in this case.

The virtual movement depicted in Figure 6 is particularly suitable for the geometric approach. The movement depicted there is a pure planar rotation. That is, considering the points of the body lying in the plane, one point remains fixed as the other points follow circular arcs centered at the fixed point. We are dealing with an infinitesimal angle of rotation $\delta\theta$, so a point at a radial distance r from this fixed point will displace a distance $r\,\delta\theta$. Furthermore, because of the smallness of $\delta\theta$, the direction of the virtual displacement of a point will be perpendicular to the radial line from the fixed point, in the sense of the rotation.

These general observations enable us to construct directly the required expressions for the virtual displacements. For instance, point B is at a distance L from the fixed point A. Thus, the virtual displacement $\delta\bar{r}_{B/O}$ is $L\,\delta\theta$ perpendicular to the radial line AB. From the fact that the virtual rotation is counterclockwise, it follows that point B displaces upward and to the left. By referring to Figure 6 for the appropriate direction cosines, we find that

$$\delta\bar{r}_{B/O} = L\,\delta\theta[-\cos(90° - \theta)\,\bar{I} + \sin(90° - \theta)\,\bar{J}]$$

$$= L\,\delta\theta[-\sin\theta\,\bar{I} + \cos\theta\,\bar{J}] \tag{5}$$

Clearly, comparable expressons for the virtual displacements of other points may also be obtained by this method.

Although this geometric approach is particularly expedient for situations where a body undergoes a pure planar rotation, it has serious shortcomings. As we have seen, the most general movement that may be imparted to a body is a combination of a translation and a rotation. Most students find the geometric approach difficult and prone to error when treating such a movement.

Let us now develop an alternative method, called the analytical method, for expressing the virtual displacements. Returning to the system and virtual movement depicted in Figure 6, we need an XYZ coordinate system whose origin is located at a point that does not move during the virtual movement of the system. The coordinate system appearing in Figure 6 fits this requirement. The position vector of the point of application of each force with respect to the origin of the coordinate system must be expressed in terms of an *arbitrary* value of the geometric parameter that is being incremented in the virtual movement. From Figure 6 we find that, for an arbitrary value of θ,

$$\bar{r}_{B/A} = L \cos \theta \, \bar{I} + L \sin \theta \, \bar{J}$$

$$\bar{r}_{C/A} = \tfrac{1}{2}\bar{r}_{B/A} \qquad \bar{r}_{A/A} = \bar{0} \tag{6}$$

The virtual displacement of each point is then found by evaluating the differential change in each position vector resulting from the infinitesimal virtual change in the geometric parameter θ. The unit vectors $\bar{I}$ and $\bar{J}$ are constants. Therefore, the differential changes of the position vectors in equations (6) are

$$\delta\bar{r}_{B/A} = \frac{d\bar{r}_{B/A}}{d\theta}\,\delta\theta = \left[\frac{d}{d\theta}(L\cos\theta)\,\bar{I} + \frac{d}{d\theta}(L\sin\theta)\,\bar{J}\right]\delta\theta$$

$$= -L\,\delta\theta\sin\theta\,\bar{I} + L\,\delta\theta\cos\theta\,\bar{J}$$

$$\delta\bar{r}_{C/A} = \frac{1}{2}\,\delta\bar{r}_{B/A} = -\frac{L}{2}\,\delta\theta\sin\theta\,\bar{I} + \frac{L}{2}\,\delta\theta\cos\theta\,\bar{J}$$

$$\delta\bar{r}_{A/A} = 0 \tag{7}$$

Note that the expression for $\delta\bar{r}_{B/A}$ is the same result as that obtained in equation (5) by the geometric approach. Also, note that if θ had a specific numerical value, this value could be substituted into equations (7) *after* the derivatives have been evaluated.

Once the virtual displacements are expressed, the virtual work of the forces is readily obtained from equation (4) by expressing each force in terms of its components. Using Figure 6 to determine the force components, the virtual work resulting from the virtual rotation $\delta\theta$ is

$$\delta U = (-N_B\bar{I}) \cdot \delta\bar{r}_{B/A} + (F\sin\theta\,\bar{I} - F\cos\theta\,\bar{J}) \cdot \delta\bar{r}_{C/A}$$

$$+ (A_x\bar{I} + A_y\bar{J}) \cdot \delta\bar{r}_{A/A}$$

$$= (-N_B)(-L\,\delta\theta\sin\theta) + (F\sin\theta)\left(-\frac{L}{2}\,\delta\theta\sin\theta\right)$$

$$+ (-F\cos\theta)\left(\frac{L}{2}\,\delta\theta\cos\theta\right) + 0$$

$$= \left[N_B L\sin\theta - \frac{FL}{2}(\sin^2\theta + \cos^2\theta)\right]\delta\theta$$

$$= \left[N_B L\sin\theta - \frac{FL}{2}\right]\delta\theta \tag{8}$$

It could have been foreseen by inspection that the reactions $\bar{A}_X$ and $\bar{A}_Y$ do no work, because point A is fixed during the virtual movement.

In the foregoing we discussed one type of virtual movement of the bar in Figure 5, specifically, the virtual rotation produced when we imagine that the support at end B is removed. Another type of virtual movement occurs if we pretend that the pin support at point A is changed to a roller, thus allowing for a virtual displacement of point A horizontally. As shown in Figure 7, this results in a vertical virtual displacement of end B.

The coordinate system in Figure 7 is different from that used in Figure 6, because we need a set of axes having an origin that is fixed during the virtual movement. However, the steps in computing the virtual displacement analytically follow the prescribed order. First we express the position vectors of the points where the forces are applied. Using θ as the independent geometric parameter, this gives

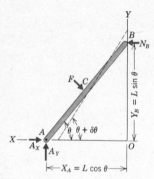

Figure 7

$$\bar{r}_{B/O} = Y_B \bar{J} = L \sin \theta \, \bar{J}$$
$$\bar{r}_{C/O} = \tfrac{1}{2} X_B \bar{I} + \tfrac{1}{2} Y_B \bar{J} = \tfrac{1}{2} L \cos \theta \, I + \tfrac{1}{2} L \sin \theta \, \bar{J} \tag{9}$$
$$\bar{r}_{A/O} = X_A \bar{I} = L \cos \theta \, \bar{I}$$

The virtual displacements are then found by computing the differentials. Thus,

$$\delta \bar{r}_{B/O} = \frac{d\bar{r}_{B/O}}{d\theta} \, \delta\theta = L \, \delta\theta \cos \theta \, \bar{J}$$

$$\delta \bar{r}_{C/O} = \frac{d\bar{r}_{C/O}}{d\theta} \, \delta\theta = \tfrac{1}{2} L(-\sin \theta \, \bar{I} + \cos \theta \, \bar{J}) \, \delta\theta \tag{10}$$

$$\delta \bar{r}_{A/O} = \frac{d\bar{r}_{A/O}}{d\theta} \, \delta\theta = -L \, \delta\theta \sin \theta \, \bar{I}$$

The corresponding virtual work is then found from equation (4) to be

$$\delta U = (N_B \bar{I}) \cdot \delta \bar{r}_{B/O} + (-F \sin \theta \, \bar{I} - F \cos \theta \, \bar{J}) \cdot \delta \bar{r}_{C/O}$$
$$\quad + (-A_X \bar{I} + A_Y \bar{J}) \cdot \delta \bar{r}_{A/O}$$
$$= 0 + \tfrac{1}{2} FL \, \delta\theta (\sin^2 \theta - \cos^2 \theta) + A_X L \, \delta\theta \sin \theta$$
$$= (-\tfrac{1}{2} FL \cos 2\theta + A_X L \sin \theta) \, \delta\theta \tag{11}$$

In this case, the fact that $\bar{N}_B$ does no work could have been anticipated without calculation, because $\bar{N}_B$ is perpendicular to the virtual displacement given to point B.

A feature of the concept of virtual movement that sometimes proves troublesome for students is that the choice of the geometric parameter to represent the movement is arbitrary. For instance, for the virtual movement of Figure 7, we could regard X_A as the geometric parameter describing the virtual movement because X_A is known if θ is known; specifically, $X_A = L \cos \theta$. The Pythagorean theorem then gives $Y_B = (L^2 - X_A{}^2)^{1/2}$, so the position vectors are

$$\bar{r}_{B/O} = Y_B \bar{J} = (L^2 - X_A{}^2)^{1/2} \bar{J}$$

$$\bar{r}_{C/O} = \tfrac{1}{2} X_A \bar{I} + \tfrac{1}{2} Y_B \bar{J} = \tfrac{1}{2} X_A \bar{I} + \tfrac{1}{2}(L^2 - X_A{}^2)^{1/2} \bar{J} \tag{12}$$

$$\bar{r}_{A/O} = X_A \bar{I}$$

The virtual displacements are then

$$\delta \bar{r}_{B/O} = \frac{d\bar{r}_{B/O}}{dX_A} \delta X_A = - \frac{X_A \, \delta X_A}{(L^2 - X_A{}^2)^{1/2}} \bar{J}$$

$$\delta \bar{r}_{C/O} = \frac{d\bar{r}_{C/O}}{dX_A} \delta X_A = \tfrac{1}{2} \, \delta X_A \, \bar{I} - \tfrac{1}{2} \frac{X_A \, \delta X_A}{(L^2 - X_A{}^2)^{1/2}} \bar{J} \tag{13}$$

$$\delta \bar{r}_{A/O} = \delta X_A \, \bar{I}$$

The corresponding virtual work may then be found from dot products, as before. This demonstrates that any convenient parameter may be chosen to describe the virtual movement.

Perhaps the most essential observation regarding the virtual work in the two types of virtual movements we considered is that when we imagine that a support is removed, the reaction force associated with this support does virtual work. This result occurs because the geometric constraint condition imposed by the support is violated in the virtual movement. In contrast, the reaction forces corresponding to supports that are unmodified during a virtual movement do no work.

For the two types of virtual movement discussed above, in each case one support was removed, so the virtual displacement could be described in terms of the infinitesimal change of one geometrical parameter. This parameter is called a *generalized coordinate*. After removal of the support, the system is said to have one *degree of freedom;* the number of degrees of freedom is defined as the number of generalized coordinates the system has. It follows that if we consider removing more than one support at a time, more than one generalized coordinate would be required. This added complication is unnecessary in statics, so it will not be pursued further.

The bar in Figure 5 is a fully constrained system, because its supports fully prevent movement. A primary goal of a statics analysis of such systems is the determination of the reaction forces. There are systems that are only partially constrained, that is, that move unless some unknown force (or forces) has the necessary value for static equilibrium. For our purposes, such systems may be analyzed from the same viewpoint as that used for fully constrained systems. The number of generalized coordinates is then the number of geometrical quantities required to describe the virtual movement of the system consistent with the existing supports. The unknown external forces required to prevent the system from moving, in effect, are then additional constraint forces.

We have seen how the concept of a virtual movement leads to the virtual work of the forces acting on the system. In the following sections we shall develop the principles that enable us to use this concept to solve statics problems. However, because the determination of virtual displacements is the key to evaluating the virtual work, the following examples and homework problems are provided to gain practice in this technique.

EXAMPLE 1

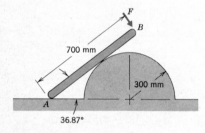

Bar AB rests on the 300-mm radius semicylinder and its lower end A rests on the floor, as shown. Determine the virtual work done by the force $\bar{F}$ at end B in a virtual movement in which the bar remains tangent to the semicylinder as end A moves horizontally.

Solution

The starting point of the solution is a sketch depicting the virtual movement.

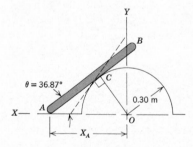

In general, we require a coordinate system whose origin is fixed in the virtual movement. The XYZ coordinate system shown in the sketch fits this requirement.

Next, we select the generalized coordinate for the virtual movement. One possible choice is the angle θ between the bar and the horizontal, which is shown in the sketch to change from its initial value of 36.87° as a result of the virtual movement. Another possible choice is the distance X_A. A way in which to choose between alternative choices for the generalized coordinate is to consider the construction of the position vectors for the points of application of the forces. For the problem at hand, the vector $\bar{r}_{B/O}$ may be obtained by replacing the 0.70-m length of bar AB

by its horizontal and vertical projections. This gives

$$\bar{r}_{B/O} = (X_A - 0.70 \cos \theta) \, \bar{I} + 0.70 \sin \theta \, \bar{J}$$

If a parameter is to serve as a generalized coordinate, it must be possible to express all other geometrical parameters that change in the virtual movement in terms of the chosen parameter. Thus, regardless of whether we choose X_A or θ, the complete description of $\bar{r}_{B/O}$ requires that we determine the relationship between these parameters. This relationship is readily determined, once it is noted that in the sketch of the system X_A forms the hypotenuse of the right traingle ACO. Thus

$$X_A = \frac{0.30}{\sin \theta} \text{ m}$$

It now appears that choosing the angle θ for the generalized coordinate gives the simpler expression for $\bar{r}_{B/O}$. Hence, for an arbitrary value of θ, we have

$$\bar{r}_{B/O} = \left(\frac{0.30}{\sin \theta} - 0.70 \cos \theta \right) \bar{I} + 0.70 \sin \theta \, \bar{J}$$

The virtual displacement of point B is obtained as a differential. This gives

$$\delta \bar{r}_{B/O} = \frac{d\bar{r}_{B/O}}{d\theta} \, \delta\theta = \left[\left(-\frac{0.30}{\sin^2 \theta} \cos \theta + 0.70 \sin \theta \right) \bar{I} \right.$$
$$\left. + 0.70 \cos \theta \, \bar{J} \right] \delta\theta$$

In the position of interest, $\theta = 36.87°$. Substituting this value into the foregoing expression for $\delta \bar{r}_{B/O}$ yields

$$\delta \bar{r}_{B/O} = \left\{ \left[-\frac{0.30}{(0.60)^2} (0.80) + 0.70(0.60) \right] \bar{I} + 0.70(0.80)\bar{J} \right\} \delta\theta$$
$$= (-0.2467\bar{I} + 0.56\bar{J}) \, \delta\theta \text{ m}$$

In the position where $\theta = 36.87°$, the force $\bar{F}$, which is perpendicular to bar AB, is given by

$$\bar{F} = F(-\sin 36.87° \, \bar{I} - \cos 36.87° \, \bar{J})$$
$$= -F(0.6\bar{I} + 0.8\bar{J}) \text{ m}$$

Hence, the virtual work done by this force is

$$\delta U = \bar{F} \cdot \delta \bar{r}_{B/O} = -F(0.6\bar{I} + 0.8\bar{J}) \cdot (-0.2467\bar{I} + 0.56\bar{J}) \, \delta\theta$$
$$= -0.300 \, F\delta\theta \text{ m-N} \equiv -0.300 \, F\delta\theta \text{ J}$$

A particular fact to bear in mind is that one choice for a generalized coordinate is better than another only in the sense that the resulting expressions for the position vectors of the points of application of the forces are simpler. However, regardless of what your choice for the generalized coordinate is, it must be used *consistently* throughout the solution.

EXAMPLE 2

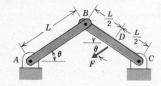

A force $\bar{F}$ is applied to the two-bar frame, as shown. Determine the virtual work done by this force in a virtual movement in which pin C is given a horizontal displacement, while the connections at pins A and B are maintained.

Solution

The following sketch illustrates the prescribed virtual movement.

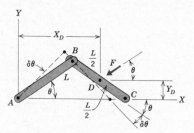

Because pin A remains fixed in the virtual movement, we select this point as the origin for the coordinate system.

As shown in the sketch, the position of the system can be established from the fact that triangle ABC is isosceles with side AC horizontal, provided that the angle θ is known. This suggests the use of θ as the generalized coordinate. Then, considering the aforementioned isosceles triangle, we see that the angle between bar BC and the X axis is also θ. Therefore, the general position vector for point D is

$$\bar{r}_{D/A} = X_D\,\bar{I} + Y_D\,\bar{J} = \frac{3L}{2}\cos\theta\,\bar{I} + \frac{L}{2}\sin\theta\,\bar{J}$$

The virtual displacement of point D is then found as a differential. This yields

$$\delta\bar{r}_{D/A} = \frac{d\bar{r}_{D/A}}{d\theta}\,\delta\theta = \left(-\frac{3L}{2}\sin\theta\,\bar{I} + \frac{L}{2}\cos\theta\,\bar{J}\right)\delta\theta$$

The angle between the force $\bar{F}$ and the horizontal is also θ, so the virtual work done by this force is

$$\delta U_F = \bar{F} \cdot \delta \bar{r}_{D/A} = F(-\cos\theta\,\bar{I} - \sin\theta\,\bar{J}) \cdot \left(-\frac{3L}{2}\sin\theta\,\bar{I}\right.$$

$$\left. + \frac{L}{2}\cos\theta\,\bar{J}\right)\delta\theta$$

$$= FL(\sin\theta\cos\theta)\,\delta\theta = \tfrac{1}{2}FL\sin 2\theta\,\delta\theta \qquad \triangleleft$$

HOMEWORK PROBLEMS

VIII.1 The mass m is supended by a cable that is attached at point C to the outer drum of the stepped pulley. A spring of stiffness k is attached at point B to the inner drum. There is no force in the spring when $\theta = 0$. Using the geometrical method, compute the virtual work done by the weight of the mass and by the spring force in a virtual rotation of the stepped pulley for an arbitrary value of θ.

VIII.2 The balance scale holds the small masses m_1 and m_2 in the position shown. Determine the work done by the weight forces acting on these masses in a virtual rotation about the pivot O. Compare this result to the moment of the weight forces about pivot O.

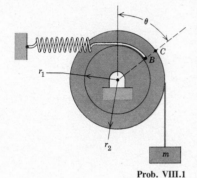

Prob. VIII.1

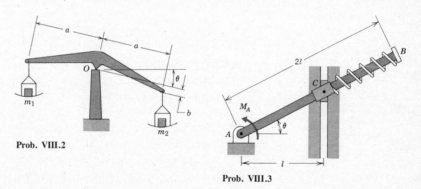

Prob. VIII.2

Prob. VIII.3

VIII.3 Collar C, which rides on bar AB, is pinned to a block that is guided by the straight groove. The spring, whose stiffness is k, is unstressed when $\theta = 0$. Determine the virtual work done by the spring in a virtual rotation of bar AB about pin A.

VIII.4 Bar AB is pinned to collars C and D, which ride on vertical rods. The lower end B is supported by the smooth ground. Determine the work done by the force $\bar{F}$ in a virtual movement that violates (a) only the constraint of the ground against the vertical movement of end B, (b) only the constraint of the left bar against horizontal movement of collar C.

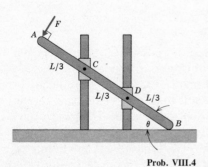

Prob. VIII.4

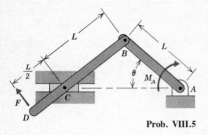

VIII.5 Block *C*, which rides in the groove, is pinned to bar *BD*. At an arbitrary angle θ the system is given a virtual movement in which all pin connections remain intact as block *C* moves parallel to the groove. Determine the virtual work done by the force $\bar{F}$ and by the couple $\bar{M}_A$.

Prob. VIII.5

VIII.6 A spring-cushioned platform supports a package of mass *m*. The spring, whose stiffness is *k*, is unstressed when $h = 800$ mm. At an arbitrary value of *h* the system is given a virtual movement in which all pin connections remain intact as roller *A* displaces horizontally. Determine the virtual work done by the spring force.

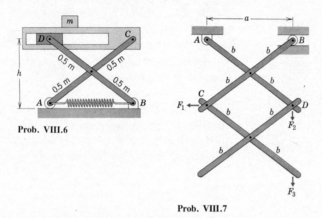

Prob. VIII.6

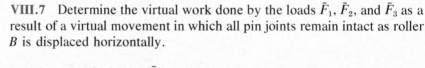

Prob. VIII.7

VIII.7 Determine the virtual work done by the loads $\bar{F}_1$, $\bar{F}_2$, and $\bar{F}_3$ as a result of a virtual movement in which all pin joints remain intact as roller *B* is displaced horizontally.

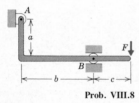

VIII.8 A vertical force $\bar{F}$ is applied to the bent bar. Determine the work done by $\bar{F}$ in a virtual movement that violates (a) only the constraint of roller *B* against vertical movement of that point, (b) only the constraint of pin *A* against vertical movement of that point.

Prob. VIII.8

C.   THE PRINCIPLE OF VIRTUAL WORK

From the preceding section we know that it is possible to select a virtual movement such that only selected unknown forces acting on the system do work. Therefore, if we could determine what the total amount of virtual work should equal, we would be able to form a relationship for the unknown forces doing work in the virtual movement. To achieve this goal, consider the body shown in Figure 8a on the following page, which is in equilibrium under the loading of an arbitrary system of forces.

We learned in Module III that the system of forces in Figure 8a is

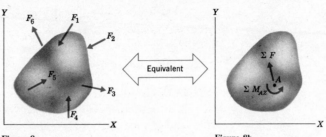

Figure 8a

Figure 8b

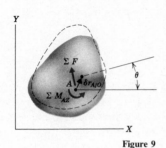

Figure 9

equivalent to the force-couple system at an arbitrary point A shown in Figure 8b. In this figure $\Sigma \bar{F}$ denotes the sum of all forces acting on the body, and ΣM_{AZ} denotes the sum of the moments of all forces about point A.

Recall that the most general infinitesimal movement of a body is the superposition of a translation corresponding to the infinitesimal displacement of a point plus an infinitesimal rotation about that point. Choosing point A for this point, a general virtual movement is as shown in Figure 9. In the translation the couple does no work, and in the rotation the force does no work. Hence, the total virtual work done by the forces acting on the body is

$$\delta U = \Sigma \bar{F} \cdot \delta \bar{r}_{A/O} + \Sigma M_{AZ} \, \delta\theta \tag{14}$$

This is a general expression for the virtual work done by an arbitrary set of forces. However, we have specified that the body is in equilibrium. The equilibrium conditions require that $\Sigma \bar{F} = 0$ and $\Sigma M_{AZ} = 0$. It then follows from equation (14) that

$$\boxed{\delta U = 0} \tag{15}$$

This simple equation is the *principle of virtual work*. As derived, it applies only to an isolated rigid body. However, the principle is also valid for systems of interconnected rigid bodies, because the total virtual work done by the forces acting on the system is the sum of the virtual work done in moving each body in the system, which is zero. Hence, the principle of virtual work may be stated as

> The virtual work δU done by all forces acting on any system in static equilibrium is zero.

The key feature of this principle is related to the fact that the type of virtual movement given to the system is at our discretion. A movement that violates only one constraint condition results in the appearance of only the associated constraint force (out of the full set of unknown reactions), in the virtual work δU. Setting $\delta U = 0$ then yields a single scalar equation for the unknown force.

For example, consider the bar of Figure 5, which is shown again in Figure 10. Recall that $\bar{F}$ is a given loading. When we wish to find $\bar{N}_B$ we imagine that the vertical wall is removed, leaving the pin at point A intact. This leads to a virtual rotation about point A, in which the reactions $\bar{A}_X$ and $\bar{A}_Y$ do no work. The resulting virtual work was found in equation (8). Setting that expression for $\delta U = 0$ yields

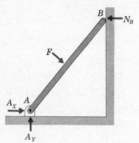

Figure 10

$$\delta U = \left(N_B L \sin \theta - \frac{FL}{2} \right) \delta \theta = 0$$

Because $\delta \theta$ is an arbitrary nonzero quantity, it must be true that

$$N_B = \frac{F}{2 \sin \theta}$$

Should we wish to determine the reaction $\bar{A}_X$, we would imagine that the constraint of pin A against horizontal movement is removed. We would then maintain the constraint of pin A against vertical movement, as well as the constraint against horizontal motion of end B imposed by the wall. In the resulting virtual movement forces $\bar{A}_Y$ and $\bar{N}_B$ do no work. The virtual work resulting from this type of virtual movement was found in equation (11). Setting that expression for $\delta U = 0$ yields

$$\delta U = (-\tfrac{1}{2} FL \cos 2\theta + A_X L \sin \theta) \, \delta \theta = 0$$

$$A_X = \frac{F \cos 2\theta}{2 \sin \theta}$$

Clearly, the preceding requires a large effort to solve some simple problems. The values of N_B and A_X for the bar can be determined from the equations of static equilibrium. In fact, summing moments about pin A would yield the value of N_B in a single equation, whereas a moment sum about the point of intersection of the horizontal line through end B and the vertical line through pin A would yield the value of A_X in a single equation. Nevertheless, even for this simple system, the principle of virtual work gives us a new viewpoint of the influence of the forces acting on the system.

The significant area of application of the principle of virtual work is for systems consisting of interconnected rigid bodies. To demonstrate this, consider the situation depicted on the next page in Figure 11a, wherein two arbitrary bodies are connected by a frictionless pin. According to Newton's third law, the interaction forces exerted be-

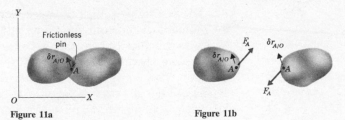

Figure 11a   Figure 11b

tween the two bodies are $\bar{F}_A$ and $-\bar{F}_A$, as shown in Figure 11b. Thus, in a virtual movement of the two bodies in which the bodies remain connected, so that the displacement of their common point of connection is the vector $\delta\bar{r}_{A/O}$, the virtual work of $\bar{F}_A$ and $-\bar{F}_A$ is

$$\delta U_{\text{pin}} = \bar{F}_A \cdot \delta\bar{r}_{A/O} + (-\bar{F}_A) \cdot \delta\bar{r}_{A/O} \equiv 0$$

Thus, frictionless pins do no work in a virtual movement in which the connected bodies remain joined.

In general, when we consider friction to have a negligible effect in the connections between rigid bodies, we are forming an *ideal* model of the system. In effect, the internal forces in ideal systems are constraint forces. For example, the pin forces $\bar{F}_A$ and $-\bar{F}_A$ in Figure 11 are the constraint forces required to ensure that the points on each body that are joined by pin A have the same virtual displacement $\delta\bar{r}_{A/O}$. As we have seen, if a virtual movement is consistent with a constraint, then the corresponding constraint force does no work.

To gain further insight into this result, and also to learn how the presence of friction affects the problem, consider the two bars in Figure 12a which are connected by a collar that may slide on bar 1. As

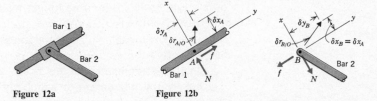

Figure 12a   Figure 12b

shown in Figure 12b, the interaction forces exerted between the two bars consist of a pair of forces $\bar{N}$ and $-\bar{N}$ normal to bar 1, and a pair of frictional forces $\bar{f}$ and $-\bar{f}$ tangent to bar 1. Also, as illustrated, there are two points we must consider, because the virtual displacement $\delta\bar{r}_{A/O}$ of point A on bar 1 will generally not be the same as $\delta\bar{r}_{B/O}$ of point B on bar 2. Note that for convenience in discussing the various vectors, an xyz coordinate system is defined, for which x is normal and y is tangent to bar 1.

The first thing we will consider about this connection is the nature of the geometrical constraint it imposes. If we were to position ourselves on bar 1, a virtual movement that is consistent with this type of connection would seem to cause the collar to displace tangentially, that is, in the y direction. This is the only possible movement of the collar relative to bar 1. Thus, although points A and B may displace by entirely different values δy_A and δy_B parallel to bar 1 in a virtual movement, the collar has the effect of making them displace by the same amount δx_A perpendicular to bar 1. This is the constraint condition. The reactions $\bar{N}$ and $-\bar{N}$ are the corresponding constraint forces, because they restrict the motion of the bars in the x direction.

Let us now compute the virtual work. Writing the virtual displacements of points A and B in terms of their x and y components, we have

$$
\begin{aligned}
\delta U_{\text{collar}} &= (N\bar{\imath} + f\bar{\jmath}) \cdot \delta\bar{r}_{A/O} + (-N\bar{\imath} - f\bar{\jmath}) \cdot \delta\bar{r}_{B/O} \\
&= (N\bar{\imath} + f\bar{\jmath}) \cdot (\delta x_A\,\bar{\imath} + \delta y_A\,\bar{\jmath}) \\
&\quad + (-N\bar{\imath} - f\bar{\jmath}) \cdot (\delta x_A\,\bar{\imath} + \delta y_B\,\bar{\jmath}) \\
&= f(\delta y_A - \delta y_B)
\end{aligned}
\tag{16}
$$

In this result the quantity $(\delta y_A - \delta y_B)$ represents the displacement of the collar relative to bar 1. We now observe that if bar 1 is smooth (the ideal model), then f is zero and the internal forces do no work. On the other hand, in a model that includes friction, the internal friction forces do work because they have the effect of resisting the movement of the collar relative to bar 1.

In general, if we apply the principle of virtual work to systems of rigid bodies where there is friction in the connections, these unknown frictional forces do work in addition to the work done by the unknown external forces. Equation (15) then no longer leads to a single equation for a single unknown. Solutions of problems pertaining to such systems are best achieved by returning to the fundamental methods of static equilibrium that we derived in earlier modules.

On the other hand, suppose that we give an ideal system of rigid bodies a virtual movement in which all of the bodies remain connected, thus satisfying the constraint conditions associated with the connections. The internal forces do not hinder such a movement, and therefore do no virtual work. They may therefore be ignored. This is one of the beneficial features of the method of virtual work in comparison to consideration of the equilibrium equations.

One qualifying statement applies to the preceding result. Specifically, in situations where a spring connects two bodies, so that the pair of forces it applies to the bodies are internal to the system, the spring forces will generally do work in a virtual movement. (The reason for this will be discussed in the section on potential energy.) Nevertheless, this does not represent an obstacle to the application of the principle of virtual work, for the force of a spring is known whenever its elongation is known.

D.   PROBLEM SOLVING

Before we proceed to utilize the principle of virtual work, it is best that we synthesize the foregoing development into a series of steps for solving problems.

1   Draw a free body diagram of the entire system in the position of interest as an aid in accounting for all forces. General reaction forces, such as those exerted by pin supports, may be broken up into any convenient set of components, for instance horizontal and vertical.

2   Draw another sketch showing the position of the system before any virtual movement, and also showing the supports of the system.

3   Determine the type of virtual movement desired. For fully constrained systems, this is done by violating only the constraint corresponding to the reaction of interest. For partially constrained systems the virtual movement should be the type the system would undergo if the external loads were not those required for static equilibrium. Show the position of the system after the virtual movement in the sketch of step 2.

4   Choose an XYZ coordinate system whose origin is located at a point that does not move during the virtual displacement. As always, the coordinate axes should have orientations that suit the geometry of the system. Show this coordinate system both in the free body diagram of step 1 and the sketch of step 2.

5   Choose any convenient geometrical parameter whose value changes during the virtual movement as the generalized coordinate.

6   Evaluate the virtual displacement of the points of application of the forces. In general, follow the analytical approach for this determination by first expressing the position vectors of the points in terms of an arbitrary value of the generalized coordinate. The required virtual displacements are then the differential changes in the position vectors resulting from an infinitesimal increase of the generalized coordinate. If the generalized coordinate has a specific value in the equilibrium position, substitute that value *after* evaluating the derivatives.

Optionally, in situations where a body is given a pure planar rotation in a virtual movement, the geometrical method may prove easy to apply. The virtual displacement of a point is then the product of the angle of the virtual rotation and the radial distance from the point to the fixed point, perpendicular to the radial line in the sense of the rotation.

7   Express the forces doing work in terms of their components, using the free body diagram as an aid. Compute the total virtual work δU as the sum

of the dot product of each force and the virtual displacement of its point of application. Equate δU to zero.

8   Solve the equation $\delta U = 0$ for the unknown force.

Clearly, the key step among those outlined above is step 6. In the illustrative problems that follow we will focus on the analytical viewpoint for constructing virtual displacements because it is useful for all types of virtual movement. Equally important, because this method is more mathematical, it requires less insight into the nature of the virtual movement of each body and therefore reduces the chance of making sign errors.

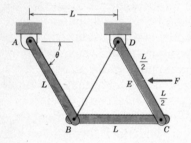

ILLUSTRATIVE PROBLEM 1

The frame shown supports the applied load $\bar{F}$. Determine the tension in cable BD using the principle of virtual work.

Solution

Step 1   Letting T denote the desired tension, the free body diagram of the system of rigid bars is as shown.

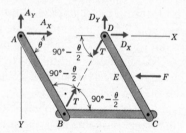

Step 2   The physical picture of the system before any virtual displacement is shown by the heavy solid lines.

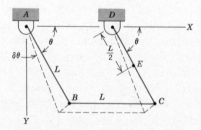

Step 3   Without the string for support, the system is only partially constrained, because the angle θ would then be a variable parameter. Thus, the virtual movement we give the system is one in which the angle θ is increased by $\delta\theta$ with all pins remaining intact, as shown by the dashed line in the sketch of step 2.

Step 4   Points A and D are both fixed during the virtual movement, so either may be chosen to be the origin. We will use the former. For convenience in representing position vectors, we choose the Y axis downward, as shown in the sketches of steps 1 and 2.

Step 5   We will use the angle θ as the generalized coordinate because we have already seen that the position of the system subsequent to the virtual movement can be described in terms of the increase in this angle.

Step 6   The only forces doing work are the tension force $\bar{T}$ at pin B and the horizontal force $\bar{F}$ at point E; the other forces appearing in the free body diagram are applied at fixed points.

We first write the position vectors for points B and E by referring to the sketch in step 2 for the dimensions. This gives

$$\bar{r}_{B/A} = L(\cos\theta\,\bar{I} + \sin\theta\,\bar{J}) \quad \bar{r}_{E/A} = \left(L + \frac{L}{2}\cos\theta\right)\bar{I} + \frac{L}{2}\sin\theta\,\bar{J}$$

The virtual displacements are then

$$\delta\bar{r}_{B/A} = \frac{d\bar{r}_{B/A}}{d\theta}\,\delta\theta = L(-\sin\theta\,\bar{I} + \cos\theta\,\bar{J})\,\delta\theta$$

$$\delta\bar{r}_{E/A} = \frac{d\bar{r}_{E/A}}{d\theta}\,\delta\theta = \frac{L}{2}(-\sin\theta\,\bar{I} + \cos\theta\,\bar{J})\,\delta\theta$$

Step 7   The force components may be obtained from the free body diagram of step 1, so the virtual work is

$$\delta U = \bar{T} \cdot \delta\bar{r}_{B/A} + \bar{F} \cdot \delta\bar{r}_{C/A}$$

$$= T\left[\cos\left(90° - \frac{\theta}{2}\right)\bar{I} - \sin\left(90° - \frac{\theta}{2}\right)\bar{J}\right] \cdot (-L\sin\theta\,\bar{I} + L\cos\theta\,\bar{J})\,\delta\theta$$

$$+ (-F\bar{I}) \cdot \left(-\frac{L}{2}\sin\theta\,\bar{I} + \frac{L}{2}\cos\theta\,\bar{J}\right)\,\delta\theta$$

$$= \left[-TL\left(\sin\frac{\theta}{2}\sin\theta + \cos\frac{\theta}{2}\cos\theta\right) + \frac{FL}{2}\sin\theta\right]\delta\theta = 0$$

Step 8   Cancelling all common factors in the equation for δU, we find that

$$TL\left(\sin\frac{\theta}{2}\sin\theta + \cos\frac{\theta}{2}\cos\theta\right) = \frac{FL}{2}\sin\theta$$

Using standard trigonometric identities this becomes

$$TL\cos\frac{\theta}{2} = \frac{FL}{2}\left(2\sin\frac{\theta}{2}\cos\frac{\theta}{2}\right)$$

Hence

$$T = F\sin\frac{\theta}{2}$$

We note in closing that step 6 could have been formulated by the geometric method. You should try it, and verify that the results are identical to those obtained above.

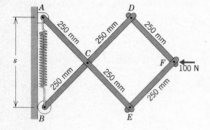

ILLUSTRATIVE PROBLEM 2

The parallelogram frame is loaded by a horizontal 100-N force. The unstretched length of the spring is 350 mm. Determine the required stiffness k of the spring if s = 400 mm in the static equilibrium position.

Solution

Step 1   The spring is extended by $s - 0.35$ m, so the tension in the spring is $k(s - 0.35)$. (For the sake of increased generality, we will substitute the given value of s later.) Thus, the free body diagram of the entire frame is as shown.

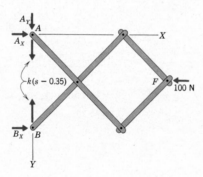

Step 2   The heavy solid lines in the sketch depict the frame before any virtual movement.

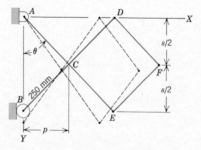

Step 3 The frame is a partially constrained system, because the distance s depends on the loads being carried by the frame. Thus the virtual displacement we give to the system is one in which the value of s is changed, with the constraints of the pins and the roller maintained. As depicted by the dashed lines in the sketch of step 2, the result of this virtual movement is that roller B moves vertically.

Step 4 The sketch of the virtual displacement shows that pin A is fixed, so it is the point chosen as the origin. The Y axis shown in the free body diagram and the sketch of the geometry is chosen downward for convenience.

Step 5 Suitable choices for the generalized coordinate are the distance s or the angle θ shown in the sketch of step 2, because the locations of points in the frame are easily described in terms of either of these variable parameters. We choose s solely because the given information gives the value of s in the equilibrium position.

Step 6 Point A is fixed, so the reaction forces there do no work. Hence, we need only construct the general position vectors for point B and F. In terms of the dimension p shown in the sketch in step 2, these vectors are

$$\bar{r}_{B/A} = s\bar{J} \qquad \bar{r}_{F/A} = 3p\bar{I} + \frac{s}{2}\bar{J}$$

Because s is the generalized coordinate, we express p in terms of s using the Pythagorean theorem. This gives

$$\bar{r}_{B/A} = s\bar{J}$$

$$\bar{r}_{F/A} = 3\left[(0.25)^2 - \left(\frac{s}{2}\right)^2\right]^{1/2}\bar{I} + \frac{s}{2}\bar{J}$$

The virtual displacements are the differential increase in the position vectors resulting from an infinitesimal increase δs. Therefore

$$\delta\bar{r}_{B/A} = \frac{d\bar{r}_{B/A}}{ds}\,\delta s = (\bar{J})\,\delta s$$

$$\delta\bar{r}_{F/A} = \frac{d\bar{r}_{F/A}}{ds}\,\delta s = \left\{3\left(\frac{1}{2}\right)\left[(0.25)^2 - \left(\frac{s}{2}\right)^2\right]^{-1/2}(-s)\bar{I} + \frac{1}{2}\bar{J}\right\}\delta s$$

Step 7 The reaction force $\bar{B}_X$ does no work because it is horizontal and point B displaces vertically (or from another viewpoint, because the constraint imposed by the roller has not been violated.) Thus, the principle of virtual work gives

$$\delta U = [-(s - 0.35)k\,\bar{J}] \cdot \delta\bar{r}_{B/A} + (-100\bar{I}) \cdot \delta\bar{r}_{F/A}$$

$$= -(s - 0.35)k\,\delta s + (-100)\left(-\frac{3}{2}s\right)\left(0.0625 - \frac{s^2}{4}\right)^{-1/2}\delta s$$

$$= \left[-(s - 0.35)k + \frac{150s}{(0.0625 - s^2/4)^{1/2}}\right]\delta s = 0\ \text{J}$$

Step 8   Cancelling the nonzero factor δs, the solution of the preceding expression is

$$k = \frac{300s}{(s - 0.35)(0.250 - s^2)^{1/2}}\ \text{N/m}$$

The value of k for the desired equilibrium position is then found by substituting $s = 0.400$ m, which yields

$$k = 8000\ \text{N/m}$$

An interesting feature of the principle of virtual work demonstrated by our analysis of this problem is that it is usually just as easy to determine the results in terms of arbitrary parameters, for instance k in terms of s, rather than for specific values.

ILLUSTRATIVE PROBLEM 3

A 2-kip load is applied to beam AB which rests on beam CD. Determine the reaction force at roller C in terms of the distance r locating the load.

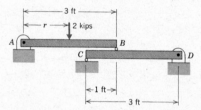

Solution

Step 1   The reaction at roller B is an internal force exerted between the beams, so the free body diagram of the system of beams is as shown.

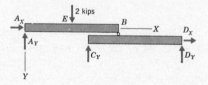

Step 2 The configuration of the system before any virtual displacement is shown as solid lines in the following sketch.

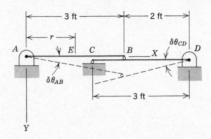

Step 3 The system of beams is fully constrained by its supports. Because it is our desire to determine the reaction at roller C, the virtual movement we give the system is one that violates only the constraint of this roller against vertical displacement. The reaction $\bar{C}_Y$ will then be the only reaction doing virtual work.

Step 4 As shown in the sketch, the result of the virtual movement we have chosen is that each bar executes a virtual rotation about its pin support. Either of the fixed pins A or D is suitable for the origin. The coordinate system we will use has an origin at point A, as shown in the sketches.

Step 5 From the sketch in step 2 we see that the position of the system after the virtual movement may be established in terms of the virtual rotation $\delta\theta_{AB}$. If this angle is known, we can locate end B. This in turn allows us to determine $\delta\theta_{CD}$ by drawing a straight line through pin D tangent to roller B. Hence, we choose θ_{AB} as the generalized coordinate.

Step 6 The virtual movement depicted in the sketch is one in which both beams execute pure rotations about their fixed ends. Because the radial line from each point on a beam to the fixed end of that beam is horizontal, all virtual displacements are vertical. For the rotations depicted in the sketch in step 2, these displacements are downward. It is therefore a straightforward matter to employ the geometrical method. We need the virtual displacements of points C and E, which are the points of application of forces doing work. Point C is on beam CD, 3 ft from the fixed end D, whereas point E is on beam AB, at a distance r from the fixed end A. Thus

$$\delta\bar{r}_{C/A} = 3\,\delta\theta_{CD}\,\bar{J}$$

$$\delta\bar{r}_{E/A} = r\,\delta\theta_{AB}\,\bar{J}$$

We now relate the angle $\delta\theta_{CD}$ to the change in the generalized coordinate θ_{AB} by noting that the roller support at point B, and hence its constraint, was not altered in producing the virtual movement. Therefore, the virtual displacement of point B at the end of beam AB must be the same as that of the point where roller B rests on beam CD. This means that

$$\delta \bar{r}_{B/A} = 3 \,\delta\theta_{AB} \,\bar{J} = 2 \,\delta\theta_{CD} \,\bar{J}$$

$$\delta\theta_{CD} = \tfrac{3}{2} \,\delta\theta_{AB}$$

Therefore,

$$\delta \bar{r}_{C/A} = 3(\tfrac{3}{2}\,\delta\theta_{AB}) \,\bar{J} = \tfrac{9}{2} \,\delta\theta_{AB} \,\bar{J}$$

$$\delta \bar{r}_{E/A} = r \,\delta\theta_{AB} \,\bar{J}$$

Step 7   The forces at points E and C are shown in the free body diagram, so the virtual work is

$$\delta U = 2\bar{J} \cdot \delta \bar{r}_{E/A} + (-C_Y \bar{J}) \cdot \delta \bar{r}_{C/A}$$

$$= 2r \,\delta\theta_{AB} - \tfrac{9}{2} C_Y \,\delta\theta_{AB} = 0 \text{ ft-kips}$$

Step 8   Cancelling the common factor $\delta\theta_{AB}$, the value of C_Y is found to be

$$C_Y = \tfrac{2}{9}(2r) = \tfrac{4}{9}r \text{ kips}$$

As an aside to this solution, we should note that the analytical method for evaluating the virtual displacements requires that the system of beams first be considered for arbitrary values of the angles θ_{AB} and θ_{CD}. This method is much more cumbersome than the geometrical method for this problem because the virtual displacements here were obvious from the sketch.

ILLUSTRATIVE PROBLEM 4

A linkage is formed by pinning collar C to bar BD. This collar may ride on the smooth horizontal guide EG. Determine the couple $\bar{M}_A$ that should be applied to bar AB to hold the linkage in the position shown when a vertical 8-kN force is applied at end D.

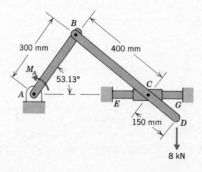

Solution

Step 1 The free body diagram of the system of bars is as shown.

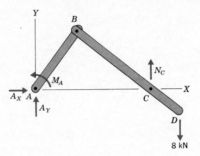

Step 2 The position of the system prior to a virtual movement is shown by solid lines in the following sketch.

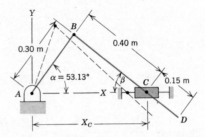

Step 3 The system is partially constrained, because the angles between the bars will change according to the loads being applied. Furthermore, we are requested to relate the values of F and M_D. Hence, the virtual movement we shall consider is one in which the constraints of the supports are unmodified, thereby preventing their reactions from doing work. As shown by the dashed lines in the sketch of step 2, the result is that bar AB executes an infinitesimal rotation about pin A, causing collar C to slide to the left.

Step 4 Pin A is the only fixed point in the system, so it is chosen as the origin for the coordinate system shown in the sketches.

Step 5 Knowledge of the angle α, the angle β, or the distance X_C would allow us to determine the other dimensions in an arbitrary position of the linkage. In view of the fact that we will need to know the virtual rotation of bar AB in order to determine the virtual work done by the couple $\bar{M}_A$, we select α to be the generalized coordinate.

Step 6 With the aid of the sketch in step 2, the position vector for point D, where the 8-kN load is applied, is

$$\bar{r}_{D/A} = (0.30 \cos \alpha + 0.55 \cos \beta)\,\bar{I} - 0.15 \sin \beta\,\bar{J}$$

It is now necessary to relate the variable angle β to the generalized coordinate α. This is done by applying the law of sines to triangle ABC, which yields

$$\frac{\sin \alpha}{0.40} = \frac{\sin \beta}{0.30} \qquad \sin \beta = 0.75 \sin \alpha$$

Replacing β by α in the expression for $\bar{r}_{D/A}$ will involve inverse trigonometric functions. Therefore, we proceed to the determination of $\delta \bar{r}_{D/A}$, bearing in mind that β is a function of α as given above. Thus

$$\delta \bar{r}_{D/A} = \frac{d\bar{r}_{D/A}}{d\alpha} \delta \alpha = [(-0.30 \sin \alpha - 0.55 \frac{d\beta}{d\alpha} \sin \beta) \bar{I} - 0.15 \frac{d\beta}{d\alpha} \cos \alpha \, \bar{J}] \delta \alpha$$

The simplest way to determine $d\beta/d\alpha$ is to differentiate the relationship between β and α implicitly. This yields

$$\frac{d}{d\alpha}(\sin \beta) \equiv \frac{d\beta}{d\alpha} \cos \beta = 0.75 \cos \alpha$$

In the position where $\alpha = 53.13°$, we have

$$\sin \beta = 0.75 \sin 53.13° = 0.6 \qquad \beta = 36.87°$$

$$\frac{d\beta}{d\alpha} \cos 36.87° = 0.75 \cos 53.13° \qquad \frac{d\beta}{d\alpha} = 0.5625$$

Thus, in the position of interest

$$\delta \bar{r}_{D/A} = \{[-0.30 \sin 53.13° - 0.55(0.5625)\sin 36.87°] \bar{I}$$

$$- 0.15(0.5625) \cos 36.87°] \bar{J}\} \delta \alpha$$

$$= (-0.4256\bar{I} - 0.0675\bar{J}) \delta \alpha \text{ m}$$

Step 7   Equation (3) gives the work done by a couple. Bar AB rotates counterclockwise with an increase in the angle α, so the counterclockwise couple $\bar{M}_A$ does positive work. Thus

$$\delta U = M_A \delta \alpha + (-8\bar{J}) \cdot (-0.4256\bar{I} - 0.0675\bar{J}) \delta \alpha$$

$$= (M_A + 0.540) \delta \alpha \text{ kN-m}$$

Step 8   When the common factor $\delta \alpha$ is cancelled, the value of M_A is found to be

$$M_A = -0.540 \text{ kN-m}$$

The negative value for M_A merely indicates that the couple applied to bar AB should be clockwise, opposite the direction in which it was assumed.
An overview of the solution we have presented shows that the com-

plications introduced by the lack of *simple* geometric relations, such as the need to evaluate derivatives implicitly, decreases the attractiveness of using the principle of virtual work for this problem. A straightforward force and moment analysis would give the same result, without such complications.

HOMEWORK PROBLEMS

VIII.9 and VIII.10   Bar *AB,* having a mass *m,* is pinned to the collar, which rides on the vertical rod. Friction is negligible. Determine the value of θ for static equilibrium.

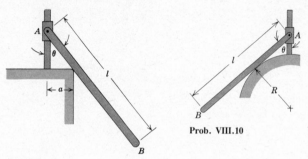

Prob. VIII.10

Prob. VIII.9

VIII.11   Bar *AB* is pinned to collars *C* and *D,* which ride on the vertical rods. The bar weighs 10 lb and the ground is smooth. Determine the reaction exerted by the vertical rod on collar *C.*

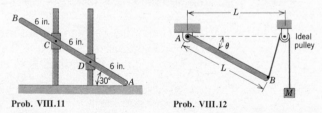

Prob. VIII.11

Prob. VIII.12

VIII.12   (a) Derive a relation between the mass *m* of bar *AB,* the mass *M* of the suspended block, and the angle θ. (b) Solve this relation for the angle θ when $M = m/4$. (c) Determine the largest value of *M* for which end *B* will not touch the pulley.

VIII.13   Collar *C,* which rides on the smooth bar *AB,* is pinned to a block that is guided by the smooth straight groove. The spring is unstressed when $\theta = 0$ and its stiffness is 5 kN/m. Determine the magnitude of the couple $\bar{M}_A$ required to maintain equilibrium in terms of θ.

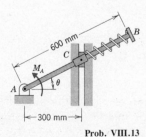

Prob. VIII.13

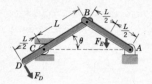

Prob. VIII.14

VIII.14 Roller C, which rides in the groove, is pinned to bar BD. Derive an expression for the force $\bar{F}_E$ required to hold the system in equilibrium against the force $\bar{F}_D$.

VIII.15 A 40-lb force is applied to arm ABC of the press punch shown in cross section. Determine the force the punch applies to the piece of sheet metal in the position shown.

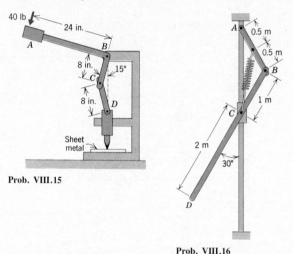

Prob. VIII.15

Prob. VIII.16

VIII.16 The mechanism shown lies in a vertical plane and is in equilibrium. Collar C rides on the smooth vertical guide, and the unstretched length of the spring is 1 m. Knowing that each of the bars has a mass of 2 kg/m, determine the stiffness of the spring.

VIII.17 The spring-cushioned platform supports a 200-kg package. The unstretched length of the spring is 700 mm. Determine the stiffness of the spring, knowing that the system is in equilibrium in the position shown.

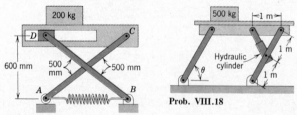

Prob. VIII.17

Prob. VIII.18

VIII.18 A 500-kg crate rests on the platform, which is supported by the mechanism shown. Determine the force in the hydraulic cylinder as a function of θ.

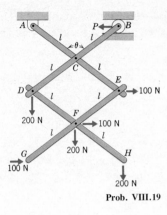

Prob. VIII.19

VIII.19 The parallelogram frame carries the loads shown. In terms of the angle θ, determine the horizontal force $\bar{P}$ that is required to maintain equilibrium.

VIII.20 The load $\bar{P}$ is always perpendicular to bar DF. The spring has stiffness k and it is unstressed when $\theta = 60°$. Derive an expression for the magnitude of $\bar{P}$ required for equilibrium at an arbitrary angle θ.

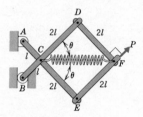

Prob. VIII.20

VIII.21 and VIII.22 Determine the reaction moment at support B of the compound beam shown.

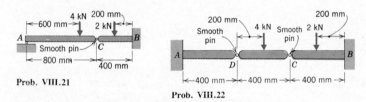

Prob. VIII.21

Prob. VIII.22

VIII.23 Determine the reaction at roller C resulting from the set of loads shown.

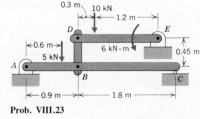

Prob. VIII.23

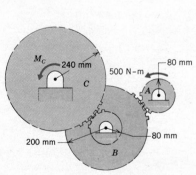

Prob. VIII.24

VIII.24 A 500 N-m couple is applied to gear A, as shown. Gear B, which transmits this couple to gear C, rides freely on its shaft. Determine the magnitude and sense of the couple $\bar{M}_C$ acting on gear C necessary for equilibrium. *Hint:* The constraint of the gear teeth requires that, in a virtual rotation of the gears, the teeth in contact on a pair of gears move the same distance circumferentially.

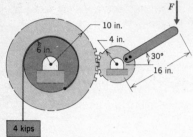

Prob. VIII.26

VIII.25   Solve Problem VIII.24 if a counterclockwise couple of 100 N-m is acting on the stepped gear B in addition to the couples on gears A and C.

VIII.26   The winch shown is used to suspend a 4-kip crate. Determine the force $\bar{F}$ required for equilibrium in the position shown. (See the hint for Problem VIII.24.)

VIII.27   The two blocks, which are interconnected by a cable passing over a pulley, are in equilibrium on the smooth inclines. Determine an expression for the mass ratio m_B/m_A. *Hint:* The length of the cable is constant in a virtual movement.

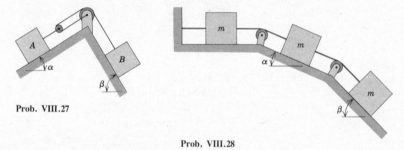

Prob. VIII.27

Prob. VIII.28

VIII.28   The three blocks having identical masses m are tied together by cables. Knowing that all contact surfaces are smooth, determine the tension in the cable that is fastened to the wall. (See the hint for Problem VIII.27.)

VIII.29   The 100-kg radioactive container is held by applying a force $\bar{P}$ to the pair of tongs. Determine the horizontal component of the force acting on the container at point F.

VIII.30   The mechanism shown is a device for clamping onto inaccessible objects. The spring has a stiffness of 2 kN and an unstretched length of 40 mm. Determine the clamping force on the object if the handles are squeezed with a pair of 10-N forces.

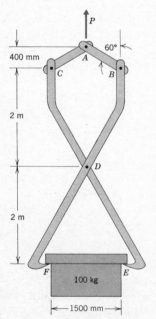

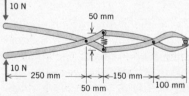

Prob. VIII.29   Prob. VIII.30

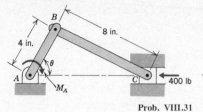

Prob. VIII.31

VIII.31 A 400-lb force is applied to the piston. Determine the couple $\bar{M}_A$ required for equilibrium when (a) $\theta = 30°$, (b) $\theta = 150°$.

VIII.32 Determine the force $\bar{F}$ required to hold the system shown in equilibrium if the couple $\bar{M}_A$ is 2 kN-m counterclockwise when (a) $\theta = 90°$, (b) $\theta = 0$, (c) $\theta = 45°$.

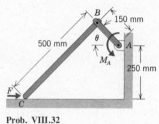

Prob. VIII.32

E. THE PRINCIPLE OF VIRTUAL WORK AND POTENTIAL ENERGY

Previous investigations were limited to the determination of the work done by a force in an infinitesimal movement of a system. This is the only consideration required in statics for the principle of virtual work. In contrast, for a system that is actually in motion, the point of application of a force will have a finite displacement. The determination of the work done in a finite movement of a system plays an important role in the study of dynamics. One feature of such a study is also relevant to statics, for it will greatly simplify the determination of the virtual work done by certain types of forces.

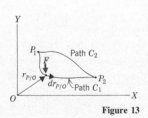

Figure 13

To investigate this matter, consider Figure 13, where we wish to investigate the work done by the force $\bar{F}$ in moving the point P from position P_1 to position P_2 in the case where the movement can take place over two different arbitrary paths C_1 and C_2.

The infinitesimal work of the force is defined as $dU = \bar{F} \cdot d\bar{r}_{P/O}$, so the total work will be the sum of all infinitesimal terms. Therefore, the work done when point P moves from position P_1 to position P_2 is

$$U_{1\rightarrow2} = \int_{r_1}^{r_2} \bar{F} \cdot d\bar{r}_{P/O} \tag{17}$$

where r_1 and r_2 denote the coordinates of points P_1 and P_2, respectively.

In general, both the magnitude and direction of $\bar{F}$ will depend on the location of point P. Furthermore, the angle between $\bar{F}$ and the infinitesimal displacement vector $d\bar{r}_{P/O}$ will depend on the orientation of the tangent to the path. We can therefore expect that the value of the work obtained from equation (17) will depend on whether path C_1 or path C_2 has been followed in the finite displacement.

Our interest in statics, however, rests in the exceptional situation where the work done does not depend on the path followed. In that case it must be true that the work $U_{1\to2}$ is a function only of the locations of the initial point P_1 and the final point P_2. Considering that the locations of the initial and final points appears only in the limits of the integral for $U_{1\to2}$, it then follows that the work will have the form

$$\boxed{U_{1\to2} = V_1 - V_2} \tag{18}$$

where V represents a scalar function whose value depends on the position of point P, so that V_1 is the value of this function at position P_1 and V_2 is the value of the function at position P_2.

The function V is called the *potential energy* of the force $\bar{F}$ and the force is said to be *conservative*. To see why, suppose that point P were to follow a closed path by moving from position P_1 to position P_2 along path C_1, and then returning to position P_1 along path C_2. According to equation (18), the work done in the second portion of the movement is

$$U_{2\to1} = V_2 - V_1 = -U_{1\to2} \tag{19}$$

This means that the total work done in following the closed path is zero, because

$$U_{1\to2\to1} = U_{1\to2} + U_{2\to1} = 0 \tag{20}$$

Equations (19) and (20) show that the work done by the force $\bar{F}$ when point P moves from one position to another is not lost. Rather, it is stored (thus, the word conservative) in the system and may be used to return point P back to its original position. In the special case where $V_2 = 0$, we have $U_{1\to2} = V_1$. Thus, we may regard the potential energy V to be the stored ability of the force to do work when point P goes to the position of zero potential energy.

Two types of conservative forces that especially interest us are the weight force acting on a body at the surface of the earth and the force of a spring. Let us consider the weight force first. Figure 14 shows a body whose center of mass G moves from position P_1 to position P_2. Note that for the purpose of computing the potential energy of the weight force shown, we need not consider what causes the movement.

To describe the vectors shown in Figure 14, we shall utilize an XYZ coordinate system whose Y axis is vertical and whose origin O is located

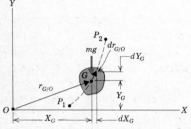

Figure 14

at an arbitrary position, as illustrated. Denoting the coordinates of point G as (X_G, Y_G), an infinitesimal displacement $d\bar{r}_{G/O}$ can be seen from Figure 14 to be equivalent to a displacement dX_G to the right and dY_G upward. That is,

$$d\bar{r}_{G/O} = dX_G \, \bar{I} + dY_G \, \bar{J}$$

The work done by the weight force as a result of this displacement is

$$dU = (-mg\bar{J}) \cdot d\bar{r}_{G/O} = -mg \, dY_G$$

Thus, the total work done by the weight force in the movement from position P_1 to position P_2 is

$$U_{1\to2} = \int_{r_{P_1/O}}^{r_{P_2/O}} (-mg \, dY_G) = -mg \, Y_G \Big|_{r_{P_1/O}}^{r_{P_2/O}}$$

$$= -mgY_2 + mgY_1 \tag{21}$$

where Y_1 and Y_2 denote the Y coordinates of points P_1 and P_2, respectively.

Equation (21) was obtained without knowing any details about the type of path followed by the center of mass, so we can conclude that gravity is a conservative force. By comparing the specific result of equation (21) to the general result for a conservative force, equation (18), we can see that

$$\boxed{V_{\text{grav}} = mgY_G} \tag{22}$$

where Y_G represents the vertical height of the center of mass G above the origin.

An alternative way of regarding the distance Y_G is to think of the X axis in Figure 14 as defining a reference elevation, called the *datum*. Then Y_G is the elevation of the center of mass above the datum. When point G is below the datum, the value of Y_G is negative, so V_{grav} is also negative. This simply means that the weight force will do negative work if the center of mass is moved back to the datum.

It is important to realize that the preceding derivation applies equally well to other forces having constant magnitude and direction, not just the weight force acting on a body near the surface of the earth. Specifically, if a force $\bar{F}$ has a constant magnitude and direction, regardless of where its point of application moves to, we can form an analogy with the gravity force by defining a line perpendicular to $\bar{F}$ to be the datum for this force. Letting h be the height of the point of application *above* the datum, that is, in the direction opposite that of $\bar{F}$, the potential energy of the force is $V_F =$

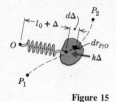

Figure 15

Fh. Hence, all forces having constant magnitude and direction are conservative. This fact will be useful for our study in the next section of the stability of the equilibrium position.

Let us now consider the force exerted by a spring. In Figure 15 there is a spring anchored at the fixed point O and attached at its other end to point P in a moving body. Again, we consider the situation where point P follows an arbitrary path as it moves from the initial position P_1 to the final position P_2.

In the situation depicted in the figure, the spring has an extension Δ beyond the unstretched length l_0. Thus, in a differential displacement $d\bar{r}_{P/O}$ in which point P moves outward from point O, the spring is extended by an additional amount $d\Delta$. Noting that the spring force applied to the body is $F = k\Delta$ directed from point P to point O, the component of the displacement $d\bar{r}_{P/O}$ perpendicular to line OP is also perpendicular to the spring force, so no work is done in this type of movement. On the other hand, the component of displacement $d\Delta$ outward along line OP is opposite from the spring force, so it results in negative work in the amount $-k\Delta$ $d\Delta$. Therefore, the total work in moving from position P_1 to position P_2 is

$$U_{1\to2} = \int_{r_{P_1/O}}^{r_{P_2/O}} (-k\Delta \; d\Delta) = -\tfrac{1}{2}k\Delta^2 \Big|_{r_{P_1/O}}^{r_{P_2/O}}$$

$$= -\tfrac{1}{2}k\Delta_2{}^2 + \tfrac{1}{2}k\Delta_1{}^2 \tag{23}$$

where Δ_1 and Δ_2 denote the extension of the spring at positions P_1 and P_2, respectively.

Equation (23) was obtained without knowledge of the path followed by point P; it follows that the spring force is conservative. Comparison of equation (23) with the general result of equation (18) then shows that

$$\boxed{V_{\mathrm{spr}} = \tfrac{1}{2}k\Delta^2} \tag{24}$$

Although this result was obtained by considering a spring that is extended, it is equally valid for springs that can sustain a compressive force $k\Delta$ due to a decrease Δ in its length. We therefore see that regardless of whether a spring is elongated or shortened, its potential energy is positive. This means that it will do positive work when the system returns to the position where the spring is undeformed.

We could go on to consider other types of conservative forces, but equations (22) and (24) will suffice for the systems we wish to consider. An important aspect of the potential energy of a force is that it is a scalar, so the potential energy of each of several conservative forces is additive as a scalar.

Let us now return to the reason why we consider the concept of potential energy in a course on statics. Suppose that we have chosen a generalized coordinate, which we will label as q. By definition, the position of all points in the system may be described in terms of the value of q. It follows that the total potential energy V of all conservative forces is a function of q; that is, $V = V(q)$. Then, with the aid of equation (18), we see that the virtual work done by the conservative forces in an infinitesimal increase δq of the generalized coordinate is

$$\delta U^{(cons)} = V(q) - V(q + \delta q)$$

$$= -\delta V(q) \equiv -\frac{dV}{dq}\delta q \tag{25}$$

We now separate the total virtual work δU into the portion $\delta U^{(cons)}$, attributable to the set of conservative forces acting on the system, and the portion $\delta U^{(nc)}$, attributable to the nonconservative forces acting on the system. Then, upon application of equation (25), the principle of virtual work becomes

$$\delta U = \delta U^{(nc)} + \delta U^{(cons)} = \delta U^{(nc)} - \frac{dV}{dq}\delta q = 0$$

$$\boxed{\delta U^{(nc)} = \frac{dV}{dq}\delta q} \tag{26}$$

Equation (26) is the principle of virtual work and energy. In words, it states that

> In the static equilibrium position the virtual work done by the nonconservative forces equals the virtual change in the potential energy of the conservative forces.

The usefulness of this principle is that we no longer have to evaluate the virtual displacements of the points of application of conservative forces. All that is necessary to account for the conservative forces is that we express their potential energy V in terms of the generalized coordinate q and then evaluate the derivative dV/dq. Thus, equation (26) is merely the result of a refinement in the method of computing the work done by a certain class of forces. Of course, all forces of uncertain nature should be regarded as nonconservative.

In the special situation where all of the forces doing work are conservative, in which case we say that we have a *conservative system*, equation (26) reduces to

$$\frac{dV}{dq} = 0 \qquad (27)$$

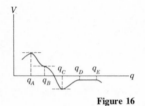

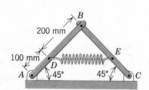

Figure 16

This is known as the *principle of stationary potential energy,* because it states that in the static equilibrium position of a conservative system, the potential energy is either a maximum or minimum value, or possibly a horizontal inflection point. A typical potential energy function for a conservative system is illustrated in Figure 16. The values q_A, q_B, q_C, and the region between q_D and q_E all correspond to values of the generalized coordinate for which the system is in static equilibrium.

EXAMPLE 3

Bars AB and BC in the system shown each have a mass of 5 kg. The spring has an unstretched length of 200 mm. Determine the value of the stiffness k for which the horizontal reaction at pin C is reduced to one-half the value that would be obtained if $k = 0$.

Solution

Aside from the fact that it is no longer necessary to compute the virtual displacement of the points of application of conservative forces, the basic method for solving problems that was developed in the last section is essentially unmodified by the utilization of the concept of potential energy. We begin with a free body diagram and a sketch of the geometry of the system.

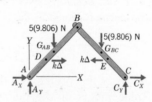

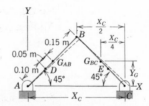

We have been requested to determine the reaction $\bar{C}_X$. Therefore, the virtual movement we give the system is one that violates the constraint of pin C against horizontal movement. The resulting virtual movement is pictured in the geometric sketch. Note that the other reaction forces do no work because the reactions $\bar{A}_X$ and $\bar{A}_Y$ are applied at a fixed point and $\bar{C}_Y$ is perpendicular to the horizontal virtual displacement of pin C. Point A is selected as the origin for the coordinate system shown because it is the only point in the system that is fixed in the virtual movement.

Suitable choices for the generalized coordinate are either the angle θ describing the angle of elevation of the bars, or else the horizontal distance X_C. For the sake of variety from previous problems, let us use the latter. The only nonconservative force doing work in the virtual displacement is the reaction $\bar{C}_X$. Thus, the only virtual displacement we need to formulate is that of point C. The position vector of this point in terms of X_C is simply

$$\bar{r}_{C/A} = X_C \bar{I}$$

so

$$\delta \bar{r}_{C/A} = \frac{d\bar{r}_{C/A}}{dX_C} \delta X_C = (\bar{I}) \, \delta X_C$$

It then follows that the work $\delta U^{(\text{nc})}$ is

$$\delta U^{(\text{nc})} = C_X \bar{I} \cdot \delta \bar{r}_{C/A} = C_X \, \delta X_C$$

We now formulate the potential energy of the conservative weight and spring forces. The X axis is a convenient datum for the potential energy of gravity, as indicated in the free body diagram. Adding the potential energy of each of the conservative forces, we have

$$V = V_{\text{grav}} + V_{\text{spr}}$$

$$= 2(5)(9.806) Y_G + \tfrac{1}{2} k \Delta^2$$

where Y_G is the elevation of the centers of mass and Δ is the elongation of the spring.

The application of equation (26) requires that we express V in terms of the generalized coordinate $q \equiv X_C$. From the sketch of the geometry we see that the horizontal distance from point C to the center of mass G_{BC} is $X_C/4$, so the Pythagorean theorem gives

$$Y_G = \left[(0.15)^2 - \left(\frac{X_C}{4} \right)^2 \right]^{1/2} \text{ m}$$

To evaluate the spring extension Δ, we subtract the unstretched length of 0.20 m from the stretched length l_{DE}. Considering the similar triangles BDE and BAC shown in the sketch, we find

$$\frac{l_{DE}}{0.20 \text{ m}} = \frac{X_C}{0.30 \text{ m}}$$

$$l_{DE} = \tfrac{2}{3} X_C$$

Thus

$$\Delta = \tfrac{2}{3}X_C - 0.20 \text{ m}$$

Upon substitution of these expressions for X_C and Δ, the potential energy becomes

$$V = 98.06\left(0.0225 - \frac{X_C{}^2}{16}\right)^{1/2} + \tfrac{1}{2}k(\tfrac{2}{3}X_C - 0.20)^2$$

We now employ the work-energy principle, equation (26). Thus

$$\delta U^{(\text{nc})} = \frac{dV}{dX_C}\,\delta X_C$$

$$C_X\,\delta X_C = \left[\frac{1}{2}(98.06)\left(0.0225 - \frac{X_C{}^2}{16}\right)^{-1/2}\left(-\frac{X_C}{8}\right)\right.$$

$$\left. + k\left(\frac{2}{3}X_C - 0.200\right)\left(\frac{2}{3}\right)\right]\delta X_C \text{ J}$$

We know that $X_C = 0.30\sqrt{2}$ m in the static equilibrium position, because triangle ABC in that case is a 45° right triangle. Therefore, after the factor δX_C is cancelled, the equilibrium equation is

$$C_X = (49.03)\left[0.0225 - \frac{0.090(2)}{16}\right]^{-1/2}\left(-\frac{0.30\sqrt{2}}{8}\right)$$

$$+ k\left[\left(\frac{2}{3}\right)0.30\sqrt{2} - 0.20\right]\left(\frac{2}{3}\right)$$

$$= -12.258 + 0.05523k \text{ N}$$

When $k = 0$ this equation gives $C_X = -12.258$ N. Thus, setting C_X equal to one-half this value, we find that

$$-\tfrac{1}{2}(12.258) = -12.258 + 0.05523k$$

$$k = 111.0 \text{ N/m}$$

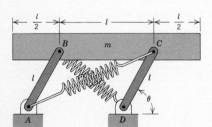

EXAMPLE 4

The table top of mass m is supported by the lightweight bars AB and CD, which are braced by two identical springs of stiffness k and unstretched length $1.20l$. Determine an expression for the value of k for which the system is in equilibrium at $\theta = 60°$.

Solution

As always, we start with a free body diagram of the system and a separate sketch describing the geometry of the system. In the latter, h is an unspecified dimension required to locate the center of mass G.

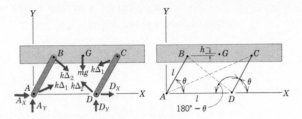

Noting that we are interested in the value of k for equilibrium, and not the reaction forces at pins A and D, the virtual movement of interest is one in which the angle θ is changed while the pin connections are maintained. Clearly, in such a movement the only forces doing work will be those of the springs and gravity, both of which are conservative. Therefore, we may employ the principle of stationary potential energy, equation (27), and it will not be necessary to evaluate virtual displacements of points.

For a generalized coordinate we choose the angle θ. The choice of the datum for the potential energy of gravity is arbitrary. The origin of the XYZ coordinate system defined in the sketch of the geometry is located at a convenient fixed point, so we shall choose the X axis for the datum. From the second sketch we see that the height of point G above the datum is $l \sin \theta + h$, so the potential energy of the gravity force is

$$V_{\text{grav}} = mg(l \sin \theta + th)$$

The potential energy of a spring is given by equation (24), so the total potential energy of both springs is

$$V_{\text{spr}} = \tfrac{1}{2}k\Delta_{AC}^2 + \tfrac{1}{2}k\Delta_{BD}^2$$

In order to employ equation (27), we will need to know the relationship between the deformation of each spring and the generalized coordinate θ. This relationship is obtained with the aid of the sketch of the geometry. Applying the law of cosines to triangle ABD, we find that the length of the spring when stretched between pins B and D is defined by

$$l_{BD}^2 = l^2 + l^2 - 2(l)(l) \cos \theta = 2l^2(1 - \cos \theta)$$

$$l_{BD} = \sqrt{2}\, l(1 - \cos \theta)^{1/2} \equiv 2l \sin \frac{\theta}{2}$$

A similar analysis of triangle ACD shows that the length of the spring joining pins A and C is

$$l_{AC} = 2l \cos \frac{\theta}{2}$$

Note that these expressions were obtained with the aid of the half-angle identities for the sine and cosine functions.

The elongation of a spring is the increase in its length beyond its unstretched value. Thus

$$\Delta_{AC} = l_{AC} - 1.20l = l\left(2 \cos \frac{\theta}{2} - 1.20\right)$$

$$\Delta_{BD} = l_{BD} - 1.20l = l\left(2 \sin \frac{\theta}{2} - 1.20\right)$$

The total potential energy of the system as a function of θ is then found to be

$$V = V_{\text{grav}} + V_{\text{spr}} = mg(h + l \sin \theta) + \frac{1}{2} kl^2\left(2 \cos \frac{\theta}{2} - 1.20\right)^2$$

$$+ \frac{1}{2} kl^2\left(2 \sin \frac{\theta}{2} - 1.20\right)^2$$

Equation (27) now requires that in the position of static equilibrium

$$\frac{dV}{d\theta} = mgl \cos \theta + kl^2\left(2 \cos \frac{\theta}{2} - 1.20\right)\left(- \sin \frac{\theta}{2}\right)$$

$$+ kl^2\left(2 \sin \frac{\theta}{2} - 1.20\right)\left(\cos \frac{\theta}{2}\right)$$

$$= mgl \cos \theta + 1.20kl^2\left(\sin \frac{\theta}{2} - \cos \frac{\theta}{2}\right) = 0$$

One possible solution of this equation is $\theta = 90°$, corresponding to the position where bars AB and CD are vertical. However, in this problem we are interested in the existence of another equilibrium position, corresponding to $\theta = 60°$. Substituting this value into the equation for $dV/d\theta$ yields

$$0.50mgl + 1.20kl^2(0.50 - 0.8660) = 0$$

$$k = 1.138\frac{mg}{l}$$

HOMEWORK PROBLEMS

VIII.33   Two identical springs, which are anchored at points A and B and attached together, have a tension T_0 when they are not supporting the block. Derive an expression for the mass m of the block that will give equilibrium at a specific value of the angle θ. Show that, for a fixed value of θ, this value of m increases if either the stiffness k or the initial tension

T_0 is increased. Explain this trend physically by considering the amount of energy stored in a spring for a given increase in its elongation.

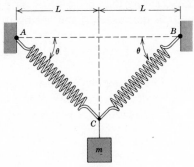

Prob. VIII.33

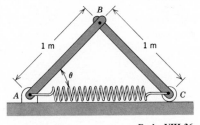

Prob. VIII.36

VIII.34 Solve Problem VIII.9 using the concept of potential energy.

VIII.35 Solve Problem VIII.13 using the concept of potential energy.

VIII.36 Identical bars *AB* and *BC,* each having a mass of 10 kg, form a linkage in the vertical plane. The spring has a stiffness of 400 N/m. Knowing that the equilibrium position corresponds to $\theta = 45°$, determine the undeformed length of the spring.

VIII.37 and VIII.38 Determine the relationship between the mass *m* of the suspended body and the angle θ for equilibrium. The identical springs have stiffness *k* and are unstressed when $\theta = 30°$.

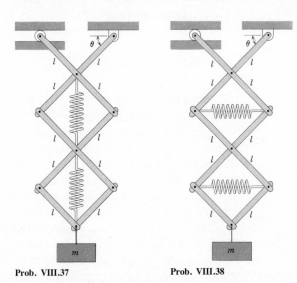

Prob. VIII.37   **Prob. VIII.38**

VIII.39 Consider the system studied in Problem VIII.16. Explain why the virtual work of the spring in the position shown is more easily formulated by treating the spring force as a nonconservative force, rather than formulating the effect of the spring in terms of potential energy.

VIII.40 Solve Problem VIII.17 using the concept of potential energy.

VIII.41 The spring has a stiffness of 50 lb/in. and an unstretched length of 20 in. Neglecting the mass of the bars, determine the horizontal component of the reaction at pin support D.

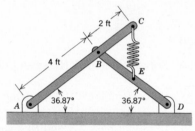

Prob. VIII.41

VIII.42 The spring, whose stiffness is 0.80 kN/m, is unstretched in the position where $\theta = 90°$. Determine the couple $\bar{M}_C$ required for equilibrium in terms of the angle θ.

VIII.43 Collar C, which rides on the smooth bar BD, is pinned to the ground. The spring, whose stiffness is 30 lb/in., is unstressed when $\theta = 22.62°$. Determine an expression for the vertical force $\bar{F}$ at point B required for equilibrium in terms of the angle θ.

VIII.44 Solve Problem VIII.20 using the concept of potential energy.

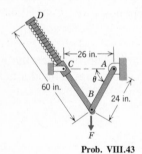

Prob. VIII.43

F. STABILITY OF THE EQUILIBRIUM POSITION

When we are dealing with partially constrained systems, it will frequently happen that several different equilibrium positions are theoretically possible. For example, in the case of the table top of Example 4, which was supported by two parallel bars, we found that equilibrium was possible with the bars in the vertical position, as well as when they were inclined at 60° from the horizontal. In such situations it may not actually be possible to maintain some of the equilibrium positions determined theoretically, for the reason that some of them may be unstable.

A very simple situation illustrating the question of the stability of the equilibrium position is the particle in Figure 17, which is placed on the smooth surface whose vertical profile is shown. At positions A, B, and C

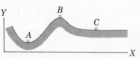

Figure 17

the tangent to the hill is horizontal. It then follows that each of these locations is a theoretical static equilibrium position for the block. Let us consider each of these.

Position A is the lowest point in a valley. After being moved away from this position by a slight disturbance, such as a gust of wind, the particle will return to the low point. This is said to be a *stable* equilibrium position. On the other hand, point B is the top of a hill. There, a small disturbance away from the top causes the particle to fall down the hill. This is an *unstable* equilibrium position. At location C the particle will remain at any point along the level portion of the hill, because there is neither a tendency to move away nor return to location C after a disturbance. This is the condition of *neutral* equilibrium.

Two observations regarding the situation in Figure 17 are significant. First, notice that in each case the only constraint acting on the particle is that of the hill, which exerts a normal force on the particle in order to prevent movement of the particle in the normal direction. In the movement satisfying this constraint, that is, in a displacement of the particle tangent to the hill, the only force doing work is gravity. Thus, the particle on the hill forms a partially constrained system that is conservative with respect to virtual movements satisfying the existing constraints. The second observation regarding the system of Figure 17 is that when the X axis is chosen as the datum for the potential energy, then the potential energy of the system is proportional to the coordinate Y; that is, $V = mgY$. If we now choose the horizontal coordinate X as the generalized coordinate, a graph of the potential energy as a function of the generalized coordinate has the same shape as the vertical profile of the hill shown in Figure 17.

The preceding discussion suggests that the type of stability of a partially constrained conservative system is determined by the properties of its potential energy function. This result can be proven, and the results are analogous to those for the particle on the hill.

Let q denote the generalized coordinate that describes the movement of a partially constrained system satisfying whatever constraints there are. Suppose that Figure 16, which is repeated here as Figure 18, represents

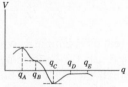

Prob. VII.18

the potential energy of such a system. When $q = q_A$, where the potential energy is a maximum (analogous to a particle at the peak of a hill), there is unstable equilibrium. If the system is given a slight disturbance from this location, the conservative forces will act to move the system farther away from q_A. At location C, where the potential energy is minimum (a particle at the bottom of a valley), the conservative forces act to return the system to this location after a disturbance. This is a stable equilibrium position. Thus we have the following criteria for conservative systems.

$$
\left. \frac{dV}{dq} \right|_{q=q_0} = 0 \qquad \left. \frac{d^2V}{dq^2} \right|_{q=q} > 0 \qquad
\begin{array}{l} q_0 \text{ represents a} \\ \text{stable equilibrium} \\ \text{position} \end{array}
$$

$$
\left. \frac{dV}{dq} \right|_{q=q_0} = 0 \qquad \left. \frac{d^2V}{dq^2} \right|_{q=q_0} < 0 \qquad
\begin{array}{l} q_0 \text{ represents an} \\ \text{unstable equilibrium} \\ \text{position} \end{array}
\tag{28}
$$

In the special case where the second derivative of the potential energy is zero, one must distinguish between the horizontal inflection point at location B in Figure 18 and the level plateau lying between locations D and E. At the former, after a disturbance that increases the value of q slightly from q_B, the conservative forces will cause the system to continue to move away from this location. Even though the system will return to this location after a slight decrease of q from q_B, a horizontal inflection point must be regarded as representing an unstable equilibrium position. In contrast, all values of q lying in the range between q_D and q_E, where the potential energy is constant, represent possible equilibrium positions $(dV/dq \equiv 0)$. Therefore, this is a region of neutral stability.

It should be noted that the stability theorems developed here apply only to conservative systems. The stability of partially constrained systems in which nonconservative forces do work is a more complicated question, being concerned with the dynamics of the system. Thus, before applying equations (28) it must be ascertained that all forces doing work are conservative.

EXAMPLE 5

Consider the system of Example 4, where a table top is supported by bars and braced by springs. (a) Determine the range of values of the spring stiffness k for which the system is in stable equilibrium at $\theta = 90°$. (b) Letting k be the value required for equilibrium at $\theta = 60°$, determine the stability of that equilibrium position.

Solution

As we saw in the solution of Example 4, the system is conservative. The derivative $dV/d\theta$ was found to be

$$\frac{dV}{d\theta} = mgl \cos\theta + 1.20kl^2\left(\sin\frac{\theta}{2} - \cos\frac{\theta}{2}\right)$$

This derivative is zero when $\theta = 90°$. It also was made to be zero at $\theta = 60°$ by setting $k = 1.138$ mg/l, which was the solution to Example 4.

To evaluate the stability we calculate the second derivative of the potential energy. This is

$$\frac{d^2V}{d\theta^2} = -mgl \sin\theta + \left(\frac{1}{2}\right)(1.20)kl^2\left(\cos\frac{\theta}{2} + \sin\frac{\theta}{2}\right)$$

Part (a) is then solved by setting $\theta = 90°$, with the result that

$$\frac{d^2V}{d\theta^2}\Big|_{\theta=90°} = -mgl + 0.8485kl^2$$

If k is too small, then $d^2V/d\theta^2$ is negative and the equilibrium is unstable. If the springs are sufficiently stiff (large k), then $d^2V/d\theta^2$ is positive, and the system is stable. The allowable range of values of k for stability is therefore found by setting

$$\frac{d^2V}{d\theta^2}\Big|_{\theta=90°} = -mgl + 0.8485kl^2 > 0$$

$$k > 1.179\frac{mg}{l} \qquad \triangleleft$$

To solve part (b) we set $\theta = 60°$ and $k = 1.138$ mg/l in the general expression for $d^2V/d\theta^2$. This gives

$$\frac{d^2V}{d\theta^2} = [-1 + 0.8485(1.138)]mgl < 0 \qquad \triangleleft$$

The second derivative is negative, so $\theta = 60°$ is an unstable equilibrium position for the given value of k.

EXAMPLE 6
The system of lightweight rigid bars are loaded by a compressive force $\bar{P}$. In the horizontal position shown, the springs of stiffness k have identical tensile forces T. Determine the maximum value of P for which this horizontal equilibrium position is stable.

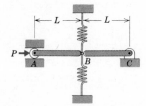

Solution
The system of bars is only partially constrained, because joint B may move up or down, causing roller A to move to the right. To assist us in

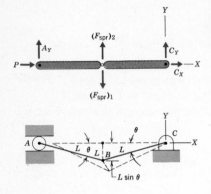

discussing the forces, we draw a free body diagram, and as an aid in locating points in general, we draw a sketch of the system in an arbitrary (in this case, nonhorizontal) position. These are shown to the left.

The angle θ can be seen from the geometric sketch to serve for describing the movement of the system consistent with the constraint conditions imposed by the supports. The equilibrium position of interest corresponds to $\theta = 0$.

In the virtual movement appearing in the geometric sketch, the only forces doing work are those of the springs and $\bar{P}$. The application of the criteria of equations (28) requires that all forces doing work are conservative ones. Assuming the force $\bar{P}$ to be constant in magnitude and direction, we may then regard it as a horizontal "weight" force. Choosing the Y axis (a line perpendicular to $\bar{P}$) as the datum for this force, the point of application A of this force is at a distance $2L \cos \theta$ above the datum, so

$$V_P = P(2L \cos \theta)$$

In order to formulate the potential energy of the springs in the form required for application of the energy principles, we need to relate their extension to the generalized coordinate θ. If we were interested in the equilibrium of the system for large values of θ, these expressions would be somewhat complicated to obtain because it would then be necessary to consider the horizontal and vertical distances locating pin B in the displaced position. However, because our interest here is in the equilibrium position at $\theta = 0$, we only need an expression for potential energy that is accurate for small values of θ.

When θ is a very small angle, pin B in the displaced position may be regarded as being a distance $L \sin \theta$ vertically below its position when $\theta = 0$. Further, for small values of θ, we may approximate $\sin \theta \approx \theta$, $\cos \theta = 1 - \theta^2/2$. Thus, the upper spring is elongated by an additional amount $L \sin \theta \approx L\theta$ in the displaced position, whereas the lower spring is shortened by the same amount. From the given information, we know that in the position $\theta = 0$ each spring has a tension T and thus an elongation T/k. Combining this elongation with the deformation introduced by a nonzero value of θ, we find that

$$V_{\text{spr}} = \frac{1}{2}k\left(\frac{T}{k} + L\theta\right)^2 + \frac{1}{2}k\left(\frac{T}{k} - L\theta\right)^2$$

$$= \frac{T^2}{k} + kL^2\theta$$

As an aside, notice that the only effect of the pretensioning T given the springs is to change the reference level of the potential energy.

We now add V_{spr} and V_P and compute the required derivatives of the total potential energy V. This gives

$$V = 2PL\left(1 - \frac{\theta^2}{2}\right) + \frac{T^2}{k} + kL^2\theta^2$$

$$\frac{dV}{d\theta} = -2PL\theta + 2kL^2\theta$$

$$\frac{d^2V}{d\theta^2} = -2PL + 2kL^2$$

The expression for $dV/d\theta$ is identically zero at $\theta = 0$, confirming that this is an equilibrium position. The requirement for stability then reveals that

$$\frac{d^2V}{d\theta^2}\Big|_{\theta=0} \equiv -2PL + kL^2 > 0$$

$$P < \tfrac{1}{2} kL$$

Any value of P exceeding this limit will cause the system to seek an equilibrium position where θ is nonzero. We cannot determine this position on the basis of the foregoing analysis, because we have limited the analysis to very small values of θ.

It is interesting to note that many aspects of the analysis are similar to those of the much more complicated problem of the buckling of structures due to compressive loads. Buckling is treated in courses on strength of materials and structures.

HOMEWORK PROBLEMS

VIII.45 Blocks A and B of mass m, resting against the smooth wall and floor, are connected by a bar of mass $2m$. The spring attached to block A has a stiffness k and is unstressed when the bar is vertical. (a) Determine the equilibrium positions of the bar. (b) Determine the range of values of k for which the equilibrium positions found in part (a) are stable.

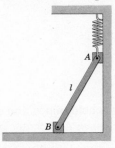

Prob. VIII.45

VIII.46   Determine the type of stability of the equilibrium position $\theta = 45°$ for the linkage of Problem VIII.36.

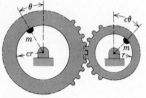

VIII.47   The two gears lying in the vertical plane carry identical eccentric masses m at radial distances that are in the same ratio c as the ratio of the radii of the gears. (a) Determine an expression for the angle θ at the equilibrium positions. (b) For the case $c = 1$, solve the equation obtained in part (a) and evaluate the stability of each possible equilibrium position. (c) Repeat part (b) for $c = 2$.

Prob. VIII.47

VIII.48   In each case illustrated, a bar of mass m is supported by two identical bars of negligible mass. For each system determine the type of stability of the equilibrium position shown.

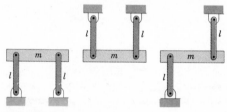

Prob. VIII.48

VIII.49   Each of the bars in the mechanism shown has a mass per unit length m. Bar BD is pinned to collar C which is supported by the spring of stiffness k and unstretched length $3l$. Determine the values of the angle θ for equilibrium and evaluate the stability of each position.

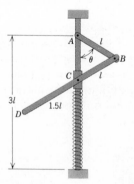

Prob. VIII.49

VIII.50   The lamp is held in position by the mechanism shown. The bulb and reflector has mass m and center of mass G, whereas the mass of the bars is negligible. The spring is unstressed in the condition corresponding to $\theta = 180°$. Determine the stiffness k required for equilibrium at $\theta = 120°$ and evaluate the stability of this equilibrium position.

Prob. VIII.50

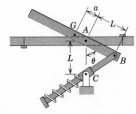

Prob. VIII.51

VIII.51 A ventilating door in a roof, having mass m and center of mass G, is hinged at point A. The angle of inclination θ, measured from the vertical, is controlled by the mechanism shown. The spring is unstressed in the condition corrresponding to $\theta = 0°$. Determine the stiffness k of the spring required for equilibrium at a specific angle θ and determine the stability of the corresponding equilibrium position.

VIII.52 Evaluate the stability of the equilibrium position of the mechanism of Problem VIII.43.

VIII.53 The mechanism shown is a component of a seismometer. When the block is directly over the pivot point B, the springs of stiffness k have equal compression forces T. Considering the mass of all elements except the block to be negligible, determine the range of values of k for which the vertical position is stable.

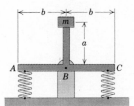

Prob. VIII.53

VIII.54 and VIII.55   A horizontal force $\bar{P}$ is applied at end A of the mechanism shown. Derive an expression for the maximum force $\bar{P}$ for which the mechanism is stable in the position shown. The springs have a stiffness k and carry a compressive force T in the position shown.

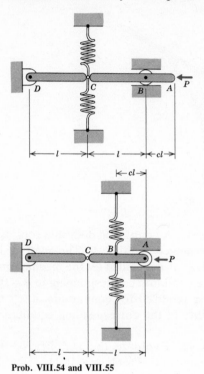

Prob. VIII.54 and VIII.55

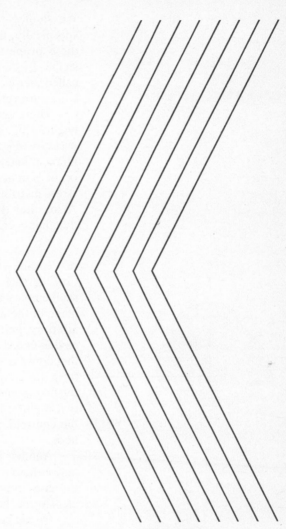

MODULE IX
AREA MOMENTS
AND PRODUCTS
OF INERTIA

We studied two important properties of a planar figure in Module V, specifically, the area and the location of the centroid. However, neither of these properties conveys much information about the geometry of the shape. In this module we will study numerical parameters, alternatively called *second moments of area* or *area moments and products of inertia*, which are related to the shape of a planar figure.

There are many applications for the inertia parameters in the field of engineering. It was shown in Module V that the effect of a hydrostatic pressure distribution on a flat surface can be described in terms of a moment and a product of inertia. Another subject where these parameters occur is in mechanics of materials, where they are needed to describe the stress distribution in beams resulting from transverse and torsional loads. Also, they have certain similarities to parameters which occur in dynamics, called mass moments of inertia.

A.   BASIC DEFINITIONS

Consider the area $\mathcal{A}$ shown in Figure 1. To locate points on this area we shall employ an *xyz* coordinate system whose *xy* plane coincides with the plane of the area. The origin O of this coordinate system is located at an arbitrary point. The area $\mathcal{A}$ is composed of infinitesimal elements $d\mathcal{A}$ having coordinates $(x, y, 0)$. Suppose that we want to define a parameter that conveys information about the "average" distance of these elements from the *x* axis. One possible parameter is $y\,d\mathcal{A}$, which is the integrand in the first moment of area with respect to the *y* coordinate. However, recall that the first moment of area provides information about the location of the centroid. Therefore, $y\,d\mathcal{A}$ will not provide the information we seek here.

Another parameter to consider is $y^2\,d\mathcal{A}$. Unlike the term $y\,d\mathcal{A}$, which may either be positive or negative, $y^2\,d\mathcal{A}$ is always positive. Therefore, $y^2\,d\mathcal{A}$ does contain information regarding the (positive) distance from an element to the *x* axis.

On the basis of these observations we may define the *area moment of inertia* (or alternatively, the *second moment of area*) about the *x* axis which we shall denote by the symbol I_x. Clearly, similar reasoning when discussing the distribution of area about the *y* axis will lead us to define the area moment of inertia about the *y* axis, I_y. The sum of an infinite number of infinitesimal terms with respect to $d\mathcal{A}$ is an integral, so the definitions are

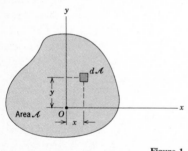

Figure 1

$$
\begin{aligned}
I_x &= \int_{\mathcal{A}} y^2\,d\mathcal{A} \\
I_y &= \int_{\mathcal{A}} x^2\,d\mathcal{A}
\end{aligned}
\tag{1}
$$

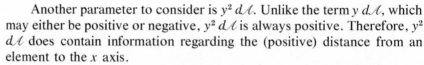

Observe that equations (1) are functions of the coordinate values squared, hence the phrase, *second moments*. The phrase *moment of inertia* about the x axis is a result of the similarity between the properties of an area as defined above and the properties of the mass distribution of a rigid body. Thus, although the word *area* is frequently omitted when referring to the moments of inertia, it is acceptable to do so only when there is no possibility of confusion with the mass properties of a rigid body.

The values of I_x and I_y are always positive, because their integrands are always positive or zero. It follows, then, that if two areas have the same shape (for instance, circular, or rectangular in a fixed proportion), the larger area will have the larger values for the moments of inertia. Also, for two different shapes having the same area, the one whose area is distributed farther from an axis will have the larger value for the moment of inertia about that axis.

The parameters we have just defined convey information about the distribution of area about the x and y axes. A third moment of inertia is concerned with the distribution of area about the z axis. Referring to Figure 1, we see that the z axis is perpendicular to the plane of the area, intersecting the origin O. Hence, the distance from this axis to an area element $d\mathcal{A}$ is $(x^2 + y^2)^{1/2}$. Squaring this distance and multiplying it by $d\mathcal{A}$, we form the integral for the (area) *polar moment of inertia J_z*.

$$J_z = \int_{\mathcal{A}} (x^2 + y^2)\, d\mathcal{A} \qquad (2)$$

The reason for using the symbol J_z rather than I_z for the polar moment of inertia is that it is conventional to do so.

The polar moment of inertia is not an independent quantity. To see this, it is only necessary to express equation (2) as a sum of integrals rather than an integral of a sum. Employing equations (1) then leads to the identity

$$J_z = \int_{\mathcal{A}} x^2\, d\mathcal{A} + \int_{\mathcal{A}} y^2\, d\mathcal{A} = I_y + I_x \qquad (3)$$

A common way to describe the moments of inertia of a shape about an axis is to give the *radius of gyration k* about that axis. This quantity is simply the square root of the ratio of the corresponding moment of inertia to the area $\mathcal{A}$. That is,

$$k_x = \sqrt{\frac{I_x}{\mathcal{A}}} \qquad k_y = \sqrt{\frac{I_y}{\mathcal{A}}} \qquad k_z = \sqrt{\frac{I_z}{\mathcal{A}}} \tag{4}$$

A radius of gyration has units of length, because the area moments of inertia have units of the fourth power of length. The radii of gyration have a simple interpretation. Suppose that we have a thin rectangular strip whose area $\mathcal{A}$ is the same as that for the general shape of interest. Let us orient this strip parallel to the x axis at a distance equal to the value of k_x, as shown in Figure 2a. Because all of the elements of area are then

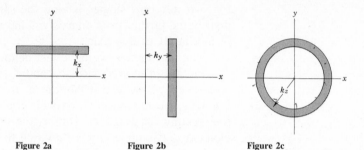

Figure 2a   Figure 2b   Figure 2c

essentially at the distance k_x from the x axis, the moment of inertia of this strip about the x axis is $k_x^2 \mathcal{A} \equiv I_x$. Similarly, the moment of inertia about the y axis of the strip in Figure 2b is identical to I_y for the general shape, and the polar moment of inertia about the z axis for the thin ring in Figure 2c is identical to J_z for the general shape. In view of equation (3), the radii of gyration are related by

$$k_z^2 = k_x^2 + k_y^2 \tag{5}$$

Having seen how the moments of inertia characterize the distribution of area about the axes, it may seem that these are the only parameters we need. This is not so, as exhibited by the two squares shown in Figure 3. For every element of area $d\mathcal{A}_1$ in the square on the left, there is a companion element $d\mathcal{A}_2$ in the right square. Both of these elements are situated at the same distances from each of the coordinate axes, so both squares have identical moments of inertia. Yet clearly, these squares differ, in the sense that they are located in different coordinate quadrants.

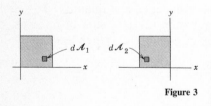

Figure 3

To describe how the area of a planar figure is situated in the coordinate quadrants, we define the area *product of inertia* as follows:

$$I_{xy} \equiv I_{yx} = \int_{\mathcal{A}} xy \, d\mathcal{A} \tag{6}$$

Because xy is positive if an element of area is located in the first or third quadrant, and negative in the second or fourth quadrant, we conclude that I_{xy} being positive means that area predominates in the first and/or third quadrants. Obviously, when I_{xy} is negative, the area predominates in the second and/or fourth quadrants.

In the important case where a shape is symmetric about one of the coordinate axes, the value of xy for an element to the left of the axis of symmetry is cancelled by the value of xy for the mirror-image element to the right. This is demonstrated in Figure 4 for a shape that is symmetric about the x axis. Thus

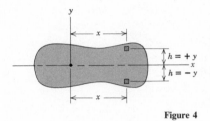

Figure 4

> Whenever a planar shape has an axis of symmetry that is either the x or y axis, then $I_{xy} \equiv 0$.

In general, when the x and y coordinate axes give a zero product of inertia, we say that they are *principal axes*. All shapes, not just symmetric ones, have principal axes. Later in this module we will see how to locate the principal axes for an arbitrary figure.

For the remainder of this module, we will consider various techniques for evaluating the area moments and products of inertia. We could go on to define higher moments of area. However, these would only be of theoretical interest, as they do not find application in the field of mechanics.

B. INTEGRATION TECHNIQUES

In view of the fact that moments and product of inertia are defined in terms of integrals, the obvious way to evaluate these properties is to evaluate the integrals. Doing this requires that we describe the position of points contained within the closed curve forming the area $\mathcal{A}$.

Recall that in Module V, when we located centroids and centers of mass by integration, we formulated the necessary integrals in terms of either Cartesian or cylindrical coordinates, whichever best suited the description of the boundaries of the body. The same is true here. However, in describing a planar area, we are concerned only with the Cartesian coordinates (x, y) or the polar coordinates (R, ϕ) locating points in the plane. A possible choice for the polar coordinates is shown in Figure 5.

For the coordinates defined in the figure, we have

$$x = R \cos \phi \qquad y = R \sin \phi$$

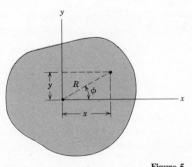

Figure 5

It should be noted that other choices for the polar coordinates are possible, for instance, by measuring the angle ϕ from a different axis.

After having made a decision to use Cartesian or polar coordinates on the basis of the shape of the curve forming the boundary of the area, we may describe the area $d\mathscr{A}$. The expressions for $d\mathscr{A}$ were derived in Module V, but for completeness we have illustrated the two basic types of elements again in Figure 6 below.

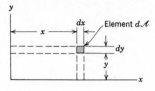
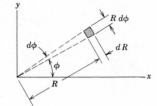

Figure 6

From Figure 6 we see that

$$d\mathscr{A} = dx\ dy \qquad \text{or} \qquad d\mathscr{A} = R\ d\phi\ dR \tag{7}$$

The following examples will demonstrate that the methods for formulating the limits of an integral extending over all of the area are the same as those we employed in locating centers of mass by integration.

EXAMPLE 1

Calculate the moments and product of inertia of the parabolic sector shown about the axes of the given coordinate system.

Solution

The parabolic curve is defined in terms of the x and y coordinates, so we will formulate the solution in terms of these coordinates. The constant k is related to the given dimensions because the points (a, b) and $(a, -b)$ are points on the parabola. Thus, we have $a = kb^2$, so $k = a/b^2$. For the formulation of the various integrals we will use the element of area shown in the diagram below.

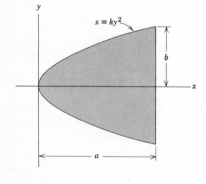

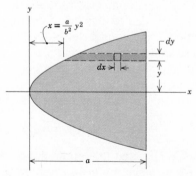

An integration over the entire area may be achieved by integrating first over the range of all x variables or all y variables, thus forming horizontal or vertical strips, respectively. As indicated in the preceding sketch, the equation for the parabola gives the value of x on the parabola for each value of y. Therefore, we will form horizontal strips by integrating x over the range $ay^2/b^2 \leq x \leq a$. All horizontal strips are then accounted for by integrating y over the range $-b \leq y \leq b$.

The parabolic sector is symmetrical about the x axis, so we have

$$I_{xy} = 0$$

Also, because the polar moment of inertia J_z may be obtained from equation (3), it is only necessary for us to evaluate the integrals in equation (1). These are

$$I_x = \int_{-b}^{b} \int_{ay^2/b^2}^{a} y^2 \, dx \, dy = \int_{-b}^{b} y^2 \left(a - \frac{ay^2}{b^2} \right) dy$$

$$= \frac{4}{15} ab^3 \qquad \qquad \triangleleft$$

$$I_y = \int_{-b}^{b} \int_{ay^2/b^2}^{a} x^2 \, dx \, dy = \tfrac{1}{3} \int_{-b}^{b} \left(a^3 - \frac{a^3 y^6}{b^6} \right) dy$$

$$= \frac{4}{7} a^3 b \qquad \qquad \triangleleft$$

From equation (3) we then find

$$J_z = I_x + I_y = \frac{4}{105} ab(7b^2 + 15a^2) \qquad \qquad \triangleleft$$

These results demonstrate a typical property of moments of inertia. Specifically, each value increases in proportion to the cube of the dimension perpendicular to the corresponding axis.

EXAMPLE 2

Calculate the moments and product of inertia of the circular sector shown with respect to the xyz coordinate system. Also, determine the corresponding radii of gyration.

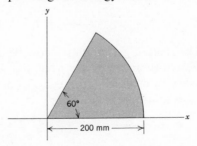

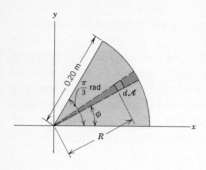

Solution

We could formulate the solution using (x, y) coordinates by writing the equation for the circular arc in terms of x and y. However, it is easier to employ polar coordinates in this problem, because points on the arc are at a constant radial distance from the origin, and equally important, points on the straight sides are at constant polar angles ϕ from the coordinate axes. Our choice for the polar coordinates is shown in the sketch.

For the coordinates defined in the sketch we have

$$x = R \cos \phi \qquad y = R \sin \phi$$

and the element of area is $d\mathcal{A} = R\, d\phi\, dR$. As indicated in the sketch, the sequence of integrations we choose will form a differential sector by integrating over $0 \le R \le 0.20$ m. All of these sectors will then be accounted for by integrating over $0 \le \phi \le \pi/3$ radians.

We rearrange the differentials in the expression for $d\mathcal{A}$ to match the order of integration, so the moments of inertia are

$$I_x = \int_0^{\pi/3} \int_0^{0.2} y^2\, R\, dR\, d\phi = \int_0^{\pi/3} \int_0^{0.2} R^3 \sin^2 \phi\, dR\, d\phi$$

$$= \tfrac{1}{4}(0.2)^4 \int_0^{\pi/3} \sin^2 \phi\, d\phi = 1.228(10^{-4})\ \text{m}^4 \qquad \triangleleft$$

$$I_y = \int_0^{\pi/3} \int_0^{0.2} x^2\, R\, dR\, d\phi = \int_0^{\pi/3} \int_0^{0.2} R^3 \cos^2 \phi\, dR\, d\phi$$

$$= \tfrac{1}{4}(0.2)^4 \int_0^{\pi/3} \cos^2 \phi\, d\phi = 2.960(10^{-4})\ \text{m}^4 \qquad \triangleleft$$

$$I_{xy} = \int_0^{\pi/3} \int_0^{0.2} xy\, R\, dR\, d\phi = \int_0^{\pi/3} \int_0^{0.2} R^3 \cos \phi \sin \phi\, dR\, d\phi$$

$$= \tfrac{1}{4}(0.2)^4 \int_0^{\pi/3} \cos \phi \sin \phi\, d\phi = 1.50(10^{-4})\ \text{m}^4 \qquad \triangleleft$$

Then, from equation (3), we find

$$J_z = I_x + I_y = 4.188(10^{-4})\ \text{m}^4 \qquad \triangleleft$$

The radii of gyration may now be determined using their definitions, equations (4). For this we need the area $\mathcal{A}$. In this problem it is not necessary to evaluate an integral to determine $\mathcal{A}$, because the sector is one-sixth of a full circle. Thus

$$\mathcal{A} = \tfrac{1}{6}\pi(0.2)^2 = 2.094(10^{-2})\ \text{m}^2$$

Hence, for the radius of gyration about the x axis we have

$$k_x = \sqrt{\frac{1.228(10^{-4})}{2.094(10^{-2})}} = 0.0766 \text{ m}$$

Similarly,

$$k_y = \sqrt{\frac{2.960(10^{-4})}{2.094(10^{-2})}} = 0.1189 \text{ m}$$

$$k_z = \sqrt{\frac{4.188(10^{-4})}{2.094(10^{-2})}} = 0.1414 \text{ m}$$

HOMEWORK PROBLEMS

IX.1–IX.6 By integration, determine the area moments of inertia I_x and I_y for the shaded area shown.

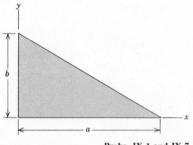

Probs. IX.1 and IX.7

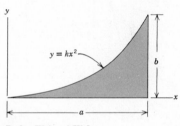

$y = kx^2$

Probs. IX.2 and IX.8

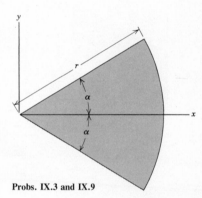

r, α, α

Probs. IX.3 and IX.9

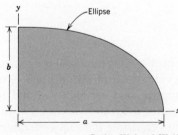

Ellipse

Probs. IX.4 and IX.10

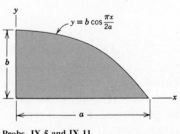

$y = b \cos \dfrac{\pi x}{2a}$

Probs. IX.5 and IX.11

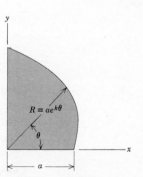

$R = ae^{k\theta}$, θ

Probs. IX.6 and IX.12

IX.7–IX.12 By integration, determine the polar area moment of inertia J_z and the product of inertia I_{xy} for the shaded area shown.

IX.13–IX.18   Determine the radius of gyration k_x of the shaded area shown.

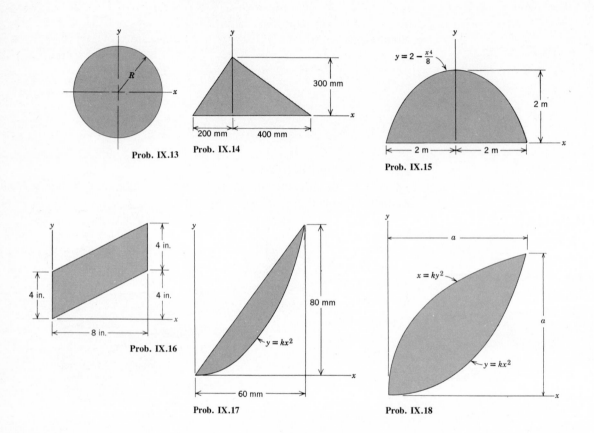

Prob. IX.13

Prob. IX.14

Prob. IX.15

Prob. IX.16

Prob. IX.17

Prob. IX.18

C.   PARALLEL AXIS THEOREMS

In the preceding examples and homework problems the area moments and product of inertia of various shapes were evaluated for a specific choice of location of the *xyz* coordinate axes. Suppose that we now wanted to know the properties of one of these shapes for a different location of the coordinate system. It would be preferable to calculate the required values by means of algebraic formulas, rather than by evaluating a new set of integrals. In this section we shall study the relationships between the moments and products of inertia for two mutually parallel coordinate systems.

Consider the two coordinate systems *xyz* and $\hat{x}\hat{y}\hat{z}$ used to describe the

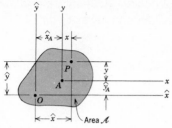

Figure 7

arbitrary shape in Figure 7. As shown in the figure, the coordinates of the origin A of the xyz coordinate system with respect to the $\hat{x}\hat{y}\hat{z}$ system are $(\hat{x}_A, \hat{y}_A, 0)$. Note that the distance between the x and $\hat{x}$ axis is $|\hat{y}_A|$ and the distance between the y and $\hat{y}$ axes is $|\hat{x}_A|$, and that the signs for these quantities are chosen in accord with the quadrant of the $\hat{x}\hat{y}\hat{z}$ system in which the origin A is located. From Figure 7, we see that the coordinates of an arbitrary point P are related by

$$x = \hat{x} - \hat{x}_A \qquad y = \hat{y} - \hat{y}_A \tag{8}$$

Let us first focus attention on determining I_x. From its definition and equations (8), we find that

$$I_x = \int_{\mathcal{A}} y^2 \, d\mathcal{A} = \int_{\mathcal{A}} (\hat{y} - \hat{y}_A)^2 \, d\mathcal{A}$$

$$= \int_{\mathcal{A}} \hat{y}^2 \, d\mathcal{A} - 2\hat{y}_A \int_{\mathcal{A}} \hat{y} \, d\mathcal{A} + (\hat{y}_A)^2 \int_{\mathcal{A}} d\mathcal{A}$$

The last of the integrals on the right side of the preceding equation is readily identified to give the area $\mathcal{A}$, and the first integral is seen to be the definition of $I_{\hat{x}}$. Recall that the integral in the second term is the first moment of area with respect to the $\hat{y}$ coordinate. Hence, *by choosing the origin O to be the centroid of the area, we can make this integral vanish.*

It is not difficult to show that similar steps for I_y and J_z yield the following set of *parallel axis theorems* for converting the moments of inertia for one coordinate system into those of the other.

$$\boxed{\begin{aligned} I_x &= I_{\hat{x}} + \mathcal{A}\,(\hat{x}_A)^2 \\ I_y &= I_{\hat{y}} + \mathcal{A}\,(\hat{y}_A)^2 \\ J_z &= J_{\hat{z}} + \mathcal{A}\,[(\hat{x}_A)^2 + (\hat{y}_A)^2] \end{aligned}} \tag{9}$$

Two important facts to bear in mind when applying equations (9) are that they are valid only if the origin of the $\hat{x}\hat{y}\hat{z}$ coordinate system coincides with the centroid of the area, and that $(\hat{x}_A, \hat{y}_A)$ are the $\hat{x}, \hat{y}$ coordinates of the origin A of the xyz system.

The interpretation of equations (9) are fairly direct. As we noted earlier, $|\hat{y}_A|$ and $|\hat{x}_A|$ represent the distances between the parallel pairs of x and y axes, respectively. Similarly, it may be seen by applying the Pythagorean theorem to Figure 7 that $[(\hat{x}_A)^2 + (\hat{y}_A)^2]^{1/2}$ is the distance between the parallel pair of z axes. Thus, as we transfer from an axis passing through the centroid of the area to another parallel axis, the moment of inertia about this new axis is obtained by adding the product of the area and the square of the distance between the axes to the old value of the moment of inertia.

With respect to the product of inertia, we follow a similar procedure to the one just presented. Substituting the coordinate transformations of equation (8) into the definition of I_{xy}, we find

$$I_{xy} = \int_{\mathcal{A}} xy \, d\mathcal{A} = \int_{\mathcal{A}} (\hat{x} - \hat{x}_A)(\hat{y} - \hat{y}_A) \, d\mathcal{A}$$

$$= \int_{\mathcal{A}} \hat{x}\hat{y} \, d\mathcal{A} - \hat{x}_A \int_{\mathcal{A}} \hat{y} \, d\mathcal{A} - \hat{y}_A \int_{\mathcal{A}} \hat{x} \, d\mathcal{A} + \hat{x}_A\hat{y}_A \int_{\mathcal{A}} d\mathcal{A}$$

The first integral is the definition of $I_{\hat{x}\hat{y}}$, whereas the second and third integrals are zero because the $\hat{x}$ and $\hat{y}$ axes pass through the centroid of the area. Hence, the parallel axis theorem for the product of inertia is

$$\boxed{I_{xy} = I_{\hat{x}\hat{y}} + \mathcal{A}\, \hat{x}_A \, \hat{y}_A} \tag{10}$$

Notice that in equations (9) the sign of $\hat{x}_A$ and $\hat{y}_A$ are not significant, because those parallel axis theorems depend only on the distance between corresponding axes. However, particular attention must be paid to the sign of $\hat{x}_A$ and $\hat{y}_A$ when applying equation (10). Again we remind you that $(\hat{x}_A, \hat{y}_A, 0)$ *are the coordinates of the origin A of the xyz system with respect to the $\hat{x}\hat{y}\hat{z}$ system whose origin is the centroid of the area.*

EXAMPLE 3
The area moments and product of inertia of the parabolic sector with respect to the *xyz* coordinate system were determined in Example 1. Determine these parameters for the *x'y'z'* system shown.

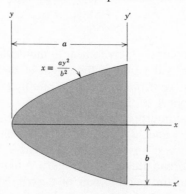

Solution

A common error is solving problems such as this is to try to transfer directly between the xyz and $x'y'z'$ coordinate systems, thereby applying the parallel axis theorems only once. This procedure is wrong because it overlooks the fact that the parallel axis theorems as we derived them are valid only when transforming between a centroidal and a noncentroidal coordinate system. Therefore, to solve the problem we must first locate the centroid of the parabolic sector. We will also need its area.

The properties of a semiparabolic section, such as that on either side of the x axis, are tabulated in Appendix B. In view of the symmetry of a parabola, we determine the centroid G to be situated as shown to the left. The area $\mathcal{A}$ is also given in Appendix B, so

$$\mathcal{A} = \tfrac{4}{3}ab$$

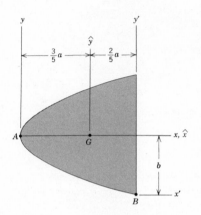

As noted earlier, we need a parallel $\hat{x}\hat{y}\hat{z}$ coordinate system whose origin is at the centroid. This is shown in the sketch, where the origins of the given coordinate systems have been labeled as points A and B. With respect to the $\hat{x}\hat{y}\hat{z}$ coordinate system, the coordinates of point A are $(-3a/5, 0, 0)$ and those of point B are $(2a/5, -b, 0)$.

We next determine the values of $I_{\hat{x}}$, $I_{\hat{y}}$, and $I_{\hat{x}\hat{y}}$ by substituting into the parallel axis theorems. Using the results of Example 1, this gives

$$I_{\hat{x}} = I_x - \mathcal{A}\,(\hat{y}_A)^2 = \tfrac{4}{15}ab^3 - (\tfrac{4}{3}ab)(0) = \tfrac{4}{15}ab^3$$

$$I_{\hat{y}} = I_y - \mathcal{A}\,(\hat{x}_A)^2 = \tfrac{4}{7}a^3b - (\tfrac{4}{3}ab)(-\tfrac{3}{5}a)^2 = \tfrac{16}{175}\,a^3b$$

$$I_{\hat{x}\hat{y}} = I_{xy} - \mathcal{A}\,\hat{x}_A\,\hat{y}_A = 0 - (\tfrac{4}{3}ab)(-\tfrac{3}{5}a)(0) = 0$$

Note that it is not necessary to apply the parallel axis theorem for $J_{\hat{z}}$ because we will be able to use equation (3) to determine $J_{z'}$.

We now apply the parallel axis theorems again in order to transfer from the $\hat{x}\hat{y}\hat{z}$ system to the $x'y'z'$ system. This yields

$$I_{x'} = I_{\hat{x}} + \mathcal{A}\,(\hat{y}_B)^2 = \tfrac{4}{15}ab^3 + (\tfrac{4}{3}ab)(-b)^2$$

$$= \tfrac{8}{5}ab^3 \qquad\qquad \triangleleft$$

$$I_{y'} = I_{\hat{y}} + \mathcal{A}\,(x_B)^2 = \tfrac{16}{175}a^3b + (\tfrac{4}{3}ab)(\tfrac{2}{5}a)^2$$

$$= \tfrac{32}{105}a^3b \qquad\qquad \triangleleft$$

$$I_{x'y'} = I_{\hat{x}\hat{y}} + \mathcal{A}\,\hat{x}_B\,\hat{y}_B = 0 + (\tfrac{4}{3}ab)(\tfrac{2}{5}a)(-b)$$

$$= -\tfrac{8}{15}a^2b^2 \qquad\qquad \triangleleft$$

We then find

$$J_{z'} = I_{x'} + I_{y'} = \tfrac{8}{105}ab(21b^2 + 4a^2) \qquad\qquad \triangleleft$$

HOMEWORK PROBLEMS

IX.19–IX.26 Determine the moment of inertia I_x, the polar moment of inertia J_z, and the product of inertia I_{xy} for the shaded area and corresponding coordinate system shown. The properties in Appendix B may be employed where necessary.

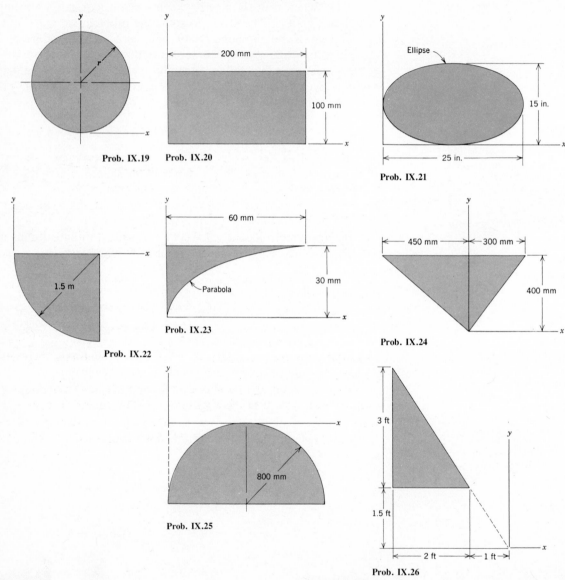

Prob. IX.19   Prob. IX.20

Prob. IX.21

Prob. IX.22

Prob. IX.23

Prob. IX.24

Prob. IX.25

Prob. IX.26

IX.27 Point C is the centroid of the shaded area. It is known that for the xyz coordinate system shown, $I_x = 1230$ in.^{4} and $J_z = 2580$ in.^{4}, whereas for the centroidal $\hat{x}\hat{y}\hat{z}$ coordinate system $J_{\hat{z}} = 1830$ in.[4]. Determine (a) the area $\mathcal{A}$, (b) the values of $I_{\hat{x}}$ and $I_{\hat{y}}$.

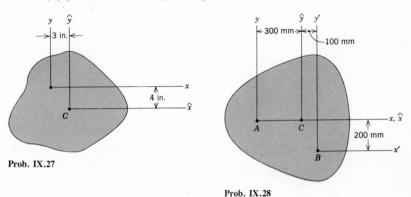

Prob. IX.27

Prob. IX.28

IX.28 Point C is the centroid of the shaded area shown. Because of symmetry it is known that the centroidal $\hat{x}\hat{y}\hat{z}$ axes are principal axes for which the products of inertia with respect to these axes is zero. It is also known that for the other coordinate systems shown, $I_y = 8.50(10^{-2})$ m[4], $I_{x'} = 3.25(10^{-2})$ m[4], and $I_{y'} = 6.50(10^{-2})$ m[4]. Determine (a) the area $\mathcal{A}$, (b) the moment of inertia $I_{\hat{x}}$, (c) the products of inertia I_{xy} and $I_{x'y'}$.

D. COMPOSITE AREAS

It is common practice to tabulate the properties of various basic shapes. Appendix B gives the area moments and product of inertia for axes having an origin at the centroid. We may then determine the properties of a planar area whose shape does not appear in Appendix B by considering it to be a composite area.

Figure 8 depicts an area that is formed from two basic shapes having areas $\mathcal{A}_1$ and $\mathcal{A}_2$. In the definitions of the inertia properties of an area, integrals extending over the total area $\mathcal{A}$ may be expressed as the sum of integrals extending over the individual areas $\mathcal{A}_1$ and $\mathcal{A}_2$. The following identities are then found.

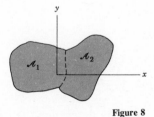

Figure 8

$$
\begin{aligned}
I_x &= (I_x)_1 + (I_x)_2 \\
I_y &= (I_y)_1 + (I_y)_2 \\
J_z &= (J_z)_1 + (J_z)_2 \\
I_{xy} &= (I_{xy})_1 + (I_{xy})_2
\end{aligned}
\tag{11}
$$

In other words the area moments and product of inertia are additive.

An aspect of equations (11) that can lead to errors is that these properties are additive only when they have been expressed with respect to the same coordinate system. Thus, noting that the properties in Appendix B are for centroidal axes of the basic shapes, it will usually be necessary to employ the parallel axis theorems before carrying out the appropriate addition indicated in equations (11).

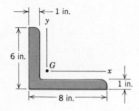

EXAMPLE 4

The cross section of an angle iron used in the construction of steel structures is shown. Neglecting the fillets (rounded-off edges), determine the values of I_x, I_y, and I_{xy} for the coordinate system shown having its origin at the centroid G of the cross section.

Solution

Let us consider the cross section to be a composite formed from the rectangular shape of each leg. First, we must determine the location of the centroid G. To do this we define a convenient $x'y'z'$ coordinate system parallel to the given system. The component rectangles, as well as the x' and y' axes, are illustrated in the sketch to the left.

The areas are

$$A_1 = 7(1) = 7 \text{ in.}^2 \qquad A_2 = 6(1) = 6 \text{ in.}^2$$

$$A = A_1 + A_2 = 13 \text{ in.}^2$$

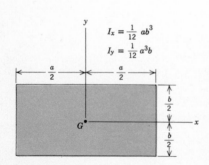

The centroid of each rectangle is at its middle, so the (x', y') coordinates of centroid G_1 are $(4.5, 0.5)$ and those of centroid G_2 are $(0.5, 3.0)$. The first moments of area with respect to the x' and y' coordinates then give

$$13(x'_G) = 7(4.5) + 6(0.5)$$

$$x'_G = 2.654 \text{ in.}$$

$$13(y'_G) = 7(0.5) + 6(3.0)$$

$$y'_G = 1.654 \text{ in.}$$

Now that we have located point G, we may determine the moments and product of inertia of the two basic shapes with respect to the centroidal xyz coordinate system. We refer to Appendix B to determine the inertia properties of each shape with respect to its own centroidal coordinate system. The properties of a rectangle are repeated in the adjacent sketch. Note that $I_{xy} = 0$ because of symmetry.

In this problem there are two rectangles, so we will need centroidal axes $\hat{x}_i \hat{y}_i \hat{z}_i$ $(i = 1, 2)$ for each, as shown in the sketch on the next page.

$$I_x = \frac{1}{12} ab^3$$

$$I_y = \frac{1}{12} a^3 b$$

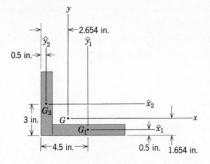

Considering rectangle 1 first, we see that in comparison with the rectangle reproduced on the previous page from Appendix B, $a = 7$ in., $b = 1$ in. In order to employ the parallel axis theorems to transfer from the $\hat{x}_1\hat{y}_1\hat{z}_1$ system to the xyz system, we need the $(\hat{x}_1, \hat{y}_1)$ coordinates of point G. These values are $(-4.50 + 2.654, 1.654 - 0.50)$ inches. Thus, using the parallel axis theorems, we have

$$(I_x)_1 = (I_{\hat{x}_1})_1 + \mathcal{A}_1(1.654 - 0.50)^2 = \tfrac{1}{12}(7)(1)^3 + 7(1.332)$$

$$= 9.91 \text{ in.}^4$$

$$(I_y)_1 = (I_{\hat{y}_1})_1 + \mathcal{A}_1(-4.50 + 2.654)^2 = \tfrac{1}{12}(1)(7)^3 + 7(3.408)$$

$$= 52.44 \text{ in.}^4$$

$$(I_{xy})_1 = (I_{\hat{x}_1\hat{y}_1})_1 + \mathcal{A}_1(-4.50 + 2.654)(1.654 - 0.50)$$

$$= 0 + 7(-2.130) = -14.91 \text{ in.}^4$$

For rectangle 2, we see that $a = 1$ in. and $b = 6$ in. The $(\hat{x}_2, \hat{y}_2)$ coordinates of point G are $(2.654 - 0.5, -3.0 + 1.654)$ inches, so the parallel axis theorems give

$$(I_x)_2 = (I_{\hat{x}_2})_2 + \mathcal{A}_2(-3.0 + 1.654)^2 = \tfrac{1}{12}(1)(6)^3 + 6(1.812)$$

$$= 28.87 \text{ in.}^4$$

$$(I_y)_2 = (I_{\hat{y}_2})_2 + \mathcal{A}_2(2.654 - 0.50)^2 = \tfrac{1}{12}(6)(1)^3 + 6(4.640)$$

$$= 28.34 \text{ in.}^4$$

$$(I_{xy})_2 = (I_{\hat{x}_2\hat{y}_2}) + \mathcal{A}_2(2.654 - 0.50)(-3.0 + 1.654)$$

$$= 0 + 6(-2.899) = -17.40 \text{ in.}^4$$

We now add corresponding inertia properties to obtain the results for the composite cross section. This yields

$$I_x = (I_x)_1 + (I_x)_2 = 38.8 \text{ in.}^4$$

$$I_y = (I_y)_1 + (I_y)_2 = 80.8 \text{ in.}^4$$

$$I_{xy} = (I_{xy})_1 + (I_{xy})_2 = -32.3 \text{ in.}^4$$

Notice that it was not necessary to calculate the polar moments of inertia of the individual rectangles, because we may use equation (3) to write

$$J_z = I_x + I_y = 119.6 \text{ in.}^4$$

HOMEWORK PROBLEMS

IX.29–IX.32   Determine the moment of inertia I_x, the product of inertia I_{xy}, and the radius of gyration k_x of the shaded area shown.

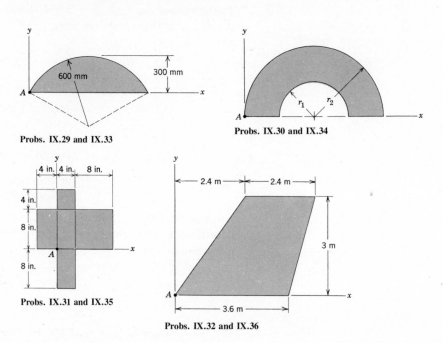

Probs. IX.29 and IX.33

Probs. IX.30 and IX.34

Probs. IX.31 and IX.35

Probs. IX.32 and IX.36

IX.33–IX.36   Determine the polar moment of inertia J_z and the radius of gyration k_z for the shaded area shown.

IX.37–IX.43   Point G is the centroid of the area shown. Determine the location of this point and the moments and product of inertia of the area with respect to the xyz coordinate system having origin G.

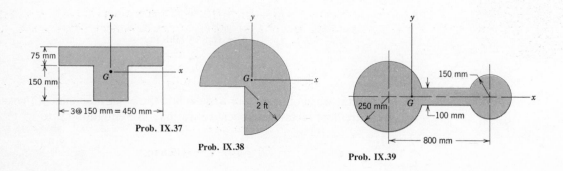

Prob. IX.37

Prob. IX.38

Prob. IX.39

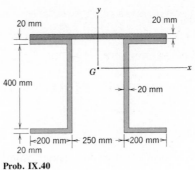

Prob. IX.40

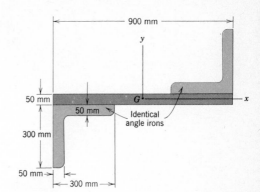

Prob. IX.41

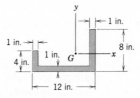

Prob. IX.42

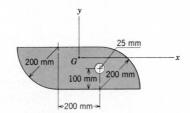

Prob. IX.43

E. ROTATION TRANSFORMATION — MOHR'S CIRCLE

The parallel axis theorems enable us to determine the moments and product of inertia corresponding to one coordinate system in terms of the properties for another parallel coordinate system. In what follows we shall determine the algebraic relationship between the inertia properties for two coordinate systems having identical origins, but different orientations. A typical situation is shown in Figure 9, where the xyz system is the one for which the moments and product of inertia are known. To proceed with the derivation we need expressions for the (x', y') coordinates in terms of the (x, y) values. These could be obtained by using trigonometry to measure various lengths in Figure 9.

An alternative method employs the fact that the position vector $\bar{r}_{P/O}$ may be described in terms of either coordinate system. Thus, we write

$$\bar{r}_{P/O} = x'\bar{i}' + y'\bar{j}' = x\bar{i} + y\bar{j}$$

Expressing the unit vectors $\bar{i}$ and $\bar{j}$ in terms of their components with respect to the $x'y'z'$ system, we find

$$\bar{i} = (\cos\theta)\,\bar{i}' - (\sin\theta)\,\bar{j}'$$

$$\bar{j} = (\sin\theta)\,\bar{i}' + (\cos\theta)\,\bar{j}'$$

Figure 9

Upon substitution of these expressions into the earlier equation for $\bar{r}_{P/O}$, we have

$$\bar{r}_{P/O} = x'\bar{i}' + y'\bar{j}' = x[(\cos\theta)\,\bar{i}' - (\sin\theta)\,\bar{j}'] + y[(\sin\theta)\,\bar{i}' + (\cos\theta)\,\bar{j}']$$

The following results are obtained when corresponding components are matched.

$$x' = x\cos\theta + y\sin\theta \qquad y' = -x\sin\theta + y\cos\theta \qquad (12)$$

We now employ this transformation in the expression for $I_{x'}$, as follows.

$$I_{x'} \equiv \int_{\mathcal{A}} (y')^2\, d\mathcal{A} = \int_{\mathcal{A}} (-x\sin\theta + y\cos\theta)^2\, d\mathcal{A}$$

$$= \cos^2\theta \int_{\mathcal{A}} y^2\, d\mathcal{A} - 2\cos\theta\sin\theta \int_{\mathcal{A}} xy\, d\mathcal{A} + \sin^2\theta \int_{\mathcal{A}} x^2\, d\mathcal{A}$$

Notice that the integrals on the right side of this expression are I_x, I_{xy}, and I_y, respectively. When similar steps are followed for $I_{y'}$ and $I_{x'y'}$, we obtain the following set of *rotation transformations* for the area moments of inertia.

$$
\boxed{
\begin{aligned}
I_{x'} &= I_x \cos^2\theta + I_y \sin^2\theta - 2\sin\theta\cos\theta\, I_{xy} \\[4pt]
I_{y'} &= I_x \sin^2\theta + I_y \cos^2\theta + 2\sin\theta\cos\theta\, I_{xy} \\[4pt]
I_{x'y'} &= I_{xy} (\cos^2\theta - \sin^2\theta) + (I_x - I_y)\sin\theta\cos\theta
\end{aligned}
}
\qquad (13)
$$

There is no need to consider $J_{z'}$; the coincidence of the z and z' axes means that $J_{z'} = J_z$. This is confirmed by the foregoing expressions, which show that

$$J_{z'} = I_{x'} + I_{y'} \equiv I_x + I_y$$

This also shows that the sum of the I_x and I_y is a constant regardless of the angular orientation of the coordinate axes.

In your later studies you will encounter transformations similar to these, for example, in mechanics of materials for stress and strain and in dynamics for mass moments of inertia.

Equations (13) are easy enough to employ. Nevertheless, by performing a few simple manipulations we may obtain a pictorial representation of the transformation that makes it unnecessary to remember formulas. To do this, we employ the following trigonometric identities.

$$\cos^2\theta = \tfrac{1}{2}(1 + \cos 2\theta) \qquad \sin^2\theta = \tfrac{1}{2}(1 - \cos 2\theta) \qquad \sin\theta\cos\theta = \tfrac{1}{2}\sin 2\theta$$

Substitution of the foregoing into equations (13) then yields

$$I_{x'} = \tfrac{1}{2}(I_x + I_y) + \tfrac{1}{2}(I_x - I_y)\cos 2\theta - I_{xy}\sin 2\theta \tag{14a}$$

$$I_{y'} = \tfrac{1}{2}(I_x + I_y) - \tfrac{1}{2}(I_x - I_y)\cos 2\theta + I_{xy}\sin 2\theta \tag{14b}$$

$$I_{x'y'} = I_{xy}\cos 2\theta + \tfrac{1}{2}(I_x - I_y)\sin 2\theta \tag{14c}$$

We next transpose the terms $\tfrac{1}{2}(I_x + I_y)$ from the right side to the left in equations (14a) and (14b). Then, squaring either of equations (14a) or (14b), and adding the result to the square of equation (14c), we obtain the following expression.

$$\left[I_{p'} - \left(\frac{I_x + I_y}{2}\right)\right]^2 + (I_{x'y'})^2 = \left(\frac{I_x - I_y}{2}\right)^2 + (I_{xy})^2 \tag{15}$$

where the symbol p' represents either x' or y', depending on which equation was squared.

The significance of this expression becomes apparent when we recall that I_x, I_y, and I_{xy} are specific known values. Therefore, equation (15) can be regarded as a relationship between $I_{x'}$ and $I_{x'y'}$, or alternatively, between $I_{y'}$ and $I_{x'y'}$. In terms of a graph with $I_{p'}$ plotted along the abscissa and $I_{x'y'}$ plotted along the ordinate, equation (15) represents a circle of radius R that is centered on the abscissa at a distance d from the origin, where

$$R = \left[\left(\frac{I_x - I_y}{2}\right)^2 + (I_{xy})^2\right]^{1/2} \tag{16}$$

$$d = \tfrac{1}{2}(I_x + I_y)$$

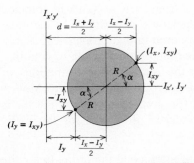

Figure 10

The parameters R and d are readily determined from a geometric construction by plotting the points (I_x, I_{xy}) and $(I_y, -I_{xy})$ and then connecting these points by a straight line. This construction and the resulting circle are illustrated in Figure 10.

When $\theta = 0°$, the x and x' axes in Figure 9 are coincident, as are the y and y' axes. Hence, the points we have plotted in Figure 10 may be considered to represent $I_{x'}$, $I_{y'}$, and $I_{x'y'}$ when $\theta = 0$. The question we must now consider is which points on the circle correspond to the values of $I_{x'}$ and $I_{x'y'}$, or $I_{y'}$ and $I_{x'y'}$, for a nonzero value of θ. To answer this question, we need the angle α. Referring to Figure 10, we see that this angle satisfies

$$R\cos \alpha = \tfrac{1}{2}(I_x - I_y) \qquad R\sin \alpha = I_{xy} \tag{17}$$

Substituting the foregoing expressions and the definition of the distance d into the basic relationships of equations (14), and making use of the trigonometric identities for the sine and cosine of the sum of two angles, we find that

$$I_{x'} = d + R(\cos \alpha \cos 2\theta - \sin \alpha \sin 2\theta) \equiv d + R \cos(\alpha + 2\theta)$$

$$I_{y'} = d - R(\cos \alpha \cos 2\theta - \sin \alpha \sin 2\theta) \equiv d - R \cos(\alpha + 2\theta)$$

$$I_{x'y'} = R(\sin \alpha \cos 2\theta + \cos \alpha \sin 2\theta) \equiv R \sin(\alpha + 2\theta)$$

As illustrated in Figure 11, these relations show that the point $(I_{x'}, I_{x'y'})$ is

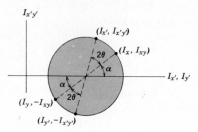

Figure 11

obtained by rotating the radial line from the point (I_x, I_{xy}) coun-terclockwise through an angle 2θ. A similar statement applies to the point $(I_{y'}, -I_{x'y'})$. It is significant that in this construction the angle 2θ is positive in the counterclockwise sense, for that is the same sense as the positive rotation from the x and y axes to the x' and y' axes originally depicted in Figure 9.

A question that bothers some students about this derivation is, "Why do we plot the point $(I_{y'}, -I_{x'y'})$, and not the point $(I_{y'}, I_{x'y'})$?" To answer this question it is only necessary to note that the x' and y' axes are separated by 90°. Thus, if we give the x' axis an additional rotation of 90° beyond its initial orientation at an angle θ from the x axis, it will coincide with the original y' axis. Hence, $I_{x'}$ for the new x' axis should be identical to $I_{y'}$ for the old y' axis. This result is assured in our geometrical sketch because angles of rotation there are twice the physical angle of rotation. This pictorial method for representing the rotation transformations of equations (13) is known as *Mohr's circle* after the engineer Otto Mohr.

In summation, the Mohr's circle for area moments of inertia is employed as follows to determine the inertia properties for the $x'y'z'$ coordinate system.

1   Using the abscissa to plot moments of inertia and the ordinate for the product of inertia, plot the two points (I_x, I_{xy}) and $(I_y, -I_{xy})$.
2   Connect the two points plotted in the previous step. The length of this line is the diameter of the Mohr's circle, and the point of intersection of this line with the abscissa is the center of the circle. Sketch the corres-ponding circle.

3   Draw a diameter in the Mohr's circle that is at an angle 2θ from the line in step 2, in the same sense as the angle of rotation θ from the x axis to the x' axis.

4   The point of intersection of the diameter just drawn with the Mohr's circle, adjacent to the point (I_x, I_{xy}) in the sense of the rotation from the x axis to the x' axis, has the coordinates $(I_{x'}, I_{x'y'})$. The point of intersection adjacent to the point $(I_y, -I_{xy})$ in the sense of the rotation is the point $(I_{y'}, -I_{x'y'})$.

Note that when constructing Mohr's circle, the points should be plotted consistently with the sign of I_{xy}. Also, there is no restriction that I_x be larger than I_y, even though this was the case in the derivation.

We will not utilize Mohr's circle to obtain a graphical solution. Rather, Mohr's circle will be employed as a device that makes it unnecessary to remember equations (13).

An important aspect of the rotation transformation is exhibited directly by Mohr's circle. Specifically, referring to Figure 11, notice that the two points where the circle intersects the abscissa correspond to orientations of the $x'y'z'$ system for which $I_{x'y'} = 0$. A coordinate system for which the product of inertia is zero is called a set of *principal axes,* and the corresponding moments of inertia are the *principal* values. Earlier, in discussing shapes having an axis of symmetry, we saw that any coordinate system containing this axis is a set of principal axes. We now see that all planar figures have a set of principal axes for any choice of the origin.

The angle of rotation $\bar{\theta}$ from the xyz axes having known properties to the principal axes is easily determined from Figure 10 or Figure 11. Recalling that the angle of rotation is one-half the angle in Mohr's circle, and considering counterclockwise rotations to be positive rotations, we see that $2\bar{\theta} = -\alpha$. (The alternative $2\bar{\theta} = 180° - \alpha$ gives the same coordinate system as the foregoing, except that the coordinate axes are labeled differently.) Using equations (17), we then have

$$\bar{\theta} = -\tfrac{1}{2} \tan^{-1}\left(\frac{2I_{xy}}{I_x - I_y}\right) \tag{18}$$

In practice, there is no need to remember equation (18), for the magnitude and sense of the angle of rotation $\bar{\theta}$ are easily determined after the Mohr's circle has been constructed.

Additional features of the rotation transformation are apparent from the Mohr's circle in Figure 11. We see that the principal moments of inertia are the maximum and minimum values of I_x and I_y that can be obtained in a rotation of the coordinate axes. Also, the axes for which the product of inertia is a maximum are at an angle of 45° ($2\theta = 90°$) from the principal axes.

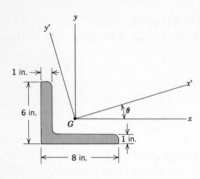

EXAMPLE 5

The moments and product of inertia of the cross section of the angle iron shown with respect to the centroidal xyz coordinate system were determined in Example 4. Determine (a) the values of $I_{x'}$, $I_{y'}$, and $I_{x'y'}$ when $\theta = 40°$, (b) the angle $\bar{\theta}$ for principal axes and the corresponding principal values of the moments of inertia.

Solution

From Example 4 we know that $I_x = 38.8$ in.[4], $I_y = 80.8$ in.[4], $I_{xy} = -32.3$ in.[4]. We plot the points (I_x, I_{xy}) and $(I_{y'}, -I_{xy})$, and label them as points A and B, respectively, for future reference. We then connect points A and B by a straight line and draw a circle having line AB as a diameter. This is shown in the following sketch.

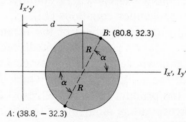

In order to locate other points on the Mohr's circle, we need the distance d, the radius R, and the angle α. From the sketch we see that

$$d = \tfrac{1}{2}(80.8 + 38.8) = 59.8$$

$$R = [(80.8 - 59.8)^2 + (32.3)^2]^{1/2} = 38.5$$

$$\alpha = \tan^{-1}\left(\frac{38.5}{80.8 - 59.8}\right) = 61.41°$$

For part (a) we are interested in the properties with respect to the $x'y'z'$ axis when $\theta = 40°$ counterclockwise, so we draw a diameter that is rotated from line AB by $2\theta = 80°$. We label this diameter as CD. In part (b) we want the principal axes, which are obtained by rotating line AB clockwise by $2\bar{\theta} = \alpha = 61.41°$, thus resulting in the diameter EF coincident with the abscissa. These diameters are illustrated in the sketch to the left.

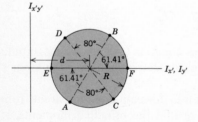

Solving part (a) first, as a result of the rotation point A goes to point C, so the coordinates of this point are $(I_{x'}, I_{x'y'})$. Similarly, point B goes to point D, so the corresponding coordinates are $(I_{y'}, -I_{x'y'})$. Using trigonometry to evaluate the distances, we then have

$$I_{x'} = d + R\cos(180° - 61.41° - 80°) = 89.9 \text{ in.}^4$$

$$I_{y'} = d - R\cos(180° - 61.41° - 80°) = 29.7 \text{ in.}^4$$

$$I_{x'y'} = -R\sin(180° - 61.41° - 80°) = -24.01 \text{ in.}^4$$

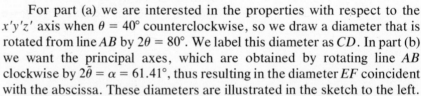

For part (b), point A goes to point E and point B goes to point F. Hence, point E has the coordinates $(I_{\bar{x}}, 0)$ and point F has the coordinates $(I_{\bar{y}}, 0)$, where $I_{\bar{x}}$ and $I_{\bar{y}}$ are the principal moments of inertia. Thus, for principal axes we have

$$\theta = \bar{\theta} = \tfrac{1}{2}(61.41°) = 30.7° \text{ clockwise}$$

$$I_{\bar{x}} = d - R = 21.3 \text{ in.}^4$$

$$I_{\bar{y}} = d + R = 98.3 \text{ in.}^4$$

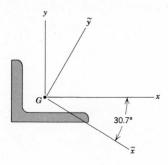

The resulting principal axes are shown in the adjacent sketch. Notice that their orientation could not be guessed on the basis of the geometry of the cross section.

HOMEWORK PROBLEMS

IX.44–IX.47 Determine the moments of inertia $I_{x'}$ and $I_{y'}$ and the product of inertia $I_{x'y'}$ for the shaded area and corresponding centroidal $x'y'z'$ axes (a) if $\theta = 36.87°$, (b) if $\theta = 135°$, (c) if $\theta = -30°$. The properties in Appendix B may be used where necessary.

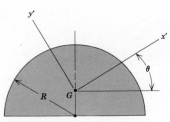

Prob. IX.44

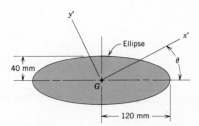

Prob. IX.45

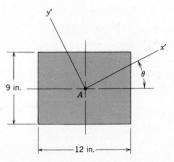

Prob. IX.46

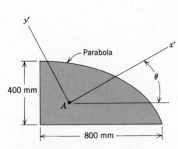

Prob. IX.47

IX.48–IX.51 Determine the moments of inertia $I_{x'}$ and $I_{y'}$ and the product of inertia $I_{x'y'}$ for the shaded area and corresponding xyz coordinate system shown (a) if $\theta = 60°$, (b) if $\theta = -53.13°$, (c) if $\theta = -135°$. The properties in Appendix B may be used where necessary.

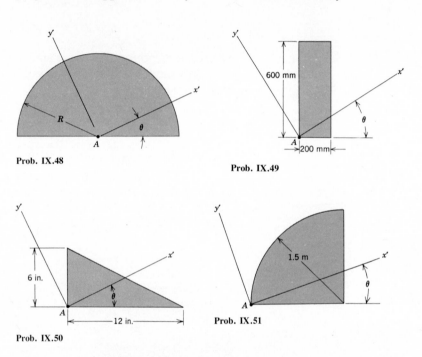

Prob. IX.48

Prob. IX.49

Prob. IX.50

Prob. IX.51

IX.52–IX.55 Determine the moments of inertia I_x I_y and the product of inertia I_{xy} for the composite shape with respect to the coordinate axes shown.

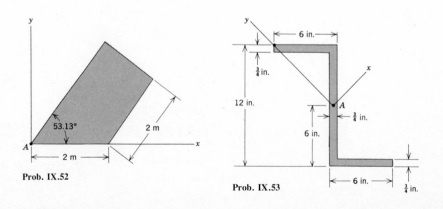

Prob. IX.52

Prob. IX.53

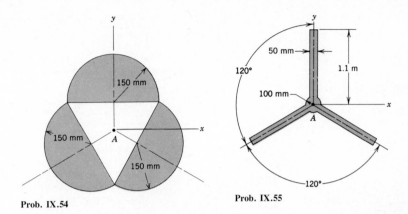

Prob. IX.54

Prob. IX.55

IX.56–IX.61   For the shape in the problem listed below, determine the orientation of the principal axes having origin A, and also determine the corresponding principal moments of inertia.

Homework Problem	IX.56	IX.57	IX.58	IX.59	IX.60	IX.61
Shape in Problem	**IX.46**	**IX.47**	**IX.49**	**IX.50**	**IX.52**	**IX.53**

IX.62–IX.67   For the shape in the problem listed below, determine the orientation of the axes having origin A for which the product of inertia has its maximum value. Also determine the moments and product of inertia for this coordinate system.

Homework Problem	IX.62	IX.63	IX.64	IX.65	IX.66	IX.67
Shape in Problem	**IX.46**	**IX.47**	**IX.49**	**IX.50**	**IX.52**	**IX.53**

IX.68   Letting $\bar{x}\bar{y}\bar{z}$ denote a set of principal axes for an arbitrary area and xyz be another set of axes having the same origin, prove that

$$I_{xy} = \sqrt{I_x I_y - I_{\bar{x}} I_{\bar{y}}}$$

APPENDIX A
SI UNITS

Table 1 Conversion factors from British units

PHYSICAL QUANTITY	U.S.-BRITISH UNIT	= SI EQUIVALENT
		BASIC UNITS
Length	1 foot (ft)	= 3.048(10^{-1}) meter (m)*
	1 inch (in.)	= 2.54(10^{-2}) meter (m)*
	1 mile (U.S. statute)	= 1.6093(10^3) meter (m)
Mass	1 slug (lb-s^2/ft)	= 1.4594(10) kilogram (kg)
	1 pound mass (lbm)	= 4.5359(10^{-1}) kilogram (kg)
		DERIVED UNITS
Acceleration	1 foot/second^2 (ft/s^2)	= 3.048(10^{-1}) meter/second^2 (m/s^2)*
	1 inch/second^2 (in./s^2)	= 2.54(10^{-2}) meter/second^2 (m/s^2)*
Area	1 foot^2 (ft^2)	= 9.2903(10^{-2}) meter^2 (m^2)
	1 inch^2 (in.^{2})	= 6.4516(10^{-2}) meter^2 (m^2)*
Density	1 slug/foot^3 (lb-s^2/ft^4)	= 5.1537(10^2) kilogram/meter^3 (kg/m^3)
	1 pound mass/foot^3 (lbm/ft^3)	= 1.6018(10) kilogram/meter^3 (kg/m^3)
Energy and Work	1 foot-pound (ft-lb)	= 1.3558 joules (J)
	1 kilowatt-hour (kW-hr)	= 3.60(10^6) joules (J)*
	1 British thermal unit (Btu)	= 1.0551(10^3) joules (J)
	(1 joule $\equiv$ 1 meter-newton)	
Force	1 pound (lb)	= 4.4482 newtons (N)
	1 kip (1000 lb)	= 4.4482(10^3) newtons (N)
	(1 newton $\equiv$ 1 kilogram-meter/second^2)	
Power	1 foot-pound/second (ft-lb-s)	= 1.3558 watt (W)
	1 horsepower (hp)	= 7.4570(10^2) watt (W)
	(1 watt $\equiv$ 1 joule/second)	
Pressure and Stress	1 pound/foot^2 (lb/ft^2)	= 4.7880(10) pascal (Pa)
	1 pound/inch^2 (lb/in.^{2})	= 6.8948(10^3) pascal (Pa)
	1 atmosphere (standard, 14.7 lb/in.^{2})	= 1.0133(10^5) pascal (Pa)
	(1 pascal $\equiv$ 1 newton/m^2)	
Speed	1 foot/second (ft/s)	= 3.048(10^{-1}) meter/second (m/s)*
	1 mile/hr	= 4.4704(10^{-1}) meter/second (m/s)
	1 mile/hr	= 1.6093 kilometer/hr (km/hr)
Volume	1 foot^3 (ft^3)	= 2.8317(10^{-2}) meter^3 (m^3)
	1 inch^3 (in.^{3})	= 1.6387(10^{-5}) meter^3 (m^3)
	1 gallon (U.S. liquid)	= 3.7854(10^{-3}) meter^3 (m^3)

*Denotes an exact factor.

Table 2   Conversion factors from "old" metric units

PHYSICAL QUANTITY	"OLD" METRIC UNIT	= SI EQUIVALENT
Energy	1 erg	$= 1.00(10^{-7})$ joule (J)*
Force	1 dyne	$= 1.00(10^{-5})$ newton (N)*
Length	1 angstrom	$= 1.00(10^{-10})$ meter (m)*
	1 micron	$= 1.00(10^{-6})$ meter (m)*
Pressure	1 bar	$= 1.00(10^{5})$ pascal (Pa)*
Volume	1 liter	$= 1.00(10^{-3})$ meter^3 (m^3)*

*Denotes an exact factor

Table 3   SI prefixes

PREFIX	SYMBOL	FACTOR BY WHICH UNIT IS MULTIPLIED
tera*	T	10^{12}
giga*	G	10^{9}
mega*	M	10^{6}
kilo*	k	10^{3}
hecto	h	10^{2}
deka	da	10
deci	d	10^{-1}
centi	c	10^{-2}
milli*	m	10^{-3}
micro*	μ	10^{-6}
nano*	n	10^{-9}
pico*	p	10^{-12}
femto*	f	10^{-15}
atto*	a	10^{-18}

*Denotes preferred prefixes.

Properties of
Geometrical Shapes APPENDIX B

Used over & over.

Use only with
thorough understanding
of the axes.

APPENDIX B
AREAS, CENTROIDS,
AND AREA MOMENTS
OF INERTIA

2ⁿᵈ moment of Area.

SHAPE	AREA	AREA MOMENT OF INERTIA FOR CENTROIDAL AXES $(J_{\hat{z}} = I_{\hat{x}} + I_{\hat{y}})$

Rectangle

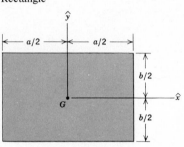

$A = ab$

$I_{\hat{x}} = \frac{1}{12}ab^3$

$I_{\hat{y}} = \frac{1}{12}a^3b$

Scalene triangle

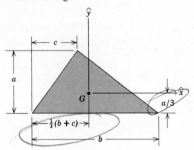

$A = \frac{1}{2}ab$

$I_{\hat{x}} = \frac{1}{36}a^3b$

$I_{\hat{y}} = \frac{1}{36}ab(b^2 + c^2 - bc)$

$I_{\hat{x}\hat{y}} = \frac{1}{72}a^2b(2c - b)$

Circle

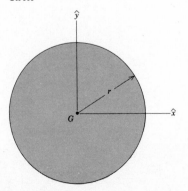

$A = \pi r^2$ $I_{\hat{x}} = I_{\hat{y}} = \frac{1}{4}\pi r^4$

Quarter circle

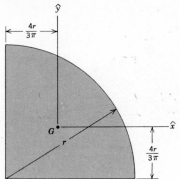

$$\mathcal{A} = \tfrac{1}{4}\pi r^2 \qquad I_{\hat{x}} = I_{\hat{y}} = \left(\frac{9\pi^2 - 64}{144\pi}\right)r^4$$

$$I_{\hat{x}\hat{y}} = \left(\frac{9\pi - 32}{72\pi}\right)r^4$$

Circular sector

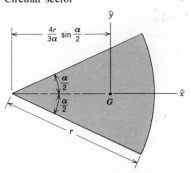

$$\mathcal{A} = \tfrac{1}{2}\alpha r^2 \qquad I_{\hat{x}} = \tfrac{1}{8}(\alpha - \sin \alpha)r^4$$

$$I_{\hat{y}} = \left[\frac{\alpha + \sin \alpha}{8}\right.$$

$$\left. - \frac{4}{9\alpha}(1 - \cos \alpha)\right]r^4$$

Circular arc

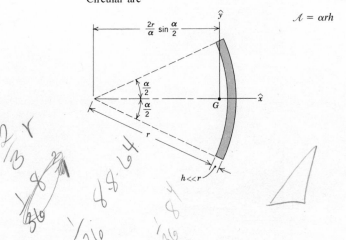

$$\mathcal{A} = \alpha r h \qquad I_{\hat{x}} = \tfrac{1}{2}(\alpha - \sin \alpha)r^3 h$$

$$I_{\hat{y}} = \left[\frac{\alpha + \sin \alpha}{2}\right.$$

$$\left. - \frac{2}{\alpha}(1 - \cos \alpha)\right]r^3 h$$

Ellipse

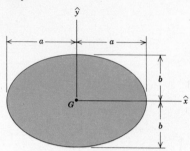

$$\mathscr{A} = \pi ab \qquad I_{\hat{x}} = \frac{\pi}{4}ab^3$$

$$I_{\hat{y}} = \frac{\pi}{4}a^3b$$

Quarter ellipse

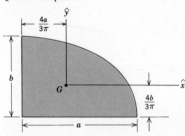

$$\mathscr{A} = \tfrac{1}{4}\pi ab \qquad I_{\hat{x}} = \left(\frac{9\pi^2 - 64}{144\pi}\right)ab^3$$

$$I_{\hat{y}} = \left(\frac{9\pi^2 - 64}{144\pi}\right)a^3b$$

$$I_{\hat{x}\hat{y}} = \left(\frac{9\pi - 32}{72\pi}\right)a^2b^2$$

Parabolic section

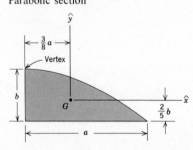

$$\mathscr{A} = \tfrac{2}{3}ab \qquad I_{\hat{x}} = \tfrac{8}{175}ab^3$$

$$I_{\hat{y}} = \tfrac{19}{480}a^3b$$

$$I_{\hat{x}\hat{y}} = -\tfrac{1}{60}a^2b^2$$

Parabolic spandrel

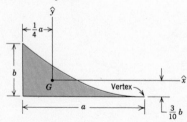

$$\mathscr{A} = \tfrac{1}{3}ab \qquad I_{\hat{x}} = \tfrac{19}{1050}ab^3$$

$$I_{\hat{y}} = \tfrac{1}{80}a^3b$$

$$I_{\hat{x}\hat{y}} = -\tfrac{1}{120}a^2b^2$$

SHAPE	VOLUME

Hemisphere

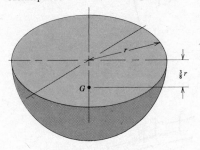

$$V = \tfrac{2}{3}\pi r^3$$

Hemispherical shell

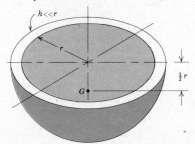

$$V = 2\pi r^2 h$$

Semicylinder

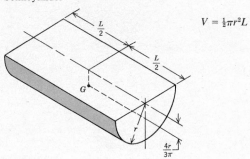

$$V = \tfrac{1}{2}\pi r^2 L$$

Semicylindrical shell

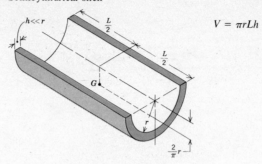

$$V = \pi r L h$$

Semicone

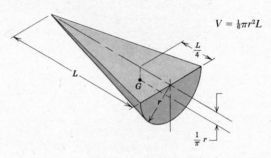

$$V = \tfrac{1}{6}\pi r^2 L$$

Semiconical shell

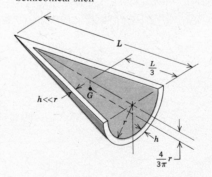

$$V = \frac{\pi}{2} r h (r^2 + L^2)^{1/2}$$

Orthogonal tetrahedron

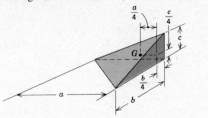

$V = \frac{1}{6}abc$

Triangular prism

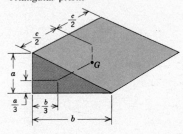

$V = \frac{1}{2}abc$

all of volume

$=$ assumed constant
densities (for mass).

APPENDIX D
DENSITIES

MATERIAL	SPECIFIC GRAVITY $(\rho/\rho_{\text{water}})$*
Aluminum	2.69
Concrete (av.)	2.40
Copper	8.91
Earth (av. wet)	1.76
(av. dry)	1.28
Glass	2.60
Iron (cast)	7.21
Lead	11.38
Mercury	13.57
Oil (av.)	0.90
Steel	7.84
Water (fresh, liquid)	1.00
(ice)	0.90
Wood (soft pine)	0.48
(hard oak)	0.80

*$\rho_{\text{water}} = 1.00(10^3)$ kg/m^3.

$\gamma_{\text{water}} = \rho_{\text{water}}g = 62.4$ lb/ft^3.

[handwritten: → x 1000 = density]

ANSWERS TO ODD-NUMBERED PROBLEMS

I.1 437 N at 11.49° above lt horiz

I.3 (a) 21.28 at 25.59° E of N
(b) 9.23 at 5° N of W
(c) 21.75 at 32.24° N of W

I.5 36.22 at 39.59° N of E

I.7 35.64 at 23.46° E of S

I.9 (a) 1 in.^{2} = 6.45(10^{-4}) m^2
(b) 1 ft^3 = 2.83(10^{-2}) m^3
(c) 1 oz = 0.278 N
(d) 1 lb/in.^{2} = 6.90(10^3) N/m^2
(e) 1 lb/ft^2 = 47.9 N/m^2
(f) 1 ft/s = 0.305 m/s
(g) 1 in./hr = 7.06(10^{-6}) m/s
(h) 1 mile/hr^2 = 1.242(10^{-4}) m/s^2

I.11 1 touchdown = 91.4 m
1 dash = 9 seconds
1 golf ball = 0.450 N

I.13 k = 150 lb/in. = 26.27 kN/m

I.15 L^2/T

I.17 (a) 1 kg − m^2/s^2
(b) 1 kg − m^2/s^3
(c) 1 J/s

II.1 F_X = 50.7 N
F_Y = 108.8 N

II.3 θ = 63.43°, F_X = 107.3, F_Y = 53.7 N; or
θ = − 116.57°, F_X = − 107.3, F_Y = − 53.7 N

II.5 (a) 45.40°
(b) 300 lb (dn)
(c) F$_{\parallel}$ = 351 lb (uphill), F$_{\perp}$ = 356 lb (dn)

II.7 (a) F_{hor} = 80 lb (rt), F_{vert} = 60 lb (up)
(b) F_{hor} = 80 lb (lt), F_{vert} = 60 lb (dn)

II.9 (a) 400 $\bar{I}$ + 600 $\bar{J}$ + 200 $\bar{K}$ m
(b) α = 57.69°, β = 36.70°
($\bar{I}$ south, $\bar{J}$ east)

II.11 $\bar{F}$ = F(0.8365 $\bar{I}$ + 0.50 $\bar{J}$ + 0.2241 $\bar{K}$)
α = 33.23°, β = 60°, γ = 77.05°

II.13 $\bar{F}$ = 60 $\bar{I}$ − 40 $\bar{J}$ + 120 $\bar{K}$
α = 64.62°, β = 106.60°, γ = 31.00°

II.15 $\bar{F}$ = −7.16 $\bar{I}$ − 3.58 $\bar{J}$ kN
$\bar{e}_{C/D}$ = −0.8944 $\bar{I}$ − 0.4472 $\bar{J}$

II.17 $\bar{e}_{B/A}$ = $\bar{K}$
$\bar{e}_{C/A}$ = −0.9487 $\bar{I}$ + 0.3167 $\bar{K}$
$\bar{e}_{D/A}$ = −0.5455 $\bar{I}$ − 0.8182 $\bar{J}$ + 0.1818 $\bar{K}$

II.19 244 N

II.21 39.8 N at 59.02° below rt horiz

II.23 5.04 kN at 57.52° below rt horiz

II.25 (a) F = 4.70 tons, α = 90°
(b) F = 1.71 tons, α = 180°

II.27 (a) 335 N
(b) 900 N

II.29 −54.71° or −145.29°

II.31 F_2 = 5 kN, F_4 = 3 kN

II.33 (a) F = 2.25 kN, β = 90°
(b) F = 3 kN, β = 90°

II.35 −1.846 $\bar{I}$ + 2.26 $\bar{J}$ − 3.02 $\bar{K}$ kN
($\bar{I}$ = $\bar{e}_{C/B}$, $\bar{K}$ up)

II.37 −49.3 $\bar{J}$ − 186.9 $\bar{K}$
($\bar{J}$ = $\bar{e}_{D/C}$, $\bar{K}$ up)

II.39 F_{AC} = 1.719 kN
F_{AD} = 1.613 kN
R = 0.957 kN

II.41 F_{AB} = 72N, F_{AC} = 664N

II.43 T_{AD} = 6.76 kN, T_{BD} = 8.65 kN, T_{CD} = 5.53 kN

II.45 −4.11(10^4) N^2

II.47 (a) 75.64°
(b) 100.66°
(c) 90°

II.49 39.88°

II.51 −0.5196 $\bar{I}$ + 0.30 $\bar{J}$ + 0.8 $\bar{K}$

II.53 (a) F_t = 5R/(h^2 + R^2)$^{1/2}$ kN
F_s = 5h/(h^2 + R^2)$^{1/2}$ kN
(b) $\bar{F}_t$ = 5R(0.3420h $\bar{I}$ + 0.9397h $\bar{J}$ + R $\bar{K}$)/(h^2 + R^2) kN
$\bar{F}_s$ = 5h(−0.3420R $\bar{I}$ − 0.9397R $\bar{J}$ + h $\bar{K}$)/(h^2 + R^2) kN

II.55 (a) F_t = 2.56 kN
F_s = 4.29 kN
(b) $\bar{F}_t$ = 1.55 $\bar{I}$ + 1.55 $\bar{J}$ + 1.31 $\bar{K}$ kN
$\bar{F}_s$ = −1.55 $\bar{I}$ − 1.55 $\bar{J}$ + 3.69 $\bar{K}$ kN

II.57 W_2 = 20 lb, W_3 = 40 lb

II.59 F = 70.3 N, β = 58.83°

II.61 N_{lt} = 0.335 N
N_{rt} = 0.490 N

II.63 N_{lt} = 0.389 N
N_{rt} = 1.074 N

II.65 cos 2θ − 2(kR/mg)(sin 2θ − sin θ) = 0

II.67 (a) $mg/(k_1 + k_2)$
(b) $k_1 + k_2$

II.69   0.1011 m

II.71 $m = 160y/(y^2 + 0.01)^{1/2}$ kg

II.73   0.384 W

II.75 $l_1 = 0.228$ m
        $l_2 = 0.150$ m

II.77   1125 lb

II.79 $T_{lt} = 628$ N
        $T_{rt} = 439$ N
        $T_{rear} = 844$ N

II.81   0.1719 m

II.83 $T_{lt} = 550$ lb
        $T_{rt} = 599$ lb

II.85 $N_{groove} = 167$ N
        $N_{block} = 1035$ N

III.1   (a) $-2\,\bar{I} - 16\,\bar{J} + 6\,\bar{K}$
        (b) $-4\,\bar{I} + 8\,\bar{J} - 3\,\bar{K}$
        (c) $-4\,\bar{I} - 2\,\bar{J} + 8\,\bar{K}$

III.3   (a) $22\,\bar{I} + 7\,\bar{J} - 2\,\bar{K}$
        (b) $\bar{I} - 2\,\bar{J} - \bar{K}$

III.7 $Y_B = -1.70$ m, $Z_C = -0.0125$ m

III.9 $X_B = 2.1 - 2Z_C$ m
        $\bar{e}_n = [0.4Z_C\,\bar{I} + (-2Z_C{}^2 + 2.4Z_C - 0.42)\,\bar{J}$
        $+ (-0.8Z_C + 0.84)\,\bar{K}]/(4Z_C{}^4 - 9.6Z_C{}^3 + 2.48Z_C{}^2$
        $- 3.36Z_C + 0.882)^{1/2}$

III.11 $M_E = 4620$ lb-in. (ccw)

III.13 $M_O = 12.99$ N-m (ccw)

III.15 $M_B = 20.49$ N-m (ccw)

III.17 $\theta = 53.13°$
        $M_C = 3.75$ kN-m (cw)

III.19 $\phi = 90°$
        $M_A = 250$ lb-ft (cw)

III.21 (a) 5.12°
        (b) 95.12°

III.23 (a) $-4.8\,\bar{I} - 8.31\,\bar{J} + 16.63\,\bar{K}$ N-m
        (b) 16.63 N-m
        (c) $r_z = 0.1386$ m
            $|\bar{r}| = 0.160$ m
        ($\bar{J}$ rt, $\bar{K}$ up)

III.25 28.3 kip-in.

III.27 (a) $87.8\,\bar{J} + 202.8\,\bar{K}$ N-m
        (b) 131.8 N-m

III.29 $M_{AB} = 0.360$ kN-m

III.31 (a) 8000 lb-ft

III.33 $F_A = 25.1$ N @ 26.57° cw from upward vert
        $F_B = 25.1$ N @ 26.57° cw from downward vert

III.35 247 N-m (ccw)

III.37 $M = 15.81$ kN-m
        $\alpha = 71.57°$
        $\beta = 161.57°$
        $\gamma = 90°$

III.39 $F = 7.071$ kN
        $\alpha = 74.05°$
        $\bar{M} = -0.8\,\bar{J} + 0.6\,\bar{K}$ kN-m
        ($\bar{J}$ rt, $\bar{K}$ up)

III.41 (a) & (b) $\bar{R} = 50\,\bar{I} - 86.6\,\bar{J}$ N-m
        (a) $\bar{M}_A = -80\,\bar{K}$ N-m
        (b) $\bar{M}_O = -85\,\bar{K}$ N-m
        ($\bar{I}$ rt, $\bar{J}$ up)

III.43 (a), (b) & (c) $\bar{R} = 1.04\,\bar{I} + 3.86\,\bar{J}$ kN
        (a) $\bar{M}_A = -0.630\,\bar{K}$ kN-m
        (b) $\bar{M}_B = 0.091\,\bar{K}$ kN-m
        (c) $\bar{M}_C = 0.721\,\bar{K}$ kN-m

III.45 $\bar{R} = -10\,\bar{I} + 14.64\,\bar{J}$ N
        $\bar{M}_O = -6\,\bar{K}$ N-m
        ($\bar{I}$ rt, $\bar{J}$ up)

III.47 $F = 267$ N
        $\bar{R} = 154\,\bar{I} - 455\,\bar{J}$ N
        ($\bar{I}$ rt, $\bar{J}$ up)

III.49 $T = 172.8$ lb
        $\bar{R} = -72\,\bar{I} - 522\,\bar{J}$ lb
        ($\bar{I}$ rt, $\bar{J}$ up)

III.51 $F_A = 1.0$ kips (up)
        $F_B = 3.0$ kips (dn)

III.53 $F_A = -4.48\,\bar{J}$ kN
        $F_B = 4.10\,\bar{I} - 8.39\,\bar{J}$ kn
        ($\bar{I}$ rt, $\bar{J}$ up)

III.55 (a) & (b) $\bar{R} = -40\,\bar{I}$ lb
        (a) $\bar{M}_A = 170\,\bar{J} + 1449\,\bar{K}$ lb-in
        (b) $\bar{M}_B = 170\,\bar{J} + 849\,\bar{K}$ lb-in
        ($\bar{J} = \bar{e}_{B/A}$, $\bar{K}$ up)

III.57 (a) $\bar{R} = -120\,\bar{K}$ N
            $\bar{M}_A = 4.50\,\bar{K}$ N-m
        (b) $M_{AD} = 4.50$ N-m
        ($\bar{I} = \bar{e}_{B/D}$, $\bar{K}$ up)

III.59 (a) & (b) $\bar{R} = -0.6\,\bar{I} + 2.5\,\bar{J} + 0.9\,\bar{K}$ kN
        (a) $\bar{M}_A = -0.78\,\bar{K}$ kN-m
        (b) $\bar{M}_B = -1.50\,\bar{I} - 0.36\,\bar{J} - 0.78\,\bar{K}$ kN-m
        ($\bar{J}$ rt, $\bar{K}$ up)

III.61 $\bar{F} = 70.7\,\bar{I} - 10.7\,\bar{K}$ N
        $\bar{M} = 28.6\,\bar{I} + 70.7\,\bar{J} - 141.4\,\bar{K}$ N-m
        ($\bar{I} = \bar{e}_{B/A}$, $\bar{J} = \bar{e}_{C/B}$)

III.63 (a) $F = 3400$ lb
        (b) $\bar{R} = -1200\,\bar{I} - 6600\,\bar{J}$ lb

$\bar{M}_A = -6000 \, \bar{J} + 216{,}000 \, \bar{K}$ lb-ft
$(\bar{I} = \bar{e}_{A/B}, \, \bar{J} = \bar{e}_{C/B})$

III.65 $R = 90$ kN (dn) @ 0.722 m from lt end

III.67 $R = 424$ N, 0.66 m above B @ 45° cw from $\bar{e}_{A/B}$

III.69 21.7 lb, 21.3 in. rt of C @ 64.03° cw from $\bar{e}_{D/C}$

III.71 $\bar{R} = 0.18 \, \bar{I} - 2.04 \, \bar{J}$ kN, 3.33 m above A
$(\bar{I}$ rt, $\bar{J}$ up)

III.73 $R = 213.6$ kN, 69.44° cw from $\bar{e}_{B/A}$, 11.30 m rt of A

III.75 3.6 in. lt and 5.4 in. below the centroid

III.77 $\bar{R} = -61 \, \bar{K}$ kN @ $1.72 \, \bar{I} + 1.23 \, \bar{J}$ m from near lt corner
$(\bar{I}$ rt, $\bar{K}$ up)

III.79 $\bar{R} = -150 \, \bar{K}$ N @ $0.373 \, \bar{I} + 0.149 \, \bar{J}$ m

III.81 40 N dn, along lt edge 0.125 m from far corner, or 40 N up, along rt edge 0.125 m from near corner

III.99 $\bar{F}_A = -0.10 \, \bar{I} + 12.05 \, \bar{J}$ kN
$(\bar{I}$ rt, $\bar{J}$ up)

III.101 0.497 kN-m

III.103 For 20° dn: $N_{rt} = 5.23$ kN, $N_{lt} = 6.75$ kN
For no slope: $N_{rt} = 6.12$ kN, $N_{lt} = 6.63$ kN

III.105 $m_B/m_A = (3 \cos \theta + \sin \theta) / (3 \cos \theta - \sin \theta)$

III.107 $T_1 = 6.83$ kN
$T_2 = 11.32$ kN
$\alpha = 83.66°$

III.109 40 lb (lt), bears on rt surface

III.111 $\bar{F}_A = 224 \, \bar{I}$ N
$\bar{F}_B = -224 \, \bar{I} + 736 \, \bar{J}$ N
$(\bar{I}$ rt, $\bar{J}$ up)

III.113 3000 lb (lt)

III.115 235 kips

III.117 832 N (T)

III.119 $F_{lt} = 0.5$ kN (up)
$F_{rt} = 0.5$ kN (dn)

III.121 $F_{lt} = 500$ lb (up)
$M_{lt} = 2000$ lb-ft (ccw)

III.123 $\bar{F}_{lt} = -0.54 \, \bar{I} + 7.07 \, \bar{J}$ kN
$\bar{F}_{rt} = 2.54 \, \bar{I} + 4.39 \, \bar{J}$ kN
$(\bar{I}$ rt, $\bar{J}$ up)

III.125 $\bar{F}_A = -158.6 \, \bar{I} + 80 \, \bar{J}$ kN
$\bar{F}_B = 118.6 \, \bar{I}$ kn
$(\bar{I}$ rt, $\bar{J}$ up)

III.127 (a) $A = 37.5$ kN (T), $B = 50$ kN (C), $C = 7.5$ kN (T)
(b) $A = 31.3$ kN (T), $B = 42.5$ kN (C),
$C = 6.3$ kN (T)

III.129 4.02 kN

III.131 $\theta - \dfrac{mgd}{kR} \sin \theta = 0$

III.133 $m_A/m_B = 2 \, [\cos \alpha \cos \beta / \cos (\alpha + \beta)] - 1$

III.135 $\theta = \sin^{-1}[(2d/L)^{1/3}]$, for $d < L/2$

III.137 (a) complete constraint: 1,2,3,5,6,9,10,11
partial constraint: 4
improper constraint: 7,8,12
(b) determinate: 1,2,6,10
indeterminate: 3,5,9,11

III.139 0.5 m

III.141 0

III.143 143.13°

III.145 $\bar{F}_A = -100 \, \bar{I} - 200 \, \bar{J} + 150 \, \bar{K}$ N
$\bar{M}_A = 100 \, \bar{I} - 72 \, \bar{J} + 156 \, \bar{K}$ N-m
$(\bar{I} = \bar{e}_{D/C}, \, \bar{J} = \bar{e}_{B/A})$

III.147 $m = 35.7$ kg
$R = 0.0756$ m
$\theta = 160.89°$

III.149 44.1 N

III.151 $T_{BD} = 160.1$ N
$T_{BE} = 98.1$ N

III.153 $T_{CE} = 15.81$ kips
$T_{BD} = 16.59$ kips
$T_{CF} = 7.91$ kips

III.155 $\bar{F}_A = -50 \, \bar{I} - 33.3 \, \bar{J}$ N
$\bar{F}_B = -66.7 \, \bar{J} - 16.7 \, \bar{K}$ N
$\bar{F}_C = 16.7 \, \bar{K}$ N
$(\bar{J}$ rt, $\bar{K}$ up)

III.157 $T_{CD} = 3000$ lb
$F_B = 0$

III.159 $\bar{N}_D = 4.25 \, \bar{I} + 7.35 \, \bar{J}$ N
$\bar{F}_A = -4.25 \, \bar{I} - 7.45 \, \bar{J} + 26.28 \, \bar{K}$ N
$\bar{F}_B = 0.10 \, \bar{J} + 12.95 \, \bar{K}$ N
$(\bar{I} = \bar{e}_{A/B}, \, \bar{K}$ up)

III.161 49.0 N *(lt)*

IV.1 $F_{AB} = 8$ kN (T)
$F_{AC} = 6$ kN (T)
$F_{BC} = 10$ kN (C)

IV.3 $F_{1-2} = 205$ lb (C)
$F_{1-3} = 220$ lb (C)
$F_{2-3} = 564$ lb (C)

IV.5 $F_{AB} = 0$
$F_{AD} = 6$ kN (C)
$F_{BC} = 5$ kN (T)
$F_{BD} = 3.61$ kN (T)
$F_{CD} = 4$ kN (C)

IV.7 $F_{AB} = 8$ kN (T)
$F_{AC} = 11.31$ kN (C)
$F_{AF} = 8$ kN (T)
$F_{BC} = 12$ kN (C)
$F_{CD} = 12$ kN (C)
$F_{CE} = 11.31$ kN (C)
$F_{CF} = 16$ kN (T)
$F_{DE} = 0$
$F_{EF} = 8$ kN (T)

IV.9 $F_{AB} = F_{CE} = F_{CF} = 0 = F_{DG}$
$F_{AG} = 29.6$ kN (C)
$F_{BC} = 11.5$ kN (C)
$F_{BF} = 23.1$ kN (T)
$F_{CD} = 11.5$ kN (C)
$F_{DE} = 10.5$ kN (C)
$F_{EF} = 10.5$ kN (C)
$F_{FG} = 29.6$ kN (C)

IV.11 $F_{1-2} = 22.5$ kN (C)
$F_{1-8} = 16.88$ kN (C)
$F_{2-3} = 3.75$ kN (C)
$F_{2-6} = 12.5$ kN (C)
$F_{2-7} = 5$ kN (C)
$F_{2-8} = 18.75$ kN (T)
$F_{3-4} = 3.75$ kN (C)
$F_{3-6} = 0$
$F_{4-5} = 5$ kN (C)
$F_{5-6} = 0$
$F_{6-7} = 11.25$ kN (T)
$F_{7-8} = 11.25$ kN (T)

IV.13 $F_{AB} = F_{DE} = 260$ lb (T)
$F_{AI} = F_{EF} = 520$ lb (C)
$F_{BC} = F_{CD} = 673$ lb (T)
$F_{BH} = F_{DG} = 312$ lb (C)
$F_{BI} = F_{DF} = 520$ lb (T)
$F_{CG} = F_{CH} = 104$ lb (T)
$F_{GH} = 667$ lb (C)
$F_{FG} = F_{HI} = 460$ lb (C)

IV.15 $F_{AB} = F_{AC} = F_{AD} = F_{AE} = F_{BF} = F_{CF} = F_{DF} = F_{EF} =$
1.382 kN (T)
$F_{BC} = F_{BE} = F_{CD} = F_{DE} = 0.833$ kN (C)

IV.17 $F_{AB} = 760$ kN (T)
$F_{AC} = 13.0$ kN (C)
$F_{AD} = F_{BD} = 72.1$ kN (C)
$F_{BC} = 65.0$ kN (C)
$F_{CD} = 78.0$ kN (T)
$F_{DE} = 101.8$ kN (T)

IV.19 $F_{AB} = 6.15$ kips (T)
$F_{AC} = 1.04$ kips (C)
$F_{AD} = 9.22$ kips (C)
$F_{BC} = 0.67$ kips (C)
$F_{BD} = 0.90$ kips (C)
$F_{CD} = 6.04$ kips (T)

IV.21 $F_{1-2} = 207$ lb (C)
$F_{1-3} = 1172$ lb (T)
$F_{1-4} = 325$ lb (T)
$F_{2-3} = 75$ lb (T)
$F_{2-4} = 207$ lb (C)
$F_{2-5} = 400$ lb (T)
$F_{3-4} = 1586$ lb (C)

IV.23 $F_{AD} = 180$ kN (T)
$F_{BD} = 39.0$ kN (C)
$F_{BE} = 36.0$ kN (C)
$F_{CD} = 58.5$ kN (C)
$F_{CE} = 10.1$ kN (C)
$F_{CF} = 45.0$ kN (C)
$F_{DE} = 15.0$ kN (T)
$F_{DF} = 22.5$ kN (T)
$F_{DG} = 90.0$ kN (T)
$F_{EF} = 4.5$ kN (C)
$F_{EG} = 39.0$ kN (C)
$F_{EH} = 9.0$ kN (C)
$F_{FG} = 58.5$ kN (C)
$F_{FH} = 10.1$ kN (T)
$F_{FI} = 0$
$F_{GH} = 15.0$ kN (T)
$F_{GI} = 22.5$ kN (T)
$F_{HI} = 13.5$ kN (C)

IV.25 $F_{3-4} = 7.5$ kips (T)
$F_{3-6} = 20.0$ kips (T)

IV.27 $F_{CD} = 32$ kN (C)
$F_{DG} = 22.6$ kN (T)

IV.29 $F_{2-9} = 0$
$F_{8-9} = 40.6$ kips (C)

IV.31   31.2 kN (T)

IV.33 $F_{2-7} = 5.66$ kN (T)
$F_{3-7} = 6.0$ kN (C)
$F_{6-7} = 13.42$ kN (C)

IV.35 $F_{4-12} = 0$
$F_{5-12} = 166.7$ N (C)

IV.37 $F_{4-9} = 1.08$ kN (C)
$F_{9-11} = 6.15$ kN (T)

IV.39 $F_{1-2} = 1667$ lb (C)
$F_{1-3} = 4800$ lb (T)
$F_{4-12} = 2692$ lb (C)
$F_{6-10} = 2115$ lb (T)

IV.41 $d = 5$ m: $F_{CD} = 67.1$ kN (C) & $F_{FG} = 35$ kN (T)
$d = 2.5$ m: no

IV.43 $F_{IJ} = 50.0$ kips (C)
$F_{IO} = 6.25$ kips (T)

IV.45 $F_{BF} = 0$
$F_{CG} = 21.3$ kN (T)

IV.47 $F_{DE} = 15.0$ kN (T)
$F_{DF} = 22.5$ kN (T)
$F_{DG} = 90.0$ kN (T)

IV.49 $F_3 = 15.56$ kN (C)
$F_4 = 13.33$ kN (C)

IV.51 $\bar{F}_A = -211\,\bar{I} - 70\,\bar{J}$ lb
$\bar{F}_C = 211\,\bar{I} + 470\,\bar{J}$ lb
($\bar{I}$ rt, $\bar{J}$ up)

IV.53 $F_{bolt} = 4P$
$F_A = 5P$ (dn on AC)

IV.55 20 N-m (ccw)

IV.57 500 lb-ft (ccw)

IV.59 $k = 418$ N/m
$\bar{F}_A = 5.7\,\bar{I} + 78.5\,\bar{J}$ N
($\bar{I}$ rt, $\bar{J}$ up)

IV.61 $F_A = 1$ kN (up)
$F_B = 5$ kN (up)
$M_B = 1.6$ kN-m (cw)

IV.63 $F_A = 2$ kN (up)
$M_A = 0.8$ kN-m (ccw)
$F_B = 4$ kN (up)
$M_B = 1.2$ kN-m (cw)

IV.65 member AC: $\bar{F}_A = 362\,\bar{I} - 43\,\bar{J}$ N
$\bar{F}_B = -696\,\bar{I} + 187\,\bar{J}$ N
$\bar{F}_C = 335\,\bar{I} - 230\,\bar{J}$ N
member CE: $\bar{F}_C = -335\,\bar{I} + 230\,\bar{J}$ N
$\bar{F}_D = -696\,\bar{I} + 387\,\bar{J}$ N
$\bar{F}_E = -362\,\bar{I} + 243\,\bar{J}$ N
($\bar{I}$ rt, $\bar{J}$ up)

IV.67 bar AB: $\bar{F}_A = 45\,\bar{I} + 38.7\,\bar{J}$ kN
$\bar{F}_B = -60\,\bar{I} + 6.3\,\bar{J}$ kN
bar BC: $\bar{F}_B = 60\,\bar{I} - 6.3\,\bar{J}$ kN
$\bar{F}_C = -75\,\bar{I} + 51.3\,\bar{J}$ kN
($\bar{I}$ rt, $\bar{J}$ up)

IV.69 $\bar{F}_A = -28.7\,\bar{I} + 100.3\,\bar{J}$ kN
$\bar{F}_D = 28.7\,\bar{I} + 49.7\,\bar{J}$ kN
($\bar{I}$ rt, $\bar{J}$ up)

IV.71 $\bar{F}_A = -268\,\bar{I} + 3000\,\bar{J}$ lb
$\bar{F}_E = 268\,\bar{I} + 5000\,\bar{J}$ lb
($\bar{I}$ rt, $\bar{J}$ up)

IV.73 (a) bar AB: $\bar{F}_A = -0.289W\,\bar{I} + 0.5\,W\,\bar{J}$
$\bar{F}_B = 0.289W\,\bar{I} + 0.5W\,\bar{J}$
bar AC: $\bar{F}_A = 0.289W\,\bar{I} + W\,\bar{J}$
$\bar{F}_C = -0.289W\,\bar{I}$
bar BC: $\bar{F}_B = -0.289W\,\bar{I} + W\,\bar{J}$
$\bar{F}_C = 0.289W\,\bar{J}$
(b) bar AB: $\bar{F}_A = -\bar{F}_B = -0.289W\,\bar{I}$
bar AC: $\bar{F}_A = -\bar{F}_C = 0.289W\,\bar{I} + 0.5W\,\bar{J}$
bar BC: $\bar{F}_B = -\bar{F}_C = -0.289W\,\bar{I} + 0.5W\,\bar{J}$
($\bar{I}$ rt, $\bar{J}$ up)

IV.75 $M_C = 3750$ N-m (cw)

IV.77 $M_2 = 2(1 + r_2/r_1)M_1$ (cw)
$M_3 = (1 + 2r_2/r_1)M_1$ (ccw)

IV.79 $\bar{F}_A = 2.67\,\bar{I} + 1.60\,\bar{J}$ kips
$\bar{F}_F = -2.67\,\bar{I} + 3.40\,\bar{J}$ kips
($\bar{I}$ rt, $\bar{J}$ up)

IV.81 $F_A = 4$ kN (up)
$F_B = 4$ kN (dn)

IV.83 67.5 N

IV.85 $\bar{F}_{plate} = -145.9\,\bar{I}$ lb
$\bar{F}_A = -173.8\,\bar{J}$ lb
$\bar{F}_E = -192.3\,\bar{I} + 192.3\,\bar{J}$ lb
($\bar{I}$ rt, $\bar{J}$ up)

IV.87 $\bar{F}_A = 8\,\bar{I} - 21.6\,\bar{J}$ kN
$\bar{F}_C = -8\,\bar{I} + 25.6\,\bar{J}$ kN
$\bar{F}_D = -1.01\,\bar{I} - 0.08\,\bar{J}$ kN
$\bar{F}_F = 1.01\,\bar{I} - 3.92\,\bar{J}$ kN
($\bar{I}$ rt, $\bar{J}$ up)

IV.89 $1.5\,P \tan\theta$ (rt)

IV.91 $F_{sp} = 27.26$ kN (C)
$\bar{F}_A = -3.12\,\bar{I} - 1.25\,\bar{J}$ kN
$\bar{F}_B = 10.18\,\bar{I} - 21.08\,\bar{J}$ kN
($\bar{I}$ rt, $\bar{J}$ up)

V.1 $\dfrac{a}{3}$ forward of rear face, $\dfrac{b}{3}$ above bottom,

$\dfrac{h}{2}$ lt of rt face

V.3 $(\tfrac{2}{3}h, 0, 0)$

V.5 $(\tfrac{3}{8}a, 0, 0)$

V.7 $\left(\tfrac{5}{16}h, -\dfrac{a}{4}, 0\right)$

V.9 $0.3619a$ on ctr line from flat face

V.11 $(0.807a, 0.540b, 0.2421c)$

V.13 $\left(\dfrac{4}{3\pi}\,a, \dfrac{4}{3\pi}\,b\right)$

V.15 $(\tfrac{4}{5}a, \tfrac{2}{7}b)$

V.17 $\left[\dfrac{(n-1)^2 a}{2n\left(n-1+\ln\dfrac{1}{n}\right)}, \dfrac{(n-1)^2 b}{2n\left(n-1+\ln\dfrac{1}{n}\right)}\right]$

V.19 $(0.574a, 0.410a)$

V.21 $\tfrac{2}{3}h$ from apex on ctr line

V.23 $\dfrac{h}{5}\left[\dfrac{3(a^2 + 4h^2)^{3/2}}{(a^2 + 4h^2)^{3/2} - a^3} - \dfrac{a^2}{2h^2}\right]$ from apex on ctr line

V.25 0.1611 m dn from top ctr

V.27 $\dfrac{4}{3\pi}\dfrac{r_2{}^2 + r_1 r_2 + r_1{}^2}{r_1 + r_2}$ up from bot ctr

V.29 7.25 in. lt, 3.25 in. up from rt tip

V.31 0.289 m rt, 0.296 m up from lower lt corner

V.33 0.01045 ft lt, 0.00523 ft up from ctr

V.35 0.125 m dn from top ctr

V.37 0.779 m from lt tip on ctr line

V.39 11.592 in. rt, 3.147 in. up from lower lt corner

V.41 0.367 in. up from ctr

V.43 1.5°

V.45 8.36° or 38.36°

V.47 4.25 m from lt apex on ctr line

V.49 (a) $m = 2.85$ kg, G is 22.2 mm up from bot ctr
(b) $m = 3.81$ kg, G is 21.0 mm up, 297 mm lt from bot rt ctr

V.51 $m = 8.58$ kg, 163.1 mm from apex on ctr line

V.53 $R = 2.4$ kN @ 0.667 m rt of A
$R_A = 1.40$ kN (up)
$R_B = 1.00$ kN (up)

V.55 couple $= \frac{1}{6} wL^2$ N-m (cw)
$R_A = \frac{1}{12} wL$ N (dn)
$R_B = \frac{1}{12} wL$ N (up)

V.57 $R = 2w_0L/\pi$ @ $L(1-2/\pi)$ rt of A
$R_A = 4w_0L/\pi^2$ (up)
$R_B = (2w_0L/\pi)(1-2/\pi)$ (up)

V.59 $R_A = 4.33$ kips (up)
$R_B = 0.67$ kips (up)

V.61 $R_A = 10.92$ kN (up)
$R_B = 15.48$ kN (up)

V.63 (a) $q_1 = 2q_2$, $R_A = \frac{1}{2} q_2L$ (dn)
(b) $q_1 = q_2$, $M_A = q_2L^2/6$ (ccw)

V.67 $\bar{R}_A = 1.571\ qr\ \bar{K}$
$\bar{M}_A = qr^2(\bar{I} + 0.571\ \bar{J})$
($\bar{J}$ rt, $\bar{K}$ up)

V.69 $\bar{R}_A = q_0r\ \bar{J}$
$\bar{M}_A = q_0r^2(0.5\ \bar{I} + 0.7854\ \bar{K})$
($\bar{I}$ rt, $\bar{J}$ up)

V.71 $\bar{R}_A = 3.93\ \bar{I}$ kN
$\bar{M}_A = 9.82\ \bar{J}$ kN-m
($\bar{J}$ rt, $\bar{K}$ up)

V.73 $\bar{R}_A = 2.50\ \bar{I}$ kN
$\bar{M}_A = 6.67\ \bar{J} - 0.83\ \bar{K}$ kN-m
($\bar{J}$ rt, $\bar{K}$ up)

V.75 $\bar{R}_A = 3.82\ \bar{I}$ kN
$\bar{M}_A = 9.54\ \bar{J} + 0.01\ \bar{K}$ kN-m
($\bar{J}$ rt, $\bar{K}$ up)

V.77 $R = 178.5$ kN (dn) @ 9.07 m from lt edge along ctr line

V.79 $R = 48.7$ kN (dn) @ 1 m rt, 1.5 m forward from pt of max p

V.81 $\bar{R} = -0.25p_0ab\ \bar{K}$ @ $2a/3\ \bar{I} + 2b/3\ \bar{J}$

V.83 $R_{front} = 280.8$ lb @ 0.5 ft up from bot on ctr line
$R_{side} = 140.4$ lb @ 0.5 ft up from bot on ctr line
$R_{bot} = 748.8$ lb @ ctr of bot

V.85 $P = 8356$ kN ($\perp$ wall) @ 8 m depth

V.87 (a) 71.3 kN
(b) 159.5 kN

V.89 (a) 71.9 kN
(b) 194.1 kN

V.91 398 lb

V.93 7.27 kN

V.95 $R = 59.96$ kN @ 11.11° below lt horiz, intersecting ctr

V.97 $\bar{R} = -499\ \bar{I} + 1082\ \bar{J}$ lb @ $1.016\ \bar{I} + 0.254\ \bar{J}$ ft

V.99 1.961 kN

V.101 2.43 kN

V.103 (a) 5611 lb
(b) 23,587 lb

V.105 $a = 0.5656R$

V.107 1: 13.15 kN @ 25.9 mm below ctr
2: 18.59 kN @ ctr
3: 18.59 kN @ 73.2 mm below ctr

V.109 (a) vertex up: $R = \frac{1}{6}\rho g\ ab(3h + 2a)$ @ $\frac{a}{2}\frac{2h + a}{3h + 2a}$ above A
vertex dn: $R = \frac{1}{6}\rho g\ ab(3h + a)$ @ $\frac{a}{2}\left(\frac{4h + a}{3h + a}\right)$ above A
(b) vertex up: $\bar{R}_A = \frac{1}{6}\rho g\ ab(3h + 2a)\ \bar{k}$
$\bar{M}_A = -\frac{1}{12}\rho g\ a^2b\ (2h + a)\ \bar{\imath}$
vertex dn: $\bar{R}_A = \frac{1}{6}\rho g\ ab\ (3h + a)\ \bar{k}$
$\bar{M}_A = -\frac{1}{12}\rho g\ a^2b\ (4h + a)\ \bar{\imath}$
($\bar{\imath}$ rt, $\bar{\jmath}$ dn)

V.111 243 N

V.113 $T_A = T_B = 77.1$ lb
$T_C = 154.1$ lb

VI.1 (a) $R_a = 0$
$\left|R_s\right| = 7.5$ kN
$\left|M_b\right| = 0.75$ kN-m
(b) $R_a = 0$
$\left|R_s\right| = 2.5$ kN
$\left|M_b\right| = 2$ kN-m

VI.3 (a) $R_a = 3$ kips
$\left|R_s\right| = 10$ kips
$\left|M_B\right| = 21$ kip-ft

(b) $R_a = 3$ kips
$\left|R_s\right| = 6$ kips
$\left|M_b\right| = 5$ kip-ft

VI.5 (a) $R_a = 0$
$\left|R_s\right| = 1.85$ kN
$\left|M_b\right| = 2.35$ kN-m
(b) $R_a = 4.8$ kN
$\left|R_s\right| = 4.0$ kN
$\left|M_b\right| = 1.6$ kN-m

VI.7 (a) $R_a = -1.73$ kips
$\left|R_s\right| = 2.63$ kips
$\left|M_b\right| = 97.0$ kip-in.
(b) $R_a = 0$
$\left|R_s\right| = 1.63$ kips
$\left|M_b\right| = 8.15$ kip-in.

VI.9 (a) $R_a = -10.33$ kN
$\left|R_s\right| = 4.10$ kN
$\left|M_b\right| = 0.459$ kN-m
(b) $R_a = 1.518$ kN
$\left|R_s\right| = 3.83$ kN
$\left|M_b\right| = 1.290$ kN-m

VI.11 (a) $R_a = 0$
$\left|R_s\right| = 2.57$ kips
$\left|M_b\right| = 15.4$ kip-ft

(b) $R_a = 0$
$\left|R_s\right| = 6$ kips
$\left|M_b\right| = 42.0$ kip-ft

VI.13 (a) $R_a = 0.333q_0R$
$\left|R_s\right| = 0.083q_0R$
$\left|M_b\right| = 0.0417q_0R^2$
(b) $R_a = -0.455q_0R$
$\left|R_s\right| = 0.12q_0R$
$\left|M_b\right| = 0.1725q_0R^2$

VI.15 $R_a = 4$ kN
$\left|R_s\right| = 5$ kN
$\left|M_b\right|$ 2.72 kN-m
$\left|M_t\right| = 0.45$ kN-m

VI.17 $R_a = 0.707\,F$
$\left|R_s\right| = 1.225\,F$
$\left|M_b\right| = FR$
$\left|M_t\right| = 0.293\,FR$

VI.19 $\left|V\right| = P$
$\left|M\right|_{max} = PL$ @ lt end

VI.21 $\left|V\right|_{max} = 3.2$ kN @ lt end
$\left|M\right|_{max} = 5.12$ kN-m @ 3.2 m from lt end

VI.23 $\left|V\right|_{max} = \frac{1}{4}\,w_0L$ @ midpt
$\left|M\right|_{max} = \frac{1}{8}\,w_0L^2$ @ lt end

VI.25 $\left|V\right|_{max} = 12.5$ lb on $8 < x < 16$ in. from lt end
$\left|M\right|_{max} = 120$ lb-in. on $16 < x < 24$ in. from lt end

VI.27 $\left|V\right|_{max} = 32$ kN @ lt end
$\left|M\right|_{max} = 192$ kN-m @ 8 m from lt end

VI.29 $\left|V\right|_{max} = 10$ kn @ midpt
$\left|M\right|_{max} = 10$ kN-m @ midpt

VI.31 $\left|V\right|_{max} = 600$ N @ 2 m from lt end
$\left|M\right|_{max} = 133.3$ N-m @ 1 m and 2 m from lt end

VI.33 $\left|V\right|_{max} = 100$ N on $0 < x < 0.1$ m from rt end
$\left|M\right|_{max} = 10$ N-m @ 0.1 m from rt end

VI.35 $\left|V\right|_{max} = 12$ kN on $0 < x < 2$ m from lt end
$\left|M\right|_{max} = 24$ kN-m on $2 < x < 6$ m from lt end

VI.37 $\left|V\right|_{max} = 20.7$ kN on $0 < x < 0.4$ m from lt end
$\left|M\right|_{max} = 14.67$ kN-m @ 1 m from lt end

VI.39 $\left|V\right|_{max} = 12.5$ lb on $8 < x < 16$ in. from lt end
$\left|M\right|_{max} = 120$ lb-in on $16 < x < 24$ in. from lt end

VI.41 $\left|V\right|_{max} = \frac{1}{3}w_0L$ @ lt end
$\left|M\right|_{max} = 0.0642w_0L^2$ @ 0.423L from lt end

VI.43 $\left|V\right|_{max} = 32$ kN @ lt end
$\left|M\right|_{max} = 192$ kN-m @ 8 m from lt end

VI.45 $\left|V\right|_{max} = 21.3$ kips @ rt end
$\left|M\right|_{max} = 15.80$ kip-ft @ 1.723 ft from lt end

VI.47 $\left|V\right|_{max} = 30$ kN @ rt end
$\left|M\right|_{max} = 60$ kN @ rt end

VI.49 (a) $q = 5$ kips/ft (dn), on $0 < x < 5$ ft
$F_1 = 10$ kips (up), @ $x = 0$
$F_2 = 5$ kips (dn), @ $x = 3$ ft
$F_3 = 25$ kips (up), @ $x = 5$ ft
$F_4 = 5$ kips (dn), @ $x = 9$ ft
(b) 9 ft
(c) $\left|M\right|_{max} = 22.5$ kip-ft @ $x = 5$ ft

VI.51 $\left|V\right|_{max} = 2650$ N @ 15 m from lt end
$\left|M\right|_{max} = 8640$ N-m @ 7.84 m from lt end

VI.53 (a) 14.40 lb
(b) 50 in.

VI.55 (a) 3.83 m
(b) $T_{AB} = 3.54$ kN
(c) $\theta_{AB} = 62.45°$

VI.57 (a) $T_{DE} = 1.18$ kN
(b) 15.49 m

VI.59 (a) $T_{min} = 1.226(10^5)$ kN
$T_{max} = 1.320(10^5)$ kN
(b) 1059 m

VI.61 $\dfrac{L_2}{h_2} = \left[\dfrac{L_1^2}{L_2^2}\left(\dfrac{L_1^2}{h_1^2} + 16\right) - 16\right]^{1/2}$

VI.63 $h = 75$ ft
$T_A = T_B = 1.061(10^5)$ lb

VI.65 (a) $T_A = 231$ kN
(b) 106.09 m

VI.67 $h = 0.333$ m
            $s = 56.30$ m

VI.69   (a) 131.7 kips
            (b) 24.93 ft
            (c) 2.41 ft

VI.71   (a) $y = \dfrac{1}{9} \dfrac{W}{T_o L^2} x^2(x + 3L)$
            (b) $\dfrac{4}{9} \dfrac{WL}{T_o}$
            (c) $(T_o^2 + W^2)^{1/2}$

VI.73 $s = 202.23$ ft
            $h = 12.99$ ft

VI.75 $0.203\, w_o d$

VI.77   (a) $0.1468\, w_o d$
            (b) $0.1288\, w_o d$

VI.79   (a) 45.46 ft
            (b) $F_{hor} = 59.6$ lb
                $F_{vert} = 50$ lb

VI.81   0.1408

VI.83   0.66 m or 74.08 m

VII.1   147.1 N

VII.3   7.01 N

VII.5   2.83 kg

VII.7   3.93 N (slipping betw A & B)

VII.9   All slide together

VII.11 0.884

VII.13 0

VII.15 (a) 0.484
            (b) $\bar{e} = -0.659\,\bar{I} + 0.706\,\bar{J} - 0.257\,\bar{K}$

VII.17 8.53°

VII.19 $74 < P < 368$ N

VII.21 287 N-m

VII.23 $L = 0.5\left(\dfrac{d}{\mu} - h\right)$

VII.25 (a) 1.439 kN (lt)
            (b) 0.746

VII.27 0.311

VII.29 0.570 kN

VII.31 210 N

VII.33 Yes

VII.35 8.83 kN (up)

VII.37 (a) 43.7 lb (lt)
            (b) 1.017

VII.39 (a) 550 N
            (b) 1450 N

VII.41 $\mu_A = 0.420$
            $\mu_C = 0.141$
            $\mu_D = 0.598$

VII.43 34.23°

VII.45 25.02°

VII.47 (a) $1.5 < d < 2$ m
            (b) $2 < d < 3$ m

VII.49 $0 < h < 3.57$ m

VII.51 (a) & (b) 235 N
            (c) 136 N

VII.53 (a) 2.67 m
            (b) & (c) 36.8 N

VII.55 (a) 4.0 m
            (b) 24.5 N
            (c) 49.0 N

VII.57 No sliding: $\beta < \tan^{-1}\mu$
            No tipping: $\beta < \tan^{-1}\dfrac{b}{h}$

VII.59 1.197 kN

VII.61 0.245 kN

VII.63 2.20 kips

VII.65 (a) 2.69 kN
            (b) 0.0145
            (c) Yes

VII.67 (a) $F_B = 1.880$ kN
            (b) $F_A = 0$, $F_C = 2$ kN (rt)

VII.69 219 lb

VII.71 1.842 N-m

VII.73 1.047 N-m

VII.75 110 N

VII.77 238 lb-in

VII.79 273 N

VII.81 8.47 N-m

VII.83 (a) 31.0 lb
            (b) 149.0 lb

VII.85 $0.4 < T < 1.071(10^5)$ N

VII.87 $T = 68.9$ N

VII.89 (a) 39.4 N-m
            (b) 33.7 N-m

VII.91 $M_A = 70.2$ N-m
            $M_B = 281$ N-m
            Slipping at A first

VII.93 $M_A = 92.1$ lb-in (cw)
            $M_B = 184.2$ lb-in (ccw)

VIII.1 $(mgr_2 - kr_1^2\, \theta)\, \delta\theta$

VIII.3 $[- kl^2\,(1 - \cos\theta)\sin\theta/\cos^2\theta]\delta\theta$

VIII.5 $[M_A - FL(4 \sin^2 \theta + 1)/2]\delta\theta$

VIII.7 $-\dfrac{a}{(4b^2 - a^2)^{1/2}} (F_2 + 2F_3)\delta a$

VIII.9 $\theta = \sin^{-1}\left[\left(\dfrac{2a}{L}\right)^{1/3}\right]$

VIII.11   26.0 lb (rt)

VIII.13 $M_A = 450 \dfrac{\sin \theta\,(1 - \cos \theta)}{\cos^3 \theta}$ N-m

VIII.15   232 lb

VIII.17   26.1 kN/m

VIII.19 $P = 150\left(1 - 6 \tan \dfrac{\theta}{2}\right)$ N

VIII.21   1.6 kN-m (cw)

VIII.23   5.11 kN (up)

VIII.25   3.45 kN-m (cw)

VIII.27 $m_B/m_A = 0.5 \sin \alpha/\sin \beta$

VIII.29   1204 N (rt)

VIII.31   (a) 1158 lb-in (cw)
(b) 442 lb-in (cw)

VIII.33 $m = \dfrac{2kL}{g}\cot^3 \theta \left[1 + \left(\dfrac{T_0}{kL} - 1\right) \sin \theta\right]$

VIII.35 $M_A = 450 \dfrac{\sin \theta\,(1 - \cos \theta)}{\cos^3 \theta}$ N-m

VIII.37 $m = (0.8\,kl/g)\,(2 \sin \theta - 1)$

VIII.41   3520 lb (lt)

VIII.43 $F = 780 \tan \theta\,[1 - 10(1252 - 1248 \cos \theta)^{-1/2}]$ lb

VIII.45   (a) & (b) $\theta = 90°$ is unstable

For $m < kl/g$, $\theta = \sin^{-1}\left(1 - \dfrac{mg}{kl}\right)$ is stable

For $m > kl/g$, $\theta = 0$ is stable
(θ is angle of elev of bar AB)

VIII.47   (a) $\sin c\theta = -\sin \theta$
(b) $\theta = 0$ is unstable
$\theta = 180°$ is stable
(c) $\theta = 0°$ is unstable

$\theta = 180°$ is unstable
$\theta = 120°$ is stable

VIII.49   If $k < 1.531\,mg$, $\theta = 0$ is stable
If $k > 1.531\,mg$, $\theta = 0$ is unstable
and $\theta = \cos^{-1}(1.531\,mg/k)$ is stable

VIII.51 $k = mg\,a/L^2$ is neutral equil

VIII.53 $K > 0.5\,mg/b^2$

VIII.55 $P = c^2\,kl$

IX.1 $I_x = ab^3/12$
$I_y = a^3b/12$

IX.3 $I_x = r^4\,(2\alpha - \sin 2\alpha)/8$
$I_y = r^4\,(2\alpha + \sin 2\alpha)/8$

IX.5 $I_x = 4ab^3/(9\pi)$
$I_y = (2/\pi^3)\,(\pi^2 - 8)a^3b$

IX.7 $J_z = ab(a^2 + b^2)/12$
$I_{xy} = a^2b^2/24$

IX.9 $J_z = R^4\alpha/2$, $I_{xy} = 0$

IX.11 $J_z = 2ab[9(\pi^2 - 8)a^2 + 2\pi^2b^2]/(9\pi^2)$
$I_{xy} = (\pi^2 - 4)a^2b^2/(8\pi^2)$

IX.13 $R/2$

IX.15   1.046 m

IX.17   37.0 mm

IX.19 $I_x = (5\pi/4)r^4$
$J_z = (3\pi/2)r^4$
$I_{xy} = 0$

IX.21 $I_x = 2.071(10^4)$ in.^{4}
$J_z = 7.823(10^4)$ in.^{4}
$I_{xy} = 2.761\,(10^4)$ in.^{4}

IX.23 $I_x = 2.94(10^{-7})$ m^4
$J_z = 5.10(10^{-7})$ m^4
$I_{xy} = 2.16(10^{-7})$ m^4

IX.25 $I_x = 0.2581$ m^4
$J_z = 1.0624$ m^4
$I_{xy} = -0.3703$ m^4

IX.27   (a) $A = 30$ in.^{2}
(b) $I_{\hat{x}} = 750$ in.^{4}
$I_{\hat{y}} = 1080$ in.^{4}

IX.29 $I_x = 6.29(10^{-3})$ m^4
$I_{xy} = 1.413(10^{-2})$ m^4
$k_x = 0.1686$ m

IX.31 $I_x = 5035$ in.^{4}
$I_{xy} = 2112$ in.^{4}
$k_x = 5.349$ in.

IX.33 $J_z = 7.89(10^{-2})$ m^4
$K_z = 0.573$ m

IX.35 $J_z = 10{,}070$ in.^{4}
$k_z = 7.56$ in.

IX.37 G is 142.5 mm above lower edge on ctr line
$I_x = 2.29(10^{-4})$ m^4
$I_y = 6.12(10^{-4})$ m^4
$I_{xy} = 0$
$J_z = 8.41(10^{-4})$ m^4

IX.39 G is 242.8 mm rt of ctr of lt circle
$I_x = 3.50(10^{-3})$ m^4
$I_y = 3.92(10^{-2})$ m^4
$I_{xy} = 0$
$J_z = 4.27(10^{-2})$ m^4

IX.41 G is at midpt
$I_x = 1.221(10^{-3})$ m^4
$I_y = 1.048(10^{-2})$ m^4
$I_{xy} = 2.58(10^{-3})$ m^4
$J_z = 1.170(10^{-2})$ m^4

IX.43 G is 285 mm rt & 100 mm below the upper lt corner
$I_x = 3.23(10^{-4})$ m^4
$I_y = 2.44(10^{-3})$ m^4
$I_{xy} = -2.28(10^{-4})$ m^4
$J_z = 2.76(10^{-3})$ m^4

IX.45 (a) $I_{x'} = 2.34(10^{-5})$ m^4
$I_{y'} = 3.69(10^{-5})$ m^4
$I_{x'y'} = -2.32(10^{-5})$ m^4
(b) $I_{x'} = I_{y'} = 3.02(10^{-5})$ m^4
$I_{x'y'} = 2.41(10^{-5})$ m^4
(c) $I_{x'} = 1.81(10^{-5})$ m^4
$I_{y'} = 4.22(10^{-5})$ m^4
$I_{x'y'} = 2.09(10^{-5})$ m^4

IX.47 (a) $I_{x'} = 6.06(10^{-3})$ m^4
$I_{y'} = 4.39(10^{-3})$ m^4
$I_{x'y'} = -3.25(10^{-3})$ m^4
(b) $I_{x'} = 3.52(10^{-3})$ m^4
$I_{y'} = 6.93(10^{-3})$ m^4
$I_{x'y'} = 2.88(10^{-3})$ m^4
(c) $I_{x'} = 2.30(10^{-3})$ m^4
$I_{y'} = 8.14(10^{-3})$ m^4
$I_{x'y'} = 1.64(10^{-3})$ m^4

IX.49 (a) $I_{x'} = 1.682(10^{-3})$ m^4
$I_{y'} = 1.4318(10^{-2})$ m^4
$I_{x'y'} = 3.74(10^{-3})$ m^4
(b) $I_{x'} = 9.66(10^{-3})$ m^4
$I_{y'} = 6.34(10^{-3})$ m^4
$I_{x'y'} = -7.15(10^{-3})$ m^4
(c) $I_{x'} = 4.40(10^{-3})$ m^4
$I_{y'} = 1.16(10^{-3})$ m^4
$I_{x'y'} = 6.40(10^{-3})$ m^4

IX.51 (a) $I_{x'} = 0.531$ m^4
$I_{y'} = 2.058$ m^4
$I_{x'y'} = -0.788$ m^4
(b) $I_{x'} = 2.391$ m^4
$I_{y'} = 0.198$ m^4
$I_{x'y'} = -0.007$ m^4
(c) $I_{x'} = 0.240$ m^4
$I_{y'} = 2.349$ m^4
$I_{x'y'} = -0.301$ m^4

IX.53 $I_x = 356.4$ in.^{4}
$I_y = 90.6$ in.^{4}
$I_{xy} = 134.1$ in.^{4}

IX.55 $I_x = I_y = 0.0334$ m^4
$I_{xy} = 0$

IX.57 $\bar{\theta} = 15.32°$ (cw)
$I_{\bar{x}} = 1.874(10^{-3})$ m^4
$I_{\bar{y}} = 8.574(10^{-3})$ m^4

IX.59 $\bar{\theta} = 20.82°$ (ccw)
$I_{\bar{x}} = 106$ in.^{4}
$I_{\bar{y}} = 974$ in.^{4}

IX.61 $\bar{x}$ is 22.37° ccw from rt horiz
$I_{\bar{x}} = 412$ in.^{4}
$I_{\bar{y}} = 34.7$ in.^{4}

IX.63 $\theta = 29.69°$ (ccw)
$I_{x'} = I_{y'} = 5.22(10^{-3})$ m^4
$I_{x'y'} = -3.35(10^{-3})$ m^4

IX.65 $\theta = 24.19°$ (cw)
$I_{x'} = I_{y'} = 540$ in.^{4}
$I_{x'y'} = 434$ in.^{4}

IX.67 x' is 22.63° cw from rt horiz
$I_{x'} = I_{y'} = 224$ in.^{4}
$I_{x'y'} = -189$ in.^{4}

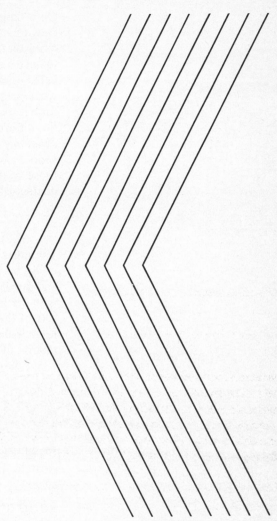

MODULE I
FUNDAMENTAL
CONCEPTS

On a daily basis we encounter physical systems, such as automobiles, elevators, and other types of machinery, in which objects are made to move by the application of forces. To use such objects, it is not necessary to have a detailed understanding of the exact manner in which the systems behave, although in some cases we do have an intuitive understanding of what is taking place. However, as engineers designing a system to perform a specific task, we need considerably more than a qualitative feel for the different ways in which systems respond to forces. In this text on *dynamics* we shall study the quantitative relationships among the responses of many types of physical systems and the forces that act on those systems. The primary new feature of the field of dynamics that distinguishes it from statics is that here systems are always in motion.

A. BASIC DEFINITIONS

The investigation of the dynamic response of a physical system can generally be broken into two parts, one treating the subject of *kinematics,* and the other the subject of *kinetics.* A kinematical study is one that deals with methods for describing the motion of a system without regard to the role of the forces causing the motion. A kinetics study is one where the relationship of the forces to the kinematical variables representing the motion of the system is considered.

A key phrase used above is *physical system.* By it we mean a collection of basic atomic elements that combine to form bodies. In the field of mechanics, for the convenience of problem solving, we tend to categorize these collections of elements into three broad groupings: particles, rigid bodies, and deformable bodies.

PARTICLE   A body whose dimensions are negligible is said to be a particle. It follows, then, that a particle occupies only a single point in space.

RIGID BODY   A body occupying more than one point in space is said to be rigid if all of the constituent elements of matter within the body are always at fixed distances from each other.

DEFORMABLE BODY   A body is said to be deformable if its constituent elements of matter experience changes in their distances from each other that are significant to the problem being investigated.

Clearly there is a certain amount of ambiguity in these definitions. For instance, is the group of molecules that compose a gas a system of particles or a deformable body? Another basic question we could ask is whether any body can correctly be considered to be rigid, because all real materials deform when forces are exerted on them.

There are no absolute answers to these questions, for the *model* of the system we form depends on what knowledge we wish to gain about the response of the system. This leads us to the next topic, which is the modeling process.

B. THE MODELING PROCESS

Our general approach in studying the response of a physical system will be first to construct a conceptual model of the system. That is, we shall consciously consider its components to be either particles, rigid bodies, or deformable bodies. We shall then employ our knowledge of kinematics to describe the motion of the chosen model, and thus of the physical system it represents. The response of the system will then be determined by relating the results of the kinematical study to the force concepts of kinetics. The general approach to the subject of dynamics can be summarized in the conceptual diagram given as Figure 1.

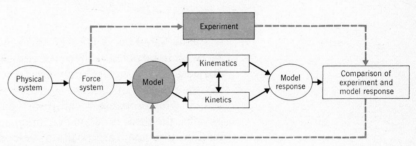

Figure 1 Dynamics study flow chart.

The dotted paths in Figure 1 illustrate a key feature of the modeling process. That is, to be sure of the validity of the model initially chosen to represent the system, it is necessary that the model display any relevant phenomena that are observed experimentally. If not, the model must be improved. Hence, modeling is an educated trial-and-error procedure. That is, the modeling process is, in part, an art based on prior experience and physical intuition. On the other hand, the modeling process is also a science based on a knowledge of the purpose of the physical system and of the available analytical methods to study the phenomena exhibited by the system.

As an example of the foregoing, consider the case of the flight of a golf ball. For a first attempt, one would tend to model the ball as a particle occupying a single point in space. As will be seen in our studies, there are some situations where such an approach results in a successful analysis; that is, the theoretical and experimental results are in good agreement. On the other hand, we shall also see that considering the golf ball to be a particle leaves one unable to explain why the ball hooks, slices, and otherwise deviates from a straight course when the wind is calm. The latter phenomena stem from the aerodynamic forces produced by the spin of the golf ball, so an accurate study of its flight requires that the ball be modeled as a rigid body, as opposed to a simple particle. Carrying these thoughts one step further, if the problem was one of determining how a golf club imparts energy to a ball, we would quickly find that the correct model would be one where the golf ball (and club) must be modeled as deformable bodies. From the foregoing example we see that the model chosen to represent the physical system is a crucial element in the solution process.

The study of the dynamics of deformable bodies is not presented in this book, because the scope of the mathematical tools necessary for this subject far exceeds that required to study the motion of particles and rigid bodies. This would be especially so if our goal was to attain the same level of comprehension and versatility in each subject area. Therefore, only those physical systems that can successfully be modeled as particles and rigid bodies are considered in this first course in dynamics.

C. FOUNDATIONS OF NEWTONIAN MECHANICS

We wish to be able to determine completely the motion of all points in the physical system we are studying. To achieve this goal we must first quantify the concept of motion.

As you probably recall from your previous courses in elementary physics, the three fundamental kinematical quantities used to describe the motion of a point are position, velocity, and acceleration. These basic terms are defined with respect to a frame of reference that describes the three-dimensional space within which the motion occurs. This frame of reference can be visualized as a set of rectangular Cartesian axes.

The laws of motion stated by Sir Isaac Newton (1666–1727) in his historic work, *Principia* (1687), are the foundation of the study of dynamics. To employ these laws we must postulate the existence of an "absolute" (fixed) reference frame. This requirement is in conflict with the concepts of the theory of relativity, which tells us that nothing is fixed in space. Fortunately, it can be shown that the laws of the theory of relativity reduce to the simpler laws of Newtonian mechanics in the special case where bodies are moving much slower than the speed of light. This is certainly the case for most systems of engineering significance. Hence, we shall assume that there is an absolute reference frame, which we shall denote as XYZ. (Throughout the text capital letters will indicate that the coordinate axes are fixed in space.) For most problems we can use the earth to define the XYZ frame. However, to set things in proper perspective, in Module VI, Section E, we shall study some effects of the motion of the earth.

Let us now return to the basic kinematical quantities. The location of a point P within an arbitrarily chosen XYZ reference frame cannot be

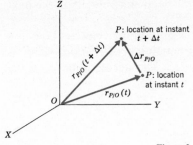

Figure 2

specified until we know both the length and orientation of the line from the origin O to point P. For example, let us locate point P in Figure 2. To describe the *position* of point P with respect to point O we employ the position vector $\bar{r}_{P/O}$ (which should be read as r of P with respect to O).

You will note that Figure 2 indicates that as point P moves in space its position vector changes, so that $\bar{r}_{P/O}$ is a function of time. The vector $\Delta\bar{r}_{P/O}$ represents the change in the vector $\bar{r}_{P/O}$ in the time interval from t to $t + \Delta t$. According to the rules for adding vectors diagrammatically, from Figure 2 we may write

$$\bar{r}_{P/O}(t) + \Delta\bar{r}_{P/O} = \bar{r}_{P/O}(t + \Delta t)$$

so that

$$\Delta\bar{r}_{P/O} = \bar{r}_{P/O}(t + \Delta t) - \bar{r}_{P/O}(t)$$

When we divide $\Delta\bar{r}_{P/O}$ by Δt, we obtain a quantity that resembles the time derivative of a scalar function $f(t)$. Recall from basic calculus that for a scalar function

$$\frac{df}{dt} \equiv \lim_{\Delta t \to 0} \frac{\Delta f}{\Delta t} = \lim_{\Delta t \to 0} \frac{f(t + \Delta t) - f(t)}{\Delta t}$$

A similar definition is used for differentiation of a vector function, where the time rate of change of the position vector $\bar{r}_{P/O}$ is called the *velocity* $\bar{v}_P$. This is the second of the three basic kinematical quantities. Thus

$$\boxed{\begin{aligned} \bar{v}_P &\equiv \frac{d}{dt}\bar{r}_{P/O} \equiv \lim_{\Delta t \to 0} \frac{\Delta\bar{r}_{P/O}}{\Delta t} \\ &= \lim_{\Delta t \to 0} \frac{\bar{r}_{P/O}(t + \Delta t) - \bar{r}_{P/O}(t)}{\Delta t} \end{aligned}} \tag{1}$$

In denoting the velocity as $\bar{v}_P$ we eliminated any indication that this quantity is defined with respect to a fixed point O. This is done to facilitate notation, and equally important, because we shall see later that the velocity of a point with respect to a fixed reference frame is independent of the location of the origin O of that frame.

Recalling that a vector describes a quantity having magnitude and direction, let us examine the velocity vector $\bar{v}_P$. The magnitude of this vector is defined as the *speed* v_P: $v_P = |\bar{v}_P|$. The speed expresses a rate of motion (e.g., 50 km/hr) and is a scalar quantity. When we are requested to determine how fast a point is moving, it is synonymous to being asked to find the speed. The direction of $\bar{v}_P$ tells us in which direction the point P is traveling. We shall have more to say about speed and velocity in the next module.

Now that we have the mathematical definition for velocity, equation (1), we may follow a similar approach in defining *acceleration*, which is the third of the basic kinematical quantities. The acceleration $\bar{a}_P$ is the time rate of change of the velocity, thus we have

$$\bar{a}_P \equiv \frac{d}{dt}\bar{v}_P \equiv \frac{d^2}{dt^2}\bar{r}_{P/O} \qquad (2)$$

The kinematical quantities of position and velocity are pieces of information that our eyes and brain are attuned to sense. Were it not for Newton's laws of motion, it would not be obvious that we should also be interested in acceleration. Because these laws are the basis for our future studies we restate them here, in modern language.

FIRST LAW   A particle will remain at rest or move with constant speed along a straight line, unless it is acted upon by a resultant force.

SECOND LAW   When a resultant force is exerted on a particle, the acceleration of that particle is parallel to the direction of the force and the magnitude of the acceleration is proportional to the magnitude of the force.

THIRD LAW   Each force exerted upon a body is the result of an interaction with another body, the forces of action and reaction being equal in magnitude, opposite in direction, and collinear.

In our studies we shall find that the first law is included in the second; it merely emphasizes that the state of motion of a particle can be changed only by the action of forces. The third law helps to define what forces are acting on a body.

The first and third laws are the only ones needed to solve problems in statics. The added feature in dynamics is the application of the second law. This law may be reworded to state that the force $\bar{F}$ acting on a particle is proportional to the acceleration $\bar{a}_P$ of the particle. As you probably know, the constant of proportionality in this relation is the mass m of the particle. Thus, we have the familiar equation

$$\bar{F} = m\bar{a}_P \qquad (3)$$

Before we can use this equation to solve kinetics problems, we must be sure that the units chosen to describe each term are mutually consistent. In the next section we consider this topic.

D. SYSTEMS OF UNITS

The quantities appearing in Newton's second law, $\bar{F} = m\bar{a}$, involve measurements of length, time, force, and mass, which we denote for brevity as $L, T, F,$ and $M,$ respectively. The units chosen for the measurement of these four dimensional quantities cannot be defined independently. They must obey the *law of dimensional homogeneity,* which requires that the physical units of all terms in an equation be identical. Stated in colloquial language, "apples cannot be equated to oranges."

In terms of the four basic measurements, the dimensions of acceleration are L/T^2. Dimensional homogeneity of equation (3) requires that

$$F = M\frac{L}{T^2}$$

In other words, we are free to choose the units of three of the four quantities, $L, T, F,$ and M. The fourth must be derived. Historically, the units by which length and time are measured are well defined, and, through the efforts of international standards committees, universally accepted.

This leaves us with the choice of defining either a standard unit of mass or of force, and then deriving the units for the undefined quantity. When the unit of mass is defined according to some standard, the system is called *absolute,* whereas a system in which the unit of force is defined is called *gravitational.* This terminology originates from the relationship between the weight W and the acceleration g of a particle that is falling freely in a vacuum at the surface of the earth. According to Newton's second law this relationship is

$$W = mg \tag{4}$$

In using equation (4) we must be aware that the values of W and g depend on the location of the particle with respect to the surface of the earth, and equally important on the choice of measurement system. Presently, the two systems that are in wide usage are the metric absolute system and the British gravitational system. In terms of the units of these systems, for most problems we shall use the value $g = 9.806$ m/s² (metric) or $g = 32.17$ ft/s² (British) as average values. These values will be looked at in considerable detail in Example 5, Module VI.

In Table 1 we summarize the commonly used systems of units and indicate how the derived units are obtained. The basic units in each system are defined with respect to standard bodies or with respect to physical phenomena. For example, a kilogram is the mass of a particular bar of platinum alloy. On the other hand, a meter is defined as 1,650,763.73 wavelengths in a vacuum of the orange-red line of krypton-86. The derived quantities in the table are the starred values.

Table 1   System of Units

UNIT/SYSTEM	LENGTH	TIME	MASS	FORCE
Metric absolute	Meter (m)	Second (s)	Kilogram (kg)	Newton (N)* = kg-m/s²
Metric gravitational	Meter (m)	Second (s)	Metric slug* = kg-s²/m	Kilogram (kg)
British absolute	Foot (ft)	Second (s)	Pound (lb)	Poundal (pdl)* = lb-ft/s²
British gravitational	Foot (ft)	Second (s)	Slug* = lb-s²/ft	Pound (lb)

The choice for systems of units presented in Table 1 is further complicated by the fact that certain multiples and submultiples of the basic units in each system are given their own names, such as the mile, which is 5280 ft, or the dyne, which is one hundred-thousandth of a newton. To simplify this situation the SI (Standard International) system is now being adopted worldwide.

Essentially, the SI system is the metric absolute system with only one unit being used to describe each type of quantity. The SI units for length, time, mass, and force are meters, seconds, kilograms, and newtons, respectively. Then, if one desires, decimal multiples and submultiples of the basic units may be indicated by the appropriate prefixes. Thus, in the SI system, length dimensions of 0.2 meters and 200 millimeters are identical and correct. On the other hand, $2(10^{-3})$ newtons, or equivalently 2 millinewtons, is the correct description for a force whose magnitude is 200 dynes. Lists of the SI units and of the preferred prefixes may be found in Appendix A.

Many of the examples and problems in this text will be presented in SI units. Unfortunately, this system is not yet universally used in the United States, where many engineers, as well as nontechnical people, still use the British gravitational system. We shall therefore present some examples and problems that are posed in both systems of units. In general, we shall solve problems in the same system of units as that used for the given information, and we shall not convert between different systems of units. Obviously, the units of any physical quantity can be changed with the aid of conversion factors. These conversion factors are numbers that are physically unity, but not numerically. For example, 25.40 mm = 1 in. A brief set of factors is given in Appendix A.

Conversion factors can be used consecutively. For instance, to change from a speed in miles per hour to a speed in meters per second, we can compute

$$1\frac{\text{mile}}{1\text{ hr}}\left(\frac{1\text{ hr}}{3600\text{ s}}\right)\left(\frac{5280\text{ ft}}{1\text{ mile}}\right)\left(\frac{12\text{ in.}}{1\text{ ft}}\right)\left(\frac{25.40\text{ mm}}{\text{in.}}\right)\left(\frac{1\text{ m}}{1000\text{ mm}}\right) = 0.4470\text{ m/s}$$

Incidentally, a conversion factor that is contained in the preceding equation is that 1 mile/hr = (88/60) ft/s. We shall use this factor frequently in solving problems having data in the British system of units. Remember, the units of all quantities in an equation must be consistent in order to satisfy the law of dimensional homogeneity.

As practice in obtaining a consistent set of units, and also to gain an intuitive understanding of the magnitude of British units relative to SI units, we present the following example.

EXAMPLE 1

Using the facts that 1 in. equals 25.40 mm and that a body weighing 1 lb on the surface of the earth has a mass of 0.4536 kg, determine the factors for converting the British gravitational units of pounds force, slugs, mass, and density (slugs/ft³) into their SI equivalents.

Solution

The conversion factors between pounds and newtons is determined by using $W = mg$. Let us compute the weight in newtons of a 0.4536-kg body. Thus, not knowing where the body was weighed, we use the average value $g = 9.806$ m/s² to obtain

$$W = (0.4536 \text{ kg})(9.806 \text{ m/s}^2) = 4.448 \text{ N}$$

In other words,

$$1 \text{ lb} = 4.448 \text{ N}$$

Then, because 1 lb-s²/ft is 1 slug, we have

$$1 \text{ slug} = \left(1\,\frac{\text{lb-s}^2}{\text{ft}}\right)\left(\frac{4.448 \text{ N}}{1 \text{ lb}}\right)\left(\frac{1 \text{ ft}}{12 \text{ in.}}\right)\left(\frac{1 \text{ in.}}{25.40 \text{ mm}}\right)\left(\frac{1000 \text{ mm}}{1 \text{ m}}\right)$$

$$= 14.593\,\frac{\text{N-s}^2}{\text{m}} = 14.593 \text{ kg}$$

To convert the units of density, we use

$$1\frac{\text{slug}}{\text{ft}^3} = \left(1\,\frac{\text{slug}}{\text{ft}^3}\right)\left(\frac{14.593 \text{ kg}}{\text{slug}}\right)\left(\frac{1 \text{ ft}}{12 \text{ in.}}\right)^3\left[\left(\frac{1 \text{ in.}}{25.40 \text{ mm}}\right)\left(\frac{1000 \text{ mm}}{1 \text{ m}}\right)\right]^3$$

$$= 515.3 \text{ kg/m}^3$$

The differences between the values obtained here and those in the table in Appendix A are attributable to the fact that we were only using four significant figures for our calculations, which is the maximum allowed by our choice for the value of g.

HOMEWORK PROBLEMS

In the following problems, t is time in seconds, x is distance, v is speed, a is the magnitude of acceleration, and m is mass.

I.1   The work done by a force is Fx and the power developed by this force is Fv. In the SI system the basic unit of work and energy is joules (J), a joule being the work done by a 1-N force in moving a body a distance of 1 m. The basic SI unit for power is a watt (W), a watt being the power developed by a 1-N force when it moves a body at a speed of 1 m/s. Determine what (a) 1 J is in terms of meters, kilograms, and seconds, (b) 1 W is in terms of meters, kilograms, and seconds, and (c) 1 W is in terms of joules and seconds.

I.2   The power developed by a force is Fv. In the SI system the basic unit of power is a watt (W), a watt being the power developed by a 1-N force when it moves a body at a speed of 1 m/s. In the British gravitational system the basic unit of power is a horsepower, a horsepower being the power developed by a 550-lb force when it moves a body at a speed of 1 ft/s. Determine (a) how many kilowatts there are in a horsepower, and (b) how many horsepower there are in a kilowatt.

I.3   Derive conversion factors for changing the following British gravitational units to their SI equivalents. (a) area: square inch; (b) volume: cubic foot; (c) force: ounce; (d) pressure: pounds per square inch; (e) pressure: pounds per square foot; (f) speed: foot per second; (g) speed: inches per hour; (h) acceleration: miles per hour squared.

I.4   In each of the following formulas c_1, c_2, etc., denote constants and θ is an angle in radians. Determine the units of these constants if the formula is to be dimensionally correct. (a) $a = c_1 v^2 / x$, (b) $\frac{1}{2} m v^2 = \frac{1}{2} c_1 x^2$, (c) $x = c_1 x_0 + c_2 v_0 + c_3 t^2$, (d) θ (degrees) $= c_1 \theta$ (radians), (e) $d\theta/dt = c_1 + c_2 t$, (f) $t = c_1 \sqrt{x}$, (g) $m c_1^2 \, d^2\theta/dt^2 = Fx$.

I.5   At a certain large university known for its athletic prowess, the following new set of physical units are in use. The basic length unit is the *touchdown*, which is 100 yd, the basic time unit is the *dash*, which is 9 s, and the basic force unit is the *golf ball*, which is 1.620 oz. Determine the conversion factors between these units and their SI equivalents.

I.6   At the university in Problem I.5 the unit of mass is the *basket*, one basket being one golf ball-dash^2/touchdown. Determine the number of kilograms in one basket.

I.7   A well-known fluid mechanics equation for the pressure drop along a pipe of length l and diameter D is

$$P_1 - P_2 = f \frac{l}{D} \, \rho \, \frac{v^2}{2}$$

where P denotes pressure, ρ is the density of the fluid, v is its velocity, and f is the resistance coefficient, which is given by the expression

$$\sqrt{f} = 2.30 \log_{10}\left(\frac{Dv}{\nu}\sqrt{f}\right) - 0.91$$

Determine the dimensions of ν, the kinematic viscosity.

I.8   A restaurant buys 1000 lb of coffee at a standard locality and also a spring scale calibrated at that locality. Determine what the coffee weighs on the spring scale in a locality where (a) $g = 32.11$ ft/s², (b) $g = 9.791$ m/s².

E.   VECTORS

In establishing the definitions for the kinematical variables of position, velocity, and acceleration, we saw that a basic tool in the study of dynamics is the description of vector quantities. As was true in the subject of statics, we shall frequently need to perform the algebraic operations of addition, subtraction, and multiplications. Equally important, in view of the relationship of the kinematical variables, we shall also be required to differentiate vectors. In what follows we shall briefly review some fundamental operations and results of vector algebra and then study some basic theorems of vector calculus.

1   Vector Algebra

A vector is any quantity that is fully specified when its magnitude and direction are known. Graphically, to depict any vector $\bar{A}$ in a diagram we draw an arrow, as we did in Figure 2 to illustrate the position vector $\bar{r}_{P/O}$. The length of this arrow is proportional to the magnitude $|\bar{A}|$ of the vector, and the orientation of the arrow shows its direction.

Multiplication by a Scalar   If q is a scalar number, then the product $q\bar{A}$ is defined to have the property that

$$|q\bar{A}| = |q||\bar{A}|$$

The resulting vector $q\bar{A}$ is parallel to $\bar{A}$, in the same sense as $\bar{A}$ if q is positive, and in the opposite sense if q is negative.

The concept of a unit vector is an important application of the multiplication by a scalar. Suppose that we multiply $\bar{A}$ by $(1/|\bar{A}|)$; the result of this operation is a vector $\bar{e}_A$, parallel to $\bar{A}$. Because the magnitude of $\bar{e}_A$ is the dimensionless number one, we call $\bar{e}_A$ a unit vector. Thus

$$\bar{e}_A = \frac{\bar{A}}{|\bar{A}|} \tag{5}$$

As you may recall from statics, this equation proves to be very useful for representing vector quantities.

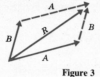

Figure 3

Addition of Vectors The basic rule for adding any two vectors $\bar{A}$ and $\bar{B}$ is the parallelogram law, according to which the tails of the two vectors are made to coincide. The resultant, $\bar{R} = \bar{A} + \bar{B} = \bar{B} + \bar{A}$, is then obtained from the parallelogram formed by these vectors, as shown in Figure 3. This figure also illustrates an equivalence rule for adding vectors, specifically, that of forming a triangle by placing the tail of one vector at the head of the other and obtaining the resultant as the third side of the triangle.

The difference of two vectors can be obtained as a corollary of the addition rule. This is so because

$$\bar{A} - \bar{B} = \bar{A} + (-\bar{B})$$

where, according to the law for multiplication by a scalar, $-\bar{B} \equiv (-1)\bar{B}$ is a vector having the same magnitude as $\bar{B}$, but opposite direction.

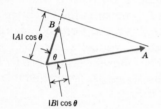

Figure 4

Dot Product of Vectors The product $\bar{A} \cdot \bar{B}$ is defined to be the product of the magnitude of one vector with the component of the other vector in the direction of the first. From Figure 4 we see that this definition can be written mathematically as

$$\bar{A} \cdot \bar{B} = \bar{B} \cdot \bar{A} \equiv |\bar{A}||\bar{B}| \cos \theta$$

where θ is the angle between the vectors $\bar{A}$ and $\bar{B}$. Notice that the dot product will be negative if $\theta > 90°$, which means that the projection of $\bar{A}$ onto $\bar{B}$ is opposite the sense of $\bar{B}$, or vice versa. Also, from this definition, $\bar{A} \cdot \bar{A} = |\bar{A}|^2$.

Cross Product of Vectors The definition of the cross product $\bar{A} \times \bar{B}$ is based on the fact that two intersecting vectors $\bar{A}$ and $\bar{B}$ define a flat plane. The cross product is defined to be a vector normal (perpendicular) to the plane of the two vectors. To specify the sense of this normal we use the "right-hand" rule, in accordance with which we curl the fingers of our right hand from the first vector to the second and define the sense of the result to be in the direction of the thumb, as shown in Figure 5.

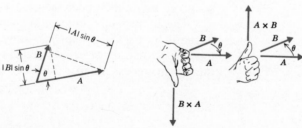

Figure 5

The definition of the magnitude of the cross product is

$$|\bar{A} \times \bar{B}| \equiv |\bar{A}| |\bar{B}| \sin \theta$$

From Figure 5 it can be seen that the definition involves the component of one vector normal to the other vector.

An important property of the cross product, illustrated in Figure 5, is that

$$\bar{B} \times \bar{A} = -\bar{A} \times \bar{B}$$

This is so because according to the right-hand rule, both products have the same magnitude but opposite direction.

Theorems of Vector Algebra The definition of the basic operations leads to rules for combining these operations that resemble those of scalar algebra. Thus

$$q(\bar{A} + \bar{B}) = q\bar{A} + q\bar{B}\,;\ q(\bar{A} \cdot \bar{B}) = (q\bar{A}) \cdot \bar{B} = \bar{A} \cdot (q\bar{B})$$

$$q(\bar{A} \times \bar{B}) = (q\bar{A}) \times \bar{B} = \bar{A} \times (q\bar{B})$$

$$\bar{C} \cdot (\bar{A} + \bar{B}) = (\bar{C} \cdot \bar{A}) + (\bar{C} \cdot \bar{B})$$

$$\bar{C} \times (\bar{A} + \bar{B}) = \bar{C} \times \bar{A} + \bar{C} \times \bar{B}$$

Rectangular Components So far we have described vectors diagrammatically by using arrows, an approach that has two serious shortcomings. First, three-dimensional vectors are difficult to represent in a diagram. Second, even for two-dimensional vectors, the addition of vectors according to the parallelogram law requires extensive use of the trigonometry of scalene triangles. To circumvent these difficulties, throughout the text we shall employ the alternative approach of representing vectors by their components with respect to a set of orthogonal (mutually perpendicular) directions.

To describe the components of a vector $\bar{A}$ with respect to a rectangular Cartesian coordinate system XYZ, we place the origin of the coordinate axes at the tail of the vector. We then drop perpendiculars (indicated by dashed lines in Figure 6) from the tip of the vector onto the three coordinate axes. From the intersections of these perpendiculars with the axes, we form the rectangular parallelipiped depicted in the figure. The lengths A_X, A_Y, and A_Z are the components of the vector $\bar{A}$ with respect to the XYZ coordinate system.

In terms of the angles α, β, γ between the vector $\bar{A}$ and the respective coordinate axes, we have

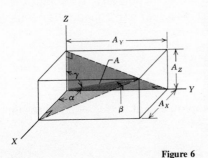

Figure 6

$$A_X = |\bar{A}| \cos \alpha \qquad A_Y = |\bar{A}| \cos \beta \qquad A_Z = |\bar{A}| \cos \gamma \qquad (6)$$

The angles α, β, γ are called the *direction angles;* the cosines of the direction angles are the *direction cosines.* Because $\bar{A}$ is a diagonal for the parallelipiped, it follows that

$$|\bar{A}| = \sqrt{A_X{}^2 + A_Y{}^2 + A_Z{}^2} \tag{7}$$

Substituting equations (6) into equation (7) yields

$$\cos^2 \alpha + \cos^2 \beta + \cos^2 \gamma = 1$$

which shows that only two of the direction angles are independent quantities.

From the foregoing we see that the components of a vector completely describe the properties of a vector, provided that we know which coordinate system is associated with the components. To specify the coordinate system we use the unit vectors $\bar{I}$, $\bar{J}$, and $\bar{K}$, which are oriented in the directions of the positive X, Y, and Z axes, respectively. Then, for example, equation (5) shows that $A_X\bar{I}$ is a vector parallel to the X axis whose magnitude is the component A_X. Figure 7, which illustrates the three component vectors, also shows that the vector sum of these components is $\bar{A}$. Hence

$$\boxed{\bar{A} = A_X\bar{I} + A_Y\bar{J} + A_Z\bar{K}} \tag{8}$$

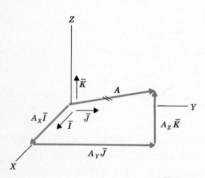

Figure 7

This is the vector component representation of $\bar{A}$.

An important result of representing vectors by their components is that the equality of vectors takes on a simpler description. Two vectors can be equal only when their components are equal. In other words,

$$\bar{A} = \bar{B} \quad \text{means} \quad A_X = B_X, \quad A_Y = B_Y, \quad A_Z = B_Z \tag{9}$$

Equation (9) is particularly useful for our studies, as it shows that a vector equation can be subdivided into the three scalar equations obtained by equating corresponding components from each side of the equation.

In general, the various vector operations are more easily performed in terms of components than from their basic definitions. To obtain the unit vector $\bar{e}_A$, we substitute equations (7) and (8) into equation (5), with the result that

$$\bar{e}_A = \frac{\bar{A}}{|\bar{A}|} = \frac{A_X\bar{I} + A_Y\bar{J} + A_Z\bar{K}}{\sqrt{A_X{}^2 + A_Y{}^2 + A_Z{}^2}} \tag{10a}$$

For an alternative form of $\bar{e}_A$, we can employ equation (6), with the result that

$$\bar{e}_A = \cos \alpha\bar{I} + \cos \beta\bar{J} + \cos \gamma\bar{K} \tag{10b}$$

The computation of the sum or difference of two vectors is also straightforward, for

$$\bar{A} \pm \bar{B} = (A_X \pm B_X)\bar{I} + (A_Y \pm B_Y)\bar{J} + (A_Z \pm B_Z)\bar{K} \tag{11}$$

By induction, equation (11) can be extended to treat more than two vectors.

The scalar and vector products are obtained by first considering the various products of the unit vectors, using the definitions of these operations. For example, for the scalar product

$$\bar{I} \cdot \bar{I} = |\bar{I}||\bar{I}| \cos 0 = 1$$

Hence, it follows that

$$\bar{I} \cdot \bar{I} = \bar{J} \cdot \bar{J} = \bar{K} \cdot \bar{K} = 1$$

$$\bar{I} \cdot \bar{J} = \bar{J} \cdot \bar{I} = \bar{J} \cdot \bar{K} = \bar{K} \cdot \bar{J} = \bar{K} \cdot \bar{I}$$

$$= \bar{I} \cdot \bar{K} = 0 \tag{12}$$

whereas, for the vector product,

$$|\bar{I} \times \bar{I}| = |\bar{I}||\bar{I}| \sin 0 = 0$$

Thus,

$$\bar{I} \times \bar{I} = \bar{J} \times \bar{J} = \bar{K} \times \bar{K} = \bar{0}$$

$$\bar{I} \times \bar{J} = -\bar{J} \times \bar{I} = \bar{K}$$

$$\bar{J} \times \bar{K} = -\bar{K} \times \bar{J} = \bar{I}$$

$$\bar{K} \times \bar{I} = -\bar{I} \times \bar{K} = \bar{J} \tag{13}$$

As an aid in obtaining the resulting direction and sign for the nonzero cross products in equation (13), we can consider the cross product of two unit vectors to be positive if the operation coincides with the clockwise order shown in Figure 8 and negative if the operation coincides with the counterclockwise order depicted there.

The products of two vectors then follows from the rules of vector algebra. Thus the scalar product is

$$\bar{A} \cdot \bar{B} = (A_X\bar{I} + A_Y\bar{J} + A_Z\bar{K}) \cdot \bar{B}$$

$$= A_X[\bar{I} \cdot (B_X\bar{I} + B_Y\bar{J} + B_Z\bar{K})]$$

$$+ A_Y[\bar{J} \cdot (B_X\bar{I} + B_Y\bar{J} + B_Z\bar{K})]$$

$$+ A_Z[\bar{K} \cdot (B_X\bar{I} + B_Y\bar{J} + B_Z\bar{K})]$$

$$= A_XB_X + A_YB_Y + A_ZB_Z \tag{14}$$

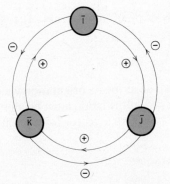

Figure 8

A similar procedure for the vector product gives

$$\bar{A} \times \bar{B} = (A_X \bar{I} + A_Y \bar{J} + A_Z \bar{K}) \times \bar{B}$$

$$= A_X[\bar{I} \times (B_X \bar{I} + B_Y \bar{J} + B_Z \bar{K}]$$

$$+ A_Y[\bar{J} \times (B_X \bar{I} + B_Y \bar{J} + B_Z \bar{K})]$$

$$+ A_Z[\bar{K} \times (B_X \bar{I} + B_Y \bar{J} + B_Z \bar{K})]$$

$$= A_X(B_Y \bar{K} - B_Z \bar{J}) + A_Y(-B_X \bar{K} + B_Z \bar{I})$$

$$+ A_Z(B_X \bar{J} - B_Y \bar{I})$$

$$= (A_Y B_Z - A_Z B_Y)\bar{I} + (A_Z B_X - A_X B_Z)\bar{J}$$

$$+ (A_X B_Y - A_Y B_X)\bar{K} \qquad (15a)$$

An alternative way of computing the cross product, which involves the evaluation of a determinant, is a convenient device for expediting calculations. This method gives

$$\bar{A} \times \bar{B} = \begin{vmatrix} \bar{I} & \bar{J} & \bar{K} \\ A_X & A_Y & A_Z \\ B_X & B_Y & B_Z \end{vmatrix} \qquad (15b)$$

To illustrate the various vector operations, let us now solve a problem that reviews the basic operations of statics.

EXAMPLE 2

The crate shown in the sketch has a mass of 50 kg and its center of mass coincides with the geometric centroid of the volume. A force $\bar{F}$ of magnitude 650 N parallel to the diagonal line BC pushes the crate at point C. Determine (a) the force-couple system acting at point A that is equivalent to the combined effect of gravity and $\bar{F}$; (b) the moment of these forces about the edge AE.

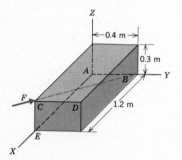

Solution

We begin by writing the forces in terms of their components. The weight force mg is downward, and we have

$$\bar{W} = -50(9.806)\bar{K} = -490.3\bar{K} \text{ N}$$

The force $\bar{F}$ has a magnitude of 650 N. To describe its orientation we use a unit vector $\bar{e}_F$, which according to the given information must be along the line from point C to point B. Because $\bar{r}_{B/C}$ is also a vector from point C to point B, equation (5) gives

$$\bar{F} = 650 \bar{e}_F = 650 \frac{\bar{r}_{B/C}}{|\bar{r}_{B/C}|}$$

The components of the vector from point C to point B are 1.2 m in the negative X direction, 0.4 m in the positive Y direction, and 0.3 m in the negative Z direction, so that

$$\bar{r}_{B/C} = -1.2\bar{I} + 0.4\bar{J} - 0.3\bar{K}$$

$$\bar{F} = 650\left[\frac{-1.2\bar{I} + 0.4\bar{J} - 0.3\bar{K}}{\sqrt{(1.2)^2 + (0.4)^2 + (0.3)^2}}\right]$$

$$= 500(-1.2\bar{I} + 0.4\bar{J} - 0.3\bar{K})$$

$$= -600\bar{I} + 200\bar{J} - 150\bar{K} \text{ N}$$

Recalling a theorem from statics, an equivalent force system acting at any point consists of a force $\bar{R}$, which is the sum of the forces in the system, and a couple $\bar{M}$, which is the moment of the force system about the chosen point. Hence

$$\bar{R} = \bar{F} + \bar{W} = -600\bar{I} + 200\bar{J} - 150\bar{K} - 490.3\bar{K}$$

$$= -600\bar{I} + 200\bar{J} - 640.3\bar{K} \text{ N}$$

The moment of a force about a point is the cross product of the position vector from that point to any point on the line of action of the force with the force. In that we want a force-couple system at point A, we compute moments about this point. Denoting the center of mass as G, we therefore have

$$\bar{M}_A = \bar{r}_{C/A} \times \bar{F} + \bar{r}_{G/A} \times \bar{W}$$

It is given that point G is at the center of the crate, so its coordinates are (0.6, 0.2, 0.15) m. Hence

$$\bar{M}_A = (1.2\bar{I} + 0.3\bar{K}) \times (-600\bar{I} + 200\bar{J} - 150\bar{K})$$

$$+ (0.6\bar{I} + 0.2\bar{J} + 0.15\bar{K}) \times (-490.3\bar{K})$$

$$= 1.2(200)\bar{K} + 1.2(-150)(-\bar{J}) + 0.3(-600)\bar{J}$$

$$+ 0.3(200)(-\bar{I}) + 0.6(-490.3)(-\bar{J}) + 0.2(-490.3)\bar{I}$$

$$= -158.1\bar{I} + 294.2\bar{J} + 240\bar{K} \text{ N-m}$$

To solve the last portion of the problem, recall that the moment about an axis is the component parallel to that axis of the moment about any point on the axis. In this problem point A is on the axis defined by edge AE, so we need the component of $\bar{M}_A$ parallel to this edge. Such a component is obtained by calculating the scalar product of $\bar{M}$ and a *unit* vector parallel to line AE, which in this case is $\bar{I}$, so

$$\bar{M}_{AE} = (\bar{M}_A \cdot \bar{I})\bar{I} = -158.1\bar{I} \text{ N-m}$$

HOMEWORK PROBLEMS

I.9   A force $\bar{F}$ of magnitude 200 N is applied at point A. Determine the moment of $\bar{F}$ about point O.

I.10   A force $\bar{R}$ of magnitude 300 N is applied at point B. Determine the moment of $\bar{R}$ about point C.

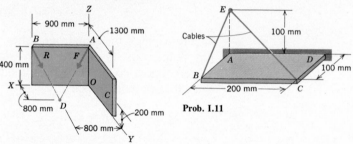

Prob. I.11

Problems I.9 and I.10

I.11   Each of the cables supporting the 500-kg flat plate has a tension of 5000 N. The weight force acts at the centroid of the plate. Determine the force-couple system acting at point D that is equivalent to the forces exerted by gravity and the supporting cables for the plate.

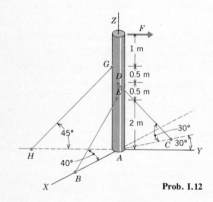

I.12   Three cables are attached to the vertical post AB, as shown, in order to support it against the 1-kN force $\bar{F}$. Determine the tension force in each of the cables if the tension in cable EB is 2 kN and the resultant moment about point A is zero.

I.13   Determine the three vertex angles θ_A, θ_B, and θ_C for the triangle ABC shown. Also determine the unit vector normal to the plane of the triangle that is oriented outward from the origin.

I.14   A 1000-N force parallel to the XY plane is applied at point D along edge AB of the triangle, as shown. Resolve this force into two force vectors, one perpendicular to the plane of triangle ABC, the other parallel to that plane.

I.15   For the two forces acting on the disk, determine the equivalent force-couple system acting about (a) point O, (b) point C.

Prob. I.12

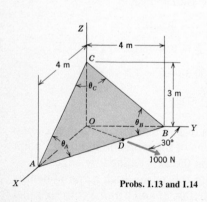

Probs. I.13 and I.14

I.16   A cable is wrapped around pulleys A and B on a block that slides on the horizontal guide bar. The tension in the cable is 50 lb. Determine (a)

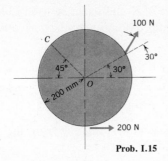

100 N

200 mm

45°

30°

30°

c

O

200 N

Prob. I.15

the components of the resultant force exerted by the cable on the pulley block assembly in directions parallel and perpendicular to the guide bar, (b) the component of the resultant cable force parallel to the line connecting the centers of the two pulleys, and (c) the resultant moment of the cable forces about the center of pulley *A*.

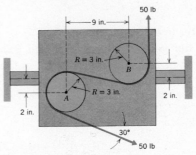

50 lb

—9 in.—

R = 3 in.

B

R = 3 in.

2 in.

A

2 in.

30°

50 lb

Prob. I.16

2   Vector Calculus

In dynamics, derivatives with respect to time are of primary importance, so we shall use *t* as the independent variable in the following discussion. Letting $\bar{A}(t)$ be any vector whose value depends on *t*, the derivative is

$$\frac{d\bar{A}}{dt} = \lim_{\Delta t \to 0} \frac{\bar{A}(t + \Delta t) - \bar{A}(t)}{\Delta t} \tag{16}$$

The following example is intended to give you a basic appreciation of the significance of equation (16), and, equally important, to emphasize the meaning of the basic concept of velocity.

EXAMPLE 3

A radar station tracks an airplane by measuring the distance *R* to the airplane and the angle of elevation θ. At a certain instant an airplane is sighted at $R = 4100$ m, $\theta = 33.7°$. After an interval of 0.75 s the airplane is at $R = 4240$ m and $\theta = 29.3°$. What is the approximate speed of the airplane and the approximate angle α at which it is diving or climbing?

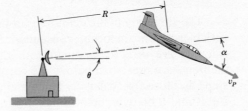

R

α

θ

v_P

Solution

Both v_P and α can be found once the velocity is known. Because the functions defining *R* and θ are unknown, simple differentiation cannot be performed. Noting that the given values of R and θ can be used to locate $\bar{r}_{P/O}(t)$ and $\bar{r}_{P/O}(t + \Delta t)$, we can form an approximation to the derivative for the velocity. Thus

$$\bar{v}_P \simeq \frac{\bar{r}_{P/O}(t + \Delta t) - \bar{r}_{P/O}(t)}{\Delta t}$$

where, in this problem, $\Delta t = 0.75$ s.

For general values of R and θ the vector $\bar{r}_{P/O}$ going from the radar station to the airplane is shown in the diagram to be

$$\bar{r}_{P/O} = R(\cos \theta \bar{I} + \sin \theta \bar{J})$$

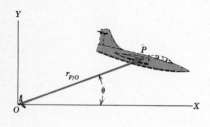

Thus

$$\bar{r}_{P/O}(t) = 4100\ (\cos 33.7°\bar{I} + \sin 33.7°\bar{J})$$

$$= 3411\bar{I} + 2275\bar{J} \text{ m}$$

$$\bar{r}_{P/O}(t + \Delta t) = 4240\ (\cos 29.3°\bar{I} + \sin 29.3°\bar{J})$$

$$= 3698\bar{I} + 2075\bar{J} \text{ m}$$

$$\bar{v}_P = \frac{1}{0.75}\ [(3698 - 3411)\bar{I} + (2075 - 2275)\bar{J}]$$

$$= 383\bar{I} - 267\bar{J}$$

Then from the following sketch we find the magnitude and direction of $\bar{v}_P$ to be

$$v_P = |\bar{v}_P| = \sqrt{(383)^2 + (267)^2}$$

$$= 466 \text{ m/s}$$

$$\alpha = \tan^{-1} \frac{267}{383} = 34.9°$$

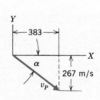

Thus the speed of the airplane is 467 m/s and it is diving at an angle of 34.9° from the horizontal.

The similarity of the definitions of the derivatives of scalar and vector variables has the important consequence that the basic theorems of vector calculus are directly analogous to those for scalar calculus. These theorems are presented below; the corresponding proofs follow the same steps as those for scalar variables and are therefore not presented here. Let $\bar{A}$ and $\bar{B}$ be two vector functions and let c be a scalar function. Then

$$\frac{d}{dt}(\bar{A} \pm \bar{B}) = \frac{d\bar{A}}{dt} \pm \frac{d\bar{B}}{dt}$$

$$\frac{d}{dt}(c\bar{A}) = \frac{dc}{dt}\bar{A} + c\frac{d\bar{A}}{dt}$$

$$\frac{d}{dt}(\bar{A} \cdot \bar{B}) = \frac{d\bar{A}}{dt} \cdot \bar{B} + \bar{A} \cdot \frac{d\bar{B}}{dt}$$

$$\frac{d}{dt}(\bar{A} \times \bar{B}) = \frac{d\bar{A}}{dt} \times \bar{B} + \bar{A} \times \frac{d\bar{B}}{dt} \qquad (17)$$

Note that in evaluating the derivative of a cross product, the order of multiplication must be maintained, because the cross product is not commutative ($\bar{A} \times \bar{B} \neq \bar{B} \times \bar{A}$).

The first and second identities in equations (17) will be of particular use to us, as they provide the basis for differentiating the component representation of a vector. Also, we shall frequently apply the *chain rule* for differentiation, because some vector $\bar{A}$ may be a function of a parameter α, whose value in turn depends on t. In such a situation,

$$\frac{d\bar{A}}{dt} = \frac{d\alpha}{dt}\frac{d\bar{A}}{d\alpha} \qquad (18)$$

Next, let us consider the process of integration, which also has the same basis for vectors as for scalars. In dynamics, we are generally concerned with definite integrals, hence

$$\bar{A} = \frac{d\bar{B}}{dt} \quad \text{means} \quad \bar{B}(t) = \bar{B}(t_0) + \int_{t_0}^{t} \bar{A}(t)\, dt \qquad (19)$$

Equations (17)–(19) will be employed extensively in the next module.

To simplify writing derivatives, we shall hereafter denote a derivative with respect to time by a dot over the quantity. For example, the basic definitions, equations (1) and (2), may now be rewritten as

$$\bar{v}_P \equiv \dot{\bar{r}}_{P/O} \qquad \bar{a}_P = \dot{\bar{v}}_P = \ddot{\bar{r}}_{P/O}$$

As can be seen, two dots over a quantity denotes the second derivative.

In closing this section we call attention to an important fact that follows from the foregoing development. Consider the representation of a vector in terms of its unit vector, as given by equation (5). Then

$$\frac{d\bar{A}}{dt} = \frac{d}{dt}(|\bar{A}|\bar{e}_A) = \left[\frac{d}{dt}|\bar{A}|\right]\bar{e}_A + |\bar{A}|\frac{d}{dt}\bar{e}_A$$

from which we see that a vector is constant only if both its magnitude is constant ($d|\bar{A}|/dt = 0$) and its orientation is constant ($d\bar{e}_A/dt = \bar{0}$).

F. PROBLEM SOLVING

The goal of a course in dynamics is the development of an understanding of the laws governing the motion of physical systems, with a primary emphasis on the application of this knowledge to solving problems. In general, engineering is the process of applying established physical laws to the solution of practical problems. Therefore, a common methodology has evolved for solving problems, although in each area the physical laws are different. The pertinent steps are detailed below.

1   Read (or formulate) the problem. If necessary draw a sketch of the system.
2   Show any known information in the sketch, or else list it. Also, list the information that is to be determined from the solution.
3   Choose a model to represent the system. For this model, write the equations corresponding to all physical principles relating the given and desired information.
4   Substitute the given information into these equations.
5   Count the number of equations and number of unknowns. An excess of unknowns usually means that either some given information was not used or a principle governing the model was not considered. In contrast, an excess of equations usually means that improper assumptions were made, or a principle employed in step 3 was incorrectly formulated, or possibly, that such a principle is not truly applicable to the problem.
6   Solve the problem.

These steps are very general. At crucial points in the text we shall present detailed steps, followed by a series of *Illustrative Problems*. The purpose of these problems is to demonstrate the applications of the procedural steps to a broad class of problems. This is in contrast to *Examples,* such as the three we have already studied, whose purpose is to demonstrate specific principles or techniques.

To conclude this module, let us consider the question that invariably occurs regarding numerical calculations, specifically, accuracy and precision. As you may have already noticed, all quantities are considered to have four significant figures. This approach is motivated by the extensive use of electronic calculators, which can easily handle numbers of such precision. We therefore have introduced $g = 9.806$ m/s^2 $= 32.17$ ft/s^2. However, as you will see, the fourth significant figure in g varies with the location of the physical system on the earth's surface. Hence, without any specification of this location, we generally have only three-place accuracy in a solution, because calculations are only as accurate as the least accurate quantity. Furthermore, three-place accuracy is more than sufficient for most engineering problems. Hence it makes sense to compute to four places and round off the final figures for the result to three places.

HOMEWORK PROBLEMS

I.17 The radar station in Example 3 tracks another aircraft. At one sighting the aircraft is at $R = 3500$ m, $\theta = 55.2°$. If this aircraft is traveling at 1836 km/hr at a climb angle of 30°, determine what the values of R and θ will be when the aircraft is sighted 0.50 s later.

I.18 At a third sighting of the aircraft in Example 3, 0.75 s after the second sighting, the aircraft is at $R = 4450$ m, $\theta = 25.6°$. Determine an approximation to the acceleration of the aircraft. Would you say that the airplane is under control?

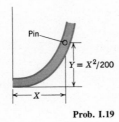

Prob. I.19

I.19 A pin in a machine follows a parabolic groove defined by $Y = X^2/200$, where X and Y are measured in millimeters. When first observed the pin is at $X = 255.2$ mm, and 2 ms later it is at $X = 239.4$ mm. Determine the approximate velocity of the pin.

I.20 An automobile is observed while traversing a level circular turn of 500-m radius. An observer at the center O of the curve initially sees the vehicle at a position where $\theta = 30°$ east of south. One second later, the car is at the position where $\theta = 33°$ east of south. What are the speed and the direction in which the vehicle is moving, approximately?

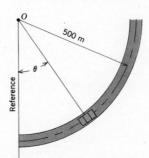

Probs. I.20 and I.21

I.21 Two seconds after the first sighting, a third observation of the automobile in Problem I.20 places it at $\theta = 36°$ east of south. Determine an approximation to the acceleration of the vehicle.

I.22 To track a UFO whose flight path does not pass over the station, a radar station measures the azimuth angle ϕ, as well as the angle of elevation θ and the distance R, as shown. Two sightings of the UFO 0.40 s apart place the object first at $R = 22,050$ ft, $\theta = 46.1°$, $\phi = 35.7°$, and then at $R = 21,130$ ft, $\theta = 47.3°$, $\phi = 32.2°$. Determine its speed and describe the direction in which it is moving.

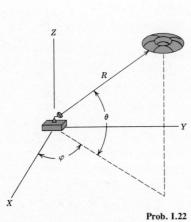

Prob. I.22

MODULE II
KINEMATICS
OF A
PARTICLE

Because a particle occupies only a single point in space, it represents the simplest model we can construct for a physical system. Thus, the study of the kinematics of a particle reduces to an investigation of methods for relating the position, velocity, and acceleration of a point.

A. BASIC KINEMATICAL PROPERTIES

One of the principal properties of the motion of a particle is its *path,* which is the line (curved or straight) the particle follows as it travels through space. To depict the path of a particle we trace the point occupied by the particle at each instant of time, thereby forming the locus of these points. A major topic for consideration in this module is the relationship between the particle path and the kinematical quantities of position, velocity, and acceleration, whose definitions we shall repeat here from Module I.

Referring to Figure 1, assume that at an instant t the particle is at a point P on its path. To serve as a frame of reference, we choose a fixed coordinate system XYZ whose origin is point O. Then the vector $\bar{r}_{P/O}$ shown in the figure describes the position of point P with respect to point O.

Having defined the position vector $\bar{r}_{P/O}$, we next define the velocity vector $\bar{v}_P$, which is the absolute velocity of point P with respect to the fixed reference frame. By definition, velocity is the time rate of change of position, so that

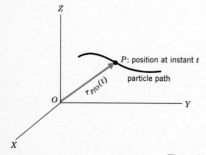

Figure 1

$$\bar{v}_P \equiv \frac{d}{dt}\bar{r}_{P/O} \equiv \dot{\bar{r}}_{P/O} \tag{1}$$

There is no need to introduce the fixed origin O into the notation for velocity because, as we shall show later, the velocity with respect to the fixed frame of reference is independent of the location of the origin.

Now that we have a mathematical definition for velocity, a similar approach may be followed in defining acceleration. Acceleration is the time rate of change of the velocity, so that

$$\bar{a}_P \equiv \dot{\bar{v}}_P \equiv \ddot{\bar{r}}_{P/O} \tag{2}$$

The three vectors $\bar{r}_{P/O}$, $\bar{v}_P$, and $\bar{a}_P$, can be described by their components with respect to the XYZ coordinate system shown in Figure 1. This coordinate system can also be used to write the equations describing the particle's path. These and other uses of a coordinate system will be discussed in the following sections. For now we merely call your attention to the fact that XYZ is a fixed rectangular Cartesian coordinate system with right-hand orientation. We shall use capital letters, such as X, Y, Z and $\bar{I}$, $\bar{J}$, $\bar{K}$, throughout the text to indicate that the axes and unit vectors of the reference frame are fixed in space.

In summary, it is important that you understand clearly the basic definitions of position, velocity, and acceleration. The next section will utilize these definitions in demonstrating different methods for describing the basic kinematical vectors. Indeed, the definitions will be of major significance throughout our study of dynamics.

B. KINEMATICAL RELATIONSHIPS

There are several methods for describing the basic kinematical properties of a particle. A fundamental task in learning to set up problems for solution is to develop the ability to identify which method is best. Here, we shall first derive the basic equations for the several methods and then see how they are used.

1 Comparison of Fixed Coordinate Systems and Path Variables

Referring to Figure 1, let us consider two ways in which we can locate the point P at a particular instant in time. One possible way of specifying the position is to give the rectangular Cartesian coordinates of the particle as functions of time, that is, to give $X(t)$, $Y(t)$, and $Z(t)$. Clearly, using these functions at any instant t we can find the corresponding values of X, Y, and Z. A plot of the values of X, Y, and Z corresponding to each instant will yield a trace of the path followed by the particle.

Now consider an alternative approach. Suppose that we are given the particle's path, and the position of the particle along its path is defined by the arc length that the particle has traveled from its initial position. The arc length is called a *path variable*. An example of such a description is to be told that a particle follows a circular path in the XY plane and that at $t = t_0$ it has traveled one-quarter of the circumference of the circle from its initial position in a clockwise direction. As we can see in Figure 2, this locates the position of the particle, provided that we know its initial position.

The fundamental difference between the Cartesian coordinate description and path variables is that in Cartesian coordinates the specification of the position at any instant is dependent on the choice of coordinate axes, and, as we shall see, is independent of knowledge of the path. The same is true for the velocity and acceleration vectors. Hence in the Cartesian coordinate approach the choice of coordinate system will change the description of the basic variables. On the other hand, when using a path variable approach all quantities, including position, are defined with respect to the path. In this method the description is essentially independent of the choice of coordinate system; the main purpose of the coordinate

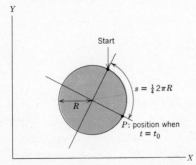

Figure 2

system is to describe the path. Because the path variables are intrinsically tied to the properties of the path, not to the fixed coordinate system, this description is sometimes referred to as *intrinsic coordinates*.

In Cartesian coordinates, and later for cylindrical coordinates, as well as many other coordinate systems that we shall not consider here, all of the variables are defined independent of knowledge of the path. They are sometimes referred to as *extrinsic coordinates*.

All mathematical descriptions of the kinematical properties of a physical system relate to one or the other of the types of coordinate systems just described. Hence, we must study them in detail. In summary, then, we wish to study both

$$\left\{ \begin{array}{c} \text{Fixed} \\ \text{coordinate} \\ \text{systems} \end{array} \right\} \leftrightarrow \left\{ \begin{array}{c} \text{Extrinsic} \\ \text{coordinates} \end{array} \right\} \leftrightarrow \left\{ \begin{array}{c} \text{Path independent} \\ \text{and coordinate} \\ \text{axes dependent} \end{array} \right\}$$

and

$$\left\{ \begin{array}{c} \text{Path} \\ \text{variables} \end{array} \right\} \leftrightarrow \left\{ \begin{array}{c} \text{Intrinsic} \\ \text{coordinates} \end{array} \right\} \leftrightarrow \left\{ \begin{array}{c} \text{Path dependent} \\ \text{and coordinate} \\ \text{axes indepedent} \end{array} \right\}$$

2   Velocity in Path Variables

We begin our study by determining the kinematical relationship for velocity in terms of path variables.

To find the velocity we must take the time derivative of the position vector. In path variables the path is given and the position vector is defined in terms of the arc length $s(t)$ that the particle has traveled. Restating equation (1), we have

$$\bar{v}_P = \dot{\bar{r}}_{P/O} = \frac{d\bar{r}_{P/O}}{dt}$$

Note that we were given s as an explicit function of t. This suggests using the chain rule for differentiation. Hence

$$\bar{v}_P = \frac{d\bar{r}_{P/O}}{dt} = \frac{d\bar{r}_{P/O}}{ds}\frac{ds}{dt} = \frac{d\bar{r}_{P/O}}{ds}\dot{s} \tag{3}$$

In order for equation (3) to be useful we must define $d\bar{r}_{P/O}/ds$. Proceeding toward this objective, in Figure 3 the vector $\Delta\bar{r}_P$ is the change in $\bar{r}_{P/O}$ when the particle moves an arc length Δs from position P. From Figure 3 we may write

$$\Delta\bar{r}_{P/O} = \bar{r}_{P/O}(s + \Delta s) - \bar{r}_{P/O}(s)$$

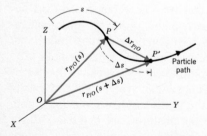

Figure 3

Geometrically, $\Delta \bar{r}_{P/O}$ is a chord for the path connecting points P and P'.

Now consider when happens as Δs becomes smaller and smaller. Because the path is a continuous line, point P' approaches P, and the chord and the tangent to the path become indistinguishable. In other words, in the limit the cord length $|\Delta \bar{r}_{P/O}|$ and arc length Δs are identical. Thus if we define $\bar{e}_t$ to be a *unit vector tangent to the path at point P* in the direction of increasing s, we have

$$\frac{d\bar{r}_{P/O}}{ds} = \lim_{\Delta s \to 0} \frac{\Delta \bar{r}_{P/O}}{\Delta s} = \lim_{\Delta s \to 0} \frac{\Delta s \bar{e}_t}{\Delta s} = \bar{e}_t$$

Returning to eqation (3), we now may write

$$\bar{v}_P = \frac{d\bar{r}_{P/O}}{ds} \dot{s} = \dot{s} \bar{e}_t \tag{4}$$

Equation (4) yields a valuable piece of information. Noting that the path we chose in Figure 3 was quite arbitrary, it must be true that the velocity is always tangent to the path.

Now let us compute the speed, which is the absolute value of the velocity:

$$v_P = |\bar{v}_P| = \sqrt{\bar{v}_P \cdot \bar{v}_P} = \dot{s}\sqrt{\bar{e}_t \cdot \bar{e}_t} = \dot{s}$$

Thus we can conclude that

$$\boxed{\begin{aligned} \bar{v}_P &= v_P \bar{e}_t \\ v_P &= \dot{s} \end{aligned}} \tag{5}$$

In summary, equations (5) say that the velocity (a vector) is tangent to the path and that the speed (a scalar) is the time rate of change of the distance traveled along the path. Also note that because $\bar{e}_t$ is tangent to the path, the direction of $\bar{e}_t$ will change as the location of point P changes, with one exception. Which? LINE

EXAMPLE 1
A locomotive follows a circular track that lies in a horizontal plane. The track has a radius of 2000 ft. If the distance the train has traveled from the start of the curve is given by $s = 40t + t^2$ (s is feet, and t is seconds), what is the velocity of the train when $t = 10$ seconds?

Solution
We are given $s(t)$. Then, knowing that $\bar{v}_P = \dot{s}\bar{e}_t$ we compute

$$\dot{s} = 40 + 2t$$

For $t = 10$ seconds, this yields $\dot{s} = 60$ ft/s, or

$$\bar{v}_P = 60\bar{e}_t \text{ ft/s}$$

The unit vector $\bar{e}_t$ is illustrated in the accompanying sketch.

As an additional exercise, let us express $\bar{v}_P$ in terms of its X and Y components. From the diagram,

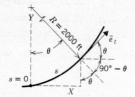

$$\theta = \frac{s}{R} \quad \text{and} \quad \bar{e}_t = \cos\theta\bar{I} + \sin\theta\bar{J}$$

When $t = 10$ seconds, $s = 40(10) + 10^2 = 500$ ft, so that

$$\theta = \frac{500}{2000} \text{ rad} = 0.25 \text{ rad} \left(\frac{180°}{\pi \text{ rad}}\right) = 14.32°$$

Thus

$$\bar{v}_P = 60 \,(\cos 14.32° \,\bar{I} + \sin 14.32°\bar{J})$$
$$= 58.13\bar{I} + 14.84\bar{J} \text{ (ft/s)}$$

3 Acceleration in Path Variables

We can now obtain a relationship for acceleration by differentiating the velocity (using the laws for vector differentiations).

$$\bar{a}_P \equiv \frac{d}{dt}(v_P\bar{e}_t) = \dot{v}_P\bar{e}_t + v_P\dot{\bar{e}}_t \tag{6}$$

In equation (6) the only term with which we are unfamiliar is $\dot{\bar{e}}_t$. Remember that $\bar{e}_t$ is tangent to the path at point P. Thus $\bar{e}_t$ is a function of the arc length s, and the chain rule gives

$$\frac{d\bar{e}_t}{dt} = \frac{d\bar{e}_t}{ds}\frac{ds}{dt} = \dot{s}\frac{d\bar{e}_t}{ds} = v_P\frac{d\bar{e}_t}{ds}$$

Placing this into equation (6), we have

$$\bar{a}_P = \dot{v}_P\bar{e}_t + v_P{}^2\frac{d\bar{e}_t}{ds} \tag{7}$$

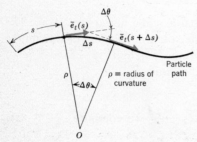

Figure 4

We now have as an unknown $d\bar{e}_t/ds$. To deal with it consider the unit vectors $\bar{e}_t(s)$ and $\bar{e}_t(s + \Delta s)$, illustrated in Figure 4.

Although we are dealing with an arbitrary smooth curve in space, without loss of generality we may consider the arc length in question to have a center of curvature O, which is defined as the center of the circle that most closely fits the path at position s. Further, for ease of visualization, consider the curve to be contained in the plane of the page. [We shall discuss the matter of center of curvature for a three-dimensional curve in Section II.B.5].

Let us denote the angle between points $\bar{e}_t(s)$ and $\bar{e}_t(s + \Delta s)$ in Figure 4 as $\Delta\theta$. We shall now slide the unit vectors to a position where their tails coincide. Hence we may form the triangle shown in Figure 5. The angle between the unit vectors is $\Delta\theta$. In Figure 5 the parameter $\Delta\theta$ appears explicitly, whereas Δs does not. Because of this, let us again use the chain rule to evaluate $d\bar{e}_t/ds$.

Figure 5

$$\frac{d\bar{e}_t}{ds} = \frac{d\bar{e}_t}{d\theta} \frac{d\theta}{ds} \tag{8}$$

Because

$$\left| \bar{e}_t(s) \right| = \left| \bar{e}_t(s + \Delta s) \right| = 1$$

the triangle in Figure 5 is an isosceles triangle. Thus we may write

$$\left| \Delta\bar{e}_t \right| = 2 \left| \bar{e}_t \right| \sin\frac{\Delta\theta}{2} = 2 \sin\frac{\Delta\theta}{2} \tag{9}$$

In equation (8) we require $d\bar{e}_t/d\theta$. Hence we divide both sides of equation (9) by $\Delta\theta$ to get

$$\frac{\left| \Delta\bar{e}_t \right|}{\Delta\theta} = \frac{\sin \Delta\theta/2}{\Delta\theta/2}$$

Now, in the limit as $\Delta\theta \to 0$, we get

$$\lim_{\Delta\theta \to 0} \frac{\left| \Delta\bar{e}_t \right|}{\Delta\theta} = \frac{d\left| \bar{e}_t \right|}{d\theta} = \lim_{\Delta\theta \to 0} \frac{\sin \Delta\theta/2}{\Delta\theta/2} \equiv 1 \tag{10}$$

Do you remember this limit from your elementary calculus course?

From Figure 5 we see graphically that as $\Delta\theta \to 0$, the vector $\Delta\bar{e}_t/\Delta\theta$ approaches the perpendicular to $\bar{e}_t(s)$. From Figure 4 we see that this perpendicular is directed toward the center of curvature. Finally, from equation (10) we see that the perpendicular has magnitude unity.

When we put these facts together we find that $d\bar{e}_t/d\theta$ is a unit vector perpendicular (normal) to $\bar{e}_t$ and directed toward the center of curvature. We denote this unit vector as $\bar{e}_n$, where the letter n is for normal; hence

$$\frac{d\bar{e}_t}{d\theta} = \bar{e}_n$$

The plane defined by $\bar{e}_t$ and $\bar{e}_n$ is called the *osculating plane*. The center of curvature, of course, also lies in the osculating plane. For a planar path the osculating plane is the plane of the path.

Let us once again return to the central theme of determining the acceleration. Equation (8) now reads:

$$\frac{d\bar{e}_t}{ds} = \frac{d\theta}{ds}\,\bar{e}_n \tag{11}$$

Inspecting equation (11), we must now express $d\theta/ds$ in a more usable form. Referring to Figure 4, you will note that as Δs becomes infinitesmal, it may be considered to form the arc of a circle of radius ρ, which is called the radius of curvature. Thus

$$\rho = \frac{ds}{d\theta}$$

Do you recognize this formula from basic calculus? From this formula equation (11) then becomes

$$\frac{d\bar{e}_t}{ds} = \frac{1}{\rho}\bar{e}_n$$

Hence equation (7) may be written in the standard form

$$\boxed{\bar{a}_P = \dot{v}_P\bar{e}_t + \frac{v_P{}^2}{\rho}\,\bar{e}_n} \tag{12}$$

Before investigating some examples involving acceleration in path variables, we should note that there are basically two classes of problems that occur in dynamics. In the simpler class of problems the given information describes the position, and the velocity and acceleration are desired. As we have seen in the derivations of the basic formulas, the velocity and acceleration are obtained through the process of differentiation.

A more complicated problem occurs when information about the acceleration is given and velocity and position are the desired quantities. A typical situation, involving equation (12), is to be given the tangential acceleration $\dot{v}_P$ in equation form. An equation that contains a derivative of one of the unknown parameters is called a *differential equation*. For example, $\dot{v}_P = t^2$ is a differential equation involving v_P and t in which v_P is called the dependent variable and t the independent variable. We shall study the topic of differential equations as it applies to dynamics problems in the next module on the equations of motion in particle motion. In this module we restrict ourselves to cases where we find velocity and/or acceleration using the differentiation process only.

EXAMPLE 2
Shown are the path of a particle and representations of the velocity and acceleration of the particle at that instant in time. Indicate if each of the conditions shown is possible or not.

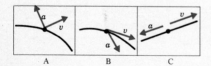

Solution

Figure A: We know that the velocity *must* be tangent to the path (equation (4)). Therefore this is not a possible situation. Also, we know that the acceleration *must* point toward the inside of the curve, for its two components are directed tangent to the path and normal to the path towards the center of curvature. Hence, this too is not possible.

Figure B: These are possible situations, from the arguments for Figure A.

Figure C: The velocity is tangent to the path. The path is a straight line, hence ρ is infinite and the only component of $\bar{a}_P$ is the one that is tangent to the path. Hence this situation is possible. What does the condition that the direction of $\bar{a}$ is opposite to the direction of $\bar{v}$ mean?

EXAMPLE 3

The diagram shown is a portion of a roller coaster track. It consists of a circle in a vertical plane of radius 150 m. Find the magnitude of the total acceleration of the roller coaster if, at the time shown, the car is moving at 50 km/hr and is gaining speed at the rate of $0.4g$ ($1g = 9.806$ m/s²).

Solution

The given information is clearly in the form of path variables (path, speed, and rate of increase of speed). Therefore select the unit vectors at P as shown. In general we know that

$$\bar{a}_P = \dot{v}_P \bar{e}_t + \frac{v_P{}^2}{\rho} \bar{e}_n$$

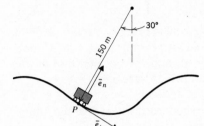

The given information is

$$v_P = 50,000 \text{ m/hr} \left(\frac{1}{3600} \frac{s}{\text{hr}}\right) = 13.889 \text{ m/s}$$

$$\dot{v}_P = 0.4(9.806) = 3.922 \text{ m/s}^2$$

Hence we compute

$$\bar{a}_P = 3.922\bar{e}_t + \frac{(13.889)^2}{150}\bar{e}_n$$

$$= 3.922\bar{e}_t + 1.286\bar{e}_n \text{ m/s}^2$$

Then the magnitude of $\bar{a}_P$ is

$$|\bar{a}_P| = \sqrt{(3.922)^2 + (1.286)^2}$$

$$= 4.13 \text{ m/s}^2$$

HOMEWORK PROBLEMS

II.1   A bicyclist whose speed is 18 miles/hr wishes to turn left. What is the minimum radius turn for which the acceleration will not exceed 5 ft/s²?

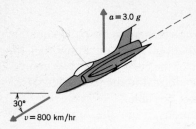

Prob. II.3

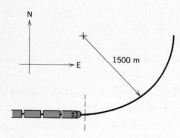

Prob. II.4

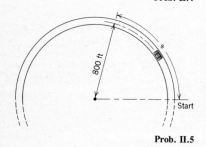

Prob. II.5

II.2 An automobile on a 40-m-radius curve has a speed of 60 km/hr. What rate of increase of this speed will result in a 50% increase in the magnitude of the acceleration of the vehicle?

II.3 At a particular instant an airplane is diving at a 30° angle with a speed of 800 km/hr. Its instantaneous acceleration is $3.0g$ upward. Determine (a) the rate of increase of the speed, (b) the unit vector normal to the path of the airplane, and (c) the radius of curvature of the path.

II.4 A freight train heading east enters a 1500-m-radius curve to the north. The arc length traveled along the curve is given by $s = 80t - t^2$ (s is in meters and t is in seconds). What is the speed, the heading in degrees with respect to the easterly direction, and the acceleration of the train after it has traveled 1200 m?

II.5 The motion of an automobile along a 800-ft-radius track is defined by $s = 6t^2 - 0.2t^3$, where t is measured in seconds and s is in feet. Determine the distance traveled, the speed, and the acceleration of the vehicle (a) when $t = 5$ seconds, (b) when $t = 15$ seconds.

II.6 Solve Problem II.5 for the instant when (a) the speed is maximum, (b) the tangential acceleration is maximum.

II.7 The constant speed of the particle spinning in a circular horizontal path at the end of a cable of length l is known to depend on the height h illustrated in the diagram according to $v^2 = v_0^2 + 2gh$. For what value of v_0 will the acceleration have a magnitude of $2v_0^2/l$ when the height is $0.8l$?

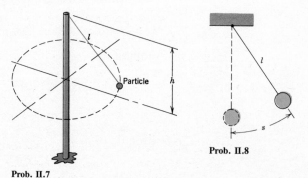

Prob. II.7

Prob. II.8

II.8 The arc length s for the pendulum in a vertical plane shown in the diagram is given by $s = (l/10) \sin \sqrt{g/l}\, t$. Determine the acceleration of the pendulum, and show the result in a sketch, (a) for $s = 0$, (b) for $s = l/10$, (c) for $s = l/20$.

4 Further Remarks on Path Variable Kinematics

Equation (12) shows that the total acceleration is the vector sum of two components that lie in the osculating plane: The tangent component is the rate of increase of the speed, $a_t = \dot{v}_P$, whereas in the normal direction the centripetal component, $a_n = v_P^2/\rho$, is directed toward the center of curvature. Symbolically the acceleration vector is often written in the form

$$\bar{a} = a_t \bar{e}_t + a_n \bar{e}_n \tag{13}$$

From this expression we can formulate the magnitude of the total acceleration as

$$|\bar{a}| = \sqrt{\bar{a} \cdot \bar{a}} = \sqrt{a_t^2 + a_n^2} = \sqrt{\dot{v}^2 + v^4/\rho^2}$$

Let us briefly turn to two special cases that frequently occur. In many problems the speed will be given as a function of the distance traveled by the particle. In other words, $v_P(s)$ will be given. Then to determine the rate of increase of the speed we can use the chain rule.

$$\dot{v}_P = \frac{dv_P}{ds}\frac{ds}{dt} = \dot{s}\,\frac{dv_P}{ds} = v_P\,\frac{dv_P}{ds} \tag{14}$$

Another special case occurs when the curved path is circular. As can be seen in Figure 6a, $\rho \equiv R$ and $\bar{e}_n$ always points toward the center of the circle. For an arbitrary path, Figure 6b, the value of ρ changes as the location of the particle changes and, in general, $\bar{e}_n$ is no longer directed toward a single point.

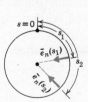

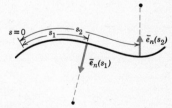

Figure 6a   Figure 6b

The process of functionally determining $\bar{e}_t$, $\bar{e}_n$, and ρ for a general three-dimensional path is a mathematical topic treated in the next section. For the case of a two-dimensional path, without loss of generality we can describe the path by selecting XYZ so as to make the particle move in the XY plane. Then the functional equation $Y = f(X)$ defines the path.

For the planar path shown in Figure 7, formulas developed in elementary calculus give

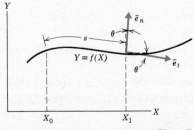

Figure 7

$$s = \int_{X_0}^{X_1}\left[1 + \left(\frac{dY}{dX}\right)^2\right]^{1/2} dX \tag{15}$$

$$\theta = \tan^{-1} \frac{dY}{ds} \tag{16}$$

$$\frac{1}{\rho} = \frac{|d^2 Y/dX^2|}{[1 + (dY/dX)^2]^{3/2}} \tag{17}$$

We should note, however, that equation (15) will usually lead to a complicated integral.

Once the value of θ is found at point s, we can, if we so desire, transform the expressions for velocity and acceleration by expressing the unit vectors $\bar{e}_t$ and $\bar{e}_n$ in terms of their components with respect to the Cartesian unit vectors $\bar{I}$ and $\bar{J}$.

As a concluding remark, we again call attention to the fact that the basic formulas developed for path variables depend on knowing either $s(t)$ and/or v_P at any position along the path. It follows that if we have knowledge about the path a particle follows and information about the distance traveled along the path (or speed), it will prove advantageous to use path variables. Without such information, as we shall see, there are better ways to proceed.

Let us now look at an example that utilizes most of the equations governing path variables for planar motion.

EXAMPLE 4

A ski jump is a curve in the vertical plane defined by $Y = 0.005X^2$, where X and Y are the horizontal and vertical coordinates of a point relative to the bottom of the ski jump measured in meters. In one trial a skier began her run at $X = 150$ m and her speed was observed to vary according to $v_P = 30 \sin(\pi s/2s_0)$ m/s, where s is the distance she has traveled and s_0 is the arc length from the start to the bottom. Determine the acceleration of the skier when she is at a horizontal distance of 50 m from the bottom of the ski jump.

Solution

The given information about the speed is expressed in terms of the parameter s, thus suggesting the path variable approach.

$$\bar{a}_P = \dot{v}_P \bar{e}_t + \frac{v_P^2}{\rho} \bar{e}_n$$

Let us begin with a sketch of the ski jump that also shows the path variable unit vectors.

We must determine the value of s_0 and also the value of s corresponding to the position where $X = 50$ m. From equation (15) we have

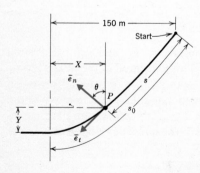

$$s_0 = \int_0^{150} \left[1 + \left(\frac{dY}{dX} \right)^2 \right]^{1/2} dX$$

$$= \int_0^{150} (1 + 0.0001X^2)^{1/2} \, dX$$

$$= 0.005\{X(100^2 + X^2)^{1/2} + 100^2 \ln[X + (100^2 + X^2)^{1/2}]\} \Big|_0^{150}$$

$$= 194.95 \text{ m}$$

(The general solution of the integral was obtained from a table of integrals.) As indicated earlier, for most paths whose shapes are common curves the evaluation of the arc length is not a trivial step. Following a similar step for the value of s at point P yields

$$s = 194.95 - \int_0^{50} \left[1 + \left(\frac{dY}{dX} \right)^2 \right]^{1/2} dX = 142.94 \text{ m}$$

The corresponding value of v_P may now be determined by direct substitution.

$$v_P = 30 \sin \left(\frac{\pi \times 142.93}{2 \times 194.95} \right) = 30 \sin(1.1517 \text{ rad})$$

$$= 30 \sin(65.98°) = 27.40 \text{ m/s}$$

Because we are given $v_P(s)$, we use the chain rule to find $\dot{v}_P(s)$.

$$\dot{v}_P(s) = v_P \frac{dv_P}{ds} = 30 \sin\left(\frac{\pi s}{2s_0} \right) \times 30 \left(\frac{\pi}{2s_0} \right) \cos \left(\frac{\pi s}{2s_0} \right)$$

Thus when $s = 142.93$ m,

$$\dot{v}_P(s) = \left(\frac{450\pi}{194.95} \right) \sin(65.98°) \cos(65.98°)$$

$$= 2.696 \text{ m/s}^2$$

Finally, we use equation (19) to find $1/\rho$.

$$\frac{1}{\rho} = \frac{|d^2Y/dX^2|}{[1 + (dY/dX)^2]^{3/2}} = \frac{0.01}{[1 + (0.01X)^2]^{3/2}}$$

$$= \frac{0.01}{[1 + 0.25]^{3/2}} = 0.00716 \, \frac{1}{\text{m}}$$

Substituting these values into the acceleration equation gives

$$\bar{a}_P = \dot{v}_P \bar{e}_t + \frac{v_P^2}{\rho} \bar{e}_n$$

$$= 2.70 \bar{e}_t + 5.38 \bar{e}_n \text{ m/s}^2$$

The solution is now complete because we have the components of acceleration with respect to a set of unit vectors that are defined in the diagram.

In the next step, which is only necessary if the problem requires it, we show how the answer may be converted to Cartesian components. This can be accomplished by expressing $\bar{e}_t$ and $\bar{e}_n$ in their Cartesian components. The diagram of the unit vectors shows that

$$\theta = \tan^{-1} \frac{dY}{dX} = \tan^{-1}(0.01X) = \tan^{-1}(0.5) = 26.57°$$

$$\bar{e}_t = -\cos\theta\bar{I} - \sin\theta\bar{J} = -0.894\bar{I} - 0.447\bar{J}$$

$$\bar{e}_n = -\sin\theta\bar{I} + \cos\theta\bar{J} = -0.447\bar{I} + 0.894\bar{J}$$

Then substituting these results into the equation for $\bar{a}_P$ gives

$$\bar{a}_P = 2.70(-0.894\bar{I} - 0.447\bar{J}) + 5.38(-0.447\bar{I} + 0.894\bar{J})$$

$$= -4.82\bar{I} + 3.60\bar{J} \text{ m/s}^2$$

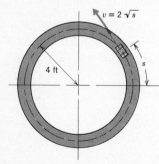

Prob. II.9

Prob. II.10

HOMEWORK PROBLEMS

II.9 A toy slot car follows a 4-ft-radius track. The speed of the toy is related to the distance traveled, s, according to $v = 2\sqrt{s}$ (s is in feet and v is in feet per second). Determine the acceleration (a) when $s = 1$ ft., (b) when $s = 6.25$ ft. Draw a sketch of the results.

II.10 The computerized controller of a high-speed passenger train slows the train according to the given graph. A train is on a 10-km curve with an initial speed $v_0 = 360$ km/hr when the brakes are applied. Determine the magnitude of the acceleration when $s = 1.5$ km.

II.11 For Problem II.10 determine the position (a) where $\dot{v}$ has its maximum magnitude, (b) where the acceleration has its maximum magnitude.

II.12 An automobile moves with a constant speed of 90 km/hr along a highway whose vertical elevation Y in meters is given by $Y = 2 \cos(\pi X/100)$, where X is the horizontal distance traveled, also in meters.

Determine the acceleration of the automobile and sketch the vector along the curve when $X = 75$ m.

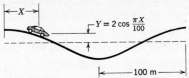

$Y = 2 \cos \dfrac{\pi X}{100}$

100 m

Prob. II.12

II.13 The brakes of the automobile in Problem II.12 are suddenly applied when $X = 75$ in, causing the speed to decrease instantaneously at the rate of $0.5g$. Determine and sketch the resulting acceleration in this position.

II.14 Arm AB rotates at a rate such that the pin follows the parabolic groove $Y = 4 - X^2$ (X and Y are in feet) at a constant speed of 3 ft/s. Determine the horizontal and vertical components of the velocity and acceleration of the pin when $X = 2$ ft.

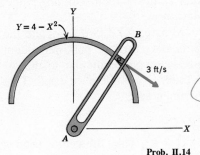

$Y = 4 - X^2$

B

3 ft/s

A

Prob. II.14

5 Path Variables in Three Dimensions

From equation (12) we see that the acceleration of a particle in space consists of two scalar components perpendicular to each other. In two dimensions it is easy to visualize that the plane that contains the vector $\bar{a}_P$ is the plane of motion. What about a particle traversing a three-dimensional curvilinear path? Let us see how the $\bar{e}_t \bar{e}_n$ plane can be described for such a general case.

A general three-dimensional path is shown in Figure 8. As Δs becomes infinitesmal, but nonzero, it can be proved that $\bar{e}_t(s)$ and $\bar{e}_t(s + \Delta s)$ lie in a common plane, but are not collinear. This plane is the *osculating plane*. To get some idea of the meaning of the osculating plane, we

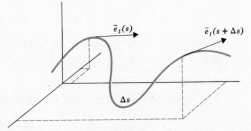

$\bar{e}_t(s)$

$\bar{e}_t(s + \Delta s)$

Δs

Figure 8

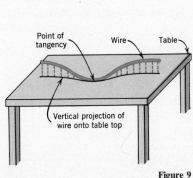

Point of tangency

Wire

Table

Vertical projection of wire onto table top

Figure 9

suggest the following. Bend a wire clothes hanger into an arbitrary shape and pick out some point on the hanger. Put the hanger into contact with a flat table (as shown in Figure 9) and rotate the wire until you find the orientation for which a small region of the wire in the vicinity of the point seems to be fully in contact with the table. You have found the osculating plane for the chosen point on the wire. Note that the center of curvature for the infinitesmal arc length ds containing the chosen point also lies in the osculating plane. As mentioned previously, if the wire is bent into a planar curve, the osculating plane for such a curve is the plane of the wire. In the special case where the wire forms a straight line, *any* plane that contains the line is the osculating plane.

In summary, at an arbitrary point on a particle's path, the unit vector $\bar{e}_n$ is perpendicular with the tangent vector $\bar{e}_t$, lies in the osculating plane, and is directed toward the center of curvature.

When we think of the Cartesian coordinates XYZ, we associate the unit vectors $\bar{I}$, $\bar{J}$, $\bar{K}$ to the system. Similarly, for the path-dependent variables $\bar{e}_t$ and $\bar{e}_n$ we can define a unit vector perpendicular to the osculating plane (the plane of $\bar{e}_t$ and $\bar{e}_n$) in order to complete the triad. That direction is called the *binormal* and is defined as

$$\bar{e}_b \equiv \bar{e}_t \times \bar{e}_n$$

(Clearly, this is directly analogous to $\bar{I} \times \bar{J} = \bar{K}$.) Thus the vector triad is $\bar{e}_t$, $\bar{e}_n$, $\bar{e}_b$.

The importance of $\bar{e}_b$ for particle motion is that when one considers Newton's second law, $\bar{F} = m\bar{a}$, from equation (12) we see that the component of $\bar{a}$ in the direction of $\bar{e}_b$ is zero. This means that we have an equation of statics to call upon in problem solving, as will be seen in Illustrative Example 3 in the next module.

In general, the determination of the path variable quantities for three dimensional motion involves more mathematical manipulations than that required for path variables in two dimensional motion. Let us look at an example to see what is involved.

EXAMPLE 5

A common way of defining a three-dimensional path is in parametric form. A helix is defined in parametric form by

$$X = A \cos \alpha \qquad Y = A \sin \alpha \qquad Z = C\alpha$$

where α is the parameter. Determine the tangent, normal, and binormal unit vectors and the radius of curvature for an arbitrary value α.

Solution

The basic equations for the desired quantities are

$$\bar{e}_t = \frac{d\bar{r}_{P/O}}{ds} \qquad \bar{e}_n = \rho \frac{d\bar{e}_t}{ds} \qquad \bar{e}_b = \bar{e}_t \times \bar{e}_n$$

The point P has coordinates (X, Y, Z) so

$$\bar{r}_{P/O} = X\bar{I} + Y\bar{J} + Z\bar{K}$$

$$= A \cos \alpha \bar{I} + A \sin \alpha \bar{J} + C\alpha \bar{K}$$

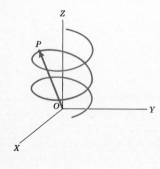

Because $\bar{r}_{P/O}$ is a function of α, we will use the chain rule with respect to α:

$$\bar{e}_t = \frac{d\bar{r}_{P/O}}{ds} = \frac{d\bar{r}_{P/O}}{d\alpha}\frac{d\alpha}{ds}$$

$$= \frac{d\alpha}{ds}[-A \sin \alpha \bar{I} + A \cos \alpha \bar{J} + C\alpha \bar{K}]$$

We do not know $(d\alpha/ds)$. To determine this quantity we use the fact that $\bar{e}_t$ must be a unit vector, so

$$|\bar{e}_t| = 1 = \frac{d\alpha}{ds}[(-A \sin \alpha)^2 + (A \cos \alpha)^2 + C^2]^{1/2}$$

$$= \frac{d\alpha}{ds}[A^2 + C^2]^{1/2}$$

Hence

$$\frac{d\alpha}{ds} = (A^2 + C^2)^{-1/2}$$

so that

$$\bar{e}_t = (A^2 + C^2)^{-1/2}(-A \sin \alpha \bar{I} + A \cos \alpha \bar{J} + C\alpha \bar{K})$$

Now, using the chain rule again,

$$\bar{e}_n = \rho\frac{d\bar{e}_t}{ds} = \rho\frac{d\bar{e}_t}{d\alpha}\frac{d\alpha}{ds}$$

$$= \rho(A^2 + C^2)^{-1/2}(-A \cos \alpha \bar{I} - A \sin \alpha \bar{J})(A^2 + C^2)^{-1/2}$$

$$= -\rho(A^2 + C^2)^{-1}(A \cos \alpha \bar{I} + A \sin \alpha \bar{J})$$

However, $\bar{e}_n$ is also a unit vector.

$$|\bar{e}_n| = 1 = \rho(A^2 + C^2)^{-1}[(A \cos \alpha)^2 + (A \sin \alpha)^2]^{1/2}$$

$$= \rho(A^2 + C^2)^{-1}A$$

This enables us to find the radius of curvature and $\bar{e}_n$.

$$\rho = \frac{A^2 + C^2}{A}$$

and

$$\bar{e}_n = -\frac{1}{A}(A \cos \alpha \bar{I} + A \sin \alpha \bar{J})$$

Finally, we find $\bar{e}_b$.

$$\bar{e}_b = \bar{e}_t \times \bar{e}_n$$

$$= -\frac{1}{A}(A^2 + C^2)^{-1/2}(-A \sin \alpha \bar{I} + A \cos \alpha \bar{J} + C\alpha \bar{K})$$
$$\times (A \cos \alpha \bar{I} + A \sin \alpha \bar{J})$$

$$= -(A^2 + C^2)^{-1/2}[-C \sin \alpha \bar{I} + C \cos \alpha \bar{J}$$
$$-A(\sin^2 \alpha + \cos^2 \alpha)\bar{K}]$$

$$= (A^2 + C^2)^{-1/2}(C \sin \alpha \bar{K} - C \cos \alpha \bar{J} + A\alpha \bar{K})$$ ◁

HOMEWORK PROBLEMS

II.15 A conical helix is defined in parametric form by

$$X = 40\alpha \cos \pi\alpha \qquad Y = 40\alpha \sin \pi\alpha \qquad Z = 30\alpha$$

where X, Y, and Z have units of millimeters. Determine the tangent, normal, and binormal unit vectors when $Z = 18$ mm.

II.16 In the position where $\alpha = 3.25$ a particle following the conical helix of Problem II.15 has a speed of 50 m/s and is gaining speed at 10 m/s². Determine its corresponding position, velocity and acceleration.

6 Fixed Rectangular Cartesian Coordinates

The underlying principle for the Cartesian coordinate description of the kinematical relationships is to express the components of vector quantities in a fixed XYZ coordinate system. The unit vectors $\bar{I}, \bar{J}, \bar{K}$ describe the positive X, Y, and Z directions. Note again that the use of capital letters for $\bar{I}, \bar{J},$ and $\bar{K}$ indicate that they correspond to the fixed XYZ system.

Figure 10a shows a point P whose coordinates are (X, Y, Z). The components of the position vector $\bar{r}_{P/O}$ parallel to the $\bar{I}, \bar{J},$ and $\bar{K}$ unit vectors are

$$\bar{r}_{p/O} = X\bar{I} + Y\bar{J} + Z\bar{K} \tag{18}$$

In Figure 10b we selected a Cartesian coordinate system having the same origin O but different orientation from the one shown in Figure 10a. Note that the values of X, Y, and Z are different, as are, of course, the $\bar{I}, \bar{J},$ and $\bar{K}$ directions. However, the resulting vector $\bar{r}_{P/O}$ represents the same physical position in space.

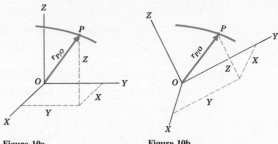

Figure 10a   **Figure 10b**

When the particle moves, the value of the X, Y, and Z components change with time. The functions $X(t)$, $Y(t)$, and $Z(t)$ define the path of the particle in parametric form, with t being the parameter. For instance, $X = Y = 0$, $X = -1/2t^2$ defines a path that is the Z axis.

The velocity is found by differentiating the position vector with respect to time. Because XYZ is a fixed system, $\bar{I}$, $\bar{J}$, and $\bar{K}$ have constant magnitude and direction, so that $\dot{\bar{I}} = \dot{\bar{J}} = \dot{\bar{K}} = \bar{0}$. Then

$$\bar{v}_P \equiv \dot{\bar{r}}_{P/O} = \dot{X}\bar{I} + X\dot{\bar{I}} + \dot{Y}\bar{J} + Y\dot{\bar{J}} + \dot{Z}\bar{K} + Z\dot{\bar{K}}$$

$$\boxed{\bar{v}_P \equiv \dot{X}\bar{I} + \dot{Y}\bar{J} + \dot{Z}\bar{K}} \tag{19}$$

The speed is the magnitude of the velocity of $\bar{v}_P$.

$$v_P = \left|\bar{v}_P\right| = \sqrt{\bar{v}_P \cdot \bar{v}_P}$$

$$= \sqrt{(\dot{X})^2 + (\dot{Y})^2 + (\dot{Z})^2}$$

Differentiation of the velocity expression yields the acceleration formula:

$$\bar{a}_P = \ddot{X}\bar{I} + \dot{X}\dot{\bar{I}} + \ddot{Y}\bar{J} + \dot{Y}\dot{\bar{J}} + \ddot{Z}\bar{K} + \dot{Z}\dot{\bar{K}}$$

$$\boxed{\bar{a}_P = \ddot{X}\bar{I} + \ddot{Y}\bar{J} + \ddot{Z}\bar{K}} \tag{20}$$

The Cartesian coordinate formulation is one alternative to path variables for determining velocity and acceleration. Depending on the way a problem is posed, one method or the other will yield the solution in a simpler and more straightforward manner.

Basically, we use Cartesian coordinates in problems where the important variables are most easily expressed as functions of the Cartesian coordinates, as in the case where quantities are given with respect to fixed directions, e.g., horizontal and vertical. For any given motion a solution by Cartesian coordinates will, of course, give the same vector resultants for velocity or acceleration as a solution by path variables.

EXAMPLE 6
A ball is thrown horizontally off the top of a building and follows a parabolic path in the vertical plane defined by $Z = 0$, $Y = -cX^2$, where Y is its elevation measured from the top of the building and X is the horizontal distance traveled by the ball. We know that the horizontal component of the velocity is constant. Find the acceleration of the particle as a function of its position.

Solution
In this problem the path is given. We are also given information about the velocity component in the (fixed) horizontal direction. These data strongly indicate that Cartesian coordinates will give an easier solution than path variables.

We begin the solution by drawing a sketch of the path. Because $Z \equiv 0$, the acceleration is

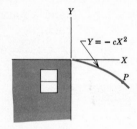

$$\bar{a}_P = \ddot{X}\bar{I} + \ddot{Y}\bar{J}$$

The component of velocity in the X direction, V_X, was given to the constant, so that $\dot{X} = V_X$, $\ddot{X} = 0$. We must now determine $\ddot{Y}$. Given that $Y = -cX^2$, the chain rule yields

$$\dot{Y} = \frac{dY}{dt} = \frac{dY}{dX}\frac{dX}{dt} = (-2cX)\dot{X} = -2cV_X X$$

Then differentiating again gives

$$\ddot{Y} = -2c\dot{V}_X X + 2cV_X \dot{X} = -2cV_X^2$$

Substituting the results for $\ddot{X}$ and $\ddot{Y}$ into the basic acceleration equation gives

$$\bar{a}_P = -2cV_X^2\bar{J} \qquad \qquad \triangleleft$$

Alternative Approach   The foregoing illustrates how the derived formulas may be utilized to solve problems. As an alternative, one can always fall back on the basic definitions to solve *any* problem as is illustrated below.

For any particle motion problem in Cartesian coordinates we can write

$$\bar{r}_{P/O} = X\bar{I} + Y\bar{J} + Z\bar{K}$$

In this case, from the given information we write

$$\bar{r}_{P/O} = X\bar{I} - cX^2\bar{J} + 0\bar{K}$$

Note that we eliminate Y rather than X because: (a) we know something about X, and (b) the mathematical representation is cleaner. Differentiating, we find the velocity.

$$\bar{v}_P = \dot{X}\bar{I} - 2cX\dot{X}\bar{J}$$

Differentiating again, we find the acceleration.

$$\bar{a}_P = \ddot{X}\bar{I} - (2c\dot{X}^2 + 2cX\ddot{X})\bar{J}$$

It is given that $\dot{X} = V_X = $ constant, therefore

$$\bar{a}_P = -2cV_X^2\bar{J}$$

EXAMPLE 7

The coordinates of a moving particle are given parametrically by $X = 7t$, $Y = 3 + t^2$, and $Z = t^3/3$ m, where t is in seconds. Find the normal and tangential components of velocity and acceleration, the radius of curvature, and the Cartesian components of the unit normal and tangential vectors when $t = 3$ s.

Solution
We were given that

$$\bar{r}_{P/O} = 7t\bar{I} + (3 + t^2)\bar{J} + \frac{t^3}{3}\bar{K}$$

Differentiating this gives

$$\bar{v}_P \equiv \dot{\bar{r}}_{P/O} = 7\bar{I} + 2t\bar{J} + t^2\bar{K} \text{ m/s}$$

from which

$$v_P = |\bar{v}_P| = \sqrt{(7)^2 + (2t)^2 + (t^2)^2} \text{ m/s}$$

This information allows us to determine $\bar{e}_t$ using the facts that the velocity is parallel to $\bar{e}_t$ and that the speed is the $\bar{e}_t$ component of $\bar{v}_P$; that is,

$$\bar{v}_P = v_P\bar{e}_t$$

so that

$$\bar{e}_t = \frac{\bar{v}_P}{v_P} = \frac{7\bar{I} + 2t\bar{J} + t^2\bar{K}}{\sqrt{(7)^2 + (2t)^2 + (t^2)^2}}$$

For $t = 3$ s, we find that

$$v_P = \sqrt{49 + 36 + 81} = 12.9 \text{ m/s}$$

and

$$\bar{e}_t = 0.543\bar{I} + 0.465\,\bar{J} + 0.698\,\bar{K}$$

What do the three components of $\bar{e}_t$ represent?

To find the path variable components of acceleration, we first find $\bar{a}_P$ in Cartesian coordinates.

$$\bar{a}_P \equiv \dot{\bar{v}}_P = 2\bar{J} + 2t\bar{K}$$

$$= 2\bar{J} + 6\bar{K} \qquad \text{when } t = 3 \text{ s}$$

But we also know that

$$\bar{a}_P = \dot{v}_P\bar{e}_t + \frac{v_P^2}{\rho}\bar{e}_n = a_t\bar{e}_t + a_n\bar{e}_n$$

Recall that the dot product of a vector and a unit vector $\bar{e}$ gives the component the vector in the direction of $\bar{e}$. Hence to find a_t we compute

$$\bar{a}_P \cdot \bar{e}_t = a_t = \dot{v}_P$$

so that

$$a_t = \frac{7\bar{I} + 2t\bar{J} + t^2\bar{K}}{\sqrt{49 + 4t^2 + t^4}} \cdot (2\bar{J} + 2t\bar{K})$$

$$= \frac{4t + 2t^3}{\sqrt{49 + 4t^2 + t^4}}$$

At $t = 3$ s, this yields

$$a_t = 5.12 \text{ m/s}^2$$

To find a_n, we match the magnitude of $\bar{a}_P$ as given by path variables and by Cartesian coordinates.

$$|\bar{a}_P| = \sqrt{a_t^2 + a_n^2} = \sqrt{(\ddot{X})^2 + (\ddot{Y})^2}$$

At $t = 3$ s, this gives

$$\sqrt{(5.12)^2 + (a_n)^2} = \sqrt{(2)^2 + (6)^2}$$

Then solving for a_n yields

$$a_n = \sqrt{4 + 36 - (5.12)^2} = 3.71 \text{ m/s}^2$$

Finally, we know that $a_n = v_P{}^2/\rho$, so that

$$\rho = \frac{v_P{}^2}{a_n} = \frac{(12.88)^2}{3.71} = 44.7 \text{ m} \qquad \text{when } t = 3 \text{ s}$$

Once we have determined a_n, we can find $\bar{e}_n$ by equating the expressions for $\bar{a}_P$ is path variables and Cartesian coordinates. When $t = 3$ s,

$$a_t \bar{e}_t + a_n \bar{e}_n = 2\bar{J} + 6\bar{K}$$

$$(5.12)\frac{1}{12.9}(7\bar{I} + 6\bar{J} + 9\bar{K}) + 3.71\bar{e}_n = 2\bar{J} + 6\bar{K}$$

Thus

$$\bar{e}_n = \frac{1}{3.71}\left[-\frac{7 \times 5.12}{12.88}\bar{I} + \left(2 - \frac{6 \times 5.12}{12.88}\right)\bar{J} \right.$$

$$\left. + \left(6 - \frac{9 \times 5.12}{12.88}\right)\bar{K}\right]$$

$$= -0.75\bar{I} - 0.10\bar{J} + 0.65\bar{K}$$

Question: How can we define the osculating plane?

HOMEWORK PROBLEMS

II.17 The planar motion of a particle is defined by $X = 4t - 3$, $Y = \sqrt{24t - 12}$, where X and Y are expressed in meters and t is in seconds. Determine the velocity and acceleration of the particle (a) when $t = 1$ s, (b) when $t = 3$ s. Show the results in a sketch of the path.

II.18 The coordinates of a particle in planar motion are defined by $X = e^{t/2}$, $Y = 16e^{-t/2}$, where X and Y are measured in feet and t is in seconds. Show that the path is a hyperbola and determine the velocity and acceleration when $t = 2$ s. Show the solution in a sketch.

II.19 An electron in a cathode-ray tube follows the path

$$Y = 0.7X, \quad Z = \sqrt{X^2 - 2500}$$

where X, Y, and Z are measured in millimeters. The value of X varies with time t according to $X = 1.5(10^6)t^4$, where t is in seconds. Determine the velocity and acceleration of the electron when $X = 150$ mm.

II.20 The roof of a building has the shape of a hyperbolic paraboloid $Z =$

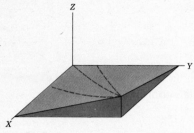

Prob. II.20

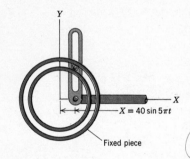

Prob. II.22

$XY/10$, where the length unit is meters. A ball rolling on the roof is observed to have the motion $X = 10 - 20t$, $Y = 10 - 20t^2$, where t is in seconds. Determine the velocity and acceleration of the ball when it reaches the edge where $X = 0$.

II.21 Determine the tangent and normal vectors and the radius of curvature of the path of the ball in Problem II.20 at the instant when the ball reaches the edge where $X = 0$. Also determine the rate of increase of the speed in this position.

II.22 The (X, Y) coordinates of a sliding pin connecting the two grooved members of the machine shown in the diagram simultaneously satisfy the relations $X^2 + Y^2 = 2500$, $X = 40 \sin 5\pi t$, where X and Y have units of millimeters and t is in seconds. Determine the velocity and acceleration of the pin when $t = 0$.

II.23 Solve Problem II.22 for the position where $t = \frac{1}{30}$ second.

II.24 For the pin of Problem II.22, determine the tangent and normal components of acceleration and the radius of curvature of the path when $t = \frac{1}{30}$ second.

7 Cylindrical Coordinates

For Cartesian coordinates we selected three independent quantities, namely the X, Y, and Z coordinates of a particle, to uniquely determine its position in the chosen reference frame. Then, in terms of the unit vectors of the coordinate system, we obtained position, velocity, and acceleration formulas for the particle. In this section we follow a similar procedure using a different set of quantities to define the position of a particle.

Suppose that we select an XYZ coordinate system, Figure 11, and wish to define the position of a particle at point P in this reference system by another method. We may begin by dropping a perpendicular point P

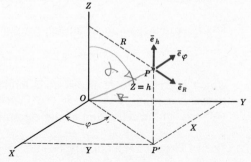

Figure 11

onto the XY plane, and calling this projected point P'. For notation, we shall denote the length of the line OP' by R, the angle between the line OP' and the X axis by ϕ, and the length of the line PP' by h. If we know the values of R, ϕ, and h, we obviously know where point P is. These quantities R, ϕ, and h are by definition *extrinsic coordinates*, because they locate the point P, even when the path is unknown.

If we wish to, we can write a unique coordinate transformation to convert between (X, Y, Z) and (R, ϕ, h) if the range of values of ϕ is limited to 2π, for example, $0 \le \phi < 2\pi$. Specifically, these transformations are

$$Z = h \qquad X = R \cos \phi \qquad Y = R \sin \phi$$

so that

$$R = \sqrt{X^2 + Y^2} \quad \text{and} \quad \phi = \tan^{-1}\frac{Y}{X}$$

The terminology we shall use is to call R the radial distance, ϕ the transverse angle, and h the axial distance. The values (R, ϕ, h) are called *cylindrical coordinates*. This name arises because if we hold R constant and vary ϕ, the line PP' follows the surface of a right circular cylinder. Note that for the special case of plane motion in the XY plane, h is zero and the remaining coordinate system (R, ϕ) is called *polar coordinates*.

Having defined the three coordinates, we now require unit vectors to describe their directions. Let $\bar{e}_R$ be the vector that points in the direction in which point P would move if the value of R were increased. Similarly, let a unit vector $\bar{e}_\phi$ point in the direction of increasing ϕ (perpendicular to $\bar{e}_R$), and let $\bar{e}_h$ point in the direction of increasing h. Thus, as we view Figure 11, $\bar{e}_R$ is perpendicular to the Z axis directed outward from point P, $\bar{e}_h$ is parallel to the Z axis, so that $\bar{e}_h \equiv \bar{K}$. Then, because $\bar{e}_R$, $\bar{e}_\phi$, and $\bar{e}_h$ are mutually perpendicular unit vectors, in accordance with the right-hand rule they are related by $\bar{e}_R \times \bar{e}_\phi = \bar{e}_h$. Note that with these definitions only $\bar{e}_h$ is a constant vector; the orientation of $\bar{e}_R$ and $\bar{e}_\phi$ will change as the angle ϕ changes.

Once the set of three independent unit vectors $\bar{e}_R$, $\bar{e}_\phi$, and $\bar{e}_h$ has been defined, it follows that we can express any vector in terms of their radial, transverse, and axial components. Using these unit vectors we shall now determine expressions for the position, velocity, and acceleration of a point.

In Figure 11 we see that the radial component of $\bar{r}_{P/O}$ is R, and the axial component of $\bar{r}_{P/O}$ is h. Note that $\bar{r}_{P/O}$ lies in the plane formed by $\bar{e}_R$ and $\bar{e}_h$. Hence there is no transverse component of position.

$$\bar{r}_{P/O} = R\bar{e}_R + h\bar{e}_h \tag{21}$$

Previously we showed that the values of R, ϕ, and h were required to define the position. Then why doesn't the value of ϕ appear in this expression for $\bar{r}_{P/O}$? The answer is that the equation for $\bar{r}_{P/O}$ really does depend on the value of ϕ, but not explicitly, because one must know ϕ in order to define (locate) the direction of $\bar{e}_R$.

Now let us find the velocity by differentiating $\bar{r}_{P/O}$ in equation (21).

$$\bar{v}_P = \dot{\bar{r}}_{P/O} = \dot{R}\bar{e}_R + R\dot{\bar{e}}_R + \dot{h}\bar{e}_h + h\dot{\bar{e}}_h \tag{22}$$

As in the case of path variables, we must determine the time derivatives of the nonconstant unit vectors. The unit vector $\bar{e}_h$ has constant direction, so $\dot{\bar{e}}_h = 0$. The unit vectors $\bar{e}_R$ and $\bar{e}_\phi$ depend on ϕ only. Using the chain rule with respect to ϕ yields

$$\dot{\bar{e}}_R \equiv \frac{d\bar{e}_R}{dt} = \frac{d\bar{e}_R}{d\phi}\frac{d\phi}{dt} = \dot{\phi}\,\frac{d\bar{e}_R}{d\phi}$$

$$\dot{\bar{e}}_\phi = \dot{\phi}\frac{d\bar{e}_\phi}{d\phi}$$

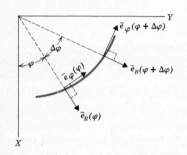

Figure 12a

Let us look down the Z axis in the negative direction and observe how $\bar{e}_R$ and $\bar{e}_\phi$ change when ϕ is increased by $\Delta\phi$, as shown in Figure 12a. The argument is exactly the same as was developed for unit vectors in path variables. First we form the difference vectors

$$\Delta\bar{e}_R = \bar{e}_R(\phi + \Delta\phi) - \bar{e}_R(\phi)$$

and

$$\Delta\bar{e}_\phi = \bar{e}_\phi(\phi + \Delta\phi) - \bar{e}_\phi(\phi)$$

Figure 12b

as shown in Figures 12b and 12c.

Because $\bar{e}_R(\phi)$ and $\bar{e}_R(\phi + \Delta\phi)$ are unit vectors, we can write

$$\left|\Delta\bar{e}_R\right| = 2\left|\bar{e}_R\right|\,\sin\frac{\Delta\phi}{2} = 2\,\sin\frac{\Delta\phi}{2}$$

Figure 12c

Dividing both sides by $\Delta\phi$ and taking the limit as $\Delta\phi \to 0$,

$$\frac{\left|d\bar{e}_R\right|}{d\phi} = \lim_{\Delta\phi \to 0}\frac{(\Delta\bar{e}_R)}{\Delta\phi} = \lim_{\Delta\phi \to 0}\frac{\sin(\Delta\phi/2)}{(\Delta\phi/2)} = 1$$

Therefore $d\bar{e}_R/d\phi$ is a unit vector. From Figure 12b we see that as $\Delta\phi$ goes to zero, $\Delta\bar{e}_R$ approaches the perpendicular to $\bar{e}_R(\phi)$. From Figure 12a we see that the perpendicular in question is in the $\bar{e}_\phi$ direction. Thus,

$$\frac{d\bar{e}_R}{dt} = \frac{d\bar{e}_R}{d\phi}\frac{d\phi}{dt} = \dot{\phi}\bar{e}_\phi \tag{23}$$

By similar reasoning we can show that

$$\frac{d\bar{e}_\phi}{dt} = \frac{d\bar{e}_\phi}{d\phi}\frac{d\phi}{dt} = -\dot{\phi}\bar{e}_R \tag{24}$$

We shall have use for equation (24) shortly. Placing equation (23) into equation (22) yields the standard form of the velocity equation,

$$\boxed{\bar{v}_P = \dot{R}\bar{e}_R + R\dot{\phi}\bar{e}_\phi + \dot{h}\bar{e}_h} \tag{25}$$

Equation (25) is the velocity equation in cylindrical coordinates. The radial component $v_R = \dot{R}$ occurs when the particle moves in the radial direction. The axial component $v_h = \dot{h}$ occurs when the particle moves in the axial direction. The transverse component $v_\phi = R\dot{\phi}$ occurs when ϕ is not a constant. We can see that, for a given value of $\dot{\phi}$, the particle will move faster for larger values of R. The quantity $\dot{\phi}$ is termed the *angular speed*.

We find the acceleration formula by differentiating the velocity (25).

$$\bar{a}_P = \dot{\bar{v}}_P = \ddot{R}\bar{e}_R + \dot{R}\dot{\bar{e}}_R + \dot{R}\dot{\phi}\bar{e}_\phi + R\ddot{\phi}\bar{e}_\phi + R\dot{\phi}\dot{\bar{e}}_\phi + \ddot{h}\bar{e}_h + \dot{h}\dot{\bar{e}}_h \tag{26}$$

Placing equations (23) and (24) into equation (26) and grouping like components gives

$$\boxed{\bar{a}_P = (\ddot{R} - R\dot{\phi}^2)\bar{e}_R + (R\ddot{\phi} + 2\dot{R}\dot{\phi})\bar{e}_\phi + \ddot{h}\bar{e}_h} \tag{27}$$

Let us look at the acceleration equation (27) by considering motion along some simple paths. If a particle moves along a radial line, then only R will change with time so that equation (27) reduces to $\bar{a}_P = \ddot{R}\bar{e}_R$. If a particle moves along a line parallel to the Z axis, only h changes with time and equation (27) reduces to $\bar{a}_P = \ddot{h}\bar{e}_h$. If a particle moves on a circular path of constant radius R parallel to the XY plane, then equation (27) reduces to

$$\bar{a}_P = -R\dot{\phi}^2\bar{e}_R + R\ddot{\phi}\bar{e}_\phi \tag{28}$$

The first term on the right-hand side of equation (28) is the centripetal component of acceleration normal to the path. The second term is the tangential component. For a general motion, equation (27) contains an additional effect. The radial acceleration is a combination arising from a varying radial distance and a centripetal-like motion. The transverse acceleration is a combination of the effect of angular acceleration and an as yet unexplained term $2\dot{R}\dot{\phi}$ which is an interaction brought about by a combination of radial and angular motion. This term is called the *Coriolis acceleration* effect, so named after G. Coriolis (1792–1843), who first disclosed its existence. In a later module we shall deal with this term at length. For now, recognize $2\dot{R}\dot{\phi}$ as the Coriolis component of acceleration, and take solace in the fact that like the other components in (27) it can be measured in field experiments.

EXAMPLE 8

A bicycle wheel rotates about its hub at a constant clockwise angular speed of 10 rad/s. A bead on a spoke approaches the hub with a constant speed of 5 m/s. Determine the acceleration of the bead as a function of the distance from the hub to the head.

Solution

The given information involves an angular speed and a rate of approach to a fixed axis. In addition, the solution desired is a function of the radial distance from that axis, and all the motion takes place in a plane. It follows quite naturally that we formulate the problem in terms of the special case of cylindrical coordinates in a plane, namely polar coordinates.

The accompanying sketch shows R and ϕ and the corresponding unit vectors for a general position. In general we know that

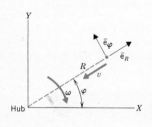

$$\bar{a}_P = (\ddot{R} - R\dot{\phi}^2)\bar{e}_R + (R\ddot{\phi} + 2\dot{R}\dot{\phi})\bar{e}_\phi$$

One important feature of the sketch is that $\bar{e}_R$ always points outward and $\bar{e}_\phi$ is perpendicular to $\bar{e}_R$ in the direction of increasing ϕ. The angular speed is the absolute value of $\dot{\phi}$, and the fact that the rotation is clockwise means that for the situation shown in the diagram ϕ is decreasing. Thus $\dot{\phi} = -10$ rad/s. It is given that the angular speed is constant, therefore $\ddot{\phi} = 0$. We are also told that the bead is approaching the hub at 5 m/s. The radial component of velocity $v_R = \dot{R}$ is the rate that a particle is moving away from the axis, so in this case $\dot{R} = -5$ m/s. Because this is a constant rate, $\ddot{R} = 0$. Thus the equation for $\bar{a}_P$ reduces to

$$\bar{a}_P = -R(10^2)\bar{e}_R + 2(-5)(-10)e_\phi$$

$$= -100R\bar{e}_R + 100\bar{e}_\phi \text{ m/s}^2$$

This answer could be converted to X and Y components by simply expressing the unit vectors $\bar{e}_R$ and $\bar{e}_\phi$ in terms of their components in the $\bar{I}$ and $\bar{J}$ directions, just as we did for the path variable unit vectors $\bar{e}_t$ and $\bar{e}_n$. However, a solution that shows the unit vectors for the descriptive method you choose and gives the magnitude of the components will usually be sufficient. Before leaving this problem, can you see what portions of the solution would have changed if we had defined ϕ as the clockwise angle?

EXAMPLE 9

A particle slides along a path that resembles a circular helix of radius R and vertical displacement $h = bt^2/2$. The velocity vector has a constant known magnitude v_P. (a) Find the acceleration. (b) Find the maximum time T for which such a motion is possible.

Solution

Part a

In general we know that

$$\bar{a}_P = (\ddot{R} - R\dot{\phi}^2)\bar{e}_R + (R\ddot{\phi} + 2\dot{R}\dot{\phi})\bar{e}_\phi + \ddot{h}\bar{e}_h$$

From the given information R is constant, so $\dot{R} = \ddot{R} = 0$. Therefore $\bar{a}_P$ reduces to

$$\bar{a}_P = -R\dot{\phi}^2\bar{e}_R + R\ddot{\phi}\bar{e}_\phi + \ddot{h}\bar{e}_h$$

We were given $h(t)$ and the fact that the speed is constant. We do not as yet know $\dot{\phi}$ and $\ddot{\phi}$. In terms of cylindrical coordinates we have

$$\bar{v}_P = \dot{R}\bar{e}_R + R\dot{\phi}\bar{e}_\phi + \dot{h}\bar{e}_h$$

$$v_P = \left|\bar{v}_P\right| = \sqrt{\bar{v}_P \cdot \bar{v}_P} = \sqrt{(\dot{R})^2 + (R\dot{\phi})^2 + (\dot{h})^2}$$

For the given motion we know that $\dot{h} = bt$, so that

$$v_P = \sqrt{(R\dot{\phi})^2 + b^2t^2}$$

Now, solving for $\dot{\phi}$ gives

$$\dot{\phi} = \frac{1}{R}\sqrt{v_p{}^2 - b^2t^2}$$

To find $\ddot{\phi}$ we differentiate this expression with respect to time.

$$\ddot{\phi} = -\frac{b^2t}{R\sqrt{v_P{}^2 - b^2t^2}}$$

We now can substitute the expressions for $\dot{\phi}$, $\ddot{\phi}$, and $\ddot{h}$ ($\ddot{h} = b$) into the equation for $\bar{a}_P$, with the result that

$$\bar{a}_P = -R\left(\frac{v_P{}^2 - b^2t^2}{R^2}\right)\bar{e}_R - \frac{b^2t}{\sqrt{v_P{}^2 - b^2t^2}}\bar{e}_\phi + b\bar{e}_n \qquad \triangleleft$$

Part b

In the solution for $\bar{a}_P$ in Part a, we saw that the transverse component of acceleration a_ϕ contains the square root of a number that decreases with time, specifically $\sqrt{v_P{}^2 - b^2t^2}$. The acceleration will cease to be a real quantity when $v_P{}^2 - b^2t^2 < 0$. Thus the largest allowable value $t = T$ is the one that gives

$$v_P{}^2 - b^2T^2 = 0$$

so that

$$T = \frac{v_P}{b}$$

The reason this situation occurs is that we required the particle to have a constant speed v_P, but the axial component $\dot{h}$ increases with time according to $\dot{h} = bt$. At $t = T\,(= v_P/b)$ we find that $\dot{h} = v_P$, so that at this instant the transverse component of velocity is zero. For $t > T$, the value of $\dot{h}$ would exceed this value of v_P, which is obviously an impossible situation because a component of a vector can never exceed the magnitude of the vector.

HOMEWORK PROBLEMS

II.25 The sliding block C is known to have the instantaneous velocity and acceleration illustrated in the diagram. Determine the corresponding values of $\dot{R}$, $\ddot{R}$, $\dot{\phi}$, and $\ddot{\phi}$ in this position.

II.26 The cylindrical coordinates (R, ϕ, h) of a particle are given by $R = 10t^2$, $\phi = 5t^3$, $h = 2t^4$, where R and h are expressed in meters, ϕ is in radians, and t is in seconds. (a) Determine the expressions for the velocity and acceleration of the particle. (b) Draw a sketch of the path.

II.27 The pendulum shown is moving to the right and slowing down. When $\theta = 15°$ it has a speed of 16 in./s and an acceleration of 20 in./s². Determine the angular speed $\dot{\theta}$ and angular acceleration $\ddot{\theta}$ at this instant.

II.28 The rotating arm AB guides pin P through the fixed spiral, which is defined by $R = 4\phi$, where R has units of inches and ϕ has units of radians. The arm has a constant counterclockwise angular speed of 16 rad/s. Determine the velocity and acceleration of the pin when $\phi = 90°$.

II.29 For the system of Problem II.28, determine the horizontal and vertical components of the velocity and acceleration when $\phi = 45°$.

II.30 When the ball is at the midpoint C of the ramp AB, it is falling at a speed of 10 ft/s and is gaining speed at the rate of 6 ft/s². For the coordinate system shown in the figure at the top of page 55, determine the cylindrical components of the position, velocity, and acceleration vectors.

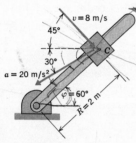

Prob. II.25

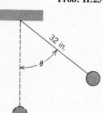

Prob. II.27

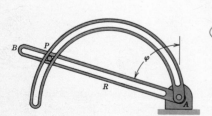

Prob. II.28

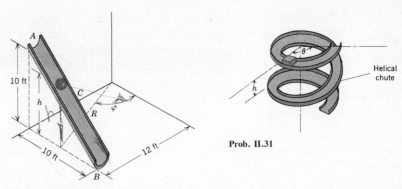

Prob. II.30

Prob. II.31

II.31 A package is descending a helical chute that is defined by $R = 0.8$ m, $h = 0.4\theta/2\pi$ m, where θ is measured in radians. The motion of the package is defined by $\theta = 0.5t + t^2$, where t is in seconds. Determine the acceleration of the package when $t = 1$ s.

II.32 The velocity and acceleration of a particle in planar motion are fully known in terms of cylindrical coordinates. (a) Show that the tangent and normal unit vectors are given by $\bar{e}_t = (v_R\bar{e}_R + v_\phi\bar{e}_\phi)/(v_R{}^2 + v_\phi{}^2)^{1/2}$ and $\bar{e}_n = (v_R\bar{e}_\phi - v_\phi\bar{e}_R)/(v_R{}^2 + v_\phi{}^2)^{1/2}$. (b) From the expressions in part (a), determine the expressions for the normal and tangential components of acceleration.

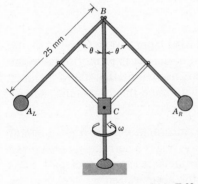

Prob. II.33

II.33 A governor consists of two heavy balls A on the end of 25-mm rods that are hinged to the vertical shaft at B. The position of the collar C on the vertical shaft may be adjusted to change the angle θ. The entire device rotates at a constant rate $\omega = 100$ rpm about the vertical. The angular speed $\dot{\theta}$ is a constant 0.2 rad/s. Determine the velocity and acceleration of each ball for $\theta = 30°$.

C. STRAIGHT AND CIRCULAR PATHS

Here we wish to consider the simple particle paths of a straight line and a circle. These paths are an important study unto themselves because of the frequency with which they occur in problems. Further, this study will help increase our understanding of the three kinematical descriptions we have previously presented.

1 Straight Path

Classically, the straight path case has the fancy title *rectilinear motion*. The simplest way of describing the motion, if the given information suits it, is to use path variables. We choose some reference point for $s = 0$ and

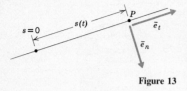

Figure 13

take $\bar{e}_t$ to be tangent to the straight line in the direction of increasing s, Figure 13. Because the straight line has an infinite radius of curvature, there is no center of curvature and we can choose $\bar{e}_n$ to be in any convenient direction perpendicular to $\bar{e}_t$, as is done in Figure 13. Then either from the figure and the fact that $\bar{e}_t$ is oriented in a fixed direction, or from the formulas derived for path variables, we find

$$\bar{r}_{P/O} = s\bar{e}_t$$

$$\bar{v}_P = \dot{\bar{r}}_{P/O} = \dot{s}\bar{e}_t$$

$$\bar{a}_P = \dot{\bar{v}}_P = \ddot{s}\bar{e}_t \tag{29}$$

from which we obtain the scalar identities

$$v_P = \dot{s} \qquad a_P = \dot{v}_P = \ddot{s} = v_P \frac{dv_P}{ds} \tag{30}$$

From equation (29) we may conclude that for rectilinear motion, the acceleration, as well as the velocity, is tangent to the path. It is for this reason that $a_P = \dot{v}_P$ in rectilinear motion, whereas for a general path we must use the full path variable equation, (13).

EXAMPLE 10

A car accelerates along a straight, level road from rest to a speed of 100 km/hr in a distance of 400 m. If the speed of the car is proportional to the distance the car has traveled, find the position along the 400-m distance where the acceleration is a maximum. Determine this maximum value.

Solution

We know that the path is a straight line. Information is also given about speed and distance traveled, specifically, $v_P = ks$, where k is a constant to be determined. This type of information indicates the use of path variables. For a straight path we know that the acceleration we seek may be written as $\bar{a}_P = \dot{v}_P\bar{e}_t$. We also know that

$$v_P = ks$$

so

$$\dot{v}_P = v_P \frac{dv_P}{ds} = v_P k = k^2 s$$

From the equation for v_P we can find k, because we were given that at $s = 400$ m,

$$v_P = 100 \ \frac{km}{hr}\left(\frac{1000 \ m}{1 \ km}\right)\left(\frac{1 \ hr}{3600 \ s}\right)$$

$$= 27.78 \ m/s$$

Hence,

$$27.78 = k(400)$$

or

$$k = 0.0695 \ 1/\text{seconds}$$

Then, substituting this value for k into the equation for $\dot{v}_P$, we find that

$$\dot{v}_P = (0.0695)^2 s \ \text{m/s}^2$$

From this we see that $\dot{v}_P$ is a maximum when $s = 400$ m and

$$\dot{v}_{P(\text{max})} = (0.0695)^2 400 = 1.932 \ \text{m/s}^2 \qquad \triangleleft$$

In the previous discussion we assumed that the information given for rectilinear motion best suited path variables. This, however, may not always be the case. For instance, we may have an inclined straight path in the vertical plane with information given about a horizontal or vertical component of velocity as a function of time, as shown in Figure 14. This information is most naturally suited for Cartesian coordinates.

To treat a problem like this we express $\bar{v}_P(t)$ by its Cartesian components,

$$\bar{v}_P(t) = v_P(t)(\cos \alpha \bar{I} + \sin \alpha \bar{K})$$

and then relate the velocity equation to the information given in the problem. Note too that if the constant angle $\alpha = 0$ or $90°$, the Cartesian unit vectors will be parallel to $\bar{e}_t$ and $\bar{e}_n$.

Similarly, we may have a situation of rectilinear motion with information that fits polar coordinates. Such a case is shown in Figure 15, where the information given for the same inclined path problem would be the angle $\theta(t)$ locating the position vector from the X axis. We then can follow the method derived for cylindrical coordinates in planar motion (changing the symbol ϕ in the formulas to θ). The appropriate unit vectors are shown in Figure 15, and equations (21), (25), and (27) may be used to express the position, velocity, and acceleration.

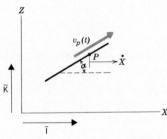

Figure 14

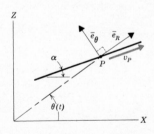

Figure 15

2 Circular Path

The first question that arises when encountering a problem containing a circular path is what method of description we should use. Those problems that are best suited for Cartesian coordinates will be quite obvious, because the given information will involve the Cartesian components of position, velocity, or acceleration. Therefore let us now consider how to determine when path variables or cylindrical coordinates are most suitable.

We have already treated the use of path variables for circular paths in Section II.B.4. As is true for all problems, the fact that the problem is

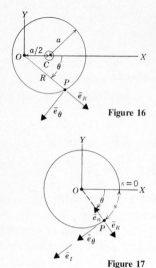

Figure 16

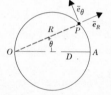

Figure 17

posed in terms of distance traveled or speed will indicate that path variables are preferable. The key for the use of cylindrical coordinates in circular motion will be that the significant parameters are radial distance from a fixed point to the particle and an angle or angular speed for this radial line. A subtle aspect of the foregoing is that the fixed point need not necessarily be the center of the circle, as is illustrated in Figure 16.

If we were to draw the path variable unit vectors $\bar{e}_t$ and $\bar{e}_n$, for point P in Figure 16, $\bar{e}_t$ would be tangent to the circle and $\bar{e}_n$ would be perpendicular to $\bar{e}_t$ pointing toward the geometric center of the circle C. Thus, in this case neither of the path variable unit vectors would be parallel to the cylindrical coordinate unit vectors. However, if in some problem the fixed point O is the center C, we have the situation shown in Figure 17, where the sense of the direction of $\bar{e}_t$ is dictated by the way the arc length s is measured. Hence, in this case, $\bar{e}_t \equiv \bar{e}_\theta$ and $\bar{e}_n \equiv -\bar{e}_R$.

EXAMPLE 11

A particle moves on the circular path shown with constant angular speed $\dot{\theta}$, where θ is defined in the sketch. Given that $v_P = v_0$ when the particle passes through point A, determine the velocity and acceleration in terms of the angle θ.

Solution

In this case the fact that we are given information regarding the angle θ, specifically that $\dot{\theta}$ is constant, indicates that we can use cylindrical coordinates. We have shown R, $\bar{e}_R$, and $\bar{e}_\theta$ in the sketch accompanying the problem. From this diagram we may write

$$\bar{r}_{P/O} = R\bar{e}_R$$

and because angle OPA is a right angle,

$$\cos \theta = \frac{R}{D} \quad \text{or} \quad R = D \cos \theta$$

As always in problem solving, we have the choice of substituting into the derived formula for velocity or differentiating the position vector using the formulas for the rates of change of the unit vectors. Usually the former approach is faster, but for variety let us follow the latter, recalling that $\dot{\bar{e}}_R = \dot{\theta}\bar{e}_\theta$, and $\dot{\bar{e}}_\theta = -\dot{\theta}\bar{e}_R$. In either approach we shall need to use the chain rule because time is given implicitly. That is, R is a function of θ and θ is a function of time.

$$\bar{v}_P = \dot{\bar{r}}_{P/O} = \frac{d}{dt}(D \cos \theta \bar{e}_R) = -(D \sin \theta)\dot{\theta}\bar{e}_R + D \cos \theta \dot{\bar{e}}_R$$

$$= \dot{\theta}D(- \sin \theta \bar{e}_R + \cos \theta \bar{e}_\theta)$$

Noting that $\dot{\theta}$ is a constant, we then find that $\ddot{\theta} = 0$. Thus when we differentiate $\bar{v}_P$ to find $\bar{a}_P$, we obtain

$$\bar{a}_P = \dot{\bar{v}}_P = \dot{\theta}D[-(\cos \theta)\dot{\theta}\bar{e}_R - \sin \theta \dot{\bar{e}}_R$$
$$- (\sin \theta)\dot{\theta}\bar{e}_\theta + \cos \theta \dot{\bar{e}}_\theta]$$
$$= -2\dot{\theta}^2 D(\cos \theta \bar{e}_R + \sin \theta \bar{e}_\theta)$$

We still must determine $\dot{\theta}$. A piece of information we have not used is that $v_P = v_0$ at point A. From our result for $\bar{v}_P$ we can form

$$v_P = |\bar{v}_P| = \dot{\theta}D\sqrt{\sin^2 \theta + \cos^2 \theta} = \dot{\theta}D$$

From this we find that v_P is constant because $\dot{\theta}$ and D are both constants. It was given that $v_P = v_0$ at point A, so we can solve for $\dot{\theta}$.

$$\dot{\theta} = \frac{v_0}{D}$$

Placing this into our result for $\bar{a}_P$ gives

$$\bar{a}_P = -\frac{2v_0{}^2}{D}(\cos \theta \bar{e}_R + \sin \theta \bar{e}_\theta)$$

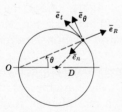

Let us check this result by using path variables. For constant speed,

$$\bar{a}_P = \frac{v_P{}^2}{\rho}\bar{e}_n = \frac{v_0{}^2}{\rho}\bar{e}_n$$

For a circle, the radius of curvature is the radius of the circle, so that $\rho = D/2$. From the sketch we have

$$\bar{e}_n = -\cos \theta \bar{e}_R - \sin \theta \bar{e}_\theta$$

Substituting for ρ and $\bar{e}_n$ in the path variable expression gives

$$\bar{a}_P = \frac{v_0{}^2}{(D/2)}(-\cos \theta \bar{e}_R - \sin \theta \bar{e}_\theta)$$
$$= -\frac{2v_0{}^2}{D}(\cos \theta \bar{e}_R + \sin \theta \bar{e}_\theta)$$

This is the same result as the one we obtained using cylindrical coordinates. If you were skeptical earlier when we said that the resulting *vectors* for velocity and acceleration are the same regardless of what kinematical description is used, you should believe it now!

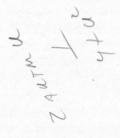

D. PROBLEM SOLVING

By now you should have a reasonable feel for the meaning of the variables and unit vectors that appear in the three kinematical descriptions presented. Problem solving using these descriptions (i.e., picking the right one) is not an art, but rather a professional decision made by recognizing the proper keys. In what follows we present a logical way of thinking to identify the keys. It is not meant for memorization, but rather as a guide while working problems. Work enough problems *correctly* and the key will become part of your thinking process.

1 Read the problem carefully and express all quantitative information in equation form, at the same time converting the information to a consistent set of units. For instance, if a problem states "After the particle has traveled 700 ft, it has a speed of 30 miles/hr," you should write down

when $s = 700$ ft,

$$v_P = 30 \, \frac{\text{miles}}{\text{hr}} \left(\frac{88 \text{ ft/s}}{60 \text{ miles/hr}} \right) = 44 \text{ ft/s}$$

2 If the path is defined but no diagram is provided, make a sketch of the path.
3 Choose path variables, Cartesian coordinates, or cylindrical coordinates. Make this choice based on which formulation most closely fits the information given in the problem and, equally important, based on the solution variables whose values are sought. Select the coordinate parameters and unit vectors and show them in the sketch. Now, in terms of these coordinates, write general geometric relationships for the lengths and angles.
4 Write down the formulas in the kinematical description you have chosen that relate the given information to the variables of the solution. After studying the formulas and the given information, from the sketch write any additional information needed to relate the system's geometry.
5 Solve the appropriate equations. In the illustrative problems, we shall refer to these steps by their appropriate numbers.

ILLUSTRATIVE PROBLEM 1

A slider moves in the vertical plane along a wire that is bent into a hyperbolic shape according to the equation $XZ = 36$ m². Determine the horizontal and vertical components of the acceleration of the slider when it is at $X = 6$ m if (a) the slider has a constant speed of 10 m/s directed as shown, and (b) the slider has a constant vertical component of velocity of 10 m/s downward.

Solution

In both parts of this problem we know the path and desire the Cartesian components of the acceleration. However, in (a) the speed is given, which suggests the

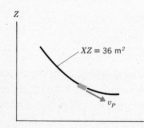

use of path variables. In (b), a Cartesian component of velocity is given, which suggests the use of Cartesian coordinates.

Part a

Step 1 $XZ = 36$ m², find a_X and a_Z when $X = 6$ m, given $v_P = 10$ m/s, and $\dot{v}_P = 0$ because v_p is constant.

Steps 2 and 3   Use path variables.

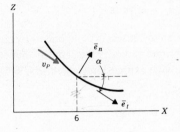

$$Z = \frac{36}{X}$$

$$\alpha = \tan^{-1}\left|\frac{dZ}{dX}\right|$$

Step 4

$$\bar{a} = \dot{v}_P\bar{e}_t + \frac{v_P^2}{\rho}\bar{e}_n$$

We know v_P and $\dot{v}_P$, but we must find the curvature $1/\rho$. This is motion in a plane, so

$$\frac{1}{\rho} = \frac{\left|d^2Z/dX^2\right|}{[1 + (dZ/dX)^2]^{3/2}}$$

Step 5

$$Z = \frac{36}{X} \qquad \frac{dZ}{dX} = \frac{-36}{X^2} \qquad \frac{d^2Z}{dX^2} = \frac{72}{X^3}$$

When $X = 6$,

$$\frac{1}{\rho} = \frac{(72/6^3)}{[1 + (36/6^2)^2]^{3/2}} = 0.118 \ 1/\text{m}$$

$$\bar{a}_P = 0\bar{e}_t + 10^2(0.1180)\bar{e}_n = 11.80\bar{e}_n \ \text{m/s}^2$$

In most cases we may consider the solution to be completed if we have shown the unit vectors in a diagram and determined the corresponding components. However, in this problem we are requested to express $\bar{a}_P$ by its $\bar{I}$ and $\bar{K}$ components. To do this, we express $\bar{e}_n$ (and $\bar{e}_t$ if it were necessary) by its Cartesian components. From the diagram (Steps 2 and 3 above) we have

$$\bar{e}_n = \cos(90° - \alpha)\bar{I} + \sin(90° - \alpha)\bar{K}$$

but

$$\alpha = \tan^{-1}\left|\frac{-36}{6^2}\right| = \tan^{-1}(1.0) = 45°$$

so that

$$\bar{e}_n = 0.707\bar{I} + 0.707\bar{K}$$

$$\bar{a}_P = 11.80 \times 0.707(\bar{I} + \bar{K}) = 8.34(\bar{I} + \bar{K})$$

Thus

$$a_X = 8.34 \text{ m/s}^2 \qquad a_Z = 8.34 \text{ m/s}^2$$

As an aside, you can verify the transformation to Cartesian components by checking to see that the magnitude of the transformed vector is the same as that of the original vector. In this problem the path variable expression yielded $|\bar{a}_P| = 11.80$, whereas the Cartesian expression gives

$$|\bar{a}_P| = \sqrt{(8.34)^2(2)} = 11.80$$

Thus, the transformation checks.

Part b

Step 1 $XZ = 36$ m², find a_X and a_Z when $X = 6$ m, given $v_Z = \dot{Z} = -10$ m/s. $\dot{Z}$ is negative because the particle is moving down and Z is decreasing, and $\ddot{Z} = 0$ because $\dot{Z}$ is constant.

Steps 2 and 3   Use Cartesian coordinates.

Step 4 $\bar{a}_P = \ddot{X}\bar{I} + \ddot{Z}\bar{K}$, $\dot{Z} = -100$ m/s², $\ddot{Z} = 0$. From the geometry of the problem we may write $X = 36/Z$. Differentiating twice yields

Step 5

$$\dot{X} = -\frac{36}{Z^2}\dot{Z} \qquad \ddot{X} = \frac{72}{Z^3}(\dot{Z})^2 - \frac{36}{Z^2}\ddot{Z}$$

Thus, when $X = 6$ m, $Z = 36/6 = 6$ m, $\ddot{X} = 72(10^2/6^3) = 33.3$ m/s² and

$$\bar{a}_P = 33.3 \text{ m/s}^2\, \bar{I}$$

Note that the difference between the accelerations in Parts a and b is substantial, even though the given information seems so similar.

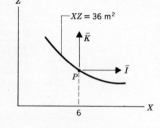

ILLUSTRATIVE PROBLEM 2

A high-speed railroad train travels along a straight track at a speed of 50 m/s. A photographer stands 10 m from the track and wishes to pan (rotate) the camera so that the camera always points directly at the locomotive. Determine what values for the rate of rotation ω of the camera and rate of increase of ω are necessary to accomplish this.

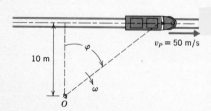

Solution

Here two methods come to mind. One obvious method to solve the problem is to express ϕ in terms of the distance traveled by the locomotive, and then differentiate the expression to find expressions for $\omega(= \dot\phi)$ and $\dot\omega$ as a function of ϕ. This approach is fine, but the derivatives will be complicated. Another approach, the one we shall follow here, does not require differentiation.

Step 1 $v_P = 50$ m/s, $\dot{v}_P = 0$ along a straight path. The distance from fixed point O (photographer's location) to the path is 10 m. Find ω and $\dot\omega$ for the radial line from point O to the locomotive.

Steps 2 and 3   Use cylindrical coordinates with $h = 0$ (plane polar coordinates), because an angular speed and angular acceleration of a radial line are desired.

Step 4

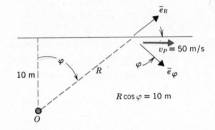

$$\bar{v}_P = \dot{R}\bar{e}_R + R\dot\phi\bar{e}_\phi$$

$$\bar{a}_P = (\ddot{R} - R\dot\phi^2)\bar{e}_R + (R\ddot\phi + 2\dot{R}\dot\phi)\bar{e}_\phi$$

$R = \dfrac{10}{\cos\phi}$

In terms of the problem variables, $\dot\phi = \omega, \ddot\phi = \dot\omega$.

Step 5   We know that $\bar{v}_P$ is tangent to the path and that $\bar{a}_P = \bar{0}$ (because $\bar{v}_P$ is constant in magnitude and direction). From the sketch we can express $\bar{v}_P$ by its polar components.

$$\bar{v}_P = 50(\sin \phi\bar{e}_R + \cos \phi\bar{e}_\phi)$$

But the polar coordinates formula gives

$$\bar{v}_P = \dot{R}\bar{e}_R + R\omega\bar{e}_\phi$$

Thus matching components yields

$$\bar{e}_R \text{ direction:} \qquad \dot{R} = 50 \sin \phi$$

$$\bar{e}_\phi \text{ direction:} \qquad R\omega = 50 \cos \phi$$

From step 3 we have $R = 10/\cos \phi$, so that

$$\dot{R} = 50 \sin \phi \text{ m/s}$$

and

$$\omega = \frac{50 \cos \phi}{R} = 5 \cos^2 \phi \text{ rad/s}$$

Next, we know that $\bar{a}_P = \bar{0}$, so that the transverse component of $\bar{a}_P$ is zero. Hence from the polar coordinate formula we have

$$a_R = R\dot{\omega} + 2\dot{R}\omega = 0$$

$$\dot{\omega} = -\frac{2\dot{R}\omega}{R} = \frac{-2 \times 50 \sin\phi \times 5 \cos^2\phi}{(10/\cos\phi)}$$

$$\dot{\omega} = -50 \sin\phi \cos^3\phi \text{ rad/s}^2$$

ILLUSTRATIVE PROBLEM 3

A race car traveled at 200 miles/hr along a straight path and then entered a 1000-ft radius curve. Immediately upon entering the curve, the driver applied his brakes and the speed decreased at a $2g$ rate ($g = 32.17$ ft/s²). Determine the maximum total acceleration of the car and where this maximum occurred. Draw a sketch of this maximum acceleration.

Solution

Step 1 Initially,

$$v = 200 \frac{\text{miles}}{\text{hr}} \left(\frac{88 \text{ ft/s}}{60 \text{ miles/hr}} \right) = 293.3 \text{ ft/s}$$

Also, $\dot{v} = -2g = -64.34$ ft/s² ($\dot{v}$ is negative, because v is decreasing). The path has a constant radius of 1000 ft, so it is a circle. Find $\left|\bar{a}_P\right|_{\max}$ and the corresponding location, and draw a sketch of $\bar{a}_P$.

Steps 2 and 3 We use path variables because we know the path and have information about the speed and rate of increase of the speed.

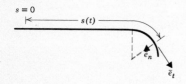

Step 4

$$\bar{a}_P = \dot{v}_P \bar{e}_t + \frac{v_P^2}{\rho} \bar{e}_n$$

Step 5 $\dot{v}_P = -64.34$ ft/s² and $\rho = 1000$ ft. Then the magnitude of the acceleration is

$$\left|\bar{a}_P\right| = \sqrt{\dot{\bar{v}}_P \cdot \dot{\bar{v}}_P} = \sqrt{(\dot{v}_P)^2 + \left(\frac{v_P^2}{\rho}\right)^2}$$

$$= \sqrt{(64.34)^2 + \left(\frac{v_P^2}{1000}\right)^2}$$

Clearly $\left|\bar{a}_P\right|$ is a maximum when v_P is a maximum. This occurs at the beginning of

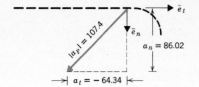

the curve, where $v_P = 293.3$ ft/s. Thus

$$a_n = \frac{v_P^2}{\rho} = \frac{(293.3)^2}{1000} = 86.02 \text{ ft/s}^2$$

$$a_t = \dot{v}_P = -64.34 \text{ ft/s}^2$$

$$\left| \bar{a}_P \right| = \sqrt{(64.34)^2 + (86.02)^2} = 107.4 \text{ ft/s}^2$$

ILLUSTRATIVE PROBLEM 4

The voltage applied to a cathode-ray tube is adjusted so that the position coordinates of an electron vary with time according to $Z = At$, $X = A \cos pt$, $Y = A \sin bt$. Show (a) that for this motion the acceleration of the electron is always parallel to the perpendicular line from the electron to the Z axis, and (b) that the magnitude of the acceleration is proportional to the length of this line.

Solution

Step 1   We are given $X(t)$, $Y(t)$, and $Z(t)$ and wish to determine some properties of $\bar{a}_P$.

Steps 2 and 3   We have the Cartesian coordinates of the electron, so clearly we shall adopt them for the kinematical description.

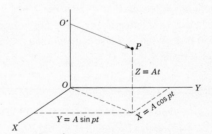

Step 4 $\bar{r}_{P/O} = X\bar{I} + Y\bar{J} + Z\bar{K}$ with $X(t)$, $Y(t)$, and $Z(t)$ given. We want to show that $\bar{a}_P$ is parallel to $\bar{r}_{P/O'}$ and that $\left| \bar{a}_P \right|$ is proportional to $\left| \bar{r}_{P/O'} \right|$. Hence the kinematical expression we want can be obtained by straightforward differentiation of the position vector, bearing in mind that $\bar{I}$, $\bar{J}$, and $\bar{K}$ are constant.

Step 5   Differentiating twice yields

$$\bar{r}_{P/O} = X\bar{I} + Y\bar{J} + Z\bar{K}$$

$$= A \cos pt\bar{I} + A \sin pt\bar{J} + At\bar{K}$$

$$\bar{v}_P = -Ap \sin pt\bar{I} + Ap \cos pt\bar{J} + A\bar{K}$$

$$\bar{a}_P = -Ap^2 \cos pt\bar{I} - Ap^2 \sin pt\bar{J}$$

Part a

From the sketch we see that

$$\bar{r}_{P/O'} = X\bar{I} + Y\bar{J}$$

$$= A \cos pt\bar{I} + A \sin pt\bar{J}$$

Therefore

$$\bar{a}_P = -p^2\bar{r}_{P/O'}$$

This shows that $\bar{a}_P$ is in the opposite direction from $\bar{r}_{P/O'}$.

Part b

Taking the absolute value of both sides of the result for $\bar{a}_P$, we have

$$|\bar{a}_P| = p^2|\bar{r}_{P/O'}|$$

and we see that the magnitude of the acceleration is proportional to $|\bar{r}_{P/O'}|$.

This last portion of the problem may seem trivial to you, but it is important because it is one of the things that differentiates the technician from the engineer: We did not leave the answer in the form of a number, but went to the further effort of interpreting the result.

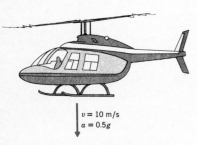

ILLUSTRATIVE PROBLEM 5

A helicopter descends at a speed of 10 m/s with an acceleration of $0.5g$. The main rotor blades are 5 m long, oriented horizontally, and are rotating at a constant rate of 450 rpm. What is the speed and magnitude of the acceleration of a point on the tip of a blade?

$v = 10$ m/s
$a = 0.5g$

Solution

Step 1 Given $v_h = 10$ m/s; $a_h = 0.5g[(9.806 \text{ m/s}^2)/1g] = 4.903 \text{ m/s}^2$, both down, $\omega_{\text{rotor}} = 450$ rpm $(2\pi \text{ rad/rev})$ $(1 \text{ min/60 s}) = 15\pi$ rad/s, $\dot{\omega} = 0$. (Note that the angular units for rotation must be radians). Find $|\bar{v}_P|$ and $|\bar{a}_P|$.

Steps 2 and 3 The motion of the axis of the rotor shaft is the same as that of the helicopter. Therefore, we know the rate of rotation of a radial line from the rotor axis to a blade tip and the motion of the tip parallel to the rotor shaft axis. This is ideal for cylindrical coordinates. If we measure h positive downward, then $\bar{e}_h$ is parallel to the velocity and acceleration of the helicopter. To produce a right-handed system we now measure positive ϕ in the direction of rotation.

Step 4

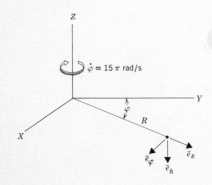

$$\bar{v}_P = \dot{R}\bar{e}_R + R\dot{\phi}\bar{e}_\phi + \dot{h}\bar{e}_h$$

$$|\bar{v}_P| = \sqrt{\bar{v}_P \cdot \bar{v}_P}$$

$$\bar{a}_P = (\ddot{R} - R\dot{\phi}^2)\bar{e}_R + (R\ddot{\phi} + 2\dot{R}\dot{\phi})\bar{e}_\phi + \ddot{h}\bar{e}_h$$

$$|\bar{a}_P| = \sqrt{\bar{a}_P \cdot \bar{a}_P}$$

$\dot{\phi} = 15\pi$ rad/s

Step 5

$$R = 5 \text{ m} \qquad \dot{R} = \ddot{R} = 0$$

$$\dot{\phi} = 15\pi \text{ rad/s}, \ \ddot{\phi} = 0$$

$$\dot{h} = 10 \text{ m/s} \qquad \ddot{h} = 4.90 \text{ m/s}^2$$

$$\bar{v}_P = 5(15\pi)\bar{e}_\phi + 10\bar{e}_h, \ |\bar{v}_P| = \sqrt{(75\pi)^2 + 10^2}$$

$$= 235.8 \text{ m/s}$$

$$\bar{a}_P = -5(15\pi)^2\bar{e}_R + 4.90\bar{e}_h$$

$$|\bar{a}_P| = \sqrt{(1125\pi^2)^2 + (4.90)^2} = 1.110(10^4) \text{ m/s}^2$$

ILLUSTRATIVE PROBLEM 6

In order to ease the shock on packages moving around a curve, a conveyor belt is designed with a transition curve in which the radius of curvature gets smaller as the package moves along the curve. The radius of curvature is given in graphical form in the diagram as a function of the distance s the package has traveled along the conveyor belt. Draw a graph of the acceleration in g's that the package undergoes as a function of s when the conveyor belt is moving at 100 in./s.

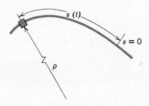
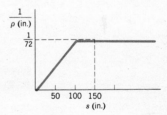

Solution

Step 1 $v_P = 100$ in./s, $\dot{v}_P = 0$, $\rho(s)$ given. Find $|\bar{a}_P|$ in units of g.

Steps 2 and 3 We are given the speed and radius of curvature as a function of distance traveled (arc length). This information fits path variables.

Step 4

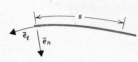

$$\bar{a}_P = \dot{v}_P\bar{e}_t + \frac{v_P^2}{\rho}\bar{e}_n$$

Step 5

$$\bar{a}_P = 0\bar{e}_t + \frac{(100)^2}{\rho}\bar{e}_n \text{ in./s}^2$$

To find $|\bar{a}_P|$ in units of g, divide by

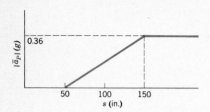

$$g = 32.17 \ \frac{\text{ft}}{\text{s}^2} \left(\frac{12 \text{ in.}}{1 \text{ ft}} \right) = 386 \text{ in./s}^2$$

Then

$$|\bar{a}_P| = \frac{10,000}{386\rho} g \qquad (\text{units for } \rho \text{ are inches})$$

This results in the graph shown.

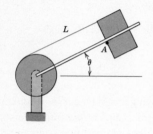

ILLUSTRATIVE PROBLEM 7

A portion of a machine consists of a block that is free to slide along a rotating rod. The block is tied by means of an inextensible string that is wrapped around a fixed drum, as shown. The length L of the string shown varies according to $L = 5 + 0.3\theta$ (L in meters and θ in radians). If θ varies according to the formula $\theta = (\pi/6)(3 - 2t^2)$, what is the acceleration of point A on the block when $t = 1$ s?

Solution

Step 1 $L = 5 + 0.3\theta$, $\theta = (\pi/6)(3 - 2t^2)$, plane motion. Find $\bar{a}_P$.

Steps 2 and 3 We are given the radial distance from a fixed point and the angle of rotation of the radial line. Therefore we can use polar coordinates, changing the symbols for these coordinates from R, ϕ to L, θ.

Step 4

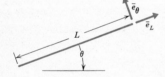

$$\bar{a}_P = (\ddot{L} - L\dot{\theta}^2)\bar{e}_L + (L\ddot{\theta} + 2\dot{L}\dot{\theta})\bar{e}_\theta$$

Step 5 We have $L(\theta)$ and $\theta(t)$, which suggests the use of the chain rule to find $\dot{L}$, $\ddot{L}$. For instance,

$$\dot{L} = \dot{\theta}\frac{dL}{d\theta}$$

A more direct alternative is to substitute $\theta(t)$ into $L(\theta)$ to find

$$L(t) = 5 + 0.3 \times \frac{\pi}{6}(3 - 2t^2)$$

Continuing with the latter,

$$\dot{L} = -0.2\pi t \qquad \ddot{L} = -0.2\pi$$

$$\dot{\theta} = -\frac{2\pi}{3}t \qquad \ddot{\theta} = -\frac{2\pi}{3}$$

When $t = 1$ s, $L = 5 + 0.05\pi = 5.157$ m, $\dot{L} = -0.2\pi$ m/s, $\ddot{L} = -0.2\pi$ m/s², $\theta = \pi/6$ rad = 30°, $\dot{\theta} = -2\pi/3$ rad/s, $\ddot{\theta} = -2\pi/3$ rad/s².

$$\bar{a}_P = a_L \bar{e}_L + a_\theta \bar{e}_\theta$$

$$\bar{a}_P = \left[-0.2\pi - 5.157\left(\frac{2\pi}{3}\right)^2 \right]\bar{e}_L + \left[5.157\left(\frac{-2\pi}{3}\right) \right.$$

$$\left. + 2(-0.2\pi)\left(-\frac{2\pi}{3}\right) \right]\bar{e}_\theta$$

$$= -23.25\bar{e}_L - 8.17\bar{e}_\theta \text{ m/s}^2$$

To convert to Cartesian components, if so desired, we can express (from the sketch) $\bar{e}_L$ and $\bar{e}_\theta$ in terms of their Cartesian components.

$$\bar{e}_L = \cos 30°\bar{I} + \sin 30°\bar{J}$$

$$\bar{e}_\theta = -\sin 30°\bar{I} + \cos 30°\bar{J}$$

$$\bar{a}_P = (a_L \cos 30° - a_\theta \sin 30°)\bar{I}$$

$$+ (a_L \sin 30° + a_\theta \cos 30°)\bar{J}$$

As we have pointed out a number of times, unless expressly asked for, either form will suffice.

ILLUSTRATIVE PROBLEM 8

A ball is thrown at an initial angle of elevation α_0 with an initial speed v_0. The horizontal and vertical distances traveled by the ball in terms of the coordinates shown are given by

$$X = (v_0 \cos \alpha_0)t \qquad Y = (v_0 \sin \alpha_0)t - \tfrac{1}{2}gt^2$$

(We shall show how to obtain this result in the next module.) Determine the radius of curvature of the path and the rate of increase of the speed at the instant of release ($t = 0$).

Solution

Step 1 $X = (v_0 \cos \alpha_0)t$, $Y = (v_0 \sin \alpha_0)t - \tfrac{1}{2}gt^2$. Find ρ and $\dot{v}_P$ when $t = 0$.

Steps 2 and 3 If we wished, we could approach this problem by a scalar method as follows. To find ρ we could determine the equation of the path, that is, Y as a function of X, and then use the formula for ρ. To find $\dot{v}_P$ we could write a formula for v_P as a function of time (using $v_P = \sqrt{(\dot{X})^2 + (\dot{Y})^2}$), and then differentiate.

Such a solution would result in a cumbersome set of calculations. We can obtain an easier solution if we note that $\dot{v}_P$ and ρ both occur in the acceleration formula for path variables. In this approach we relate the acceleration of the particle in Cartesian coordinates to the description in path variables. Let us pursue this method.

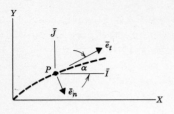

Step 4

$$\bar{v}_P = \dot{X}\bar{I} + \dot{Y}\bar{J} = v_P\bar{e}_t$$

$$\bar{a}_P = \ddot{X}\bar{I} + \ddot{Y}\bar{J} = \dot{v}_P\bar{e}_t + \frac{v_P^2}{\rho}\bar{e}_n$$

Step 5

$$\dot{X} = v_0 \cos \alpha_0 \qquad \dot{Y} = v_0 \sin \alpha_0 - gt$$

$$\ddot{X} = 0 \qquad \ddot{Y} = -g$$

Thus

$$\bar{v}_P = v_0 \cos \alpha_0 \bar{I} + (v_0 \sin \alpha_0 - gt)\bar{J}$$

$$\bar{a}_P = -g\bar{J}$$

$$v_P = |\bar{v}_P| = \sqrt{(v_0 \cos \alpha_0)^2 + (v_0 \sin \alpha_0 - gt)^2}$$

$$= \sqrt{v_0^2 - 2v_0 gt \sin \alpha_0 + g^2 t^2}$$

We can now find the tangent and normal components of $\bar{a}_P$ with the aid of the sketch shown above.

$$\bar{J} = \sin \alpha \bar{e}_t - \cos \alpha \bar{e}_n$$

$$\bar{a}_P = -g(\sin \alpha \bar{e}_t - \cos \alpha \bar{e}_n) = -g \sin \alpha \bar{e}_t + g \cos \alpha \bar{e}_n$$

$$= \dot{v}_P \bar{e}_t + \frac{v_P^2}{\rho}\bar{e}_n$$

Thus, by equating components,

$$\bar{e}_t \text{ direction:} \qquad \dot{v}_P = -g \sin \alpha$$

$$\bar{e}_n \text{ direction:} \qquad \frac{v_P^2}{\rho} = g \cos \alpha$$

At the initial position $\alpha = \alpha_0$ and $v_P = v_0$; the results are

$$\dot{v}_P = -g \sin \alpha_0 \qquad \rho = \frac{v_0^2}{g \cos \alpha_0}$$

If the results at a general instant were desired, the angle α could be found by expressing $\bar{e}_t$ by its Cartesian components, that is, $\bar{e}_t = \cos \alpha \bar{I} + \sin \alpha \bar{J}$, so that

$$\bar{v}_P = v_P \bar{e}_t = v_P \cos \alpha \bar{I} + v_P \sin \alpha \bar{J} = \dot{X}\bar{I} + \dot{Y}\bar{J}$$

$$\tan \alpha = \frac{v_P \sin \alpha}{v_P \cos \alpha} = \frac{\dot{Y}}{\dot{X}} = \frac{v_0 \sin \alpha_0 - gt}{v_0 \cos \alpha_0}$$

This expression for α can be checked by setting $t = 0$, from which we obtain an identity for $\tan \alpha_0$. Noting that the value of v_P can be found from the derived result, in general we have

$$\dot{v}_P = -g \sin \alpha$$

$$\rho = \frac{v_P^2}{g \cos \alpha}$$

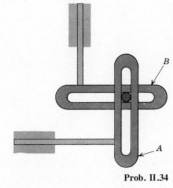

Prob. II.34

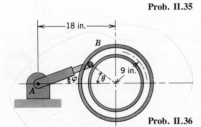

Prob. II.35

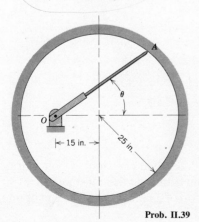

Prob. II.36

Prob. II.39

HOMEWORK PROBLEMS

II.34   A pin slides in two yokes as shown. Yoke A has a velocity of 50 mm/s to the right and an acceleration of 20 mm/s² to the left whereas yoke B has a constant downward velocity of 10 mm/s. For the position shown determine the radius of curvature of the pin's path and sketch where the center of curvature is located.

II.35   A *lazy tongs* consisting of bars 1 m long is pivoted at a fixed point O. The pin A is constrained to move along OA. Find the velocity and acceleration of pin A when $\theta = 47°$ and $\dot{\theta}$ is a constant 3 rad/s clockwise. *Hint:* Express the distance from point O to point A in terms of the angle θ.

II.36   A hydraulic piston inside arm AB causes the length of the arm to increase at the constant rate of 3 in./s. Determine (a) the velocity and acceleration of the pin where $\theta = 90°$, (b) the corresponding angular velocity $\dot{\phi}$ and the angular acceleration $\ddot{\phi}$ of arm AB.

II.37   Solve Problem II.36 for the position where $\theta = 120°$.

II.38   The acceleration of a satellite in a circular orbit of radius R around the earth is $g(R_e/R)^2$, where $R_e = 6370$ km is the radius of the earth and $g = 9.806$ m/s². Determine the value of R for which the satellite is in a synchronous orbit, which is an orbit that requires 1 day for a complete revolution. Also determine the corresponding speed of the satellite. (Consider a day to be 24 hr.)

II.39   The angle of rotation of arm OA about the eccentric pivot point O is defined by $\theta = 200t$, where θ is in radians and t is in seconds. A spring within the telescoping arm keeps end A in contact with the circular hole. Determine the velocity and acceleration of point A when $\theta = 90°$.

II.40   Solve Problem II.39 for the instant when $\theta = 30°$.

II.41   At the instant shown the telescoping passenger ramp is 45 m long and is being extended at the constant rate of 0.5 m/s. What is the constant rotation rate $\dot{\phi}$ that will produce an acceleration of 1 m/s² at end A at this instant? What is the corresponding speed of end A?

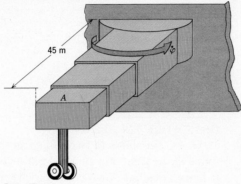

45 m

A

Prob. II.41

II.42   It is known that the speed of a rocket sled on a straight track is related to the distance it travels by $v = 2000s - 500s^2$, where v is in kilometers per hour and s is in km. Determine (a) the maximum speed and the location where it occurs, (b) the maximum acceleration and the location where it occurs, (c) the location where the sled comes to rest and the acceleration in that position.

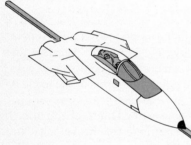

Prob. II.42

II.43   A block slides along the bent road $ABCD$ such that the distance traveled is given by $s = 30t + 5t^2$, where s has units of inches and t is seconds. Determine the acceleration of the block as a function of t. Draw a graph showing the results for each acceleration component.

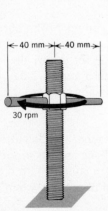

← 40 mm →← 40 mm →

30 rpm

Prob. II.44

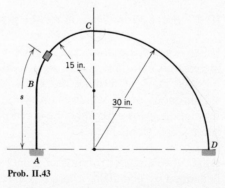

C

15 in.

B

s

30 in.

A D

Prob. II.43

II.44   Two bars are welded to a nut that advances along a bolt whose pitch is 4 mm. The 80-mm bar-nut system rotates at the constant rate of 30 rpm. What is the velocity and acceleration of a point at the free end of one of the bars?

II.45 A pipe rotates about the vertical axis at 30 rpm as water flows through it at 20 m/s. Determine the velocity, acceleration, and radius of curvature of the path of the particle of fluid that is at $R = 0.5$ m from the vertical axis.

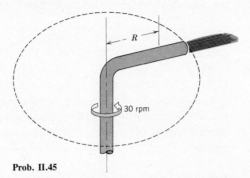

Prob. II.45

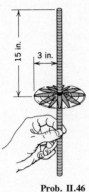

Prob. II.46

II.46 The propeller on the toy shown is pushed along a threaded shaft so that it flies into the air while rotating. The pitch of the thread causes the propeller to revolve once every tenth of an inch. If the hub of the propeller is moved along the shaft with a speed t^2 in./s, determine (a) how long it takes the hub to move the length of the shaft, and (b) the magnitude of the acceleration of the tip of the propeller blade when the propeller leaves the shaft.

II.47 A spring of unstretched length l is attached to the block. Functionally, the speed of the block is

$$v = \pm \sqrt{v_0^2 - K_1 x^2 - \tfrac{1}{2}K_2 x^4}$$

where the sign for v is chosen to fit the direction of the motion. Determine the acceleration of the block as a function of x.

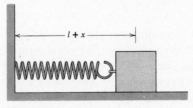

Prob. II.47

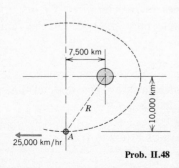

Prob. II.48

II.48 The acceleration of a satellite in an elliptical orbit has a magnitude of $g(R_e/R)^2$, where $g = 9.806$ m/s² and $R_e = 6370$ km is the radius of the earth. Also, the acceleration is always directed toward the center of the earth. At point A the satellite has a speed of 25,000 km/hr. What is the rate of change of the speed and the radius of curvature of the path at this location?

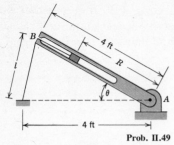

Prob. II.49

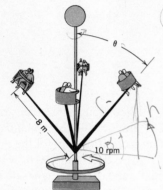

Prob. II.51

II.49 The block is pulled outward along the slot in the rotating arm AB by the cable whose length is 3 ft. The arm AB has a constant angular speed of 4 rad/s. Determine the speed and magnitude of the acceleration of the block when $\theta = 36.87°$. *Hint:* The rate of increase of the distances R and l are identical because the cable has a constant length.

II.50 Determine the horizontal and vertical components of the velocity and acceleration of the block in Problem II.49 when $\theta = 30°$.

II.51 The amusement park ride rotates about its vertical axis at the constant rate of 10 rpm as the arms swing in and out. At the instant when $\theta = 30°$ the arms are swinging outward at $\dot{\theta} = 0.3$ rad/s and this rate is decreasing at 0.1 rad/s². Determine the acceleration of the cabin at this instant.

II.52 The pin is guided through the fixed straight groove AB by arm CD, which moves horizontally to the right according to $X = 200(1 - e^{-t/2})$ mm. Determine the speed and acceleration of the pin along groove AB in the position where $X = 100$ mm.

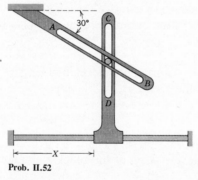

Prob. II.52

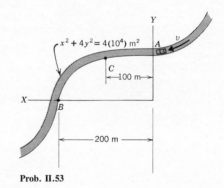

Prob. II.53

II.53 The diagram shows an elliptic curve forming a portion of an automobile race track. For maximum performance the driver operates the vehicle such that the normal component of acceleration is constant at $1.2g$. Given that $x^2 + 4y^2 = 4(10^4)$ m², determine the speed of the vehicle at points A, B, and C.

II.54 An observer at point O on the ground follows an airplane in level flight at a speed of 900 km/hr. For the position where $\phi = 0$, determine the rates of change of the angle ϕ and the distance d.

II.55 Determine $\ddot{d}$ and $\ddot{\phi}$ for the system of Problem II.54.

II.56 A block is pushed up the circular hill; its horizontal component of

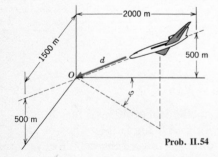

Prob. II.54

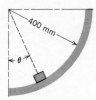

Prob. II.56

velocity is constant at 200 mm/s. Determine the velocity and acceleration of the block when $\theta = 0$ and $\theta = 60°$.

II.57   A cycloidal path is defined in parametric form by $X = 0.5(\alpha - \sin \alpha)$ ft, $Y = -0.5 \cos \alpha$ ft, where α is the parameter. In a certain case it is known that the motion of a particle following this path is given by $\alpha = 4t^2$. Determine expressions for the velocity and acceleration of this point.

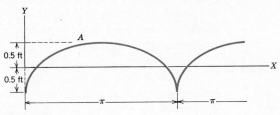

Prob. II.57

II.58   For the motion in Problem II.57, determine the radius of curvature of the path and the rate of change of the speed when the particle is at point A.

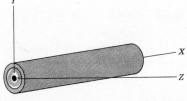

Prob. II.59

II.59   A compression wave in the air contained in the hollow tube causes the particles of air to follow a rectilinear motion in the X direction defined by $v = A \sin[\pi(X - ct)/L]$, where A and c are constant. Determine an expression for the acceleration of a particle of air.

II.60   The rectangular Cartesian coordinates of a particle are given by $X = 200 \sin \pi t/10$, $Y = 200 \cos \pi t/10$, $Z = 400 \cos \pi t/20$, where X, Y, and Z are expressed in millimeters and t is in seconds. Determine the velocity and acceleration vectors and show that the path lies on the surface of a circular cylinder.

II.61   For the motion in Problem II.60, determine expressions for the speed, the rate of increase of the speed, and the radius of curvature, all as functions of time.

II.62   By transforming the unit vectors, express the solutions to Problem II.61 in terms of cylindrical coordinates for which the Z axis is the axial direction. From this result determine the corresponding values of the angular velocity and angular acceleration.

II.63   The particle at the end of the flexible bar follows an elliptic path whose equation is

$$R = \frac{ab}{[b^2 + (a^2 - b^2) \cos^2 \phi]^{1/2}}$$

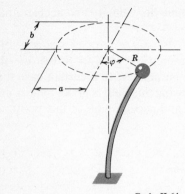

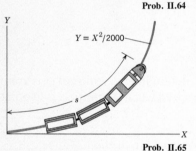

It is known that the angular velocity for the motion is given by $\dot{\phi} = K/R^2$. Find an expression for the acceleration of the particle in an arbitrary position.

Prob. II.64

II.64 Along a portion of a railroad line the transition curve from a straight track to a curve is defined by $Y = X^2/2000$, where X and Y are in meters. Measuring the distance s from the position where the locomotive is at $X = 0$, the motion of the train is defined by $s = t^2/10$, where s is the arc length in meters and t is in seconds. What is the speed, the rate of change of the speed, and the magnitude of the acceleration of the locomotive when $X = 1000$ m?

$Y = X^2/2000$

Prob. II.65

II.65 For sufficiently large values of the radial distance R from the centerline of a *vortex* in a fluid, the cylindrical components of the velocity of a particle of fluid are $v_R = 0$, $v_\phi = A/R$, $v_Z = 0$, where A is a constant. Determine an expression for the acceleration of a particle. What is the path of a particle of fluid?

II.66 An important type of flow in the study of fluids is the *doublet,* for which the velocity components in cylindrical coordinates are $v_R = (A \cos \phi)/R^2$, $v_\phi = (A \sin \phi)/R^2$, $v_Z = 0$, where A is a constant. Determine an expression for the acceleration of a particle.

E. RELATIVE MOTION

1 Translating Frame of Reference

The kinematical equations developed so far describe how we see the motion of a particle when our own position is a fixed point in space, that is, when the origin of the reference frame is a fixed point. Clearly, if we were to view the same particle motion from the window of a moving vehicle, we would not be viewing it from a fixed frame of reference. However, in order to convey our observations to someone not in the vehicle with us, we must be able to relate these observations to those viewed from a fixed point in space. This can be accomplished by the concept of relative motion. Such a situation is depicted in Figure 18.

The XYZ frame is a fixed coordinate system. The location of particle P with respect to the fixed point O is $\bar{r}_{P/O}$. Let point Q denote the position of an observer in a moving vehicle. Hence the observer's position with respect to the fixed point O is $\bar{r}_{Q/O}$.

We now choose a second reference frame xyz whose origin is the moving point Q. To aid in describing the kinematical properties that point P has relative to the observer, we ascribe fixed directions to the x, y, and z coordinate axes. That is, these axes are chosen to remain parallel to their initial orientation. Hence, xyz is a reference frame executing *translational*

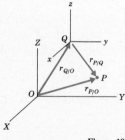

Figure 18

motion. Our convention, throughout this text, will be to use lowercase letters to indicate that a reference frame is moving.

From Figure 18 we see that $\bar{r}_{P/Q}$ (read r of P relative to Q) is the position of point P as seen by the observer at point Q, and further, we can write

$$\bar{r}_{P/O} = \bar{r}_{Q/O} + \bar{r}_{P/Q} \tag{31}$$

We shall refer to any variable that is observed from a fixed point as an absolute quantity. Thus, equation (31) states that the absolute position of point P is the sum of the absolute position of point Q and the relative position of point P with respect to point Q.

The next kinematical quantity of interest is velocity. The time derivative of the relative position vector $\bar{r}_{P/Q}$ is the relative velocity $\bar{v}_{P/Q}$ of point P with respect to point Q. That is,

$$\frac{d}{dt}(\bar{r}_{P/Q}) = \dot{\bar{r}}_{P/Q} = \bar{v}_{P/Q} \tag{32}$$

Then, differentiating equation (31), we have the following relationship between the absolute and relative velocities:

$$\dot{\bar{r}}_{P/O} = \dot{\bar{r}}_{Q/O} + \dot{\bar{r}}_{P/Q}$$

$$\bar{v}_P = \bar{v}_Q + \bar{v}_{P/Q} \tag{33}$$

In a similar manner we can reason that the acceleration of point P as seen by the observer at Q must be

$$\ddot{\bar{r}}_{P/Q} = \dot{\bar{v}}_{P/Q} = \bar{a}_{P/Q} \tag{34}$$

Then, by differentiating equation (33), we find that

$$\bar{a}_P = \bar{a}_Q + \bar{a}_{P/Q} \tag{35}$$

Notice the similarity in the form of equations (31), (33), and (35). In their present form they are symbolic. As will be seen in the examples that follow, when combined with our knowledge of kinematics with respect to a fixed point, these equations become a valuable computational tool. One immediate result arises in the special case where Q is also a fixed point, for then $\bar{v}_Q = \bar{a}_Q = 0$. This means that

$$\bar{v}_P = \bar{v}_{P/O} \quad \text{and} \quad \bar{a}_P = \bar{a}_{P/Q}$$

thus proving that the absolute velocity and acceleration are independent of the location of the fixed point of observation.

Note that in using a translating reference frame we are still describing motions, be they absolute or relative, in terms of fixed directions. In Module VI we shall relax this restriction and allow the moving frame to rotate. This will enable us to create a method for handling a broader class of problems.

Recall that vector components can be added only when all the vectors are written with respect to the same set of unit vectors. When using a moving reference frame *xyz,* dynamicists tend to perform computations by describing all vectors in terms of their $(\bar{\imath}, \bar{\jmath}, \bar{k})$ components with respect to *xyz*. We shall adhere to this procedure.

EXAMPLE 12
A rowboat has a speed of 8 ft/s with respect to a river that is flowing with a constant speed of 4 ft/s. What is the direction of the relative velocity vector, and what is the absolute speed of the boat if it is following a straight path from point A to point B?

Solution
To understand the motion of the rowboat, suppose that the river was at rest. Then the velocity of the boat relative to the river would be the true velocity of the boat. Here, the river is flowing. Hence the boat must head in a somewhat cross-stream direction to maintain its absolute direction along AB.

Using equation (33), in the terms of our problem,

$$\bar{v}_{\text{boat}} = \bar{v}_{\text{water}} + \bar{v}_{\text{boat/water}}$$

The given information is that $\bar{v}_{\text{water}}$ is 4 ft/s downstream and that $\bar{v}_{\text{boat/water}}$ has a magnitude of 8 ft/s. Also, $\bar{v}_{\text{boat}}$ is tangent to line $AB,$ because this is the true path of the boat. To assist in describing these vectors, we draw a sketch and define the axes of the *xyz* coordinate system. The angle α in this sketch, which defines the orientation of $\bar{v}_{\text{boat/water}}$, also describes the orientation of the longitudinal axis of the boat.

From this sketch we have

$$\bar{v}_{\text{boat}} = v_{\text{boat}} (\sin 15°\, \bar{\imath} + \cos 15°\bar{\jmath})$$

Because the axes of the moving and fixed reference frames are parallel, we have

$$\bar{v}_{\text{water}} = 4\bar{I} \equiv 4\bar{\imath}$$

$$\bar{v}_{\text{boat/water}} = 8(-\sin \alpha\bar{\imath} + \cos \alpha\bar{\jmath})$$

Thus, the relative velocity equation gives

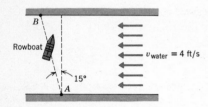

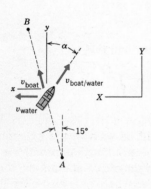

$$v_{\text{boat}}(\sin 15°\bar{\imath} + \cos 15°\bar{\jmath}) = 4\bar{\imath} + 8(-\sin\alpha\bar{\imath} + \cos\alpha\bar{\jmath})$$

Equating like components in this vector equation yields

$$\bar{\imath}\text{ component:} \quad v_{\text{boat}} \sin 15° = 4 - 8\sin\alpha$$

$$\bar{\jmath}\text{ component:} \quad v_{\text{boat}} \cos 15° = 8\cos\alpha$$

Hence

$$8\sin\alpha = 4 - 0.2588v_{\text{boat}}$$

$$8\cos\alpha = 0.9659v_{\text{boat}}$$

$\tan\alpha = \dfrac{4 - .2588 v_{boat}}{.9654 v_{boat}}$

Squaring each of these equations and adding yields

$$8^2(\sin^2\alpha + \cos^2\alpha) \equiv 64 = (4 - 0.2588v_{\text{boat}})^2 + (0.9659v_{\text{boat}})^2$$

$$64 = 16 - 2.070v_{\text{boat}} + v_{\text{boat}}^2$$

$$v_{\text{boat}} = 8.040, \ -5.970$$

The positive value of v_{boat} is the only one with meaning in this problem (the negative value corresponds to motion from point B to point A).

Substituting v_{boat} into the scalar equations then gives

$$\sin\alpha = \frac{4 - 0.2588(8.040)}{8} = 0.2399$$

$$\cos\alpha = \frac{0.9659(8.040)}{8} = 0.9707$$

Thus, because $\sin\alpha$ and $\cos\alpha$ are both positive, α is positive and we find

$$\alpha = 13.89°$$

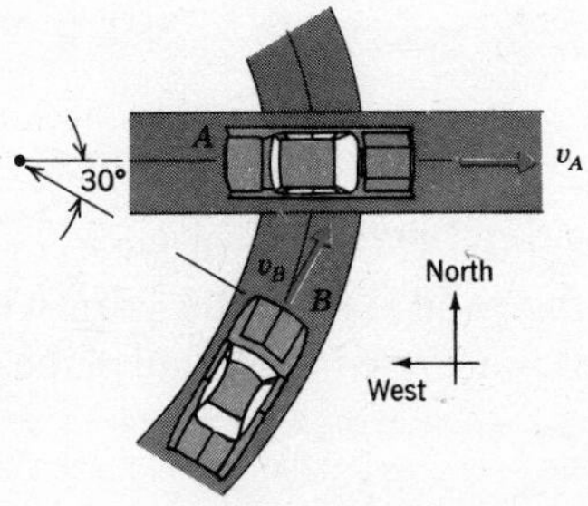

EXAMPLE 13

At the instant depicted in the diagram, car A is on the east-west overpass and has a constant speed of 36 km/hr whereas car B has a constant speed of 72 km/hr along the highway. The radius of curvature of the highway is 100 m. Determine the velocity and acceleration of car A as viewed from car B.

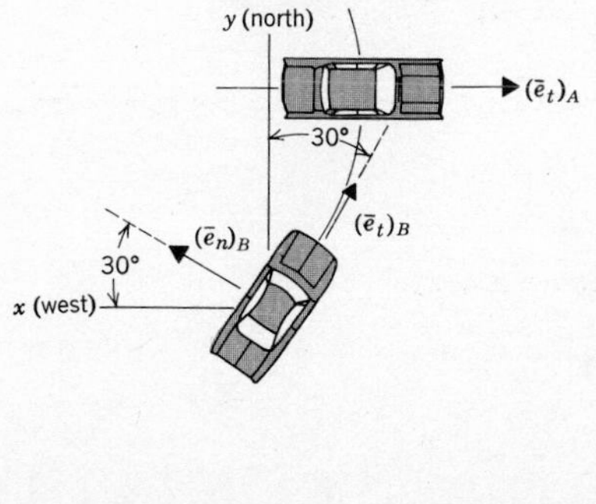

Solution

OK.

This problem requires the determination of the motion of car A relative to car B, so we place the origin of the moving reference frame at car B. Then, because we know the speed along its path for each car, we describe the absolute velocities and accelerations by using path variables.

Car A is following a straight path at a constant speed of 36 km/hr = 10 m/s, whereas car B has a constant speed of 72 km/hr = 20 m/s along the curved road. Thus

$$\bar{v}_A = 10(\bar{e}_t)_A \qquad \bar{v}_B = 20(\bar{e}_t)_B \text{ m/s}^2$$

$$\bar{a}_A = \bar{0} \qquad \bar{a}_B = \frac{(20)^2}{100}(\bar{e}_n)_B = 4(\bar{e}_n)_B \text{ m/s}^2$$

Before we can substitute these vectors into the relative motion equations, we must express them in terms of their components with respect to the *xyz* reference frame. Following the standard procedure, we transform the unit vectors. From the sketch we see that

$$(\bar{e}_t)_A = -\bar{\imath} \quad \text{and} \quad (\bar{e}_t)_B = -\sin 30°\bar{\imath} + \cos 30°\bar{\jmath}$$

$$(\bar{e}_n)_B = \cos 30°\bar{\imath} + \sin 30°\bar{\jmath}$$

Thus

$$\bar{v}_A = -10\bar{\imath} \qquad \bar{v}_B = 20(-0.5\bar{\imath} + 0.8660\bar{\jmath}) = -10.0\bar{\imath} + 17.32\bar{\jmath}$$

$$\bar{a}_A = \bar{0} \qquad \bar{a}_B = 4(0.8660\bar{\imath} + 0.50\bar{\jmath}) = 3.464\bar{\imath} + 2\bar{\jmath}$$

Finally, from equations (33) and (35) we find that

$$\bar{v}_{A/B} = \bar{v}_A - \bar{v}_B = -17.32\bar{\jmath} \text{ m/s}^2$$

and

$$\bar{a}_{A/B} = \bar{a}_A - \bar{a}_B = -3.464\bar{\imath} - 2.0\bar{\jmath} \text{ m/s}^2$$

Thus we see that, to an observer in car B, at this instant car A seems to be moving south at 17.32 m/s, with acceleration components of 3.464 m/s^2 east and 2.0 m/s^2 south.

HOMEWORK PROBLEMS

II.67   The nose of a small airplane is oriented 15° south of west and the airplane is traveling west. Knowing that the wind is from the southwest at 100 km/hr, determine the airspeed and groundspeed of the airplane.

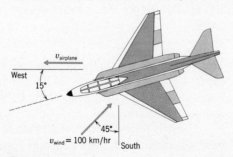

Prob. II.67

Prob. II.68

II.68 Passengers looking out the side window of an automobile traveling on a level road at a speed v_0 observe rain falling at an angle α forward from the vertical, whereas an observer at rest sees it falling at a forward angle β. Knowing that the absolute speed of a raindrop is 25 m/s, obtain an expression for α in terms of v_0 and β.

II.69 Two automobiles are following straight and level roads as they approach an intersection. Vehicle A has a speed of 100 km/hr and vehicle B has a speed of 60 km/hr, both constant. Determine the values of $\dot{R}$ and $\ddot{R}$ for the position shown. *Hint:* Evaluate the relative motion and express it in terms of polar coordinates having origin at vehicle B.

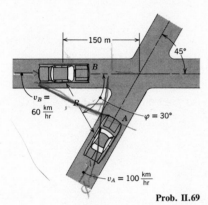

Prob. II.69

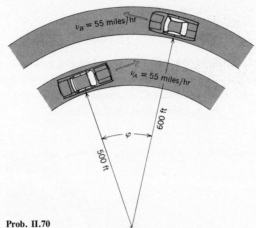

Prob. II.70

II.70 Two automobiles are traveling in opposite directions along a divided highway. Both vehicles have a speed of 55 miles/hr. Determine the velocity and acceleration of vehicle A with respect to vehicle B when the angle of separation $\phi = 0°$.

II.71 Solve Problem II.70 for $\phi = 30°$.

II.72 The pin P is constrained to follow the grooves in the two sliding arms AB and CD. In the position depicted in the diagram, arm AB has a constant speed of 80 mm/s, whereas arm CD has a speed of 40 mm/s and is gaining speed at the rate of 10 mm/s². For this position determine (a) the velocity of the pin, (b) the acceleration of the pin.

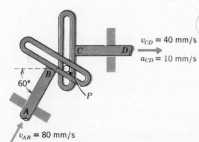

Prob. II.72

II.73 In order to avoid a collision, ships A and B maneuver in circular paths. At the instant shown ship A has a speed of 4 m/s and is decelerating at the rate of 0.4 m/s², whereas ship B has a constant speed of 8 m/s. What is the relative position, velocity, and acceleration of ship B with respect to ship A at this instant?

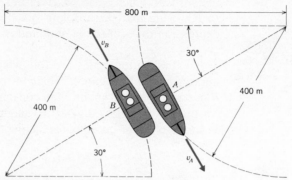

Prob. II.73

II.74   A pin at the end of the 3-ft-long arm AB follows the circular groove in arm CD, which is moving to the right with a constant speed of 60 ft/s. For the instant when $\phi = 45°$, determine (a) the velocity of the pin, and (b) the acceleration of the pin.

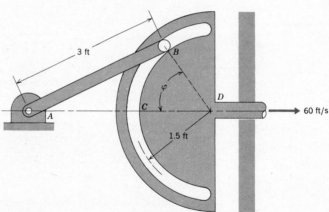

Prob. II.74

II.75   An airplane that can fly at v km/hr in still air is to travel to a point $\bar{r}$ $= R(\cos \theta \bar{I} + \sin \theta \bar{J})$ km from its starting point at the origin in a wind of velocity $v_w \bar{I}$ km/hr. Determine
(a) the direction in which the plane should be headed,
(b) the ground speed v_0 of the airplane,
(c) the time it will take to make the flight,
(d) the wind speed $v_{\max}$ for which it becomes impossible for the flight to be made in a straight-line path.

II.76 In the position shown, airplane A, which is diving in a vertical circle, has a speed of 200 miles/hr and is gaining speed at the rate of 5 ft/s². It is being observed by airplane B, which is flying in a straight path at a constant speed of 300 miles/hr. Determine the velocity of airplane A with respect to airplane B.

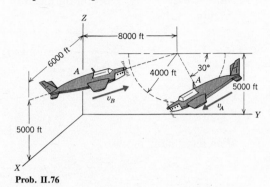

Prob. II.76

II.77 In Problem II.76, determine the acceleration of airplane B with respect to airplane A.

2 Cable-Pulley Systems

When cables and pulleys are utilized to connect bodies, we are concerned with the motion of more than one point, just as we were in the preceding study on relative motion. In rigid body mechanics cables are generally modeled as inextensible, massless bodies. Hence the bodies connected to the cable are constrained to move with respect to each other in a way that can easily be determined. As will be seen, this is accomplished by writing an equation for the constrained length of the cable.

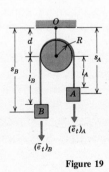

Figure 19

Consider the two masses A and B in Figure 19 which are connected by an inextensible cable. The distances s_A and s_B locate the masses with respect to the fixed reference point O. The length of the cable is

$$l = l_A + \pi R + l_B$$

where πR is the length of the cable in contact with the pulley. From Figure 19 we may write

$$l_A = s_A - C_A \quad \text{and} \quad l_B = s_B - C_B$$

where C_A and C_B are constants whose magnitudes depend on the fixed distance d and the dimensions of the masses. Replacing l_A and l_B in the equation for l gives

$$l = s_A - C_A + \pi R + s_B - C_B$$

or

$$s_A + s_B = C$$

where C is a combined single constant whose value will prove to be of no importance to us. Differentiating this relationship with respect to time yields

$$\frac{d}{dt}(s_A + s_B) \equiv \dot{s}_A + \dot{s}_B = 0$$

and

$$\ddot{s}_A + \ddot{s}_B = 0$$

For rectilinear motion,

$$\bar{v}_A = \dot{s}_A(\bar{e}_t)_A \qquad \bar{v}_B = \dot{s}_B(\bar{e}_t)_B$$

and

$$\bar{a}_A = \ddot{s}_A(\bar{e}_t)_A \qquad \bar{a}_B = \ddot{s}_B(\bar{e}_t)_B$$

where *the unit vectors,* shown in Figure 19, *are in the direction of increasing s* by definition. Thus, expressing the results in terms of the motion of block A, we have the following constraint equations

$$\bar{v}_A = \dot{s}_A(\bar{e}_t)_A \quad \text{and} \quad \bar{v}_B = -\dot{s}_A(\bar{e}_t)_B$$
$$\bar{a}_A = \ddot{s}_A(\bar{e}_t)_A \quad \text{and} \quad \bar{a}_B = -\ddot{s}_A(\bar{e}_t)_B$$

For the system of Figure 19, $(\bar{e}_t)_B \equiv (\bar{e}_t)_A$. Thus, these constraint equations state that blocks A and B have the same speed and magnitude of acceleration, but the velocity and acceleration of mass B are opposite to those of mass A.

For an arbitrary cable-pulley system we may generalize this approach for obtaining the constraint equations as follows. First, locate each of the connected bodies by its distance from a fixed point of reference. Then sum the length of each free portion of the cable, ignoring any constant values, and equate this sum to a constant. Finally, differentiate this sum to obtain the speed and magnitude of acceleration relationships.

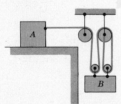

EXAMPLE 14
Block A is moving to the left at 4 ft/s and decelerating at the rate of 12 ft/s². Determine the velocity and acceleration of block B.

Solution
We begin with a sketch in which we define the distance s_A and s_B for the

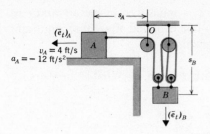

location of the blocks. Remember that these distances must be measured from a fixed point. We have chosen point O.

In terms of the unit vectors shown in the sketch,

$$\bar{v}_A = \dot{s}_A(\bar{e}_t)_A = 4(\bar{e}_t)_A \text{ ft/s} \qquad \bar{v}_B = \dot{s}_B(\bar{e}_t)_B$$

$$\bar{a}_A = \ddot{s}_A(\bar{e}_t)_A = -12(\bar{e}_t)_A \text{ ft/s}^2 \qquad \bar{a}_B = \ddot{s}_B(\bar{e}_t)_B$$

We now construct an equation for the length of the cable. There are four vertical lengths of cable and one horizontal length, so

$$4s_B + s_A = C$$

Differentiating gives

$$4\dot{s}_B + \dot{s}_A = 0 \qquad 4\ddot{s}_B + \ddot{s}_A = 0$$

Then, because the preceding equations for velocity and acceleration give $\dot{s}_A = 4$ ft/s, $\ddot{s}_A = -12$ ft/s^2, we find

$$\dot{s}_B = -1 \text{ ft/s} \qquad \ddot{s}_B = 3 \text{ ft/s}^2$$

Thus

$$\bar{v}_B = -1(\bar{e}_t)_B \text{ ft/s} \qquad \bar{a}_B = 3(\bar{e}_t)_B \text{ ft/s}^2$$

HOMEWORK PROBLEMS

II.78 Block A is given an acceleration of $0.5g$ to the right. Determine (a) the acceleration of block B and (b) the acceleration of block B with respect to block A.

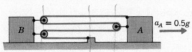

Prob. II.78

II.79 The tow truck A pulls the automobile B out of a ravine by driving forward. Starting from the position shown, the distance the truck moves

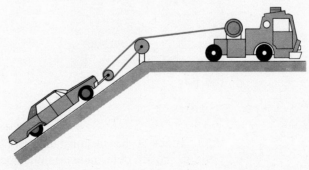

Prob. II.79

is given by $s = t^2/2 - t^3/100$, where s is measured in meters and t in seconds. Determine how far the automobile is pulled after 10 seconds and its velocity at the end of this time interval.

II.80   Block B is moving downward with a speed v_B. What is the velocity of block A?

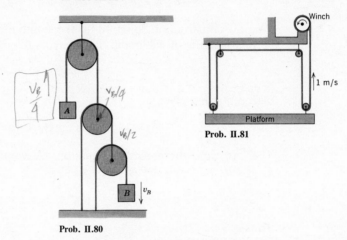

Prob. II.80

Prob. II.81

II.81   The electrically powered winch pulls the cable upward at the constant speed of 1 m/s. What is the velocity of the platform?

II.82   Two alternative designs for an elevator are shown in the diagram. In design A the pulley attached to the motor pulls the cable in with an absolute speed v, whereas in design B the speed v with which the cable is pulled in is with respect to the elevator. Which elevator is faster, and by what percentage is it faster?

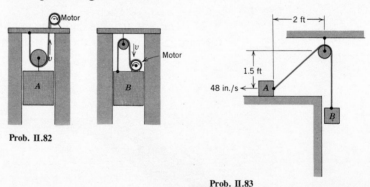

Prob. II.82

Prob. II.83

II.83   Block A has a constant velocity of 48 in./s to the left. In the position shown, determine (a) the velocity of block B, and (b) the acceleration of block B. (Note: The radius of the pulley may be neglected.)

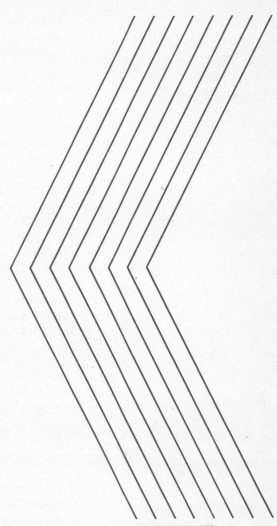

MODULE III
KINETICS OF
A PARTICLE

In this module we shall apply our knowledge of the kinematics of a particle to determine how the forces exerted upon the particle affect its motion. The foundation for this study is Newton's second law of motion.

Letting $\Sigma \bar{F}$ denote the resultant of the set of forces acting on a particle P, we may write

$$\Sigma \, \bar{F} = m\bar{a}_P \tag{1}$$

When employing this relationship we shall express the forces and accelerations in terms of their components with respect to either Cartesian coordinates, cylindrical coordinates, or path variables. As one of our objectives, we shall learn how to recognize which of these coordinate descriptions is best suited for the formulation and solution of particular classes of dynamics problems.

Before continuing, it will prove beneficial to place Newton's two other laws of motion in proper perspective. The first law is embodied in the second law, reaffirming the fact that $\bar{a}_P = \bar{0}$ when $\Sigma \, \bar{F} = \bar{0}$. In effect, by stating this law separately, we are considering the study of statics to be independent of the study of dynamics. The third law proves most useful when we model physical systems as interacting, but independent, bodies. For such instances the action and reaction will be needed to represent the interaction forces correctly.

The foregoing brings to mind the concept of a free body diagram, in which we isolate a body from its surroundings, showing all forces exerted on the body of interest. This diagram serves two purposes. Drawing the diagram helps us to understand how the forces affect the body, and the resulting diagram aids us to account consistently for all the forces. Because the free body diagram is a primary tool in solving statics problems, we shall not dwell on its construction here.

An important feature that arises in analyzing the motion of a system concerns the restrictions, called *constraints,* that are often placed on the motion. In the case of particle motion, the most common constraint is that the particle must follow a specific path. Such constraint conditions can be thought of as being imposed by the application of constraint forces. For example, consider the particle in Figure 1a, which is moving under the influence of a given force $\bar{F}_1$ along a smooth curved path in a *horizontal plane.* The free body diagram of the particle when it is at a general point P must show the constraint force $\bar{N}$ that is normal to the path. This force has the effect of restraining the velocity of the particle to be tangent to the path. Without this constraint force the particle would follow a different path from the prescribed one. Thus, Figure 1b is the free body diagram for this problem.

For ease in reading Figure 1 we have accentuated the arrows depicting forces to differentiate them from the arrows that describe unit vectors. Also note that the forces normal to the plane of motion, in this case the

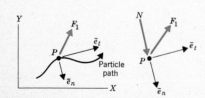

Figure 1a   **Figure 1b**

vertical forces, are not shown in Figure 1b. Included in these forces is the constraint force necessary to prevent motion (due to gravity) perpendicular to the plane. If we were interested in this force, we would draw a free body diagram showing the system as it is seen when viewed horizontally.

In the kinematics study in Module II, we concentrated on systems where we knew the position and/or velocity of a particle; the acceleration was obtained by differentiation of the position vector. No problem involved the determination of position and velocity from given information regarding the acceleration. Our reason for presenting these two seemingly similar types of problems separately is motivated by the fact that there are essentially two broad classes of practical problems in dynamics.

The first class involves situations where the motion of the physical system is given and we seek the forces producing this motion. For a particle, this reduces to a kinematical study using the methods in Module II to determine $\bar{a}_P$. Then, a free body diagram and $\Sigma \bar{F} = m\bar{a}_P$ will yield the necessary information about the forces.

In the second class of problems, the forces exerted on a physical system are given and we seek to determine the system's motion. For the case of a particle, we may determine its acceleration from $\bar{a}_P = \Sigma \bar{F}/m$. It is then necessary to follow an integration process (the opposite of differentiation) to determine the velocity and position vectors. (Clearly, a third class of problems is a combination of the foregoing.)

In this module we shall study how to determine the motion of a particle under the action of a given force system. In order to understand this process clearly, we shall first develop procedures for problem solving by considering the case of rectilinear (straight-line) motion. We shall then study a spectrum of problems in curvilinear motion, treating situations where forces are given, where the motion is specified, and where combinations of the two exist.

A. EQUATIONS OF RECTILINEAR MOTION

We begin the study of the motion of a particle along a straight path by drawing a free body diagram that shows *all* forces acting on the particle. To complete the sketch we also need a set of unit vectors, which of course involves a decision as to which kinematical formulation we should select to describe the motion. As always, this decision depends on how the information given in the problem is presented. Let us assume that the given forces are functions of the path variable quantities, which are the distance s (traveled along the path), the speed v, and the time t. With this type of given information we normally use path variables with unit vectors $\bar{e}_t$, $\bar{e}_n$, and $\bar{e}_b$ when setting up the free body diagram. Recall that $\bar{e}_b$ is perpendicular to the plane of $\bar{e}_t$ and $\bar{e}_n$ and is defined by $\bar{e}_t \times \bar{e}_n = \bar{e}_b$. Thus,

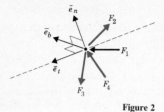

Figure 2

for the motion of a particle along a straight path, letting $\bar{F}_1$, $\bar{F}_2$, etc., denote the set of forces acting on the particle, we have the general free body diagram of Figure 2. Note that $\bar{e}_t$ is shown tangent to the path and that the plane formed by $\bar{e}_n$ and $\bar{e}_b$ is perpendicular to the path.

We now wish to apply Newton's second law to this problem. To do this we resolve all vectors into their components with respect to the directions of the unit vectors. The resultant force $\Sigma \bar{F}$ is the sum of the vectors $\bar{F}_i$,

$$\Sigma \bar{F} = \Sigma \bar{F}_i \equiv (\Sigma F_t)\bar{e}_t + (\Sigma F_n)\bar{e}_n + (\Sigma F_b)\bar{e}_{\mathrm{b}} \tag{2}$$

where ΣF_t, ΣF_n, and ΣF_b are the sum of the force components in the tangent and two normal directions, respectively. From kinematics we know that for rectilinear motion

$$\bar{a}_P = \dot{v}_P \bar{e}_t \tag{3}$$

Thus, because two vectors are equal only when their components are equal, the vector relation $\Sigma \bar{F} = m\bar{a}_P$ reduces to the scalar equations

$$\boxed{\begin{aligned} \Sigma F_t &= m\dot{v}_P \\ \Sigma F_n &= 0 \\ \Sigma F_b &= 0 \end{aligned}} \tag{4}$$

Equations (4) are the basic equations for rectilinear motion. The components perpendicular to the direction of motion express the fact that there must be a balance of forces in these directions, whereas the tangential component relates the forces to the rate of increase of the speed.

Equations (4) may be used to determine the forces for a known value of $\dot{v}_P$ or to determine $\dot{v}_P$ for a known set of forces. In the first case we have three algebraic equations for the force components, which can be solved by conventional methods. Here we shall concentrate on the latter case, in that it gives rise to an interesting situation. Let us assume that we know the functional dependence of the forces. Then, when we substitute the functions describing these forces into equations (4), we obtain a *differential equation,* that is, an equation relating the derivative of a function to other functions. The solution of this equation will tell us how the speed is related to the elapsed time and to the distance traveled and, equally important, how the distance traveled depends on the time. These data are defined by the functions $v_P(t)$, $v_P(s)$, and $s(t)$.

It will soon be obvious to you that we cannot determine these characteristics of the motion unless we know what state the particle was in prior to the application of the forces. (Remember, we require a force to alter a state of motion.) To define the initial state, we need *initial conditions,* which are the time t_0 when the motion was initiated, the distance s_0 describing the initial location of the particle, and the initial speed v_0.

B. SEPARATION OF VARIABLES

In general, there are many methods, both analytical and computerized, for solving the differential equation of motion obtained from equation (4). An effective approach is the method of separating variables. This method, which is applicable in situations where the differential equation contains only two variables, aims at forming integrable terms. In what follows we shall consider situations where this method can be employed.

1 Force as a Function of Time

In this case we have $\dot{v}_P = f(t)$, where $f(t) \equiv \Sigma\, F_t/m$ represents a known function of time. Then the differential equation given by the first of equations (4) can be placed into a form suitable for separating variables by using $\dot{v}_P \equiv dv_P/dt$. Hence equation (4) becomes

$$\dot{v}_P = \frac{dv_P}{dt} = f(t) \tag{5}$$

The term "separating variables" refers to the process of manipulating the differential equation so that only one variable appears on each side of the equation. To obtain this result for equation (5), we multiply through by dt, which gives

$$dv_P = f(t)\, dt$$

Knowing $f(t)$, this equation can be integrated. For dynamics problems with given initial conditions, a definite integral can be formed. We therefore have

$$\int_{v_0}^{v_P} dv_P \equiv v_P - v_0 = \int_{t_0}^{t} f(t)\, dt \tag{6}$$

Notice that the lower limits in the integrals are the initial values of the variables, and the upper limits are the values associated with point P. As an aside, remember that a valuable aid in the integration process is standard tables of integrals.

The foregoing equation yields the solution for $v_P(t)$, which can then be integrated to determine $s(t)$. Toward this end, we write

$$\dot{s} = \frac{ds}{dt} = v_P(t)$$

altar

Separating variables and integrating yields

$$\int_{s_0}^{s} ds \equiv s - s_0 = \int_{t_0}^{t} v_P(t)\, dt \tag{7}$$

Finally, if we desire to relate v_P to s, we can treat t as a parameter that is to be eliminated between equations (6) and (7).

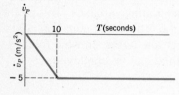

EXAMPLE 1

An aircraft lands with a speed of 200 km/hr and the pilot then gradually brakes the aircraft, producing the deceleration rate shown in the graph. How far will the aircraft travel prior to coming to rest?

Solution

The given initial conditions for $t_0 = 0$ are

$$v_0 = 200 \frac{km}{hr}\left(\frac{1\ hr}{3600\ s}\right)\left(\frac{1000\ m}{km}\right) = 55.56\ m/s$$

$$s_0 = 0$$

The graph shows that v_P is defined by two functions. For $0 \le t \le 10$ seconds, we have

$$\dot{v}_P = -\tfrac{1}{2}t\ m/s^2$$

and for $t \ge 10$ seconds, we have

$$\dot{v}_P = -5\ m/s^2$$

We first solve for the motion during the initial time period. Thus, for $0 \le t \le 10$ seconds, the acceleration is

$$\dot{v}_P \equiv \frac{dv_P}{dt} = -\tfrac{1}{2}t$$

Upon separating variables and integrating, this yields

$$\int_{55.56}^{v_P} dv_P \equiv v_P - 55.56 = -\tfrac{1}{2}\int_0^t t\, dt = -\tfrac{1}{4}t^2$$

$$v_P = 55.56 - \tfrac{1}{4}t^2\ m/s$$

To find $s(t)$, we write

$$\dot{s} \equiv \frac{ds}{dt} \equiv v_P = 55.56 - \tfrac{1}{4}t^2$$

$$\int_0^s ds \equiv s = \int_0^t (55.56 - \tfrac{1}{4}t^2)\, ds$$

$$s = 55.56t - \tfrac{1}{12}t^3\ m$$

We can now determine if the aircraft will come to rest within this 10 second time interval by substituting $t = 10$ seconds. This gives

$$v_P(10) = 30.36 \text{ m/s}$$

$$s(10) = 472.3 \text{ m}$$

Hence the aircraft is still moving at the end of the first time period. We next consider the time interval $t \geq 10$ seconds. The initial conditions for this time interval are the values of v_P and s at the end of the initial time interval. Repeating the foregoing steps,

$$\dot{v}_P \equiv \frac{dv_P}{dt} = -5$$

$$\int_{30.56}^{v_P} dv_P \equiv v_P - 30.56 = -5 \int_{10}^{t} dt = -5(t - 10)$$

$$v_P = 80.56 - 5t \text{ m/s}$$

Now, to find $s(t)$ we write

$$\dot{s} \equiv \frac{ds}{dt} \equiv v_P = 80.56 - 5t$$

$$\int_{472.3}^{s} ds \equiv s - 472.3 = \int_{10}^{t} (80.56 - 5t) \, dt$$

$$s - 472.3 = 80.56(t - 10) - 2.5(t^2 - 10^2)$$

$$= -83.3 + 80.56t - 2.5t^2 \text{ m}$$

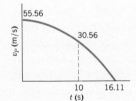

The value of t at which the aircraft comes to rest is obtained by setting $v_P = 0$. Thus

$$0 = 80.56 - 5t, \text{ so} \qquad t = 16.11 \text{ seconds}$$

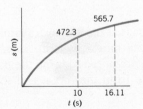

Finally, we can now obtain the distance required for the airplane to come to rest.

$$s = -83.3 + 80.56(16.11) - 2.5(16.11)^2 = 565.7 \text{ m}$$

If we were interested in how the motion varies with time, we could present the results for $v_P(t)$ and $s(t)$ graphically as shown in the diagram.

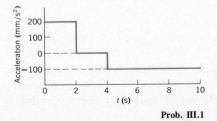

Prob. III.1

HOMEWORK PROBLEMS

III.1 A particle in rectilinear motion is given the acceleration plotted in the diagram. When $t = 0$ the particle is at rest at the origin. Sketch diagrams showing how the speed and distance traveled vary with time.

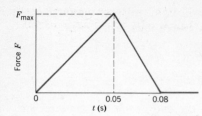

Prob. III.2

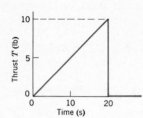

Prob. III.3

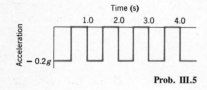

Prob. III.5

III.2 Experimental measurement of the contact force F exerted between a golf ball and a club indicates that the force-time plot is as shown. However, because of an error the scale factor for the force was lost, so the value of F_{max} is unknown. If the velocity of the ball after the impact was 70 m/s, determine the value of F_{max} in terms of the mass m of the ball.

III.3 The thrust T of a toy rocket that follows a straight vertical path is shown in the diagram. The toy weighs 2 lb and rests on its platform until the thrust is sufficient to launch it. Determine the maximum speed of the toy and the maximum height it attains.

III.4 The acceleration of a particle starting from rest along a straight line is known to follow the relation

$$a = k_n t^n$$

where k_n is a constant that depends on the nonnegative integer n. It is also known that the distance $s = l$ when $t = \tau$ seconds. Determine (a) an expression for k_n in terms of l, τ, and n, and (b) the speed of the particle when $t = \tau$. Does this speed increase or decrease with increasing values of n?

III.5 In order to prevent an automobile from skidding on ice when coming to a stop, it is common practice to "pump" the brakes. A driver followed this procedure, resulting in the acceleration-time plot shown. (a) Letting v_n and s_n denote the speed and distance traveled after n seconds, derive expressions for v_{n+1} and s_{n+1} in terms of v_n and s_n. (b) Use the results of part (a) to determine how long it takes to bring the automobile to rest from an initial speed of 20 miles/hr, and also determine the corresponding distance traveled.

III.6 A particle is rectilinear motion passes the origin traveling at 400 m/s when it is given a deceleration of $50t$ mm/s², where t is in units of seconds. Determine (a) the length of time required to bring the particle to rest, and (b) the distance of the particle from the origin and the actual distance it has traveled after 6 seconds.

2 Force as a Function of Position
When force is a function of position, $\dot{v}_P = \Sigma\, F_t/m = f(s)$, we cannot separate variables by writing $\dot{v}_P = dv_P/dt$, in that we would then have three variables (v_P, s, and t) in the differential equation. The s variable

must be retained because it appears in the force term. Hence, we shall eliminate time from the equation. This is achieved by employing the chain rule, specifically,

$$\dot{v}_P \equiv \frac{dv_P}{dt} \equiv \frac{ds}{dt}\frac{dv_P}{ds} \equiv v_P\frac{dv_P}{ds} = f(s)$$

We now can separate variables and integrate. Thus

$$\int_{v_0}^{v_P} v_P\, dv_P \equiv \tfrac{1}{2}(v_P{}^2 - v_0{}^2) = \int_{s_0}^{s} f(s)\, ds \tag{8}$$

Equation (8) yields the function $v_P(s)$. To determine how the distance s depends on time, we write

$$\dot{s} \equiv \frac{ds}{dt} = v_P(s)$$

We then separate variables and integrate, with the result that

$$\int_{s_0}^{s} \frac{ds}{v_P(s)} = \int_{t_0}^{t} dt \equiv t - t_0 \tag{9}$$

Upon evaluating the integral on the left side of equation (9), we can solve for the value of s at any instant t. If the value of v_P at a specific instant is required, we can substitute the result for s at this instant into equation (8) and solve for the corresponding value of v_P.

EXAMPLE 2
A cable is attached to a 0.5-kg block that slides over the smooth rigid horizontal rod AB. The diagram shown depicts the vertical plane. The cable tension is a constant value T. At point C the speed of the block is 6 m/s to the left and at point D the speed is 1 m/s to the left. Determine the value of T.

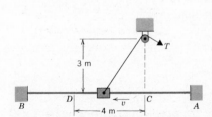

Solution
In order to relate the speed of the block at points C and D, we sketch a free body diagram of the block at a general position. (Do not forget the normal force N that constrains the block to follow rod AB.)
We let s be the distance the block has traveled from point C and show $\bar{e}_t$ in the direction of increasing s.
Summing forces in the normal direction, we have

$$\Sigma F_n = N + T \sin\alpha - 4.903 = 0.5a_n = 0$$

Summing forces in the tangential direction yields

$$\Sigma F_t = -T\cos\alpha = 0.5a_t = 0.5\dot{v}_P$$

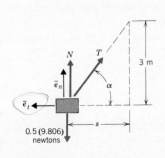

The first equation gives

$$N = 4.903 - T \sin \alpha$$

In this problem the expression for N is not needed. The second equation shows that the acceleration is a function of the angle α, but from the right triangle illustrated in the free body diagram we find that

$$\cos \alpha = \frac{s}{\sqrt{s^2 + 3^2}}$$

Hence, the tangential force equation will give $\dot{v}_P$ as a function of s. We therefore write

$$\dot{v}_P \equiv v_P \frac{dv_P}{ds} = -\frac{T}{(0.5)} \frac{s}{\sqrt{s^2 + 9}}$$

The initial conditions are that $s = 0$ and $v_P = 6$ m/s when $t = 0$. Therefore, separating variables and integrating yields

$$\int_6^{v_P} v_P \, dv_P = -2T \int_0^s \frac{s \, ds}{\sqrt{s^2 + 9}}$$

The integrand on the right side has the form $\int x^{-1/2} \, dx$. Hence

$$\tfrac{1}{2}(v_P{}^2 - 6^2) = -2T[\sqrt{s^2 + 9} - \sqrt{9}]$$
$$v_P{}^2 = 36 - 4T[\sqrt{s^2 + 9} - 3]$$

The foregoing defines $v_P(s)$ for a given value of T. To find T we use the fact that $v_P = 1$ m/s when $s = 4$ m, which yields

$$1^2 = 36 - 4T[\sqrt{4^2 + 3^2} - 3]$$

$$T = \frac{35}{4(2)} = 4.375 \text{ newtons}$$

Before leaving this problem, let us note the problem confronting us if we wish to follow the analogous steps to those in equation (9). The form of the function $v_P(s)$ that was obtained above would not allow for a closed-form evaluation of the integral on the left side of this equation. For such a case the integral could be evaluated numerically with the aid of a computer.

HOMEWORK PROBLEMS

III.7   The 10-kg block is pulled upward from the rest position shown by a

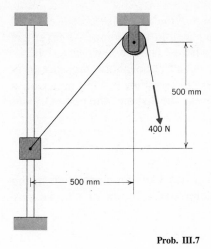

Prob. III.7

constant cable tension of 400 N. Determine the velocity of the block when it reaches the elevation of the pulley (a) if friction is negligible, (b) if the coefficient of friction between the block and its guide is $\mu = 0.1$.

III.8 A steel block weighing 800 lb is to be picked up by an electromagnet whose attractive force in pounds is given by the inverse square formula $F = 3(10^4)/s^2$ lb, where s is the distance in feet. The block was initially at $s = 5$ ft. Determine the velocity with which the block strikes the electromagnet.

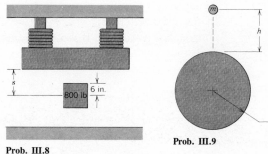

Prob. III.8

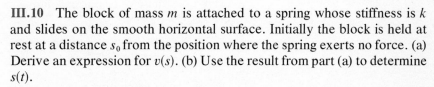

Prob. III.9

III.9 The inverse square law for the gravitational attraction exerted on a particle of mass m is

$$F = \frac{mg}{(1 + h/R_e)^2}$$

where $g = 32.17$ ft/s², h is the altitude of the particle above the surface of the earth, and $R_e = 2.091(10^7)$ ft is the mean radius of the earth. In a hypothetical situation, a particle is released from rest at a height $h = 10,000$ ft. Neglecting air resistance, determine the velocity of the particle when it hits the earth. Compare this result to the value obtained by considering the gravity force to be simply mg.

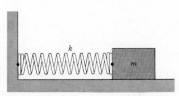

Prob. III.10

III.10 The block of mass m is attached to a spring whose stiffness is k and slides on the smooth horizontal surface. Initially the block is held at rest at a distance s_0 from the position where the spring exerts no force. (a) Derive an expression for $v(s)$. (b) Use the result from part (a) to determine $s(t)$.

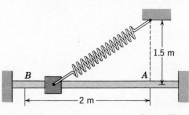

Prob. III.11

III.11 A 0.25-kg collar slides over a horizontal rod under the action of a linear spring whose constant is 20 N/m. There is no force in the spring when the collar is at point A. In a certain experiment the collar is released from rest at point B. Determine the velocity of the collar when it passes point A. Friction is negligible.

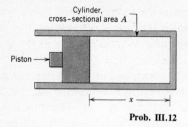

Cylinder,
cross-sectional area A

Piston →

x

Prob. III.12

III.12 The 0.5-kg piston slides within a cylinder whose cross-sectional area $A = 5000$ mm². The gas pressure p within the cylinder is inversely proportional to the volume of gas Ax, where x is as shown. Also, when $x = 50$ mm, the pressure p equals the constant atmospheric pressure $p_0 = 1.0133(10^5)$ pascal. If the piston passes the position $x = 50$ mm with a velocity to the right of 500 mm/s, determine the distance the piston travels before coming to rest.

3   Force as a Function of Speed
In the case where $\dot{v}_P = \Sigma\, F_t/m = f(v_P)$, we have an alternative in the approaches available. If we desire to determine $v_P(t)$ and/or $s(t)$, we may write

$$\dot{v}_P \equiv \frac{dv_P}{dt} = f(v_P)$$

Then separating variables and integrating yields

$$\int_{v_0}^{v_P} \frac{dv_P}{f(v_P)} = \int_{t_0}^{t} dt \equiv t - t_0 \qquad (10)$$

After evaluating the integral on the left side of equation (10), we can solve for $v_P(t)$. To determine $s(t)$, we write

$$\dot{s} \equiv \frac{ds}{dt} = v_P(t)$$

so that

$$\int_{s_0}^{s} ds \equiv s - s_0 = \int_{t_0}^{t} v_P(t)\, dt \qquad (11)$$

The alternative method involves the situation where we desire to know the value of v_P corresponding to a value of s. Here the chain rules yields

$$\dot{v}_P = \frac{dv_P}{dt} = \frac{ds}{dt}\frac{dv_P}{ds} = v_P \frac{dv_P}{ds} = f(v_P)$$

which, upon separating variables and integrating, gives

$$\int_{v_0}^{v_P} \frac{v_P\, dv_P}{f(v_P)} = \int_{s_0}^{s} ds \equiv s - s_0 \qquad (12)$$

Note that we now have two ways in which $s(t)$ can be evaluated. The first is to follow equation (11), whereas the second is to treat v_P as a parameter that is to be eliminated between equations (10) and (12).

$\tan^{-1}(0.1)$

EXAMPLE 3

The brakes of an automobile whose mass is 1500 kg are released and the automobile rolls down a 10% grade. The air resistance is kv_P^2, where $k = 2.5$ newtons-seconds²/meter², and the rolling resistance of the tires is 0.01 N, where N is the normal force exerted by the ground. Determine (a) the speed of the automobile after it has rolled 100 m, and (b) the length of time required for the car to reach this position.

Solution

As shown in the free body diagram, the forces acting on the automobile are the frictional air and rolling resistances (opposing the velocity of the automobile), the weight force, and the normal force exerted by the ground.

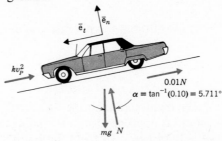

$\bar{e}_t$ $\bar{e}_n$

kv_P^2

$0.01N$

$\alpha = \tan^{-1}(0.10) = 5.711°$

mg N

Summing forces in the $\bar{e}_n$ direction yields

$$\Sigma F_n = N - mg \cos \alpha = 0$$

whereas the force sum in the direction of $\bar{e}_t$ gives

$$\Sigma F_t = mg \sin \alpha - kv_P^2 - 0.01N = m\dot{v}_P$$

Using the expression for N from the first equation to solve the second equation for $\dot{v}_P$, we obtain

$$\dot{v}_P = g(\sin \alpha - 0.01 \cos \alpha) - \frac{k}{m} v_P^2$$

This can be written in the more convenient form

$$\dot{v}_P = \frac{k}{m}(c^2 - v_P^2), \text{where} \quad c = \sqrt{\frac{mg}{k}(\sin \alpha - 0.01 \cos \alpha)}$$

The distance s is measured from the position where the car is released, so that the initial conditions are $v_P = 0$ and $s = 0$ when $t = 0$. To determine the speed at a specified position, namely $s = 100$ m, we write

$$\dot{v}_P \equiv v_P \frac{dv_P}{ds} = \frac{k}{m} (c^2 - v_P^2)$$

Upon separating variables and integrating, we get

$$\int_0^{v_P} \frac{v_P \, dv_P}{c^2 - v_P{}^2} = \frac{k}{m} \int_0^s ds$$

From this result v_P may be obtained as follows:

$$-\tfrac{1}{2} \ln(c^2 - v_P{}^2) + \tfrac{1}{2} \ln(c^2) = \frac{k}{m} s$$

$$c^2 - v_P{}^2 = c^2 \exp\!\left(\frac{-2ks}{m}\right)$$

$$v_P = c\left[1 - \exp\!\left(\frac{-2ks}{m}\right)\right]^{1/2}$$

Now, substituting the parameters and letting $s = 100$ m yields

$$c = 22.95 \text{ m/s}$$

$$v_P = 12.22 \text{ m/s} \qquad \triangleleft$$

Having determined the value of v_P corresponding to $s = 100$ m, the most direct approach for determining the time required to reach this position is to obtain $v_P(t)$ from the original differential equation. To achieve this we write

$$\dot{v}_P \equiv \frac{dv_P}{dt} = \frac{k}{m}(c^2 - v_P{}^2)$$

$$\int_0^{v_P} \frac{dv_P}{(c^2 - v_P{}^2)} = \frac{k}{m} \int_0^t dt \equiv \frac{k}{m} t$$

With the aid of a table of integrals we find that

$$t = \frac{1}{2c} \frac{m}{k} \ln\!\left(\frac{c + v_P}{c - v_P}\right)$$

Hence, for $v_P = 12.22$ m/s, we get

$$t = 15.52 \text{ seconds} \qquad \triangleleft$$

For completeness, the derived equations relating v_P to t, and v_P to s, are presented here graphically. If we desired to relate s to t, the solution for v_P as a function of s could be substituted into the time equation.

The dashed lines in the graph show that v_P approaches a maximum value as s and $t \to \infty$. This maximum value, which is called the *terminal velocity*, can be determined without actually solving the differential equation of motion by simply setting $\dot{v}_P = 0$. Thus, for this problem we have

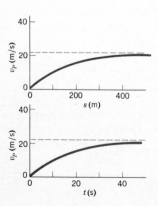

$$\dot{v}_P = 0 = \frac{k}{m}\left[c^2 - (v_P)^2_{\text{terminal}}\right]$$

$$(v_P)_{\text{terminal}} = c = 22.95 \text{ m/s}$$

HOMEWORK PROBLEMS

III.13 The viscous friction between a 30° inclined plane coated by a layer of lubricant and a 10-lb block is given by $f = 0.25v$, where f is expressed in pounds and v is in feet per second. The block is released from rest. Determine (a) the time required for the block to attain a speed of 10 ft/s, and (b) the corresponding distance traveled by the block.

Prob. III.13

III.14 The frictional resistance on a ball bearing falling through oil is know to be proportional to the speed of the ball bearing, $f = kv$. It is also known that the terminal speed is 800 mm/s. The ball bearing is released from rest. Determine (a) the time required for the ball bearing to reach one-half its terminal speed, and (b) the speed of the ball bearing at twice the elapsed time found in part (a).

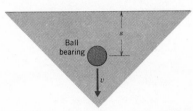

Prob. III.14

III.15 A block of mass m sliding on a horizontal surface hits a shock absorbing device (called a *linear dashpot*), which slows the block by exerting a retarding force $f = kv$, where v is the speed of the block and k is a constant. Letting s be the distance the dashpot is compressed, derive expressions for $v(t)$, $v(s)$, and $s(t)$ if $v = v_0$ and $s = 0$ initially.

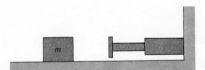

Prob. III.15

III.16 The air resistance on a 200-g ball in free flight is given by $f = 2(10^{-4})v^2$, where f is in newtons and v is in meters per second. If the ball is dropped from rest 500 m above the ground, determine the speed at which it hits the ground. What percentage of the terminal speed is the result?

III.17 A 200-g ball is thrown vertically upward from the ground with a speed of 25 m/s. The air resistance is $f = 2(10^{-4})v^2$, where f is in newtons and v is in meters per second. Determine (a) the maximum height attained by the ball, (b) the speed of the ball when it returns to the ground, (c) the total elapsed time.

III.18 The resistance of the water to the motion of a $5(10^4)$-ton ship is known to be proportional to the square of the speed. It is also known that the propellers must produce a thrust of a quarter of a million pounds to give the ship a speed of 40 ft/s. The ship is traveling at 40 ft/s when the engine is turned off and the ship slows to a speed of 10 ft/s. Determine (a) the distance the ship travels, and (b) the elapsed time.

4 Force as a Constant

If the forces acting on a particle in rectilinear motion have constant magnitude and direction, then equation (9) shows that $\dot{v}_P$ is constant. We could consider this case to be a special case of any of the three preceding ones. However, let us treat it separately because this situation arises quite frequently. We first determine v_P as a function of time by writing $\dot{v}_P \equiv dv_P/dt$, separating variables and integrating as follows.

$$\int_{v_0}^{v_P} dv_P = \dot{v}_P \int_{t_0}^{t} dt$$

$$v_P - v_0 = \dot{v}(t - t_0)$$

$$v_P = v_0 + \dot{v}_P(t - t_0) \tag{13}$$

To find $s(t)$ we then have

$$v_P \equiv \frac{ds}{dt} = v_0 + \dot{v}_P(t - t_0)$$

$$\int_{s_0}^{s} ds = \int_{t_0}^{t} [v_0 + \dot{v}_P(t - t_0)]\, dt$$

$$s - s_0 =$$

$$s = s_0 + v_0(t - t_0) + \tfrac{1}{2}\dot{v}_P(t - t_0)^2 \tag{14}$$

Finally, we find $v_P(s)$ by using the chain rule.

$$v_P \frac{dv_P}{ds} = \dot{v}_P$$

$$\int_{v_0}^{v_P} v_P\, dv_P = \dot{v}_P \int_{s_0}^{s} ds$$

$$(v_P)^2 = (v_0)^2 + 2\dot{v}_P(s - s_0) \tag{15}$$

Equations (13)–(15) are commonly employed in elementary physics courses, but their derivation is sufficiently easy to make it unnecessary to memorize them. However, if you should choose to memorize them, take care not to apply these formulas in situations where $\dot{v}_P$ is not constant.

EXAMPLE 4

The two blocks shown in the sketch are released from rest. It is observed that block A has a downward velocity of 6 ft/s after falling 2 ft. Knowing that block A weighs 50 lb, determine the weight of block B. Each pulley has negligible mass.

Solution

We draw a free body diagram of each block, and also of the pulley D, because it supports block B. In drawing these diagrams we use the fact that the tension in a cable is unchanged as it passes over ideal (massless and frictionless) pulleys.

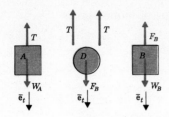

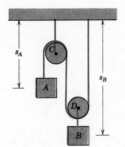

We have used the same tangent unit vector for each free body diagram, so we may relate the motion of the blocks by expressing their distances from a fixed datum. Thus, following the steps discussed in Module II, we draw a sketch of the pulley system that shows the various distances.

The length of the cable is constant, so that

$$2s_B + s_A = \text{constant}$$

Differentiation of this expression gives

$$2\dot{s}_B + \dot{s}_A = 0 \quad \text{and} \quad 2\ddot{s}_B + \ddot{s}_A = 0$$

so

$$v_B = -\tfrac{1}{2}v_A \quad \text{and} \quad \dot{v}_B = -\tfrac{1}{2}\dot{v}_A$$

These relationships may now be used in the equations of motion. Summing forces in the $\bar{e}_t$ direction for each block yields

$$\text{Block } A: \quad \Sigma F_t = W_A - T = \frac{W_A}{g}\dot{v}_A$$

$$\text{Block } B: \quad \Sigma F_t = W_B - F_B = \frac{W_B}{g}\dot{v}_B = -\tfrac{1}{2}\frac{W_B}{g}\dot{v}_A$$

The value F_B is determined by summing the forces on massless pulley D. Thus,

$$\text{Pulley } D: \quad \Sigma F_t = F_B - 2T = \frac{W_{\text{pulley}}}{g}\dot{v}_{\text{pulley}} = 0$$

negligible mass, hence can't no acceleration.

As a result, the equations of motion become

$$W_A - T = \frac{W_A}{g}\dot{v}_A$$

$$W_B - 2T = -\tfrac{1}{2}\frac{W_B}{g}\dot{v}_A$$

There are three unknowns in the foregoing two equations (W_A was given). The third relation needed is obtained from the given information about the motion of block A, which states that the speed of this block increases from zero to 6 ft/s as the block travels 2 ft. Because the equations of motion show that $\dot{v}_A$ is a constant, we may use equation (15) to get

$$v_A^2 = v_0^2 + 2\dot{v}_A(s - s_0)$$

$$6^2 = 0 + 2\dot{v}_A(2)$$

$$\dot{v}_A = 9.0 \text{ ft/s}^2$$

(handwritten notes in margins:)

$S = S_0 + V_0 t + \tfrac{1}{2}at^2$

$D = \tfrac{1}{2}at^2$

$6(t) + \tfrac{1}{2}at^2$

$\dfrac{dv}{dt} = \dfrac{dv}{ds}v$

$\dfrac{6}{2}(v)$

$\displaystyle\int v_f \cdot v_i = at$

$\dfrac{dv}{ds}\dfrac{ds}{dt}$

Substituting this value into the equations of motion gives

$$W_A - T = \frac{W_A}{g}(9.0)$$

$$W_B - 2T = -\frac{W_B}{g}(4.5)$$

from which we find

$$T = W_A\left(1 - \frac{9.0}{g}\right)$$

$$W_B - 2W_A\left(1 - \frac{9.0}{g}\right) = -\frac{W_B}{g}(4.5)$$

$$W_B = 2W_A\left(\frac{g-9}{g+4.5}\right) = 63.2 \text{ lb}$$

In conclusion, note that the tension force is not equal to the weight of the block A, for only in the case of static equilibrium is $T = W_A$. The difference is attributable to the inertia of the block.

HOMEWORK PROBLEMS

III.19 A rock is thrown downward from a bridge with a speed of 5 m/s and hits the water 4 s later. Determine (a) the height of the bridge and (b) the speed with which the rock hits the water.

III.20 A 3000-lb automobile accelerates at a constant rate from rest to 60 miles/hr in 10 s. The rolling resistance attributable to the undriven wheels is 20 lb. What is the tractive force exerted by the drive wheels on the ground during the acceleration? The air resistance is negligible.

III.21 A rider on an open elevator that is descending at 3 m/s throws a ball vertically upward with a speed of 9 m/s relative to the elevator. How long will it be before that ball returns to the platform, and how far will the platform have descended in this time interval?

III.22 The 100-lb box was initially descending at 8 ft/s. What constant force $\bar{F}$ must be applied to the 20-lb counterweight to bring the box to rest in 4 s? How far does the counterweight move in this time interval?

III.23 The driver of an automobile traveling 40 miles/hr sees a traffic light 1500 ft ahead turn red. The light is timed to be red for 15 s and green

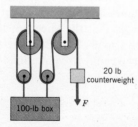

20 lb counterweight

100-lb box

F

Prob. III.22

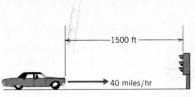

Prob. III.23

for 30 s. (a) What constant acceleration will enable the automobile to pass the light at the instant it turns green, and what is the corresponding speed as it passes the light? (b) What is the maximum constant deceleration the automobile can have and still pass the light in the green cycle? Also, find the speed of the automobile as it passes the light for this deceleration rate.

III.24   A motorist in an automobile traveling 72 km/hr at a distance of 60 m behind a truck going at the same speed wishes to pass the truck. In order to accomplish this maneuver safely, the automobile must accelerate and reach a position 60 m in front of the truck in no more than 30 s. (a) Determine the minimum constant acceleration of the automobile that is safe. (b) Determine the corresponding speed of the car at the end of the passing maneuver.

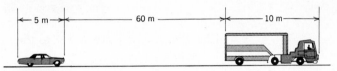

Prob. III.24

5   Force as a General Function

Force is a general function when the tangential component of the resultant force depends on more than one of the kinematic variables t, s, and v_P. Then, in the most general case, we find $\dot{v}_P = \Sigma F_t/m = f(t, s, v_P)$. This equation is not easy to solve, for we *cannot* use the method of separating variables to reduce the problem to a series of differential equations in first derivatives. Instead, we must form a second-order differential equation by using $v_P \equiv \dot{s}$ to obtain

$$\ddot{s} = f(t, s, \dot{s}) \tag{16}$$

Equations of this type can be solved, but in many cases the solution is possible only with the aid of an electronic computer. In certain situations equation (16) may form a linear differential equation, methods for the solution of which are the subject of the section on vibrations in Module VIII. However, obtaining solutions for the case of general functions $f(t, s, \dot{s})$ is beyond the scope of this text.

C.   EQUATIONS OF CURVILINEAR MOTION

With some basic techniques for solving differential equations established, we can now proceed to study the motion of a particle as it follows a curved path. Scalar equations of motion can be obtained from the components of Newton's second law, equation (1). To do this we must first choose the unit vectors associated with a particular coordinate system.

This choice is an important facet of the process of problem solving, in that a poor choice can unnecessarily complicate the scalar equations whose solutions we require. Consequently, we shall now discuss the form of the equations of motion that arise when each of the kinematical descriptions (path variables, rectangular Cartesian coordinates, and cylindrical coordinates) is considered.

1   Path Variables

The path variable approach can be employed when the particle path is known. Letting s be the arc length traversed by the particle, we know that

$$\bar{v}_P = \dot{s}\bar{e}_t \quad \text{and} \quad \bar{a}_P = \dot{v}_P \bar{e}_t + \frac{v_P{}^2}{\rho}\bar{e}_n$$

Equating components of $\Sigma \bar{F}$ to their corresponding components of $m\bar{a}_P$, we obtain

$$\Sigma F_t = ma_t \equiv m\dot{v}_P \tag{17}$$

$$\Sigma F_n = ma_n \equiv m\frac{v_P{}^2}{\rho}$$

where ΣF_t and ΣF_n are the force components in the direction of the tangent unit vector $\bar{e}_t$ and normal unit vector $\bar{e}_n$, respectively. Equations (17) are sufficient for treating planar problems. However, as we saw in the case of rectilinear motion, an added equation is available for solving three-dimensional problems. In such a case we use the fact that there is no acceleration in the $\bar{e}_b$ (binormal) direction. Recalling that the binormal is perpendicular to the osculating plane formed by $\bar{e}_t$ and $\bar{e}_n$, there is no acceleration in that direction. Hence, there is a force balance in that direction, as there was in equations (4). The importance of this fact will be demonstrated in Example 6.

The similarity of the first of equations (4) to the first of equations (17) suggests that path variables should be employed in situations where the particle is following a given path *and* the significant kinematical quantities for describing the forces and motion are t, s, and/or v_P. We shall see that situations that fail to meet these two criteria can be treated more readily by using the other kinematical descriptions.

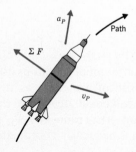

EXAMPLE 5

A rocket is programmed to gain speed as it moves along the path shown. Is the depicted situation possible?

Solution

An obvious error in the diagram is that $\bar{F}$ and $\bar{a}_P$ are not parallel, which violates Newton's second law. Also, we know that

$$\bar{v}_P = v_P \bar{e}_t$$

so that the velocity vector must be tangent to the path. Finally, we can determine the general orientation of the acceleration vector using

$$\bar{a}_P = \dot{v}_P \bar{e}_t + \frac{v_P^2}{\rho} \bar{e}_n$$

From the given information we know that v_P is increasing, so that $\dot{v}_P$ must be positive. The centripetal acceleration is directed toward the center of curvature. Hence the correct situation is as shown below.

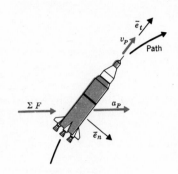

EXAMPLE 6

A 3000-lb automobile is traveling at 54 miles/hr along a level highway whose radius of curvature is 500 ft. The highway is banked at 10° from the horizontal. What are the normal and frictional forces that must be exerted between the tires and the road if the automobile is to remain in its lane?

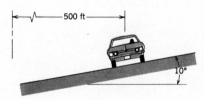

Solution

Let us begin by assuming that the automobile would skid outward if there were no friction, in which case the friction force $\bar{f}$ is inward. Recall that for two surfaces that are not sliding over each other, the only restriction on the friction force is that it must be less than the coefficient of static friction times the normal force. Thus, the free body diagram is as shown at left.

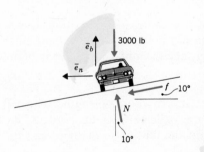

We choose path variables for the formulation, noting that the automobile is following a circular path in the horizontal plane. Thus $\bar{e}_n$ is horizontal, oriented toward the center of curvature, and $\bar{e}_t$ is perpendicular to the plane of the free body diagram. We must therefore define a unit vector $\bar{e}_b$ to describe the vertical direction.

From the given information we have

$$v_P = 54 \frac{\text{miles}}{\text{hr}} \times \left(\frac{88 \text{ ft/s}}{60 \text{ miles/hr}} \right) = 79.20 \text{ ft/s}$$

$$\dot{v}_P = 0 \qquad \rho = 500 \text{ ft}$$

Thus, the acceleration of the automobile is

$$\bar{a}_P = \dot{v}_P \bar{e}_t + \frac{v_P^2}{\rho} \bar{e}_n = 12.545 \bar{e}_n \text{ ft/s}^2$$

The force sums in the directions illustrated in the free body diagram then yield, with $g = 32.17 \text{ ft/s}^2$,

$$\Sigma F_n = N \sin 10° + f \cos 10° = ma_n = \frac{3000}{32.17}(12.545)$$

$$\Sigma F_b = N \cos 10° - f \sin 10° - 3000 = 0$$

The solution of these equations is

$$N = 3158 \text{ lb}$$

$$f = 631 \text{ lb}$$

Because f is positive, our initial assumption that the vehicle would slide outward in the absence of friction was correct. What would we do if this assumption were wrong?

HOMEWORK PROBLEMS

III.25　The longitudinal axis of a rocket ship is aligned tangent to the path, at an angle of 30° from the vertical. The thrust is 1,000,000 lb and the weight of the rocket is 200,000 lb. If the speed at this instant is 4000 ft/s, determine (a) the rate at which the rocket is gaining speed, and (b) the radius of curvature of the path.

Prob. III.26

III.26　A 50-g sphere is initially held by two cables in the position shown. The left cable is cut. Determine the tension in the right cable just before and just after the cable is cut. Also determine the initial acceleration of the sphere.

III.27　A small sphere of mass m is attached to a cord of length l. The ball rotates in a horizontal circle for which the angle between the cord and the vertical axis is θ. Derive expressions for the tension in the cord and the constant speed of the ball in terms of l and θ.

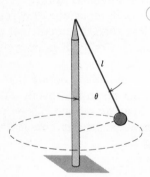

Prob. III.27

III.28　The driver of a miniature go-cart wants to perform a stunt by driving along a circular track that is banked at 90°, so that the surface of the track is vertical. The top speed of the car is 45 miles/hr and the coefficient of friction between the tires and the ground is 0.8. What is the maximum radius r that the track can have?

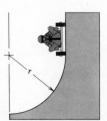

Prob. III.28

III.29　An automobile is slowed at a constant rate from 108 km/hr to 72 km/hr in a distance of 600 m as the vehicle follows a level curve of 500-m radius. What are the magnitude and angle relative to the road of the horizontal force exerted between the automobile and the ground at a position 400 m from the location where the braking was begun?

III.30   The 0.5-kg collar is pulled down the curved rod by a spring whose stiffness is 20 N/m and whose unstretched length is 500 mm. Friction is negligible. If the collar has a speed of 10 m/s in the position shown, determine the corresponding acceleration of the collar and the force the collar exerts on the rod.

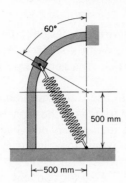

Prob. III.30

2   Rectangular Cartesian Coordinates

A general feature of extrinsic (path-independent) coordinates is that they may be applied to problems where the particle path is to be determined, as well as to situations where the particle path is given. The specific decision to use rectangular Cartesian coordinates is usually based on noticing that the forces *and* acceleration are most easily described by components in fixed directions, for example, horizontal or vertical.

Let (X, Y, Z) denote the rectangular coordinates of a particle. The acceleration of the particle is

$$\bar{a}_P = \ddot{X}\bar{I} + \ddot{Y}\bar{J} + \ddot{Z}\bar{K}$$

Hence, matching the components of $\Sigma \bar{F}$ and $m\bar{a}_P$, as required by equation (1), yields

$$\boxed{\begin{aligned} \Sigma F_X &= m\ddot{X} \\ \Sigma F_Y &= m\ddot{Y} \\ \Sigma F_Z &= m\ddot{Z} \end{aligned}}$$

(18)

The following example demonstrates a common application of equation (18).

EXAMPLE 7

Firefighters are trying to put out a fire on the roof of a warehouse that is 21 m tall. To do this they set up their hose 15 m from the base of the building,

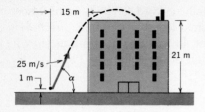

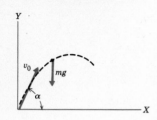

holding the nozzle 1 m above the ground. They aim the hose at an angle of elevation α, so as just to clear the edge of the roof. The speed of the water as it leaves the nozzle of the hose is 25 m/s. Neglecting the effects of air resistance, (a) determine the value of the angle α, and (b) locate where the water will land on the level roof.

Solution

Let us first determine the general relationships for the motion of a particle in freefall near the earth's surface. We begin by drawing a free body diagram of the particle in a general position, using rectangular Cartesian coordinates.

The decision to use Cartesian coordinates is based primarily on the fact that the weight force is in the (constant) vertical direction, and also on the fact that the particle path is not given. The latter makes path variables unsuitable. For convenience, the origin of the XYZ coordinate system was chosen to coincide with a known point, in this case the launching point of the particle. Also for convenience the XY plane was chosen such that the initial velocity vector $\bar{v}_0$ is contained in it.

In general, for the chosen coordinate system, we have

$$\bar{a}_P = \ddot{X}\bar{I} + \ddot{Y}\bar{J} + \ddot{Z}\bar{K}$$

The initial velocity and position vectors are

$$\bar{v}_0 = v_0(\cos \alpha \bar{I} + \sin \alpha \bar{I}) \qquad \bar{r}_0 = \bar{0}$$

Summing force components yields

$$\Sigma F_x = 0 = ma_x = m\ddot{X}$$

$$\Sigma F_y = -mg = ma_y = m\ddot{Y}$$

$$\Sigma F_z = 0 = m\ddot{Z}$$

Thus, the acceleration components reduce to

$$\ddot{X} = 0 \qquad \ddot{Y} = -g \qquad \ddot{Z} = 0$$

In words, the acceleration downward is constant and there is no component of acceleration in either horizontal direction.

We now integrate the equations of motion, using equations (13)–(15) for constant acceleration, which give

$$X = X_0 + \dot{X}_0 t$$

$$Y = Y_0 + \dot{Y}_0 t - \tfrac{1}{2}gt^2$$

$$Z = Z_0 + \dot{Z}_0 t$$

In the foregoing X_0, Y_0, Z_0, and $\dot{X}_0$, $\dot{Y}_0$, $\dot{Z}_0$ represent the components of the initial position and velocity vectors, $\bar{r}_0$ and $\bar{v}_0$, respectively. Therefore, using the initial conditions, we have

$$X = (v_0 \cos \alpha)t \qquad Y = (v_0 \sin \alpha)t - \tfrac{1}{2}gt^2 \qquad Z = 0$$

Because $Z = 0$, we can conclude that the particle path is planar. In Homework Problem III.33 we shall eliminate t between the X and Y equations, to find that Y is a quadratic function of X. Hence, the path is a parabola. However, this knowledge is not essential for solving the problem at hand.

To determine α we use the fact that the edge of the roof is located at $X = 15$ m, $Y = 20$ m. Thus, upon substituting $v_0 = 25$ m/s and $g = 9.806$ m/s^2, we find

$$15 = (25 \cos \alpha)t$$

$$20 = (25 \sin \alpha)t - \tfrac{1}{2}(9.806)t^2$$

Solving the first equation for t and placing it into the second, we get

$$t = \frac{0.6}{\cos \alpha}$$

$$20 = (25 \sin \alpha)\left(\frac{0.6}{\cos \alpha}\right) - 4.903\left(\frac{0.6}{\cos \alpha}\right)^2$$

Using the trigonometric identity, $\sin \alpha = (1 - \cos^2 \alpha)^{1/2}$, the latter equation becomes

$$20 = 15(1 - \cos^2 \alpha)^{1/2}\left(\frac{1}{\cos \alpha}\right) - 1.7651\left(\frac{1}{\cos^2 \alpha}\right)$$

Simplifying this expression yields

$$625.0 \cos^4 \alpha - 154.40 \cos^2 \alpha + 3.116 = 0$$

We now have a quadratic equation for $\cos^2 \alpha$. The resulting values of the angle are $\alpha = 64.69°$ and $\alpha = 81.44°$; the latter value corresponds to the case where the stream hits the corner of the roof on the way down. Hence, the solution to part (a) is

$$\alpha = 64.69°$$

Now part (b) can be solved by substituting this value for α into the equations for $X(t)$ and $Y(t)$. From the sketch provided with the problem, we see that we need the larger value of X corresponding to $Y = 20$ m. Thus

$$X = (25 \cos 64.69°)t$$

$$20 = (25 \sin 64.69°)t - \tfrac{1}{2}(9.806)t^2$$

From this we find that a particle will be at the elevation of the roof at both 1.194 s and 3.415 s after leaving the nozzle. The first value of time corresponds to the position at the near corner of the roof. Using the second value of time, $t = 3.415$ s, we get

$$X = (25 \cos 64.69°)(3.415) = 37.28 \text{ m}$$

from which we see that the stream of water will fall on the roof 22.28 m from the near corner.

HOMEWORK PROBLEMS

III.31   A basketball is thrown from the balcony of a building that is 4 m above the ground. It is thrown at an angle of elevation of 45° with an initial speed of 12 m/s. Determine the maximum height attained by the ball and the location where the ball strikes the ground.

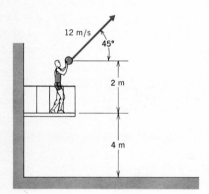

Prob. III.31

III.32   A pitching machine can throw baseballs at any angle of elevation with a speed of 120 ft/s. A ball player standing 180 ft away catches the ball at the same height above the level ground as that from which it was thrown. Neglecting air resistance, determine (a) the two possible angles of elevation for the throw, and (b) the final speed of the ball and the time duration of the flight for both possible angles.

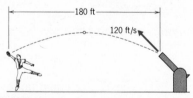

Prob. III.32

III.33   Use the equations derived in Example 7 to show that the path of a particle in free flight is a parabola. Then derive an expression for the velocity of the particle in terms of the horizontal distance it has traveled.

III.34   A ball is projected from a hill with a speed v_0 at an angle of elevation α. Determine the distance d from the launch point to the location where the ball hits the hill (a) using the xyz coordinate system shown, and (b) using the $x'y'z'$ coordinate system shown.

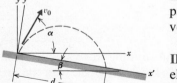

Prob. III.34

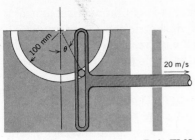

Prob. III.35

III.35 A 200-g pin is pulled through a circular groove by the grooved vertical arm, which is moving to the right at the steady speed of 20 m/s. Determine the normal force exerted by each groove on the pin in the position where $\theta = 36.87°$.

III.36 A 1-oz steel ball bearing in free flight passes through a magnetic field that exerts a constant horizontal pull of 2 oz in the X direction. Initially, the ball bearing is at a point 1 ft above the ground and its velocity is $20\bar{J}$ ft/s. Determine (a) the location of point B where the ball bearing hits the XY plane, and (b) the velocity of the ball bearing when it reaches point B.

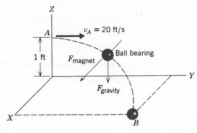

Prob. III.36

3 Cylindrical Coordinates

The acceleration of a particle in cylindrical coordinates is

$$\bar{a}_P = (\ddot{R} - R\dot{\phi}^2)\bar{e}_R + (R\ddot{\phi} + 2\dot{R}\dot{\phi})\bar{e}_\phi + \ddot{h}\bar{e}_h$$

We follow the approach of the previous cases by matching the radial, transverse, and axial components of $\Sigma \bar{F}$ to those of $m\bar{a}_P$. Hence, the scalar equations of motion are

$$
\begin{aligned}
\Sigma F_R &= m(\ddot{R} - R\dot{\phi}^2) \\
\Sigma F_\phi &= m(R\ddot{\phi} + 2\dot{R}\dot{\phi}) \\
\Sigma F_h &= m\ddot{h}
\end{aligned}
\tag{19}
$$

As is true for Cartesian coordinates, cylindrical coordinates can be applied in cases where the particle path is not given. Key factors in choosing this kinematical description are that the forces are described with respect to a fixed axis or point, or equally important, the angle of rotation and radial distance to a fixed axis (or point) are the significant kinematical quantities.

EXAMPLE 8

A sharpshooter is tracking a moving target. In so doing, the rifle barrel is

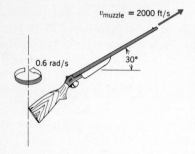

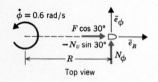

Top view

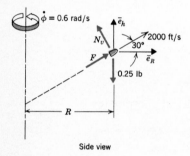

Side view

elevated 30° above the horizontal and rotated about the vertical axis at a constant rate of 0.6 rad/s. A bullet weighing 4 oz leaves the barrel with a muzzle velocity of 2000 ft/s. Determine the horizontal component of the force the bullet exerts on the barrel immediately before leaving the barrel.

Solution

The forces acting on the bullet are its weight (0.25 lb), the components N_ϕ and N_v of the contact force perpendicular to the barrel in the horizontal and vertical planes, and the explosive force F tangent to the barrel. The value of N_ϕ represents the solution of the problem. To illustrate this three-dimensional situation let us draw free body diagrams in side and top views.

Because the rate of rotation about a fixed axis is given, we choose cylindrical coordinates defined with respect to this axis. It follows that $\bar{e}_R$ is perpendicular to the axis. To describe the motion of the bullet, we note that the muzzle velocity, which is tangent to the rifle barrel, represents the radial and axial components of the total velocity of the bullet. Thus

$$\bar{v}_B = \dot{R}\bar{e}_R + R\dot{\phi}\bar{e}_\phi + \dot{h}\bar{e}_h$$

$$= 2000(\cos 30°\bar{e}_R + \sin 30°\bar{e}_h) + 0.6R\bar{e}_\phi$$

Therefore, by equating coefficients of the unit vectors, we find

$$\dot{R} = 2000 \cos 30° = 1732 \text{ ft/s}$$

$$\dot{h} = 2000 \sin 30° = 1000 \text{ ft/s}$$

Also, notice that the length of the rifle is not given, so we cannot determine the value of R. (We continue, hoping that this will not affect the computations for the force N_h.)

We next determine the acceleration of the bullet. The values of $\dot{R}$, $\dot{\phi}$, and $\dot{h}$ are known; it is also known that $\ddot{\phi} = 0$ because $\dot{\phi}$ is constant. Hence

$$\bar{a}_B = (\ddot{R} - R\dot{\phi}^2)\bar{e}_R + (R\ddot{\phi} + 2\dot{R}\dot{\phi})\bar{e}_\phi + \ddot{h}\bar{e}_h$$

$$= [\ddot{R} - R(0.6)^2]\bar{e}_R + [0 + 2(1732)(0.6)]\bar{e}_\phi + \ddot{h}\bar{e}_h$$

The value of N_ϕ is determined directly from the force sum in the transverse direction, which gives

$$\Sigma F_\phi = N_\phi = ma_\phi = \left(\frac{0.25}{32.17}\right)(2)(1732)(0.6)$$

$$N_\phi = 16.15 \text{ lb}$$

Does this force assist or oppose the rotation of the rifle barrel?

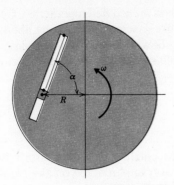

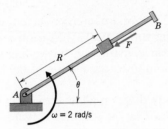

Prob. III.37

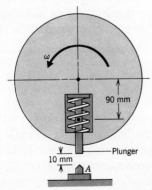

Prob. III.41

HOMEWORK PROBLEMS

III.37   A block is held in position within a groove in a rotating disk by a cable. Derive expressions for the tension force in the cable and the normal force exerted by the groove on the block in terms of the constant angular speed ω of the disk and the dimensions R and α shown. The motion occurs in the horizontal plane.

III.38   The 2-kg block slides over the rigid arm AB, which is rotating in the vertical plane at 2 rad/s counterclockwise. A radial force F gives the block a constant inward speed of 1 m/s. When $\theta = 30°$, the radial distance R is 400 mm. Determine the corresponding value of F if the coefficient of friction is 0.10.

Prob. III.38

III.39   Passengers on an amusement park ride sit in seats that are suspended by 25-ft-long cables from the horizontal arm AB, as shown. Arm AB rotates about the vertical axis at a constant angular speed ω. If $\theta = 36.87°$, determine the corresponding value of ω.

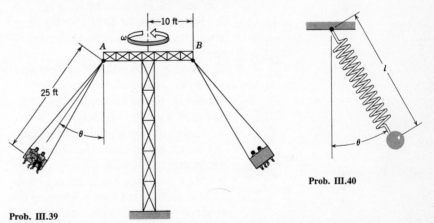

Prob. III.39

Prob. III.40

III.40   An 8-oz sphere is attached to a spring whose stiffness is 10 lb/ft as it swings in the vertical plane. The unstretched length of the spring is 18 in. When $\theta = 30°$ it is observed that $l = 20$ in. and that the velocity of the sphere is 80 in./s horizontally to the right. Determine the corresponding values of $\ddot{\theta}$ and $\ddot{l}$.

III.41   The speed governer shown in cross section consists of a 200-g plunger that is restrained by a spring whose stiffness is 30 N/mm. When the angular speed is zero the compressive force in the spring is 200 N and the plunger is in the position shown. Determine the constant angular speed ω that will cause the plunger to move the 10 mm required to trip switch A.

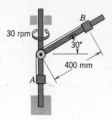

30 rpm

B

30°

400 mm

A

Prob. III.42

III.42 Collars A and B, each having mass m, are tied together by a cable as they slide over their guides. The entire assembly is rotating about the vertical axis at 30 rpm. In the position shown block A is descending at 4 m/s. Determine the acceleration of collar B at this instant. Friction is negligible.

D. PROBLEM SOLVING

The steps for formulating and solving the equations of motion for a body when it is modeled as a particle are probably apparent to you as a result of the eight examples we have considered. Nevertheless, let us present them formally.

1 Draw a free body diagram. If several bodies are interconnected, draw a free body diagram of each body, being sure to satisfy Newton's third law of action and reaction. As an aid in constructing these diagrams, remember that each constraint (restriction) upon the motion of the body can only be imposed by a force.

2 Write down or depict in the free body diagram all given information about the forces and kinematical variables. List the quantities required for the solution.

3 Choose a coordinate system for describing vector components, and show the associated unit vectors in the free body diagram. This choice should be based on suiting the kinematical quantities and the forces associated with the system.

4 Following the kinematical methods in Module II, express the acceleration of each body in terms of the coordinate system chosen in the preceding step.

5 Sum the force components in the direction of each unit vector and equate each term to the corresponding component of $m\bar{a}_P$. Check that the number of algebraic and differential equations match the number of unknowns. Common reasons for these two numbers being unequal are a force that was not properly accounted for, a given quantity that was not used, or a kinematical constraint that was not considered.

6 Solve the equations of motion.

ILLUSTRATIVE PROBLEM 1

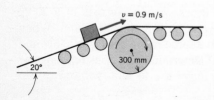

$v = 0.9$ m/s

300 mm

20°

A conveyor belt moves with a constant speed of 0.9 m/s up a 20° incline and passes over an idler wheel whose radius is 300 mm. The coefficient of static friction between a package and the conveyor belt is 0.4. Determine the maximum deceleration rate that can be imparted to the belt without causing packages to slip as they pass over the idler wheel. How does this result compare to the maximum deceleraton rate for which packages will not slip when they are on the incline?

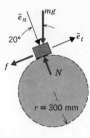

Solution

Step 1   The critical position in regard to slipping of the packages is the location where the packages first begin to follow the idler, for it is at this position along the idler that the component of gravity opposite the direction of motion is largest. The friction force $\bar{f}$ is oriented as shown in the free body diagram, because this force opposes the direction in which the package would slip if the deceleration rate were too large.

Note that if we set $r = \infty$, the diagram then describes the situation when the package is approaching the idler on the incline.

Step 2   Given $v_P = 0.9$ m/s and $\mu_s = 0.4$, find the maximum value of $\dot{v}_P$ for the condition of no slipping at the idler and on the incline.

Step 3   As shown in the free body diagram, we chose to use path variables based on the fact that the speed is the principal kinematical quantity in this problem.

Step 4   The acceleration of the package is easily shown to be

$$\bar{a}_P = \dot{v}_P \bar{e}_t + \frac{v_P{}^2}{\rho}\, \bar{e}_n = \dot{v}_P \bar{e}_t + 2.7\bar{e}_n \text{ m/s}^2$$

Step 5   The maximum deceleration is obtained when $f = \mu_s N$. Thus, the equations of motion are

$$\Sigma F_t = -0.4N - mg \sin 20° = ma_t = m\dot{v}_P$$

$$\Sigma F_n = mg \cos 20° - N = ma_n = m(2.7)$$

Considering m as a parameter, we can solve these equations for N and $\dot{v}_P$.

Step 6   Using $g = 9.806$ m/s², we find

$$N = m(9.806 \cos 20° - 2.7) = 6.515m \text{ newtons}$$

$$\dot{v}_P = -0.4\,\frac{N}{m} - 9.806 \sin 20° = -5.960 \text{ m/s}^2$$

Next we consider the situation when the package is on the inclined ramp. Here, the maximum deceleration rate is easily determined, because the free body diagram for this situation is the same as the one shown earlier. Then, because this is rectilinear motion, we set the centripetal acceleration to zero, from which it follows that

$$N = m(9.806 \cos 20°)$$

$$\dot{v}_P = -0.4\frac{N}{m} - 9.806 \sin 20° = -7.040 \text{ m/s}^2$$

The maximum deceleration rate is smaller for the position on the roller because the centripetal acceleration at this location decreases the value of the normal force, which, in turn, has the effect of lowering the friction force.

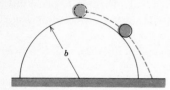

ILLUSTRATIVE PROBLEM 2

A small ball bearing moves downward over the surface of a smooth semicircular cylinder after being released from rest at the top. Determine the location at which the ball bearing will cease to be in contact with the cylinder.

Solution

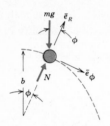

Step 1 The ball bearing loses contact with the cylinder when the normal force $\bar{N}$ becomes zero. Therefore, we draw a free body diagram with the aim of determining this force.

Step 2 We must find the value of the angle ϕ defined in the free body diagram at which $N = 0$, knowing that $\bar{v} = \bar{0}$ when $\phi = 0$.

Step 3 Because the ball bearing follows a circular path, we could use path variables to formulate the equations of motion. However, as we wish to relate the force $\bar{N}$ to the angle ϕ, a more direct approach is to utilize polar coordinates, as illustrated in the free body diagram.

Step 4 For polar coordinates we have

$$\bar{a}_P = (\ddot{R} - R\dot{\phi}^2)\bar{e}_R + (R\ddot{\phi} + 2\dot{R}\dot{\phi})\bar{e}_\phi$$

For the coordinate system chosen, $R = b$. However, the only information given about ϕ is that $\dot{\phi} = 0$ when $\phi = 0$. Therefore, we have

$$\bar{a}_P = -b\dot{\phi}^2\bar{e}_R + b\ddot{\phi}\bar{e}_\phi$$

Step 5 Newton's second law gives

$$\Sigma F_R = N - mg \cos \phi = ma_R = -mb\dot{\phi}^2$$

$$\Sigma F_\phi = mg \sin \phi = ma_\phi = mb\ddot{\phi}$$

The second of these two equations is a differential equation for $\phi(t)$, whereas the first serves as an algebraic equation that gives N. We therefore proceed to the solution of these equations.

Step 6 We want to find the value of ϕ for which $N = 0$. Hence, set $N = 0$ in the first equation of motion to obtain

$$0 - mg \cos \phi = mb\dot{\phi}^2$$

In other words, $N = 0$ when $\dot{\phi}^2 = (g/b) \cos \phi$. We must now use the second equation of motion to determine how $\dot{\phi}$ depends on ϕ. To achieve this we write

$$\ddot{\phi} = \frac{g}{b}\sin\phi$$

Notice that this differential equation is identical in form to that obtained in rectilinear motion when acceleration is a function of position. To separate variables we write

$$\ddot{\phi} \equiv \frac{d}{dt}\dot{\phi} \equiv \frac{d\phi}{dt}\frac{d}{d\phi}\dot{\phi} \equiv \dot{\phi}\frac{d\dot{\phi}}{d\phi} = \frac{g}{b}\sin\phi$$

Using the initial condition that $\dot{\phi} = 0$ when $\phi = 0$, we have

$$\int_0^\phi \dot{\phi}\,d\dot{\phi} = \frac{g}{b}\int_0^\phi \sin\phi\,d\phi$$

$$\tfrac{1}{2}\dot{\phi}^2 = \frac{g}{b}(1 - \cos\phi)$$

By substituting the expression for the value of $\dot{\phi}^2$ at which $N = 0$ into this equation, we find

$$\tfrac{1}{2}\frac{g}{b}\cos\phi = \frac{g}{b}(1 - \cos\phi)$$

$$\cos\phi = \tfrac{2}{3}, \text{ so } \quad \phi = 48.18° \qquad \triangleleft$$

When the ball bearing reaches this position it flies off tangent to the cylinder and follows a free-fall trajectory until it hits the ground. We mention in closing that the expression for the value of $\dot{\phi}^2$ at any angle ϕ can more easily be determined by energy principles. This approach, which will be presented later in this module, avoids the need to solve a differential equation.

ILLUSTRATIVE PROBLEM 3

A horizontal force $\bar{F}$ pushes a 5-kg block up a smooth parabolic hill whose shape is defined by $Y = 4X^2$, where X and Y are measured in meters. Determine how $\bar{F}$ must vary with the location of the block in order for the horizontal component of the velocity of the block to be a constant value c.

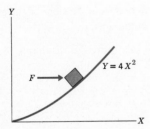

Solution

Step 1 There is no friction force because the hill is smooth. Thus, we have the following free body diagram.

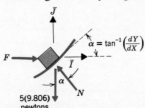

Note that the normal force $\bar{N}$ is perpendicular to the local tangent to the parabola.

Step 2 We are told that the equation of the path is $Y = 4X^2$ and that $\dot{X} = c$, $\ddot{X} = 0$. We wish to determine the value of F.

Step 3 As indicated in the free body diagram, we shall use the XY coordinate system to formulate the equations of motion. This approach is chosen because the given kinematical information involves the velocity component in the X direction.

Step 4 In terms of rectangular Cartesian components, the acceleration of the block is

$$\bar{a}_P = \ddot{X}\bar{I} + \ddot{Y}\bar{J} + \ddot{Z}\bar{K}$$

but for planar motion $\ddot{Z} = 0$. Furthermore, because we know that $\dot{X} = c$ and $\ddot{X} = 0$, we can determine $\ddot{Y}$ by differentiating the equation for Y twice with respect to time. Thus

$$\dot{Y} \equiv \frac{dY}{dt} = \frac{d}{dt}(4X^2) = 8X\dot{X} = 8cX$$

$$\ddot{Y} = 8c\dot{X} = 8c^2$$

Therefore

$$\bar{a}_P = 8c^2\bar{J}$$

Step 5 Referring to the free body diagram, we have

$$\Sigma F_X = F - N \sin \alpha = ma_X = 0$$

$$\Sigma F_Y = N \cos \alpha - 5(9.806) = ma_Y = 5(8c^2)$$

Because the value of the angle α can be determined from the slope of the parabola, the two equations of motion have only two unknowns.

Step 6 We solve the first equation for N, and substitute this result into the second equation, with the following results.

$$N = \frac{F}{\sin \alpha} \qquad \left(\frac{F}{\sin \alpha}\right) \cos \alpha = 49.03 + 40c^2$$

$$F = (49.03 + 40c^2) \tan \alpha \text{ newtons}$$

Noting that

$$\tan \alpha = \frac{dY}{dX} = 8X$$

we find that

$$F = (392.2 + 320c^2) X \text{ newtons}$$

You probably did not expect it to be necessary to increase the force $\bar{F}$ in proportion to the horizontal distance traveled by the block. Can you give a physical explanation of why a constant value of F does not produce the desired motion?

ILLUSTRATIVE PROBLEM 4

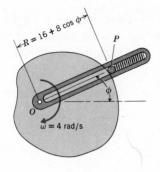

The slotted arm rotates in a horizontal plane at a constant clockwise rate $\omega = 4$ rad/s as it guides the pin P over the fixed cam. The cam is defined by $R = 16 + 8 \cos \phi$, where R is the distance in inches from the fixed point O to a point on the perimeter of the cam. The compression force in the spring is 3.0 lb when $\phi = 0°$ and 1.0 lb when $\phi = 180°$. Calculate the value of the contact force exerted between the 12-oz pin and the cam for the position where $\phi = 60°$.

Solution

Step 1   There are three forces acting on the pin in the horizontal plane: the spring force $\bar{S}$ which acts radially inward, a force $\bar{F}$ exerted by the slotted arm which is pependicular to the walls of the groove, and a normal force $\bar{N}$ exerted by the cam, perpendicular to the local tangent to the cam. Because the value of N is desired when $\phi = 60°$, we draw a free body diagram of the pin in that position.

Step 2   It is given that $\omega = 4$ rad/s and $\dot{\omega} = 0$. Also, it is known that $S = 3$ lb when $\phi = 0$ and $S = 1$ lb when $\phi = 180°$, and that the weight of the pin is $\frac{12}{16}$ lb. We wish to find the value of N when $\phi = 60°$.

Step 3   It is logical to use cylindrical coordinates whose origin is point O, because the given kinematical information is the angular speed of the radial line from point O to the pin. Recalling that the unit vector $\bar{e}_\phi$ is oriented in the direction of increasing ϕ, the unit vectors are as shown in the free body diagram.

Step 4   For cylindrical coordinates we have

$$\bar{a}_P = (\ddot{R} - R\dot{\phi}^2)\bar{e}_R + (R\ddot{\phi} + 2\dot{R}\dot{\phi})\bar{e}_\phi + \ddot{h}\bar{e}_h$$

For planar motion $\ddot{h} = 0$, and for clockwise rotation ϕ is decreasing, so that

$$\dot{\phi} = -\omega = -4 \text{ rad/s} \qquad \ddot{\phi} = -\dot{\omega} = 0$$

The distance R can be determined by using the given equation defining the shape of the cam. Thus

$$R = 16 + 8 \cos \phi$$

$$\dot{R} = -8\dot{\phi} \sin \phi$$

$$\ddot{R} = -8\dot{\phi}^2 \cos \phi - 8\ddot{\phi} \sin \phi$$

Solving for these values at the position where $\phi = 60°$ yields

$$R = 20.0 \text{ in.} \qquad \dot{R} = 27.71 \text{ in./s} \qquad \ddot{R} = -64.0 \text{ in./s}^2$$

Therefore

$$\bar{a}_P = [-64.0 - 20(4)^2]\bar{e}_R + [0 + 2(27.71)(-4)]\bar{e}_\phi$$

$$= -384.0\bar{e}_R - 221.7\bar{e}_\phi \ \text{in./s}^2$$

Another quantity that is related to the kinematics of the motion is the angle α between the normal force $\bar{N}$ and the radial line, as shown in the free body diagram. The accompanying sketch shows that α is also the angle between the local tangent to the cam and the transverse line.

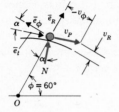

Hence, to determine α we can use the fact that the velocity $\bar{v}_P$ is tangent to the path. From the sketch we have

$$\alpha = \tan^{-1}\left(-\frac{v_R}{v_\phi}\right)$$

Note that the negative sign in this equation arises because the vector $\bar{v}_P$ illustrated in the sketch has a negative transverse component. Thus, evaluating $\bar{v}_P$ at $\phi = 60°$, we find

$$\bar{v}_P = \dot{R}\bar{e}_R + R\dot{\phi}\bar{e}_\phi = 27.71\bar{e}_R + 20.0(-4)\bar{e}_\phi$$

$$\alpha = \tan^{-1}\left(\frac{27.71}{80.0}\right) = 19.10°$$

Step 5   Referring to the free body diagram in step 1, the equations of motion are

$$\Sigma F_R = N\cos\alpha - S = \frac{0.75}{g}a_R = \frac{0.75}{g}(-384.0)$$

$$\Sigma F_\phi = N\sin\alpha + F = \frac{0.75}{g}a_\phi = \frac{0.75}{g}(-221.7)$$

where $g = 386.0 \ \text{in./s}^2$, because the units of length are inches.

These two equations have three unknowns (N, S, and F), so we need another equation. This is obtained by using the given information about the spring force. For a linearly elastic spring $S = k\Delta$, where Δ is the amount by which the length of the spring has been decreased from the unstretched value. Let Δ_0 be this value at $\phi = 0$. Then, because $R = (16+8)$ in. at $\phi = 0$, $R = 20.0$ in. at $\phi = 60°$, and $R = (16 - 8)$ in. at $\phi = 180°$, it follows that

$$S_0 = 3 \ \text{lb} = k\Delta_0$$

$$S_{60} = k[\Delta_0 - (24.0 - 20.0)]$$

$$S_{180} = 1 \ \text{lb} = k[\Delta_0 - (24.0 - 8.0)]$$

Solving these equations, we find

$$1 = k\Delta_0 - 16k = 3 - 16k, \text{ so } \quad k = 0.125 \text{ lb/in.}$$

$$\Delta_0 = 3/k = 24.0 \text{ in.}$$

$$S_{60} \doteq 0.125[24.0 - (24.0 - 20.0)] = 2.50 \text{ lb}$$

Step 6 We may now find N at $\phi = 60°$ by solving the equation for ΣF_R in the preceding step, using the value of α found in step 4.

$$N = \frac{1}{\cos 19.10°}[2.50 - \frac{0.75}{386.0}(384.0)] = 1.856 \text{ lb}$$

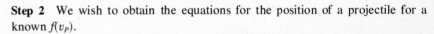

ILLUSTRATIVE PROBLEM 5

The magnitude of the air resistance on a projectile is known to be a specific function of the speed of the projectile, $f(v_P)$. For example, in many situations $f(v_P) = kv_P{}^2$. Determine the equations describing the motion of a projectile near the surface of the earth whose solution would yield the location of the projectile at any instant. The initial velocity of the projectile is v_0 at an angle of elevation θ.

Solution

Step 1 The air resistance is a frictional force opposite the velocity of the projectile. Because the path is planar in the absence of air resistance, it follows that the motion here is also planar. Hence, the free body diagram is as shown below.

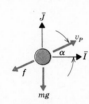

Step 2 We wish to obtain the equations for the position of a projectile for a known $f(v_P)$.

Step 3 As indicated in the free body diagram, we choose rectangular Cartesian coordinates for the formulation. To a certain extent this choice is by default, in that we cannot employ path variables because the path is unknown, whereas polar coordinates are clearly not applicable. Also, the chosen approach is consistent with that employed in Example 7 to treat projectile motion in the absence of air resistance.

Step 4 For rectangular Cartesian coordinates,

$$\bar{a}_P = \ddot{X}\bar{I} + \ddot{Y}\bar{J} + 0\bar{K}$$

We also need the angle α between the velocity vector and the horizontal line in order to compute the sum of the forces. Toward this end, we draw a sketch of the velocity vector $\bar{v}_P = \dot{X}\bar{I} + \dot{Y}\bar{J}$.

From this sketch we see that

$$\tan \alpha = \frac{\dot{Y}}{\dot{X}} \qquad \sin \alpha = \frac{\dot{Y}}{\sqrt{\dot{X}^2 + \dot{Y}^2}} \qquad \cos \alpha = \frac{\dot{X}}{\sqrt{\dot{X}^2 + \dot{Y}^2}}$$

Step 5   Using the free body diagram to sum forces, we have

$$m\ddot{X} = -[\,f(v_P)]\cos\alpha$$

$$m\ddot{Y} = -mg - [\,f\,(v_P)]\sin\alpha$$

We now express all terms in these equations by their Cartesian coordinate equivalents. Thus

$$\ddot{X} + \frac{1}{m}[f(\sqrt{\dot{X}^2 + \dot{Y}^2})\,]\,\frac{\dot{X}}{\sqrt{\dot{X}^2 + \dot{Y}^2}} = 0$$

$$\ddot{Y} + \frac{1}{m}[f(\sqrt{\dot{X}^2 + \dot{Y}^2})\,]\,\frac{\dot{Y}}{\sqrt{\dot{X}^2 + \dot{Y}^2}} = -g$$

Accompanying these equations are the initial conditions that

$$\bar{r}_{P/O}(t = 0) = X_0\bar{I} + Y_0\bar{J} \text{ and } \bar{v}_P(t = 0) = v_0(\cos\theta\bar{I} + \sin\theta\bar{J})$$

where X_0 and Y_0 are the initial coordinates of the projectile.

The foregoing equations are, in general, complicated differential equations whose solutions are intertwined. As a specific illustration, consider the case when $f(v_P) = kv_P^2$. We then have

$$f(\sqrt{\dot{X}^2 + \dot{Y}^2}) = k[\sqrt{\dot{X}^2 + \dot{Y}^2}]^2 = k(\dot{X}^2 + \dot{Y}^2)$$

so that the equations of motion become

$$\ddot{X} + \frac{k}{m}\dot{X}\sqrt{\dot{X}^2 + \dot{Y}^2} = 0$$

$$\ddot{Y} + \frac{k}{m}\dot{Y}\sqrt{\dot{X}^2 + \dot{Y}^2} = -g$$

Clearly, the solution of these equations is beyond the scope of an elementary course. In fact, accurate solutions are possible only with the aid of an electronic computer. However, if the air resistance were proportional to the speed, we could solve the resulting equations of motion. This is the subject of Homework Problem III.82.

HOMEWORK PROBLEMS

III.43   A 3-kg block moving down the inclined rod is brought to a stop by a cord that is held horizontally. The tension in the cord is 100 N and the coefficient of friction between the block and the rod is 0.4. If the block was initially sliding at 2 m/s, determine the distance the block slides before coming to rest.

III.44   A pendulum composed of a 2-lb sphere and a 3-ft-long cable is mounted from a post on a cart whose weight is 4 lb. If the cart is pulled by

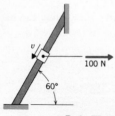

Prob. III.43

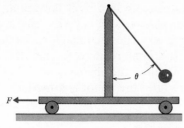

Prob. III.44

Crate

Prob. III.45

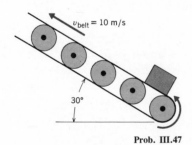
$v_{belt} = 10$ m/s

30°

Prob. III.47

a horizontal force $\bar{F}$, determine an expression for the constant angle θ between the cable of the pendulum and the vertical direction.

III.45   A crate weighing 1 ton rests on the flat bed of a 5-ton truck traveling at 55 miles/hr. The coefficients of static and kinetic friction between the crate and the truck are $\mu_s = 0.50$ and $\mu_k = 0.45$, respectively. Determine the minimum length of time in which the truck may be brought to rest without causing the crate to shift relative to the truck. What is the total braking force that the tires must produce for this deceleration?

III.46   The truck in Problem III.45 decelerates at 20 ft/s². Determine the deceleration rate of the crate and the required braking force for this deceleration.

III.47   A package is at rest when it is placed on a conveyor belt that is moving up a 30° incline at a constant speed of 10 m/s. The coefficient of kinetic friction between the package and the belt is 0.75. Determine how far a package will move up the incline before it ceases to slip relative to the conveyor belt. Also determine the corresponding elapsed time.

III.48   Each car of a commuter train is self-propelled and has a mass of 10,000 kg. The maximum forward traction force from the electric motors in each car is 10,000 N and the maximum braking force is 20,000 N. What is the minimum length of time in which the train can depart from one station and come to a stop at the next station 6 km away without exceeding a speed 90 km/hr? What is the average speed of the train between stops?

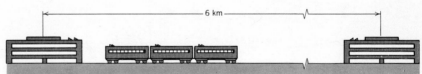
6 km

Prob. III.48

III.49   A locomotive pulls a two-car train with a constant acceleration from rest to 20 m/s in 1 min along a level track. The mass of each car is 10,000 kg and the mass of the locomotive is 20,000 kg. The frictional resistance on each car and the locomotive attributable to the undriven wheels is 1000 N each. Determine the traction force of the locomotive and the force exerted within the couplers between the cars during the acceleration.

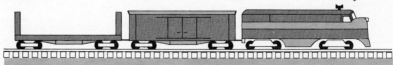

Prob. III.49

III.50 A 1000-kg motorboat is pointed upstream in a river flowing at 3 m/s and the thrust of the outboard motor is 400 N, with the result that the boat holds its position relative to the riverbank. The frictional force of the water is proportional to the speed of the boat relative to the water. Determine (a) the true speed of the boat 30 s after the engine is turned off, and (b) the distance the boat will drift 30 s after the motor is turned off.

Prob. III.50

III.51 The 70-kg sign painter applies a tensile force of 220 N to the free end of the cable. Determine (a) the acceleration of the painter, and (b) the force exerted between the painter and the seat.

Prob. III.51

III.52 Each block has a mass of 20 kg. What force $\bar{F}$ must be applied to block A to accelerate block B at 5 m/s²? What is the corresponding tension force in the cable? Friction is negligible.

III.53 The weights of the blocks are $W_A = 10$ lb, $W_B = 5$ lb, and $W_C = 20$ lb. The coefficient of friction for blocks A and B is $\mu = 0.20$. Determine how far block C will descend in 4 s, starting from rest.

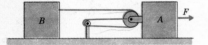

Prob. III.52

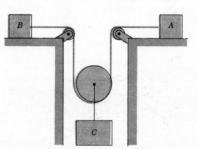

Prob. III.53

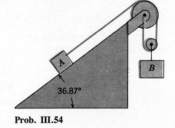

Prob. III.54

III.54 When $t = 0$, block A is moving down the incline at 2 m/s. Block A has a mass of 4 kg, block B has a mass of 8 kg, and the coefficient of friction between the block and the incline is $\mu = 0.2$. How far will block A move down the incline before coming to rest?

III.55 The crate is raised by a tensile force $\bar{F}$, which pulls the end of the cord downward at the constant speed v_0. Derive an expression for the value of F in terms of the speed v_0, the mass of the crate m, and the dimensions h and L.

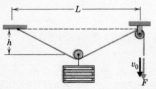

Prob. III.55

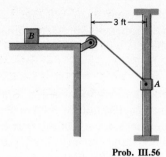

Prob. III.56

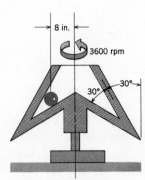

Prob. III.58

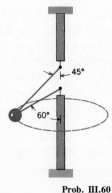

Prob. III.60

III.56 Block *A* weighs 3 lb and block *B* weighs 6 lb. The two blocks are tied together by a cord and slide with negligible resistance from friction. Block *A* is released at the elevation of block *B* with the cord taut. Derive the differential equation of motion governing the distance block *A* descends.

III.57 A 150-m-radius turn in a highway is designed such that 120 km/hr is the maximum safe speed of an automobile whose tires have a coefficient of friction of 0.40 for sliding (sideways) on the road. What is the angle of banking for the road?

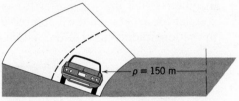

Prob. III.57

III.58 Hollow plastic spheres weighing 2 oz are tested within the centrifuge device whose cross section is shown. Determine the normal forces exerted between a sphere and each of the smooth walls for a constant rate of rotation of 3600 rpm.

III.59 A small ball bearing follows a horizontal circle as it rolls at a constant speed *v* on the interior of a hemispherical bowl of radius *r*. Determine the relationship between *v* and the depth *h* of the ball below the rim of the bowl.

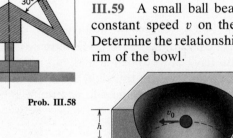

Prob. III.59

III.60 Two cords are attached to a 1-kg sphere that revolves in a horizontal circle of 500-mm radius at a constant speed of 2.5 m/s. Determine the tension force in each cord.

III.61 For what range of values of the speed of the sphere in Problem III.60 will both cords be taut?

III.62 Two blocks rest on a turntable that is rotating about the vertical axis at a constant angular speed ω. The blocks are connected by an

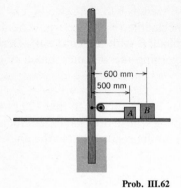

Prob. III.62

inextensible cable. The masses of blocks A and B are 1.0 and 1.5 kg, respectively, and the coefficient of friction is 0.30. Determine the maximum angular speed for which the blocks will not slide. Which way will the blocks slide if ω exceeds this value?

III.63 A train follows a level unbanked curve of 5-km radius. Starting from rest, the locomotive has a constant acceleration of 0.25 m/s². The mass of the locomotive is $2(10^4)$ kg and the mass of each of the four cars it is pulling is $2(10^5)$ kg. Also, the frictional resistance on each car is (10^4) N. Determine the components of the total reaction force between the locomotive and the track after it has traveled 1000 m. Assume that the coupler force between the locomotive and the cars being pulled is tangent to the longitudinal axis of the locomotive.

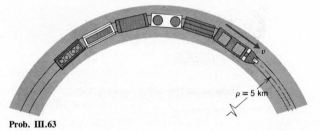

Prob. III.63

III.64 Solve Problem III.63 if the track is banked at 10°.

III.65 An automobile is traveling at a constant speed over a hill. At the top the radius of curvature is 500 ft. (a) Determine the maximum speed the automobile can have in this positon and still remain in contact with the road. (b) If the automobile is traveling at 60 miles/hr and the brakes are suddenly applied at the top of the hill, determine the maximum possible deceleration rate. The coefficient of friction between the tires and the ground is 0.70.

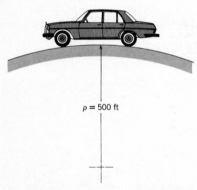

Prob. III.65

III.66 The vertical elevation of a straight road is defined by $Y = 2\sin(\pi X/20)$, where X and Y are measured in meters. Determine the maximum deceleration rate of an automobile traveling at 45 km/hr at (a) point A, (b) point B, (c) point C. The coefficient of friction between the tires and the ground is 0.60.

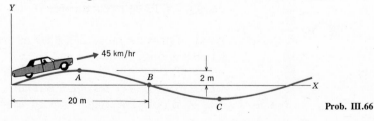

Prob. III.66

III.67   The curve of a level unbanked entrance ramp on a highway is defined by $Y = X^2/1600$, where X and Y are expressed in feet. A 3000-lb automobile, starting from rest at point A, accelerates at a constant rate and passes point B with a speed of 50 miles/hr. What are the components of the frictional force between the tires and the ground as the automobile passes point B? *Hint:* First determine the distance the automobile travels between points A and B.

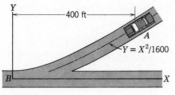

Prob. III.67

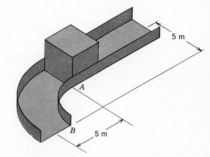

Prob. III.68

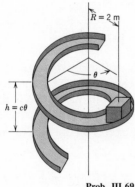

**Prob. III.69
and Prob. III.70**

III.68   A 20-kg block of ice slides within a chute. Section AB forms a level 90° turn of 5-m radius. The viscous friction force between the ice and the metal chute is given by $f = 0.2v$ N, where v is the speed in meters per second. At position A the ice block is entering the turn at a speed of 10 m/s. Determine (a) the maximum and (b) the minimum normal forces exerted by the block of ice on the side wall of the chute in section AB.

III.69   A package descends a spiral chute that is defined by $R = 2$ and $h = c\theta$, where R and c are expressed in meters and θ is in radians. There is no friction. Determine the speed of the package after it has fallen 4 m with (a) $c = 1$ m, and (b) $c = 2$ m. Also determine the time required for the descent.

III.70   A package descends a spiral chute defined by $R = 2$, $h = c\theta$, where R and c are measured in meters and θ is in radians. Because of friction the package has a constant speed of 4 m/s. The coefficient of friction between the package and the side wall is the same value μ as that between the package and the floor of the chute. What is the value of μ?

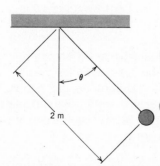

Prob. III.71

III.71   A pendulum is formed from a 400-g sphere at the end of a 2-m cable. If the sphere is released from rest at $\theta = 60°$, determine the tension in the cable (a) when $\theta = 0°$, (b) when $\theta = 30°$.

III.72   A satellite following an ellipitical orbit about the earth has the

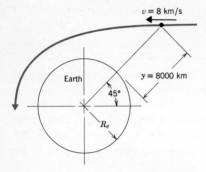

$v = 8$ km/s

Earth

$45°$

$y = 8000$ km

R_e

Prob. III.72

instantaneous position and velocity shown. The inverse square law for gravitation is $F = mgR_e^2/(R_e + y)^2$, where $R_e = 6730$ km is the mean radius of the earth. Determine (a) the value of $\ddot{y}$, (b) the radius of curvature of the orbit at this location.

III.73 A particle follows a spiral in the horizontal plane defined by $R = 200\pi/\theta$, where R is in millimeters and θ is in radians. The tangential force $\bar{F}_t$ is varied with the angle θ so as to maintain $\dot{\theta}$ constant at 4 rad/s. Determine the tangential and normal forces acting on the particle when $\theta = 120°$.

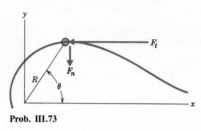

Prob. III.73

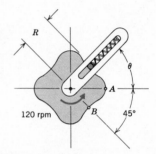

Prob. III.74

III.74 Arm AB rotates counterclockwise at the constant rate of 120 rpm as it guides an 8-oz pin over the fixed cam shaft. The equation of the camshaft is $R = 10 + 2 \cos 4\theta$, where R is in inches and θ is in radians. The maximum compressive force in the spring is 8 lb and the minimum force is 4 lb. What is the force that the pin exerts on the cam (a) at point A, (b) at point B?

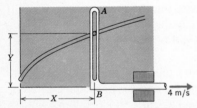

Prob. III.75

III.75 Guide AB moves to the right at a constant speed of 4 m/s as it guides the 250-g pin through the fixed parabolic groove lying in the horizontal plane. The equation of the groove is $X = Y^2$, where X and Y have units of meters. Determine the force the pin exerts on guide AB in terms of the distance X.

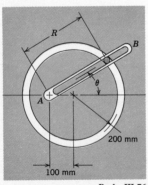

Prob. III.76

III.76 The eccentric arm AB rotates clockwise at the steady rate of 180 rpm, causing the 100-g pin to follow the fixed circular groove whose radius is 200 mm. Friction is negligible. Determine the normal force exerted between arm AB and the pin when $\theta = 45°$. Which side of the groove in arm AB is in contact with the pin? *Hint:* Use the law of cosines to express the radial distance R in terms of θ.

III.77 The hollow tube rotates about a vertical axis with a constant

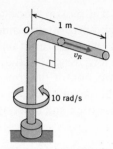

Prob. III.77

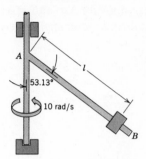

Prob. III.78

angular speed of 10 rad/s, as a rubber ball rolls inside the tube. (a) Derive the equation defining the distance of the ball from the axis of rotation. (b) If a ball is introduced at the origin with a radial speed of 500 mm/s, determine the radial speed and the absolute velocity of the ball as it emerges from the tube.

III.78 The collar slides over bar *AB*, which is rotating about the vertical axis at the constant rate of 10 rad/s. Friction is negligible. (a) Derive the equation governing the distance *l* of the collar from point *A*. (b) If the collar was held at $l = 1$ ft and then released, determine the absolute velocity of the collar at $l = 3$ ft.

III.79 A rider on an open platform that is descending at the constant speed of 3 m/s throws a ball. Relative to the platform, the ball's initial velocity is horizontal at 12 m/s. The ground is 10 m below the location where the ball is thrown. (a) Where does the ball hit the ground? (b) How long after the ball hits the ground does the platform reach ground level?

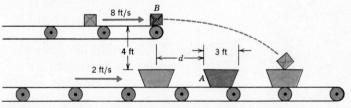

Prob. III.79

III.80 Mail parcels are projected from a horizontal conveyor belt moving at 8 ft/s and are to be caught by bins running on a parallel conveyor belt 4 ft below. The speed of this conveyor belt is 2 ft/s. In order to catch package *B* in bin *A*, what are the range of values for the distance *d* that the bin *A* should be ahead of package *B* when the package is projected from the upper conveyor belt?

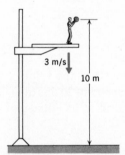

Prob. III.80

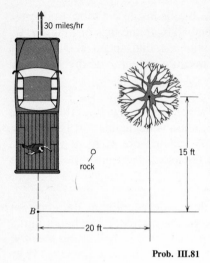

15 ft

B

|← 20 ft →|

Prob. III.81

III.81 A person standing on the bed of a pickup truck traveling at 30 miles/hr on a straight road throws a rock horizontally, aiming the throw perpendicular to the velocity of the truck, with the result that the rock hits a tree at point A, as shown in the top view. Knowing that the rock was thrown at position B, 15 ft before the location of the tree, determine (a) the true initial velocity of the rock, (b) the location of the rider when the rock hits the tree, and (c) the distance the rock had fallen when it hits the tree.

III.82 The air resistance on a slow-moving body is essentially proportional to its velocity, that is, $\bar{f} = -k\bar{v}_P$. Use the equations derived in Illustrative Problem III.5 to derive expressions for $X(t)$ and $Y(t)$ for a body in free flight. The object had an initial speed v_0 at an angle of elevation α.

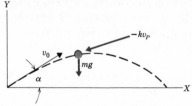

Prob. III.82

III.83 In Problem III.82 the value of k for a 200-g ball is 0.05 N-s/m. The ball has an initial speed of 20 m/s at an angle of elevation of 45°. Determine where the ball returns to its initial elevation and compare this result to the result for $k = 0$.

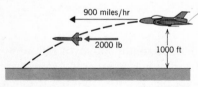

Prob. III.84

III.84 The longitudinal axis of a 1000-lb guided missile remains horizontal as the engine produces a constant thrust of 2000 lb. If the missile was launched from an aircraft traveling at 900 miles/hr in level flight at an altitude of 1000 ft, determine (a) the horizontal distance the missile will travel before it hits the ground, (b) the velocity of the missile when it hits the ground. Neglect air resistance.

III.85 A steel ball bearing rolls on a 30° incline. Its initial speed at point A was 5 m/s horizontally, as shown. Determine (a) the length of time required for the ball bearing to reach point B at the base of the incline, and (b) the distance d locating point B.

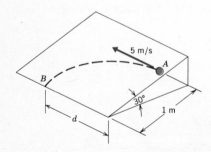

Prob. III.85

III.86 An electron is introduced at rest midway between the two parallel charged plates shown, which are separated by a distance D. The force on the electron, whose mass is m, is eV/D, where e is the electron charge and V is the voltage difference. When $t = 0$ the voltage is varied sinusoidally according to $V = V_0 \cos \omega t$. (a) What is the maximum value of V_0 for

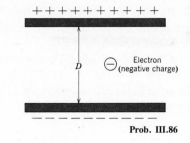

Prob. III.86

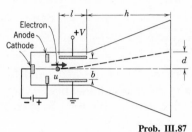

Prob. III.87

which the electron does not hit either charged plate? (b) What is the resulting position of the electron as a function of time for this maximum value of V_0?

III.87 The gravitational force on an electron of mass m and charge e in a cathode-ray tube is negligible, so the path of an electron is essentially a straight line except when it passes between the charged plates. The electron is attracted to the positively charged plate by a force eV/b perpendicular to the plates, where V is the voltage difference and b is the distance between the plates. When an electron is emitted from the cathode and passes through the anode, its speed is u. Derive an expression for the deflection d of the electron for a constant voltage $V = V_0$.

III.88 Solve Problem III.87 if the voltage is varied according to $V = V_0(ut/l)$.

E. NEWTON'S EQUATION OF MOTION — SYSTEM OF PARTICLES

In all of the examples and problems discussed thus far in this text we have treated some large bodies, such as automobiles and airplanes, as if they were particles. To understand why this may be done, and also to derive an important result for our later studies, let us consider a general system of particles.

A primary property of such a system is the position of its center of mass. Therefore, let us start by reviewing how this point is located. In Figure 3a we see a system of N particles, in which gravity acts in the negative Z direction.

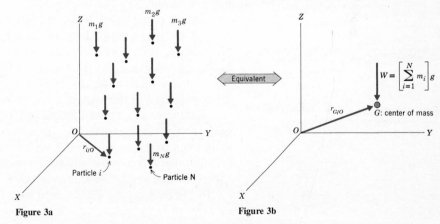

Figure 3a   Figure 3b

The center of mass is found by noting that the resultant of the gravitational forces in Figure 3a is equivalent to the total weight force acting at the center of mass G, as illustrated in Figure 3b. Recall that in statics it was shown that equivalence of the moments of the force systems in the figures leads to an equation relating the *first moments of mass*.

To detail this result we denote the coordinates of the ith particle by (X_i, Y_i, Z_i) and the coordinates of the center of mass by (X_G, Y_G, Z_G). Then, for the orientation of the gravity force in Figures 3, moment equivalence about the X and Y axes gives, respectively,

$$\left(\sum_{i=1}^{N} m_i\right) X_G = \sum_{i=1}^{N} m_i X_i$$

$$\left(\sum_{i=1}^{N} m_i\right) Y_G = \sum_{i=1}^{N} m_i Y_i$$

These two equations may be used to determine the values of X_G and Y_G. To determine Z_G we can imagine realigning the gravitational field in Figures 3; for instance, let gravity act in the negative Y direction. In such a case, moment equivalence about the X axis requires

$$\left(\sum_{i=1}^{N} m_i\right) Z_G = \sum_{i=1}^{N} m_i Z_i$$

These relations may be summarized in vector form by

$$\boxed{m\bar{r}_{G/O} = \sum_{i=1}^{N} m_i \bar{r}_{i/O}} \tag{20}$$

where m is the total mass of the body and the vectors are as defined in Figures 3a and 3b.

Before considering Newton's laws for the system of particles, let us categorize the individual forces acting on each particle in the system according to their source. We denote by $\bar{F}_i$ the resultant force on particle i from sources external to the system. In addition to the external forces, there will be interaction forces $\bar{f}_{ij}$ $(\bar{i} \neq \bar{j})$ on each particle i due to every other particle j in the system.

Examples of external forces are gravity and contact forces exerted between the earth and the particles being studied. Interaction forces can arise from particles within the system that are in contact, and from the gravitational attractions between particles. A simple case is illustrated in Figure 4 for a system of three particles. Notice that each pair of interaction forces satisfies Newton's third law. Extending Figure 4 to the case of a system of N particles, we see that Newton's second law for particle i is

$$\bar{F}_i + \sum_{\substack{j=1 \\ \bar{i} \neq j}}^{N} \bar{f}_{ij} = m_i \bar{a}_i = m_i \ddot{\bar{r}}_{i/A} \tag{21}$$

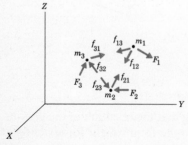

Figure 4

where A is an arbitrary point in the system.

Let us now compute the total resultant of all forces. As can be seen from the system of Figure 4, the forces $\bar{f}_{ij}$ and $\bar{f}_{ji}$ associated with any two particles i and j cancel each other. This occurs because the forces are equal in magnitude but opposite in direction. Therefore, when we sum equation (21) for each particle, we find

$$\sum_{i=1}^{N} \bar{F}_i = \sum_{i=1}^{N} m_i \ddot{\bar{r}}_{i/A} = \frac{d^2}{dt^2} \sum_{i=1}^{N} m_i \bar{r}_{i/A}$$

This expression gains greater significance when we note that the summation in the right side defines the center of mass of the system according to equation (20). Hence

$$\sum_{i=1}^{N} \bar{F}_i = \frac{d^2}{dt^2} (m \bar{r}_{G/A}) = m \ddot{\bar{r}}_{G/A}$$

$$\boxed{\sum_{i=1}^{N} \bar{F}_i = m \bar{a}_G} \qquad (22)$$

In words, equation (22) states that the acceleration of the center of mass is affected only by the total external force, and, equally important, it is unaffected by the interaction between the particles of the system.

The primary reason for presenting equation (22) at this point in our studies is the direct analogy between this equation and Newton's second law. This shows that *when we model the collection of atoms composing a physical body as a single particle, we are studying the motion of the center of mass of the body*.

Equation (22) is essential to the study of kinetics of rigid bodies, which we shall begin in Module V. We shall also use it later in this module to investigate interacting particles further.

F. WORK-ENERGY RELATIONS

We have encountered a number of problems where the application of Newton's laws resulted in differential equations relating the acceleration of a particle in a general position to the forces. The work-energy principle, which follows, represents a standard integral of such differential equations. Basically, it relates the motion of a particle at two different positions along its path.

Because the work-energy principle circumvents the need to solve a differential equation, it is a powerful tool for solving problems. However, as we shall see, the work-energy relations cannot be applied in all situations. Therefore, they complement, but do not replace, the formulation and solution of equations of motion.

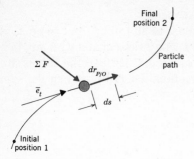

Figure 5

1 Work Performed by Forces

Let us follow the particle in Figure 5 as it travels an infinitesimal arc length ds along its path under the action of the resultant force $\Sigma \bar{F}$.

We define the *work dU* performed by this resultant force in moving the particle this infinitesimal amount as the product of the component of the force in the direction of motion and the infinitesimal distance traveled. Mathematically, this definition is

$$dU = \Sigma F_t \, ds$$

Thus, the total work done by the resultant force in moving the particle from point 1 to point 2, as indicated in Figure 5, is

$$U_{1 \to 2} \equiv \int_{s_1}^{s_2} \Sigma F_t \, ds \tag{23a}$$

The evaluation of this expression requires that we express the resultant force in terms of its tangent and normal components. However, it may be preferable in some problems to write $\bar{F}$ in terms of a different set of unit vectors. Then, a more general definition of the work done can be obtained by referring to the infinitesimal change in the position vector of the particle $d\bar{r}_{P/O}$, as illustrated in Figure 5. Specifically, we have

$$U_{1 \to 2} = \int_{r_1}^{r_2} \Sigma \bar{F} \cdot d\bar{r}_{P/O} \tag{23b}$$

where r_1 and r_2 denote the coordinates of the initial and final positions of point P.

Before considering some examples of the evaluation of the work done by forces in moving a particle, we can make a few general observations about this determination.

1   Notice that the work done by a force is a scalar quantity, so the total work performed by a system of forces is the scalar sum of the work done by each force in the system.

2   Any force that is normal to the direction of motion, such as a reaction force, does no work because it has no component tangent to the direction of motion.

3   In order to compute the work done by a force, its component in the direction of motion *must be a function only of the position* of the particle.

The last statement is particularly significant, because it states that we should *not* try to apply the work-energy principle to problems where the forces doing work depend on either time or the particle velocity. An example of such a force is the air resistance in Illustrative Problem 5.

EXAMPLE 9

The 4-kg block shown in the sketch is pulled up the smooth incline by applying a 40-N force to the end of the cable. Compute the total work done in moving the block from position A to position B on the incline.

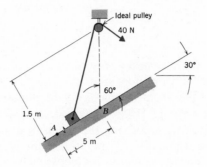

Solution

To see which forces do work in moving the block, we draw a free body diagram of the block in a general position at a distance s from point A. From this diagram we see that work is done by gravity and the cable force, but not by the normal force $\bar{N}$.

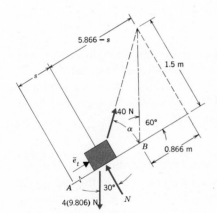

To compute the work done by gravity, we note that the component of gravity in the direction of motion is the constant value $-4g \sin 30°$. Thus

$$
\begin{aligned}
(U_{A \to B})_{\text{gravity}} &= \int_{s_A=0}^{s_B=5} (-4g \sin 30°) \, ds \\
&= -4(9.806)(0.50)(S_B - S_A) \\
&= -98.06 \text{ N-m} \\
&\equiv -98.06 \text{ joules}
\end{aligned}
$$

Note that the SI units of work are joules (J), a joule being defined as a newton-meter. Also, the work done by gravity in this problem is negative because it is opposite to the direction of motion.

To determine the work done by the cable force, we can express the component of $\bar{F}$ tangent to the incline in terms of the angle α, which gives

$$
(U_{A \to B})_{\text{cable}} = \int_0^5 (40 \cos \alpha) \, ds
$$

Then, from the dashed triangle in the free body diagram we find

$$
\cos \alpha = \frac{5.866 - s}{\sqrt{(1.5)^2 + (5.866 - s)^2}}
$$

so that

$$(U_{A \to B})_{\text{cable}} = 40 \int_0^5 \frac{(5.866 - s)}{\sqrt{2.25 + (5.866 - s)^2}} \, ds$$

$$= -40[\sqrt{2.25 + (5.866 - s)^2}] \Big|_{s=0}^{s=5}$$

$$= -40[1.732 - 6.055] = 172.91 \ J$$

An alternative way to obtain the work done by the cable is to consider the work done in pulling the end of the cable. Note that the length of the cable from the block to the pulley, when the block is at point A, is $((5.866)^2 + (1.5)^2)^{1/2} = 6.055$ m, and that it is $1.5 \sin 60° = 1.732$ m at point B. Hence, in moving the block from point P to point B, the end of the cable is pulled 4.323 m, so that the work done is $40(4.323) = 172.91$ J.

The solution is now completed by adding the individual work terms to obtain

$$W_{A \to B} = 172.91 - 98.06 = 74.85 \ J \qquad \triangle$$

EXAMPLE 10
One end of a linear elastic spring whose unstretched length is L_0 is attached to a fixed point O, and the other end is attached to collar C which slides on a smooth curved bar, as shown in the diagram. The spring and collar are coplanar. Determine the work done by the spring force when the collar C is moved from point A to point B.

Solution
We formulate the solution in terms of cylindrical coordinates having origin at point O because the force in the spring depends on the deformed length of the spring. Also, the spring force has only a radial component. Therefore, if the spring is elongated by an amount Δ from its undeformed length, we have the following free body diagram of the collar.

We use equation (23b) to compute the work done by the spring. The infinitesimal change in the position vector is the vector sum of the radial movement dR and the transverse movement $Rd\phi$ of the collar C. Thus, assuming that the spring force is inward,

$$U_{1 \to 2} = \int_{r_1}^{r_2} \bar{F}_{sp} \cdot d\bar{r}_{P/O}$$

$$= \int_{r_1}^{r_2} (-k\Delta \bar{e}_R) \cdot (dR\bar{e}_R + Rd\phi \bar{e}_\phi)$$

$$= -\int_{R_1}^{R_2} k\Delta \, dR$$

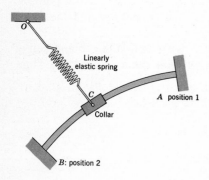

O

Linearly
elastic spring

C

Collar

A position 1

B: position 2

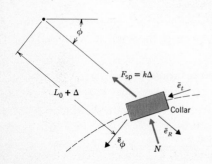

ϕ

$F_{sp} = k\Delta$

$\bar{e}_t$

$L_0 + \Delta$

Collar

$\bar{e}_R$

$\bar{e}_\phi$

N

Notice that the spring does work only if its length is changed. Then, because the value of dR is the amount of additional elongation in the spring, $dR \equiv d\Delta$, so

$$U_{1\to2} = -k \int_{\Delta_1}^{\Delta_2} \Delta d\Delta \equiv \tfrac{1}{2}k(\Delta_1{}^2 - \Delta_2{}^2)$$

We can gain added insight to the meaning of this result by examining a graph of the force exerted by the spring as a function of its elongation, as shown below.

When the motion of the particle results in increasing the elongation of the spring from Δ_1 to Δ_2, the force applied by the spring is opposite the motion of the particle, so that

$$U_{1\to2} = -\int_{\Delta_1}^{\Delta_2} F_{sp}\, d\Delta = \tfrac{1}{2}k(\Delta_1{}^2 - \Delta_2{}^2)$$

This integral is the lightly shaded trapezoidal area in the foregoing sketch. Hence

$$U_{1\to2} = -\tfrac{1}{2}[(F_{sp})_1 + (F_{sp})_2](\Delta_2 - \Delta_1)$$
$$= -\tfrac{1}{2}[k\Delta_1 + k\Delta_2](\Delta_2 - \Delta_1) = \tfrac{1}{2}k(\Delta_1{}^2 - \Delta_2{}^2)$$

This, of course, is the same result as was obtained earlier by integration.

It should be noted that the result obtained in this example shows that the value of the work depends only on the magnitude of the initial and final deformation, and not on whether these deformations are positive (elongation) or negative (compression).

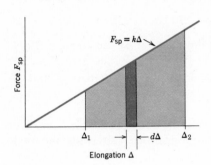

HOMEWORK PROBLEMS

III.89 As the 50-kg package descends the incline, it is caught by a bumper (spring) at point A. The stiffness of the spring, which is initially held at 50 N compression by the two tensioned cables, is 100 N/m. The coefficient of friction is 0.20. Determine the work done by all the forces acting on the package if the package compresses the spring 10 mm.

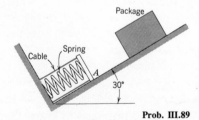

Prob. III.89

III.90 The throttle of a 600-lb motorcycle carrying a 175-lb rider is held steady, resulting in a constant traction force of 150 lb. Neglecting friction, determine the total work done by all the forces in the system when the motorcyclist moves from point A to point B. The center of mass for the motorcycle and rider is 3 ft above the ground.

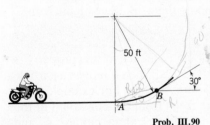

Prob. III.90

III.91 A ball is thrown from the ground at 80 ft/s at an angle of elevation

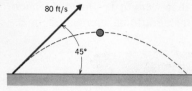

Prob. III.91

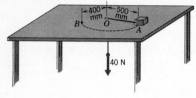

Prob. III.92

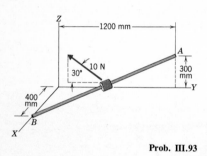

Prob. III.93

of 45°. Determine the work done by gravity as the ball moves from its initial location to the highest point in its trajectory.

III.92 A 2-kg block is tied to a cable as it slides on a smooth table in the horizontal plane. The cable is passed through a hole in the table at point *O,* and a 40-N force is applied to its free end. This force causes the block to spiral inward, from point *A* where $R = 500$ mm, to point *B* where $R = 400$ mm. Determine the work done by the cable (a) by considering the force applied to the block, (b) by considering the force applied to the free end of the cable.

III.93 A 0.5-kg collar is pushed along the rigid bar by the constant 10-N force shown. Determine the work done by all the forces acting on the block moving the block from position *A* to position *B*. The *Z* axis is vertical, and friction is negligible.

III.94 A constant tension force of 100 lb at the end of the cable raises the suspended 40-lb crate 1.75 ft from the position shown. Determine the total work done in raising the crate.

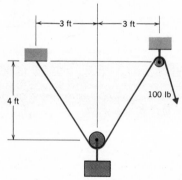

Prob. III.94

2 Potential Energy and Conservative Forces

A subtlety contained in Example 10 is that we did not need to know the curve formed by the bar in order to compute the work done by the spring. Any force that has the characteristic of doing the same amount of work in moving a particle between two points, independent of the path the particle follows, is called a *conservative force*. A consequence of this property is that the total work done by a conservative force when it moves a particle along a closed path back to its starting point is zero. For example, suppose that the spring in Example 10 were freed from the constraint of the curved bar and went from position 1 to position 2 along *any* arc C_1 and returned along any other arc C_2, as shown in Figure 6.

Then, from the result of Example 10 we may write

$$U_{1\to2} = \tfrac{1}{2}k\Delta_1{}^2 - \tfrac{1}{2}k\Delta_2{}^2$$

so that interchanging the subscripts gives

$$U_{2\to1} = \tfrac{1}{2}k\Delta_2{}^2 - \tfrac{1}{2}k\Delta_1{}^2 = -U_{1\to2}$$

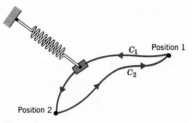

Figure 6

Hence the total work performed by the spring in following the closed path is

$$U_{1\to2\to1} = U_{1\to2} + U_{2\to1} \equiv 0 \tag{24}$$

Thus, if we suppose that the spring force does negative work in moving a particle from point 1 to point 2 because it opposes the motion, this work is not lost from the system; it is stored within the spring and can be used to return the spring to position 1.

Because the work done by a conservative force depends only on the initial and final positions of the particle, it follows that the work can be expressed as the difference in the value of a function that depends only on the position of the particle. That is, for such forces we can create *potential energy* functions. A potential energy function, V, is defined to be a position-dependent function that satisfies

$$U_{1\to2} = V_1 - V_2 \tag{25}$$

To understand this further, consider equations (24) and (25) in the special case where the potential energy at position 1 is zero. Then

$$U_{2\to1} = -U_{1\to2} = V_2$$

Thus, potential energy may be regarded as a measure of the work a conservative force must perform to return the particle to the reference position of zero potential energy.

From the foregoing it is readily verified that in the case of a linear spring,

$$V_{sp} = \tfrac{1}{2}k\Delta^2 \qquad (26)$$

Note that for all cases (compression or extension) V_{sp} is positive if $\Delta \neq 0$.

In addition to the linearly elastic spring whose potential energy is given by equation (26), another conservative force that is frequently encountered is gravity. The determination of the potential energy of gravity is the topic of the next example.

EXAMPLE 11

Determine the potential energy of the gravity force acting on (a) a ball that we have thrown from the surface of the earth, and (b) an artificial satellite that is in orbit about the earth.

Solution

Part a

In this case we may neglect the curvature of the surface of the earth and thus consider the vertical direction to be fixed. This results in the following free body diagram.

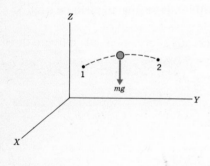

To determine the work done by the gravity force, we use equation (23b) and express all quantities in terms of rectangular Cartesian coordinates. Thus,

$$d\bar{r}_{P/O} = dX\bar{I} + dY\bar{J} + dZ\bar{K}$$

$$\bar{F} = -mg\bar{K}$$

$$U_{1\to2} = \int_{r_1}^{r_2} \bar{F} \cdot d\bar{r}_{P/O} = -mg \int_{Z_1}^{Z_2} dZ$$

$$= mgZ_1 - mgZ_2$$

Comparing this result with equation (25), we have $V_1 = mgZ_1$, $V_2 = mgZ_2$. Hence

$$V_{gr} = mgZ \qquad (27)$$

where Z is the elevation of the body above an *arbitrary* reference position, which we call the *datum*.

Part b

In this case we must account for the fact that the gravitational force is always oriented toward the center of the earth. Also, we must use the inverse square law to describe the magnitude of the force. This law is

$$F_{gr} = G\frac{Mm}{R^2}$$

where G is the universal gravitational constant, M is the mass of the earth, and R is the distance from the body to the center of the earth. The value of GM may be estimated by noting that $F_{gr} = mg$ at the earth's surface. Substituting for F_{gr} in the inverse square law yields

$$GM = gR_e^2$$

where $R_e = 3960$ miles $= 6370$ km is the radius of the earth when the earth is considered to be a sphere.

We select cylindrical coordinates with origin at the center of the earth, because it facilitates the description of $\bar{F}_{gr}$. This results in the free body diagram shown. The sketch implies that the satellite follows a planar path; this fact will be established when we study orbital motion in Module VIII.

We now use equation (23b) to compute the work done by $\bar{F}_{gr}$. Thus

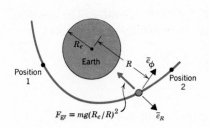

$$d\bar{r}_{P/O} = dR\bar{e}_R + Rd\phi\bar{e}_\phi$$

$$\bar{F}_{gr} = -mg\frac{R_e^2}{R^2}\bar{e}_R$$

$$U_{1\to2} = \int_{r_1}^{r_2} \bar{F}_{gr} \cdot d\bar{r}_{P/O}$$

$$= -mgR_e^2\int_{R_1}^{R_2}\frac{dR}{R^2} = mgR_e^2\left(\frac{1}{R_2} - \frac{1}{R_1}\right)$$

Comparing this result with equation (25), we have $V_1 = -mgR_e^2/R_1$, $V_2 = -mgR_e^2/R_2$, so that

$$\boxed{V_{gr} = -mg\frac{R_e^2}{R}} \tag{28}$$

This result does not seem to resemble that obtained in part a, even though both results describe the same physical phenomenon. To treat this matter let us specialize equation (28) to the case of motion very close to the earth's surface. Letting $R = R_e + Z$, where Z is the elevation above the surface, an infinite series expansion yields

$$V = -mg\frac{R_e^2}{R_e + Z} = -mg(R_e - Z + \frac{Z^2}{R_e} - \dots)$$

$$= -mgR_e + mgZ - \dots$$

Thus, as long as $Z/R_e \ll 1$, the series converges rapidly and the two expressions for the potential energy differ only by a constant. In general, the addition of an arbitrary constant to the result for the potential energy of any conservative force has no effect, for the work done by a conservative force only depends on the *change* in the value of the potential energy. An additive constant does not affect the difference of the potential energy values.

A more mathematical treatment of potential energy can be established by proving that a conservative force is defined by the *gradient* of its potential energy. For rectangular Cartesian coordinates this operation is given by

$$\bar{F} = -\bar{\nabla}V \equiv -\frac{\partial V}{\partial X}\bar{I} - \frac{\partial V}{\partial Y}\bar{J} - \frac{\partial V}{\partial Z}\bar{K}$$

where the symbol $\bar{\nabla}$ is called the ''grad'' or ''del'' operator. The gradient formulation has many useful applications in the study of mechanics, particularly fluid mechanics.

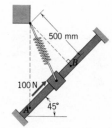

Prob. III.95

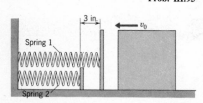

Prob. III.96

HOMEWORK PROBLEMS

III.95 A 1-kg collar slides over the inclined guide under the action of a 100-N tangential force and an elastic spring whose stiffness is 200 N/m. The spring's unstretched length is 400 mm. The diagram depicts the vertical plane. Determine the total work done by all the forces acting on the collar in moving the collar from point A to point B.

III.96 A variable-stiffness bumper for packages on a level ramp is composed of two springs. As shown, the unstretched length of spring 1, of stiffness 10 lb/in., is 3 in. greater than the length of spring 2, of stiffness 20 lb/in. In terms of the distance x that spring 1 is compressed by a package, derive an expression for the total potential energy stored in the springs (a) if $x < 3$ in., (b) if $x > 3$ in.

III.97 The elevation of a 5-kg object above the surface of the earth is increased by 1 km. Determine the increase in the potential energy of the gravitational force acting on this body (a) if the object was initially at the surface of the earth, (b) if the object was initially 6370 km above the earth's surface.

III.98 A type of slingshot is formed by stretching a rubber band horizontally between pegs A and B with an intial tension force $\bar{F}_0$. An object is pulled by the horizontal force $\bar{P}$ and held at point D. The stiffness of the rubber band is k, (a) Derive an expression for the potential energy V_{rb} in

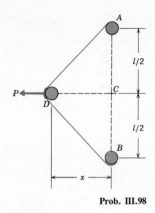

Prob. III.98

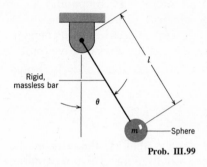

Rigid, massless bar

θ

m — Sphere

Prob. III.99

terms of k, F_0, and the dimensions x and l shown. (b) How much work is done by the force $\bar{P}$ in moving the object from point C to point D? (c) When the rubber band is released and the object is projected past point C, how much work does the rubber band do?

III.99 Derive an expression for the potential energy V_{gr} of the gravitational force acting on the sphere in terms of the mass m, the length l, and the angle θ. Plot the results for $-180° \leq \theta \leq 180°$. What is the significance of the locations where V_{gr} is a maximum or a minimum?

III.100 The tensile force in a nonlinear spring is given by $F = k_1\Delta + k_2\Delta^3$, where Δ is the elongation. As shown, the spring is said to be of the hardening or softening type according to whether k_2 is positive or negative. (a) Derive an expression for the potential energy V of such a spring. (b) For the case where it is of the softening type, determine the values of Δ where $F = 0$ and where $V = 0$. Why don't the two coincide?

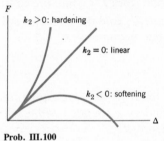

Prob. III.100

3 General Principle of Work and Energy

The work done by the resultant force in moving a particle is related by Newton's second law to the kinematic variables describing the motion of the particle. We begin by using the path variables description to relate the work done by the resultant force moving along arc length ds to the motion variables. This gives

$$\Sigma F_t \, ds = m\dot{v}_P \, ds$$

From the chain rule we have $\dot{v}_P \equiv v_P \, dv_P/ds$, so the foregoing becomes

$$\Sigma F_t \, ds = mv_P \, dv_P$$

Integrating this expression from its initial position to its final position along the path yields

$$\int_{s_1}^{s_2} \Sigma F_t \, ds = m \int_{v_1}^{v_2} v_P \, dv_P \equiv \tfrac{1}{2}m(v_2^2 - v_1^2) \qquad (29)$$

III $A - F \, 1 \, \& \, 2$

From equation (23a) we see that the left side of equation (29) is the total work done by the forces in moving the particle from position 1 to position 2. We now define *kinetic energy,* denoted by the symbol T, to be

$$T = \tfrac{1}{2}mv_P{}^2 \equiv \tfrac{1}{2}m\bar{v}_P \cdot \bar{v}_P \tag{30}$$

so equation (29) may be rewritten as

$$T_2 - T_1 = U_{1\to2} \tag{31}$$.

In other words, the change in the kinetic energy of the particle equals the total work done by the forces acting on the particle. Note that the work-energy equation (31) is a scalar equation.

Applying equation (31) to solve problems requires that, for each problem, we compute the work done by all forces. The computations can be considerably reduced if we use the concept of potential energy to evaluate the work done by conservative forces. Therefore, let us decompose the total work $U_{1\to2}$ into the contributions of the conservative forces, denoted as $U_{1\to2}^{(c)}$, and the nonconservative forces, denoted as $U_{1\to2}^{(nc)}$. Equation (31) then becomes

$$T_2 - T_1 = U_{1\to2}^{(c)} + U_{1\to2}^{(nc)}$$

Now let V *be the total potential energy,* that is, the sum of the potential energy associated with each conservative force. From equation (25) we then have

$$T_2 - T_1 = V_1 - V_2 + U_{1\to2}^{(nc)}$$

Rearranging the terms in this expression yields

$$T_2 + V_2 = T_1 + V_1 + U_{1\to2}^{(nc)} \tag{32}$$

This is the form of the work-energy relation that we shall employ to solve problems. It can be utilized to relate the speed of the particle at any two positions, provided that the work done by the nonconservative forces can be evaluated. The significance of the latter requirement will be obvious in the second of the following examples. However, before we study some examples, let us consider the implication of this work-energy equation.

First, if all forces acting on the particle are nonconservative, then there is no potential energy and equation (32) reduced to equation (31). On the other hand, if all of the forces are conservative, we then have

$$T_2 + V_2 = T_1 + V_1 \tag{33}$$

This means that the sum of potential and kinetic energy must remain constant in the motion. Equation (33) is usually referred to as the principle of *conservation of mechanical energy.*

There are other forms of energy in addition to mechanical energy, such as heat and acoustic energy. When we account for these, we find that equation (32) is equivalent to the first law of thermodynamics, which states that the sum of all forms of energy in a system is constant. For example, the friction force on a sliding body always opposes the motion, and therefore does negative work. This results in a dissipation in the value of $T + V$, where the lost mechanical energy is manifested mainly in the production of heat.

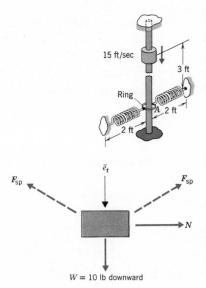

EXAMPLE 12

The 10-lb block shown in the sketch slides over the smooth vertical rod until it is caught at point A by a ring that is attached to two linearly elastic springs. When the springs are in position A they are each elongated by 6 in. and the tensile force within them is 40 lb. Knowing that the block has a velocity of 15 ft/s downward when the block is 3 ft above point A, determine the maximum distance the block will fall below point A.

Solution

We begin with a free body diagram of the block. The spring forces are depicted by dashed arrows in order to indicate that these forces act in only a portion of the motion.

From this diagram we see that the only forces doing work are the gravity and spring forces. ($\bar{N}$ is always perpendicular to the motion; in fact, $\bar{N} \equiv 0$ in this problem.) We therefore apply the work-energy equation (32), using the concept of potential energy to describe the work done by the springs and gravity. Thus

$$T_2 + V_2 = T_1 + V_1 + U_{1 \rightarrow 2}^{(nc)}$$

$$V = V_{gr} + V_{sp}, \quad U_{1 \rightarrow 2}^{(nc)} = 0$$

We are assuming that there is no loss of mechanical energy when the block collides with the ring. This assumption allows us to relate directly the motion of the block at the initial and final positions, without being concerned about the collision process.

We find the properties of the springs from the information given in the problem, which states that $F_{sp} = 40$ lb and $\Delta = 6$ in. $= \frac{1}{2}$ ft at position A. For a linear spring, $F_{sp} = k\Delta$, so in this case

$$k = \frac{40}{\frac{1}{2}} = 80 \text{ lb/ft}$$

To determine the elongation of each spring at any position other than point A, we need to know the unstretched lengths. Noting that the springs are each 2 ft long at point A where they are elongated by $\frac{1}{2}$ ft, we have

$$L_0 = 2.0 - 0.5 = 1.5 \text{ ft}$$

We can now evaluate the system's potential energy. To do this let s be the distance the block descends below point A and choose point A to be the datum for gravity, as shown in the following sketch. Considering that there are two springs, equations (23) and (24) for potential energy give

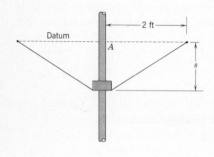

$$V \doteq 2 \left(\tfrac{1}{2}k\Delta^2\right) + mgh = 80\Delta^2 + 10h$$

At the initial position the block is 3 ft above the datum ($h = +3$ ft) and $\Delta = 0.5$ ft, hence

$$V_1 = 80(0.5)^2 + 10(3) = 50 \text{ lb-ft}$$

From the sketch we see that the length of the spring in the final position is $(s^2 + 2^2)^{1/2}$, so that

$$\Delta_2 = \sqrt{s^2 + 2^2} - L_0 = \sqrt{s^2 + 4} - 1.5$$

Then, because the block is below the datum in the final position, $h = -s$, so that

$$V_2 = 80(\sqrt{s^2 + 4} - 1.5)^2 - 10s$$

The kinetic energy in the initial position is

$$T_1 = \tfrac{1}{2}\left(\frac{10}{32.17}\right)(15)^2 = 34.97 \text{ lb-ft}$$

In the final position $T_2 = 0$ because the value of s is a maximum when the block comes to rest ($\dot{s} = 0$). Thus the work-energy equation gives

$$0 + 80(\sqrt{s^2 + 4} - 1.5)^2 - 10s = 34.97 + 50$$

Removing the radical sign from this expression results in a quartic polynomial for s. The real roots of this polynomial are

$$s = 1.707 \text{ ft} \qquad s = -1.400 \text{ ft}$$

The positive answer is the solution we seek. The negative value of s represents the maximum elevation above point A to which the block would rebound if it were actually attached to the springs.

EXAMPLE 13

A skier descends a 40° slope and enters a circular transition to a level grade at a speed of 50 km/hr. The transition has a radius of 140 m. Determine the speed of the skier at the bottom of the transition curve (a) when friction is neglected, and (b) when friction between the snow and skies is accounted for. The coefficient of sliding friction is $\mu = 0.10$.

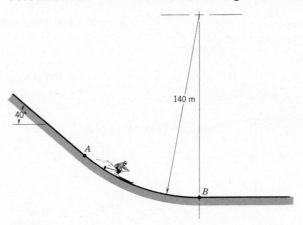

Solution

In case a the friction force on the skier is negligible, whereas in case b the frictional force obviously opposes the velocity. Thus a free body diagram that includes both cases is the following.

Case a For this case we set $f \equiv 0$, and determine the individual terms in the work-energy equation (32). At point A the speed is 50 km/hr, so that the kinetic energies at the initial and final positions are

$$T_1 = \tfrac{1}{2}m\left[(50)\,\frac{1000}{3600}\right]^2 = 96.45(m) \text{ J} \qquad T_2 = \tfrac{1}{2}mv_B^2$$

Turning to the free body diagram, we see that the normal force $\bar{N}$

does no work and that the only other force is gravity, which is conservative. Hence, $U_{1\to2} \equiv 0$, so that equation (32) reduces to the conservation of energy equation (33). Let us choose the end of the slope, point B as the datum elevation. To compute the elevation of the skier at point A, we sketch the geometry of the slope, considering the skier as a particle.

From this sketch we see that

$$h = 140 - 140 \cos 40°$$

so that

$$V_1 = mg(140)(1 - \cos 40°) = 32.75mg$$

$$V_2 = 0$$

Equation (33) now gives

$$\tfrac{1}{2}mv_B{}^2 + 0 = 96.45m + 32.75mg$$

or

$$v_B = 28.90 \text{ m/s} = 104.0 \text{ km/hr}$$

Case b In this case $f = \mu N$. Thus, in order to compute the work done by the friction force we must determine the value of N. Using the path variable unit vectors shown in the free body diagram, the acceleration of the skier is

$$\bar{a}_P = \dot{v}\bar{e}_t + \frac{v^2}{140}\bar{e}_n$$

so that

$$\Sigma F_n = N - mg \cos \theta = m\,\frac{v^2}{140}$$

$$N = m\left(g \cos \theta + \frac{v^2}{140}\right)$$

For the work done by the friction force we therefore have

$$U_{1\to2}^{\text{(nc)}} = -\int_{s_A}^{s_B} \mu N\,ds = -\mu m \int_{s_A}^{s_B}\left(g \cos \theta + \frac{v^2}{140}\right) ds$$

The evaluation of this integral requires that we know how the speed depends on the value of s. Because this information is unknown, and in fact represents the solution we desire, we must use a different method to solve the problem. Remember that at the outset we indicated that the work-energy relations cannot be used in situations where forces doing work depend on the speed of the particle. The real purpose of case b, then, is to present a situation where the method of work-energy is not an effective tool.

To complete the problem we return to the free body diagram and write the equations of motion.

$$\Sigma F_t = mg \sin \theta - \mu N = m\dot{v}$$

$$\Sigma F_n = N - mg \cos \theta = m \frac{v^2}{R}$$

We leave it as a homework problem to show that the solution to these equations is

$$v^2 = Ae^{2\,\mu\theta} + \frac{2gR}{1 + 4\mu^2} [3\mu \sin \theta + (1 - 2\mu^2) \cos \theta]$$

where the constant A may be found from the initial conditions: $v = 50$ km/hr when $\theta = 40°$. Carrying out the computations yields $v = 74.64$ km/hr at the bottom point B. Comparing this to the result in part a, we see the sizable effect of friction in this problem.

HOMEWORK PROBLEMS

III.101   Solve Problem III.43 using the work-energy principle.

III.102   A 3000-lb automobile descends a 5% grade at 45 miles/hr. The coefficient of friction between the tires and the ground is 0.60. What is the minimum distance required to reduce the speed (a) to 15 miles/hr, (b) to zero. Neglect the resistance of the air.

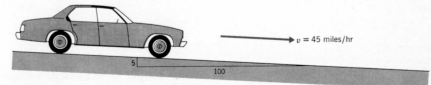

Prob. III.102

III.103   The sliding collar in Problem III.93 is at rest at point A. Determine its speed when it reaches point B.

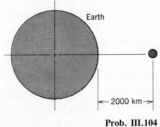

Prob. III.104

III.104   An object is at rest with respect to the center of the earth at an altitude of 2000 km above the earth's surface. Determine the speed at which it will hit the earth if it falls, neglecting air resistance.

III.105   In Problem III.90, the motorcycle starts from rest at point A. Determine the speed of the motorcycle when it reaches point B.

III.106   The maximum speed of a bicyclist on level ground is 10 m/s. What is the largest radius R of a vertical loop that the rider may safely negotiate without pedaling the bicycle? The center of mass of the bicyclist and bicycle is 1 m above the ground.

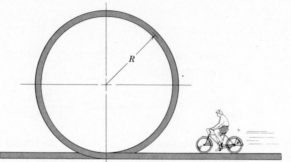

Prob. III.106

III.107   At $\theta = 10°$ the pendulum is returning to its static equilibrium position at 2.5 m/s. Determine (a) the speed of the sphere when it passes the vertical position, (b) the maximum value of the angle θ during the motion.

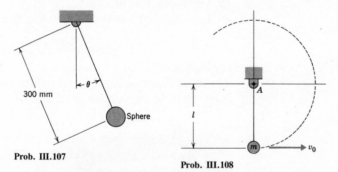

Prob. III.107

Prob. III.108

III.108   A simple pendulum of mass m and length l is given an initial velocity v_0 to the right at its lowest point. Determine the minimum value of v_0 necessary to cause the mass to follow a full circle if the mass is connected to the pivot A (a) by a massless rigid bar, (b) by a cord.

Prob. III.109

III.109   A simple pendulum of mass m and length l is released from rest in the horizontal position. The cord hits peg B, which is directly below the pivot A, and begins to wrap around it. Determine the smallest value of the distance d for which the mass m will follow a complete circle about peg B.

III.110 A daredevil toboggan rider proposes to descend the ice-covered slope from point A, fly into the air at point B, and then land on the ledge C when the toboggan is at the top of its trajectory. The take-off angle at point B is 30°. Neglecting friction and air resistance, determine (a) the speed v_B of the toboggan at take-off, (b) the dimensions h and d where the ledge C should be located, (c) the landing speed v_C.

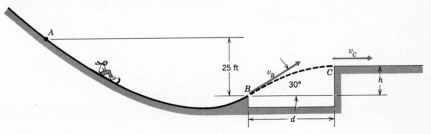

Prob. III.110

III.111 Verify by substitution that the expression for $v(\theta)$ in case (b) of Example 13 satisfies the equations of motion given there. Then use this result to drive an expression for the work done by the friction force as a function of θ.

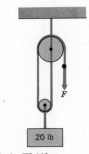

III.112 What constant tension force $\bar{F}$ must be applied to the end of the cable to give the 20-lb block a downward velocity of 2 ft/s after the block descends 4 ft from rest?

III.113 Solve Problem III.7a using the work-energy principle.

III.114 The 40-lb crate is raised from point A where it is at rest by a constant tensile force of 35 lb. Determine (a) the speed of the package after it has been raised 1.75 ft, and (b) the maximum height the crate will attain.

Prob. III.112

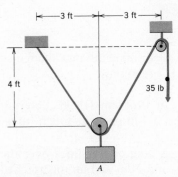

Prob. III.114

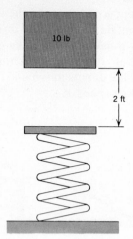

III.115 A 10-lb block is dropped from a height of 2 ft onto a spring whose stiffness is 2 lb/in. Determine the maximum deformation of the spring and the maximum compressive force it sustains.

III.116 A 10-kg package slides down from point A and is brought to rest at point B by a spring whose stiffness is 5 kN/m. Neglecting frction, determine the maximum compression of the spring.

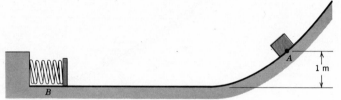

Prob. III.115   **Prob. III.116**

III.117 The 50-kg package descends the incline and is caught by the spring bumper. The stiffness of the spring is 10 kN/m and the spring is precompressed by the two tensioned cables, which apply a force of 50 N. The coefficient of friction is 0.20. If the package was initially 2 m from the bumper when it was released from rest, determine the maximum compression force in the spring.

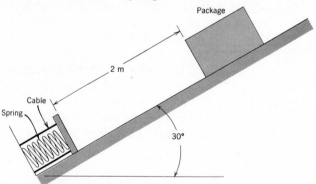

Prob. III.117

III.118 Determine the maximum distance to which the package in Problem III.117 rebounds.

III.119 In Problem III.96, determine the maximum compression of spring 1 if a 4-lb package hits it (a) at 50 in./s, (b) at 200 in./s.

III.120 A 2-kg collar slides over the rigid guide AB. The spring has a stiffness of 5 N/m and an unstretched length of 100 mm. If the collar is released from rest at point A, what is its speed when it reaches point B?

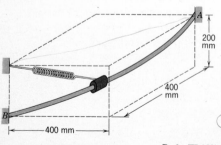

Prob. III.120

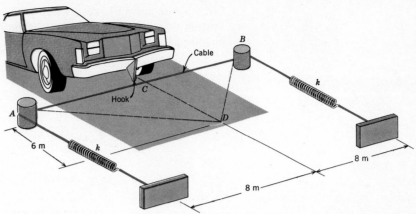

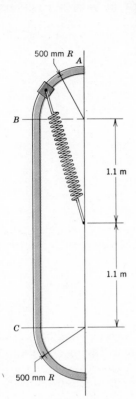

500 mm *R*

1.1 m

1.1 m

500 mm *R*

Prob. III.122

Prob. III.121

III.121   The diagram shows a schematic view of an arresting gear to slow automobiles in a crash test, in which a hook on the vehicle is caught at point *C*. When the cable is stretched between the vertical posts *A* and *B*, the tension within the springs is 10,000 N. The stiffness of each spring is $k = 2(10^4)$ N/m. After hitting the cable, the hook on the automobile moves a distance of 6 m, coming to rest at point *D*. Determine the speed of the automobile at point *C*. The mass of the vehicle is 1200 kg, and the vehicle is rolling freely.

III.122   The 0.5-kg collar is pulled down the smooth curved rod by gravity and a spring whose stiffness is 20 N/m. At point *A* the collar is at rest and the tension force in the spring is 16 N. Determine the force exerted by the collar on the rod when the collar is on the curved segment (a) at point *B*, (b) at point *C*.

III.123   An elastic cord whose stiffness is 24 lb/ft and whose unstretched length is 16 in. is attached to a 2-lb collar. The collar rides on the circular guide *ABCD* in a horizontal plane. When the collar is at point *A* it has a velocity of 10 ft/s to the left. Determine (a) the maximum speed of the collar, (b) whether the collar reaches point *D*, and if so, its speed at this location. *Hint:* Recall that a cord cannot sustain a compressive force.

III.124   Determine the minimum speed that the collar in Problem III.122 should have at point *A* in order to reach point *D*.

III.125   The 5-kg sphere is attached to arm *AB*, which may be considered to be massless and rigid. The stiffness of the spring is 800 N/m and the

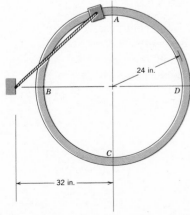

24 in.

32 in.

Prob. III.123

force in the spring is zero when arm AB is horizontal. Knowing that the system was released from rest in the horizontal position, determine the speed of the block when the arm falls to the vertical position.

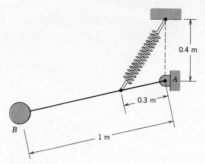

Prob. III.125

4 Work and Energy for a System of Particles
The work-energy equation for a system of N particles is obtained by summing the corresponding equations for each particle. Let T be the total kinetic energy of the system. Then

$$T = \tfrac{1}{2} \sum_{i=1}^{N} m_i v_i^2 \tag{34}$$

Now redefine V to be the total potential energy of all conservative forces acting on the particles of the system. The foregoing definitions lead to the result that *equation (32) is also the work-energy relation for a system of particles.* That this result follows so readily stems from the fact that the work done by the individual forces is additive as scalars.

EXAMPLE 14
The system shown in the sketch is released from rest. Determine the velocity of both blocks when block B has descended 1.5 m. The masses of the blocks are $m_A = 12$ kg, $m_B = 6$ kg, and the coefficient of kinetic friction between block A and the horizontal surface is 0.20.

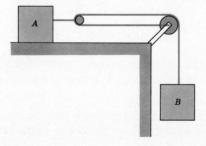

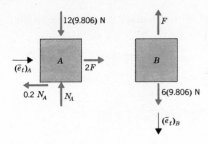

Solution

To relate the energy at the initial and final positions, we begin by drawing free body diagrams of both blocks.

An analysis of the cable-pulley system shows that block A moves 0.75 m to the right when block B descends 1.50 m, and also that $v_A = v_B/2$. The displacement information is used to express the work done by the forces acting on both bodies. Because each of the tangential components of force is constant, the total work is

$$U_{1 \to 2} = [6(9.806) - F] (1.50) + [2F - 0.2N_A] (0.75) = 88.25 - 0.15N_A \text{ J}$$

Notice that we chose not to use the fact that gravity is a conservative force, because for this comparatively uncomplicated problem the work done by this force is simple to determine. In essence, we are ignoring equation (32) and using the more fundamental relationship, equation (31).

The evaluation of the work term is completed by determining N_A. The equilibrium equation for block A normal to its direction of motion yields

$$N_A = 12(9.806) = 117.67 \text{ newtons}$$

so that

$$U_{1 \to 2} = 88.25 - 0.15(117.67) = 70.60 \text{ J}$$

Finally, the total kinetic energy is

$$T = \tfrac{1}{2}m_A v_A{}^2 + \tfrac{1}{2}m_B v_B{}^2 = 6v_A{}^2 + 3v_B{}^2$$

Because the system is initially at rest and $v_A = v_B/2$, we have

$$T_1 = 0 \qquad T_2 = 6\left[\frac{(v_B)_2}{2}\right]^2 + 3(v_B)_2{}^2 = 4.5(v_B)_2{}^2$$

Thus from equation (30) we obtain

$$T_2 = T_1 + U_{1 \to 2} \qquad 4.5(v_B)_2{}^2 = 0 + 70.60$$

$$(v_B)_2 = 3.96 \text{ m/s}$$

A significant feature of this system, common to *all* systems in which cables and pulleys are considered to have ideal properties (massless, inextensible, and frictionless), is that the cable tension does not contribute to the total work done in moving the connected bodies, so that the work done by the cable forces may be ignored. This subtlety is displayed in the formulation of $U_{1 \to 2}$ in the foregoing example.

HOMEWORK PROBLEMS

III.126 Determine the force $\bar{F}$ that must be applied to the 20-kg block A

to make the 20-kg block B attain a speed of 4 m/s after moving 2 m to the right from rest. Neglect friction.

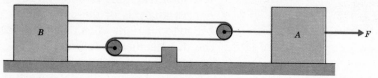

Prob. III.126

III.127 Solve Problem III.126 if the coefficient of friction between the block and the ground is 0.20.

III.128 When released from rest, block A attains a speed of 1 m/s after rising 2 m. Knowing that the coefficient of friction between block B and the incline is 0.40 and that the mass of block B is 5 kg, determine the mass of block A.

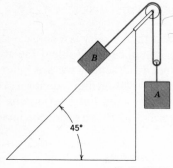

Prob. III.128

III.129 Two spheres A and B are attached to opposite ends of the rigid bar, which pivots about point C. The masses of the spheres are $m_A = 2$ kg, $m_B = 1$ kg, and the mass of the bar is negligible. The bar is released from rest in the horizontal position. Determine the angular velocity of the bar when it passes the vertical position.

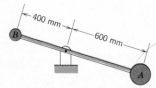

Prob. III.129

III.130 A 4-oz pin is guided through the circular groove by the grooved vertical arm, which is pulled to the right by a 2-lb force. The arm weighs 1 lb. The pin starts from rest at point A. Determine the speed of the pin when it has traveled 8 in. to the right.

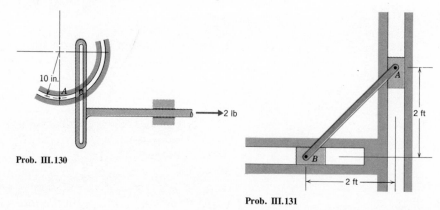

Prob. III.130

Prob. III.131

III.131 Sliders A and B are connected by a rigid bar of negligible mass in a vertical plane. If block A is released from rest in the position shown, determine its velocity after it falls (a) 2 ft, (b) 4 ft. Both blocks weigh 2 lb, and friction is negligible.

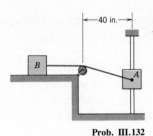

←40 in.→

B

A

Prob. III.132

III.132 Block A is released from rest at the same elevation as block B with the cable taut. Neglecting friction, determine the velocity of block B after block A has fallen 30 in. Both blocks weigh 1 lb.

III.133 Solve Problem III.132 for the case where there is frictional resistance to the motion of block B, the coefficient of friction being 0.20.

G. POWER

The concept of power is intimately tied to the energy concepts we have just studied. Basically, the *mechanical power* supplied to a system by a force is the time rate at which this force does work; that is,

$$\text{Power} = \frac{dU}{dt} \tag{35}$$

A more specific relation for power is obtained by considering the definition of the work performed by a force in an infinitesimal displacement, which is

$$dU = \bar{F} \cdot d\bar{r}_{P/O}$$

Dividing this expression by dt yields

$$\frac{dU}{dt} = \bar{F} \cdot \frac{d\bar{r}_{P/O}}{dt}$$

$$\text{Power} = \bar{F} \cdot \bar{v}_P \tag{36}$$

The unit of power in the SI system is a watt (W), which is equal to 1 J/s, which is 1 N-m/s. In the British system the basic unit is a horsepower (hp) one horsepower being defined as 550 ft-lb/s. (One horsepower is 776 W.)

For a different viewpoint of the power developed by a force, let us consider equation (32), rewritten as follows:

$$U_{1\to2}^{(nc)} = (T_2 - T_1) + (V_2 - V_1)$$

In the situation where position 2 is different from position 1 by a differential amount, this equation becomes

$$dU^{(nc)} = dT + dV$$

Dividing this result by dt gives

$$\text{Power}^{(nc)} \equiv \frac{d}{dt}(T + V) \tag{37}$$

Thus, we see that the total power supplied to a system by a nonconservative force equals the rate of increase of the potential and kinetic energy of the system.

A common type of nonconservative force is friction, which generally does negative work. Thus, the power associated with a friction force is negative, representing a dissipation of energy essentially in the form of heat. In this book we are concerned only with mechanical energy; the science of thermodynamics is concerned with all forms of energy and the efficiency with which this energy is used.

EXAMPLE 15

The power output of the motor of an electric car is constant over its entire speed range for a fixed setting of the motor control. The maximum power that can be delivered to the drive wheels is 60 hp. The automobile weighs 2500 lb and the air resistance in pounds is $0.012v^2$, where v is in units of feet per second. Consider the rolling resistance to be negligible. Determine the maximum speed of the vehicle and the minimum time required to accelerate the car from rest to a speed of 60 miles/hr, both on level ground.

Solution

In drawing a free body diagram we must remember that the friction force $\bar{f}$ on the drive wheels pushes the automobile forward.

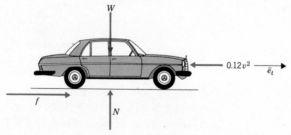

The power of the motor is transmitted to the drive wheels. The friction force $\bar{f}$ is in the direction of motion. Therefore, using power $= fv$, for maximum acceleration

$$f = \frac{60(550)}{v} = \frac{33,000}{v} \text{ lb}$$

We now sum forces in the tangential direction.

$$\Sigma F_t = f - 0.012v^2 = m\dot{v} = \frac{2500}{32.17}\dot{v}$$

$$\frac{33,000}{v} - 0.012v^2 = 77.71\dot{v}$$

This is a differential equation defining acceleration as a function of position. To determine the maximum speed we set $\dot{v} = 0$ and find

$$\frac{33,000}{v_{max}} - 0.012v^2_{max} = 0$$

$$v_{max} = 140.10 \text{ ft/s} = 95.5 \text{ miles/hr}$$

The solution to the second part of the problem is more complicated, because it requires that we solve the differential equation of motion. To determine the time required to attain a speed of 60 miles/hr (= 88 ft/s), we write $\dot{v} = dv/dt$ and separate variables.

$$\frac{77.71 \, dv}{(33,000/v) - 0.012v^2} = dt$$

Using the initial conditions we obtain the following integral

$$\frac{77.71}{0.012}\int_0^{88} \frac{v \, dv}{2.75(10^6) - v^3} = \int_0^t dt \equiv t$$

From a table of integrals we find

$$t = \frac{77.71}{0.012}\frac{1}{3k}\left[\frac{1}{2}\ln\left(\frac{k^2 - kv + v^2}{k^2 + 2kv + v^2}\right) + \sqrt{3}\tan^{-1}\left(\frac{2v - k}{k\sqrt{3}}\right)\right]_0^{88}$$

where

$$k = [2.75(10^6)]^{1/3} = 140.10$$

Evaluating this expression for t yields

$$t = 8.33 \text{ seconds}$$

It may seem unlikely to you that an automobile whose top speed is only 95 miles/hr should be able to accelerate this rapidly. The result is obtained because the electric motor is capable of high power output at low speed. A gasoline engine does not have this characteristic. In fact, it is for this reason that automobiles with gasoline engines require transmissions.

HOMEWORK PROBLEMS

III.134 A 30,000-kg truck has a constant speed of 40 km/hr up a 10% grade. The air resistance is $F = 40.0v^2$, where F is in newtons and v is the speed of the vehicle in meters per second. It is known that 75% of the power produced by the engine is actually used to propel the truck. Determine the power produced by the engine.

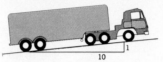

Prob. III.134

III.135   In Problem III.134, what acceleration will the truck have when it reaches level ground if its engine continues to produce the same power?

III.136   Once a motorboat exceeds 3 m/s, its hull begins to *plane* over the water. It is known that the water resistance is then given by $F_w = k_1/v^2 + k_2 v^2$, where v is the speed of the boat. It is also known that 5 kW are required to propel the boat at 15 m/s, and that a speed of 5 m/s requires the minimum power output from the engine. (a) Determine the values of the coefficients k_1 and k_2. (b) At what speed is the water resistance a minimum? Why isn't this speed the same as that for which the power is a minimum?

III.137   Four identical diesel-electric locomotives weighing 30 tons each accelerate a freight train from rest. There are 150 cars in the train, the average weight of each car being 60 tons. The rolling resistance on each car is 0.2% of its weight, and air resistance is negligible. Each locomotive is producing 2000 hp of usable mechanical power. Determine (a) the top speed of the train, (b) the time required to attain a speed of 50 ft/s.

Prob. III.137

III.138   An electrically powered winch is utilized to move the 50-kg crate with a constant acceleration from rest to 5 m/s in a distance of 10 m. The inclined ramp consists of a series of roller bearings, so friction is negligible. Determine (a) the maximum horsepower produced by the motor, and (b) the horsepower produced in the position where the crate has moved 5 m from its initial position.

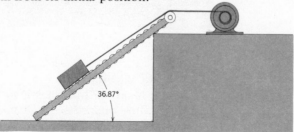

36.87°

Prob. III.138

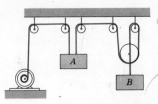

Prob. III.139 and Prob. III.140

III.139 The electric motor raises block A at the constant speed $v_A = 6$ m/s. Both blocks have a mass of 2 kg. Determine the power required of the motor.

III.140 The motor is switched on when block A is descending at 6 m/s, immediately producing 200 watts to slow the system. Determine the deceleration rate of block A when the motor is turned on.

III.141 The electric motor pulls the cable in at the constant rate v. Derive an expression for the power required from the motor in terms of v, the mass m of the package being raised, and the dimensions l and d shown.

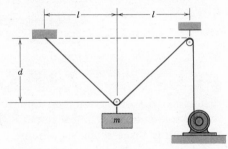

Prob. III.141

III.142 The escalator carries people upward at 4 ft/s. What is the average power that the motors must furnish in order to transport 45 people simultaneously, if the average weight of a person is 170 lb? Also, determine how many people are being transported per minute.

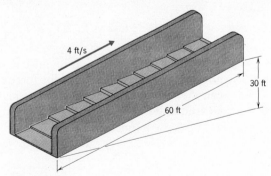

Prob. III.142

H. LINEAR IMPULSE-MOMENTUM RELATIONS

The momentum principles that we shall study are additional standard integrals of the differential equations of motion. In contrast to the energy principle, which treats situations where forces vary with the position of the particle, we *consider the momentum of the particle when the forces depend only on time.*

The derivation is straightforward. We begin with Newton's second law and note that when the forces depend only on t, we can separate variables in the vector equation by writing $\bar{a}_P \equiv \dot{\bar{v}}_P = d\bar{v}_P/dt$. Thus

$$m \, d\bar{v}_P = \Sigma \bar{F} \, dt$$

Integrating over an arbitrary time interval from t_1 to t_2 yields

$$m \int_{\bar{v}_1}^{\bar{v}_2} d\bar{v}_P = \int_{t_1}^{t_2} \Sigma \bar{F} \, dt$$

$$\boxed{m\bar{v}_2 - m\bar{v}_1 = \int_{t_1}^{t_2} \Sigma \bar{F} \, dt} \qquad (38)$$

where $\bar{v}_1$ and $\bar{v}_2$ are the initial and final velocities of the particle.

The quantity $m\bar{v}_P$ is the *linear momentum* of the particle. The integral in the right side of equation (38) is called the *linear impulse* of the force resultant. In words, equation (38) states that the change in the linear momentum equals the linear impulse. Notice that this is a vector relation representing three scalar equations.

Let us now consider how we can apply equation (38) in conjunction with the several kinematical descriptions. In the case of rectangular Cartesian coordinates, we have

$$\Sigma \bar{F} = \Sigma F_X \bar{I} + \Sigma F_Y \bar{J} + \Sigma F_Z \bar{K}$$

Then, because the unit vectors are constant, the linear impulse is

$$\int_{t_1}^{t_2} \Sigma \bar{F} \, dt = \int_{t_1}^{t_2} \Sigma F_X \, dt \, \bar{I} + \int_{t_1}^{t_2} \Sigma F_Y \, dt \, \bar{J} + \int_{t_1}^{t_2} \Sigma F_Z \, dt \, \bar{K}$$

This tells us that the linear impulse may be evaluated whenever we know how the XYZ components of the forces vary with time. If we so desire, we can then obtain scalar equations equating corresponding components in equation (38). For instance, the $\bar{I}$ component gives

$$m(v_X)_2 - m(v_X)_1 = \int_{t_1}^{t_2} \Sigma F_X \, dt$$

We should note that this result can also be obtained with equal ease from Newton's second law whenever we know how ΣF_X depends on time by writing

$$\Sigma F_X = ma_X \equiv m\ddot{X} = m\dot{v}_X$$

Separating variables and integrating then gives the same result as that obtained in the momentum development.

Let us now consider what is involved in evaluating the linear impulse in situations where we are using either path variables or cylindrical coordinates. The unit vectors in these coordinate descriptions vary, so they cannot be treated as constants in the integral for the impulse. For example, in the case of path variables we have

$$\int_{t_1}^{t_2} \Sigma \, \bar{F} \, dt = \int_{t_1}^{t_2} (\Sigma \, F_t \, \bar{e}_t) \, dt + \int_{t_1}^{t_2} (\Sigma \, F_n \, \bar{e}_n) \, dt$$

For a general motion these integrals can be evaluated only by resolving the unit vectors into their $(\bar{I}, \bar{J}, \bar{K})$ components. Therefore, we make the general statement that the principle of linear momentum and impulse should always be formulated in terms of rectangular Cartesian coordinates.

One situation where a momentum formulation is very useful occurs when we are interested only in the average value of the forces in a time interval, because in that case the linear impulse is

$$\int_{t_1}^{t_2} \Sigma \, \bar{F} \, dt = (\Sigma \, \bar{F}_{\text{average}})(t_2 - t_1) \tag{39}$$

Other applications of the impulse-momentum principles are for systems of interacting particles and for impulsive forces, such as those exerted between colliding bodies. These are the next topics we shall study.

1 Interacting Particles

Let us begin with the equation for the acceleration of the center of mass G of a system of N particles, equation (22),

$$\sum_{i=1}^{N} \bar{F}_i = m\bar{a}_G$$

where $\Sigma \, \bar{F}_i$ is the resultant of the external forces acting on the system. From equation (2) we know that the position of the center of mass is defined by

$$m\bar{r}_{G/O} = \sum_{i=1}^{N} m_i \, \bar{r}_{i/O}$$

Therefore

$$m\bar{a}_G \equiv m\ddot{\bar{r}}_{G/O} = \sum_{i=1}^{N} m_i \, \ddot{\bar{r}}_{i/O} = \sum_{i=1}^{N} \frac{d}{dt}(m_i \, \dot{\bar{r}}_{i/O})$$

$$= \sum_{i=1}^{N} \frac{d}{dt}(m_i \bar{v}_i)$$

Assuming that we know how the external forces vary with time, separation of variables yields

$$\sum_{i=1}^{N} d(m_i \bar{v}_i) = \sum_{i=1}^{N} \bar{F}_i \, dt$$

We now integrate over the time interval $t_1 \le t \le t_2$ to obtain

$$\sum_{i=1}^{N} \int_{(\bar{v}_i)_1}^{(\bar{v}_i)_2} d(m_i \bar{v}_i) = \sum_{i=1}^{N} \int_{t_2}^{t_1} \bar{F}_i \, dt$$

$$\sum_{i=1}^{N} m_i (\bar{v}_i)_2 - \sum_{i=1}^{N} m_i (\bar{v}_i)_1 = \sum_{i=1}^{N} \int_{t_1}^{t_2} \bar{F}_i \, dt \tag{40}$$

In words, equation (40) shows that the total linear momentum is changed only by the impulse of the external forces.

Occasionally, we shall encounter a situation where the external forces exert a negligible impulse in comparison to the impulse of the interaction forces. In such a case equation (40) yields

$$\sum_{i=1}^{N} m_i (\bar{v}_i)_2 = \sum_{i=1}^{N} m_i (\bar{v}_i)_1$$

In conventional terminology this is the principle of *conservation of linear momentum*.

The primary feature of equation (40) in general is that it provides a relationship for the motion of a system that does not require knowledge of the interactions occurring within the system. In addition to its application to conventional mechanical systems, we shall employ this formulation in Module VIII to treat some problems in fluid mechanics.

EXAMPLE 16

A truck towing a trailer starts from rest and accelerates along level ground to a speed of 40 km/hr in 20 s. The truck has a mass of 5000 kg and the trailer has a mass of 15,000 kg. The frictional resistances on the truck and the trailer attributable to the undriven wheels are 200 N each. Determine the average traction force produced by the drive wheels of the truck during the acceleration and determine the average pull exerted by the truck on the trailer.

Solution

We draw free body diagrams of the truck and of the trailer, letting $\bar{H}$ be the horizontal pull force exerted by the truck on the trailer and $\bar{F}$ be the tractive force.

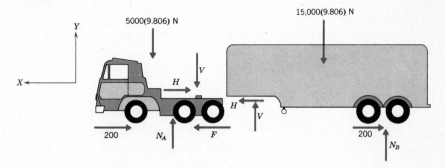

As you can see, the interaction between the truck and the trailer also involves the vertical force $\bar{V}$ needed to establish equilibrium for the trailer. The determination of the value of this force is beyond the scope of this module, because such a determination requires that we first study the kinetics of rigid bodies.

We begin by considering the truck and trailer as a system in order to eliminate the unknown interaction force. We refer to the free body diagrams to obtain the external forces acting on the system and use equation (39) to formulate the impulse in terms of average forces. Hence

$$(m_{\text{truck}} + m_{\text{trailer}})(\bar{v}_2 - \bar{v}_1) = \Sigma \bar{F}_{\text{average}}(\Delta t)$$

$$20{,}000[40\left(\frac{1{,}000}{3{,}600}\right)\bar{I} - \bar{0}] = \{(F - 200 - 200)\bar{I}$$

$$+ [-(15{,}000 + 5{,}000)(9.806)$$

$$+ N_A + N_B]\bar{J}\}(20)$$

Equating the $\bar{I}$ components in this equation gives

$$20{,}000(40)\left(\frac{10}{36}\right) = (F - 400)(20)$$

$$F = 1.151(10^4) \text{ N}$$

To determine the force $\bar{H}$ we consider either the truck or the trailer individually, so that the effect of this force does not cancel in the interaction. Treating the trailer, we write

$$15{,}000[40\left(\frac{1{,}000}{3{,}600}\right)\bar{I} - \bar{0}] = \{(H - 200)\bar{I} + [N_B + V - 15{,}000(9.806)]\bar{J}\}(20)$$

Equating the $\bar{I}$ components yields

$$15,000(40)\left(\frac{10}{36}\right) = (H - 200)(20)$$

$$H = 8530 \text{ N}$$

Can you explain why H and F are not equal?

HOMEWORK PROBLEMS

III.143 With the aid of a catapult, a 20-ton aircraft is launched in an easterly direction from the deck of an aircraft carrier. The wind is calm and the aircraft has an airspeed of 150 miles/hr when launched, attained after being accelerated for 4 s. If the engines provide 40,000 lb of thrust, determine the force provided by the catapult, (a) if the aircraft carrier is at rest, (b) if the aircraft carrier is heading east at a constant speed of 30 miles/hr.

III.144 A three-car commuter train slows from an initial speed of 72 km/hr on a level track. Each car has a mass of (10^4) kg, and the coefficient of friction between the wheels and the track is 0.20. Determine the minimum amount of time required to stop the train and the corresponding force within the couplers (a) if the brakes in all cars are functioning, (b) if the brakes in the middle car fail. Air resistance is negligible.

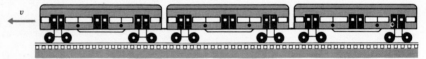

Prob. III.144

III.145 The 1500-kg automobile accelerates from rest and attains a speed of 5 m/s relative to the 6000-kg barge in 5 s. Determine the average force in the mooring cable that is required to hold the barge stationary and the average traction force exerted by the tires on the barge.

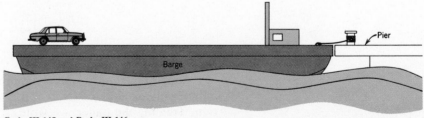

Prob. III.145 and Prob. III.146

III.146 The 1500-kg automobile starts from rest and attains a speed of 5 m/s relative to the 6000-kg barge in 5 s. Because someone forgot to attach the mooring cable, the barge moves away from the pier. Determine (a) the velocity of the barge after 5 s, (b) the average traction force exerted by the tires on the barge, and (c) the distance the barge moves in this 5-s interval.

III.147 The 20-lb block slides down the smooth ramp of the 100-lb cart. The system was at rest initially. Determine (a) the velocity of the block and of the cart when the box reaches the bottom of the ramp, and (b) the distance the cart shifts during this motion.

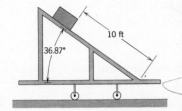

Prob. III.147

III.148 A 40-lb piece of luggage has a speed of 8 ft/s when it is projected from the conveyor belt onto the ramp. It then slides down the smooth ramp and onto the 80-lb cart. The coefficient of friction between the piece of luggage and the cart is 0.50. Determine (a) the velocity of the cart when the luggage comes to rest on it, (b) the time required for the luggage to come to rest on the cart, and (c) the total distance the luggage slides. Neglect rolling resistance.

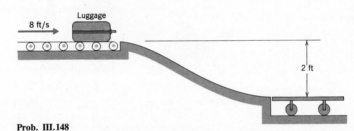

Prob. III.148

III.149 As the 1000-kg automobile descends the hill, the total friction force $\bar{f}$ exerted between the tires and the ground varies as shown, where a negative value represents braking. The initial speed of the car was 72 km/hr. Determine the speed of the car (a) after 15 s, (b) after 30 s. Neglect air resistance.

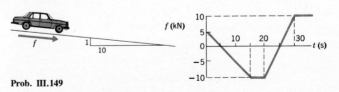

Prob. III.149

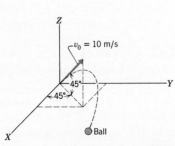

Prob. III.150

III.150 A 0.5-kg ball is thrown at 10 m/s as shown. The XY plane is horizontal. The ball is subjected to a constant 1-N force from a wind that is blowing in the negative Y direction. How long will it be before the ball is traveling parallel to the XZ plane? Determine the velocity of the ball at this time.

III.151 The initial and final velocities of a 200-g particle over a 1-min time interval are $\bar{v}_1 = 6\bar{I} - 6\bar{J} + 3\bar{K}$ m/s and $\bar{v}_2 = -8\bar{I} + 9\bar{J} + 12\bar{K}$ m/s, respectively. (a) Determine the average force acting on the particle. (b) If the force on the particle is known to vary with time according to $\bar{F} = C_1\bar{I} + C_2\bar{J} + C_3t^2\bar{K}$, where t is in seconds and $\bar{F}$ is in newtons, determine the maximum magnitude of $\bar{F}$ in the 1-min interval.

2 Impulsive Forces

To investigate impulsive forces, let us consider the example shown in Figure 7a where a block initially held at rest on an incline by friction is struck by a hammer at instant t_1. The force $\bar{F}$ resulting from the hammer blow is displayed in Figure 7b. We say that the force $\bar{F}$ is an *impulsive force* if it significantly increases the speed of the block in the small time interval from t_1 to t_2. More specifically, an impulsive force acts over a small time duration and imparts an acceleration of many g's to a body.

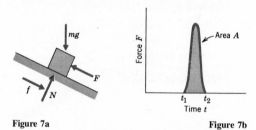

Figure 7a   **Figure 7b**

From Figure 7b we can determine the magnitude of the impulse of $\bar{F}$, because $\int F\ dt$ is the area A under the F–t curve. We shall assume that during the time interval $t_1 \leq t \leq t_2$ the hammer force exceeds the friction force $\bar{f}$ by a significant amount, so that we can approximate the velocity of the block at t_2 by the linear impulse-momentum relation

$$m(v_2\bar{I} - \bar{0}) = \int_{t_1}^{t_2} F(t)\ dt\ \bar{I} = A\bar{I}$$

$$v_2 = \frac{A}{m}$$

After the hammer blow is completed, the block moves up the incline, decelerating uniformly due to the effects of gravity and friction. Thus the graph of $v(t)$ is as shown in Figure 8.

From the shape of the graph we observed that a reasonable approximation to the velocity history of the block can be obtained by considering the time instants t_1 and t_2 to be identical, which is equivalent to considering the velocity of the particle to change instantaneously. From Figure 8

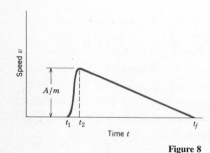

Figure 8

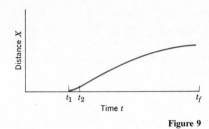

Figure 9

we can construct a graph of $X(t)$, because for rectilinear motion along the X axis,

$$X(t) = X(0) + \int_0^t v(t) \, dt$$

The qualitative character of $X(t)$, depicted in Figure 9, is obtained by considering the integral to be the area under the $v(t)$ graph. Notice that the value of X is shown close to zero at instant t_2, which is consistent with the approximation that the position of the block does not change during the period of the hammer blow.

The foregoing assertions about the action of a particle acted upon by an impulsive force can be generalized as follows:

1   The change in the velocity of a particle

$$\Delta \bar{v} = \frac{1}{m} \int_{t_1}^{t_2} \bar{F}(t) \, dt$$

occurs, in essence, instantaneously.

2   The change in position of the particle during the time interval the impulsive force acts is negligible.

3   All nonimpulsive forces are negligible compared to the impulsive force during the time interval in which the impulsive force acts.

EXAMPLE 17

The last stage of a guided missile shuts down when the system attains an orbital speed of 27,000 km/hr. To separate the 3000-kg payload from the expended 2000-kg booster, six explosive bolts are fired. The explosives in each bolt generate 400 J of energy, 40% of which converts to mechanical energy. The charges are placed such that the payload continues to move parallel to the initial velocity. Determine the speed of the payload and of the booster immediately after the firing of the explosive bolts and the separation distance between the two bodies 1 min after the firing.

Solution

We begin with free body diagrams of the payload and of the booster. Note that the impulsive interaction force $\bar{R}$ generated by the explosion is depicted parallel to the direction of the original velocity in order to maintain this direction. Also, we are using the standard value of g because no information is given about the altitude.

Neglecting the gravity force in comparison to the explosive force, the principle of conservation of linear momentum for the system yields

$$3000(7500 + \Delta v_P) + 2000(7500 + \Delta v_B) = 5000(7500) \text{ kg-m/s}$$

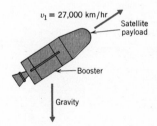

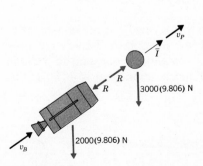

or

$$3000 \, \Delta v_P + 2000 \, \Delta v_B = 0$$

where 7500 m/s is the given initial speed and Δv_P and Δv_B represent the increase in the speeds of the payload and booster, respectively.

It is given that 40% of the 6(400) J produced by the explosion is converted to mechanical energy. Because we neglect any change in position during the short duration of the explosion, the work-energy equation (32) yields

$$T_2 = T_1 + 0.40(2400)$$

$$\tfrac{1}{2}(3,000)(7,500 + \Delta v_P)^2 + \tfrac{1}{2}(2,000)(7,500 + \Delta v_B)^2$$

$$= \tfrac{1}{2}(5,000)(7,500)^2 + 960$$

This expression simplifies to

$$1,500[15,000 \, \Delta v_P + (\Delta v_P)^2] + 1,000[15,000 \, \Delta v_B + (\Delta v_B)^2] = 960$$

We now have two equations for the two unknown increments in speed, which we solve as follows.

$$\Delta v_B = -\frac{3,000}{2,000} \Delta v_P = -1.5 \, \Delta v_P$$

$$1,500[15,000 \, \Delta v_P + (\Delta v_P)^2] + 1,000[15,000(-1.5 \, \Delta v_P)$$

$$+ (1.5 \, \Delta v_P)^2] = 960$$

$$3,750(\Delta v_P)^2 = 960$$

$$\Delta v_P = 0.5060 \text{ m/s}$$

$$\Delta v_B = -1.5(0.5060) = -0.7590 \text{ m/s}$$

Hence the speeds are v (initial) $+ \Delta v$, or

$$(v_P)_2 = 7500 + 0.5060 \text{ m/s} \qquad \triangleleft$$

$$(v_B)_2 = 7500 - 0.7590 \text{ m/s} \qquad \triangleleft$$

To determine the distance between the payload and the booster after separation, we consider their relative motions. Both bodies are only under the influence of gravity, which is essentially constant in a 1-min interval. They therefore have the same acceleration, that is

$$\bar{a}_{P/B} = \bar{a}_P - \bar{a}_B = \bar{0}$$

or

$$\bar{v}_{P/B} = \bar{v}_P - \bar{v}_B = (0.5060 + 0.7590)\bar{I}$$

$$= 1.2650\bar{I} \text{ m/s}$$

Considering the two bodies to be adjacent to each other at the end of the explosion, we find for $t = 60$ s,

$$\bar{r}_{P/B} = 1.2650(t)\bar{I} = 1.265(60)\bar{I}$$

$$= 75.4\bar{I} \text{ meters}$$

or, in words the payload is 75.4 m ahead of the booster at the end of 1 min.

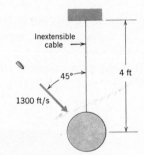

Prob. III.152

Prob. III.153

HOMEWORK PROBLEMS

III.152 The initial and final velocities of a 5-oz baseball being hit by a bat are shown. Knowing that the ball and bat are in contact for 0.05 s, determine the average force exerted between the ball and the bat.

III.153 A 2-oz bullet traveling at 1300 ft/s in the direction shown hits and is embedded in the 8-lb sphere. The impact is completed in 0.05 s. Determine (a) the average increase in the tension of the cable during the impact, (b) the speed of the sphere immediately after the impact, and (c) the height to which the sphere rises after the impact.

III.154 The velocities of a 20-lb block of wood and a 1.5-oz bullet immediately before they collide are shown. The bullet is embedded in the block 0.01 s after the first contact. Determine the final velocity of the block of wood and the average force exerted on the block by the bullet during the collison.

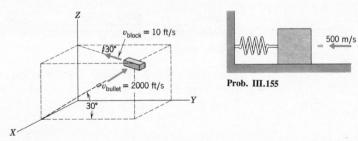

Prob. III.154

Prob. III.155

III.155 A 50-g bullet traveling at 500 m/s hits the 5-kg block and embeds in it. The block is then slowed by the spring whose stiffness is 100 N/mm. Determine the maximum distance the spring is compressed.

III.156 A schematic view of a cannon and its recoil mechanism is shown. A 20-kg shell has a muzzle velocity of 500 m/s. The shell emerges 1 ms after being fired. The mass of the barrel and its movable support is 600 kg. Also, the spring has a stiffness of 1000 kN/m. Determine (a) the average increase in the reaction force on the rails during the firing, (b) the recoil velocity of the barrel, and (c) the maximum distance that the spring is compressed.

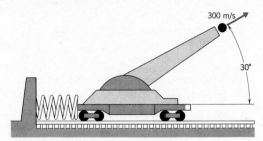

Prob. III.156

III.157 Different weight bullets are fired from a rifle with the same kinetic energy of 2500 J. Each bullet emerges from the barrel 20 ms after being fired. Determine the average force required to hold the rifle stationary for (a) a 25-g bullet, (b) a 50-g bullet.

III.158 Three identical $3(10^4)$-kg freight cars are to be coupled. The coupling process is completed in 0.50 s. Car A is at rest and the coupled cars B and C are traveling at 1 m/s. Determine the speed of the cars after the coupling and the average force in each of the two couplers during the coupling.

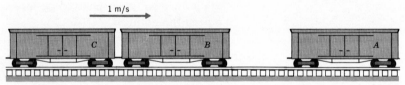

Prob. III.158

III.159 Solve Problem III.158 if cars B and C are uncoupled. Car C is sufficiently far behind car B to allow cars A and B to couple before hitting car C.

3 Central Impact

A common way in which impulsive forces are generated is through the collision of bodies. Figure 10a depicts a general condition of a collision *(impact)* of two spheres, where $(\bar{v}_A)_1$ and $(\bar{v}_B)_1$ are the velocities of the spheres immediately before impact, and $(\bar{v}_A)_2$ and $(\bar{v}_B)_2$ are the velocities immediately after the impact. To study this system we choose a coordinate system xyz such that the x axis is normal to the plane of contact (the yz plane) of the spheres, as shown in the figure.

The situation depicted in Figure 10a is called *central impact*. This is the case when the centers of mass of both bodies and the point of contact

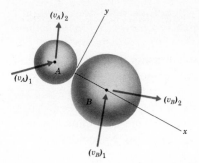

Figure 10a

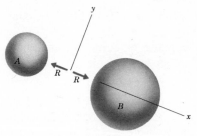

Figure 10b

are all situated along a common line normal to the plane of contact. The case where these three points are not all situated along the normal is called *eccentric impact*. The latter is a topic in rigid body kinetics, treated in Module V.

We assume that the spheres are perfectly smooth, so that the only interaction force between the bodies is the normal force $\bar{R}$ illustrated in Figure 10b. At an instant when time equals t_1, the bodies first contact and begin to deform under the influence of this force. After attaining their maximum deformation, at time t_{max}, the bodies begin to regain their shapes, until ultimately they lose contact at time t_2. To a certain extent the collision process resembles the sudden compression and release of a spring, but it is much more complex.

In order to avoid the necessity of studying the deformation characteristics of the bodies, let us consider the results of a number of experiments in which we measure the final velocities of the two spheres for many different initial velocities. We find in these experiments that the ratio of the initial and final relative velocity components in the x direction (normal to the plane of contact) is essentially constant for a range of velocities of engineering significance. In order for the particles to collide and then separate, we must have

$$(v_{Ax})_1 > (v_{Bx})_1$$

$$(v_{Ax})_2 \le (v_{Bx})_2$$

Therefore, the experiments tell us that

$$\epsilon = -\frac{(v_{Bx})_2 - (v_{Ax})_2}{(v_{Bx})_1 - (v_{Ax})_1} \tag{41}$$

where ϵ is a positive constant called the *coefficient of restitution*.

Equation (41) also has meaning in regard to the interaction force $\bar{R}$. Let us define the period of deformation by $t_1 \le t \le t_{max}$ and the period of restitution by $t_{max} \le t \le t_2$. We can then show that

$$\epsilon = \frac{\displaystyle\int_{t_{max}}^{t_2} R \, dt}{\displaystyle\int_{t_1}^{t_{max}} R \, dt} \tag{42}$$

In other words, the ratio of the impulses of the contact force in the two time intervals is the coefficient of restitution.

A collision generally involves a loss of mechanical energy, the energy normally being expended in the form of heat and sound. Only when $\epsilon = 1$, which is the case of a *perfectly elastic impact,* is there conservation of mechanical energy. The case $\epsilon = 0$, which is a *perfectly plastic impact,* represents a maximum loss of energy.

Equation (41) represents only one relationship for the velocity components $(v_{Ax})_2$ and $(v_{Bx})_2$. Clearly we require another relationship to determine the velocities of both bodies after the collision. Toward this end, we use the free body diagram of Figure 10b. Because the interaction force $\bar{R}$ is the only significant force, the other relationship is obtained by using conservation of linear momentum in the x direction. Specifically,

$$m_A(v_{Ax})_2 + m_B(v_{Bx})_2 = m_A(v_{Ax})_1 + m_B(v_{Bx})_1 \tag{43}$$

Solving this equation simultaneously with equation (41) yields the x components of the velocities.

The components of velocity in the plane of contact remain unchanged, because the impact does not produce any significant forces acting tangent to the contact plane. Hence

$$(v_{Ay})_2 = (v_{Ay})_1 \qquad (v_{By})_2 = (v_{By})_1 \tag{44}$$

$$(v_{Az})_2 = (v_{Az})_1 \qquad (v_{Bz})_2 = (v_{Bz})_1$$

We should note the limitations of the foregoing development. Equation (41), for the coefficient of restitution, is applicable to all cases of central impact. On the other hand, equations (43) and (44) were obtained for the special case where the impulsive interaction force $\bar{R}$ was the only force, Figure 10b. In some problems the bodies will be restrained by impulsive external forces that may invalidate either equation (43) or equation (44). Situations where this occurs are featured in some of the homework problems that follow.

Finally, consider the case when a small body collides with the earth. In such a situation the enormous mass of the earth results in a negligible change in the earth's velocity. Therefore, to study such problems we consider the earth to be fixed, $\bar{v}_e \equiv \bar{0}$, and determine the motion of the small body from equations (41) and (44).

EXAMPLE 18

Disk A on an air hockey game board collides with disk B, which is initially at rest. The disks are of identical mass and the coefficient of restitution is ϵ. Determine (a) the velocities with which the disks rebound from the collision, and (b) the percentage of the initial kinetic energy that is dissipated in the collision.

Solution

The study of the collision process requires that we first establish the contact plane. Therefore, we make a sketch that shows the two disks in contact.

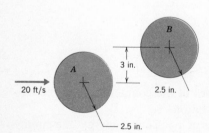

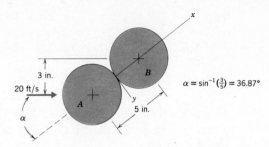

Choosing the x axis normal to the contact plane as shown, the initial velocities are

$$(\bar{v}_B)_1 = \bar{0}$$

$$(\bar{v}_A)_1 = 20(\cos \alpha \bar{i} + \sin \alpha \bar{j}) = 16\bar{i} + 12\bar{j} \text{ ft/s}$$

Because the contact force is in the x direction, the velocity components in the y direction are unchanged. Hence

$$(v_{By})_2 = 0 \qquad (v_{Ay})_2 = 12$$

The motion in the x direction can be determined from the equation for the coefficient of restitution and from conservation of linear momentum. Thus, for $m_A = m_B = m$,

$$\epsilon = -\frac{(v_{Bx})_2 - (v_{Ax})_2}{0 - 16}$$

$$m(v_{Ax})_2 + m(v_{Bx})_2 = m(16) + 0 \longleftarrow \text{DISK } B \text{ INITIALLY AT REST}$$

Solving these equations simultaneously gives

$$(v_{Ax})_2 = 8(1 - \epsilon) \quad (v_{Bx})_2 = 8(1 + \epsilon)$$

Therefore the final velocities are

$$(\bar{v}_A)_2 = 8(1 - \epsilon)\bar{i} + 12\bar{j} \text{ ft/s}$$

$$(\bar{v}_B)_2 = 8(1 + \epsilon)\bar{i} \text{ ft/s}$$

These vectors are illustrated in the following sketch.

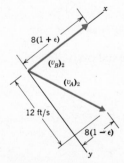

The decrease in kinetic energy is now computed as

$$T_1 = \tfrac{1}{2}m_A(v_A)_1{}^2 + \tfrac{1}{2}m_B(v_B)_1{}^2 = \tfrac{1}{2}m(20)^2 = \tfrac{1}{2}m(400)$$

$$T_2 = \tfrac{1}{2}m_A(v_A)_2{}^2 + \tfrac{1}{2}m_B(v_B)_2{}^2$$

$$= \tfrac{1}{2}m[8^2(1 - \epsilon)^2 + 12^2 + 8^2(1 + \epsilon)^2]$$

$$= \tfrac{1}{2}m[272 + 128\epsilon^2]$$

Thus

$$\Delta T(\%) = \frac{T_2 - T_1}{T_1}(100) = \frac{272 + 128\epsilon^2 - 400}{400}(100)$$

$$= -32(1 - \epsilon^2) \%$$

From this result we can deduce that 32% of the kinetic energy is dissipated in a perfectly plastic collision and no energy is lost in a perfectly elastic collision.

In closing, we call attention to a common misconception about a perfectly plastic collision; that is, that the two bodies are joined after the collision. This need not be so, as can be seen in this problem by setting $\epsilon = 0$ in $(\bar{v}_A)_2$ and $(\bar{v}_B)_2$.

EXAMPLE 19

A steel ball bearing is dropped from a height of 1.60 m onto a steel surface that is inclined at an angle of 15°. The coefficient of restitution for the collision is 0.60. Determine the maximum height to which the ball bearing will rebound.

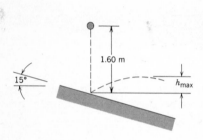

Solution

There are three time intervals involved in this problem. The first is the free-fall time of the ball bearing until it hits the steel surface. Then there is a short time interval in which the collision occurs. Finally, the ball bearing rebounds, once again under the influence of gravity.

For the initial free fall we have rectilinear motion from rest with constant acceleration g. Hence the ball bearing impacts the surface at the speed of

$$v_1 = \sqrt{v_0^2 + 2gh} = \sqrt{0 + 2(9.806)(1.60)} = 5.602 \text{ m/s}$$

To study the collision we sketch the impact, showing the x axis normal to the contact plane.

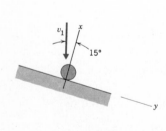

From this diagram we may write

$$\bar{v}_1 = 5.602(-\cos 15°\bar{i} + \sin 15°\bar{j})$$

$$= -5.411\bar{i} + 1.450\bar{j} \text{ m/s}$$

Treating the steel surface as a massive fixed body, from equation (41) we have

$$\epsilon = 0.60 = -\frac{(v_x)_2}{(v_x)_1} = -\frac{(v_x)_2}{(-5.411)}$$

$$(v_x)_2 = 3.247 \text{ m/s}$$

The velocity components tangent to the contact plane are unchanged, so

$$(v_y)_2 = (v_y)_1 = 1.450$$

Therefore

$$\bar{v}_2 = 3.247\bar{i} + 1.450\,\bar{j} \text{ m/s}$$

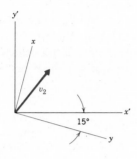

We next study the motion after the collision by computing the horizontal and vertical components of $\bar{v}_2$. This computation is conveniently performed by choosing another coordinate system $x'y'z'$ in which x' is horizontal and y' is vertical, as shown in the sketch, and then transforming the unit vectors. (In essence, we are setting up a new problem and hence choosing a more convenient set of coordinate axes.) From this diagram we have

$$\bar{i} = \sin 15°\,\bar{i}' + \cos 15°\,\bar{j}' = 0.2588\bar{i}' + 0.9659\bar{j}'$$

$$\bar{j} = \cos 15°\,\bar{i}' - \sin 15°\,\bar{j}' = 0.9659\bar{i}' - 0.2588\bar{j}'$$

Thus

$$\bar{v}_2 = 3.247(0.2588\bar{i}' + 0.9659\bar{j}') + 1.450(0.9659\,\bar{i}' - 0.2588\bar{j}')$$

$$= 2.241\bar{i}' + 2.761\bar{j}' \text{ m/s}$$

In this problem we require the value of h_{max}, which is the maximum upward distance traveled after the collision. The motion in the vertical direction involves a constant downward deceleration g with the initial upward speed of 2.761 m/s. Thus, with h as the vertical distance, we have

$$(v_{y'})^2 = (2.761)^2 - 2(9.806)h$$

The value of h is a maximum when the velocity in the vertical direction is zero, which results in

$$h_{max} = \frac{(2.761)^2}{2(9.806)} = 0.389 \text{ m}$$

HOMEWORK PROBLEMS

III.160 Two automobiles collide head-on. Vehicle A has a mass of 1200 kg and an initial speed of 3 m/s, whereas vehicle B has a mass of 1500 kg and an initial speed of 2 m/s. The coefficient of restitution is 0.20. Determine (a) the velocities of the two vehicles following the collision, and (b) the amount of mechanical energy dissipated by the collision.

Prob. III.160

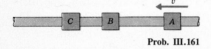

Prob. III.161

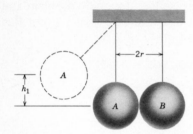

Prob. III.162

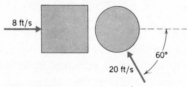

Prob. III.163

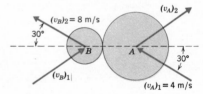

Prob. III.164

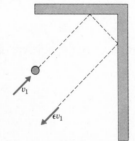

Prob. III.165

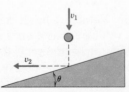

Prob. III.166

III.161 Three identical collars slide over a smooth horizontal rod. Initially, block A is moving to the left at speed v and the other blocks are at rest. Determine how many collisions will occur and the velocity of each block after all collisions are over. Let the coefficient of restitution between the collars be ϵ, where $0 < \epsilon < 1$.

III.162 Spheres A and B have identical mass m and radius r. The coefficient of restitution is ϵ. If sphere A is released from a height h_1, determine the height to which the spheres rebound after colliding.

III.163 The velocities of a 2-lb block and a 1-lb disk sliding on a horizontal table are shown. Determine their velocities after they collide, if the coefficient of restitution is 0.40.

III.164 Two disks sliding over a horizontal surface collide in the position shown. The masses of the disks are $m_A = 800$-g and $m_B = 200$-g, and the coefficient of restitution is 0.80. As shown in the diagram, the initial velocity of disk B and the final velocity of disk A are unknown. Determine these unknown quantities.

III.165 Show that, in the absence of friction, a ball rolling on a horizontal plane will emerge from a corner such that its final velocity $\bar{v}_2$ is $-\epsilon\bar{v}_1$, where $\bar{v}_1$ is the initial velocity. Also explain what would happen in the case where $\epsilon = 0$.

III.166 A ball is dropped vertically onto the inclined plane, hitting it at a speed v_1. It is desired to have the ball rebound horizontally. Derive expressions for the required angle θ for the incline and the corresponding rebound speed v_2 in terms of the coefficient of restitution ϵ.

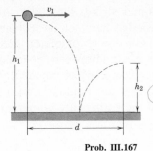

Prob. III.167

III.167 A ball bearing is thrown horizontally at a speed v_1 at a distance h_1 above the ground. Derive expressions for the distances d and h_2 locating the highest point attained by the ball bearing after it hits the ground. Let the coefficient of restitution be ϵ.

III.168 The pendulum is released from rest in the horizontal position and strikes the inclined surface when it is in the vertical position. Determine the velocity of the sphere as it rebounds from the collision if the sphere is connected to the pivot by (a) a cable, (b) a rigid, massless bar.

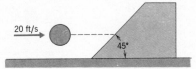

Prob. III.169

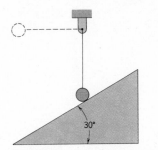

Prob. III.168

III.169 A 4-oz ball bearing hits the 2-lb steel angle block with a horizontal velocity of 20 ft/s. Assuming that all surfaces are smooth, determine the velocities of both bodies after the collision. The coefficient of restitution is 0.75.

III.170 The unit vector normal to the shaded plane is $\bar{e} = (2\bar{I} + 2\bar{J} + \bar{K})/3$. The velocity of a ball before it hits the plane is $\bar{v} = -40\bar{I} - 30K$ ft/s, and the coefficient of restitution is 0.80. Determine the velocity of the ball after it hits the plane.

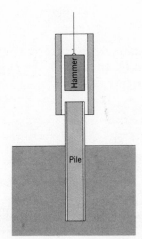

Prob. III.171

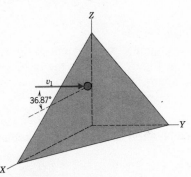

Prob. III.170

III.171 The 900-kg hammer of a pile driver falls from a height of 2 m above a 150-kg pile. The collision between the hammer and the pile is perfectly plastic, driving the pile 150 mm into the ground. Determine the average resistance of the ground to the motion of the pile.

III.72 The 4-kg head of a sledge hammer strikes the 20-kg anvil at 10 m/s. The mass of the handle of the sledge hammer is negligible. The coefficient of restitution is 0.50 and the collision is completed in 20 ms. (a) Determine the average impulsive force exerted between the hammer and the anvil if the stiffness of the two springs supporting the anvil is $k = 20$ kN/m each. (b) Solve part (a) if the springs are rigid ($k = \infty$). (c) Determine the maximum additional distance the springs are compressed after the collision in case (a).

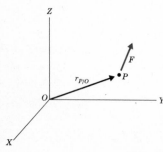

I. ANGULAR IMPULSE-MOMENTUM RELATIONS

The relationships to be developed in this section are additional standard integrals of the differential equations of motion involving the linear momentum of a particle. However, in contrast to the linear impulse-momentum principles, the result obtained here will be relationships useful for systems that are rotating about an axis or a point. In addition to their applications to particle systems, the principles we derive will be fundamental to further studies, such as the kinetics of rigid bodies. As was done previously, we shall first study the motion of a single particle, and then consider a system of particles.

1 Single Particle

Let us compute the moment of the resultant force $\bar{F}$ exerted on the particle P in Figure 11 about the fixed point O. In general, the moment of a force about an arbitrary point may be written as

$$\bar{M}_O \equiv \bar{r}_{P/O} \times \bar{F}$$

Using Newton's second law to replace $\bar{F}$ gives

$$\bar{M}_O = \bar{r}_{P/O} \times m\bar{a}_P$$

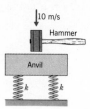

Figure 11

We can now manipulate this equation by using the fact that $\bar{a}_P = \dot{\bar{v}}_P = \ddot{\bar{r}}_{P/O}$, and applying the rule for the time derivative of a cross product, as follows.

For central force motion;

$R_1 v_1 = R_2 v_2 = $ *equ* $n\,48$

$$\bar{M}_O = \bar{r}_{P/O} \times m\dot{\bar{v}}_P \equiv \frac{d}{dt}(\bar{r}_{P/O} \times m\bar{v}_P) - \dot{\bar{r}}_{P/O} \times m\bar{v}_P$$

$$= \frac{d}{dt}(\bar{r}_{P/O} \times m\bar{v}_P) - \bar{v}_P \times m\bar{v}_P$$

$$= \frac{d}{dt}(\bar{r}_{P/O} \times m\bar{v}_P)$$

The quantity $\bar{r}_{P/O} \times m\bar{v}_P$ is called the *angular momentum* with respect to point O and is denoted by the symbol $\bar{H}_O$. As can be seen from this definition, $\bar{H}_O$ is the moment of the linear momentum vector about point O. In general then, we have

$$\bar{H}_O = \bar{r}_{P/O} \times m\bar{v}_P$$

$$\bar{M}_O = \dot{\bar{H}}_O \tag{45}$$

The derivation of the angular momentum relationship for a single particle can now be completed by assuming that we know how the moment $\bar{M}_O$ varies with time. Knowing this, we can separate variables in the second of equations (45) to obtain

$$\int_{(\bar{H}_O)_1}^{(\bar{H}_O)_2} d\bar{H}_O = \int_{t_1}^{t_2} \bar{M}_O\, dt$$

$$(\bar{H}_O)_2 = (\bar{H}_O)_1 + \int_{t_1}^{t_2} \bar{M}_O\, dt \tag{46}$$

Note the similarity to equation (38) for linear momentum. Hence it makes sense to call $\int \bar{M}_O\, dt$ the *angular impulse.* In situations where the resultant force exerts no moment about the origin, equation (46) tells us that $(\bar{H}_O)_1 = (\bar{H}_O)_2$, which is the statement of the principle of *conservation of angular momentum.*

To gain further insight about the angular momentum principles let us examine one scalar component of the vector relationships given in equations (45) and (46). Let the direction of the chosen component be the axial direction $\bar{e}_h = \bar{K}$ of the set of cylindrical coordinates shown in Figure 12.

We begin by writing the vectors in terms of their cylindrical coordinate components. Thus

$$\bar{r}_{P/O} = R\bar{e}_R + h\bar{e}_h$$

$$\bar{v}_P = \dot{R}\bar{e}_R + R\dot{\phi}\,\bar{e}_P + \dot{h}\bar{e}_h$$

$$\bar{F} = F_R\,\bar{e}_R + F_\phi\bar{e}_\phi + F_h\,\bar{e}_h$$

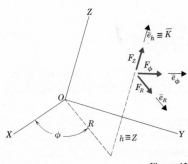

Figure 12

Now we compute

$$\bar{H}_O = \bar{r}_{P/O} \times m\bar{v}_P = m[-hR\dot{\phi}\ \bar{e}_R + (h\ \dot{R} - R\dot{h})\bar{e}_\phi + R^2\dot{\phi}\ \bar{e}_h]$$

$$\bar{M}_O = \bar{r}_{P/O} \times \bar{F} = -hF_\phi\ \bar{e}_R + (hF_R - RF_h)\bar{e}_\phi + RF_\phi\ \bar{e}_h$$

As stated above, our interest is with the axial components of $\bar{H}_O$ and $\bar{M}_O$. Noting that $\bar{e}_h \equiv \bar{K}$, these components are

$$(H_O)_Z = mR^2\dot{\phi}, \qquad (M_O)_Z = RF_\phi$$

Consider first the axial component equation (45), that is
$$(M_O)_Z = (\dot{H}_O)_Z$$

Upon substitution of the expressions for $(H_O)_Z$ and $(M_O)_Z$, this becomes

$$RF_\phi = \frac{d}{dt}(mR^2\dot{\phi})$$

Performing the differentiation yields

$$RF_\phi = m(2R\dot{R}\dot{\phi} + R^2\ddot{\phi})$$

The interesting aspect of this result is that, when a factor R is cancelled, we have $F_\phi = ma_\phi$. In other words, the axial component of equation (45) is identical to the transverse component of Newton's second law. We therefore can conclude that equation (45) does not represent a significantly new viewpoint for treating the motion of a particle.

The utility of the concept of angular momentum lies in the application of the impulse-momentum equation (46), not the basic equation (45). Using the foregoing results to express the axial components of the vectors in equation (46) gives

$$(H_O)_{Z_2} = (H_O)_{Z_1} + \int_{t_1}^{t_2} (M_O)_Z\ dt$$

or

$$mR_2{}^2\dot{\phi}_2 = mR_1{}^2\dot{\phi}_1 + \int_{t_1}^{t_2} RF_\phi\ dt \tag{47}$$

The application of equation (47) requires that we know how the product RF_ϕ varies with time, in order to evaluate the angular impulse. Such situations occur infrequently. However, there is an important class of physical problems that have the special property, $F_\phi \equiv 0$. Equation (47) then tells us that $R^2\dot{\phi} = (H_O)_Z/m$ is constant in the motion, which is the statement of conservation of angular momentum about the Z axis.

A common but not unique case in which $F_\phi = 0$ arises when the resultant force acting on a particle is always directed toward or away from a fixed point. We refer to the force in such a situation as a *central force*, as typified by the problem of orbital motion of a space vehicle under the influence of the gravitational attraction of the earth. This problem is treated separately in Module VIII because of its practical importance.

Summarizing the foregoing discussion, we can make the following generalization. The concept of angular momentum proves important to study the motion of a *single particle* only when the particle moves about an axis under the influence of a *central force*, in which case we have *conservation of angular momentum* about that axis. Denoting the axis as Z, equation (47) then yields

$$(H_O)_Z = R_2{}^2\dot{\phi}_2 = R_1{}^2\dot{\phi}_1 \tag{48}$$

or equivalently

$$(H_O)_Z = R_2(v_\phi)_2 = R_1(v_\phi)_1$$

EXAMPLE 20

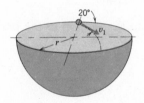

A particle slides down the smooth interior surface of a hemisphere of radius r. It was released from the rim with a velocity $\bar{v}_1$ directed at an angle of 20° with the horizontal, as shown. Determine the magnitude of v_1 that will result in the lowest position attained by the particle being an elevation of $0.8r$ below the rim.

Solution

The motion will consist of a spiralling motion about the vertical axis of the hemisphere. Hence we choose cylindrical coordinates, defined with respect to this axis. Because the surface is smooth, none of the forces acting on the particle have transverse components, as indicated in the free body diagram.

Neither of the forces acting on the particle exert a moment about the Z axis. Therefore, there is conservation of angular momentum about this axis. Also, noting that only gravity does work, we can use the work-energy principle to relate the initial and final positions.

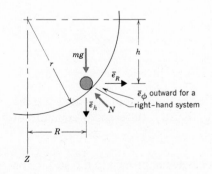

The velocity of the particle is always tangent to the hemispherical surface. Thus, at the rim the velocity is in the vertical plane, so that

$$\bar{v}_1 = v_1(\sin 20° \, \bar{e}_h + \cos 20° \, \bar{e}_\phi)$$

The specific position we wish to consider is $h = 0.8r$, where $\dot{h} \equiv v_h = 0$. This means that $v_R = 0$ also, because

$$R = \sqrt{r^2 - h^2} \qquad v_R = \dot{R} = -\frac{h}{\sqrt{r^2 - h^2}}$$

Thus $\bar{v}_2$ is in the transverse direction.

$$\bar{v}_2 = v_2\bar{e}_\phi$$

Hence, using equation (48) to describe the conservation of angular momentum, we write

$$R_2(v_\phi)_2 = R_1(v_\phi)_1$$

For the spherical surface $h^2 + R^2 = r^2$, so $R_1 = r$ and $R_2 = 0.6r$. Hence

$$0.6r(v_2) = r(v_1 \cos 20°)$$

$$v_2 = \frac{\cos 20°}{0.6} v_1 = 1.5662v_1$$

This is one relation between the initial and final speeds. As was mentioned previously, we can also use the work-energy principle to relate v_1 and v_2. Let us arbitrarily choose the elevation of the rim as the datum for the potential energy of gravity, in which case

$$V_1 = 0 \qquad V_2 = mg(-h_2) = -0.8mgr$$

Then, because only gravity does work in the motion of the particle, we have

$$T_2 + V_2 = T_1 + V_1$$

$$\tfrac{1}{2}mv_2{}^2 - 0.8mgr = \tfrac{1}{2}mv_1{}^2 + 0$$

$$v_1{}^2 = v_2{}^2 - 1.6gr$$

We now have two equations in the two unknowns v_1 and v_2. Solving for v_1 yields

$$v_1{}^2 = (1.5662v_1)^2 - 1.6gr$$

$$v_1 = 1.049 \sqrt{gr}$$

HOMEWORK PROBLEMS

III.173 Use the relationship between the moment about point O and the corresponding angular momentum to derive the differential equation of motion for the angle θ.

III.174 An artificial satellite makes its closest approach to the surface of the earth (perigee) in an elliptical orbit. It has an altitude of 400 km and a

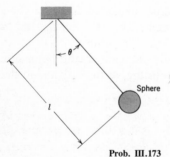

Prob. III.173

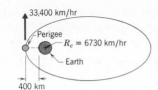

Prob. III.174

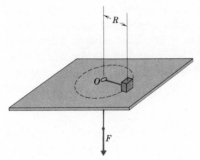

Prob. III.175

speed of 33,400 km/hr. Determine the altitude and speed of the satellite when it is farthest from the earth (apogee).

III.175 A cable is attached to an 8-oz block that may slide on a smooth horizontal surface. The cable is then passed through a hole at point O and a tensile force $\bar{F}$ is applied to it. Initially, the block is following a circular path at $R = 1$ ft corresponding to $F = 12$ lb. (a) Determine the initial speed of the block. (b) If the value of F is suddenly decreased to a constant value of 4 lb, determine the velocity of the block after it has spiralled outward to $R = 1.5$ ft.

III.176 In Problem III.175, the tensile force is decreased from 12 lb to some unknown value of F, with the result that the maximum distance attained by the block in its spiralling motion about point O is $R = 2$ ft. Determine (a) the velocity of the block at $R = 2$ ft, and (b) the corresponding value of F. (c) Do the answers to parts (a) and (b) satisfy circular motion for the block at a 2-ft radius?

III.177 A cable that is attached to a mooring post at point A is thrown to the person in the rowboat B. When the cable is caught the rowboat has the position and velocity shown. The person immediately pulls on the cable with a force of 100 N, in order to approach the dock. Determine the velocity of the rowboat after the cable is pulled in by 1.5 m. The combined mass of the passenger and the boat is 120 kg, and the water resistance is negligible.

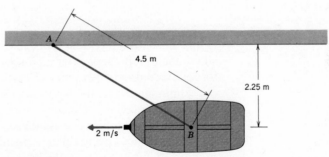

Prob. III.177

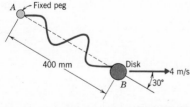

Prob. III.178

III.178 A 200-g disk on a smooth horizontal plane is tied to a rubberband whose opposite end is tied to a peg at point A. The disk is set into motion at point B, as shown. The stiffness of the rubberband is 8 N/m and the rubberband is 600 mm long when it is slack. Determine the maximum distance between the disk and end A of the rubberband.

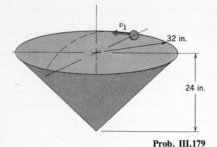

Prob. III.179

III.179 A ball bearing is projected tangentially along the rim of the cone shown at a speed v_1. At its lowest point the ball bearing is 12 in. above the apex of the cone. Determine (a) the initial speed v_1 of the ball bearing, and (b) the speed of the ball bearing at its lowest point. (c) Is the solution to part (b) the same as the speed for circular motion on the interior surface of the cone at that elevation?

III.180 An inextensible cable is passed through the hollow vertical post and then tied to a 0.5-kg sphere. The sphere is made to revolve in a horizontal circle for which $l_1 = 2$ m and $\theta_1 = 30°$. The cable is then drawn in very slowly until the sphere follows a horizontal circle for which $\theta_2 = 60°$. Determine (a) the initial speed v_1 of the sphere, (b) the final speed v_2, and (c) the amount of work done by the cable in pulling the sphere inward.

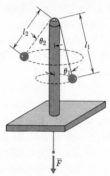

Prob. III.180

2 System of Particles

Figure 13a shows a typical set of forces acting on a three-particle system, where the $\bar{F}_i$ represent external forces and the $\bar{f}_{ij}$ represent the interaction forces exerted between pairs of particles.

Summing moments about an arbitrary point A, we find that the moment of each force $\bar{f}_{ij}$ cancels the moment of the corresponding force $\bar{f}_{ji}$ because the two forces are collinear, equal in magnitude, and opposite in direction. Hence, using relative position vectors such as those shown in Figure 13b, we can extend this analysis to conclude that the total moment about point A for a system of N particles is

$$\Sigma \bar{M}_A \equiv \sum_{i=1}^{N} \bar{r}_{i/A} \times (\bar{F}_i + \sum_{j=1}^{N} \bar{f}_{ij}) = \sum_{i=1}^{N} \bar{r}_{i/A} \times \bar{F}_i \qquad (49)$$

In other words, *the total moment acting on the system is the sum of the moments of the external forces.* This statement is basic to the study of statics.

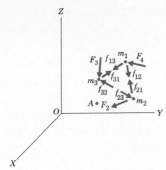

Figure 13a

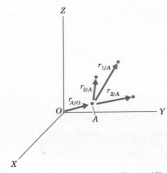

Figure 13b

Newton's second law for the ith particle is

$$\bar{F}_i + \sum_{i=1}^{N} \bar{f}_{ij} = m_i \bar{a}_i$$

Substituting this expression into equation (49) yields

$$\Sigma \, \bar{M}_A = \sum_{i=1}^{N} \bar{r}_{i/A} \times (m_i \bar{a}_i) \qquad (50)$$

This expression can be placed into a more convenient form by using the concept of relative motion with respect to point A. From Figure 13b we can write

$$\bar{r}_{i/O} = \bar{r}_{A/O} + \bar{r}_{i/A}$$

so that

$$\bar{v}_i = \bar{v}_A + \bar{v}_{i/A}$$

$$\bar{a}_i = \bar{a}_A + \bar{a}_{i/A}$$

Therefore, equation (50) becomes

$$\Sigma \, \bar{M}_A = \sum_{i=1}^{N} \bar{r}_{i/A} \times [m_i(\bar{a}_A + \bar{a}_{i/A})]$$

$$= \left(\sum_{i=1}^{N} m_i \bar{r}_{i/A} \right) \times \bar{a}_A + \sum_{i=1}^{N} \bar{r}_{i/A} \times m_i \bar{a}_{i/A} \qquad (51)$$

The first term on the right-hand side of equation (51) involves the definition of the location of the center of mass G of the system with respect to point A [refer to equation (20)]. Hence

$$\left(\sum_{i=1}^{N} m_i \bar{r}_{i/A} \right) \times \bar{a}_A \equiv \bar{r}_{G/A} \times m\bar{a}_A$$

We now rewrite the last term in equation (51) as the moment of a momentum vector by using the rule for the derivative of a vector product. This yields

$$\bar{r}_{i/A} \times m_i \bar{a}_{i/A} = \bar{r}_{i/A} \times [\frac{d}{dt}(m_i \bar{v}_{i/A})]$$

$$= \frac{d}{dt}(\bar{r}_{i/A} \times m_i \bar{v}_{i/A}) - \frac{d}{dt}(\bar{r}_{i/A}) \times m_i \bar{v}_{i/A}$$

$$= \frac{d}{dt}(\bar{r}_{i/A} \times m_i \bar{v}_{i/A}) - \bar{v}_{i/A} \times m_i \bar{v}_{i/A}$$

$$= \frac{d}{dt}(\bar{r}_{i/A} \times m_i \bar{v}_{i/A})$$

Using the results obtained above, equation (51) reduces to

$$\Sigma \, \bar{M}_A = \bar{r}_{G/A} \times m\bar{a}_A + \frac{d}{dt}\left(\sum_{i=1}^{N} \bar{r}_{i/A} \times m_i \bar{v}_{i/A}\right) \tag{52}$$

Let us consider the last term in equation (52) more carefully. If we think of $m_i \bar{v}_{i/A}$ as being the linear *momentum of particle i relative to point A*, then $\bar{r}_{i/A} \times m_i \bar{v}_{i/A}$ is the moment about point A of the momentum of particle i relative to point A. Therefore, let the vector $\bar{H}_A$ denote the *angular momentum of the system of particles relative to point A;* this vector is defined as

$$\bar{H}_A = \sum_{i=1}^{N} \bar{r}_{i/A} \times m_i \bar{v}_{i/A} \tag{53}$$

Thus, if point A is an arbitrary point, the moment equation is

$$\Sigma \, \bar{M}_A = \bar{r}_{G/A} \times m\bar{a}_A + \dot{\bar{H}}_A$$

A significant difference between this equation and the simple relation $\bar{M}_O = \dot{\bar{H}}_O$ for a single particle is that, if point A is arbitrarily chosen, it is necessary to determine the acceleration of this point in addition to determining the rate of change of the angular momentum $\bar{H}_A$.

Let us investigate the conditions required for this equation to reduce to the simpler form

$$\Sigma \, \bar{M}_A = \dot{\bar{H}}_A \tag{54}$$

For this to occur, point A must fit one of three special cases that reduces the term $\bar{r}_{G/A} \times m\bar{a}_A$ to zero.

1   If point A is fixed, or has constant velocity, then A is an inertial reference point and $\bar{a}_A \equiv 0$.
2   If point A is the center of mass, then $\bar{r}_{G/A} \equiv 0$.
3   If point A is accelerating directly toward or away from the center of mass, then $\bar{a}_A$ is parallel to $\bar{r}_{G/A}$ and their cross product vanishes.

Restricting ourselves to formulating the angular momentum with respect to one of these special points in no way limits the scope of the problems we can investigate.

We now complete the derivation of angular momentum principles by considering the situation where the moment of the forces is a known function of time. For such a case equation (54) can be integrated to obtain

$$(\bar{H}_A)_2 = (\bar{H}_A)_1 + \int_{t_1}^{t_2} \Sigma \, \bar{M}_A \, dt \tag{55}$$

This is the angular impulse-momentum equation for a system of particles.

As was true for a single particle, the concept of angular momentum for a system of independent particles is useful for problems involving conservation of angular momentum with respect to a specific axis. Equation (55) then shows that it is only necessary that the external forces exert no moment about the axis to have a constant value for the corresponding component of $\bar{H}_A$. Another important use of the concept of angular momentum is for treating the motion of rigid bodies. It will be shown in Modules V and VII that equation (54) represents one of the fundamental equations of motion for rigid bodies.

EXAMPLE 21

Blocks A and B, which are tied together by a cord, slide without friction along the horizontal arms of the cross bar shown while the entire system rotates freely about the vertical axis at the angular speed ω. Blocks A and B have masses of 2.0 kg and 0.5 kg, respectively, whereas the mass of the crossbar and the cord connecting the blocks is negligible. When block A is 600 mm from the vertical axis, it is moving outward with a speed of 40 mm/s and the angular speed of the system is $\omega = 0.5$ rad/s. Determine the value of $\dot{\omega}$ in this position, and also determine whether the radial speed of block A is increasing or decreasing.

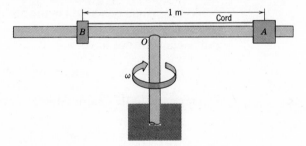

Solution

The information desired in this problem involves acceleration quantities, so we begin with free body diagrams of each block which will enable us to apply Newton's second law. Because the motion of each block is in the horizontal plane, the vertical component of the reaction force on each block merely counterbalances gravity. Hence, we only need show top views of the system illustrating the forces that act in the horizontal plane.

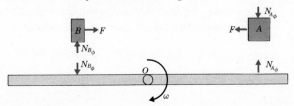

The transverse component of the reaction forces $\bar{N}_A$ and $\bar{N}_B$ are a result of the interaction of the blocks and the crossbar. If, instead of considering the two blocks and the crossbar individually, as we have done in drawing the free body diagrams, we consider them to form a system that also contains the cord connecting the two blocks, then all forces in the free body diagrams are internal to the system. It then follows that the angular momentum of this system about the vertical axis is constant, because the crossbar is rotating freely.

When we compute the total angular momentum of this system, we find that it is simply the sum of the values for the two blocks, because the other bodies have negligible mass. Hence, letting the Z axis coincide with the vertical axis of rotation, we have

$$(H_O)_Z = m_A R_A{}^2 \omega + m_B R_B{}^2 \omega$$

In order to obtain a relation for $\dot{\omega}$, we must use the derivative form of the angular momentum equation, that is, equation (55).

$$(M_O)_Z \equiv 0 = \frac{d}{dt}(H_O)_Z$$

$$= 2(m_A R_A \dot{R}_A + m_B R_B \dot{R}_B)\omega + (m_A R_A{}^2 + m_B R_B{}^2)\dot{\omega}$$

At the instant described in the problem, we know that $R_A = 0.6\,\text{m}$, $\dot{R}_A = +0.4\,\text{m/s}$, and $\omega = 0.5\,\text{rad/s}$. From the constraint that the length of the cord is constant we have

$$R_B = 1.0 - R_A = 0.4\,\text{m}$$

We differentiate this expression and use the given value of $\dot{R}_A$ to obtain

$$\dot{R}_B = -\dot{R}_A = -0.4\,\text{m/s}$$

Differentiating again yields $\ddot{R}_B = -\ddot{R}_A$. We now solve the above equations for $(M_O)_Z$ to find

$$\dot{\omega} = -\frac{2(m_A R_A \dot{R}_A + m_B R_B \dot{R}_B)}{m_A R_A{}^2 + m_B R_B{}^2}\,\omega = -0.50\,\text{rad/s}^2 \qquad \triangle$$

Here the negative sign means that the angular speed is decreasing.

Having considered the overall motion of the system, we are still free to consider the individual motion of each block. In view of the free body diagrams, the equations of motion for block A are

$$\Sigma F_R = -F = m_A(\ddot{R}_A - R_A \omega^2) = 2[\ddot{R}_A - 0.6(0.5)^2]$$

$$\Sigma F_\phi = N_{A\phi} = m_A(R_A \dot{\omega} + 2\dot{R}_A \omega) = 2[0.6(-0.5) + 2(0.4)(0.5)]$$

Because $\ddot{R}_B = -\ddot{R}_A$, the corresponding equations for block B are

$$\Sigma F_R = -F = m_B(\ddot{R}_B - R_B\omega^2) = 0.5[-\ddot{R}_A - 0.4(0.5)^2]$$

$$\Sigma F_\phi = N_{B\phi} = m_B(R_B\dot{\omega} + 2\dot{R}_B\omega)$$

$$= 0.5[0.4(-0.5) + 2(-0.4)(0.5)]$$

The solution of these equations is

$$N_{A\phi} = 0.20 \text{ newtons} \qquad N_{B\phi} = -0.30 \text{ newtons}$$

$$\ddot{R}_A = 0.10 \text{ m/s}^2$$

$$F = 0.10 \text{ newtons}$$

We conclude that the radial speed of block A is increasing becauce $\ddot{R}_A$ is positive. Finally, can you explain why the reaction forces above satisfy the equation $R_A N_{A\phi} + R_B N_{B\phi} = 0$?

HOMEWORK PROBLEMS

III.181 Two identical 1-kg disks of negligible diameter are fastened to the rigid bar AC, which is free to pivot in the vertical plane about point C. The system is at rest when disk A is hit by a 100-g bullet that is traveling at 600 m/s as shown. The bullet becomes embedded in the disk. Determine the angular velocity of the system immediately after the collision is completed.

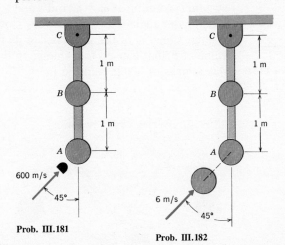

Prob. III.181

Prob. III.182

III.182 Two identical 1-kg disks of negligible diameter are fastened to the rigid bar AC, which is free to pivot in the vertical plane about point C. The system is at rest when disk A is hit by another 1-kg disk that is

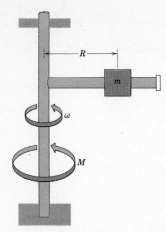

Prob. III.183

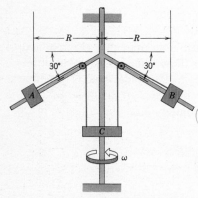

Prob. III.184 and III.185

traveling at 6 m/s, as shown. The coefficient of restitution for the collision is 0.50. Determine the angular velocity of bar AC after the collision is completed.

III.183 The collar of mass m can slide over the smooth rigid crossbar as the bar rotates about the vertical axis. The mass of the bar is negligible. A constant counterclockwise torque M is applied to the bar, causing it to rotate, and also causing the collar to slide along it. Initially, the system is at rest and $R = R_0$. Derive a single differential equation governing the radial distance R in terms of time.

III.184 Each of the three collars weighs 4 lb. The bar assembly on which they ride has negligible mass and rotates freely about the vertical axis. Initially, collars A and B are at rest relative to the bars at $R = 9$ in., and the angular speed of the assembly is $\omega = 2$ rad/s. Determine the velocity of each of the collars when $R = 18$ in.

III.185 Each of the three collars weighs 4 lb. The bar assembly on which they ride has negligible mass and rotates freely about the vertical axis. When $R = 15$ in., the collars are sliding outward relative to the bars at a speed of 9 in./s, with $\omega = 1$ rad/s. Determine the angular acceleration $\dot{\omega}$ and the acceleration of collar C at this instant.

III.186 Arm AB of the angular speed governor can be moved up and down by the force $\bar{F}$, causing the angle θ to change. In one case the system is rotating at $\omega = 80$ rad/s and $\theta = 60°$, when this force is changed causing the angle θ to change, very slowly, to $\theta = 30°$. Determine (a) the angular speed ω of the governor in its final position, and (b) the average value of F that caused this change in the value of θ. Each of the spheres has a mass of 400-g, and the mass of the rigid bars is negligible.

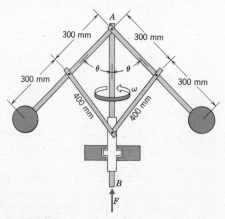

Prob. III.186

III.187 The assembly rotates in the vertical plane about the pivot at point A. The mass of each of the four disks is m, and the mass of the rigid bars is negligible. Determine the system's differential equation of motion in terms of the angle θ.

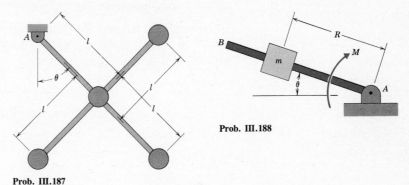

Prob. III.188

Prob. III.187

III.188 A constant clockwise couple M is applied to the bar AB, causing the system to rotate in the vertical plane about the pin at point A. This causes the block of mass m to slide along the bar. Considering bar AB to be smooth and massless, derive the differential equations of motion for the radial distance R and the angle θ.

J. SUMMARY OF PARTICLE DYNAMICS

The fundamental tool for a complete investigation of the motion of particle systems is Newton's second law expressed in terms of the appropriate kinematical description. Here we concentrated on path variables, rectangular Cartesian coordinates, and cylindrical coordinates for the kinematical description. These methods are sufficient for problems where only a force or an acceleration value are to be determined.

In some problems we must relate the initial and final velocity parameters of the system. One method for achieving this is to derive and solve the differential equation of motion by formulating Newton's second law for a general position. There are many techniques for solving differential equations of motion, the most elementary being the method of separating variables.

This approach for relating the initial and final velocities is complemented by the principles of work and energy, linear impulse and momentum, and angular impulse and momentum. All of these principles are standard solutions of the differential equations of motion. The key features to look for in applying these principles are summarized below.

1 Energy considerations are generally useful when the forces depend only on the position of the particle.

2   Linear momentum is suitable to situations where the rectangular components of the forces are functions of time, particularly, for impulsive forces.

3   Angular momentum is suitable to situations where bodies are rotating about a specific axis.

Finally, it should be noted that these principles may be utilized jointly to solve a problem, or individually to treat separate intervals in the motion of a body.

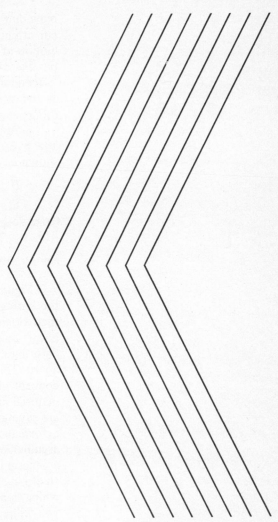

MODULE IV
KINEMATICS OF A
RIGID BODY IN
PLANAR MOTION

Now that we have completed a study of the dynamics of particles, we turn our attention to rigid bodies. Our aim in this module is to describe the motion of a rigid body when each point of the body is confined to move parallel to a prescribed plane. Hence, this module has the classic title, "kinematics of rigid bodies in plane motion," where kinematics is defined as the study of the motion of a physical system without regard to the forces causing the motion.

Simply stated, a rigid body is a collection of particles that are always at a fixed distance from each other. Consider the two-particle systems shown in Figure 1.

Figure 1   Two-particle systems.

Figure 1a shows the two particles connected by an elastic spring. If the spring is extended or shortened, the particles will no longer be at their original distance from each other; hence the system is not a rigid body. In Figure 1b, if we neglect the secondary (small) effects of the deformation of the steel bar, the particles are always at a fixed distance apart. Thus, we regard the bar to be a rigid body. Clearly, then, the concept of a rigid body is a mathematical modeling tool in that no real material can remain undeformed under the influence of forces. What we are saying is that for the class of problems we shall study, the microscopic deformation effects will not influence the results. Obviously the foregoing arguments can be extended to a system of particles (atoms) that compose a general body. Incidentally, if we did not neglect the deformation of the steel bar, we would be concerned with the dynamics of flexible bodies, which is an important topic (an integral part of the design process) that you will encounter in your later years of study.

Now let us direct attention to the phrase "plane motion." For the most part we shall use the plane of the paper to represent the plane of motion. A simple example of such motion is a disk rolling in a straight line. More generally, a rigid body executes plane motion if the path of each point in the body lies wholly in a plane, with the planes for all points being parallel to each other. An example of this is a body that is constrained to rotate about a fixed shaft.

Before we develop the basic kinematical model for the general motion of a rigid body, let us look more closely at the categories of motion we shall consider.

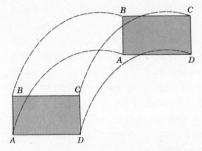

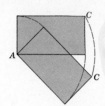

A. TYPES OF MOTION

1 Translation

When a body moves in a manner that results in every line in the body always being parallel to its original position, the body is said to be translating. Figure 2 shows a typical translation motion for a body. Note that in translation the points in the body can move in a curvilinear path. It is a common misconception to think that translation is equivalent to rectilinear motion for each point in the body.

Figure 2

2 Pure Rotation

If a rigid body moves in a manner such that one point in the body is fixed in space, then the body is said to be in pure rotation. Thus, in Figure 3 the box is executing a pure rotation about point A about an axis passing through A perpendicular to the plane of the diagram. An example of a pure rotation is the rotor of an electric motor.

Figure 3

3 General Motion

If the motion of a body is not a translation or a pure rotation, the body is in general motion. Later, we shall show that general motion consists of a combination of translation and rotation. An illustration of general motion for the box of Figure 2 is shown in Figure 4. An automobile traveling over a roadway with varying radius of curvature is in general motion.

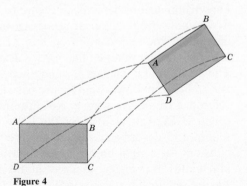

Figure 4

B. KINEMATICAL EQUATIONS OF PLANE MOTION

The important kinematic variables for rigid body motion are the same as those for particle motion, namely position, velocity, and acceleration.

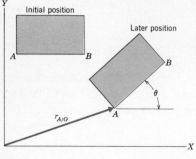

Figure 5

However, now we must be able to describe these variables for any point in the body.

To help in our thinking about the information needed to describe the position of all points in a body, let us start by considering the rigid box shown in Figure 5. Initially the edge AB of the box is parallel to the X axis. Because the distance between all points of the body is constant, we can determine the position of the box at a later time if we know the dimensions of the box and the location of a line (for instance, AB) of the box at that time. Line AB is a rigid line of known length. Hence, the only information required to describe the position of line AB is a position vector of any point on the line with respect to point O, for instance, $\bar{r}_{A/O}$, and the angle θ through which line AB was rotated.

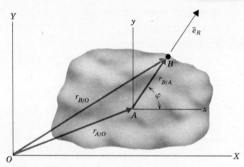

Figure 6

Let us generalize these concepts by considering the arbitrary body of known dimensions depicted in Figure 6. We know the position of the body if we know the location of a line AB in the body. This position can be defined by locating a point on line AB, for instance point A, and then locating another point on line AB, such as point B, with respect to point A. The position of point A is $\bar{r}_{A/O}$ and the position of point B with respect to A is $\bar{r}_{B/A}$. For convenience, in Figure 6 we introduced an auxiliary translating reference frame xyz whose axes are always parallel to XYZ, and whose origin is fixed at point A of the body. Then at any instant, all that is required to describe $\bar{r}_{B/A}$ is the angle φ, because the magnitude of $\bar{r}_{B/A}$ is constant for a rigid body. Reflecting on the tools we learned in particle kinematics, it is logical that we should describe $\bar{r}_{B/A}$ is polar coordinates. Let us define the polar coordinates with respect to the translating xyz reference frame. Thus

$$\bar{r}_{B/A} = \left|\bar{r}_{B/A}\right|\bar{e}_R$$

From Figure 6 we have the simple vector relationship

$$\bar{r}_{B/O} = \bar{r}_{A/O} + \bar{r}_{B/A} \tag{1}$$

We now have a formula to define the location of *any* point in the rigid body (in this case point B), which can be used when we know the absolute position of some other arbitrary point in the body (in this case point A), the distance from point A to point B, and the angle associated with the radial unit vector $\bar{e}_R$.

Let us now consider the matter of velocities. Velocity is the time derivative of the position vector. Therefore

$$\frac{d\bar{r}_{B/O}}{dt} \equiv \dot{\bar{r}}_{B/O} = \dot{\bar{r}}_{A/O} + \dot{\bar{r}}_{B/A} \tag{2}$$

But $\dot{\bar{r}}_{B/O} \equiv \bar{v}_B$ and $\dot{\bar{r}}_{A/O} \equiv \bar{v}_A$. Equation (2) may now be written as

$$\bar{v}_B = \bar{v}_A + \bar{v}_{B/A} \tag{3}$$

where $\bar{v}_{B/A}$ is the velocity of point B with respect to point A. An alternative way of considering $\bar{v}_{B/A}$ is to think of yourself as being stationed on the xyz coordinate system at point A (oblivious to the fact that this coordinate system is translating) watching point B move, in which case you would observe the velocity of B with respect to A.

We now require a way to describe the term $\bar{v}_{B/A}$ mathematically. Consider the two configurations shown in Figure 7. They show a body at some initial instant t and a slightly later time $t + \Delta t$.

Because $|\bar{r}_{B/A}|$ is a constant, we have

$$\bar{v}_{B/A} = \frac{d}{dt}(\bar{r}_{B/A}) = \frac{d}{dt}(|\bar{r}_{B/A}|\bar{e}_R) = |\bar{r}_{B/A}|\frac{d\bar{e}_R}{dt}$$

$$= |\bar{r}_{B/A}|\dot{\bar{e}}_R$$

In the time interval Δt, the unit vector $\bar{e}_R$ has rotated through an angle $\Delta\varphi$. Thus using the chain rule and the results from cylindrical coordinates for the derivative of the unit vector $\bar{e}_R$, we have

$$\frac{d\bar{e}_R}{dt} = \frac{d\bar{e}_R}{d\varphi}\frac{d\varphi}{dt} = \bar{e}_\varphi\dot{\varphi}$$

so that the equation for $v_{B/A}$ may now be written as

$$\bar{v}_{B/A} = |\bar{r}_{B/A}|\dot{\varphi}\bar{e}_\varphi \tag{4}$$

Equation (4) will prove to be useful. Hence, let us write this pure rotational effect in a more general form. If the plane of the page is the plane of motion, then for the rotational effect associated with $\varphi(t)$, it is easy to visualize the rotation as taking place about an axis perpendicular to the page. With this in mind let us define the angular velocity of line AB by a vector whose magnitude is $\dot{\varphi}$ and whose direction is normal to the

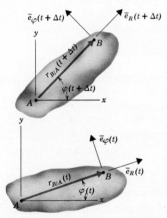

Figure 7

plane of motion. The sense of this normal is according to the right-hand rule, with the fingers of the right hand curling in the direction of increasing φ. Thus, the vector is in the direction of both the Z and z axes; that is,

$$\bar{\omega} \equiv \dot{\varphi}\bar{K} = \dot{\varphi}\bar{k} \tag{5}$$

The unit vectors of cylindrical coordinates are $\bar{e}_R, \bar{e}_\varphi, \bar{e}_h (\equiv \bar{k})$, from which we have

$$\bar{k} \times \bar{e}_R = \bar{e}_\varphi$$

This allows us to write equation (4) as

$$\bar{v}_{B/A} = \dot{\varphi}|\bar{r}_{B/A}|(\bar{k} \times \bar{e}_R) = \dot{\varphi}\bar{k} \times |\bar{r}_{B/A}|\bar{e}_R$$

or more simply,

$$\bar{v}_{B/A} = \bar{\omega} \times \bar{r}_{B/A} \tag{6}$$

Then from equation (3) we have the resulting vector equation for the velocity of any point on a rigid body in general motion:

$$\bar{v}_B = \bar{v}_A + \bar{\omega} \times \bar{r}_{B/A} \tag{7}$$

The acceleration equation may now be obtained by differentiating the velocity equation.

$$\dot{\bar{v}}_P = \dot{\bar{v}}_A + \dot{\bar{v}}_{B/A}$$
$$= \dot{\bar{v}}_A + \dot{\bar{\omega}} \times \bar{r}_{B/A} + \bar{\omega} \times \dot{\bar{r}}_{B/A}$$
$$= \dot{\bar{v}}_A + \dot{\bar{\omega}} \times \bar{r}_{B/A} + \bar{\omega} \times \bar{v}_{B/A} \tag{8}$$

The quantity $\dot{\bar{\omega}}$ is called the *angular acceleration* $\bar{\alpha}$ of the line *AB*. Because $\bar{\omega}$ is always normal to the plane of motion, only its magnitude can vary in planar motion. Thus

$$\bar{\alpha} \equiv \dot{\bar{\omega}} = \ddot{\varphi}\bar{k} \tag{9}$$

The time rate of change of velocity is acceleration. Hence substitution of equation (7) into equation (9) yields

$$\bar{a}_B = \bar{a}_A + \bar{a}_{B/A}$$
$$= \bar{a}_A + \bar{\alpha} \times \bar{r}_{B/A} + \bar{\omega} \times (\bar{\omega} \times \bar{r}_{B/A}) \tag{10}$$

Incidentally, in three-dimensional motion, the angular velocity can also change its direction. This becomes a major complication in such problems. However, the basic equations for rigid body motion in three dimensions will have the same form as those given by equations (7) and (10).

We used $\bar{\omega}$ and $\bar{\alpha}$ as properties of a line AB in the development of equations (7) and (10) for rigid body motion. Let us now show that at any instant of time $\bar{\omega}$ and $\bar{\alpha}$ are properties of the entire rigid body, so that the equations hold for any line in the rigid body.

Consider the rigid body shown in Figure 8. Arbitrarily choose two intersecting lines AB and CD that are fixed in the rigid body. Now measure their angular orientation with respect to a fixed line, say the $+X$ axis. Let γ be the fixed angle between lines AB and CD. We now have

$$\phi = \psi + \gamma \tag{11}$$

The time derivative of equation (11) gives

$$\dot{\phi} = \dot{\psi} + \dot{\gamma}$$

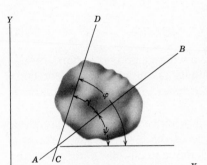

Figure 8

However, because γ is constant, $\dot{\gamma} \equiv 0$. Therefore

$$\dot{\phi} = \dot{\psi} \equiv \omega$$

and we conclude that at any given instant ω is the rate of rotation of all lines in a rigid body. Differentiating again, we have

$$\ddot{\phi} = \ddot{\psi} \equiv \alpha$$

from which we conclude that the angular acceleration is also a property of all lines in a rigid body.

The application of the basic kinematical equations (1), (7), and (10) requires that we express the vectors in terms of their components. These components may be referred to the axes of either the fixed XYZ or moving xyz reference frames. Consistent with the procedure we adopted for treating relative motion in Module Two, we shall use the latter. Hence, the unit vectors will be $(\bar{i}, \bar{j}, \bar{k})$. Recalling that vector components do not depend on the location of the origin of the reference frame, we shall specify only the orientation of the axes when solving kinematics problems. In the next module, in the process of formulating the kinetics principles for rigid bodies in planar motion, we shall develop rules governing the location of the origin.

C. PHYSICAL INTERPRETATION OF KINEMATICAL EQUATIONS

In order to understand better the physical significance of equations (7) and (10), let us use cylindrical coordinates to represent the vectors, rather than the rectangular Cartesian representation that we shall employ when solving problems. Thus, because

$$\bar{r}_{B/A} = |\bar{r}_{B/A}|\bar{e}_R, \quad \bar{\omega} = \dot{\varphi}\bar{k}, \quad \text{and} \quad \bar{\alpha} = \ddot{\varphi}\bar{k}$$

we have

$$\bar{v}_B = \bar{v}_A + \dot{\varphi}|\bar{r}_{B/A}|\bar{e}_\varphi \qquad (12)$$

$$\bar{a}_B = \bar{a}_A + \ddot{\varphi}|\bar{r}_{B/A}|\bar{e}_\varphi - \dot{\varphi}^2|\bar{r}_{B/A}|\bar{e}_R \qquad (13)$$

First, suppose that we had a problem where a body is in pure translation. Then $\dot{\varphi} = \ddot{\varphi} \equiv 0$, and our intuition tells us that all points in the body should have the same velocity and acceleration. This expectation is confirmed by equations (12) and (13).

Next, suppose that we had a problem where a body is in pure rotation about an axis perpendicular to the plane of the body passing through a point A. Then $\bar{v}_A = \bar{a}_A \equiv \bar{0}$ and all points move in circular paths about the axis of rotation. Note that the angular velocity and angular acceleration are parallel to the axis of rotation. For this case equation (12) tells us that the velocity of a point is tangent to its circular path. The acceleration is seen to have two components for this case. One is tangent to the path and the other is directed toward the center of curvature of the circular path. These properties are, of course, identical to those obtained when considering particle motion on a curved path.

In the case of general plane motion, we can regard $\bar{v}_A$ and $\bar{a}_A$ as translational effects with respect to the motion of point A and the remaining terms may be regarded as being caused by a rotation about an axis through point A normal to the plane of motion. Thus general motion is a superposition of translational and pure rotational motion. This is known as *Chasles' theorem*.

Note that if we had replaced point A with a different point, say point C, the translational portion of the motion would be that of point C, and the rotation would now be about a normal axis through C. However, as shown above, for either point A or point C, the angular velocity $\bar{\omega} \equiv \dot{\varphi}\bar{k}$ is the same, because $\bar{\omega}$ is a property of the body, and not of the points whose motion is being described.

EXAMPLE 1

Bar OA rotates in a plane about the fixed point O. At the instant when $\theta = 30°$, the angular velocity is 4 rad/s clockwise and the angular acceleration is 6 rad/s counterclockwise. Find (a) the velocity and (b) the acceleration of point A at this instant.

Solution
Part a
In general we know that

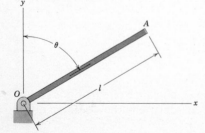

$$\bar{v}_A = \bar{v}_O + \bar{\omega} \times \bar{r}_{A/O}$$

Using the given coordinate system and the given information, we know that at the desired instant $\theta = 30°$, $\bar{\omega} = -4\bar{k}$ (note that the negative sign arises because of the right-hand rule), and $\bar{\alpha} = 6\bar{k}$. From the geometry of the figure, for $\theta = 30°$ we have

$$\bar{r}_{A/O} = l(\sin \theta \bar{\imath} + \cos \theta \bar{\jmath}) = l(0.5\bar{\imath} + 0.866\bar{\jmath})$$

Also, $\bar{v}_O = \bar{0}$ because point O is fixed.

The rigid body velocity formula then gives

$$\bar{v}_A = \bar{0} + (-4\bar{k}) \times l(0.5\bar{\imath} + 0.866\bar{\jmath})$$

$$= l(3.46\bar{\imath} - 2\bar{\jmath})$$

Part b

In general we know that

$$\bar{a}_A = \bar{a}_O + \bar{\alpha} \times \bar{r}_{A/O} + \bar{\omega} \times (\bar{\omega} \times \bar{r}_{A/O})$$

From the given information we have $\bar{\alpha} = 6\bar{k}$. We also know $\bar{\omega}$ and $\bar{r}_{A/O}$, and in part a we determined $\bar{\omega} \times \bar{r}_{A/O}$. Thus the acceleration formula yields

$$\bar{a}_A = \bar{0} + (+6\bar{k}) \times l(0.5\bar{\imath} + 0.866\bar{\jmath})$$

$$+ (-4\bar{k}) \times l(3.46\bar{\imath} - 2\bar{\jmath})$$

$$= l(-5.20\bar{\imath} + 3\bar{\jmath}) + l(-8\bar{\imath} - 13.84\bar{\jmath})$$

$$= l(-13.20\bar{\imath} - 10.84\bar{\jmath})$$

Alternative Method

In working any kinematical problem one can always revert to the fundamental definitions. Namely, the velocity is the time derivative of the position vector. As is true for any function, we must differentiate the general position vector, not the one evaluated at a specific instant. Hence

$$\bar{r}_{A/O} = l(\sin \theta \bar{\imath} + \cos \theta \bar{\jmath})$$

As the coordinate axes have fixed directions, for the velocity we find

$$\bar{v}_A = \frac{d}{dt}(\bar{r}_{A/O}) \equiv \dot{\bar{r}}_{A/O} = \frac{d}{dt}[l(\sin \theta \bar{\imath} + \cos \theta \bar{\jmath})]$$

$$= l\dot{\theta}(\cos \theta \bar{\imath} - \sin \theta \bar{\jmath})$$

From the given information, at the position of interest $\theta = 30°$ and $\dot{\theta} = 4$ (note that in this example, clockwise rotation means θ is increasing so that $\dot{\theta}$ is positive). Thus, we have

$$\bar{v}_A = l(3.46\bar{\imath} - 2\bar{\jmath}) \qquad \text{when } \theta = 30°$$

Next, we find the acceleration from the time derivative of the velocity. To do this we must be certain to differentiate the general expression for velocity. Hence

$$\bar{a}_A = \frac{d}{dt}(\bar{v}_A) = \frac{d}{dt}[l\dot{\theta}(\cos\theta\,\bar{\imath} - \sin\theta\,\bar{\jmath})]$$

$$= l\ddot{\theta}(\cos\theta\,\bar{\imath} - \sin\theta\,\bar{\jmath}) + l\dot{\theta}^2(-\sin\theta\,\bar{\imath} - \cos\theta\,\bar{\jmath})$$

Using the given information that $\theta = 30°$, $\dot{\theta} = 4$, and $\ddot{\theta} \equiv -6$ (counterclockwise angular acceleration means that $\dot{\theta}$ is decreasing), we have

$$\bar{a}_A = l(-5.20\bar{\imath} + 3\bar{\jmath}) + l(-8\bar{\imath} - 13.84\bar{\jmath})$$

$$= l(-13.20\bar{\imath} - 10.84\bar{\jmath}) \qquad \text{when } \theta = 30°$$

The key point in this alternative method is that one must work symbolically. For instance, if we inadvertently attempted to differentiate the velocity vector evaluated at $\theta = 30°$ to find the acceleration, we would have erroneously obtained zero acceleration, because we then would not have differentiated the vector velocity function.

Which method is easier? It doesn't really matter; however, our experience is that most people feel more comfortable using the derived formulas to solve kinematical problems, rather than performing differentiations according to the alternative method.

EXAMPLE 2

The steel fire door shown in the diagram is supported by two parallel arms AB and CD, each 1.5 m long. If the equal arms AB and CD are each rotating at the constant rate of $\omega = 10$ rad/s, what is the velocity and acceleration of the center of mass G of the door?

Solution

Points A, C, and G are part of the same body (the door), so we can use the rigid body equations. Because points A and C have the same motion, we can use either point. Let us use point A.

$$\bar{v}_G = \bar{v}_A + \bar{\omega}_{\text{door}} \times \bar{r}_{G/A}$$

$$\bar{a}_G = \bar{a}_A + \bar{\alpha}_{\text{door}} \times \bar{r}_{G/A} + \bar{\omega}_{\text{door}} \times (\bar{\omega}_{\text{door}} \times \bar{r}_{G/A})$$

However, the polygon formed by $ABDC$ is a parallelogram, so that line AC is always horizontal and the door only translates. Thus $\bar{\omega}_{\text{door}} \equiv \bar{0}$ and

$$\bar{v}_G = \bar{v}_A \qquad \bar{a}_G = \bar{a}_A$$

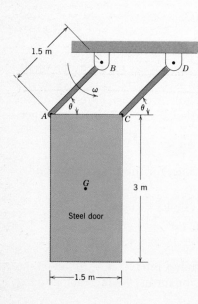

Now, the problem of finding $\bar{v}_A$ and $\bar{a}_A$ is similar to that in the previous problem, because arm AB is in pure rotation about point B.

$$\bar{v}_A = \bar{v}_B + \bar{\omega}_{AB} \times \bar{r}_{A/B}$$

$$\bar{a}_A = \bar{a}_B + \bar{\alpha}_{AB} \times \bar{r}_{A/B} + \bar{\omega}_{AB} \times (\bar{\omega}_{AB} \times \bar{r}_{A/B})$$

As shown in the sketch, the coordinate system we choose for describing the vector components has the x axis horizontal and the y axis vertical. We are given $\bar{v}_B = \bar{a}_B = \bar{0}$, $\bar{\omega}_{A/B} = 10\bar{k}$, $\bar{\alpha}_{AB} = \bar{0}$. From the sketch we have

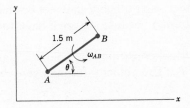

$$\bar{r}_{A/B} = 1.5(-\cos\theta\,\bar{i} - \sin\theta\,\bar{j})$$

Thus,

$$\bar{v}_G = \bar{v}_A = \bar{0} + 10\bar{k} \times 1.5(-\cos\theta\,\bar{i} - \sin\theta\,\bar{j})$$

$$= 15(-\cos\theta\,\bar{j} + \sin\theta\,\bar{i}) \text{ m/s}$$

$$\bar{a}_G = \bar{a}_A = \bar{0} + \bar{0} + 10\bar{k} \times [10\bar{k} \times 1.5(-\cos\theta\,\bar{i} - \sin\theta\,\bar{j})]$$

$$= 150(\cos\theta\,\bar{i} + \sin\theta\,\bar{j}) \text{ m/s}^2$$

An important thing to note in this example is that the system was composed of three rigid bodies. Hence we had the possibility of having three different angular velocities and three different angular accelerations. It was the geometrical configuration of the system that reduced those quantities to simply one nonzero angular velocity.

EXAMPLE 3

A portion of a machine consists of a bar AB, 750 mm long, with smooth sliders on both ends. The bar moves in the vertical plane such that slider A follows a vertical guide and slider B follows a horizontal guide, as shown in the diagram. At the instant when $\theta = 30°$, slider A is driven downward with a speed of 250 mm/s and this downward velocity is decreasing at the rate of 50 mm/s². Find the velocity and acceleration of end B at this instant.

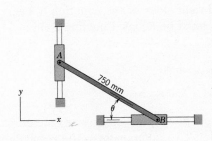

Solution

The approach we shall follow is to use the rigid body motion formulas, expressing all quantities by their vector components at the instant of interest. First we select coordinate axes xy, as shown in the diagram. With respect to these axes $\bar{v}_A = -250\bar{j}$ mm/s, $\bar{a}_A = 50\bar{j}$ mm/s² (because the downward velocity is decreasing, the acceleration is upward). The problem is to find $\bar{v}_B$ and $\bar{a}_B$. We cannot simply use the velocity and acceleration formulas for bar AB because we do not know $\bar{\omega}_{AB}$ and $\bar{\alpha}_{AB}$. There is, however, an additional requirement we have not as yet utilized, which is that slider B can only move horizontally.

The requirements that slider A moves vertically and slider B moves horizontally are called *geometrical constraints*, because they are a result of the geometrical design of the system. The importance of geometrical constraints is that they are conditions the designer can impose on a system to obtain particular types of motion from the system, that is, translation, rotation, or specialized general motion. In general, if we overlook a problem's geometrical constraint condition, we probably shall not be able to solve the problem (even in general terms) because we shall have too many unknowns.

With the knowledge of the geometrical constraint the solution may now be completed. We first determine $\bar{v}_B$. The velocity formula gives

$$\bar{v}_B = \bar{v}_A + \bar{\omega}_{AB} \times \bar{r}_{B/A}$$

The geometrical constraint requires that $\bar{v}_B = v_B \bar{\imath}$, where the speed v_B is unknown. Because this is plane motion, $\bar{\omega}_{AB} = \omega_{AB}\bar{k}$. ($\omega_{AB}$ may be either a positive or negative number. By the right-hand rule, if it is positive the rotation is counterclockwise.) Then from the diagram with $\theta = 30°$, we have

$$\bar{r}_{B/A} = 750(\cos 30° \, \bar{\imath} - \sin 30° \, \bar{\jmath})$$

Thus, we find

$$v_B \bar{\imath} = -250\bar{\jmath} + \omega_{AB}\bar{k} \times 750(\cos 30° \, \bar{\imath} - \sin 30° \, \bar{\jmath})$$

$$= 750\omega_{AB} \sin 30° \, \bar{\imath} + (-250 + 750\omega_{AB} \cos 30°)\bar{\jmath}$$

In order for the vectors on the left and right sides of the equation to be equal, their corresponding components must be equal. Equating components gives

$\bar{\imath}$ direction: $v_B = 750\omega_{AB} \sin 30°$

$\bar{\jmath}$ direction: $0 = -250 + 750\omega_{AB} \cos 30°$

Note that we have obtained two scalar equations. This was to be expected because in planar motion the velocity has two components. The results are

$$\omega_{AB} = \frac{250}{750 \cos 30°}$$

$$v_B = 750\left(\frac{250}{750 \cos 30°}\right) \sin 30° = 250 \tan 30°$$

$$\bar{\omega}_{AB} = 0.385\bar{k} \text{ rad/s} \qquad \bar{v}_B = 0.144\bar{\imath} \text{ m/s}$$

Note that in addition to using the solution for $\bar{\omega}_{AB}$ to find $\bar{v}_B$, we shall need $\bar{\omega}_{AB}$ in the solution for $\bar{a}_B$.

$$\bar{a}_B = \bar{a}_A + \bar{\alpha}_{AB} \times \bar{r}_{B/A} + \bar{\omega}_{AB} \times (\bar{\omega}_{AB} \times \bar{r}_{B/A})$$

We were given $\bar{a}_A$, and we have determined $\bar{\omega}_{AB}$ and $\bar{r}_{B/A}$. The geometrical constraint on slider B requires that $\bar{a}_B = a_B \bar{i}$, and because this is plane motion $\bar{\alpha}_{AB} = \alpha_{AB} \bar{k}$. Thus the acceleration formula yields

$$a_B \bar{i} = 50 \bar{j} + \alpha_{AB} \bar{k} \times 750(\cos 30° \, \bar{i} - \sin 30° \, \bar{j})$$

$$+ \; 0.385 \bar{k} \times [0.385 \bar{k} \times 750(\cos 30° \, \bar{i} - \sin 30° \, \bar{j})]$$

$$= (750 \alpha_{AB} \sin 30° - 111.11 \cos 30°) \bar{i}$$

$$+ \; (50 + 750 \alpha_{AB} \cos 30° + 111.11 \sin 30°) \bar{j}$$

Equating components then gives

$\bar{i}$ direction: $\qquad a_B = 375.0 \alpha_{AB} - 96.22$

$\bar{j}$ direction: $\qquad 0 = 50 + 649.52 \alpha_{AB} + 55.55$

Thus the solutions are

$$\bar{\alpha}_{AB} = -0.1625 \text{ rad/s}^2$$

$$\bar{a}_B = -0.1572 \text{ m/s}^2$$

Before leaving this problem, we should consider a geometrical interpretation of the solution steps presented. In solving for both velocity and acceleration we expressed the general motion of bar AB as a translation with the motion of slider A plus a rotation about the pivot of slider A (with the angular motion of bar AB). Then we noted that the geometrical constraint on slider B requires that the combination of the motions of slider B associated with the translation and rotation must result in slider B having horizontal motion.

HOMEWORK PROBLEMS

IV.1 The angle of rotation of the reel of a tape recorder is defined by $\theta = 1.2t + 0.01 \sin 30\pi t$, where θ is measured in radians and t in seconds. Determine the horizontal and vertical components of the acceleration of point A on the rim of the reel when (a) $t = 0$, (b) $t = \frac{1}{60}$ s, and (c) $t = \frac{1}{45}$ s.

Prob. IV.1

IV.2 The truck shown in the sketch is at rest. The tailgate of the truck is lowered such that the vertical component of the velocity of its free end is

Prob. IV.2

constant at 5 m/s. Determine the angular velocity and angular acceleration of the tailgate for the instant when $\theta = \tan^{-1}(\frac{3}{4})$.

IV.3 The acceleration of slider A is 2 m/s² down to the right, and its velocity is 1 m/s up to the left as the bar passes the horizontal. For this position determine (a) the velocity of slider B, (b) the angular acceleration of the bar, and (c) the acceleration of slider B.

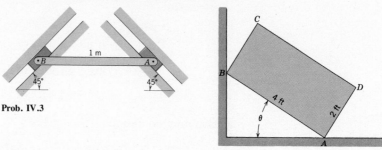

Prob. IV.3

Prob. IV.4

IV.4 A box resting against a wall begins to slide. In the position where $\theta = 30°$ the angular velocity is 3.0 rad/s and the angular acceleration is 1.0 rad/s², both counterclockwise. Compute the velocities and accelerations of corners A and C at this instant.

IV.5 In the position shown collar A at the end of a 1.30-m rod is moving to the left at 300 mm/s and is gaining speed at the rate of 100 mm/s². Determine the velocity and acceleration of collar B at this instant.

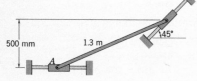

Prob. IV.5

IV.6 The rigid link ABC is a portion of the control linkage in an automotive carburetor. At the instant shown, the control rod BE has a velocity of 5 in./s and an acceleration of 2 in./s², both to the left. Determine the velocity and acceleration of point C at this instant.

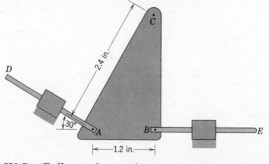

Prob. IV.6

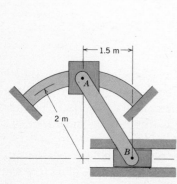

Prob. IV.7

IV.7 Collar A is moving to the right at the constant speed of 10 m/s. Determine the angular velocity and angular acceleration of rod AB in the position shown.

IV.8 Observation of the motion of the circular disk reveals the following information about the instantaneous velocity of points A, B, and C: $v_{Ax} =$ 20 m/s, $v_{By} = -15$ m/s, $v_{Cx} = 25$ m/s. Determine the angular velocity of the disk and the velocity of the midpoint O.

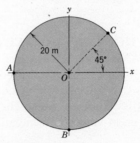

Prob. IV.8

IV.9 Bars AB and BC rotate in the vertical plane such that the angles γ and β are identical. Bar AB is rotating clockwise at the constant rate of 20 rev/min. Determine the acceleration of point C when $\gamma = 30°$.

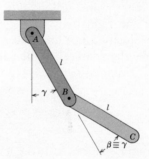

Prob. IV.9

IV.10 Slider A moves along rod OB and is also constrained to move in the vertical slot. Given that the rod is rotating with angular velocity ω and angular acceleration α as shown, derive expressions for the velocity and acceleration of the slider. *Hint:* Write the position vector for point A and differentiate.

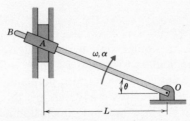

Prob. IV.10

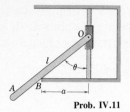

IV.11　The collar O slides along the vertical rod with constant velocity v. Determine expressions for the velocity and acceleration of point A on the link OA. *Hint:* Write the position vector for point O and differentiate.

Prob. IV.11

D.　APPLICATIONS OF KINEMATICAL EQUATIONS AND SPECIAL CASES

Before looking at some illustrative problems that demonstrate how we can systematically apply the kinematical equations to real systems, let us briefly study three special cases. The results of these discussions will prove to be of great value in formulating many problems.

1　Instantaneous Center of Zero Velocity

In this section we shall consider a rigid body undergoing general planar motion. We shall show that if we can find a point of the body that has zero velocity, it will enable us to simplify greatly the determination of the velocities of other points of the body.

Consider the rigid body of Figure 9. Let us assume that we can measure the velocities at A and B as a function of time. Hence, $\bar{v}_A$ and $\bar{v}_B$ are known. We then can relate the velocity of either point A or B to any other point C. From equation (7) we have

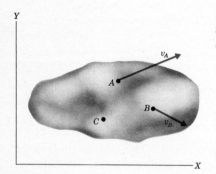

Figure 9

$$\bar{v}_A = \bar{v}_C + \bar{\omega} \times \bar{r}_{A/C} \tag{14}$$

$$\bar{v}_B = \bar{v}_C + \bar{\omega} \times \bar{r}_{B/C} \tag{15}$$

The problem now reduces to two questions. What are the conditions for $\bar{v}_C = \bar{0}$, and how can point C be found? From equations (14) and (15) we see that for $\bar{v}_c = \bar{0}$,

$$\bar{v}_A = \bar{\omega} \times \bar{r}_{A/C} \qquad \bar{v}_B = \bar{\omega} \times \bar{r}_{B/C} \tag{16}$$

From the definition of the cross product we know that $\bar{\omega} \times \bar{r}$ is perpendicular to both $\bar{\omega}$ and $\bar{r}$. Because $\bar{\omega}$ is perpendicular to the plane of motion, it must therefore be true that $\bar{\omega} \times \bar{r}$ is perpendicular to $\bar{r}$, and lies in the XY plane.

In other words, from equation (16) we see that the position vector $\bar{r}_{A/C}$ from the point of zero velocity to point A must be perpendicular to $\bar{v}_A$. Of course, a similar statement is true for point B. When we construct the perpendiculars to $\bar{v}_A$ and $\bar{v}_B$, they will intersect at point C, which is the only point common to both perpendiculars.

In Figure 10 we have shown this construction for the rigid body of Figure 9. The important conclusion is that because the velocity formulas

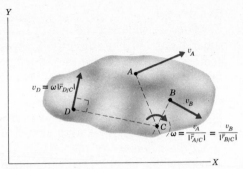

Figure 10

in equation (16) resemble those for a body in pure rotation, for the purposes of computing *velocity only,* we can think of the body as rotating about point C. Thus if we let D be any point in the body, we can say that $\bar{v}_D$ is perpendicular to $\bar{r}_{D/C}$ in the sense of the rotation of the body. This is shown in Figure 10.

Point C has zero velocity only at the instant for which we have drawn the diagram, because at a later instant the location of points A and B, as well as the velocities of these points, will have changed. The velocity formulas, equations (16), resemble those for a body in pure rotation about point C. Therefore, we call point C an *instantaneous center of zero velocity.* This is frequently shortened to *instant center.*

These statements are true for a general planar motion of a rigid body. If the body is in pure rotational motion, the instant center is the fixed center of rotation. On the other hand, if the body is in pure translation, $\bar{v}_A = \bar{v}_B$ so that the perpendiculars to $\bar{v}_A$ and $\bar{v}_B$ will never intersect; thus there is no instant center for a body in pure translation. Also note that frequently the instant center is not situated on the body. In such cases we can, if we wish, think of the instant center as lying on an imaginary extension of the body. Finally, we emphasize that although the instantaneous center can be located graphically, it can be located more exactly by trigonometry and geometry.

An obvious question to ask now is, can we use the instant center for determining accelerations? With the exception of pure rotation of a rigid body, the instant center is not fixed. Therefore the various points of the rigid body do not follow circular paths, and hence in general we *cannot* use the acceleration equations for circular motion about the instant center.

Our approach for acceleration problems will be to use the vector equation (10). However, because we need to know the angular velocity of the body before we can use equation (10), we shall find that the instant center is a convenient aid.

EXAMPLE 4

Consider the slider mechanism of Example 3. In that example the speed of slider A was 250 mm/s directed downward when $\theta = 30°$. Using the instant center concept, solve for the velocity of slider B when $\theta = 30°$, and show that the solution is the same as that obtained previously.

Solution

Sequentially, the steps in the solution are as follows: We begin an instant center solution by making a sketch that shows the known information about the velocities of the system. In this problem we know $\bar{v}_A$, and we also know that point B is constrained to move horizontally. With regard to $\bar{v}_B$, we need only know that it is horizontal in order to draw the appropriate perpendicular. Eventually the sense of $\bar{\omega}$ will yield the sense of $\bar{v}_B$, which at this stage is an unknown. The point of intersection of the perpendiculars to $\bar{v}_A$ and $\bar{v}_B$ is the instant center. The value of the angular speed ω is found by dividing the known speed v_A by the distance from point A to the instant center. The direction of $\bar{\omega}$ is determined by pretending that point A is actually in circular motion about the instant center. We then can find the magnitude and direction of $\bar{v}_B$ by pretending that point B is also in circular motion about the instant center. All of these statements are illustrated in the following sketch.

In this diagram ω is shown counterclockwise because point A is pictured as moving in a counterclockwise circle about the instant center, point C. Then, because ω is counterclockwise, circular motion for point B dictates that $\bar{v}_B$ is to the right. All that remains is to determine the lengths from C to A and B, respectively. We see that angle ACB is a right angle, so

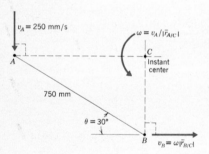

$$\left|\bar{r}_{A/C}\right| = 750 \cos 30° \qquad \left|\bar{r}_{B/C}\right| = 750 \sin 30°$$

Thus

$$\omega = \frac{250}{750 \cos 30°} = 0.384 \text{ rad/s} \qquad \text{(c.c.w.)}$$

$$v_B = \left(\frac{250}{750 \cos 30°}\right) 750 \sin 30° = 0.144 \text{ m/s} \qquad \text{(to the right)}$$

These results are, of course, the same as those obtained by the two methods used in Example 3.

Among the three methods for solving for velocities in rigid body motion, the instant center method is usually the fastest and easiest. Some people dislike it because it is more geometrical and involves more visualization. The alternative method of differentiating the general position vector becomes progressively more difficult as the system becomes more complicated. The primary advantage of using the vector equation (7) for the determination of the velocity is that the method of solving for accelerations is similar to that used in solving for velocities. The decision of which approach to use is yours. However, *do not* forget that the instant center can be used *only* in the velocity portion of the solution.

Additional Comments   As we have noted several times, the instant center changes location with time. A curve that is sometimes of interest is the locus of the points in space that the instant center occupies, called the *space centrode*. For the system of Example 4 we can determine the space centrode by finding the coordinates of the instant center X_C and Y_C for an arbitrary θ. We see from the sketch that the coordinates of the instant center with respect to XYZ are

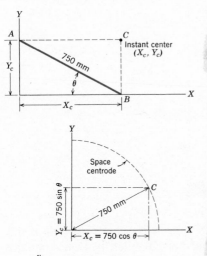

$$X_C = 750 \cos \theta \qquad Y_C = 750 \sin \theta \qquad Z_C = 0$$

Then, because $X_C{}^2 + Y_C{}^2 = (750)^2$ and the range of values of θ is from $0°$ to $90°$, we find that the space centrode is a quarter circle, centered at the origin, of radius 750 mm.

A less obvious curve that we could seek is the locus of points on the body (or its imaginary extension) that are the instant centers, called the *body centrode*. To find the body centrode in Example 4, we resketch the instant center diagram and determine the coordinates x and y that define the location of the instant center with respect to the body. We know that $|\vec{r}_{A/C}| = 750 \cos \theta$. Then by trigonometry we have

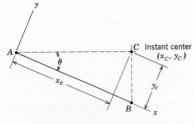

$$x_C = |\vec{r}_{A/C}| \cos \theta = 750 \cos^2 \theta = 375.0(1 + \cos 2\theta)$$

$$y_C = |\vec{r}_{A/C}| \sin \theta = 750 \sin \theta \cos \theta = 375.0 \sin 2\theta$$

From this wee see that $(x_C - 375)^2 + y_C{}^2 = (375)^2$. Also, because θ varies between $0°$ and $90°$, x is in the range of values from 0 to 750 mm. Thus, a sketch showing the locus of values of x_C and y_C is a semicircle, centered at the point $(x = 375.0, y = 0)$ of radius 375 mm.

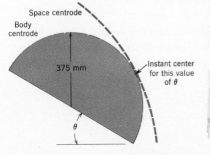

The principal significance of the body and space centrodes is that we can reconstruct the motion of the system by fixing the body centrode to the body and fixing the space centrode in space and allowing the body centrode to roll over the space centrode. These concepts prove of great value in the general area of computer-aided design of mechanisms.

HOMEWORK PROBLEMS

IV.12   At the instant shown, slider A is moving to the left with a speed of 7.5 m/s. Use the method of instant centers to determine the angular velocity of the rod and the velocity of slider B in this position.

IV.13   Determine the velocity of collar B in Problem IV.5 using the method of instant centers.

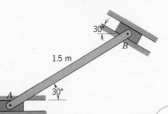

Prob. IV.12

IV.14 Two people are moving a table. Knowing that $\bar{v}_A$ is 15 in./s parallel to the diagonal line AC and that $\bar{v}_C$ is parallel to edge BC, as shown, determine the velocity that the person at corner B imparts to the table.

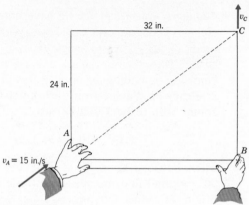

32 in.

24 in.

$v_A = 15$ in./s

Prob. IV.14

IV.15 The forward landing gear of an aircraft is being retracted at the rate of 15 rad/min. The aircraft has a speed of 120 miles/hr. Determine the location of the instantaneous center of zero velocity of the landing gear. Does the result depend on the angle θ?

$v = 120$ miles/hr

2 ft

Prob. IV.15

IV.16 A hammer is rotated such that the angle γ between the hammer and the forearm is identical to the angle β between the forearm and the horizontal. When $\beta = 45°$ the angular speed of the forearm is 2.0 rad/s. Determine the corresponding location of the instant center of the hammer and the velocity of point D on the hammer head.

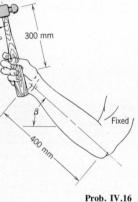

50 mm

D

300 mm

β

Fixed

400 mm

Prob. IV.16

IV.17 The stepped pulley is pinned at its center O to the sliding block. Two cords are wrapped around the pulley and pulled with constant velocities, as shown in the diagram. Determine the angular velocity of the pulley and the velocity of the block.

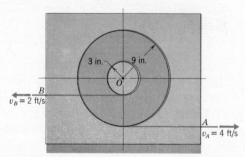

3 in.   9 in.

O

B

$v_B = 2$ ft/s

A

$v_A = 4$ ft/s

Prob. IV.17

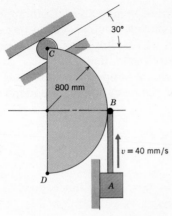

Prob. IV.18

IV.18 The hydraulic piston at A is being extended at the rate of 40 mm/s. Determine the angular velocity of the semicircular plate and the velocity of point D at the instant shown.

IV.19 The triangular plate is supported by two taut inextensible cables. In the position shown, cable CD has an angular velocity of 12 rad/s counterclockwise. Determine (a) the corresponding angular velocity of the plate, and (b) the velocity of the centroid G.

IV.20 The hydraulic cylinder actuates the motion of rod AB by moving the follower D to the right with a speed of 2.0 m/s. In the position where $\alpha = \tan^{-1}(\frac{4}{3})$, determine the angular velocity of the rod and the velocity of point A.

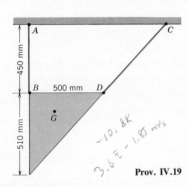

Prov. IV.19

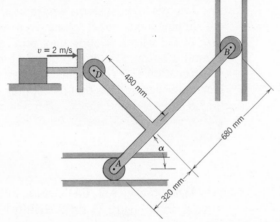

Prob. IV.20

2 Rolling Without Slipping

In this section we shall consider the case where the curved surfaces of two bodies are in contact with each other without being physically joined. Thus, as the bodies rotate, their point of contact changes. This type of motion is called rolling. We have illustrated a general rolling situation in Figure 11. Let us draw an x axis parallel to the plane of contact and a y axis perpendicular to that plane. To distinguish between the points of contact on each body, we shall denote the point on body 1 as point C_1 and the point on body 2 as point C_2. Because the bodies are rigid they cannot penetrate each other, so the velocity components of points C_1 and C_2 in the y direction must be identical.

Now let us focus on the expression "without slipping." We have

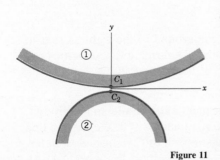

Figure 11

reasoned that the two surfaces can only move relative to each other along the tangent plane (parallel to the x axis). Such a motion is called *slipping* or *sliding*. If there is no slipping, point C_1 and C_2 will also have the same velocity components parallel to the tangent plane, that is, in the x direction. We therefore conclude that the constraint of no slip means that the x and y components of the velocities of the points of contact are identical, so that

$$\boxed{\bar{v}_{C_1} = \bar{v}_{C_2}} \qquad (17)$$

In the special case where one of the bodies, for instance body 2, is stationary, the no slip condition (17) reduces to $\bar{v}_{C_1} = \bar{0}$. For this case it then follows that the contact point is the instant center for body 1.

Two common systems whose motion fits the expression rolling without slipping are a wheel rolling over the ground and a set of gears. In the case of the wheel, friction may prevent the wheel from sliding over the ground. Clearly, the teeth of a gear limit sliding.

Equation (17) represents a constraint on the velocities. Now let us look at the restrictions that no slipping imposes upon the accelerations. If the no slip condition applies, the two points will follow each other in the direction parallel to the plane of contact. Thus the condition of no slip requires that

$$(a_{C_1})_x = (a_{C_2})_x \qquad (18)$$

As the bodies roll, the location of the point of contact changes. Thus if we redraw Figure 11 for a slightly later instant, points C_1 and C_2 will have separated. Therefore the y components of the accelerations of points C_1 and C_2 need not be the same.

Finally, it will be seen in the examples that follow that in the special case where the rolling bodies have circular surfaces, we shall be able to solve problems without recourse to equation (18). However, for general surface geometries we would need to use both equations (17) and (18).

EXAMPLE 5
A wheel of diameter 90 mm rolls over the ground to the right so that its center O has a constant speed of 900 mm/s. Determine the velocities and the accelerations of points A, B, and C at the instant shown.

Solution
We know that the center, O, follows a straight path parallel to the ground. Thus, we are given $\bar{v}_O = 900$ mm/s to the right and $\bar{a}_O = \bar{0}$. We wish to find the motion of points A, B, and C. To accomplish this we must first determine the angular velocity and angular acceleration of the wheel. As al-

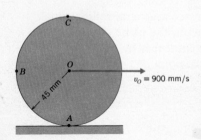

ways, we begin with angular velocity. Because contact point A is not slipping over the ground and the ground is not moving, the velocity condition for no slip gives

$$\bar{v}_A = \bar{0} \quad \checkmark$$

Thus point A is the instant center. The instant center method gives

$$\omega = \frac{v_O}{\left|\bar{r}_{B/A}\right|} \qquad v_B = \omega\left|\bar{r}_{B/A}\right| \qquad v_C = \omega\left|\bar{r}_{C/A}\right|$$

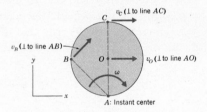

From the sketch we have $\left|\bar{r}_{O/A}\right| = 45$ mm, $\left|\bar{r}_{B/A}\right| = 45/\cos 45° = 45\sqrt{2}$ mm, $\left|\bar{r}_{C/A}\right| = 90$ mm. Also from the sketch we see that $\bar{v}_C$ must be parallel to the ground and that $\bar{v}_B$ must be at an upward 45° angle. Thus

$$\omega = \frac{900}{45} = 20 \text{ rad/s}$$

$$\bar{v}_B = (20)45\sqrt{2}(\cos 45° \, \bar{\imath} + \sin 45° \, \bar{\jmath})$$

$$= 900(\bar{\imath} + \bar{\jmath}) \text{ mm/s}$$

$$\bar{v}_C = (20) \, 90\bar{\imath} = 1800\bar{\imath} \text{ mm/s} = 1.8\bar{\imath} \text{ m/s}$$

There are two approaches to finding the angular acceleration. One which is valid for a surface of any shape, consists of noting that the x direction shown in the sketch is parallel to the contact plane. Because the ground is at rest, the no slip constraint requires that

$$(a_A)_x = 0$$

However, points A and O are two points on the wheel, so

$$\bar{a}_A = \bar{a}_O + \bar{\alpha} \times \bar{r}_{A/O} + \bar{\omega} \times (\bar{\omega} \times \bar{r}_{A/O})$$

From the velocity solution we know that $\bar{\omega} = -20\bar{k}$ (the negative sign occurs because $\bar{\omega}$ is clockwise). Also, $\bar{r}_{A/O} = -45\bar{\jmath}$ and $\bar{\alpha} = \alpha\bar{k}$ for plane motion. Thus

$$\bar{a}_A = \bar{0} + 45\alpha\bar{\imath} + 45(20)^2\bar{\jmath}$$

Taking the x component of $\bar{a}_A$ and equating it to zero gives

$$0 = 45\alpha$$

$$\bar{\alpha} = \bar{0}$$

Thus

$$\bar{a}_A = (45)400\bar{j} = 18{,}000\bar{j} \text{ mm/s}^2 = 18\bar{j} \text{ m/s}^2 \qquad \triangleleft$$

The alternative approach for finding α is considerably easier, but it is *only* valid for a *circular* surface. In this approach we note that regardless of the position of the wheel, the sketch for the instant center is unchanged. Thus, it is always true that $\omega = v_O/R$, where R is the radius of the wheel. We can differentiate this formula to find

$$\alpha = \dot{\omega} = \frac{\dot{v}_O}{R}$$

In this problem v_O is constant, so $\dot{v}_O = 0$ and $\alpha = 0$.

Once we have found $\bar{\alpha}$, we can continue on to find $\bar{a}_B$ and $\bar{a}_C$.

$$\bar{a}_B = \bar{a}_O + \bar{\alpha} \times \bar{r}_{B/O} + \bar{\omega} \times (\bar{\omega} \times \bar{r}_{B/O})$$

$$\bar{a}_C = \bar{a}_O + \bar{\alpha} \times \bar{r}_{C/O} + \bar{\omega} \times (\bar{\omega} \times \bar{r}_{C/O})$$

From the diagram we have

$$\bar{r}_{B/O} = -45\bar{i} \qquad \bar{r}_{A/O} = 45\bar{j}$$

so that

$$\bar{a}_B = \bar{0} + \bar{0} + 45(20)^2\bar{i} = 18{,}000\bar{i} \text{ mm/s}^2 = 18\bar{i} \text{ m/s}^2 \qquad \triangleleft$$

$$\bar{a}_C = \bar{0} + \bar{0} - 45(20)^2\bar{j} = -18{,}000\bar{j} \text{ mm/s}^2 = 18\bar{j} \text{ m/s}^2 \qquad \triangleleft$$

Finally, because rolling wheels are an integral part of a multitude of physical systems, let us discuss two more aspects of the general problem of rolling without slipping that relate to the problem we just solved. First, in the derivations we saw that the general motion of a body can be considered to be the superposition of a translation (having the motion of any point on the body) and a pure rotation (about that same point). In the case of the wheel in this problem, we can say that the motion is a superposition of a translation (to the right) of the center point O and a rotation (clockwise) about point O. The effect of the constraint condition of no slipping is to force the rate of rotation to be such that the contact point A has no velocity and only has an acceleration perpendicular to the ground.

With regard to the other aspect we wish to mention, suppose that we drew the path of point A in space as the wheel rolls. You may remember that this path is a cycloid. The cusps (e.g., point O) in the cycloidal path are the points in space where point A is in contact with the ground. As point A approaches a cusp in its path, its velocity is downward and its speed is decreasing. After passing the cusp the velocity has reversed direction and the speed increases, thus at the cusp, point A has zero velocity and its acceleration is upward. The foregoing is an alternative way to describe the "no slip" constraint condition.

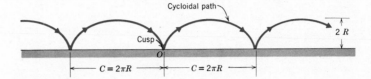

EXAMPLE 6

A portion of the steering gear of an automobile consists of the rack and pinion set shown in the sketch. The pinion rotates about a fixed axis passing through point A, and the rack is constrained to translate only. If the pinion is rotating at 0.5 rev/s clockwise and this rate is increasing at 4 rev/s², determine the velocity and acceleration of the rack.

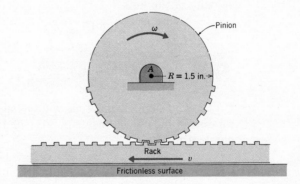

Solution

First we note that for computational purposes the angular units must be changed to radians.

$$\omega = (0.5)2\pi = \pi \text{ rad/s}$$

$$\alpha = (4)2\pi = 8\pi \text{ rad/s}^2$$

The motion of any point on the pinion is known because point A is on the axis of rotation and is fixed. We can therefore find the motion of the contact point on the pinion, and then use of the no slip condition will give the motion of the rack. Point A is at rest, so it is always the instant center.

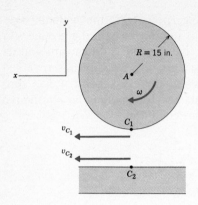

$$v_{C_1} = \omega|\bar{r}_{C_1/A}| = \omega R = 1.5\pi \text{ in./s}$$

$$\bar{v}_{C_1} = \omega R\bar{i} = 1.5\pi\bar{i}$$

For no slipping $\bar{v}_{C_1} = \bar{v}_{C_2}$ but $\bar{v}_{C_2} = \bar{v}_{rack}$. Thus $\bar{v}_{rack} = 1.5\pi\bar{i}$ in./s. To find $\bar{a}_{rack}$ we could use the no slip condition $(a_{rack})_x = (a_{C_1})_x$ and subsitute into the acceleration formula, but in this problem — because we have a circular surface — it is easier to note that $v_{rack} = \omega R$ for all positions of the system. Hence

$$\dot{v}_{rack} = \alpha R = (8\pi)1.5 = 12\pi \text{ in./s}^2$$

or vectorially,

$$\bar{a}_{rack} = \dot{\bar{v}}_{rack} = \dot{v}_{rack}\bar{i} = 12\pi\bar{i} \text{ in./s}^2$$

HOMEWORK PROBLEMS

IV.21 A cylindrical tank of radius 400 mm is rolling backward without slipping of the flatbed of a truck with a speed of 10 m/s relative to the truck. The truck has a forward speed of 20 m/s. Determine the angular velocity of the tank.

Prob. IV.21

IV.22 The stepped disk rolls without slipping on a set of rails. The disk is rotating at the constant rate of 30 rev/min. Determine the acceleration of points A and B.

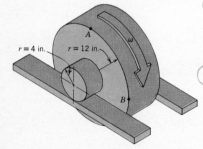

Prob. IV.22

IV.23 The stepped drum B rolls without slipping. Drum A has an angular velocity of 20 rad/s clockwise and an angular acceleration of 20 rad/s² clockwise. Determine (a) the angular velocity of body B, (b) the angular acceleration of body B, (c) the velocity of point C, (d) the acceleration of point O, and (e) the acceleration of point C.

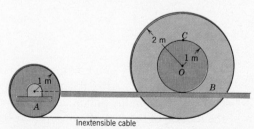

Prob. IV.23

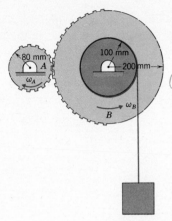

IV.24 A hoist consists of a motor driven gear A which rotates the larger stepped gear B. Determine (a) the gear ratio ω_B/ω_A, and (b) the velocity of the block when $\omega_A = 300$ rev/min.

IV.25 A wheel rolling in a circular path has an angular velocity $\bar{\omega}$ and an angular acceleration $\bar{\alpha}$, both clockwise. There is no slipping at the contact point. Determine the velocity and acceleration of the center point O.

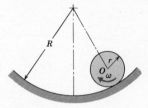

Prob. IV.26

Prob. IV.25

Prob. IV.24

IV.26 A box is placed on rollers and given an acceleration of 4 ft/s² to the right. There is no slippage between the rollers and the crate and between the rollers and the ground. At a specific instant it is known that the total acceleration of the point on the perimeter of a roller that is in contact with the crate is 5 ft/s². Determine the velocity and acceleration of the centers of the rollers at this instant. The radius of the roller is 250 mm.

IV.27 At the instant shown, the elevator is descending at the rate of 4 m/s and is slowing down at 0.5 m/s². The cable does not slip over the pulley. Determine (a) the angular velocity and angular acceleration of the large upper pulley, and (b) the velocity and acceleration of the counterweight.

IV.28 Determine the speed ratio v_A/v_B using the method of instant centers. Compare this solution to that obtained by imposing the kinematical constraint that the lengths of all cables are constant. The cables do not slip as they pass over the pulleys.

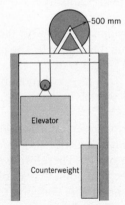

Prob. IV.27

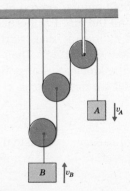

Prob. IV.28

IV.29 The semicircular cylinder rolls without slipping. When $\theta = 45°$ the angular velocity is $\bar{\omega}$ and the angular acceleration is $\bar{\alpha}$, both counterclockwise. Determine the velocity and acceleration of the centroid G at this instant.

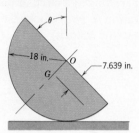

Prob. IV.29

3   Mechanical Linkages

A mechanical linkage is a mechanism by which motion and power may be transferred. For example, the crankshaft and connecting rod in Figure 12 transform the rectilinear motion of the piston into a rotary motion of the crankshaft. In a linkage, bodies are interconnected so that they cannot move independently of each other. In this module we shall consider only the case where they are connected by pins. Other types of connections will be studied in Module VI.

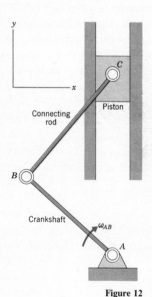

Figure 12

Let us assume that we know the angular velocity $\bar{\omega}$ of the drive shaft AB in Figure 12, and for simplicity assume that $\bar{\alpha}_{AB} = \bar{0}$. The problem is to determine the velocity and acceleration of the piston.

The method of solution differs little from the work already presented. However, as a good summary of what we have already done, and for completeness, let us discuss the analysis.

The motion of rod AB is defined by the given information. That must be our starting point. The piston can only translate, so all points on the piston undergo the same motion as the pin C. Therefore we shall seek information about point C on rod BC, because point C is a point common to the piston and the pin. Furthermore, point B is common to both rods. Thus, in the solution we shall start at A and determine the velocity of point B, considering rod AB. Then we shall consider rod BC using our knowledge of the kinematics of point B.

For link AB we may write

$$\bar{v}_B = \bar{v}_A + \bar{\omega}_{AB} \times \bar{r}_{B/A}$$

Point A is stationary, because it is on the axis of rotation of the crankshaft. Therefore $\bar{v}_A \equiv \bar{0}$. We now have

$$\bar{v}_B = \bar{\omega}_{AB} \times \bar{r}_{B/A}$$

Now, considering $\bar{v}_B$ to be known, we may express the velocity of point C using the velocity equation for the connecting rod.

$$\bar{v}_C = \bar{v}_B + \bar{\omega}_{BC} \times \bar{r}_{C/B}$$

Then substituting for $\bar{v}_B$ gives

$$\bar{v}_C = \bar{\omega}_{AB} \times \bar{r}_{B/A} + \bar{\omega}_{BC} \times \bar{r}_{C/B} \tag{19}$$

We are dealing with the motion of a point in a plane. Hence equation (19) represents two scalar equations that are obtained by matching the x and y components of the resulting vectors. How many unknowns are there in equation (19)? At this juncture, we do not know $\bar{v}_C$ and $\bar{\omega}_{BC}$. However, we know that we have plane motion, so $\bar{\omega}_{BC} = \omega_{BC}\bar{k}$ and $\bar{v}_C$ can have only two components. Thus we have three unknown scalar quantities, one more than the number of equations presently available. To get the additional equation we use the fact that the piston is constrained to follow the walls of the cylinder, so that $\bar{v}_C = v_C\bar{j}$ (a negative value for v_C will tell us that the piston is moving downward). Now we have two equations and two unknowns, so we can solve for ω_{BC} and v_C.

The acceleration problem may be solved in the same manner, after having first solved the velocity problem. We proceed through the linkage from point A to point B to point C using the relative acceleration equation for each link. For these computations we need $\bar{\omega}_{BC}$, which was determined in the solution to the velocity problem. We then impose the conditions that the system is constrained to move in a plane and that the acceleration of the piston is constrained to be a rectilinear motion in the vertical direction.

We close this discussion by noting that alternatively, the instant center method can be used to find the angular velocity of the crankshaft and the velocity of the piston. However, the vector equation approach must be used for the determination of acceleration. This will be illustrated in the next example.

EXAMPLE 7

The wheel in the linkage shown in the sketch is rotating in a clockwise direction at the constant rate of 40 rev/min. Determine the angular speed and angular acceleration of bar AB for the position shown.

Solution

From the given information we find $\omega_{CD} = (40)(2\pi/60) = (4\pi/3)$ rad/s and $\alpha_{CD} = 0$. Because point D is fixed, we can determine the velocity and acceleration of any point on the wheel. A study of the sketch reveals that we can proceed through the linkage system from point D to point A.

Let us begin by using instant centers to solve the velocity problem.

(The vector approach would work just as well.) Note that because point A is fixed, it is the center of rotation of bar AB. Similarly, point D is the center of rotation of the wheel. It is important that you realize that each rigid body in the system has its own instant center. We now draw a velocity diagram of the linkage.

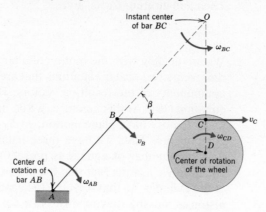

The instant center for bar BC was located as follows. The velocity of point B must be perpendicular to bar AB because point A is its instant center. Hence the instant center for BC must be along AB or its extension. Similarly, the velocity of point C must be perpendicular to line DC because point D is the instant center of the wheel. Hence, the instant center for BC must be along DC or its extension. It then follows that the instant center of bar BC is at point D, where lines AB and CD intersect. Recall that *each* rigid body has its own instant center. Before we can find ω_{AB} we must first determine the lengths of several lines. Because line OB is collinear with line AB, we know that

$$\beta = \tan^{-1} \tfrac{12}{5}$$

Then, because $\left| \bar{r}_{B/C} \right| = 100$ mm, we find

$$\left| \bar{r}_{C/O} \right| = 100 \tan \beta = 240 \text{ mm}$$

$$\left| \bar{r}_{B/O} \right| = 100 \cos \beta = 260 \text{ mm}$$

From the diagram we have

$$v_C = \left| \bar{r}_{C/D} \right| \omega_{CD} = (60)\frac{4\pi}{3} = 80\pi \text{ mm/s to the right}$$

$$= \left| \bar{r}_{C/O} \right| \omega_{BC} = 240\omega_{BC}$$

Solving for ω_{BC} gives

$$\omega_{BC} = \frac{80\pi}{240} = \frac{\pi}{3} \text{ rad/s counterclockwise}$$

Point B is a common point to bars AB and BC. Thus we find v_B using the instant center O and center of rotation A.

$$v_B = |\bar{r}_{B/O}|\omega_{BC} = (260)\frac{\pi}{3}$$

$$= |\bar{r}_{B/A}|\omega_{AB} = 130\omega_{AB}$$

Solving for ω_{AB} gives

$$\omega_{AB} = \frac{(260)\pi/3}{130} = \frac{2\pi}{3} \text{ rad/s clockwise}$$

We can now solve the acceleration problem using the vector approach. (Instant centers are not valid for accelerations). Starting from fixed point D, $\bar{a}_D = \bar{0}$, we have

$$\bar{a}_C = \bar{0} + \bar{\alpha}_{CD} \times \bar{r}_{C/D} + \bar{\omega}_{CD} \times (\bar{\omega}_{CD} \times \bar{r}_{C/D})$$
$$\bar{a}_B = \bar{a}_C + \bar{\alpha}_{BC} \times \bar{r}_{B/C} + \bar{\omega}_{BC} \times (\bar{\omega}_{BC} \times \bar{r}_{B/C})$$
$$\bar{a}_A = \bar{a}_B + \bar{\alpha}_{AB} \times \bar{r}_{A/B} + \bar{\omega}_{AB} \times (\bar{\omega}_{AB} \times \bar{r}_{A/B})$$

Because point A is fixed, we must also require that $\bar{a}_A = \bar{0}$.

Before we can solve the foregoing equations we must select a coordinate system for describing vector components. Because we have the horizontal and vertical distances for the locations of the points, we use xyz coordinates as shown in the first diagram of the problem. From this diagram we find

$$\bar{r}_{C/D} = 60\bar{j}$$

$$\bar{r}_{B/C} = -100\bar{i}$$

$$\bar{r}_{A/B} = -50\bar{i} - 120\bar{j}$$

From the velocity solution and the fact that for plane motion $\bar{\omega}$ and $\bar{\alpha}$ are perpendicular to the plane, we have (using the right-hand rule)

$$\bar{\omega}_{CD} = \frac{-4\pi}{3}\bar{k} \qquad \bar{\omega}_{BC} = \frac{\pi}{3}\bar{k} \qquad \bar{\omega}_{AB} = \frac{-2\pi}{3}\bar{k} \text{ rad/s}$$

$$\bar{\alpha}_{CD} = \bar{0} \qquad \bar{\alpha}_{BC} = \alpha_{BC}\bar{k} \qquad \bar{\alpha}_{AB} = \alpha_{AB}\bar{k}$$

Substituting these vectors into the acceleration equations gives

$$\bar{a}_C = -\left(\frac{4\pi}{3}\right)^2 60\bar{j} = -\frac{960}{9}\pi^2\bar{j}$$

$$\bar{a}_B = -\frac{960}{9}\pi^2\bar{j} - 100\alpha_{BC}\,\bar{j} + \left(\frac{\pi}{3}\right)^2 100\bar{i}$$

$$= \frac{100}{9}\pi^2\bar{i} - \left(\frac{960}{9}\pi^2 + 100\alpha_{BC}\right)\bar{j}$$

$$\bar{a}_A = \bar{0} = \frac{100}{9}\pi^2\bar{i} - \left(\frac{960}{9}\pi^2 + 100\alpha_{BC}\right)\bar{j}$$

$$+ \alpha_{AB}(-50\bar{j} + 120\bar{i}) + \left(\frac{2\pi}{3}\right)^2(50\bar{i} + 120\bar{j})$$

$$= \left(\frac{100}{9}\pi^2 + 120\alpha_{AB} + \frac{200}{9}\pi^2\right)\bar{i}$$
$$+ \left(-\frac{960}{9}\pi^2 + 100\alpha_{BC} - 50\alpha_{AB} + \frac{480}{9}\pi^2\right)\bar{j}$$

We may now obtain the scalar equations for α_{AB} and α_{BC} by matching components on the left- and right-hand sides of the equation for $\bar{a}_A$. Therefore

$$\bar{i} \text{ direction:} \qquad 0 = \frac{300}{9}\pi^2 + 120\alpha_{AB}$$

$$\bar{j} \text{ direction:} \qquad 0 = -\frac{480}{9}\pi^2 + 100\alpha_{BC} - 50\alpha_{AB}$$

Using these equations the angular accelerations are found to be

$$\bar{\alpha}_{AB} = 2.74\bar{k} \text{ rad/s}^2 \qquad \bar{\alpha}_{BC} = 6.63\bar{k} \text{ rad/s}^2$$

HOMEWORK PROBLEMS

IV.30 Each system shown is a four-bar linkage (the fixed line *AD* being the fourth bar). In the position shown, bar *AB* is rotating clockwise at 100 rev/min. Determine the corresponding angular velocity of bar *CD*.

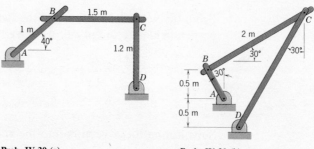

Prob. IV.30 (a)   Prob. IV.30 (b)

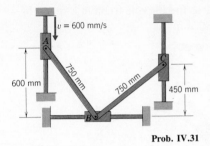

$v = 600$ mm/s

600 mm

750 mm

750 mm

450 mm

Prob. IV.31

IV.31 Slider A has a downward velocity of 600 mm/s. Determine the angular velocity of rod BC and the velocity of slider C for the position shown in the sketch.

IV.32 A motor rotates the inner gear in the planetary gear transmission at a rate of 3600 rev/min clockwise. (a) Determine the angular velocity of the outer gear when the spider arms are held fixed. (b) Determine the angular velocity of the spider arms when the outer gear is held fixed.

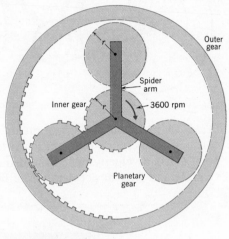

Outer gear

Spider arm

Inner gear

3600 rpm

Planetary gear

Prob. IV.32

IV.33 The disk rotates at a constant angular velocity $\bar{\omega}$ clockwise. (a) Determine the angular velocity of bar BC as a function of r, l, ω, and θ. (b) Determine the ratio v_C/v_B as a function of θ for the case when $(l/r) = 2$.

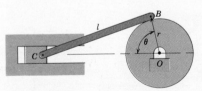

Prob. IV.33

IV.34 End A of rod AB is given a constant velocity of 10 ft/s to the left. The wheel to which its other end is pinned rolls without slipping. Determine the angular velocity of the wheel and the velocity of the center point O (a) when $\theta = 0°$, and (b) when $\theta = 90°$.

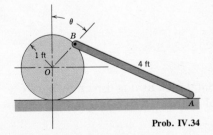

θ

B

1 ft

O

4 ft

A

Prob. IV.34

IV.35 The driven gear B and the rear wheel of the bicycle shown in the sketch are mounted on the same shaft, with the result that their angular velocities are identical. The diameter of the drive gear A is twice that of gear B. What rate of rotation ω_A will cause the bicycle to have a speed of 4 m/s? The diameter of each wheel is 700 mm.

Prob. IV.35

IV.36   A hydraulically driven piston moves point C to the right with a constant acceleration of 4.5 m/s². At the instant depicted in the diagram, point C has a velocity of 15 m/s, also to the right. Determine the angular velocity and angular acceleration of both rods at this instant.

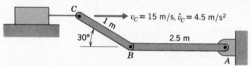

Prob. IV.36

IV.37   In the position shown, the angular velocity of bar AB is the constant value $\bar{\omega}$. Determine the angular velocities and angular accelerations of bars BC and CD.

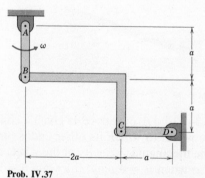

Prob. IV.37

IV.38   Gear A meshes with the horizontal rack as it is driven by bar BC. When $\theta = 60°$, slider C has a velocity of 10 ft/s to the right and its speed is decreasing at the rate of 4 ft/s². Determine the corresponding velocity and acceleration of the center of the gear.

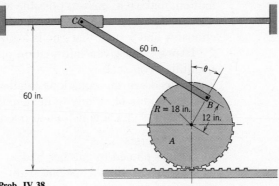

Prob. IV.38

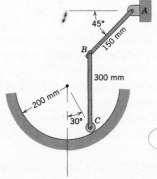

Prob. IV.41

IV.39   Determine the acceleration of point C in Problem IV.30 (a) if for the position shown α_{AB} is 2 rad/s² counterclockwise.

IV.40   Determine the acceleration of point C in Problem IV.30 (b) if for the position shown α_{AB} is 2 rad/s² clockwise.

IV.41   Bar AB is rotating at the constant rate of 3 rad/s counterclockwise. Determine the values of the angular velocity and angular acceleration of bar BC for the position shown. The radius of the roller at C is negligible.

E.   PROBLEM SOLVING

As was done in previous modules, let us develop a series of logical steps that lead to a systematized approach to problem solving. Again, we emphasize that this is strictly to help organize your thoughts. If these steps are just memorized the result will be that you will not truly understand the process of problem solving. As you gain proficiency you will intuitively perform many of the steps delineated below.

In this module we dealt with kinematics of rigid bodies in plane motion. Within the framework of problems in this topic we must always solve a velocity problem. Remember, even if we are only required to find an acceleration, the values of angular and linear velocities were shown to be an integral part of the solution for acceleration. With this in mind we have our starting point, the velocity problem.

In order to formulate any problem properly, we require a pictoral representation from which we can work.

1 The Velocity Problem

1   Draw a simple sketch of the system in the position of interest. Choose an xyz coordinate system such that x and y, which are in the plane of motion, make it as easy as possible to express the position of points in the system. For instance, if horizontal and vertical dimensions are given, choose x and y such that one is horizontal and the other is vertical.

2   Express the given information in the form of equations and list the quantities you seek.

3   Write down the equations relating the velocities of each of the constrained points of a body to the angular velocity of the body. Do this for each body. Typical constrained points to look for are

a)   fixed points
b)   points pinned to another body
c)   points to follow a specific path
d)   points of contact for bodies that roll

4   Write down the mathematical restrictions imposed on *all* constrained points. For instance, if a point A must follow a straight line in the x direction, write down $\bar{v}_A = v_A \bar{i}$.

5   Express all position vectors and angular velocity vectors by their xyz components. Remember that for plane motion $\bar{\omega} = \pm \omega \bar{k}$.

6   Substitute the constraint conditions from step 4 and the position vectors and angular velocity vectors from step 5 into the velocity equations from step 3. Calculate the cross products.

7   Collect components in the x and y directions. Count the number of scalar equations you have. Count the number of scalar unknowns in these equations. If the two numbers are not equal, you have probably forgotten to use a given piece of information or forgotten to utilize a constraint condition that the system must obey.

8   Solve the equations.

9   After determining the velocity of any point on the body and the angular velocity of the body, the velocity of unconstrained points may be found from the relative velocity equation.

   Alternative to (3–9): The method of instant centers may be used in lieu of the vector equations. This method uses the fact that the velocity $\bar{v}_P$ of any point P on the body must be perpendicular to the line from point P to the instant center C of the body and that $v_P = \omega |\bar{r}_{P/C}|$.

2 The Acceleration Problem

10–16   If accelerations must be determined, repeat steps 3 through 9, replacing the word velocity by the word acceleration. In these steps consider the solutions for velocities as given information. Remember that instant centers are not valid for determining acceleration.

We shall now investigate a series of illustrative problems, calling attention to the application of these steps as we proceed through the solutions.

ILLUSTRATIVE PROBLEM 1

Two movers lowered a cabinet from the window of an apartment building. They each let their cables out at the same constant rate of 3 ft/s when, as the cabinet neared the ground, they simultaneously tightened their grips to slow it down. In so doing the cable at A was immediately given an upward acceleration of 5 ft/s² and the cable at B was given an upward acceleration of 1 ft/s². Determine the acceleration of the center of the cabinet C and the angular acceleration of the cabinet at the instant the movers tightened their grips.

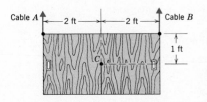

Solution

Step 1 Draw a sketch labeling all important points and showing xyz.

Step 2 Given $\bar{v}_A = \bar{v}_B = -3\bar{j}$ ft/s, $\bar{a}_A = 5\bar{j}$ ft/s², and $\bar{a}_B = \bar{j}$ ft/s². Find the acceleration of point C. Our thinking at this point is that the sudden change in the motion of the cables will *tend* to cause a rotation of the cabinet as shown. However, we also realize that the time interval during which the movers tightened their grips is so brief that the cabinet will not have time to alter its initial velocity appreciably, so that we should expect the cabinet to have no angular velocity at this instant. If this seems strange to you, think of other situations where the derivative of a quantity undergoes a sudden change. For instance, suppose that you are in a car traveling at 90 km/hr and suddenly apply the brakes. What is your velocity at time zero when you activate the brakes?

Step 3 We know that the formula we shall eventually use for the acceleration requires that we know $\bar{\omega}$, the angular velocity of the cabinet. Let us prove that $\bar{\omega} = \bar{0}$. To do this we relate the velocities of the two points about which we have information. Clearly, in this problem they are points A and B.

$$\bar{v}_B = \bar{v}_A + \bar{\omega} \times \bar{r}_{B/A}$$

Step 4 The constraints we know are that at the instant the cables were seized the velocities of A and B were equal and directed downward. Hence

$$\bar{v}_A = \bar{v}_B = -3\bar{j} \text{ ft/s}$$

Step 5 We know that $\bar{\omega} = -\omega\bar{k}$ for the coordinate system chosen and assumed sense of rotation. The position vector we require in step 3 in

$$\bar{r}_{B/A} = 4\,\bar{i} \text{ ft}$$

Step 6 We may now return to the formula in step 3 and write

$$-3\bar{j} = -3\bar{j} + (-\omega\bar{k}) \times 4\bar{i}$$

$$= -3\bar{j} - 4\omega\bar{j}$$

Step 7   Match components.

$$\bar{j} \text{ direction:} \qquad -3 = -3 - 4\omega$$

Step 8   We have one equation and one unknown, from which

$$\omega \equiv 0$$

Hopefully our foreshadowing made this an expected result. However, it does not follow that the angular acceleration is zero at time zero.

Step 9   We were not asked to find $\bar{v}_C$, but because $\omega = 0$, $\bar{v}_C = \bar{v}_B = \bar{v}_A = -3\bar{j}$.

Step 10   We now wish to determine the acceleration of point C. Begin by repeating step 3. Points A and B are constrained, so we write

$$\bar{a}_B = \bar{a}_A + \bar{\alpha} \times \bar{r}_{B/A} + \bar{\omega} \times (\bar{\omega} \times \bar{r}_{B/A})$$

Because $\bar{\omega} = \bar{0}$ at the initial instant, the equation reduces to

$$\bar{a}_B = \bar{a}_A \times \bar{\alpha} \times \bar{r}_{B/A}$$

Step 11   The constraints given in the problem are, at time zero,

$$\bar{a}_A = 5\bar{j} \quad \text{and} \quad \bar{a}_B = 1\bar{j} \text{ ft/s}^2$$

Step 12

$$\bar{\alpha} = -\alpha\bar{k}$$

$$\bar{r}_{B/A} = 4\bar{j} \text{ ft}$$

Step 13   Place the vectors in steps 11 and 12 into the formula in step 10.

$$\bar{j} = 5\bar{j} - \alpha\bar{k} \times 4\bar{i} = 5\bar{j} - 4\alpha\bar{j}$$

Step 14   Match components.

$$\bar{j} \text{ direction:} \qquad 1 = 5 - 4\alpha$$

Step 15   We have one equation and one unknown, therefore

$$\alpha = 1$$

$$\bar{\alpha} = -1\bar{k} \text{ rad/s}^2$$

Step 16

$$\bar{a}_C = \bar{a}_A + \bar{\alpha} \times \bar{r}_{C/A}$$

$$= 5\bar{j} + (-\bar{k}) \times (2\bar{i} - \bar{j})$$

$$= 5\bar{j} - 2\bar{j} - \bar{i}$$

$$= 3\bar{j} - \bar{i} \text{ ft/s}^2$$

ILLUSTRATIVE PROBLEM 2

A slider mechanism of length L slides with its ends constrained to move in slots that make an angle θ with one another. (a) Prove that when the link is perpendicular to one slot, the velocity of the slider in contact with the other slot vanishes. (b) Show that when the link makes equal angles with the slots, the speeds of the two sliders are the same. (c) Find the speed for part (b) in terms of l, θ, and ω_{AB}.

Solution

Step 1 Because we require two different geometric configurations, let us sketch the system in terms of a general angle α.

Step 2 In terms of the angle α, let us restate the problem. For part (a), show that $v_A \equiv 0$ when $\alpha = 90°$. For part (b), show that $|\bar{v}_A| = |\bar{v}_B|$ when $\alpha = (90° - \theta) + (90° - \alpha)$, that is, when $\alpha = 90° - \theta/2$. Part (c) remains the same, namely show that $|\bar{v}_A| \equiv |\bar{v}_B|$ for the condition of part (b).

Step 3 Clearly there are two points of interest in this problem, and both are constrained. We shall solve the problem using the vector approach. Hence we write

$$\bar{v}_B = \bar{v}_A + \bar{\omega}_{AB} \times \bar{r}_{B/A}$$

Step 4 Points A and B are constrained to move in the slots shown. Hence, assuming $\bar{v}_B$ to be to the left, we may write

$$\bar{v}_A = v_A(\cos \theta \, \bar{i} + \sin \theta \, \bar{j})$$

$$\bar{v}_B = -v_B \bar{i}$$

Step 5 We require the position vector $\bar{r}_{B/A}$.

$$\bar{r}_{B/A} = L(-\cos \alpha \, \bar{i} + \sin \alpha \, \bar{j})$$

The motion takes place in the xy plane. Therefore

$$\bar{\omega}_{AB} = \omega_{AB}\bar{k}$$

Step 6   Place the vector relationships determined in steps 4 and 5 into the velocity formula shown in step 3.

$$-v_B \bar{\imath} = v_A(\cos\theta\,\bar{\imath} + \sin\theta\,\bar{\jmath}) + \omega_{AB}\bar{k} \times L(-\cos\alpha\,\bar{\imath} + \sin\alpha\,\bar{\jmath})$$

$$= v_A(\cos\theta\,\bar{\imath} + \sin\theta\,\bar{\jmath}) - \omega_{AB}L(\sin\alpha\,\bar{\imath} + \cos\alpha\,\bar{\jmath})$$

Step 7   These lead to the scalar equations

$$\bar{\imath}\ \text{direction:}\quad -v_B = v_A\cos\theta - \omega_{AB}L\sin\alpha$$

$$\bar{\jmath}\ \text{direction:}\quad 0 = v_A\sin\theta - \omega_{AB}L\cos\alpha$$

Note that for a problem containing numbers we would count equations and unknowns at this point. Here we have two general equations and must apply the conditions of the specific problem before looking for a solution.

Step 8   Part (a) may be looked upon as, find v_A when $\alpha = 90°$. From the second equation of step 7 we have

$$0 = v_A\sin\theta$$

Because θ is a fixed nonzero angle, we conclude that $v_A = 0$, as was to be shown.

Parts (b) and (c) require that we find v_A and v_B when $\alpha = 90° - \theta/2$. From the second equation in step 7 we have

$$v_A\sin\theta = \omega_{AB}L\cos\left(90° - \frac{\theta}{2}\right)$$

$$\equiv \omega_{AB}L\sin\frac{\theta}{2}$$

Recall the trigonometric identity

$$\sin 2\beta = 2\sin\beta\cos\beta$$

Therefore, letting $2\beta = \theta$, we may write

$$2v_A\sin\frac{\theta}{2}\cos\frac{\theta}{2} = \omega_{AB}L\sin\frac{\theta}{2}$$

or

$$v_A = \frac{\omega_{AB}L}{2\cos(\theta/2)}$$

which is the solution to part (c).

Place this into the first equation in step 7 with $\alpha = 90° - \theta/2$.

$$-v_B = v_A \cos \theta - 2v_A \cos \frac{\theta}{2} \sin \left(90° - \frac{\theta}{2}\right)$$

$$\equiv v_A \cos \theta - 2v_A \left(\cos \frac{\theta}{2}\right)^2$$

We now make use of another trigonometric identity,

$$\cos 2\beta = \cos^2 \beta - \sin^2 \beta = 2 \cos^2 \beta - 1$$

Letting $2\beta = \theta$ again, we find

$$-v_B = v_A \left[2\left(\cos \frac{\theta}{2}\right)^2 - 1\right] - 2v_A \cos \left(\frac{\theta}{2}\right)^2 = -v_A$$

This is the identity we were asked to show in part (b).

3 Alternative Methods

There are two alternate methods that are equally valid to use for a problem such as this. One is, of course, the method of instant centers. The other, a bit more subtle, is to write a position vector, say $\bar{r}_{A/C}$, and differentiate.

ILLUSTRATIVE PROBLEM 3

A winch is being used to move a large spool of inextensible electrical transmission wire. The wire from the spool is wrapped around the drum of the winch and the spool rolls along the ground without slipping. At the instant shown, the winch has an angular speed of 1 rad/s and an angular acceleration of 2 rad/s², both clockwise. Find (a) the velocity of the center of the wire spool, (b) the rate at which the cord is being wound up or let out from the spool, and (c) the acceleration of the center of the wire spool.

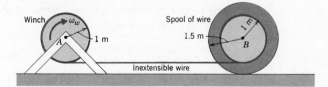

Solution

Intuitively, we expect the spool of wire to be pulled to the left. This is useful information to know, but we shall not require it, for the analysis will yield the proper direction of motion even if we begin with an incorrect assumption. Hence, assuming the motion to be to the left, if the contact point on the spool does not slip, the spool will rotate counterclockwise. You may be wondering how the given information could state as a fact that the spool does not slip. The answer to this question is that it is a result that we shall establish in the next module. There, our study of the effects of the system's forces will tell us whether or not slipping occurs. Here, we accept the given premise of rolling without slipping.

We shall solve the problem by the vector approach.

Step 1 Draw a sketch labeling any important points and showing xyz.

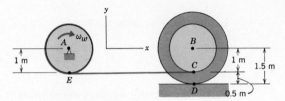

Step 2 Given $\omega_w = 1$ rad/s, $\alpha_w = 2$ rad/s². Find (a) $\bar{v}_B$, (b) rate of winding or unwinding of cord, (c) $\bar{a}_B$.

Part (b) is unlike anything we have encountered thus far. To motivate our thinking, let us imagine a few hypothetical situations. First, if center point B and point C have the same velocity, the spool will have no rotation and the wire will neither wind nor unwind. However, if there is no rotation, slipping will occur. Hence, by the constraint of the given information this cannot be. Second, if point C is approaching the winch faster than point B, then the spool will be rotating clockwise and the cord will unwind. Third, if point C is approaching the winch slower than point B, then the spool will rotate counterclockwise and the cord will wind around the drum.

From the foregoing we conclude that the rate of unwinding is determined from the difference between $\bar{v}_C$ and $\bar{v}_B$. But

$$\bar{v}_C - \bar{v}_B = \bar{v}_{C/B} = \bar{\omega}_s \times \bar{r}_{C/B}$$

so that we need to find the angular velocity of the spool, $\bar{\omega}_s$.

Step 3 For the winch we know that point A is fixed and that point E is the contact point between the winch and the wire. We therefore write

$$\bar{v}_E = \bar{v}_A + \bar{\omega}_w \times \bar{r}_{E/A}$$

For the spool we know that point D is the contact point with the ground and that point C is the contact point between the wire and the spool. Thus,

$$\bar{v}_C = \bar{v}_D + \bar{\omega}_s \times \bar{r}_{C/D}$$

Note that point B is an unconstrained point, so we do not consider it in this step.

Step 4 The wire is inextensible, and because it does not slip over either drum, we have the constraint condition

$$\bar{v}_C = \bar{v}_E$$

Point A is fixed, so $\bar{v}_A = \bar{0}$, and the ground is stationary, so the no slip condition gives $\bar{v}_D = \bar{0}$.

Step 5 We know $\bar{\omega}_w = -\omega_w \bar{k} = -1\bar{k}$ rad/s. For plane motion $\bar{\omega}_s = \omega_s \bar{k}$. The position vectors in step 3 are

$$\bar{r}_{E/A} = -1\bar{j} \qquad \bar{r}_{C/D} = 0.5\bar{j} \text{ m}$$

Step 6 Substitute steps 4 and 5 into step 3, leaving $\bar{\omega}_w$ in terms of the algebraic rate ω_w, in lieu of substituting its current numerical value. The reason for this will be obvious shortly.

$$\bar{v}_E = \bar{0} - \omega_w \bar{k} \times -\bar{j} = -\omega_w \bar{i}$$

$$= \bar{v}_C = \bar{0} + \omega_s \bar{k} \times 0.5\bar{j} = -0.5\omega_s \bar{i}$$

Step 7 Match components.

$$\bar{i} \text{ direction:} \qquad -\omega_w = -0.5\omega_s$$

We have one equation and one unknown.

Step 8 Solve the equation.

$$\omega_s = \frac{\omega_w}{0.5} = 2\omega_w = 2 \text{ rad/s} \qquad \text{at this instant}$$

Therefore, at this instant,

$$\bar{\omega}_s = 2\bar{k} \text{ rad/s}$$

We now can discuss the reason for leaving ω_w as an algebraic variable. Generally in problems of circular wheels that are rolling without slipping, the geometry of the system does not change as time goes by. We shall therefore be able to differentiate the general formula for the angular speed(s) to obtain a relationship for angular acceleration(s). This may enable us to skip some or all of steps 10–16. For example, in this problem the cord is always horizontal, so the diagram in step 1 applies at all instants of time. Therefore at this instant,

$$\alpha_s = \dot{\omega}_s = 2\dot{\omega}_w = 2\alpha_w = 4 \text{ rad/s}^2$$

so that

$$\bar{\alpha}_s = 4\bar{k} \text{ rad/s}^2$$

Step 9 For part (a) we want $\bar{v}_B$. Clearly, we wish to relate B to a point in the spool of known velocity, which is point D. Therefore

$$\bar{v}_B = \bar{v}_D + \bar{\omega}_s \times \bar{r}_{B/D} = \bar{0} + \omega_s \bar{k} \times 1.5\bar{j}$$

$$= -1.5\omega_s \bar{i}$$

$$= -3\bar{i} \text{ m/s} \qquad \text{at this instant}$$

For part (b) we have

$$\bar{v}_{C/B} = \bar{\omega}_s \times \bar{r}_{C/B} = 2\bar{k} \times -1\bar{j} = 2\bar{i} \text{ m/s}$$

From this result and the answer to part (a), we conclude that point B is moving to the *left* at 2 m/s faster than point C. Therefore the cord is being wound up at the rate of 2 m/s. What conclusion would you reach if the result was $\bar{v}_{C/B} = -2\bar{i}$?

Steps 10–16 From part (a) we have the general equation

$$\bar{v}_B = -1.5\omega_s \bar{i}$$

Because this equation is always true, we can differentiate it. Therefore

$$\bar{a}_B = \dot{\bar{v}}_B = -1.5\dot{\omega}_s \bar{i} = -1.5\,\alpha_s \bar{i}$$

From step 8 we know that $\alpha_s = 4$ rad/s², so

$$\bar{a}_B = -6\bar{i} \text{ m/s}^2$$

ILLUSTRATIVE PROBLEM 4

Blocks at both ends of rigid bar AB slide in grooves. At the instant shown, block B is moving upward at a constant speed of 2.5 m/s. Determine the acceleration of the midpoint C of bar AB for this position.

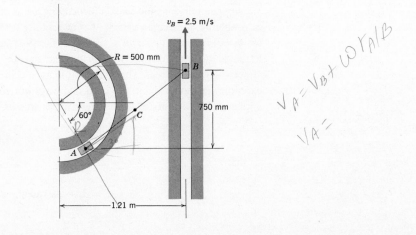

Solution

Step 1

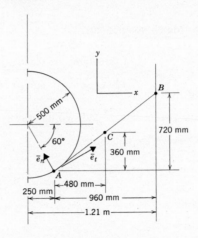

Step 2   Given $\bar{v}_B = 2.5\bar{j}$ m/s, $\bar{a}_B = \bar{0}$. Find $\bar{a}_C$.

Step 3   We shall solve the problem using the vector equation approach. Point B must follow the vertical groove, and point A must follow the circular one. Thus

$$\bar{v}_A = \bar{v}_B + \bar{\omega}_{AB} \times \bar{r}_{A/B}$$

Step 4   We were given $\bar{v}_B = 2.5\bar{j}$. From path variables we know that $\bar{v}_A$ is tangent to the circle, so that

$$\bar{v}_A = v_A\bar{e}_t$$

The unit vectors $\bar{e}_t$ and $\bar{e}_n$ are shown in the sketch of step 1. From the given geometry we may write

$$\bar{e}_t = \cos 30°\,\bar{i} + \sin 30°\,\bar{j} = 0.866\bar{i} + 0.5\bar{j}$$

$$\bar{e}_n = -\sin 30°\,\bar{i} + \cos 30°\,\bar{j} = -0.5\bar{i} + 0.866\bar{j}$$

The unit vectors $\bar{e}_t$ and $\bar{e}_n$ are expressed by their components in the x and y directions because we shall need the components of $\bar{v}_A$ and $\bar{a}_A$ in these directions. Thus

$$\bar{v}_A = v_A(0.866\bar{i} + 0.5\bar{j})$$

Step 5 $\bar{\omega}_{AB} = \omega_{AB}\bar{k}$ (ω_{AB} is positive if counterclockwise).

$$\bar{r}_{A/B} = -0.96\bar{i} - 0.72\bar{j} \text{ m}$$

Step 6   Substitute the vectors from steps 4 and 5 into the velocity equation in step 3.

$$v_A(0.886\bar{i} + 0.5\bar{j}) = 2.50\bar{j} + \omega_{AB}\bar{k} \times (-0.96\bar{i} - 0.72\bar{j})$$

$$= 0.72\omega_{AB}\,\bar{i} + (2.50 - 0.96\omega_{AB})\bar{j}$$

Step 7 Match components.

$$\bar{i} \text{ direction:} \qquad 0.866v_A = 0.72\omega_{AB}$$

$$\bar{j} \text{ direction:} \qquad 0.5v_A \;\; = 2.50 - 0.96\omega_{AB}$$

We have two equations and two unknowns.

Step 8 Solve the equations.

$$v_A = \frac{0.72}{0.866}\,\omega_{AB} = 0.831\omega_{AB}$$

$$(0.5)0.831\omega_{AB} = 2.50 - 0.96\omega_{AB}$$

$$1.375\omega_{AB} = 2.50$$

$$\omega_{AB} = \frac{2.50}{1.375} = 1.818 \text{ rad/s}$$

$$v_A = 1.511 \text{ m/s}$$

Step 9 We are not asked for $\bar{v}_C$, so we do not compute it.

Step 10 We repeat the process for accelerations. Points A and B are constrained, so we write

$$\bar{a}_A = \bar{a}_B + \bar{\alpha}_{AB} \times \bar{r}_{A/B} + \bar{\omega}_{AB} \times (\bar{\omega}_{AB} \times \bar{r}_{A/B})$$

Step 11 We were given $\bar{a}_B = \bar{0}$. From path variables we know that

$$\bar{a}_A = \dot{v}_A\bar{e}_t + \frac{v_A^{\,2}}{R}\,\bar{e}_n$$

From step 3 we know $\bar{e}_t$ and $\bar{e}_n$, and we know v_A from step 8. Thus

$$\bar{a}_A = \dot{v}_A(0.866\bar{i} + 0.5\bar{j}) + \frac{(1.511)^2}{0.50}(-0.5\bar{i} + 0.866\bar{j})$$

$$= (0.866\dot{v}_A - 2.283)\bar{i} + (0.5\dot{v}_A + 3.954)\bar{j}$$

Step 12 The only additional vector needed is

$$\bar{\alpha}_{AB} = \alpha_{AB}\bar{k}$$

Step 13 Substitute the vectors from steps 5, 11, and 12, and the value of ω_{AB} from step 8 into the acceleration equation in step 10.

$$(0.866\ddot{v}_A - 2.283)\bar{i} + (0.5\ddot{v}_A + 3.954)\bar{j}$$

$$= \bar{0} + \alpha_{AB}\bar{k} \times (-0.96\bar{i} - 0.72\bar{j})$$

$$+ 1.818\bar{k} \times [1.818\bar{k} \times (-0.96\bar{i} - 0.72\bar{j})]$$

$$= \alpha_{AB}(-0.96\bar{j} + 0.72\bar{i})$$

$$+ (1.818)^2(0.96\bar{i} + 0.72\bar{j})$$

$$= (0.72\alpha_{AB} + 3.173)\bar{i} + (-0.96\alpha_{AB} + 2.38)\bar{j}$$

Step 14 Match components.

$\bar{i}$ direction: $\qquad\qquad 0.866\ddot{v}_A - 2.283 = 0.72\alpha_{AB} + 3.173$

$\bar{j}$ direction: $\qquad\qquad 0.5\ddot{v}_A + 3.954 = -0.96\alpha_{AB} + 2.38$

We have two equations and two unknowns.

Step 15 Solve the equations. To find α_{AB}, multiply the first equation of step 14 by 0.5 and the second by 0.866.

$$0.5(0.866\ddot{v}_A - 2.283) = 0.5(0.72\alpha_{AB} + 3.173)$$

$$0.866(0.5\ddot{v}_A + 3.954) = 0.866(-0.96\alpha_{AB} + 2.38)$$

Now we subtract the second from the first.

$$-4.566 = 1.1914\alpha_{AB} - 0.4746$$

$$\alpha_{AB} = -3.434 \text{ rad/s}^2$$

Then solving for $\ddot{v}_A$ gives

$$\ddot{v}_A = 3.444 \text{ m/s}^2$$

Step 16 We now can find the acceleration of the unconstrained point C by relating it to either point A or point B. In this problem point B is the easier choice because it has no acceleration.

$$\bar{a}_C = \bar{a}_B + \bar{\alpha}_{AB} \times \bar{r}_{C/B} + \bar{\omega}_{AB} \times (\bar{\omega}_{AB} \times \bar{r}_{C/B})$$

From the diagram in step 1,

$$\bar{r}_{C/B} = -0.48\bar{i} - 0.36\bar{j}$$

Then substituting the results for $\bar{\omega}_{AB}$ and $\bar{\alpha}_{AB}$ gives

$$\bar{a}_C = \bar{0} - 3.434\bar{k} \times (-0.48\bar{i} - 0.36\bar{j}) + 1.818\bar{k}$$

$$\times [1.818\bar{k} \times (-0.48\bar{i} - 0.36\bar{j})]$$

$$= 1.648\bar{j} - 1.236\bar{i} + 1.587\bar{i} + 1.190\bar{j}$$

$$= 0.350\bar{i} + 0.284\bar{j} \text{ m/s}^2$$

To reinforce your confidence, we should note that this is one of the more complicated problems you can expect to see in the kinematics of rigid bodies in planar motion. If you followed all of the development, you have established a good level of understanding. If some of the steps gave you trouble, you should go back and review the corresponding material in the text.

HOMEWORK PROBLEMS

IV.42 The circular cam is mounted eccentrically on a fixed axis of rotation passing through point O. Express the velocity and acceleration of the follower in terms of the angle θ. *Hint:* Express the distance y in terms of the angle and the constant angular speed ω θ.

IV.43 The block is moving to the right at a constant speed of 0.5 m/s. Determine the angular velocity and angular acceleration of the 400-mm rod as a function of θ.

IV.44 The lower rack has a constant velocity of 6 in./s to the right. The upper rack has an acceleration of 8 in./s² to the right, and an instantaneous velocity of 2 in./s to the right. Determine (a) the angular velocity and angular acceleration of the gear, (b) the acceleration of point C.

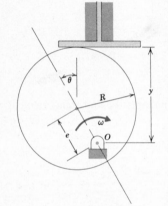

Prob. IV.42

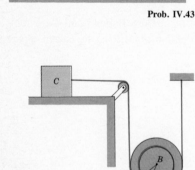

Prob. IV.43

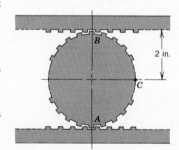

Prob. IV.44

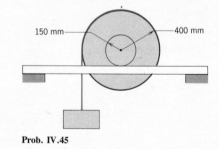

Prob. IV.45

IV.45 The cable is unwrapped from the rim of the stepped disk as the block is lowered. The angular velocity of the disk is 12 rad/s and the angular acceleration is 6 rad/s², both counterclockwise. Assuming that there is no slippage and that the cable remains vertical, determine (a) the velocity of the block, and (b) the acceleration of the block.

IV.46 Block A has fallen 1 ft from rest with a contant acceleration of 16 ft/s². There is no slippage of the cables as they pass over the stepped pulley B. For this instant, determine (a) the angular velocity of pulley B, (b) the relative velocity of block A with respect to the center of pulley B, and (c) the velocity of block C.

Prob. IV.46

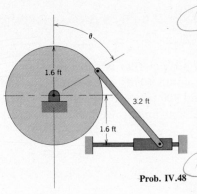

1.6 ft

3.2 ft

1.6 ft

θ

Prob. IV.48

IV.47 For the system of Problem IV.46, determine (a) the angular acceleration of pulley B, (b) the relative acceleration of block A with respect to the center of pulley B, and (c) the acceleration of block C.

IV.48 The angle of rotation of the flywheel is given by $\theta = 30t + 180t^2$, where θ is measured in degrees and t in seconds. Determine (a) the velocity of the piston when $t = 1$ s, (b) the acceleration of the piston when $t = 1$ s.

IV.49 Solve Problem IV.48 for $t = \frac{1}{2}$ s.

IV.50 Solve Problem IV.48 for $t = \frac{1}{3}$ s.

IV.51 The angular velocity of bar AB is 5 rad/s. Determine the velocity of slider C at the instant depicted in the diagram.

IV.52 If bar AB in Problem IV.51 has an angular deceleration of 2 rad/s² in the position illustrated, and all other parameters are as given, determine the acceleration of slider C.

IV.53 Solve Illustrative Problem 2 by writing the appropriate position vector and differentiating.

IV.54 In the position shown, slider A has a velocity of 20 ft/s to the left. Determine the angular velocities of bars AB and BC.

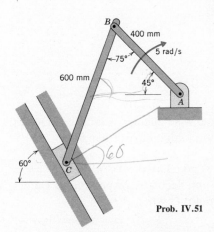

B

400 mm

5 rad/s

75°

600 mm

45°

A

60°

C

Prob. IV.51

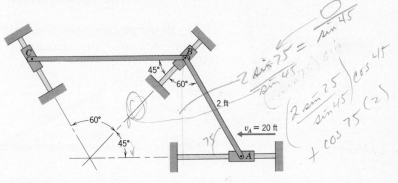

45°

60°

60°

2 ft

$v_A = 20$ ft

45°

A

Prob. IV.54

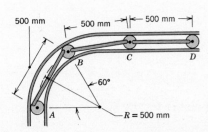

500 mm

500 mm

500 mm

B

C

D

60°

A

$R = 500$ mm

Prob. IV.56

IV.55 In Problem IV.54 the acceleration of slider A in the given position is 40 ft/s² to the right. Determine the angular accelerations of both bars.

IV.56 A side view of three identical panels from a garage door is shown in the diagram. The bottom is being raised with a speed of 7.50 m/s. For the position depicted, determine the speed of (a) roller B, (b) roller D.

IV.57   Determine the acceleration of roller D for the system of Problem IV.56.

IV.58   A portion of the mechanism of a typewriter is shown in the diagram. In this position the downward components of the velocity and acceleration of the key are 200 mm/s and 400 mm/s², respectively. Determine the corresponding values of the velocity and acceleration of point E.

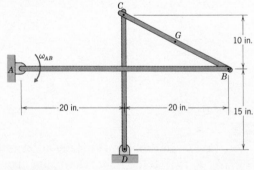

Prob. IV.58

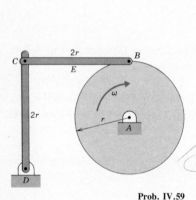

Prob. IV.59

IV.59   The flywheel has a constant clockwise angular velocity $\bar{\omega}$. Determine the velocity and acceleration of the midpoint E of bar BC.

IV.60   The angle of rotation of bar AB is given by $\varphi = (\pi/4)(1 - \cos 5\pi t)$, where the units of φ and t are radians and seconds, respectively. Determine the velocity and acceleration of point C when $\varphi = 0°$.

IV.61   In the position shown bar AB is rotating at a constant rate of 120 rev/min. Determine the velocity of the midpoint G of bar BC at this instant.

Prob. IV.60

Prob. IV.61

IV.62   Determine the acceleration of point G for the system of Problem IV.61.

IV.63 The diagram shows a possible mechanism for a log-cutting saw. The flywheel rotates at the constant rate of 60 rev/min. Determine the velocity and acceleration of the lowest point on the perimeter of the circular saw blade when $\theta = 0°$.

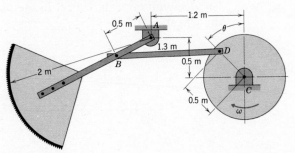

Prob. IV.63

IV.64 Arm AB in the escapement mechanism rotates clockwise at the constant rate of 6 rev/min. For the position shown, determine (a) the velocity and acceleration of the ratchet tooth E, and (b) the angular velocity and angular acceleration of the gear.

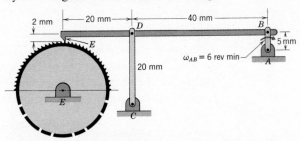

Prob. IV.64

IV.65 Two wheels which are interconnected by rod AC, roll along the ground without slipping. Wheel A has a constant velocity of 5 m/s to the right. Determine the velocity and acceleration of wheel B in the position where $\theta = 0$.

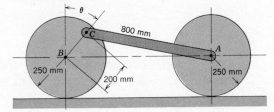

Prob. IV.65

IV.66 For the system of Problem IV.65, determine the velocity and acceleration of wheel B in the position where $\theta = 120°$.

IV.67 The machine shown in the sketch is used to test the wear of railroad wheels and tracks. The flywheel has a constant counterclockwise angular velocity of 600 rev/min. There is no slippage between the railroad wheel and the track. Determine (a) the angular velocity and (b) the angular acceleration of the wheel in the illustrated position. The diameter of the wheels is 1.2 m.

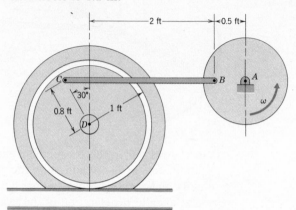

Prob. IV.67

IV.68 A truck is being driven up a 20° incline. The speedometer, which measures the speed of the rear wheel, indicates a constant 8 km/hr, and the wheels are not skidding. Determine the values of the angular velocities of the front and rear wheels and of the vehicle body for the position illustrated in the diagram. The diameter of the wheels is 0.9 m.

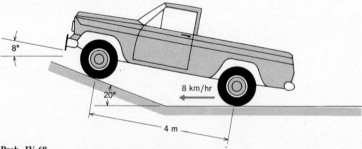

Prob. IV.68

IV.69 Determine the angular accelerations of the front and rear wheels and of the vehicle body for the truck in Problem IV.68.

IV.70 The slider has a constant velocity of 0.7 m/s to the right, thus causing gear A to move over the fixed gear B. Determine the angular velocity of gear A in this position.

IV.71 Determine the angular acceleration of gear A in Problem IV.70.

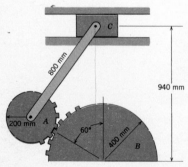

Prob. IV.70

MODULE V
KINETICS OF A
RIGID BODY IN
PLANAR MOTION

In the study of the kinematics of a rigid body undergoing planar motion, we learned that the position of a rigid body can be described by three quantities: Two variables locate a point in the rigid body at any instant, whereas the third gives the angle of rotation of the body from a reference position. The relationship between the foregoing three parameters and the force system acting on a rigid body will be fully explored in this module on kinetics.

Here we shall restrict our attention to the broad class of problems where the significant forces are parallel to the plane of motion, and to rigid bodies that have a plane of symmetry that is parallel to the plane of motion. Problems that do not fit one or both of these specifications should be treated by the methods of Module VII, on three-dimensional motion.

A kinetics study must consider the inertial effect associated with the motion of every particle in the system. In essence, a rigid body is a continuum of particles of infinitesimal mass dm that are in contact with their neighboring particles. In the field of strength of materials it is shown that the contact forces between the particles are actually stress resultants. In the study of the dynamics of rigid bodies these forces are treated as the constraint forces required to maintain the particles of the body at a fixed distance from each other. The constraint forces are internal to the system of particles composing the rigid body. Hence, they have no effect on the overall motion of the body. This was a key result emphasized in the treatment of a system of particles in Module III.

A. EQUATIONS OF MOTION

We know that the acceleration of the center of mass G of a system of particles can be determined from the basic relation

$$\Sigma \bar{F} = m\bar{a}_G \tag{1}$$

Because the center of mass of a rigid body has a fixed position relative to the body, the use of equation (1) requires that we locate the center of mass of the body. Then we can apply the kinematical formulas of Module II for the motion of this point.

An important facet of equation (1) is its similarity to Newton's second law, $\bar{F} = m\bar{a}$, for a *single particle*. It follows that equation (1) is the means by which we determine the motion of a point in the body.

Recall that when a body was modeled as a single particle in Module III, we implicitly considered the angular motion of the body to be insignificant. We know that the angular motion of a rigid body is related to the relative motion between points in the body. Also, we know that the angular momentum of a system of particles can be described in terms of the relative velocity of the particles. Hence, we shall employ the angular momentum principle to study the rotation of a rigid body.

The simplest form for the basic relationship between the angular momentum of a system of particles and the resultant moment of the force system is

$$\Sigma \, \bar{M}_A = \dot{H}_A \tag{2}$$

where point A is either (a) the center of mass, (b) a fixed point, or (c) a point accelerating directly toward or away from the center of mass. In treating conditions (b) and (c) we shall, for convenience, now impose the additional restriction that *point A must be fixed to the body.* This will enable us to use the relative motion concepts of the previous module.

Referring to the rigid body shown in Figure 1, where m_i denotes the element of mass located at point i, the angular momentum vector $\bar{H}_A$ in equation (2) is defined by

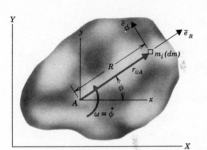

Figure 1

$$\bar{H}_A = \sum_{i=1}^{N} \bar{r}_{i/A} \times m_i \bar{v}_{i/A} \tag{3}$$

If we consider the rigid body of Figure 1 to be composed of an infinite number of infinitesimal masses $dm = m_i$, then the summation in equation (3) becomes an integral over all of the masses d/m in the volume $\mathcal{V}$ of the body. Hence

$$\bar{H}_A = \int_{\mathcal{V}} (\bar{r}_{i/A} \times \bar{v}_{i/A}) \, dm \tag{4}$$

To describe the position and velocity of the differential element dm at point i relative to point A, we introduce a translating reference frame system xyz whose origin is placed coincident with point A. As postulated earlier, the body has a plane of symmetry parallel to the plane of motion. Let us choose the xy plane to coincide with this symmetry plane, so that the z axis is normal to the plane of motion in accordance with the right-hand rule. Then, by defining cylindrical coordinates with respect to this reference frame, as shown in Figure 1, we have

$$\bar{r}_{i/A} = R\bar{e}_R$$

To obtain the relative velocity we apply the velocity relationship for two points in a rigid body undergoing planar motion, which gives

$$\bar{v}_i = \bar{v}_A + \bar{\omega} \times \bar{r}_{i/A}$$

Thus

$$\bar{v}_{i/A} \equiv \bar{v}_i - \bar{v}_A = \bar{\omega} \times \bar{r}_{i/A}$$

Because $\bar{e}_h \equiv \bar{k}$ for the chosen reference frame,

$$\bar{\omega} = \omega \bar{e}_h$$

$$\bar{\omega} \times \bar{r}_{i/A} = \omega R \bar{e}_\phi$$

so that

$$\bar{r}_{i/A} \times \bar{v}_{i/A} = \omega R^2 \bar{e}_h = \omega R^2 \bar{k} = R^2 \bar{\omega}$$

We now substitute this result into equation (4) for the angular momentum, noting that the angular velocity $\bar{\omega}$ is a function of time only, so it may be removed from the integral over the volume of the body. Hence we have

$$\bar{H}_A = \int_{\mathcal{V}} \bar{r}_{i/A} \times \bar{v}_{i/A} \, dm = \bar{\omega} \int_{\mathcal{V}} R^2 \, dm$$

$$\boxed{\bar{H}_A = I_{zz} \bar{\omega}} \tag{5}$$

where, in view of the basic relationship between the polar coordinate R and the Cartesian coordinates x and y (evident from Figure 1),

$$\boxed{I_{zz} = \int_{\mathcal{V}} R^2 \, dm = \int_{\mathcal{V}} (x^2 + y^2) \, dm} \tag{6}$$

The quantity I_{zz} is called the *second moment of mass*, or alternatively, the *moment of inertia* of the body about the z axis. Remember that the z axis is perpendicular to the plane of motion, passing through the point A that was selected for the moment sum. A significant property of the moment of inertia I_{zz}, as derived above, is that it is a constant for the rigid body. This is so because we required point A to be a point that is fixed relative to the body, thus making the radial distance R to each mass element constant. We shall consider further properties of, and computational procedures for, moments of inertia in the next section. As an aside, the double subscript in I_{zz} has no significance for the analysis of problems in planar kinetics; it is introduced here in order to present a notation that is consistant with that used for studying the kinetics of rigid bodies in three-dimensional motion.

Returning now to the derivation of the equations of motion, by differentiating equation (5) with respect to time we may determine $\dot{\bar{H}}_A$. Thus, noting that the time derivative of the angular velocity $\bar{\omega}$ is the angular acceleration $\bar{\alpha}$, and that I_{zz} is a constant, differentiating equation (5) yields

$$\dot{\bar{H}}_A = I_{zz} \dot{\bar{\omega}} = I_{zz} \bar{\alpha}$$

Hence, equation (2) becomes

$$\boxed{\Sigma \, \bar{M}_A = I_{zz}\bar{\alpha}} \tag{7}$$

Equation (7) represents the third of the three scalar equations necessary to describe the three independent position variables of a rigid body in planar motion. Two equations are obtained from the planar components of the force equation (1), and the third equation is obtained by determining the moment of the planar force system about point A and applying equation (7).

An important feature of equation (7) is the similarity of its component form to the equation for rectilinear motion of a particle that we studied in Module Three. Thus, in situations where we wish to relate the angle of rotation ϕ, the angular speed ω, and time, we need only obtain a general expression for the moment of the force system and apply the identity $\alpha = \dot{\omega} = \ddot{\phi}$. This will allow us to form a differential equation of motion for the angle ϕ.

Before we can write the equations of motion for a specific problem, we must establish procedures for the determinations of the location of the center of mass and of the moment of inertia. These topics are the subjects of the next section. We shall close this section with a brief review regarding the computation of the moment for a planar force system.

Consider the force $\bar{F}$ shown in Figure 2. The moment $\bar{M}_A$ of the force $\bar{F}$ about point A may be determined by the vector operation

$$\bar{M}_A = \bar{r}_{P/A} \times \bar{F}$$

where point P is a point on the line of action of $\bar{F}$. In terms of the components of $\bar{r}_{P/A}$ illustrated in Figure 2, this moment is

$$\bar{M}_A = (x_P \, \bar{i} + y_P \, \bar{j}) \times (F_x \, \bar{i} + F_y \, \bar{j})$$

$$= (x_P F_y - y_P F_x)\bar{k}$$

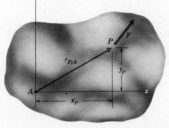

Figure 2

This result can also be obtained directly by multiplying each force component by its lever arm about point A, the latter quantity being the perpendicular distance from point A to the line of action of the force component. To obtain the proper sign we note that, for the axes shown, the z axis is outward. Hence, in accord with the right-hand rule, counterclockwise moments are positive. For planar problems this approach usually proves as efficient as the cross product.

B. INERTIA PROPERTIES

Let us begin by considering the physical meaning of the center of mass and moment of inertia. The center of mass of a body may be determined experimentally by suspending it by a cable from two points. For consis-

Figure 3a

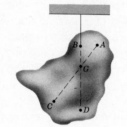

Figure 3b

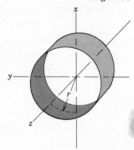

Figure 4

tency with the concepts of planar motion, let the body be symmetric with respect to the xy plane, and let points A and B be two points that coincide with this plane, as shown in Figure 3a. When suspended from point A, the body assumes a specific orientation with respect to the direction of the earth's gravitational field, the vertical line AC being the line of action of the weight force. When the body is suspended from point B, it assumes a different orientation with respect to the direction of the earth's gravitational field. This line of action of the weight force is labeled BD in Figure 3b. The point of intersection of lines AC and BD is called the *mass center, G*, or synonymously, the *center of gravity*. If the body is suspended from still another point, the vertical line of action would also intersect point G.

From the foregoing experiment we deduce that we can locate the center of mass for a specific body by considering any two orientations of the body with respect to the earth's gravitational field. The orientations we shall employ, for computation purposes, will always be 90° apart.

A moment of inertia (I_{zz}) conveys information regarding the distribution of the mass of a body about the z axis. For instance, for the thin circular ring in Figure 4 each element of mass is essentially situated along the perimeter. Hence the distance from the z axis to each mass element is the radius r of the ring, so that

$$I_{zz} = \int_{\mathscr{V}}(x^2 + y^2)\, dm = r^2\int_{\mathscr{V}} dm = mr^2$$

Clearly, if two rings have the same mass, then the one with the larger radius has the larger moment of inertia.

Dividing the value of I_{zz} by the mass and then computing the square root of that quotient, we obtain the *radius of gyration* about the z axis,

$$k_z = \sqrt{\frac{I_{zz}}{m}} \tag{8}$$

Because $k_z = r$ for the ring in Figure 4, we can think of the radius of gyration of any body as the radius of a ring having the same inertia properties as the body under consideration. Note that the integrand in equation (6) is always positive, so that I_{zz} will always be positive and hence k_z is a real number.

An understandable confusion arises between the concepts of *mass moment of inertia* and *second moment of area*. The second moment of area plays an important role in fields such as strength of materials, because it arises in the derivation for the determination of the stress distribution within materials. In our study of inertia properties we are concerned with the *polar* second moment of area. To avoid confusion with

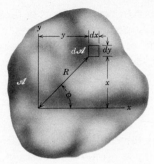

Figure 5

the symbols for mass properties, we shall denote this quantity by J_z. In what follows we shall compare I_{zz} and J_z.

For an area $\mathscr{A}$, shown in Figure 5, the definition of J_z is

$$J_z \equiv \int_{\mathscr{A}} (x^2 + y^2)\, d\mathscr{A} \tag{9}$$

The similarity of this quantity to the moment of inertia I_{zz} is obvious, because J_z measures the distribution of the area about the z axis.

Now let us consider a three-dimensional body whose outline in the xy plane is the area $\mathscr{A}$ of Figure 5. Let the thickness of the mass element dm in the z direction be h. In order to study the motion of this body by the methods of planar kinetics, we shall limit the analysis to cases where the (mass) density ρ depends only on the location of the differential mass elements with the xy plane. It then follows that

$$dm = \rho\, d\mathscr{V} = \rho h\, d\mathscr{A} \tag{10}$$

and hence

$$m = \int_{\mathscr{V}} dm = \int_{\mathscr{A}} \rho h\, d\mathscr{A} \tag{11}$$

$$I_{zz} = \int_{\mathscr{V}} (x^2 + y^2)\, dm = \int_{\mathscr{A}} (x^2 + y^2)\, \rho h\, d\mathscr{A} \tag{12}$$

Note that ρ and h were left in the integrand in the foregoing equations. If the body has an arbitrary shape and an arbitrary mass distribution, then ρ and h are not constant, and hence cannot be factored out of integrals that extend over the area $\mathscr{A}$. However, in the case when the body is *homogeneous,* so that ρ is constant, and when it has a constant thickness h, we then find that

$$m = \rho h \int_{\mathscr{A}} d\mathscr{A} = \rho h \mathscr{A} \tag{13}$$

$$I_{zz} = \rho h \int_{\mathscr{A}} (x^2 + y^2)\, d\mathscr{A} \equiv \rho h J_z \tag{14}$$

Equation (13) is the familiar result for the mass of a body with constant thickness, and equation (14) shows that in this special case the moment of inertia I_{zz} is proportional to the polar moment of area J_z. A more subtle point to note is that only in this special case of constant ρ and h does the center of mass of the body *always* coincide with the centroid of the area $\mathscr{A}$.

In general, unless we are given the value of the moment of inertia or the radius of gyration, the determination of I_{zz} becomes part of the task in the solution of problems. Also, with the exception of those cases where the geometry is such that we locate the center of mass by inspection, locating this point becomes part of our task. We shall now study some techniques for obtaining these quantities.

1 Inertia Properties by Integration

The fundamental method for determining the center of mass and moment of inertia is based on the definitions of these properties in terms of integrals.

To describe the position of the mass element we may use either rectangular Cartesian or polar coordinates. Remember that the two are interchangeable, because we can always express the transformation between the coordinate systems. For example, for the coordinate systems of Figure 5 we have

$$x = R \cos \phi \qquad y = R \sin \phi$$

Similarly, the cross-sectional area $d\mathcal{A}$ expressed in terms of Cartesian coordinates is

$$d\mathcal{A} = dx \, dy$$

whereas in terms of polar coordinates it is

$$d\mathcal{A} = R \, dR \, d\theta$$

We generally base the choice of coordinate system on whether the curve defining the perimeter of the area $\mathcal{A}$ is more easily expressed in terms of Cartesian or polar coordinates.

If the mass of the body is not given, this value may be obtained by evaluating the integral in equation (11). With the total mass determined, we can then locate the center of mass by using the result of the experiment we described in Figure 3 for determining its location. When the y axis in Figure 5 is vertical, the gravitational force on each element of mass is $(dm)g$ parallel to the y axis, so the moment about the z axis for each element is $x(dm)g$. Denoting the coordinates of the center of mass as (x_G, y_G), the total moment of the gravity force is $x_G mg$ because the resultant gravity force acts at this point. Equating this moment to the total moment of the force on each element, we obtain the *first moment of mass* with respect to the x coordinate,

$$mx_G = \int_{\mathcal{A}} x(\rho h \, d\mathcal{A}) \tag{15a}$$

Similarly, the first moment of mass with respect to the y coordinate is obtained if we imagine that gravity is aligned with the x axis. This yields

$$my_G = \int_{\mathcal{A}} y(\rho h \, d\mathcal{A}) \qquad (15b)$$

After evaluating the integrals in these expressions, we may solve for the coordinates (x_G, y_G). Conversely, equations (15) demonstrate that when the center of mass is situated on one of the coordinate axes, the first moment of mass with respect to the other coordinate is zero. This fact will prove useful for later derivations.

There is little theoretical discussion necessary to employ equation (12) to determine I_{zz}. Let us now study three examples that review the techniques of multiple integration.

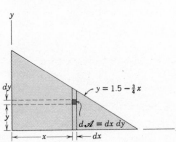

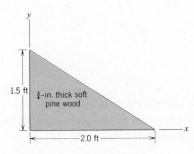

EXAMPLE 1

A $\frac{3}{4}$-in. sheet of soft pine wood is cut into the triangular shape shown in the diagram. Using the methods of integration, locate the center of mass of the sheet and then determine the value of I_{zz} for the coordinate axes shown.

Solution

Organizing our thinking, we must first determine the mass of the system, then the coordinates of the mass center, and finally I_{zz}. Because the edges of this body are all straight, the decision to formulate the solution in terms of rectangular Cartesian coordinates is obvious. We begin by describing the boundaries of the body as functions of x and y, as is done in the sketch to the left. (We used the straight-line formula, $y = mx + b$, to describe the hypotenuse.)

As the sketch indicates, we shall first form a vertical differential strip by integrating in the y direction from $y = 0$ to $y = 1.5 - 3x/4$. Then all strips will be summed by integrating from $x = 0$ to $x = 2$ ft. The thickness of the sheet is constant at $h = 3/48$ ft, and the density of soft pine is $\rho = 0.933$ lb-s²/ft⁴. Thus the mass of the body is

$$m = \rho h \int_0^2 \int_0^{1.5-3x/4} dy \, dx = \rho h \int_0^2 \left(1.5 - \frac{3x}{4}\right) dx$$

$$= \rho h[1.5(2) - \tfrac{3}{8}(2)^2] = 0.08743 \text{ lb-s}^2/\text{ft}$$

To locate the center of mass, we use the first moment of mass, equations (15), which gives

$$mx_G = \rho \int_0^2 \int_0^{1.5-3x/4} x \, dy \, dx = \rho h \int_0^2 x\left(1.5 - \frac{3x}{4}\right) dx$$

$$= 0.05828 \text{ lb-s}^2$$

$$my_G = \rho h \int_0^2 \int_0^{1.5-3x/4} y \, dy \, dx = \rho h \int_0^2 \frac{1}{2}\left(1.5 - \frac{3x}{4}\right)^2 dx$$

$$= 0.04371 \text{ lb-s}^2$$

Solving for x_G and y_G, we obtain

$$(x_G, y_G) = (0.667, 0.500) \text{ ft}$$

Finally, using equation (12), the moment of inertia is

$$I_{zz} = \rho h \int_0^2 \int_0^{1.5-3x/4} (x^2 + y^2) \, dy \, dx$$

$$= \rho h \int_0^2 \left[x^2\left(1.5 - \frac{3x}{4}\right) + \frac{1}{3}\left(1.5 - \frac{3x}{4}\right)^3\right] dx$$

$$= 0.0911 \text{ lb-s}^2\text{-ft}$$

EXAMPLE 2

A thin wire having a constant cross-sectional area A is bent into the form of a circular arc, as shown in the sketch. Determine (a) the center of mass and, (b) the value of I_{zz} for the coordinate axes shown.

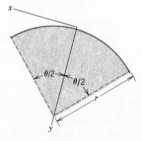

Solution

Because the wire is thin, we can consider its mass to be distributed along the centerline of the cross section, so that the differential mass element dm is defined by the infinitesimal arc length ds. The circular arc is easily described in terms of the polar coordinates defined in the following sketch.

$$\sin^2 = 1 - \cos^2$$

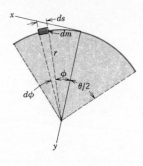

For this set of polar coordinates we have

$$x = r \sin \phi \qquad y = r - r \cos \phi \qquad ds = r \, d\phi$$

Then, because dm has cross-sectional area A and length ds, it follows that

$$dm = \rho A \, ds = \rho A r \, d\phi$$

In order to locate the center of mass, we first require the total mass, which is

$$m = \int dm = \rho A r \int_{-\theta/2}^{\theta/2} d\phi = \rho A r \theta$$

From the symmetry property of the system we note that the center of gravity is situated along the y axis, so $x_G = 0$. Therefore, we need only determine the first moment of mass with respect to the y coordinate. Using the polar coordinate expression for the value of y then gives

$$m y_G = \int y \, dm = \rho A r \int_{-\theta/2}^{\theta/2} r(1 - \cos \phi) \, d\phi$$

$$= \rho A r^2 \left(\theta - 2 \sin \frac{\theta}{2} \right)$$

Dividing by m results in

$$y_G = \frac{\rho A r^2 (\theta - 2 \sin \theta/2)}{\rho A r \theta} = r \left(1 - \frac{2}{\theta} \sin \frac{\theta}{2} \right)$$

or

$$(x_G, y_G) = \left[0, \, r \left(1 - \frac{2}{\theta} \sin \frac{\theta}{2} \right) \right]$$

Lastly, we compute the moment of inertia as follows.

$$I_{zz} = \int (x^2 + y^2) \, dm = \rho A r \int_{-\theta/2}^{\theta/2} [r^2 \sin^2 \phi + r^2 (1 - \cos \phi)^2] \, d\phi$$

$$= 2 \rho A r^3 \left(\theta - 2 \sin \frac{\theta}{2} \right)$$

To eliminate the density ρ, we may solve for ρ in the result for the total mass, which gives

$$\rho = \frac{m}{A r \theta}$$

so that

$$I_{zz} = 2\left(\frac{m}{Ar\theta}\right)Ar^3\left(\theta - 2\sin\frac{\theta}{2}\right)$$

$$= 2mr^2\left(1 - \frac{2}{\theta}\sin\frac{\theta}{2}\right)$$

EXAMPLE 3

Determine, by integration, the moment of inertia of a homogeneous sphere of radius a about an axis passing through the center of the sphere.

Solution

The problem requires the determination of I_{zz}, where the z axis passes through the center of the sphere. Hence, we locate the origin of the xyz system at the center. This also allows us to utilize the symmetry properties of a sphere. The shape of a sphere makes it logical to use polar coordinates to locate the differential mass elements. The significant variables are shown in the diagram below.

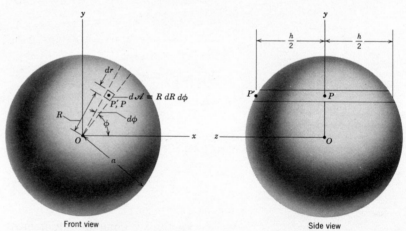

Front view   Side view

In this sketch the point P locates where the mass element intersects the xy plane and point P' locates where the mass element intersects the surface of the sphere. The length of the line OP is $R = \sqrt{x^2 + y^2}$, and the length of the line OP' is the radius a. Thus, because the mass element is perpendicular to the xy plane, the Pythagorean theorem gives

$$\frac{h}{2} = \sqrt{a^2 - R^2}$$

To obtain the contribution of all mass elements we first integrate over the range of ϕ from 0 to 2π, and then integrate over the range of R from 0 to a. Noting that ρ is a constant for a homogeneous body, we have

$$I_{zz} = \rho \int_0^a \int_0^{2\pi} R^2 h(R\ dR\ d\phi) = 2\rho \int_0^a \int_0^{2\pi} R^3\sqrt{a^2 - R^2}\ d\phi\ dR$$

$$= 2\rho(2\pi) \int_0^a R^3\sqrt{a^2 - R^2}\ dR$$

$$= \frac{8}{15}\pi\rho a^5$$

The last integral was evaluated with the aid of a table of integrals. In order to eliminate ρ, we note that the mass of a sphere is

$$m = \rho\mathcal{V} = \frac{4}{3}\pi\rho a^3$$

so that

$$\rho = \frac{3}{4\pi}\frac{m}{a^3}$$

Hence

$$I_{zz} = \frac{2}{5}ma^2$$

Each of the foregoing examples contained a body that can be termed an elementary shape. Any mechanical system can be approximated by a combination of elementary shapes. Some can be duplicated exactly. We shall concern ourselves with the latter. In such cases it is very useful to compute the centroidal and inertia properties of the shapes and compile them for future use. We present a typical standard table of properties in Appendix B.

In the next section we shall develop a formula that is of great value in inertia computations, especially when we wish to use the standard table of properties.

HOMEWORK PROBLEMS

By direct integration, determine the location of the center of mass and I_{zz} for each of the homogeneous bodies (of mass m) shown.

V.1 Slender Rod

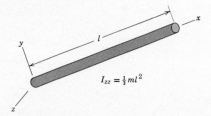

$$I_{zz} = \tfrac{1}{3}ml^2$$

Prob. V.1

V.2 Rectangular plate

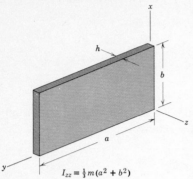

$$I_{zz} = \tfrac{1}{3} m(a^2 + b^2)$$

Prob. V.2

V.3 Semicircular plate

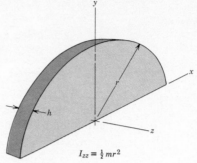

$$I_{zz} = \tfrac{1}{2} mr^2$$

Prob. V.3

V.4 Semicircular wire

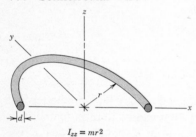

$$I_{zz} = mr^2$$

Prob. V.4

V.5 Thin circular sector

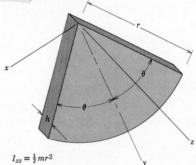

$$I_{zz} = \tfrac{1}{2} mr^2$$

Prob. V.5

V.6 Thin parabolic spandrel

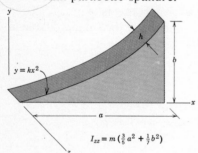

$y = kx^2$

$$I_{zz} = m(\tfrac{3}{5} a^2 + \tfrac{1}{7} b^2)$$

Prob. V.6

V.7 Solid circular cylinder

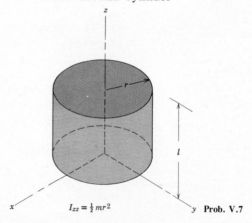

$$I_{zz} = \tfrac{1}{2} mr^2$$ **Prob. V.7**

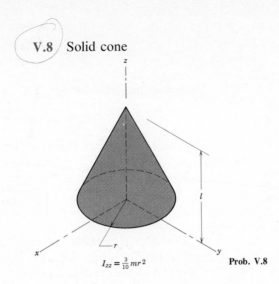

V.8 Solid cone

$$I_{zz} = \tfrac{3}{10} mr^2$$

Prob. V.8

2 The Parallel Axis Theorem

Standard tables of properties generally give the moment of inertia of a body about only one set of axes, usually axes that pass through the center of mass. Suppose that we know $I_{\hat{z}\hat{z}}$, where the $\hat{z}$ axis is perpendicular to the plane of motion of a rigid body and passes through its center of mass, as shown in Figure 6. It is our desire to establish the relation of this moment of inertia to I_{zz}, where the z axis is also perpendicular to the plane of motion (hence it is parallel to the $\hat{z}$ axis). Note that in Figure 6 we selected the two coordinate systems to contain mutually parallel axes. This proves to be an important convenience in the work to follow.

Letting the xy and $\hat{x}\hat{y}$ planes coincide, from Figure 6 we may write

$$x = \hat{x} - \hat{x}_A \qquad y = \hat{y} - \hat{y}_A$$

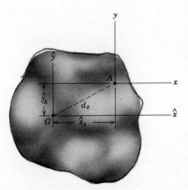

Figure 6

where $(\hat{x}_A, \hat{y}_A, 0)$ are the coordinates of the origin A of the xyz system with respect to the $\hat{x}\hat{y}\hat{z}$ system. Introducing this transformation into equation (6) gives

$$I_{zz} = \int_{\mathscr{V}} (x^2 + y^2)\, dm = \int_{\mathscr{V}} [(\hat{x} - \hat{x}_A)^2 + (\hat{y} - \hat{y}_A)^2]\, dm$$

$$= \int_{\mathscr{V}} (\hat{x}^2 + \hat{y}^2)\, dm - 2\hat{x}_A \int_{\mathscr{V}} \hat{x}\, dm - 2\hat{y}_A \int_{\mathscr{V}} \hat{y}\, dm$$

$$+ [(\hat{x}_A)^2 + (\hat{y}_A)^2] \int_{\mathscr{V}} dm \qquad (16)$$

As the origin of the $\hat{x}\hat{y}\hat{z}$ system is the center of mass, it follows that the first moments of mass with respect to this coordinate system are zero; that is,

$$\int_{\mathscr{V}} \hat{x}\, dm = \int_{\mathscr{V}} \hat{y}\, dm = 0$$

Further, denoting the distance between the z and $\hat{z}$ axes as d_z, we have

$$d_z = \sqrt{(\hat{x}_A)^2 + (\hat{y}_A)^2}$$

so that equation (16) gives the following form for the *parallel axis theorem:*

$$\boxed{I_{zz} = I_{\hat{z}\hat{z}} + md_z{}^2} \tag{17}$$

The application of equation (17) is straightforward, provided that you remember that the *$\hat{z}$ axis must pass through the center of mass.* If it does not, then the first moments of mass do not vanish and equation (17) is not valid.

As indicated earlier, in applications it is more expedient to determine the inertia properties by using the table of properties in Appendix B rather than the methods of integration. The parallel axis theorem is a powerful tool in this regard because it allows us to determine the moment of inertia about an axis other than the one referred to in the tables.

HOMEWORK PROBLEMS

Using the parallel axis theorem, and the answers given to the problems in the previous section, find the moment of inertia, I_{zz}, about the mass center for each of the following problems.

V.9   Problem V.1

V.10   Problem V.2

V.11   Problem V.3

V.12   Problem V.4

V.13   Problem V.5

V.14   Problem V.6

For the following problems find I_{zz} about an axis parallel to the given z axis containing the point $y = r$.

V.15   Problem V.7

V.16   Problem V.8

3   Composite Bodies

As mentioned earlier, it is very useful to employ tabulated properties when we consider a body to be a composite shape, that is, a shape composed of several basic shapes. Such a body is shown in Figure 7 where the areas $\mathcal{A}_1$ and $\mathcal{A}_2$ represent two elementary geometric shapes forming the outline of the body of interest.

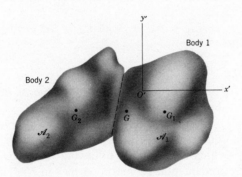

Figure 7

Any integral extending over the entire area $\mathcal{A}$ is the sum of the integrals over the areas $\mathcal{A}_1$ and $\mathcal{A}_2$. Thus, equation (11) for the mass becomes

$$m = \int_{\mathcal{A}_1} \rho h \, d\mathcal{A} + \int_{\mathcal{A}_2} \rho h \, d\mathcal{A} \quad m_1 + m_2 \tag{18}$$

This is the obvious result that the mass of the body is the sum of the mass of each component. Conversely the mass of a composite body, in the form of a body 1 from which a body 2 has been removed, can be found by *subtracting m_2 from m_2.*

We require an arbitrary reference frame to locate the center of mass G of the composite body. This is the primed frame shown in Figure 7; the prime emphasizes its arbitrary selection. In practice, the origin O' is selected as any convenient point for computing first moments of mass, such as the center of mass of either body 1 or 2. With the reference frame selected, the center of mass G is obtained by taking the first moment of mass with respect to both the x' and y' axes. Thus, considering the area $\mathcal{A}$ in terms of its component shapes, and using equation (15), we find

$$m x'_G \equiv \int_{\mathcal{A}} x'(\rho h \, d\mathcal{A}) = \int_{\mathcal{A}_1} x'(\rho h \, d\mathcal{A}) + \int_{\mathcal{A}_2} x'(\rho h \, d\mathcal{A})$$

$$= m_1 x'_{G_1} + m_2 x'_{G_2} \tag{19a}$$

Similarly,

$$my_G' = m_1 y_{G_1}' + m_2 y_{G_2}' \tag{19b}$$

where (x_{G_1}', y_{G_1}') and (x_{G_2}', y_{G_2}') are the $x'y'$ coordinates of the mass centers of bodies 1 and 2, respectively. Solution of equations (19) yields the values of x_G' and y_G', locating the center of mass of the composite body.

The final task is the determination of the moment of inertia. In terms of *any* coordinate system xyz whose z axis is perpendicular to the plane of motion, we find from equation (12) that

$$I_{zz} = \int_{\mathcal{A}} (x^2 + y^2)\, \rho h\, d\mathcal{A} = \int_{\mathcal{A}_1} (x^2 + y^2)\, \rho h\, d\mathcal{A} + \int_{\mathcal{A}_2} (x^2 + y^2)\, \rho h\, d\mathcal{A}$$

$$= (I_{zz})_1 + (I_{zz})_2 \tag{20}$$

This result tells us that the moments of inertia of the component bodies about the *same* axis are additive.

When we are solving kinetics problems, the z axis we shall employ in equation (20) is the one passing through a particular point, the selection of which we discussed following equation (2). The moments of inertia for the fundamental geometric shapes in Appendix B may not be given about an axis that coincides with this z axis. In such cases it will be necessary first to apply the parallel axis theorem (17) to determine the value of I_{zz} for each component of the composite body before applying equation (20).

EXAMPLE 4
The aluminum body shown in the sketch has a thickness of 80 mm. Determine the mass of the body, the location of its center of mass, and its moment of inertia about an axis perpendicular to the plane of the diagram through the center of mass.

Solution

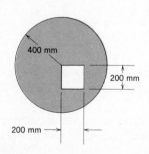

The body has a composite shape formed by removing a square parallelepiped, call it body 2, from a circular cylinder, call it body 1. Thus, because the density of aluminum is

$$\rho = 2.691(10^3)\ \text{kg/m}^3 = 2.691(10^{-6})\ \text{kg/mm}^3$$

we have

$$m_1 = \rho \mathcal{V}_1 = 2.691(10^{-6})\pi(400)^2(80) = 108.21\ \text{kg}$$

$$m_2 = \rho \mathcal{V}_2 = 2.691(10^{-6})(200)^2(80) = 8.61\ \text{kg}$$

$$m = m_1 - m_2 = 99.60\ \text{kg}$$

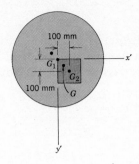

We next select a reference frame $x'y'z'$ for computing first moments. The center of the circle seems a convenient origin for this system, and it makes sense to orient the x' and y' axes to be parallel to the edges of the cut-out area, as shown in the sketch. (We shall generally find that selecting a coordinate system that is parallel to the one depicted in the tables of Appendix B will be most efficient for computations.) Because of the symmetry of the body, we can see that the center of mass G is situated along the 45° line connecting points G_1 and G_2, so it must be true that $x'_G = y'_G$. (This observation is merely for the purpose of eliminating some computations.)

Because the mass m_2 is subtracted from the mass m_1, the first moment of mass with respect to the x' coordinate is

$$mx'_G = m_1 x'_{G_1} - m_2 x'_{G_2}$$

$$= 108.21(0) - 8.61(100)$$

$$= -861 \text{ kg-mm}$$

Thus

$$x'_G = y'_G = \frac{-861}{99.60} = -8.64 \text{ mm} \qquad \triangle$$

The tabulated values of moment of inertia for both the cylinder and the parallelepiped are for axes $\hat{z}$ that pass through the respective centers of mass. We shall therefore need first to apply the parallel axis theorem to obtain the moments of inertia about an axis passing through point G. Knowing the location of point G, we now find the distances from points G_1 and G_2 to the center of mass G to be

$$(d_z)_1 = \sqrt{2}(8.64) = 12.22 \text{ mm}$$

$$(d_z)_2 = \sqrt{2}(100 + 8.64) = 153.64 \text{ mm}$$

Using the tabulated moments of inertia, we then find

$$(I_{zz})_1 = \tfrac{1}{2}m_1 r^2 + m_1 (d_z)_1^2$$

$$= 108.21[\tfrac{1}{2}(400)^2 + (12.22)^2]$$

$$= 8.673(10^6) \text{ kg-mm}^2 = 8.673 \text{ kg-m}^2$$

$$(I_{zz})_2 = \tfrac{1}{12}m_2(a^2 + b^2) + m_2(d_z)_2^2$$

$$= 8.61\{\tfrac{1}{12}[(200)^2 + (200)^2] + (153.64)^2\}$$

$$= 2.61(10^5) \text{ kg-mm}^2 = 0.261 \text{ kg-m}^2$$

$$I_{zz} = (I_{zz})_1 - (I_{zz})_2 = 8.41 \text{ kg-m}^2 \qquad \triangle$$

HOMEWORK PROBLEMS

In the following problems use the information in Appendix B to determine the location of the mass center of the composite body and the mass moment of inertia $I_{\hat{z}\hat{z}}$, where $\hat{z}$ is the axis passing through the mass center that is parallel to the z axis shown.

V.17 The steel plate shown below.

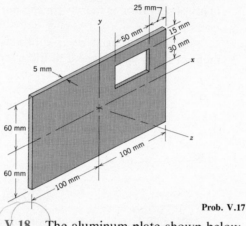

Prob. V.17

V.18 The aluminum plate shown below.

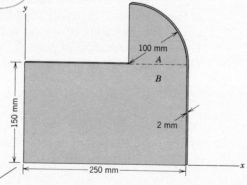

Prob. V.18

V.19 Solve Problem V.18 for the case where plate A is aluminum and plate B is steel.

V.20 The aluminum cylinder and cone shown to the left.

V.21 Solve Problem V.20 for the case where the cylinder is steel and the cone is aluminum.

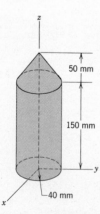

Prob. V.20

V.22 The steel assembly shown below.

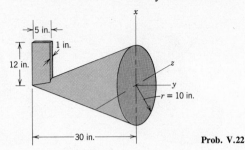

Prob. V.22

V.23 Solve Problem V.22 for the case where the cone is steel and the plate is aluminum.

V.24 The aluminum wires, 0.20 in.² cross-sectional area, shown below.

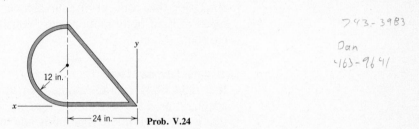

Prob. V.24

C. THREE TYPES OF RIGID BODY MOTION

Now that we have the ability to evaluate the inertial properties of a body, let us return to the consideration of the equations of motion (1) and (7). Our aim in this section is to gain further understanding of the significance of these equations. With this as our goal, let us consider an interpretation of equations (1) and (7) that builds on concepts developed in the study of the statics of rigid bodies.

According to the principle of equivalence of force systems, the effect on a rigid body of an arbitrary force system can be represented by an equivalent force-couple system acting at any point P. In this equivalent system, the force that is applied at point P is the resultant of the forces acting on the body, and the torque of the couple is the total moment of all the forces about point P.

Figure 8a, on the next page, shows a rigid body subjected to an arbitrary planar force system, where $\hat{x}\hat{y}\hat{z}$ is the coordinate system whose origin is the center of mass G. We use the center of mass G as the point P for the equivalent force-couple system because it is always a valid point for summing moments in the formulation of the dynamics problem. Thus,

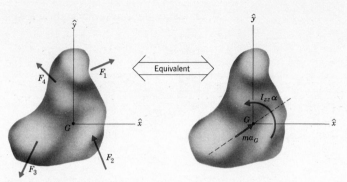

Figure 8A and Figure 8b

equations (1) and (7) tell us that the force system in Figure 8a is equivalent to a force $m\bar{a}_G$ applied at point G and a couple $I_{zz}\bar{\alpha}$, as illustrated in Figure 8b. Using this concept of equivalence, let us examine each of the three types of rigid body motion: pure translation, pure rotation, and general motion.

1   Translational Motion

If a rigid body is translating, then the angular velocity $\bar{\omega}$ and angular acceleration $\bar{\alpha}$ are both zero. From our knowledge of kinematics it follows that every point in the rigid body has the same acceleration as the center of mass. In this case, the equivalent force system in Figure 8a reduces to that shown in Figure 9.

From Figure 9 we can deduce that if we had computed the moment of the force system about any point other than the center of mass, such as point P (in Figure 9), then, in general, we would have to equate the resulting moment to the moment about point P of the equivalent force $m\bar{a}_G$. An exception occurs if point P lies on the indicated dashed line, for then the vector $m\bar{a}_G$ passes through point P. However, there will be much less chance for error if we make it standard practice to sum moments about the center of mass if the body is translating. The equations of motion in this case reduce to

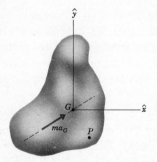

Figure 9

$$\Sigma \, \bar{F} = m\bar{a}_G \qquad \Sigma \, \bar{M}_G = \bar{0} \qquad (21)$$

2   Rotational Motion

A body undergoing pure rotation in a plane has an axis perpendicular to the plane of motion that is fixed. It then follows that the center of mass is in circular motion about the axis of rotation. In terms of polar coordinates, defined with respect to point O on the fixed axis, we have

$$\bar{a}_G = -R\omega^2\bar{e}_R + R\alpha\bar{e}_\phi$$

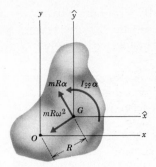

Figure 10

Using this expression for $\bar{a}_G$, the general equivalent force system in Figure 8b reduces to that shown in Figure 10.

From Figure 10 we note that the equivalent force $-mR\omega^2\bar{e}_R$ exerts no moment about point O. Therefore, when we compute $\Sigma\,\bar{M}_O$ for the actual force system and equate this to the sum of the moments about point O of the equivalent force system, we find that

$$\Sigma\,\bar{M}_O = I_{\hat{z}\hat{z}}\bar{\alpha} + mR\alpha(R)\bar{k} = (I_{\hat{z}\hat{z}} + mR^2)\alpha\bar{k}$$

In view of the parallel axis theorem, equation (17), this reduces to

$$\Sigma\,\bar{M}_O = I_{zz}\alpha\bar{K} = I_{zz}\bar{\alpha}$$

This, of course, is the same result as equation (7) for the case where we sum moments about a fixed point (thus demonstrating why a fixed point is an allowable point for summing moments). Notice that if we had computed the moment about any point other than the fixed point O or center of mass G, we would not have the simple relationship of equation (7).

The question of whether we should compute the moment of the force system about point O or point G is easily answered when we consider the forces acting on the system. In order to have a fixed axis of rotation, there must be constraint forces applied to this axis, such as the reaction forces produced by a pin joint. It follows then, that in order to eliminate these (unknown) constraint forces, we solve problems involving bodies executing pure rotational motion by summing moments about the fixed point O. The resulting equations of motion are therefore

$$\Sigma\,\bar{F} = m\bar{a}_G \qquad \Sigma\bar{M}_O = I_{zz}\bar{\alpha} \tag{22}$$

where the z axis is coincident with the fixed axis of rotation.

3 General Motion

For this type of motion, which is a superposition of translational and rotational effects, no point in the body is fixed. Also, whether or not there is any point in the body whose acceleration is directly toward or away from the center of mass depends on the geometry and constraints of the system. This can be illustrated by the simple problem of a circular wheel rolling without slipping down an incline.

If the wheel is balanced so that the center of mass G coincides with the geometric center O, then we have the kinematical situation illustrated in Figure 11a. For the no slip condition, point C of the body, which is in contact with the incline, must accelerate directly toward point O. Thus, if points O and G coincide, then point C is an allowable point for applying equation (17). In contrast, if the wheel is unbalanced, so that the center of mass is at a distance l from the geometric center O, we then have the

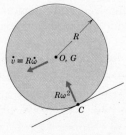

Figure 11a

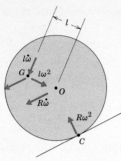

Figure 11b

kinematical situation shown in Figure 11b. From this Figure we conclude that the contact point C does not accelerate directly toward point G, except for the special cases that arise at instants when points G, O, and C are collinear.

To have an approach that is valid in all situations, we therefore solve problems in general motion by computing moments about the center of mass G. Then, the general equations of motion are

$$\Sigma \, \bar{F} = m\bar{a}_G \qquad \Sigma \, \bar{M}_G = I_{zz}\bar{\alpha} \tag{23}$$

where the z axis is perpendicular to the plane of motion, passing through point G.

D. PROBLEM SOLVING

We did not present examples using the equations of motion prior to this section for the reason that the procedures to use them are fairly straightforward. As we proceed, note that the steps we present are essentially a synthesis of the methods for treating force systems (developed in statics) and the methods of kinematics (developed in Module IV).

1   Determine the mass and location of the center of mass G of each body in the system for which such information is not given.

2   Draw a free body diagram of each individual body in the system, making use of Newton's law of action and reaction where appropriate. Remember that each constraint on the motion of a system can only be imposed by a corresponding force or moment. The free body diagram(s) should also show the location of the center of mass, even when the gravitational force is perpendicular to the plane of motion.

3   Select an xyz reference frame for each body. For bodies in translation or general motion, the origin A of this frame should be the center of mass of the body, whereas for bodies that are rotating about a fixed axis the origin should be situated on that axis. The x and y axes should be parallel to the plane of motion and their orientation should, where possible, fit the geometry of the physical system. In problems where there is more than one body, the xyz reference frame for each body should be parallel to each other, making it easier to equate various vectors.

4   Show all given values of forces in the free body diagram and write down any other given information in equation form, using the xyz frame to describe vector components.

5   Following the methods developed in Module IV for treating the kinematics of rigid bodies in planar motion, relate the angular acceleration and the x and y components of the acceleration of the center of mass. Be certain to satisfy all of the constraints imposed on the motion of the system.

6 Calculate I_{zz} for any body that has rotational motion.

7 Obtain the scalar equations of motion for each body by equating the x and y components of $\Sigma \bar{F} = m\bar{a}_G$ and the z component of $\Sigma \bar{M}_A = I_{zz}\bar{\alpha}$. In following this step, substitute the results of step 5.

8 Count the number of scalar equations and the number of scalar unknowns to be determined. If the two numbers do not match, you probably have either improperly employed the law of action and reaction, or else forgotten a constraint force or the corresponding kinematical constraint condition.

9 Solve the scalar equations of motion. Then check that the solution satisfies the initial assumptions. For example, if you have assumed that two surfaces do not slip relative to each other, check to be sure that the friction force f and normal force N satisfy the condition $|f| < \mu_s |N|$.

ILLUSTRATIVE PROBLEM 1

The automobile shown accelerates at a constant rate on a level road. It starts from rest and reaches the speed of 70 km/hr in a distance of 200 m. Determine the reactive force at the hitch of the 400-kg trailer during the motion. Neglect friction.

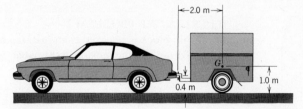

Solution

Step 1 The location of the center of mass G of the trailer is given.

Steps 2, 3, and 4 Because the reactive force at the hitch can be determined only when it appears as an external force, we isolate the trailer in a free body diagram. Note that we did not isolate the automobile, because we have no information about this body. It is reasonable to consider the hitch as a pin, hence it exerts horizontal and vertical reactions. The choice for the coordinate system is now obvious.

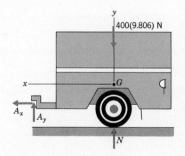

Step 5 The system is in pure translation, so $\omega = \alpha = 0$. There is no motion in the y direction, hence $v_y = a_y = 0$. For constant acceleration in the x direction we have

$$2a_x s_x = (v_x)_2^2 - (v_x)_1^2 = \left(70\frac{1000}{3600}\right)^2 - 0 = 378.1$$

$$a_x = \frac{378.1}{2(200)} = 0.9453 \text{ m/s}^2$$

Step 6   Because $\alpha \equiv 0$, I_{zz} is not needed.

Step 7

$$\Sigma F_x = A_x = m(a_G)_x = 400(0.9453)$$

$$\Sigma F_y = -400(9.806) + A_y + N = m(a_G)_y = 0$$

$$\Sigma M_{Gz} = (2A_y + 0.6A_x) = 0$$

Step 8   We have three scalar equations in the three scalar unknowns A_x, A_y, and N.

Step 9

$\bar{\imath}$ direction: $A_x = 400(0.9453) = 378.1$ newtons

$\bar{\jmath}$ direction: $A_y + N = 3922$

moment: $A_y = -0.3A_x = -113.4$ newtons

Note that we only needed the first and third equations to solve for the components of the reactive force.

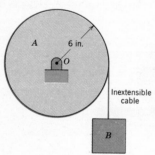

ILLUSTRATIVE PROBLEM 2

The 100-lb block is released from rest. Compute the angular velocity of the 60-lb uniform drum and the tension in the cable: (a) 5 s after release, (b) after the block has fallen 20 ft.

Solution

Step 1   We must cut the cable to determine its tension. Hence, we construct two free body diagrams. We can treat the block as a particle and assume that the center of mass of the drum coincides with the shaft supporting the drum.

Steps 2, 3, and 4   The shaft can apply horizontal and vertical reactions to support the drum, resulting in the free body diagram shown. Note that even if we were not interested in the tension force, we would have to treat the block and drum separately because jointly they do not form a rigid body.

Step 5   The acceleration of the 100-lb weight must be the same as the tangential acceleration of the point P on the circumference of the drum. Therefore,

$$-a_B \bar{\jmath} = -r\alpha_A \bar{\jmath} = -0.5\alpha_A \bar{\jmath} \text{ ft/s}^2$$

Step 6   From Appendix B, for body A we have

$$I_{zz} = \frac{1}{2} mr^2 = \frac{1}{2}\frac{60}{32.17}\left(\frac{1}{2}\right)^2 = 0.2331 \text{ lb-s}^2\text{-ft}$$

The moment of inertia of the block is not required because we are treating it as a particle.

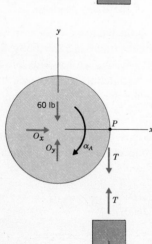

Step 7 The force sum for the drum will involve the unknown reactions at point A. As we are not interested in these forces, we only sum moments for the drum. Hence, for body A we have

$$\Sigma\, M_{Gz} = -T(0.5) = 0.2331(-\alpha_A)$$

For body B we sum forces in the y direction and use the result of step 5. This gives

$$\Sigma\, F_y = T - 100 = \frac{100}{32.17}(-a_B) = -\frac{100}{32.17}(0.5\,\alpha_A)$$

Step 8 We have two scalar equations in the two scalar unknowns T and α_A.

Step 9 The solution of the equations of motion in step 7 is

$$\alpha_A = 49.49 \text{ rad/s}^2$$

$$T = 23.07 \text{ lb}$$

The value of T is constant. We obtain the desired angular velocities by applying the formulas for constant acceleration. Hence, when $t = 5$ s for part (a), we have

$$(\omega_A)_2 = (\omega_A)_1 + \alpha_A t$$

$$= 0 + 49.49(5)$$

or

$$\bar{\omega}_A = -247.5\bar{k} \text{ rad/s}$$

For part (b) we first determine the angle of rotation corresponding to a 20-ft drop of the block.

$$\theta \equiv \frac{20}{R} = \frac{20}{0.5} = 40 \text{ rad}$$

Then using

$$(\omega_A)_2{}^2 = (\omega_A)_1{}^2 + 2\alpha_A\theta$$

for $\theta = 40$ rad we find

$$(\omega_A)_2{}^2 = 0 + 2(49.49)40$$

or

$$\bar{\omega}_A = -62.92\bar{k} \text{ rad/s}$$

ILLUSTRATIVE PROBLEM 3

The uniform plate shown is released from its rest position in a vertical plane by cutting cable BC. Determine the forces in cables BD and AE at the instant of release.

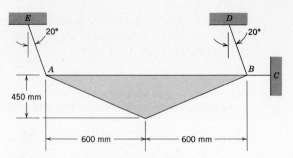

Solution

Step 1 The center of mass of a triangular plate is at a distance from the base that is one-third of the height.

Step 2 Because we want the initial values of the cable forces, we draw the free body diagram for the instant of release.

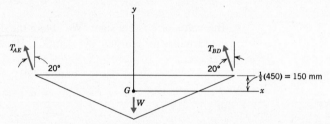

Steps 3 and 4 Edge AB will always be horizontal, so the plate translates. Hence, the origin of the xyz coordinate system must be at the center of mass. The chosen orientation of xyz expedites the description of the force components and their points of application. As the weight of the plate is not given, call it W.

Step 5 For translational motion $\bar{\omega} = \bar{\alpha} = \bar{0}$. Because the plate is released from rest, $\bar{v}_G = \bar{0}$, but $\bar{a}_G \neq \bar{0}$. The value of $\bar{a}_G$ is not arbitrary, because the acceleration of all points in a translating body are identical and points A and B are constrained by the cables to follow a circular motion. It is convenient to use path variables to determine $\bar{a}_A$.

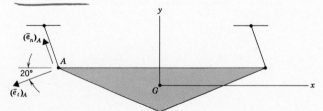

In terms of the path variable unit vectors with $v = 0$ we have

$$\bar{a}_G = \bar{a}_A = \dot{v}(\bar{e}_t)_A + \frac{v^2}{\rho}\,(\bar{e}_n)_A = \dot{v}(\bar{e}_t)_A$$

However, we shall need the x and y components of $\bar{a}_G$ for the force sum. Thus, transforming $(\bar{e}_t)_A$ yields

$$\bar{a}_G = \dot{v}(-\cos 20° \, \bar{i} - \sin 20° \, \bar{j})$$

Step 6 $\alpha \equiv 0$ so I_{zz} is not needed.

Step 7 Referring to the free body diagram in step 2, the equations of motion are

$$\Sigma F_x = -T_{AE} \sin 20° - T_{BD} \sin 20° = m(a_G)_x$$

$$= \frac{W}{g} (-\dot{v} \cos 20°)$$

$$\Sigma F_y = T_{AE} \cos 20° + T_{BD} \cos 20° - W = m(a_G)_y$$

$$= \frac{W}{g} (-\dot{v} \sin 20°)$$

$$\Sigma M_{Gz} = T_{AE} \sin 20° \, (0.15) - T_{AE} \cos 20° \, (0.60)$$

$$+ T_{BD} \sin 20° \, (0.15) + T_{BD} \cos 20° \, (0.60)$$

$$= I_{zz}\alpha = 0$$

Step 8 We have three scalar equations for the three scalar unknowns T_{AE}, T_{BD}, and $\dot{v}$.

Step 9 Solving the moment equation for T_{BD} and substituting this result into the two force equations (which are then solved simultaneously) yields

$$\dot{v} = 0.3214g$$

$$T_{AE} = 0.4816W \qquad T_{BD} = 0.4012W \qquad \triangleleft$$

ILLUSTRATIVE PROBLEM 4

A 30-kg uniform rod is sliding along the floor and wall. The coefficient of friction between all surfaces is 0.3. Determine the reactions at ends A and B for the instant shown, if end A has a velocity of 3 m/s to the left at this instant.

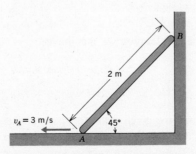

Solution

Step 1   The location of the center of mass G is known by symmetry.

Steps 2, 3, and 4   The friction forces are opposite the sliding motion. The rod is in general motion, so we place the origin of xyz at point G.

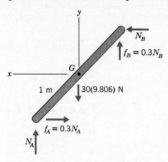

Step 5   Ends A and B are constrained to move horizontally and vertically, respectively. Hence, we can relate $\bar{a}_G$ to $\bar{\alpha}$ for the constrained motion. Following the methods established in Module Four, we first determine the angular velocity using the method of instant centers.

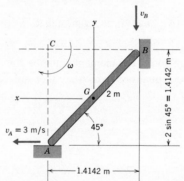

From the sketch, point C is the instant center, so

$$\omega = \frac{v_A}{2 \sin 45°} = \frac{3}{1.4142} = 2.121 \text{ rad/s} \qquad \text{clockwise}$$

$$\bar{\omega} = -2.121\bar{k}$$

For the acceleration we then have

$$\bar{a}_G = \bar{a}_A + \bar{\alpha} \times \bar{r}_{G/A} + \bar{\omega} \times (\bar{\omega} \times \bar{r}_{G/A})$$

$$= \bar{a}_B + \bar{\alpha} \times \bar{r}_{G/B} + \bar{\omega} \times (\bar{\omega} \times \bar{r}_{G/B})$$

Note that the given value for v_A is an instantaneous value and is not a constant. (An additional constraint force would be necessary to have no acceleration for this point.) The individual vectors are

$$\bar{a}_A = a_A\,\bar{i} \qquad \bar{a}_B = -a_B\,\bar{j} \qquad \bar{\alpha} = \alpha\bar{k}$$

$$\bar{r}_{G/A} = 0.7071(-\bar{i} + \bar{j}) \qquad \bar{r}_{G/B} = 0.7071(\bar{i} - \bar{j})$$

The acceleration equations then give

$$\bar{a}_G = [a_A - 0.7071\alpha + (2.121)^2(0.7071)]\bar{i}$$

$$+ [-0.7071\alpha - (2.121)^2(0.7071)]\,\bar{j}$$

$$= [0.7071\alpha - (2.121)^2(0.7071)]\bar{i}$$

$$+ [-a_B + 0.7071\alpha + (2.121)^2(0.7071)]\,\bar{j}$$

Equating corresponding components gives

$$(a_G)_x = a_A - 0.7071\alpha + 3.181 = 0.7071\alpha - 3.181$$

$$(a_G)_y = -0.7071\alpha - 3.181 = -a_B + 0.7071\alpha + 3.181$$

We are not interested in the values of a_A and a_B, so we write

$$(a_G)_x = 0.7071\alpha - 3.181 \qquad (a_G)_y = -0.7071\alpha - 3.181$$

Step 6   We obtain I_{zz} for the rod from Appendix B.

$$I_{zz} = \tfrac{1}{12}ml^2 = \tfrac{1}{12}(30)(2)^2 = 10 \text{ kg-m}^2$$

Step 7   From the free body diagram we have

$$\Sigma\,F_x = N_B - 0.3N_A = m(a_G)_x = 30(0.7071\alpha - 3.181)$$

$$\Sigma\,F_y = 0.3N_B + N_A - 30(9.806) = m(a_G)_y$$

$$= 30(-0.7071\alpha - 3.181)$$

$$\Sigma\,M_{Gz} = N_A(0.7071) - 0.3N_A\,(0.7071) - N_B(0.7071)$$

$$- 0.3N_B(0.7071) = I_{zz}\alpha = 10\alpha$$

Note that in the moment equation, clockwise moments are positive because the z axis is into the plane of the diagram.

Step 8   We have three equations for the three unknowns, N_A, N_B, and α.

Step 9   Solving the three equations simultaneously for N_A and N_B yields

$$N_A = 112.3 \text{ newtons} \qquad N_B = 19.03 \text{ newtons}$$

Hence, the reactions are

$$\bar{R}_A = -0.3N_A\bar{i} + N_A\,\bar{j} = 112.3\bar{i} - 33.69\bar{j} \text{ newtons}$$

$$\bar{R}_B = N_B\bar{i} + 0.3N_B\,\bar{j} = 19.03\bar{i} + 5.709\bar{j} \text{ newtons}$$

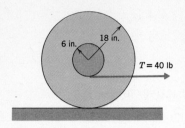

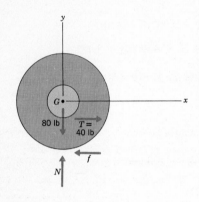

ILLUSTRATIVE PROBLEM 5

The diagram shows the cross section of an 80-lb spool of wire. The radius of gyration of the spool is 10 in. Knowing that the coefficients of static and kinetic friction are $\mu_s = 0.25$ and $\mu_k = 0.20$, respectively, determine the acceleration of the mass center G for a wire tension $T = 40$ lb.

Solution

Step 1 The location of the center of mass is known by symmetry.

Steps 2 and 3 We do not know whether there is slippage at the point of contact, so we *cannot* assume that the friction force $f = \mu N$. Hence, we consider f to be an unknown.

Regardless of whether or not there is slippage, the disk is in general motion, so the origin of the xyz axes is selected to coincide with the center of mass.

Step 4 The radius of gyration is $k_z = 10$ in. and the given value for T is shown in the free body diagram.

Step 5 The center of mass follows a horizontal path and its acceleration will probably be to the right, so

$$\bar{a}_G = a_G \bar{\imath}$$

For planar motion we know that $\bar{\alpha} = \alpha \bar{k}$, but we have two alternative kinematical hypotheses to consider. The constraint of no slippage gives $a_G = -R\alpha = -1.5\alpha$ (see Module Four), where the negative sign occurs because, in this case, negative α corresponds to clockwise angular acceleration and positive a_G. On the other hand, if there is slippage, then there is no constraint condition to relate a_G and α. However, for the latter case we know that $f = \mu_k N$, opposite the sliding motion.

Step 6 We use the given value of the radius of gyration to determine I_{zz}.

$$I_{zz} = mk_z{}^2 = \frac{80}{32.17}\left(\frac{10}{12}\right)^2 = 1.7269 \text{ lb-s}^2\text{-ft}$$

Step 7

$$\Sigma F_x = 40 - f = m(a_G)_x = \frac{80}{32.17} a_G$$

$$\Sigma F_y = N - 80 = m(a_G)_y = 0$$

$$\Sigma M_{Gz} = 40\left(\frac{6}{12}\right) - f\left(\frac{18}{12}\right) = I_{zz}\alpha = 1.7269\alpha$$

Steps 8 and 9 The three equations of motion have four unknowns: f, N, a_G, and α. For the fourth relation we assume one of the kinematical possibilities; if we are wrong in this assumption we shall find a contradiction. Let us assume that there is no slipping. Then $a_G = -1.5\alpha$ and the equations of motion reduce to

$$40 - f = 2.487a_G$$

$$N - 80 = 0$$

$$20 - 1.5f = 1.7269\left(-\frac{a_G}{1.5}\right) = -1.151a_G$$

The solution to these equations is

$$a_G = 8.194 \text{ ft/s}^2 \qquad f = 19.62 \text{ lb} \qquad N = 80 \text{ lb}$$

We now can check the assumption of no slippage, by ascertaining if $|\bar{f}| \leq \mu_s N$, where the equality sign is the maximum friction force that may be sustained without slipping. Using the given value for μ_s we have

$$\mu_s N = 0.25(80) = 20 > f$$

Hence the spool does not slip and the foregoing solution is correct. Thus

$$a_G = 8.194 \text{ ft/s}^2$$

Note the procedure we used to solve the problem. Whenever we are not sure whether surfaces slip over each other, we begin by assuming no slippage. If the required friction force exceeds $\mu_s N$, then we know that slipping will occur. In that case the friction force will be $\mu_k N$ in the same direction as the friction force obtained from the assumption of no slipping. Also, the values of a_G and α are then independent variables. The next problem features a situation where we know that slipping occurs.

ILLUSTRATIVE PROBLEM 6

The driver of an automobile causes the rear drive wheels to skid as the car accelerates. The coefficient of kinetic friction is 0.40. Assuming that the drive wheels have a constant angular velocity of 600 rev/min and that the automobile starts from rest as it moves to the left, determine the acceleration of the vehicle during the time interval when the rear wheels are skidding and also find the duration of this time interval. The mass of the vehicle is 1250 kg. The location of its center of mass C is shown in the diagram. The wheels have a mass of 15 kg, a radius of 400 mm, and a radius of gyration of 300 mm.

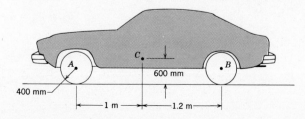

Solution

There are three types of bodies in this system: the chassis, the two rear drive wheels, and the two front wheels. If we wished to neglect the mass of the wheels, we could simply draw a free body diagram of the entire system, determine the acceleration of the vehicle, and then seek the condition where the no slip condition for a wheel, $v = \omega R$, is satisfied. However, as we are given the properties of the wheels, let us include the effect of their mass as we model the system.

Step 1 The given diagram locates the center of mass of all bodies.

Steps 2, 3, and 4 In order to accelerate to the left, the wheels must be turning counterclockwise. Hence, the friction force on the skidding drive wheels is to the left. The front wheels are not slipping, so this friction force is unknown, but it is probably to the right. We must not forget to include the counterclockwise couple exerted by the drive train on the rear wheels and to satisfy Newton's third law for any forces that correspond to interactions between the chassis and the wheels.

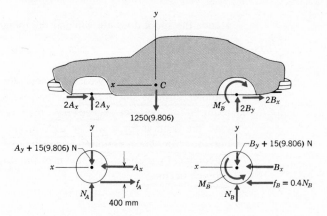

Note that the coordinate systems for each body are mutually parallel. The only given information not appearing in the free body diagram is that

$$\bar{\omega}_B = -600\left(\frac{2\pi}{60}\right)\bar{k} = -20\pi\bar{k} \text{ rad/s} \qquad \bar{\alpha}_B = \bar{0}$$

$$(k_z)_A = (k_z)_B = 300 \text{ mm}$$

Step 5 The centers of mass are all moving to the left at the same speed. Letting v denote that speed, we have

$$\bar{a}_A = \bar{a}_B = \bar{a}_G = \dot{v}\bar{\imath}$$

The chassis is translating, so

$$\bar{\alpha}_{chassis} \equiv \bar{0}$$

Finally, because the front wheels do not skid,

$$\bar{\alpha}_A = -\frac{\dot{v}}{R} \bar{k} = -\frac{\dot{v}}{0.4} \bar{k}$$

Step 6 The value of I_{zz} for the rotating wheels about their respective z axes is

$$I_{zz} = mk_z^2 = 15(0.3)^2 = 1.350 \text{ kg-m}^2$$

Step 7 Referring to the free body diagram for the forces, the equations of motion are

Chassis:

$$\Sigma F_x = -2A_x - 2B_x = m(a_G)_x = 1250\dot{v}$$

$$\Sigma F_y = 2A_y + 2B_y - 1250(9.806) = m(a_G)_y = 0$$

$$\Sigma M_{Cz} = 2A_y(1.0) - 2A_x(0.2) - 2B_y(1.2) - 2B_x(0.2) + 2M_B = 0$$

Wheel A:

$$\Sigma F_x = A_x - f_A = m_A(a_A)_x = 15\dot{v}$$

$$\Sigma F_y = N_A - A_y - 15(9.806) = m_A(a_A)_y = 0$$

$$\Sigma M_{Az} = -f_A(0.4) = I_{zz}\alpha_A = 1.350\left(-\frac{\dot{v}}{0.4}\right)$$

Wheel B:

$$\Sigma F_x = B_x + 0.4N_B = m_B(a_B)_x = 15\dot{v}$$

$$\Sigma F_y = N_B - B_y - 15(9.806) = m_B(a_B)_y = 0$$

$$\Sigma M_{Bz} = 0.4N_B(0.4) - M_B = I_{zz}\alpha_B = 0$$

Step 8 The nine equations in step 7 contain nine unknowns: A_x, A_y, B_x, B_y, M_B, f_A, N_A, N_B, and $\dot{v}$.

Step 9 The equations are not as complicated as they might appear. We can use the six equations governing the motion of the wheels to express N_A, f_A, N_B, M_B, A_x, and B_x in terms of A_y, B_y, and $\dot{v}$. Then subsituting these expressions into the equations for the chassis and solving gives

$$\dot{v} = 1.803 \text{ m/s}^2$$

The values of the forces are not pertinent to the problem, because we are interested only in the motion of the system. To determine the desired time inter-

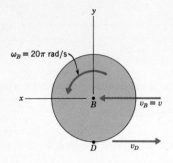

val, we examine the motion of the point of contact D between a rear wheel and the ground, as sketched to the left.

The kinematical formula for velocity gives

$$\bar{v}_D = -v_D\bar{i} = \bar{v}_B + \bar{\omega}_B \times \bar{r}_{D/B} = v\bar{i} - 0.4(20\pi)\bar{i}$$

$$v_D = 8\pi - v \text{ m/s}$$

Slipping ceases to occur when $v_D = 0$, for in that case there is no relative motion between the contact surfaces. Hence, for no slipping $v = 8\pi$ m/s. Then to determine time we use the result for $\dot{v}$ to obtain

$$v = 8\pi = \dot{v}t = 1.803t$$

$$t = 13.94 \text{ s}$$

Once the skidding ceases, it will not occur again as long as the controls of the vehicle are not changed.

HOMEWORK PROBLEMS

V.25   The truck is accelerating at 1 m/s² as the tailgate is held in the position shown by two cables (only one is shown in the side view). The tailgate has a mass of 15 kg. What is the tension force in each cable?

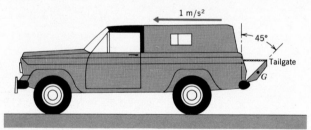

Prob. V.25

V.26 A person is trying to balance a stick in a tilted position. Determine the horizontal acceleration of this person's hand for which the angle θ is constant at 10°.

Prob. V.26

V.27   A homogeneous cylinder of mass m and radius r is pulled along a rough surface as shown by a truck whose acceleration is $g/10$. If the cylinder is sliding over the ground without rotating, determine the coefficient of friction and the force the truck applies to the two cables (only one is shown in the side view).

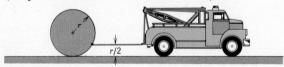

Prob. V.27

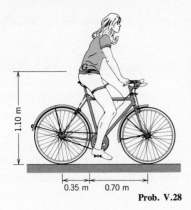

1.10 m

0.35 m   0.70 m

Prob. V.28

V.28   A cyclist moving at a speed of 20 km/hr applies the front wheel brakes only. The cyclist wishes to avoid being thrown over the handle bars. Determine the shortest safe stopping distance for which the rear wheel will not lose contact with the ground.

V.29   An automobile coasts freely down a 10% grade hill. Determine (a) the normal reactions between the road and tires, and (b) the acceleration of the 1500-kg vehicle. The mass of the wheels is negligible, as is rolling resistance.

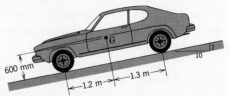

600 mm   1.2 m   1.3 m   10

Prob. V.29

V.30   The automobile in Problem V.29 is accelerated on a level road where the coefficient of static friction between the tires and car is 0.65. Determine the maximum acceleration possible if (a) it has rear wheel drive, (b) it has front wheel drive, (c) it has four wheel drive.

V.31   A 2-ton coal car is pulled up a ramp. Neglecting the mass and friction of the wheels, determine the minimum cable tension at which one set of wheels loses contact with the ramp. Also determine the corresponding acceleration of the coal car.

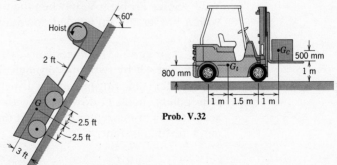

60°

Hoist

2 ft

G

2.5 ft

2.5 ft

3 ft

Prob. V.31

800 mm G_t G_c 500 mm

1 m

1 m   1.5 m   1 m

Prob. V.32

V.32   The 1000-kg forklift truck is carrying an 800-kg crate. The brakes are applied at all four wheels in order to bring the truck to rest from an initial speed of 5 m/s to the right. Determine the shortest distance in which the truck may be safely stopped without causing the truck to tilt forward or causing the crate to slide. The coefficient of static friction between the fork and the crate is 0.4.

V.33   The 100-kg refrigerator is placed on casters and then pushed across a floor by the 1000-N force shown. Determine the limiting values for the

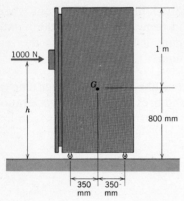

Prob. V.33

height h at which this force can be applied without causing the refrigerator to tip over. Friction is negligible.

V.34 Solve Problem V.33 for the case where the casters are locked and the coefficient of kinetic friction is 0.30.

V.35 In Problem V.33, replace the 1000-N force by a horizontal force $\bar{F}$ and consider the case where the casters are locked. The coefficient of friction for sliding between each caster and the ground is μ. Determine the range of values of the height h for which tipping will not occur. Is there any value of h for which the refrigerator will not tip over regardless of the magnitude of $\bar{F}$?

V.36 A truck traveling at a speed of 45 miles/hr brakes to a stop. Determine the shortest stopping distance for the truck, such that the 200-lb crate will neither tilt nor slip. The coefficient of friction between the crate and the truck is 0.25.

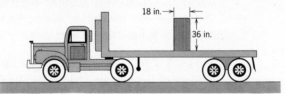

Prob. V.36

V.37 Solve Problem V.36 when the truck is initially (a) headed up a road with a 10% grade, (b) headed down a road with a 10% grade.

V.38 A 40-kg solid homogeneous cylinder is connected to the rigid vertical support AB by two links of negligible weight. In the position shown, the cylinder is moving upward at a speed of 6 m/s and the hydraulic piston D is pulling point C downward with a force of 700 N. Determine the acceleration of the center of mass of the cylinder and the reaction force at pin A for this instant.

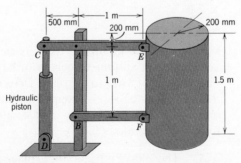

Prob. V.38

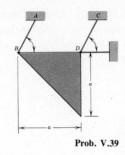

Prob. V.39

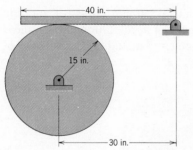

Prob. V.41

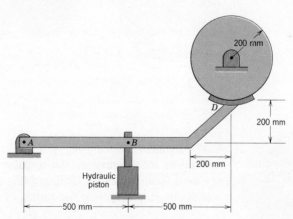

Prob. V.42

V.39 A 10-kg homogeneous plate is set into motion by severing cable *BE*. Determine the acceleration of point *D* and the tension in cables *AB* and *DC* at this instant.

V.40 The torque of an electric motor directly drives the platter of a record turntable. The platter weighs 6 lb and its radius of gyration is 10 in. What is the constant value of the torque *T* for which the platter attains a speed of $33\frac{1}{3}$ rev/min in 1 rev after starting from rest?

V.41 In Problem V.40, the torque of the electric motor varies with the angular speed ω according to the given graph. What is the value of the initial torque *T* for which the platter attains a speed of $33\frac{1}{3}$ rev/min in 1 rev after starting from rest?

V.42 The flywheel weighs 150 lb and has a radius of gyration of 10 in. It is rotating at 400 rev/min when the 100-lb beam is brought into contact with the rim. Determine how long it will take for the flywheel to come to rest, given that the coefficient of friction is 0.25.

V.43 The hydraulic cylinder *C* is used to brake the 800-kg flywheel whose radius of gyration is 160 mm. The coefficient of friction between the brake shoe *D* and the rim is 0.60. Determine the force that the hydraulic cylinder B must apply in order to bring the flywheel to rest in 25 rev from an initial angular velocity of 600 rev/min (a) clockwise, (b) counterclockwise.

Prob. V.43

V.44 A dynamometer is used to simulate the inertial resistance of an automobile. To do this the vehicle is held stationary and the drive wheels cause the drum of radius *R* to rotate clockwise. Letting m_a denote the mass of the automobile, derive an expression for the moment of inertia of the drum about its axis of rotation for which the tire traction required to impart an angular acceleration to the drum is the same as that required to actually accelerate the vehicle on level ground.

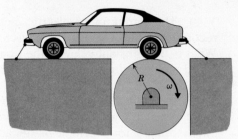

Prob. V.44

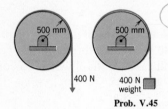

Prob. V.45

V.45 Determine the angular acceleration of the 100-kg pulley (a) if a 400-N force is applied to the end of the cable, (b) if a block whose weight is 400 N is suspended from the end of the cable.

V.46 The 20-kg double pulley shown has a 250-mm radius of gyration and is initially at rest. Determine (a) the angular acceleration of the pulley, (b) the angular velocity of the pulley when $t = 3$ s, and (c) the velocity of the 30-kg mass when it has moved 1 m.

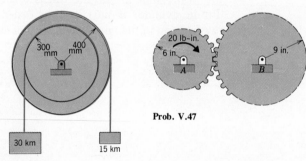

Prob. V.47

Prob. V.46

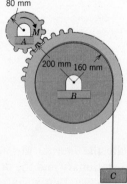

Prob. V.48

V.47 It takes 5 s for gear A, which has an external torque of 20 lb-in. applied to it, to increase the angular speed of gear B from 100 to 300 rev/min. Gear B weighs 45 lb and has a radius of gyration of 7 in. Gear A weighs 10 lb. Determine the radius of gyration of gear A.

V.48 A hoist consists of a motor driven gear A which rotates the larger stepped gear B. The masses of the gears are $m_A = 1$ kg, $m_B = 8$ kg, and their radii of gyration are $k_A = 60$ mm, $k_B = 150$ mm. A torque $\bar{M}$ is applied to gear A, with the result that the 12-kg block C is accelerated to a speed of 4 m/s from rest after rising 1 m. What is the value of M?

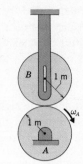

V.49 Disks A and B have identical mass m. Initially, disk A is rotating clockwise at a rate of 300 rev/min and disk B is at rest. Disk B is then gently released and brought into contact with disk A. The coefficient of friction between the two disks is 0.25, and both disks rotate freely about their shafts. Determine the time required for the disks to attain the same angular speed and thus cease to slip over each other.

Prob. V.49

V.50 In Problem V.49, a clockwise couple $\bar{M}$ is applied to disk A to maintain its angular speed at 300 rev/min. Determine (a) the required value of M, (b) the time required for the disks to cease to slip over each other.

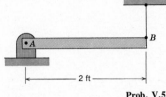

V.51 The homogeneous bar weighs 15 lb. The cord at end B breaks. Determine the angular acceleration of the bar and the reaction at the support A at this instant.

2 ft

Prob. V.51

V.52 Solve Problem V.51 for the case where the support at end A exerts a constant frictional couple of 2 lb-ft opposing the motion of the bar.

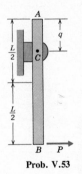

V.53 A horizontal force $\bar{P}$ is applied at end B of the uniform rod of mass m. The bar is initially at rest in the position shown. If the distance $q = L/4$, determine (a) the angular acceleration of the rod in this position, (b) the components of the reaction force at pin C for this position.

Prob. V.53

V.54 A horizontal force $\bar{P}$ is applied at end B of the uniform rod of mass m in Problem V.53. The bar is at rest in the position shown. Determine the value of q for which the horizontal component of the reaction force at pin C is zero in this position, and also determine the corresponding angular acceleration.

V.55 The equilateral triangular plate of mass m pivots about pin A. (a) Derive a differential equation of motion for the angle θ for the triangular plate. (b) The angle ϕ for the simple pendulum formed from a cable and a particle of mass m shown obeys the same equation of motion as that found in part (a). What is the length l of the cable?

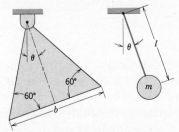

Prob. V.55

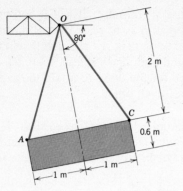

Prob. V.56

V.56   A 200-kg homogeneous block is suspended by cables OA and OC which are attached to the stationary crane at point O. In the position shown, the box is descending and the center of mass has a speed of 2 m/s. Determine the acceleration of the center of mass in this position and the tensile forces in the cables.

V.57   The acceleration of plate A is 12 ft/s² to the right and the acceleration of plate B is 4 ft/s² to the left. Assuming that the surfaces are not slipping over each other, determine the friction forces exerted by the plates on the 20-lb cylinder C.

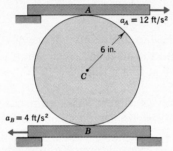

Prob. V.57

V.58   Racks A and B, having mass of 30 kg and 60 kg, respectively, are constrained by their smooth guides to move horizontally. The stepped gear has a mass of 80 kg and a radius of gyration of 500 mm. A torque of 5 kN-m is applied clockwise to the gear. Determine the acceleration of the center of the gear and of the racks.

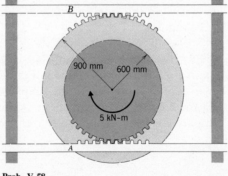

Prob. V.58

V.59   The 300-kg gear and drum assembly has a radius of gyration of 700 mm. The 1500-N force causes the gear to roll on the 100-kg rack, with negligible friction between the rack and the horizontal surface. Determine (a) the tension in the cable, (b) the length of cable wound on or off the drum in 5 s starting from rest.

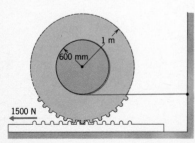

Prob. V.59

V.60 The 4-lb stepped pulley in the sketch is pinned at its center O to the 8-lb sliding block. The tension forces in the cables are $T_A = 10$ lb, $T_B = 5$ lb. The radius of gyration of the pulley is 6 in., and friction is negligible. Determine the angular acceleration of the pulley and the acceleration of the block.

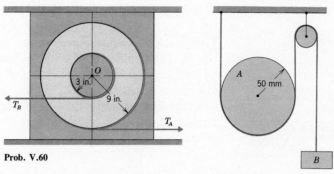

Prob. V.60

Prob. V.61

V.61 The 30-kg wheel A has a radius of gyration of 40 mm. Determine (a) the acceleration of the center of the wheel when the 20-kg counterweight B is released from rest, (b) the speed of the counterweight after it has fallen 2 m.

V.62 The stepped pulley B weighs 10 lb and its radius of gyration is 8 in., whereas both blocks weigh 20 lb. Determine the velocity of block A 2 s after the system is released from rest. The coefficient of friction between block C and the horizontal surface is 0.3.

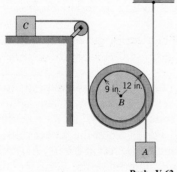

Prob. V.62

V.63 The 60-kg drum and axle assembly has a 300-mm radius of gyration. The counterweight has a mass of 40 kg. Determine (a) the acceleration of the axle assembly and of the counterweight, (b) the velocity of the counterweight 2 s after it is released from rest.

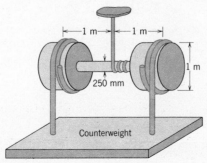

Prob. V.63

V.64   Solve Problem V.63 if the counterweight has a mass of 10 kg and all other parameters are unchanged.

V.65   A uniform disk of mass m and radius r rolls without slipping down the incline as shown. Determine (a) the acceleration of the center of the disk, (b) the value of the coefficient of friction μ corresponding to the case of impending slippage.

Prob. V.65

V.66   A 1000-kg thin pipe rests on the flatbed of a truck in the position shown. The coefficient of static friction between the pipe and the horizontal surface is 0.30. The truck accelerates from rest. Determine (a) the maximum acceleration the truck can have for which the pipe will roll without slipping until it falls off the back of the truck, (b) for the acceleration found in part (a), the distance the truck will have traveled when the pipe falls off.

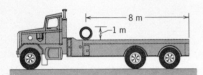

Prob. V.66

V.67   The 2.5-kg homogeneous cylinder rolls without slipping down the curved surface. In the position where $\theta = 36.87°$, the speed of the mass center is 4 m/s. Determine the acceleration of the cylinder in this position and the components of the reaction forces acting on the cylinder.

V.68   Derive the differential equation of motion for the angle θ for the 2.5 kg homogeneous cylinder in Problem V.67.

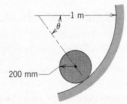

Prob. V.67

V.69   A 5-lb wheel of 8-in. radius was imperfectly manufactured, so its mass center G is 2 in. from its geometric center C, as shown. The radius of gyration with respect to an axis through the center C is 7 in. As a result of the imperfection, the velocity of point C varies when the wheel rolls. In the position where $\theta = 0$, point C is moving to the left at 2 in./s. Assuming that the wheel is rolling without slipping, determine the corresponding angular acceleration of the wheel and the acceleration of point C.

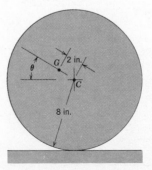

Prob. V.69

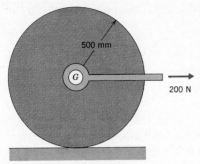

500 mm

G

200 N

Prob. V.71

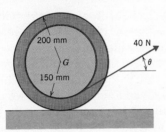

200 mm

G

150 mm

40 N

θ

Prob. V.73

V.70 Solve Problem V.69 if the given velocity for point C corresponds to the position where $\theta = 30°$.

V.71 The 100-kg roller is initially at rest and is acted upon by a 200-N force as shown. The body rolls without slipping. Determine (a) the velocity of the mass center after 6 s, (b) the friction force required to prevent slipping.

V.72 Solve Illustrative Problem 5 for a wire tension of 60 lbs.

V.73 The stepped pulley has a mass of 10 kg and a radius of gyration of 180 mm, and the coefficients of friction between the pulley and the horizontal surface are $\mu_s = 0.40$, $\mu_k = 0.35$. The angle θ is held constant as the end of the cable is pulled by a 40-N force. Determine the acceleration of the center of mass G and the angular acceleration of the disk if $\theta = 90°$.

V.74 Solve Problem V.73 (a) if $\theta = 75°$, (b) if $\theta = 45°$.

V.75 When the thin hoop of radius R and weight W was placed on the floor, its center O was at rest and it was rotating at an angular velocity $\bar{\omega}_1$. Letting the coefficient of kinetic friction be μ_k, derive expressions for (a) the speed of point O at the instant when the hoop ceases to slip relative to the ground, (b) the time duration during which slipping occurs.

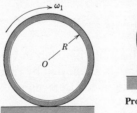

ω_1

R

O

O

25 ft/s

Prob. V.76

Prob. V.75

V.76 When projected onto the floor, the 16-lb bowling ball is moving horizontally at 25 ft/s and is not rotating. The coefficient of kinetic friction is 0.20. Determine (a) the speed of the ball when it begins to roll on the ground without slipping, (b) the distance traveled and the elapsed time for the slipping phase of the motion. Regard the bowling ball as a homogeneous 1-ft-diameter sphere.

V.77 A 12-kg uniform rod is released from rest in the position shown. Determine the required coefficient of static friction to prevent slipping at

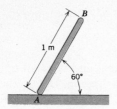

Prob. V.77

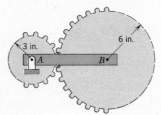

Prob. V.79

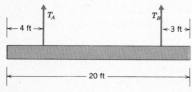

Prob. V.80

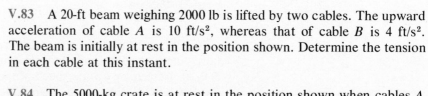

Prob. V.83

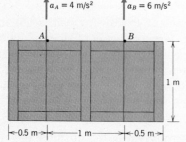

Prob. V.84

point A. In which direction would point A slip if the actual coefficient of friction were less than this value?

V.78 The coefficients of static and kinetic friction between the 12-kg uniform bar in Problem V.77 and the ground are $\mu_s = 0.27$, $\mu_k = 0.25$. Determine the angular acceleration of the rod.

V.79 The semicylinder of mass m and radius R is released from rest in the position shown. Determine (a) the minimum value of the coefficient of friction for which the semicylinder will roll without slipping, (b) the corresponding initial angular acceleration.

V.80 Gear A is stationary. Gear B weighs 10 lb and has a radius of gyration of 5 in. The uniform bar AB weighs 3 lb. If the system is released from rest in the position shown, determine (a) the initial angular acceleration of the bar, and (b) the initial acceleration of the mass center of gear B.

V.81 The 250-kg marble slab is being moved by placing it on two cylindrical rollers and applying the 150-N towing force. Each cylinder has a mass of 25 kg. Determine the acceleration of the marble slab. Assume that there is no slippage.

Prob. V.81

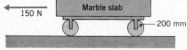
Prob. V.82

V.82 The 250-kg marble slab is being moved by placing it on a lightweight frame that is supported by two 25-kg cylinders, as shown. A 150-N towing force is applied. Determine the acceleration of the slab.

V.83 A 20-ft beam weighing 2000 lb is lifted by two cables. The upward acceleration of cable A is 10 ft/s², whereas that of cable B is 4 ft/s². The beam is initially at rest in the position shown. Determine the tension in each cable at this instant.

V.84 The 5000-kg crate is at rest in the position shown when cables A and B are given upward accelerations of 4 m/s² and 6 m/s², respectively. Determine the tension in each cable. The crate may be regarded as a homogeneous body.

V.85 To correct the course of a missile, it is proposed to increase the thrust in engine A. The mass of the missile is (10^5) kg and each engine has a thrust of $2(10^3)$ kN when the increase is to occur. It is desired to rotate the missile by 1° in 1 s. Determine (a) the required increase in the thrust of engine A, (b) the acceleration of the center of mass G immediately after the thrust of engine A is increased. The missile may be approximated as a thin uniform bar 60 m long.

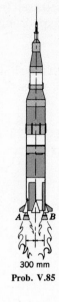

300 mm

Prob. V.85

V.86 and V.87 The 4-kg homogeneous plate, which is suspended as shown, is initially at rest. Determine the initial acceleration of the mass center when the connection at point B breaks suddenly.

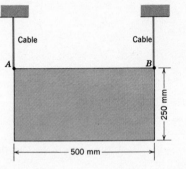

Prob. V.86

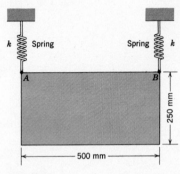

Prob. V.87

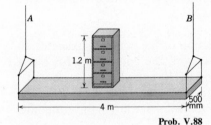

Prob. V.88

V.88 An 80-kg file cabinet is placed on the 80-kg scaffold. Cable B suddenly breaks when the system is at rest. Assuming that the cabinet does not move relative to the scaffold, determine the initial acceleration of the cabinet when it is (a) at the center of the scaffold, (b) at end B. Consider the cabinet to be a slender bar and the scaffold to be a uniform plate.

V.89 Point A on the 20-lb bracket has a constant speed of 10 in./s. For this position determine the tension in the cables and the couple $\bar{M}_G$ required for the motion.

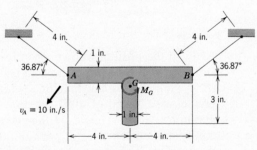

Prob. V.89

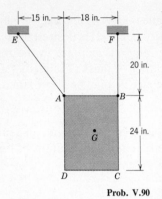

Prob. V.90

V.90 The 20-lb homogeneous plate is released from rest in the position shown. Determine (a) the tension in the supporting cables AE and BF, (b) the acceleration of the center of mass.

V.91 Solve Problem V.90 if, in the position shown, the plate is moving to the left and the speed of the center of mass G is 5 ft/s.

V.92 The 500-kg box is released from rest in the position shown. The coefficient of friction between all surfaces is 0.20. Determine (a) the angular acceleration of the crate and (b) the reactions at A and B.

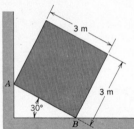

Prob. V.92

V.93 Solve Problem V.92 if in the position shown point B is moving to the right at 1 m/s.

V.94 The 100-g rigid link ABC is a portion of the control linkage of an automobile carburetor. At the instant shown, the control rod BE has a velocity of 20 mm/s and an acceleration of 10 mm/s², both to the left. Determine the value of the force F that is pushing control rod BE. The mass of both control rods is negligible and the linkage lies in the horizontal plane.

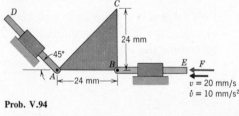

Prob. V.94

Prob. V.95

V.95 In the position shown, the power stroke of a one-cylinder engine applies a 5000-N force to the 1-kg piston head. The 2-kg connecting rod AB is pinned to crankshaft BC, which is rotating at a constant counterclockwise speed of 4000 rev/min. For the position where $\theta = 90°$, determine (a) the couple $\bar{M}$ that is required to hold the speed of the crankshaft constant for this position, (b) the reactions at pins A and B for this position. The linkage lies in the horizontal plane and the mass of the piston is negligible.

V.96 Solve Problem V.95 for $\theta = 45°$.

V.97 In the position shown, point A is moving to the left at 3 m/s and decelerating at 6 m/s². The mass of the wheel is 10 kg and its radius of gyration about point A is 80 mm. What is the value of the force F acting on the piston C if the mass of the piston and bar BC are negligible? Consider point B to be on the circumference of the wheel and assume that the wheel does not slip.

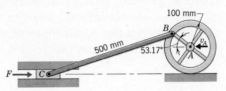

Prob. V.97

V.98 Solve Problem V.97 if the piston has a mass of 1 kg and the mass of bar BC is 3 kg.

V.99 The flywheel is rotated clockwise at a constant angular speed ω by a torque $\bar{M}$. The mass of each bar is m. Determine the required value of M for the instant shown.

V.100 Two identical 80-kg cylinders are interconnected by rod AC whose mass is 10 kg. The system is at rest in the position shown when the 400-N force is applied at point A. Determine the acceleration of each cylinder at this instant. Assume that there is no slipping.

V.101 Solve Problem V.100 if cylinder A is moving to the left at 5 m/s in the position shown when the 400-N force is applied.

V.102 The A-frame shown consists of two identical 100-lb bars that are 6 ft long. Cable AB suddenly breaks. Determine (a) the initial angular acceleration for each member and, (b) the initial reactions at points C and E.

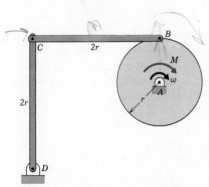

Prob. V.99

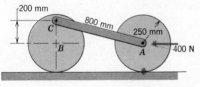

Prob. V.100

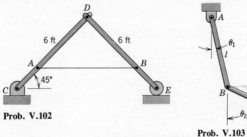

Prob. V.102

Prob. V.103

V.103 Bars AB and BC have the same mass m and length l. Derive two differential equations of motion relating the value of θ_1 and θ_2 and their derivatives to the mass m and length l.

E. ENERGY AND MOMENTUM PRINCIPLES

Our study of energy and momentum principles for particle motion showed that these principles greatly facilitate problem solving for certain classes of problems, such as those involving conservative or impulsive forces. These principles are, in essence, special forms of the differential equations that arise when using Newton's second law.

In this section we shall adapt the energy and momentum principles for a system of particles, derived in Module Three, to the special case where the particles form a rigid body in planar motion. As in the previous sections, we shall restrict the study to the case of bodies that have a plane of symmetry parallel to the plane of motion, reserving the more general case for Module Seven on three-dimensional motion.

1 Work and Energy

Recalling the symbolism for the work-energy formulation for a system of particles, T is the total kinetic energy of the system, V is the potential energy of all conservative forces acting on the system, and $U_{1 \to 2}^{\text{(nc)}}$ is the total work done by the nonconservative forces acting on the system. In terms of these symbols, the work-energy equation is

$$T_2 + V_2 = T_1 + V_1 + U_{1 \to 2}^{\text{(nc)}} \tag{24}$$

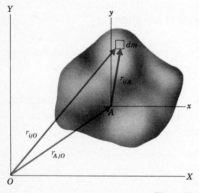

Figure 12

Let us begin the study of this equation by formulating the kinetic energy for a rigid body. In Figure 12 a rigid body is modeled as a continuum of infinitesimal particles dm whose position with respect to the origin of a fixed reference frame is $\bar{r}_{i/O}$. Denoting the velocity of each particle as $\bar{v}_i \equiv \dot{\bar{r}}_{i/O}$, the kinetic energy of the body, which is the sum of the kinetic energy of all the particles in the system, may be written as an integral over the volume of the body.

$$T = \tfrac{1}{2} \int_{\mathcal{V}} \bar{v}_i \cdot \bar{v}_i \, dm \tag{25}$$

where the mass m is a function of the volume $\mathcal{V}$ occupied by the body. Our objective now is to place equation (25) in a more useful form for computational purposes. We begin by choosing a translating reference frame xyz whose origin coincides with an arbitrary point A in the body. Then, using the vectors shown in Figure 12, for a rigid body,

$$\bar{r}_{i/O} = \bar{r}_{A/O} + \bar{r}_{i/A}$$

$$\bar{v}_i = \bar{v}_A + \bar{\omega} \times \bar{r}_{i/A}$$

Hence, equation (25) becomes

$$T = \tfrac{1}{2} \int_{\mathcal{V}} [(\bar{v}_A + \bar{\omega} \times \bar{r}_{i/A}) \cdot (\bar{v}_A + \bar{\omega} \times \bar{r}_{i/A})] \, dm$$

$$= \tfrac{1}{2} \int_{\mathcal{V}} [\bar{v}_A \cdot \bar{v}_A + 2\bar{v}_A \cdot (\bar{\omega} \times \bar{r}_{i/A})$$

$$+ (\bar{\omega} \times \bar{r}_{i/A}) \cdot (\bar{\omega} \times \bar{r}_{i/A})] \, dm$$

$$= \tfrac{1}{2}(v_A)^2 \int_{\mathcal{V}} dm + \bar{v}_A \cdot \left[\bar{\omega} \times \int_{\mathcal{V}} \bar{r}_{i/A} \, dm \right]$$

$$+ \tfrac{1}{2} \int_{\mathcal{V}} [\bar{\omega} \times \bar{r}_{i/A}) \cdot (\bar{\omega} \times \bar{r}_{i/A})] \, dm \qquad (26)$$

The first integral in equation (26) gives the mass m of the rigid body. To investigate the meaning of the other two integrals, let us resolve $\bar{r}_{i/A}$ into its components. Referring once again to Figure 12,

$$\bar{r}_{i/A} = x\bar{\imath} + y\bar{\jmath}$$

Also, for planar motion, $\bar{\omega} = \omega\bar{k}$, so the second and third integrals in equation (26) now become

$$\int_{\mathcal{V}} \bar{r}_{i/A} \, dm = \int_{\mathcal{V}} x \, dm \, \bar{\imath} + \int_{\mathcal{V}} y \, dm \, \bar{\jmath}$$

$$\int_{\mathcal{V}} [(\bar{\omega} \times \bar{r}_{i/A}) \cdot (\bar{\omega} \times \bar{r}_{i/A})] \, dm$$

$$= \omega^2 \int_{\mathcal{V}} [(x\bar{\jmath} - y\bar{\imath}) \cdot (x\bar{\jmath} - y\bar{\imath})] \, dm$$

$$= \omega^2 \int_{\mathcal{V}} (x^2 + y^2) \, dm$$

The integrals occuring in the first equation are the first moments of mass. Hence, referring to equations (15), this term becomes

$$\int_{\mathcal{V}} \bar{r}_{i/A} \, dm = mx_G \, \bar{\imath} + my_G \, \bar{\jmath} = m\bar{r}_{G/A}$$

Next, we note from equation (12) that the last integral from equation (26) is the moment of inertia. Hence

$$\int_{\mathcal{V}} [(\bar{\omega} \times \bar{r}_{i/A}) \cdot (\bar{\omega} \times \bar{r}_{i/A})] \, dm = \omega^2 I_{zz}$$

Substituting these results into equation (26) yields

$$T = \tfrac{1}{2}mv_A{}^2 + \bar{v}_A \cdot (\bar{\omega} \times m\bar{r}_{G/A}) + \tfrac{1}{2}I_{zz}\omega^2 \qquad (27)$$

This is the result when point A is arbitrarily chosen. The formula is considerably simplified when the *origin A is located at the center of mass G,* in which case $\bar{r}_{G/A} = \bar{0}$ and equation (27) reduces to

$$T = \tfrac{1}{2}mv_G{}^2 + \tfrac{1}{2}I_{zz}\omega^2 \qquad (28a)$$

In effect, when using equation (28a), we are considering the motion of the body to be a superposition of a translation following the center of mass G plus a rotation about point G. In the language of dynamics, the kinetic energy of translation is $\tfrac{1}{2}mv_G{}^2$ and the kinetic energy of rotation is $\tfrac{1}{2}I_{zz}\omega^2$. Clearly, the latter term is zero for a translating body and for motion of a particle.

We shall use equation (28a) to describe rigid bodies that are in translation or in general motion. The third type of planar motion is a pure rotation about a fixed axis. For such cases we *locate the origin A at the fixed axis.* Then, setting $\bar{v}_A = 0$ in equation (27) yields

$$T = \tfrac{1}{2}I_{zz}\omega^2 \qquad (28b)$$

This shows that the body has only rotational kinetic energy for a coordinate system whose origin coincides with the axis of rotation.

Having now derived the formula for the kinetic energy of a rigid body, let us examine the other terms of the work-energy principle, equation (24). The potential energy V is determined in the same manner as was done for particles. Each conservative force may be considered individually, as their potential energies are additive. We may also use this procedure for the nonconservative forces, tracing the path of the point of application of each force and computing the work of each force individually.

However, in some situations the points of application may not have simple motions. Let us derive an alternative method for computing the work term for such cases.

Consider the force $\bar{F}$ in Figure 13 which is applied at point P in the rigid body. The work done by the force $\bar{F}$ in changing the position of point P with respect to the fixed origin is

$$U_{1 \to 2} = \int_{(r_{P/O})_1}^{(r_{P/O})_2} \bar{F} \cdot d\bar{r}_{P/O}$$

where the limits in the integral denote the coordinates of the initial and final positions of point P.

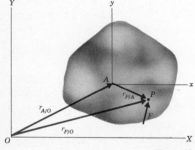

Figure 13

We may now relate the motion of point P to that of the origin A.

Using the kinematic formula relating the velocity of two points in a rigid body gives

$$\bar{v}_P \equiv \frac{d\bar{r}_{P/O}}{dt} = \bar{v}_A + \bar{\omega} \times \bar{r}_{P/A} \equiv \frac{d\bar{r}_{A/O}}{dt} + \frac{d\bar{\theta}}{dt} \times \bar{r}_{P/A}$$

Upon multiplying both sides by dt, we obtain

$$d\bar{r}_{P/O} = d\bar{r}_{A/O} + d\bar{\theta} \times \bar{r}_{P/A} \tag{29}$$

This expression shows that an infinitesimal displacement of point A causes point P to displace in the same manner, whereas an infinitesimal rotation $d\bar{\theta}$ causes P to move perpendicular to the line from point A to point P, as indicated by the cross product term.

We now substitute equation (29) into the expression for the work to obtain

$$U_{1 \to 2} = \int_{(r_{P/O})_1}^{(r_{P/O})_2} \bar{F} \cdot (d\bar{r}_{A/O} + d\bar{\theta} \times \bar{r}_{P/A})$$

Employing the readily verified identity

$$\bar{F} \cdot (d\bar{\theta} \times \bar{r}_{P/A}) \equiv (\bar{r}_{P/A} \times \bar{F}) \cdot d\bar{\theta}$$

we break the expression for the work into two terms

$$U_{1 \to 2} = \int_{(r_{A/O})_1}^{(r_{A/O})_2} \bar{F} \cdot d\bar{r}_{A/O} + \int_{\theta_1}^{\theta_2} (\bar{r}_{P/A} \times \bar{F}) \cdot d\bar{\theta} \tag{30}$$

Recall from statics that the given force $\bar{F}$ acting at point P is equivalent to a force $\bar{F}$ acting at point A and a couple $\bar{M}_A$ that is the moment of the given force about point A; $\bar{M}_A = \bar{r}_{P/A} \times \bar{F}$. Equation (30) is a direct consequence of this fact; the first term gives the work done by the equivalent force in translating the body and the second term gives the work done by the equivalent couple in rotating the body. Therefore, if we are dealing with a system of nonconservative forces, we may resolve the forces into their equivalent force and couple, $\Sigma \bar{F}^{(\text{nc})}$ and $\Sigma \bar{M}_A^{(\text{nc})}$. The work done is then given by

$$U_{1 \to 2}^{(\text{nc})} = \int_{(r_{A/O})_1}^{(r_{A/O})_2} \Sigma \bar{F}^{(\text{nc})} \cdot d\bar{r}_{A/O} + \int_{\theta_1}^{\theta_2} \Sigma \bar{M}_A^{(\text{nc})} \cdot d\bar{\theta} \tag{31}$$

An important consequence of equation (31) is that the forces internal to a rigid body do no work when the body moves. This is so because

Newton's third law tells us that all internal forces cancel each other in computing the total force and total moment.

In summary, we have found that the work-energy principle, equation (24), can be applied in conjunction with either of the kinetic energy expressions of equations (28) to relate the angular and linear velocity of a rigid body when it occupies different positions. The work of the nonconservative forces can be determined either by the same methods as that used in particle motion, or alternatively when it is difficult to follow the point of application of the force, by applying equation (30) or (31). Equally important, in those cases where there is *any* uncertainty as to whether a force is conservative or nonconservative, it is usually best to treat it as a nonconservative force, rather than pausing to determine whether it is indeed conservative. Remember that the concept of the potential energy of a conservative force was derived in Module Three merely as an aid in determining the work done by the force.

Finally, recall that it is sometimes impossible to evaluate the work done by a force until the motion of the system has been determined. In such situations we resort to formulating and solving the differential equations of motion.

EXAMPLE 5

A 200-N force is applied to the free end of cable wrapped around the drum of the stepped disk, as shown. The disk has a mass of 40 kg and a radius of gyration of 150 mm. Knowing that the disk starts from rest and rolls without slipping, determine its speed after moving 5 m up the incline.

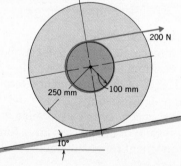

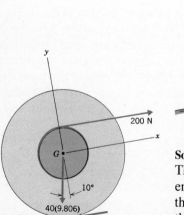

Solution

The key to recognizing that this problem is ideally suited for the work-energy principle is that the speed is required at a known position. Because the ultimate success of this approach requires that we be able to evaluate the work term, we begin with a free body diagram to assist in this determination.

The origin of the *xyz* coordinate system is located at the mass center *G* because the disk is in general motion. The weight force is a conservative force, so we shall account for it in the potential energy term, although it would not be incorrect to ignore the conservative nature of this force. We shall determine $U_{1\rightarrow 2}^{(\text{nc})}$ by applying equation (31), because the remaining forces are applied to different points of the body at each instant; that is, different points of the disk come into contact with the ground and the 200-N force as the disk rolls. Hence, from the free body diagram we have

$$\Sigma \, \bar{F}^{(\text{nc})} = (200 - f)\bar{\imath} + N\bar{\jmath} \text{ N}$$

$$\Sigma \, \bar{M}_C^{(\text{nc})} = [-f(0.25) - 200(0.10)]\bar{k} \text{ N-m}$$

The negative signs in the moment arise because clockwise moments are the direction of the negative *z* axis, according to the right-hand rule.

In order to compute the work, we next describe the differential displacement of point *G* and the differential rotation. For this system point *G* moves parallel to the incline in the *x* direction. Letting *ds* be the increment of distance traveled,

$$d\bar{r}_{G/O} = ds \, \bar{\imath}$$

The disk does not slip. Hence, this displacement will cause the disk to rotate clockwise by an amount $ds/R = ds/0.25$ rad, as sketched here. Hence

$$d\bar{\theta} = -\frac{ds}{0.25}\bar{k}$$

The initial value of *s* is zero and we want the velocity when $s = 5$ m. Therefore, equation (31) for the work gives

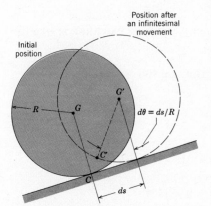

Initial position

Position after an infinitesimal movement

$d\theta = ds/R$

$$U_{1\rightarrow 2}^{(\text{nc})} = \int_{(r_{G/O})_1}^{(r_{G/O})_2} \Sigma \, \bar{F}^{(\text{nc})} \cdot d\bar{r}_{G/O} + \int_{\theta_1}^{\theta_2} \Sigma \, \bar{M}_G^{(\text{nc})} \cdot d\bar{\theta}$$

$$= \int_{s=0}^{s=5} [(200 - f)\bar{\imath} + N\bar{\jmath}] \cdot ds \, \bar{\imath}$$

$$- \int_{s=0}^{s=5} (0.25f + 20)\bar{k} \cdot (-\frac{ds}{0.25}\bar{k})$$

$$= \int_0^5 (200 - f) \, ds + \int_0^5 (f + 80) \, ds$$

$$= (200 - f)5 + (f + 80)5 = 1400 \text{ J}$$

There are several noteworthy aspects to this result. That the normal force does no work is not surprising, because it is perpendicular to the direction of motion. However, the fact that the friction force does no work at first seems to conflict with the notion that friction retards motion. The friction force in this system does no work because the contact surfaces are not moving relative to each other. However, friction does slow the disk in that it tends to make the disk rotate as well as translate. Finally, note that the work done by the cable could have (alternatively) been obtained by multiplying the cable tension by the distance the end of the cable moves.

The solution can now be completed by writing the initial and final values of the kinetic and potential energies. The disk starts from rest, so choosing the initial elevation as the datum for gravitational potential energy gives

$$T_1 = 0 \qquad V_1 = 0$$

In the final position the center of mass has a speed of v and the no slip condition gives $\omega = v/R = v/0.25$. From equation (28a) and the given value for the radius of gyration, we write

$$T_2 = \tfrac{1}{2}mv^2 + \tfrac{1}{2}I_{zz}\omega^2 = \tfrac{1}{2}(40)v^2 + \tfrac{1}{2}(40)(0.15)^2\left(\frac{v}{0.25}\right)^2$$

$$= 27.20v^2$$

When the disk moves 5 m up the incline, its vertical elevation above the datum is 5 sin 10°, so

$$V_2 = mgh = 40(9.806)(5 \sin 10°)$$

$$= 340.6 \text{ J}$$

Hence, equation (24) gives

$$T_2 + V_2 = T_1 + V_1 + U^{(nc)}_{1\rightarrow2}$$

$$27.20v^2 + 340.6 = 0 + 0 + 1400$$

$$v = 6.24 \text{ m/s}$$

EXAMPLE 6

The system shown in the sketch is used to lower a 100-lb homogeneous package. The horizontal bar weighs 25 lb and the other bars weigh 10 lb. The spring has a stiffness of 50 lb/in. and an unstretched length of 9 in. When $\theta = 53.13°$ the package is descending with a speed of 10 ft/s. What constant horizontal force $\bar{F}$ is required to bring the system to rest at $\theta = 0°$?

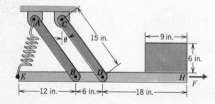

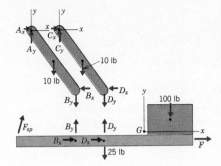

Solution

We must find the value of F that produces a specific change in the velocity of the system between the initial and final positions. This is the key that suggests employing the work-energy principle. As an aid in understanding the role of the forces, we draw a free body diagram of each body, taking care that interaction forces between bodies satisfy Newton's third law. The bar EH and the package move in unison, and thus they may be treated as one body.

Note that we selected the coordinate systems for the bodies to be parallel to each other. The origins of the coordinate systems were chosen to coincide with the fixed axes of rotation for bars AB and CD, and with the center of mass of the combination of the package and bar EH.

The weight and spring forces are conservative. The reactions at points A and C are applied at fixed points, so they do no work. The work done by each interaction force in moving one body is canceled by the work done by its counterpart in moving the other body by the same amount. Hence, the only nonconservative force doing work in moving the system is $\bar{F}$. The work done by this force, and also the system's potential energy, is determined with the aid of sketches of the system in the initial and final positions.

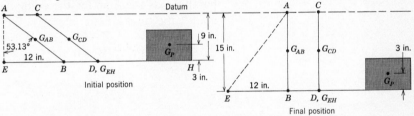

The sketches show that bar EH moves to the left by 12 in. in the motion. Because $\bar{F} = F\bar{\imath}$, it follows that

$$U^{(nc)}_{1\rightarrow 2} = -F(12.0) \text{ in.-lb}$$

Let us choose the elevation of pins A and C as the datum for gravity. In the initial position the centers of mass of bars AB and CD are 4.5 in. below the datum, whereas the corresponding points for bar EH and the package are 9 in. and 6 in. below the datum, respectively. The length of the spring connecting points A and E is 9 in., so the spring is unstretched. Thus

$$V_1 = (V_{\text{gravity}} + V_{\text{spring}})_1 = 10(-4.5) + 10(-4.5)$$

$$+ 100(-6) + 25(-9) + 0 = -915.0 \text{ in.-lb}$$

In the final position the length of the spring is $\sqrt{(15)^2 + (12)^2} = 19.209$ in. Referring to the sketch for the locations of the center of mass, we have

$$V_2 = (V_{\text{gravity}} + V_{\text{spring}})_2 = 10(-7.5) + 10(-7.5)$$
$$+ 100(-12) + 25(-15) + \tfrac{1}{2}(50)(19.209 - 9.0)^2$$
$$= 880.6 \text{ in.-lb}$$

The last terms we require are those for the kinetic energy. We want the system to come to rest at $\theta = 0$. Hence, the value of T_2 is zero. Before we can determine T_1, the velocities of the bars must be related to the given initial speed for the package, for which a velocity sketch is now required. Remember that bar EH is translating.

From this sketch we see that

$$v_D = v_B = v_{\text{package}} = 120 \text{ in./s}$$

$$\omega_{AB} = \omega_{CD} = \frac{v_B}{15} = 8 \text{ rad/s}$$

The coordinate systems in the free body diagram require that we use equation (28a) for the kinetic energy of the translating body and equation (28b) for the rotating bars. With the assistance of Appendix B and the parallel axis theorem for the computation of I_{zz} with respect to points A and C, the total kinetic energy is found to be

$$T_1 = \tfrac{1}{2}(I_{zz})_{AB}\,\omega_{AB}{}^2 + \tfrac{1}{2}(I_{zz})_{CD}\,\omega_{CD}{}^2 + \tfrac{1}{2}(m_{EH} + m_{\text{package}})v_P{}^2$$

$$= \tfrac{1}{2}(2)\tfrac{1}{3}\left[\frac{10}{32.17(12)}(15)^2\right](8)^2 + \tfrac{1}{2}\left[\frac{125}{32.17(12)}\right](120)^2 = 2456 \text{ in.-lb}$$

Finally, we substitute all terms into the work-energy principle and solve for F.

$$T_2 + V_2 = T_1 + V_1 + U^{(\text{nc})}_{1 \to 2}$$

$$0 + 730.6 = 2456 - 1065.0 - 12F$$

$$F = 55.03 \text{ lb}$$

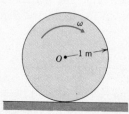

Prob. V.104

HOMEWORK PROBLEMS

V.104 The 50-kg cylinder is slipping as it rolls. It has a clockwise angular velocity of 120 rev/min and its center O has a speed of 54 km/hr. Determine the cylinder's kinetic energy when (a) the translational velocity is to the left, (b) the translational velocity is to the right.

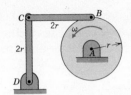

Prob. V.105

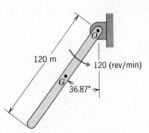

Prob. V.107

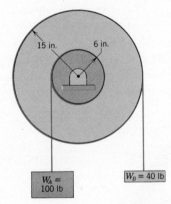

Prob. V.109

V.105 The mass of each of the uniform bars is m and that of the flywheel disk is $8\,m$. In the position shown the disk is rotating at angular speed ω counterclockwise. Determine the kinetic energy of each bar and of the disk.

V.106 For the mechanism of Problem V.105, derive an expression for the couple $\bar{M}$ applied to the disk that will bring the system to rest in 1 rev of the disk.

V.107 In the position shown, the 5-kg uniform rod is rotating counterclockwise with an angular speed of 120 rev/min. Determine the kinetic energy of the bar (a) using equation (28a), (b) using equation (28b).

V.108 For the initial motion of the bar given in Problem V.107, determine the reactions at pin O when the bar reaches the vertical position.

V.109 The 50-lb pulley has a radius of gyration of 12 in. and is rotating at $\omega = 240$ rev/min clockwise. Determine the total kinetic energy of the system.

V.110 Initially ω for the system in Problem V.109 is zero and the system is in static equilibrium. While in this position a 10-lb mass is added to block B. Determine the velocity of each block when the pulley has made 2 rev.

V.111 A 2000-kg turbine engine has a radius of gyration of 1 m. When shut down while operating at 1500 rev/min, it requires 4000 rev to come to rest. Determine the average magnitude of the frictional torque caused by the bearings and air resistance.

V.112 The 10-kg cylinder B is placed in contact with a conveyor belt traveling at 40 m/s. The cylinder is initially at rest and the coefficient of friction between the cylinder and belt is 0.25. Determine the number of revolutions required for the cylinder to attain a constant angular velocity.

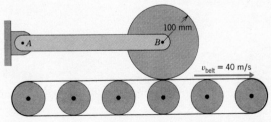

Prob. V.112

V.113 By pushing on point B with a 1000-N force, the hydraulic piston

C brakes the 800-kg flywheel whose radius of gyration is 150 mm. Initially the flywheel is rotating at 550 rev/min clockwise. The coefficient of friction between the rim and brake shoe *D* is 0.60. Determine the number of revolutions required for the flywheel to stop.

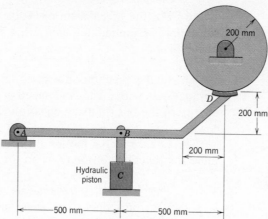

Prob. V.113

V.114 The identical inner gears *A* and *B* each have a mass of 2 kg and a radius of gyration of 100 mm. The radius of gyration of the 4-kg outer gear is 400 mm. A torque of 200 N-m is applied to gear *A*. Determine the angular velocity of gear *B* when gear *C* has made 1 rev starting from rest.

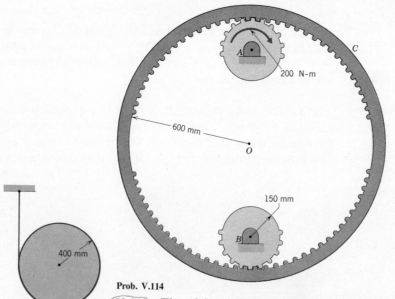

Prob. V.114

Prob. V.115

V.115 The 10-kg cylinder is released from rest. Determine the velocity of its mass center after it has made 2 rev.

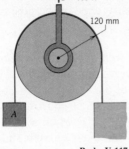

Prob. V.117

V.116   For the system of Problem V.115, determine how far the cylinder must drop to attain a speed of 20 m/s.

V.117   The 5-kg wheel of radius of gyration 80 mm is hoisted by a constant force P-400 N. Assuming that the cord does not slip on the wheel, determine the distance the 10-kg mass A must be raised, starting from rest, to attain a speed of 15 m/s.

V.118   For the system of Problem V.117, determine the force $\bar{P}$ required for the mass A to have a speed of 10 m/s when it has fallen 5 m starting from rest.

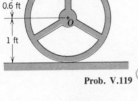

Prob. V.119

V.119   The 10-lb force is applied to pin P on the 40-lb wheel when it is at rest in the position shown. The radius of gyration of the wheel is 0.9 ft. If the force remains horizontal as the wheel rolls without slipping, determine the velocity of the center of the wheel (a) after the wheel has undergone one-half a revolution, (b) after the center of the wheel has moved 4 ft.

V.120   The solid homogeneous semicylinder is released from rest in the position shown. Determine its angular speed when its mass center is at its lowest point. Assume that slipping does not occur.

Prob. V.120

V.121   A 15-lb sphere of 1-ft radius rolls without slipping inside a curved surface of 4-ft radius. Knowing that the sphere is released from rest at $\theta = 60°$, determine (a) the velocity of the center of the sphere as it passes $\theta = 0°$, (b) the magnitude of the normal reaction at that instant.

Prob. V.121

V.122   Solve Problem V.121 for the instant when the sphere is at $\theta = 30°$.

V.123   A solid homogeneous cylinder A of radius r rolls down a fixed cylindrical surface of radius R that is sufficiently rough to prevent slipping. If cylinder A starts from rest in the position of unstable equilibrium, Q, determine the angle θ where it will lose contact with the cylindrical surface.

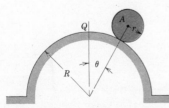

Prob. V.123

V.124   Determine the minimum initial angular speed the 2-kg slender rod

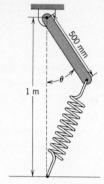

Prob. V.124

must have when $\theta = 0$ to just attain a position of $\theta = 90°$. The stiffness of the spring is 20 N/m and the spring is unstretched in the position where $\theta = 0$.

V.125 The 20-lb rod AB is constrained to move in the vertical plane by block A, which follows the horizontal groove, and block B, which follows the vertical one. The unstretched length of the spring is 5 in., and $\theta = 0$ is the static equilibrium position. If the block is released from rest when $\theta = 45°$, determine the velocity of the slider block B when $\theta = 0°$, (a) if the slider blocks have negligible mass, (b) if the mass of each slider block is 2 lb.

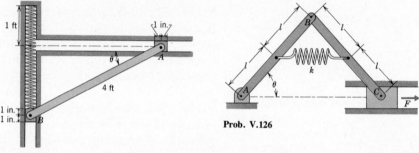

Prob. V.125

Prob. V.126

V.126 The system shown in the diagram moves in the horizontal plane. The stiffness of the spring is k and the spring is unstretched when $\theta = 60°$. Each bar has a weight W, and the weight of the piston is negligible. Determine the constant force $\bar{F}$ that, starting the system from rest at $\theta = 60°$, gives the piston a speed v to the right at $\theta = 30°$.

V.127 A 60-N vertical force is applied at point B when the system is at rest in the position where $\theta = 60°$. The mass of the rods is 2 kg each and that of the disk is 4 kg. The spring has a stiffness of 50 N/m and an unstretched length of 1 m. Assuming that the disk rolls without slipping, determine the angular velocity of each rod when $\theta = 0$.

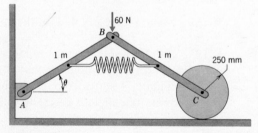

Prob. V.127

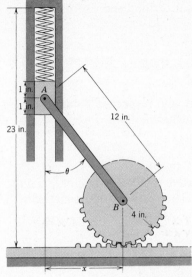

Prob. V.129

V.128 For the system of Problem V.127, determine the velocity of point C when $\theta = 30°$.

V.129 The mechanism shown lies in a horizontal plane. Bar AB weighs 10 lb and gear B weighs 5 lb and has a 3-in. radius of gyration. The spring has a stiffness of 20 lb/ft and an unstretched length of 12 in. If the system was at rest at $\theta = 0$ when gear B was given a very slight push to the right, determine (a) the maximum speed of point B and the corresponding value of x, (b) the maximum value of x attained by gear B in the motion. Neglect the mass of block A.

V.130 The 3-kg bar AB and the 4-kg bar BC form a linkage in the vertical plane. Bar AB is rotating clockwise at 180 rev/min in the position shown. What constant force $\bar{F}$ should be applied to the 1-kg collar C to bring the system to rest in the position where the bars are horizontal? Friction is negligible.

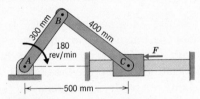

Prob. V.130

2   Impulse and Momentum

The linear impulse-momentum principle for a system of particles is the integral of Newton's second law for the total system. This principle (derived in Module Three) in equation form is

$$m(\bar{v}_G)_2 = m(\bar{v}_G)_1 + \int_{t_1}^{t_2} \Sigma\, \bar{F}\, dt \qquad (32)$$

Equation (32) is equivalent to saying that the change in the linear momentum $m\bar{v}_G$ equals the total impulse $\int_{t_1}^{t_2} \Sigma\, \bar{F}\ dt$ of the forces acting on the system.

The principle of angular impulse and momentum has similar significance, in that it is the integral of the moment equation for a system of particles. Letting point A be a fixed point or the center of mass of the system, we have

$$(\bar{H}_A)_2 = (\bar{H}_A)_1 + \int_{t_1}^{t_2} \Sigma \, \bar{M}_A \, dt \tag{33}$$

where $\bar{H}_A$ is the total angular momentum with respect to point A. Equation (5) showed that for a rigid body in planar motion,

$$\bar{H}_A = I_{zz}\bar{\omega}$$

where the origin of the xyz coordinate system is located at point A and the z axis is normal to the plane of motion. Then $\bar{\omega} = \omega \bar{k}$ and equation (33) becomes

$$r \times (ma)dt \quad \Rightarrow \quad r \times mv$$
$$r \times F) dt$$

$$\boxed{I_{zz}\omega_2 = I_{zz}\omega_1 + \int_{t_1}^{t_2} \Sigma \, M_{Az} \, dt} \tag{34}$$

Clearly, equations (32) and (34) can be employed only if we can evaluate the impulse term; that is, we must know how $\Sigma \, \bar{F}$ and $\Sigma \, \bar{M}_{Az}$ vary with time. On the other hand, if these dependencies are known, it is a simple matter to integrate the basic differential equations of motion for a rigid body to relate its velocity at two instants of time, as we did in Illustrative Problem 2, rather than employing the momentum principles. However, equations (32) and (34) are of particular importance in the case of impulsive forces, that is, very large forces acting over a short interval of time. As we saw in treating forces of this type acting on a particle, the major effect of any impulsive force is described by its impulse, and not by its actual variation with time.

EXAMPLE 7

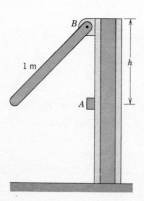

The 20-kg bar is the moving component of an impact machine used to test the strength of the object at point A. In a particular test the bar is released from rest in the horizontal position, strikes the object, and rebounds from the collision at a speed of 3 rad/s. The time of the collision was measured as 0.02 s. Determine the value of the distance h that minimizes the reactions at pin B during the collision, and also find the corresponding average value of the impact force.

Solution

In order to evaluate the impact, we must know the angular speed of the bar as it strikes the object. Because the given information defines the motion of the bar at another position, we can employ the work-energy principle to determine the angular velocity at impact. Hence, we must first solve a work-energy problem to obtain sufficient information to study the impact problem. A free body diagram of the falling bar is drawn below.

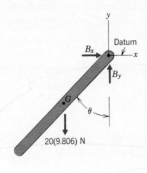

The reactions at point B act at a fixed point and therefore do no work. The only other force is gravity, which is conservative, so $U_{1\to2}^{(nc)} = 0$. Letting the elevation of point B be the datum for gravity, in the initial position $\theta = 90°$, so $V_1 = 0$. In the final position $\theta = 0°$, so the center of mass is 0.5 m below the datum, and

$$V_2 = 20(9.806)(-0.50) = -98.06 \text{ J}$$

The origin of the xyz coordinate system was placed at point B because the bar is in pure rotation about the z axis through this point. Hence, with respect to the point B,

$$T = \tfrac{1}{2}I_{zz}\omega^2 = \tfrac{1}{2}(\tfrac{1}{3}ml^2)\omega^2 = \tfrac{1}{2}[\tfrac{1}{3}(20)\,(1.0)^2]\omega^2 = 3.333\omega^2$$

Because the bar is released from rest, $T_1 = 0$. Thus

$$T_2 + V_2 = T_1 + V_1 + U_{1\to2}^{(nc)}$$

$$3.333\omega^2 - 98.06 = 0 + 0 + 0$$

$$\omega = 5.424 \text{ rad/s}$$

We can now study the impact problem. When treating an impulsive force, we can neglect any change in the position of the system. Hence, we draw a free body diagram of the bar in the vertical (impact) position.

The linear impulse-momentum relation is

$$m(\bar{v}_G)_2 = m(\bar{v}_G)_1 + \int_{t_1}^{t_2} [(B_x - F)\bar{\imath} + (B_y - 196.12)\bar{\jmath}]\, dt$$

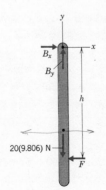

For circular motion the speed of v_G is 0.5ω. Also, $\bar{v}_G$ is to the right initially and to the left at a known speed when the bar rebounds. Hence, using these values, the foregoing equation takes the form

$$20[-0.5(3.0)]\bar{\imath} = 20[0.5(5.424)]\bar{\imath} + [(B_x - F)\bar{\imath} + (B_y - 196.12)\bar{\jmath}](0.02) \text{ kg-m/s}$$

Equating the components of this equation yields

$$B_x - F = -\frac{1}{0.02}[30.0 + 54.24] = -4212 \text{ N}$$

and

$$(B_y - 196.12)0.02 = 0$$

from which $B_y = 196.12$ N.

We are left with one equation in two unknowns. However, the angular impulse-momentum principle may also be applied. We do so with respect to the fixed point B in order to eliminate the reactions from the moment term. The value of I_{zz} was determined in the first portion of the problem. Because the z axis is outward, counterclockwise rotations and moments are positive according to the right-hand rule. Referring to the free body diagram, we see that the only nonzero moment is that of the impact force $\bar{F}$. Hence

$$I_{zz}\omega_2 = I_{zz}\omega_1 + \int_{t_1}^{t_2} \Sigma M_{Bz}\, dt$$

$$6.667(-3.0) = 6.667(5.424) - Fh(0.02)$$

This gives $Fh = 2808$ kg-m^2/s.

We now have two equations for the values of B_x and F in terms of h. Solving them yields

$$F = \frac{2808}{h} \qquad B_x = \frac{2808}{h} - 4212$$

The reaction at point B is minimized when the magnitude of B_x has its smallest value. Setting $B_x = 0$ yields

$$h = \frac{2808}{4212} = 0.667 \text{ m}$$

The other component of the reaction, B_y, is, of course, the weight force. The corresponding value of F is

$$F = 4212 \text{ N}$$

The location on a rotating body where the reaction to an impulsive force is minimized is called the *center of percussion*.

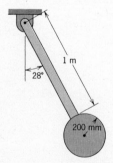

Prob. V.131

HOMEWORK PROBLEMS

V.131 The pendulum consists of a 2-kg rod rigidly attached to a 4-kg sphere. At the instant shown it is rotating at 240 rev/min clockwise. Determine (a) the linear momentum, and (b) the angular momentum of the pendulum.

V.132 The 600-kg turbine of a jet engine is rotating at 4000 rev/min when the engine is shut down. The radius of gyration of the turbine is 700 mm. Friction in the bearings produces a torque of 500 N-m. Determine the time required for the engine to come to rest.

V.133 Solve Problem V.42 using the momentum principles.

V.134 The 50-lb stepped pulley whose radius of gyration is 27 in. is released from rest. Use the linear and angular impulse-momentum principles to determine the velocity of its mass center 10 s later.

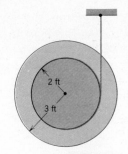
Prob. V.134

V.135 A homogeneous cylinder and a homogeneous sphere having the same radius roll without slipping down an incline starting from rest. Use the linear and angular impulse-momentum principles to derive expressions for the velocity of the center of each body after t s. Which body rolls faster?

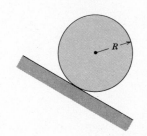
Prob. V.135

V.136 Two cables are wrapped in opposite directions around the drum of the wire spool. The radius of gyration of the spool is 500 mm. If a constant tensile force of 40 N is applied to one of the cables as shown, determine the velocity of the disk's mass center 5 s after starting from rest. Assume that the drum does not slip on the ground. The mass of the spool is 10 kg.

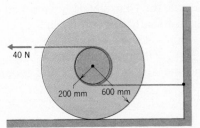
Prob. V.136

V.137 The 20-kg flywheel A, whose radius of gyration is 250 mm, and the 40-kg disk B are connected by means of a belt that does not slip over the contacting surfaces. The flywheel is driven by a torque $M = 100t^{1/2}$ where M is in newton-meters and t is in seconds. Determine the time required for the disk to attain an angular velocity of 100 rad/s.

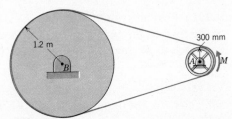
Prob. V.137

V.138 A 120-lb ice skater spins with knees bent and arms outstretched at an angular speed of 12 rev/min. In this position the skater has an 18-in. radius of gyration about the axis of rotation. When standing the radius of gyration is 6 in. and the mass center rises by 1 ft. Neglecting friction from the ice, determine (a) the angular velocity of the skater in the upright position, (b) the work done by the skater's muscles in going from the kneeling to the upright position.

V.139 A 50-kg crate is hoisted up the incline as shown. The torque applied by the 5-kg hoisting drum is $M = 800t^{1/3}$, where M is in newton-meters and t is in seconds. The radius of the drum is 600 mm and the radius of gyration is 500 mm. The coefficient of friction between the crate and the plane is 0.20. Knowing that the system was at rest when $t = 0$, determine the velocity of the block when $t = 4$ s.

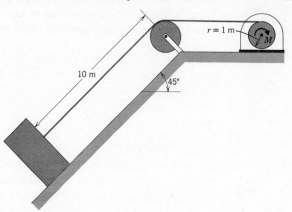

Prob. V.139

V.140 A 100-g cylindrical block slides inside the 1-kg, slender cylindrical tube, which rotates freely about the vertical shaft. Friction is negligible. Initially, the system is rotating at 20 rev/min, as shown, and the block is stationary with respect to the tube at $R = 250$ mm. The block is then released and slides outward. Determine the angular speed of the tube when the block reaches $R = 400$ mm.

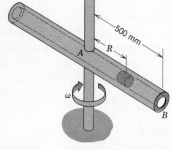

Prob. V.140

V.141 For the system in Problem V.140 determine (a) the angular speed of the tube as the block emerges from the open end B, (b) the velocity of the block as it emerges.

V.142 The 2-kg bar is initially translating at 10 m/s, as shown, when it is subjected to the impulsive force $\bar{F}$, whose average magnitude over a 0.10-s interval is 500 N. If $h = 250$ mm, determine the velocity of the center of mass and the angular velocity of the bar immediately after the impact.

V.143 Immediately after the application of the impulsive force $\bar{F}$ in Problem V.142, point A is observed to be the instantaneous axis of rotation. What are the corresponding values of h and of the angular velocity after the impact?

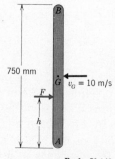

Prob. V.142

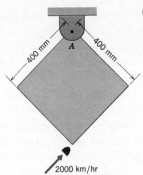

Prob. V.144

V.144 A 50-g bullet traveling at 2000 km/hr in the direction shown hits the 10-kg square plate. The bullet becomes embedded, coming to rest within the plate 0.002 s after the initial contact. Determine (a) the angular velocity of the plate immediately after the impact, (b) the impulsive reaction at pin A.

V.145 The 20-lb wooden block is initially at rest when it is struck by a 2-oz bullet that is traveling at 1500 ft/s in the direction shown. The bullet becomes embedded, coming to rest with respect to the block 0.004 s after the initial contact. Determine (a) the velocity of the center of mass of the block immediately after the impact, (b) the average tensile force within each cable during the impact.

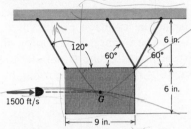

Prob. V.145

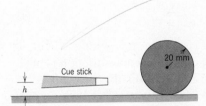

Prob. V.146

V.146 The 300-g billiard ball of 20-mm radius is struck by a cue stick that exerts an average force of 600 N horizontally over a 0.005-s interval. If h = 10 mm, determine the velocity of the center of the ball and the angular velocity of the ball immediately after the impact.

V.147 Immediately after being hit by the cue stick, the billiard ball in Problem V.146 is rolling without slipping. Determine (a) the height h for the cue stick, (b) the velocity of the ball after the impact.

3 Eccentric Impact

Recall that Example 7 considered a collision between bodies in which the normal to the plane of contact did not coincide with the line connecting the mass centers of the two bodies. This is called an *eccentric impact*. In that example the motion of the system after the collision was given. Our aim here is to learn how to determine the motion of a system when it rebounds from a collision.

When we studied the collision of particles in Module III (for which the line of impact coincides with the line connecting the mass centers), we were able to solve problems by using the concept of conservation of linear momentum in conjunction with the coefficient of restitution, ϵ, which was a measure of the loss of energy during the collision. In

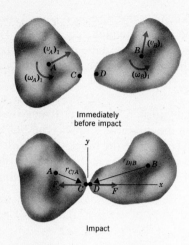

Immediately
before impact

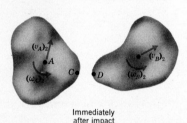

Impact

Immediately
after impact

Figure 14

essence, we recognized that particles are not really rigid and that energy can be lost due to deformation during impact. The problem for rigid bodies is quite similar, but it is complicated slightly by the fact that the velocity at the point of impact is not the same as the velocity of the mass center.

Consider the general type of collision depicted in Figure 14. For convenience we employ an xyz coordinate system having an origin at the point of contact with the y axis tangent to the plane of contact and the x axis along the line of impact. In the figure points A and B are the centers of mass of the bodies, and points C and D are the points on each body that come into contact. In the following we shall establish procedures for determining the values of the kinematical variables $\bar{v}_A$, $\bar{\omega}_A$, $\bar{v}_B$, and $\bar{\omega}_B$ after the impact in terms of their values before the collision.

The actual collision process is quite complicated. Fortunately, we have the experimental observation that the component of the relative velocity of points C and D parallel to the line of impact is related to the coefficient of restitution, according to

$$\epsilon = -\left[\frac{(v_{Dx})_2 - (v_{Cx})_2}{(v_{Dx})_1 - (v_{Cx})_1}\right] \tag{35}$$

As an aside, this relationship applies to collisions where friction is negligible, and is accurate only for a limited range of relative velocities.

Equation (35) relates the velocities of the points of contact on each body, but we are interested in the velocities of the centers of mass. We therefore employ the relative velocity relations for each of the rigid bodies:

$$\bar{v}_C = \bar{v}_A + \bar{\omega}_A \times \bar{r}_{C/A} \qquad \bar{v}_D = \bar{v}_B + \bar{\omega}_B \times \bar{r}_{D/B} \tag{36}$$

When we use equations (36) to evaluate the x components of $\bar{v}_A$ and $\bar{v}_C$, equation (35) then represents one relation for the unknowns. Other relationships are obtained from the linear and/or angular impulse-momentum principles, which, of course, involves the impulse of the contact force $\bar{F}$ exerted between the two bodies. As shown in Figure 14, this force is directed along the line of impact.

To see how these equations are used, suppose that the bodies in Figure 14 are unconstrained in their motion, so that the contact force $\bar{F}$ is the only impulsive force acting on the body. In such a situation, we apply the linear and angular momentum principles for each body, the latter with respect to the centers of mass. There are x and y components of linear momentum and a z component of angular momentum. This gives three scalar equations of momentum for each body. Thus, after using equation (35), we have a total of seven scalar equations. The seven scalar unknowns are the x and y components of $\bar{v}_A$ and $\bar{v}_B$, the z components of $\bar{\omega}_A$ and $\bar{\omega}_B$, and the impulse of the contact force $\bar{F}$. Hence, we have a well-posed problem.

In contrast, when a body rotates about a fixed axis, there are impulsive reaction forces, as was seen in Example 7. To eliminate these reactions in the impact problem, we formulate the angular impulse-momentum relation with respect to the point on the fixed axis. The linear impulse-momentum relation for the body is then discarded, as it merely defines the impulse of the reaction.

EXAMPLE 8

A 1000-kg crate is dropped by parachute, hitting the ground in the position shown. At the instant of impact the crate is in rectilinear translation, downward, with a speed of 20 m/s. Knowing that the coefficient of restitution is 0.20, determine the motion of the crate at the end of the impact. The radius of gyration of the crate about its center of mass G is 0.80 m.

Solution

The forces exerted by the parachute and the weight force are negligible compared to the impulsive force generated by the collision. Choosing the x axis normal to the plane of contact, we sketch a free body diagram for the crate.

The earth is massive enough to be considered immovable. From the given information we know that

$$(\bar{v}_G)_1 = -20\bar{i} \text{ m/s} \qquad (\bar{\omega})_1 = \bar{0}$$

In order to use the equation for the coefficient of restitution, we shall need

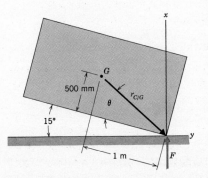

the general expression for v_{Cx} in terms of $\bar{v}_G$ and $\bar{\omega}$. Hence

$$\bar{v}_C = \bar{v}_G + \bar{\omega} \times \bar{r}_{C/G}$$

Because we want the x component of this equation, $\bar{r}_{C/G}$ must be expressed in terms of its x and y components. To do this we refer to the free body diagram, where we see that

$$\theta = \tan^{-1}\frac{0.5}{1.0} = 26.57°$$

$$\left|\bar{r}_{C/G}\right| = \sqrt{(1.0)^2 + (0.5)^2} = 1.1180 \text{ m}$$

from which it follows that

$$\bar{r}_{C/G} = \left|\bar{r}_{C/G}\right|[-\sin(\theta + 15°)\bar{\imath} + \cos(\theta + 15°)\bar{\jmath}]$$

$$= -0.7418\bar{\imath} + 0.8365\bar{\jmath} \text{ m}$$

Hence, because $\bar{\omega} = \omega\bar{k}$ for planar motion,

$$\bar{v}_C = \bar{v}_G + \omega\bar{k} \times (-0.7418\bar{\imath} + 0.8365\bar{\jmath})$$

$$v_{Cx} = v_{Gx} - 0.8365\omega$$

Thus, setting the velocity of the earth to zero, the coefficient of restitution gives

$$\epsilon = 0.2 = -\frac{(v_{Cx})_2}{(v_{Cx})_1} = -\frac{(v_{Gx})_2 - 0.8365\omega_2}{-20 - 0.8365(0)}$$

so

$$(v_{Gx})_2 - 0.8365\omega_2 = 4.0$$

The linear impulse-momentum equation for the 1000-kg crate is

$$1000(\bar{v}_G)_2 = 1000(\bar{v}_G)_1 + \int_{t_1}^{t_2} (F\bar{\imath})\ dt$$

Hence, equating components and lettering $L \equiv \int_{t_1}^{t_2} F\ dt$, we obtain

$\bar{\imath}$ direction: $1000(v_{Gx})_2 = 1000(-20) + L$

$\bar{\jmath}$ direction: $1000(v_{Gy})_2 = 1000(0)$

The crate is in general motion, so we formulate the angular momentum principle with respect to a $\hat{x}\hat{y}\hat{z}$ coordinate system, parallel to xyz, with origin at the center of mass G. From the given radius of gyration we have

$$I_{\hat{z}\hat{z}} = 1000(0.8)^2 = 640 \text{ kg-m}^2$$

The moment of the force $\bar{F}$ about point G is

$$\Sigma \, \bar{M}_G = \bar{r}_{C/G} \times \bar{F} = (-0.7418\bar{i} + 0.8365\bar{j}) \times F\bar{i}$$

$$= -0.8365F\bar{k}$$

Hence the resulting angular impulse-momentum equation is

$$I_{\hat{z}\hat{z}}\omega_2 = I_{\hat{z}\hat{z}}\omega_1 + \int_{t_1}^{t_2} \Sigma \, M_{Gz} \, dt$$

$$640\omega_2 = 640(0) - 0.8365 \int_{t_1}^{t_2} F \, dt$$

$$= -0.8365L$$

We now have four equations (one from ϵ, two from linear momentum, and one from angular momentum) for the four scalar unknowns $(v_{Gx})_2$, $(v_{Gy})_2$, ω_2, and L. The solution of these equations is

$$(v_{Gx})_2 = -8.53 \text{ m/s} \qquad (v_{Gy})_2 = 0 \qquad \triangleleft$$

$$\omega_2 = -14.99 \text{ rad/s} \qquad \triangleleft$$

Hence, after the collision the center of mass moves downward at a slower rate and the crate is rotating counterclockwise.

HOMEWORK PROBLEMS

V.148 The 2-kg slender rod is hit by a 150-g ball traveling at 20 m/s in the direction shown. The coefficient of restitution is 0.80. Determine the angular velocity of the bar and the velocity of the ball immediately after the collision.

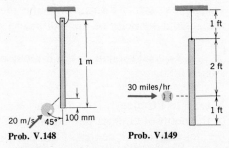

Prob. V.148   **Prob. V.149**

V.149 A 5.50-oz baseball travelling horizontally at 30 miles/hr hits the 4-lb wooden bar. The coefficient of restitution for the collision is 0.50. Determine the velocity of both ends of the bar immediately after the collision.

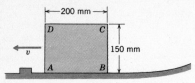

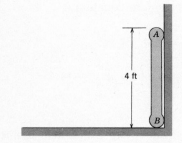

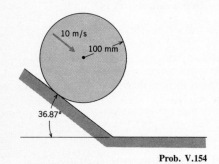

Prob. V.150

Prob. V.152

Prob. V.154

Prob. V.156

V.150 A 10-kg package on an assembly line slides along the curved chute in the vertical plane until it hits a small cleat at point A. If the speed of the package just before it hits the cleat is $v = 5$ m/s, determine the angular velocity of the package immediately after the collision. Assume that the impact is perfectly plastic.

V.151 What is the maximum speed v at which the package in Problem V.150 can hit the cleat without tipping over onto edge AD?

V.152 The 100-lb slender beam is initially at rest in the vertical position when the lower end B is given a very slight push to the left, causing the bar to fall. The floor and wall are both smooth. Determine the impulse of the reaction forces at both ends of the bar as end A hits the ground. The impact is perfectly plastic. Confine the analysis to the first impact.

V.153 In Problem V.152, the coefficient of restitution is 0.40. Determine the velocity of both ends of the beam immediately after the first impact.

V.154 The 1-kg homogeneous sphere is rolling without slipping down the incline at 10 m/s when it hits the change in grade. Assuming that the collision with the horizontal surface is perfectly plastic, determine the velocity of the center of the sphere and the angular velocity immediately after the impact. Does the sphere roll without slipping immediately after the impact?

V.155 Solve Problem V.154 if the coefficient of restitution for the collision with the horizontal surface is 0.90.

V.156 A 30-g ball bearing is rolling without slipping at 20 m/s on the horizontal surface when it hits the 5-mm ledge. The corner of the ledge is slightly rounded, so the ball bearing may slip relative to this edge. The collision is perfectly plastic. What is the velocity of the center of mass and the angular velocity of the ball bearing immediately after the collision?

Solve Problem V.156 if the coefficient of restitution for the collision is 0.80.

V.158 In Problem V.156, the corner of the ledge is sharp, so the velocity of the point of contact immediately after the collision is zero. Determine the velocity of the center of the ball bearing after the collision. *Hint:* In order to bring the point of contact to rest, an impulsive force tangent to the contact plane at the corner is required.

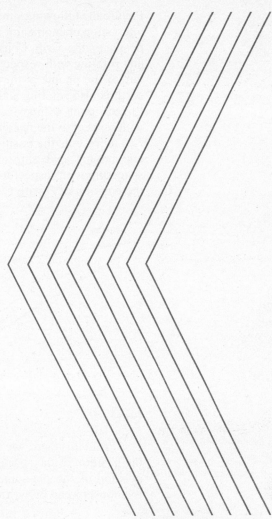

MODULE VI
KINEMATICS IN A
MOVING REFERENCE
FRAME

Picture the following situation. You are in a spaceship and you observe an interesting phenomenon whose location you wish to describe via radio to people on the earth. Considering the fact that your spaceship is translating and rotating with respect to the earth, what information should you convey to the people at the earth station? If you stop to think about it, the earth is also rotating and translating. To solve the dilemma let us pick a "fixed" point O in space and radio back the position of the phenomenon with respect to the fixed point. This situation is shown in Figure 1, where $\bar{r}_{P/O'}$ represents the position of the phenomenon at point P with respect to our moving spaceship at point O', $\bar{r}_{O'/O}$ represents the position of the spaceship with respect to a fixed point (a distant star), and $\bar{r}_{P/O}$ represents the position of P with respect to the "fixed" star.

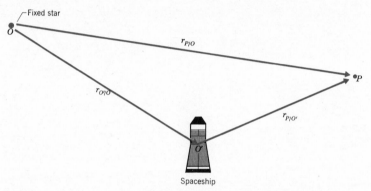

Figure 1

From this figure we see that we can describe the vector $\bar{r}_{P/O}$ by giving the position of the spaceship with respect to the fixed star ($\bar{r}_{O'/O}$) and the position of the phenomenon with respect to the spaceship ($\bar{r}_{P/O'}$). This relationship can be written in vector form as

$$\bar{r}_{P/O} = \bar{r}_{O'/O} + \bar{r}_{P/O'} \tag{1}$$

Furthermore, because you are in the spaceship you will probably choose to use the spaceship as your frame of reference for describing $\bar{r}_{P/O'}$, for instance by giving the distances that the phenomenon is ahead of, to the right of, and above the vehicle.

Whereas the preceding argument discussed a system of advanced technology, we could have made the same arguments with respect to the moving parts of a machine. In either case the new task before us is clear: to describe the motion of a particle with respect to a moving reference frame. The kinematical tools that will emerge from the development to follow are essential to the solution of three-dimensional dynamics problems. Equally important, when applied to two-dimensional problems, in many cases they can make a previously complex analysis quite trivial.

A.   REFERENCE FRAMES AND ANGULAR VELOCITY

Consider the rectangular coordinate frame xyz shown in Figure 2, which is free to translate and rotate with respect to the fixed Cartesian coordinate system XYZ. The vector $\bar{r}_{O'/O}$ locates the origin of the moving system. Clearly the components of $\bar{r}_{O'/O}$ may be written as functions of either the unit vectors of the fixed system $(\bar{I}, \bar{J}, \bar{K})$ or the moving system $(\bar{\imath}, \bar{\jmath}, \bar{k})$. Thus describing the motion of the origin O' of the moving system will not present a significant problem, for we are already familiar with the kinematics of the absolute motion of a particle (Module II).

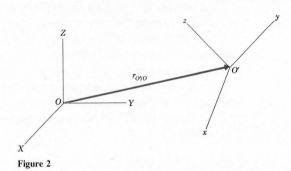

Figure 2

Let us now consider a particle whose position vector $\bar{r}_{P/O'}$ is given with respect to the xyz system, as seen in Figure 3. Writing this given vector in terms of the moving reference frame, we have

$$\bar{r}_{P/O'} = x\bar{\imath} + y\bar{\jmath} + z\bar{k} \tag{2}$$

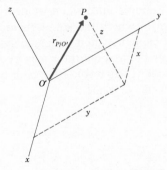

Figure 3

Because the frame of reference we shall use is rotating, we need to describe its rotational characteristics. To do this, let us consider an infinitesimal rotation that the system undergoes during the infinitesimal time interval from t to $t + dt$. The resulting rotation may be described in terms of the infinitesimal rotations $d\theta_x$, $d\theta_y$, and $d\theta_z$, where $d\theta_x$ is a rotation about an axis that is parallel to the x axis at time t, and so on. (By an infinitesimal rotation we mean that the magnitude of the angle of rotation in radians squared is negligible when compared to the angle itself.)

A rotation about an axis can be described in vector form by the right-hand rule. Our intuition tell us that the total (infinitesimal) rotation vector can be obtained by vector addition of the individual rotations, in which case

$$d\bar{\theta} = d\theta_x\,\bar{i} + d\theta_y\,\bar{j} + d\theta_z\,\bar{k} \tag{3}$$

The basic definition of angular velocity is that $\bar{\omega}$ is the rate of change of the angle of rotation directed in accordance with the right-hand rule. Therefore

$$\bar{\omega} \equiv \frac{d\bar{\theta}}{dt} = \frac{d\theta_x}{dt}\,\bar{i} + \frac{d\theta_y}{dt}\,\bar{j} + \frac{d\theta_z}{dt}\,\bar{k} = \omega_x\,\bar{i} + \omega_y\,\bar{j} + \omega_z\,\bar{k} \tag{4}$$

The direction of the vector $\bar{\omega}$ describes the orientation of the axis of rotation of the reference frame xyz at the instant t. The magnitude of $\bar{\omega}$ is the instantaneous rate of rotation of the xyz frame about this axis.

The implication of equation (3) is that after the xyz frame undergoes a set of infinitesimal rotations, its resulting orientation is independent of the order of the rotations. Knowing that, it follows that the angular velocity $\bar{\omega}$ of the moving reference system is obtained by vector addition of the rates or rotation about the various axes, as stated by equation (4). This result is proven rigorously in Section A.2. In Section A.1 we demonstrate that in the case of finite (large) rotations, the final orientation *does* depend on the order of rotation.

Sections A.1 and A.2 are presented for those who desire a rigorous treatment. If you choose, without loss of continuity, you can accept equations (3) and (4) on the basis of the foregoing discussion and proceed directly to Section B.

1   Finite Rotations

It is a simple matter to demonstrate by example that, when we deal with finite rotations, we must know the order in which the rotations are imposed. Figures 4a and 4b illustrate what happens when rectangle $OABC$, which initially lies in the xy plane, is rotated first about the x axis by $-90°$ (Figure 4a), and then about the z axis by $90°$ (Figure 4b). The result of the two rotations is that the rectangle lies in the yz plane.

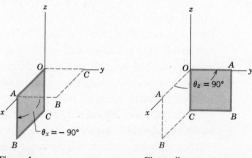

Figure 4a   **Figure 4b**

Alternatively, the rotation about the z axis could have been imposed first, followed by the rotation about the x axis. This is illustrated in Figures 5a and 5b. In this case the position of the rectangle is the xz plane. Therefore we can conclude that, in general, the order of rotation is important when considering finite rotations.

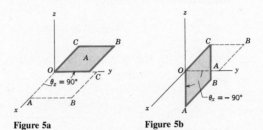

Figure 5a   Figure 5b

2 Infinitesimal Rotations and Unit Vectors

Let us now consider the case of infinitesimal rotations by considering how an arbitrary unit vector

$$\bar{e} = e_x\,\bar{\imath} + e_y\,\bar{\jmath} + e_z\,\bar{k}$$

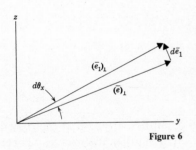

Figure 6

is changed by successive rotations $d\theta_x$, $d\theta_y$, and $d\theta_z$. Figure 6 is a view looking down the x axis at the result of the rotation $d\theta_x$. In this rotation the vector $\bar{e}$ changes by the amount $d\bar{e}_1$ to become the vector $\bar{e}_1$. The vectors $(\bar{e})_\perp$ and $(\bar{e}_1)_\perp$ in Figure 6 are the components of $\bar{e}$ and $\bar{e}_1$, respectively, which are perpendicular to the x axis, and hence tangent to the yz plane. Thus

$$(\bar{e})_\perp = e_y\,\bar{\jmath} + e_z\,\bar{k} = \bar{e} - e_x\,\bar{\imath} \tag{5}$$

In the limit as the value of $d\theta_x$ goes to zero, we see from Figure 6 that $d\bar{e}_1$, which is parallel to the yz plane, becomes perpendicular to $(\bar{e})_\perp$. Furthermore, because the chord of an infinitesimal circular sector is equal to the arc length, we have

$$\left|d\bar{e}_1\right| = \left|(\bar{e})_\perp\right|\,d\theta_x$$

Thus, in vector form,

$$d\bar{e}_1 = (d\theta_x\,\bar{\imath}) \times (\bar{e})_\perp$$

Now, in view of equation (5), we have

$$d\bar{e}_1 = (d\theta_x\,\bar{\imath}) \times (\bar{e} - e_x\,\bar{\imath}) \equiv (d\theta_x\,\bar{\imath}) \times \bar{e}$$

Therefore, we may find $\bar{e}_1$ by adding $d\bar{e}_1$ to $\bar{e}$, yielding

$$\bar{e}_1 = \bar{e} + d\bar{e}_1 = \bar{e} + (d\theta_x \, \bar{\imath}) \times \bar{e} \tag{6}$$

We shall now operate on $\bar{e}_1$ by performing the rotation $d\theta_y$, which results in $\bar{e}_1$ changing by $d\bar{e}_2$ to become $\bar{e}_2$. By following similar steps to those used in obtaining equation (6), it is easy to show that

$$\bar{e}_2 = \bar{e}_1 + d\bar{e}_2 = \bar{e}_1 + (d\theta_y \, \bar{\jmath}) \times \bar{e}_1 \tag{7}$$

Finally, with the rotation $d\theta_z$ the vector $\bar{e}_2$ changes by $d\bar{e}_3$ to become the vector $\bar{e}_3$, according to

$$\bar{e}_3 = \bar{e}_2 + d\bar{e}_3 = \bar{e}_2 + (d\theta_z \, \bar{k}) \times \bar{e}_2 \tag{8}$$

Equations (6)–(8) detail how a unit vector $\bar{e}$ becomes a vector $\bar{e}_3$. If we substitute equation (6) into equation (7) and then substitute equation (7) into equation (8), we shall obtain an expression for $\bar{e}_3$ in terms of $\bar{e}$. The first of these two steps gives

$$\bar{e}_2 = \bar{e} + (d\theta_x \, \bar{\imath}) \times \bar{e} + (d\theta_y \, \bar{\jmath}) \times [\bar{e} + (d\theta_x \, \bar{\imath}) \times \bar{e}]$$

and the second yields

$$\begin{aligned}
\bar{e}_3 = {}& \bar{e} + (d\theta_x \, \bar{\imath}) \times \bar{e} + (d\theta_y \, \bar{\jmath}) \times \bar{e} \\
& + (d\theta_y \, \bar{\jmath}) \times [(d\theta_x \, \bar{\imath}) \times \bar{e})] + (d\theta_z \, \bar{k}) \times \{\bar{e} \\
& + (d\theta_x \, \bar{\imath}) \times \bar{e} + (d\theta_y \, \bar{\jmath}) \times \bar{e} \\
& + d\theta_y \, \bar{\jmath} \times [(d\theta_x \, \bar{\imath}) \times \bar{e}]\}
\end{aligned}$$

Recall that the concept of infinitesimal quantities means that their products are negligible in comparison with the quantities themselves. Therefore, when we neglect products of $d\theta_x$, $d\theta_y$, and $d\theta_z$, we find that

$$\begin{aligned}
\bar{e}_3 &= \bar{e} + (d\theta_x \, \bar{\imath}) \times \bar{e} + (d\theta_y \, \bar{\jmath}) \times \bar{e} + (d\theta_z \, \bar{k}) \times \bar{e} \\
&= \bar{e} + (d\theta_x \, \bar{\imath} + d\theta_y \, \bar{\jmath} + d\theta_z \, \bar{k}) \times \bar{e} \\
&= \bar{e} + d\bar{\theta} \times \bar{e}
\end{aligned}$$

This equation tells us that the total infinitesimal change in the unit vector $\bar{e}$ is

$$d\bar{e} = \bar{e}_3 - \bar{e} = d\bar{\theta} \times \bar{e} \tag{9}$$

From this we can conclude that only the resultant rotation vector $d\bar{\theta}$ is significant to infinitesimal rotations.

B. RELATIVE VELOCITY—DIFFERENTIATION IN A MOVING REFERENCE FRAME

In Section A.2 we proved that the change of a unit vector $\bar{e}$ due to an arbitrary infinitesimal rotation $d\bar{\theta}$ is given by

$$d\bar{e} = d\bar{\theta} \times \bar{e} \tag{10}$$

Equation (10) simply states that in an infinitesimal rotation $d\bar{\theta}$ the unit vector $\bar{e}$ rotates by an angle $d\bar{\theta}$ about an instantaneous axis of rotation that is parallel to the direction of $d\bar{\theta}$.

It follows from equation (10) that, as a unit vector rotates, its rate of change is given by

$$\frac{d\bar{e}}{dt} \equiv \dot{\bar{e}} = \frac{d\bar{\theta}}{dt} \times \bar{e}$$

$$\boxed{\dot{\bar{e}} = \bar{\omega} \times \bar{e}} \tag{11}$$

Equation (11) provides a method for describing the rotational characteristics of a moving reference frame. We shall now utilize this result to study the motion of a point P. In Figure 3 we selected an xyz system that is rotating and translating. The origin of the xyz system is at point O', and the vector $\bar{r}_{P/O'}$ locates the position of point P within this system. The velocity $\bar{v}_{P/O'}$ of point P with respect to origin O' accounts for the total rate of change of the position vector, according to

$$\bar{v}_{P/O'} \equiv \frac{d}{dt}(\bar{r}_{P/O'})$$

Differentiating the component form of $\bar{r}_{P/O'}$, equation (2), we find that

$$\bar{v}_{P/O'} = \dot{x}\bar{i} + x\dot{\bar{i}} + \dot{y}\bar{j} + y\dot{\bar{j}} + \dot{z}\bar{k} + z\dot{\bar{k}} \tag{12}$$

To evaluate the derivatives of the unit vectors that occur in this equation, *let $\bar{\omega}$ be the angular velocity of the xyz frame* (and hence of the unit vectors $\bar{i}, \bar{j}, \bar{k}$). Thus, using equation (10) we may write

$$\dot{\bar{i}} = \bar{\omega} \times \bar{i} \qquad \dot{\bar{j}} = \bar{\omega} \times \bar{j} \qquad \dot{\bar{k}} = \bar{\omega} \times \bar{k}$$

With these results equation (11) becomes

$$\bar{v}_{P/O'} = \frac{d}{dt}(\bar{r}_{P/O'}) = \dot{x}\bar{i} + \dot{y}\bar{j} + \dot{z}\bar{k} + x(\bar{\omega} \times \bar{i}) + y(\bar{\omega} \times \bar{j}) + z(\bar{\omega} \times \bar{k})$$

$$= \dot{x}\bar{i} + \dot{y}\bar{j} + \dot{z}\bar{k} + \bar{\omega} \times (x\bar{i} + y\bar{j} + z\bar{k})$$

$$= (\dot{x}\bar{i} + \dot{y}\bar{j} + \dot{z}\bar{k}) + \bar{\omega} \times \bar{r}_{P/O'}$$

We shall now define a new process of differentiation of a vector using the symbol $\delta(\ )/\delta t$. This differentiation process will account only for the rate of change of the scalar components of a vector. It ignores the fact that the unit vectors for the components are not constant. In terms of this definition,

$$\frac{\delta}{\delta t}(\bar{r}_{P/O'}) = \dot{x}\bar{\imath} + \dot{y}\bar{\imath} + \dot{z}\bar{k} \tag{13}$$

Therefore, $\delta(\bar{r}_{P/O'})/\delta t$ is the rate of change of $\bar{r}_{P/O'}$ that someone would observe if that person did not realize that the reference frame was rotating. In other words, $\delta(\bar{r}_{P/O'})/\delta t$ *is the velocity of point P relative to the moving reference frame.* For notational convenience let us denote this term by

$$\frac{\delta}{\delta t}(\bar{r}_{P/O'}) \equiv (\bar{v}_{P/O'})_{\text{rel}}$$

This allows us to write the velocity of P with respect to O' as

$$\bar{v}_{P/O'} = \frac{d}{dt}(\bar{r}_{P/O'}) = (\bar{v}_{P/O'})_{\text{rel}} + \bar{\omega} \times \bar{r}_{P/O'} \tag{14}$$

In words, this equation states that the velocity of point P with respect to the origin of the moving reference frame xyz (point O') is the velocity of point P relative to xyz, $(\bar{v}_{P/O'})_{\text{rel}}$, plus an additional term $\bar{\omega} \times \bar{r}_{P/O'}$ that accounts for the velocity caused by the angular motion $\bar{\omega}$ of the moving reference system.

Equation (14) describes the rate of change of the vector $\bar{r}_{P/O'}$. It follows then that if we consider any vector, for instance $\bar{A}$, we may write

$$\boxed{\frac{d\bar{A}}{dt} = \frac{\delta\bar{A}}{\delta t} + \bar{\omega} \times \bar{A}} \tag{15}$$

C. ABSOLUTE VELOCITY

Let us return to the example of Figure 1, where the information we wished to transmit was the position of P with respect to the fixed star at location O. Thus, repeating equation (1), we have

$$\bar{r}_{P/O} = \bar{r}_{O'/O} + \bar{r}_{P/O'} \tag{1}$$

so that the position of the phenomenon in space ($\bar{r}_{P/O}$) may be described by knowing the position of the spaceship ($\bar{r}_{O'/O}$) and the position of the phenomenon with respect to the spaceship ($\bar{r}_{P/O'}$).

In order to describe this phenomenon to the people on the earth, we shall need to describe its velocity with respect to the fixed star. We shall refer to this quantity as the *absolute velocity* $\bar{v}_P$. In view of the basic definition of velocity, we have

$$\bar{v}_P = \frac{d}{dt}(\bar{r}_{P/O})$$

so that differentiating equation (1) gives

$$\bar{v}_P = \frac{d}{dt}(\bar{r}_{O'/O}) + \frac{d}{dt}(\bar{r}_{P/O'}) = \bar{v}_{O'} + \bar{v}_{P/O'}$$

By using equation (14) we obtain the final result,

$$\boxed{\bar{v}_P = \bar{v}_{O'} + (\bar{v}_{P/O'})_{\mathrm{rel}} + \bar{\omega} \times \bar{r}_{P/O'}} \qquad (16)$$

From equation (16) we see that the absolute velocity of point P equals the absolute velocity of the origin of the moving reference frame plus the velocity of point P relative to the moving reference frame plus the cross product of the angular velocity $\bar{\omega}$ of the moving reference frame and the relative position vector $\bar{r}_{P/O'}$. In the simple case of a reference frame that is only translating, $\bar{\omega} \equiv \bar{0}$ and $\bar{v}_{P/O'} = (\bar{v}_{P/O'})_{\mathrm{rel}}$. Equation (16) then becomes the relative motion equation that we studied in Module II, specifically,

$$\bar{v}_P = \bar{v}_{O'} + \bar{v}_{P/O'}$$

On the other hand, if we attach an *xyz* frame to a rigid body, then the coordinates of any point P of the body with respect to the *xyz* frame are constant, so that $(\bar{v}_{P/O'})_{\mathrm{rel}} \equiv \bar{0}$. For this case equation (16) becomes

$$\bar{v}_P = \bar{v}_{O'} + \bar{\omega}_{\mathrm{body}} \times \bar{r}_{P/O'} \qquad (17)$$

This result, of course, is the same as the velocity formula that we obtained for rigid bodies in planar motion (Module Four). The new feature is that we have now shown the result to be true for three dimensional motion, where the axis of rotation of the body can have a variable direction.

Before considering some examples that utilize equation (16), let us discuss criteria to be employed in the selection of a moving reference frame. Basically, we have to make three decisions regarding the *xyz* frame:

1   Where should its origin O' be located?
2   What angular velocity should it have?
3   What orientation should it have at the instant of interest?

These questions can be answered by studying equation (16). First, we note that the absolute velocity of point O' is required; hence the origin should have a fairly simple type of motion. For example, a fixed point or a point following a planar path would be ideal. The decision regarding the angular velocity $\bar{\omega}$ to be ascribed to the moving frame is based on the fact that another term we need to compute is the velocity of point P relative to the moving reference frame. Hence we look for a reference frame whose rotation makes it comparatively easy to determine $(\bar{v}_{P/O'})_{\text{rel}}$.

Finally, to select the instantaneous orientation of the xyz frame, we note that we shall require the components of various vectors. These components must *all* be with respect to the same coordinate system in order to perform the operations of vector addition and multiplication. Obviously, we can use components with respect to either the fixed unit vectors $\bar{I}, \bar{J}, \bar{K}$, or the moving unit vectors $\bar{i}, \bar{j}, \bar{k}$. As a general rule, dynamicists tend to use $\bar{i}, \bar{j}, \bar{k}$ components. We shall therefore select the orientation of the xyz frame to be such that the components of the vectors are easy to describe.

EXAMPLE 1

Turbine rotors and impeller pumps, which have fluid flowing over the vanes or blades, lead to important and interesting kinematical studies. Let us idealize such a problem. A disk of radius R rotates with constant angular velocity $\bar{\omega}$ about a fixed axis. A straight vane of length l is welded rigidly to the disk, and the angle between the vane and a radial line is θ, as shown in the sketch. If a particle of fluid slides outward along the vane at a speed u relative to the vane, determine the velocity of the particle as it flies off the end of the vane.

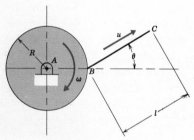

Solution

Because the axis of the disk is fixed, point A, which is on this axis of rotation, can serve as the fixed point of reference. The axes XYZ can be oriented with the Z axis parallel to the axis of rotation. We must now choose the moving reference frame. There are several possibilities for its origin: (a) on the axis of rotation, that is, point A, (b) at the junction of the disk and the vane, that is, point B, (c) at the end of the vane, that is, point C on the vane, (d) traveling with the particle P, that is, at point C at this instant, but moving with the particle. Let us study each of these possibilities.

Approach a

Select xyz to have origin O' at point A and let xyz rotate with the disk. We select the orientation of xyz to facilitate the writing of the expressions for the relative position and velocity vectors. This situation is depicted in the sketch.

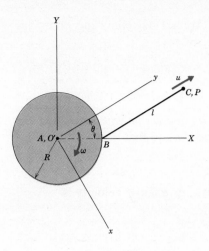

Because the origin of xyz is the fixed point A we have, at this instant,

$$\bar{r}_{O'/A} = \bar{0}$$

$$\bar{r}_{P/O'} \equiv \bar{r}_{P/A} = (R \sin \theta)\bar{i} + (R \cos \theta + l)\bar{j}$$

To determine the velocity we first obtain the individual terms in the velocity equation (16) and then perform the vector algebra. Noting that the origin O' is fixed, we have

$$\bar{v}_{O'} = \bar{0}$$

With respect to the rotating reference frame, the particle seems to be only sliding along the vane at speed u. Thus, the velocity of the particle relative to the moving reference frame is rectilinear motion parallel to the vane, so that

$$(\bar{v}_{P/O'})_{\text{rel}} = u\,\bar{j}$$

Because xyz is attached to the disk, the angular velocity according to the right-hand rule is

$$\bar{\omega} = \omega(-\bar{K}) = -\omega\bar{k}$$

Note that we have expressed $\bar{\omega}$ in terms of both XYZ and xyz components. This is done in order to anticipate a step in the acceleration problem, which we shall consider in the next section. However, when we substitute $\bar{\omega}$ and the other vectors into the velocity equation, we must use only one coordinate system for components. Thus

$$\bar{v}_P = \bar{v}_{O'} + (\bar{v}_{P/O'})_{\text{rel}} + \bar{\omega} \times \bar{r}_{P/O'}$$

$$= \bar{0} + u\,\bar{j} + (-\omega\bar{k}) \times [R \sin \theta\,\bar{i} + (R \cos \theta + l)\,\bar{j}]$$

$$= \omega(R \cos \theta + l)\bar{i} + (u - \omega R \sin \theta)\,\bar{j}$$

Approach b

Let the origin O' be point B, and let xyz rotate with the disk. Again, we select the orientation of xyz to make it as easy as possible to describe the vector components, as shown in the sketch.

For this coordinate system,

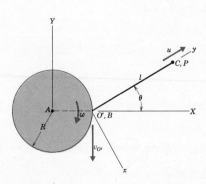

$$\bar{r}_{P/O'} = l\,\bar{j}$$

In this approach origin O' is moving. Its velocity may be derived in one of two ways. We can either note that point O' is following a circular path centered at point A, so that its velocity is ωR tangent to the path, or alternatively that point O' is a point on the rigid disk, in which case we can employ the rigid body equation (17). Let us follow the latter approach. Then

$$\bar{v}_{O'} = \bar{v}_A + \bar{\omega}_{\text{disk}} \times \bar{r}_{O'/A} = \bar{0} + (-\omega\bar{k}) \times [R(\cos\theta\,\bar{j} + \sin\theta\,\bar{i})]$$

$$= \omega(\cos\,\bar{i} - \sin\theta\,\bar{j})$$

We next determine the relative velocity by noting that with respect to xyz the motion is parallel to the y axis at speed u, so that

$$(\bar{v}_{P/O'})_{\text{rel}} = u\,\bar{j}$$

Finally, because xyz rotates with the disk, its angular velocity is

$$\bar{\omega} = -\omega\bar{K} = -\omega\bar{k}$$

Thus

$$\bar{v}_P = \bar{v}_{O'} + (\bar{v}_{P/O'})_{\text{rel}} + \bar{\omega} \times \bar{r}_{P/O'}$$

$$= \omega R(\cos\theta\,\bar{i} - \sin\theta\,\bar{j}) + u\,\bar{j} + (-\omega\bar{k}) \times (l\,\bar{j})$$

$$= \omega(R\cos\theta + l)\bar{i} + (u - \omega R\sin\theta)\bar{j} \qquad \triangleleft$$

Because the xyz system we have utilized in this approach is parallel to the one we used in Approach a, the components of $\bar{v}_P$ should be identical for each approach. A comparison shows that this is true.

Approach c

Let the origin O' be at the tip of the vane, so that the particle is at point O' at this instant. Let xyz rotate with the vane and orient xyz parallel to the reference frames used in the previous approaches. It follows, then, that, with respect to the xyz axes, the particle is moving at speed u parallel to the vane.

The tip of the vane is a point in the rigid body (consisting of the disk and the vane). Therefore, from the sketch we find

$$\bar{r}_{P/O'} = \bar{0}$$

$$\bar{v}_{O'} = \bar{v}_A + \bar{\omega}_{\text{disk}} \times \bar{r}_{A/O'}$$

Because xyz in this approach is parallel to xyz in Approach a, we can use the components of $\bar{r}_{O'/A}$ that we determined there. Thus the velocity of the origin of the moving system is

$$\bar{v}_{O'} = \bar{0} + (-\omega\bar{k}) \times [R\sin\theta\,\bar{i} + (R\cos\theta + l)\bar{j}]$$

$$= \omega(R\cos\theta + l)\bar{i} - \omega R\sin\theta\,\bar{j}$$

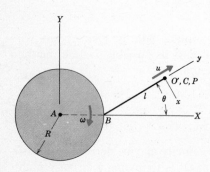

Also, we know that

$$(\bar{v}_{P/O'})_{\text{rel}} = u\,\bar{j}$$

and

$$\bar{\omega} = -\omega\bar{K} = -\omega\bar{k}$$

so that

$$\bar{v}_P = \bar{v}_{O'} + (\bar{v}_{P/O'})_{\text{rel}} + \bar{\omega} \times \bar{r}_{P/O'}$$

$$= \omega(R\cos\theta + l)\bar{i} - \omega R\sin\theta\,\bar{j} + u\,\bar{j} + \bar{0}$$

$$= \omega(R\cos\theta + l)\bar{i} + (u - \omega R\sin\theta)\bar{j}$$

Approach d

Let the origin of the moving reference frame follow the motion of the particle, so that the condition $\bar{r}_{P/O'} = \bar{0}$ always holds. It then follows that the relative velocity $(\bar{v}_{P/O'})_{\text{rel}} = \bar{0}$. Thus, regardless of the angular velocity of xyz, the velocity equation reduces to

$$\bar{v}_P = \bar{v}_{O'} + (\bar{v}_{P/O'})_{\text{rel}} + \bar{\omega} \times \bar{r}_{P/O'}$$

$$= \bar{v}_{O'} + \bar{0} + \bar{0}$$

In other words, ''The answer is the answer.'' In this case the use of a moving reference frame does not aid the solution for $\bar{v}_P$.

Let us summarize the foregoing by commenting on the advantages and disadvantages of each approach.

Clearly, the last approach is of no use. We should *never* let the origin of the moving reference frame follow the point in which we are interested. In Approach c the method we used for determining the velocity of the origin may not have been obvious to you, and the alternative of using the concepts of circular motion leads to some complications due to the geometry of the system.

The solution was obtained fairly directly in both Approaches a and b. However, of the two, in Approach b the origin O' was a moving point whose velocity was easy to compute, and each term in the velocity equation had components that were very simple to write. This turns out to be true in general. That is, if there is a point intermediate to the moving point P and the fixed point O, and if the kinematical description of this point is fairly direct, we shall use such a point as the origin of the moving reference system.

EXAMPLE 2

A centrifuge is being used to study the tolerance of human beings to motion. It consists of a main arm that rotates about the A–A axis at an angular speed ω_1, as shown in the sketch, and a cockpit that may rotate with respect to the main arm at an angular speed ω_2 about the C–C axis. The individual system rotations are computer controlled, and the head of the person being tested is stationary with respect to the cockpit seat.

Consider a test where $\omega_1 = 60$ rad/min, $\omega_2 = 90$ rad/min, the rider is 40 ft from the A–A axis, and the rider's eyes are 3 ft from the C–C axis. Determine the absolute velocity of the rider's eyes at the instant when the rider is in the horizontal position shown below.

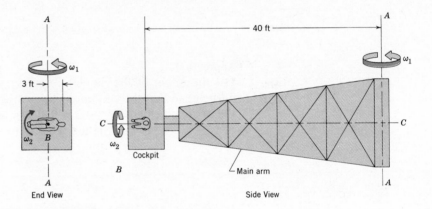

Solution

In the previous example we discussed the matter of the choice of the origin of the moving reference frame. From the experience we gained there we see that for this problem, either the moving point B (which is on the rotating axis C–C and therefore follows a circular path) or a fixed point on the A–A axis would make a reasonable choice for the origin O'. Further, in Example 1 we saw that the moving point is preferable. Therefore, we shall let the origin O' coincide with B, so that for circular motion $v_{O'} = \omega_1(40)$ directed outward from the plane of the diagram. Note that for this step we could have used equation (17) because point O' is a point on the rigid main arm.

Now let us turn our attention to the question of what angular velocity $\bar{\omega}$ we should ascribe to reference frame xyz. The reference frame can be attached either to the main arm or to the cockpit. In what follows we shall work both approaches. Regardless of which approach you select, as part of your solution you *must state* the body to which you have attached your reference frame.

Approach a

Let xyz be attached to the main arm. In order to facilitate expressing vector components at the instant when the rider is horizontal, orient the xyz axes so that one axis is parallel to the line from point B to the rider's eyes and one axis is parallel to axis C–C along the arm. The resulting fixed and moving reference systems are shown in the sketch.

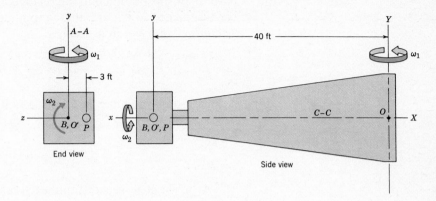

As we can see from the sketch, at this instant

$$\bar{r}_{P/O'} = -3\bar{k} \text{ ft}$$

Also, because $\bar{v}_{O'}$ is outward when viewed from the side, we have

$$\bar{v}_{O'} = 40\omega_1(-\bar{k}) = -40(60 \text{ rad/min})(1 \text{ min/60 s})\bar{k}$$

$$= -40\bar{k} \text{ ft/s}$$

To determine the velocity of the rider's eyes relative to the moving reference frame, we note that the cockpit, and hence the rider, is in pure rotation about axis C–C relative to the main arm. The xyz frame is fixed to the main arm, so point P is following a circular path. Therefore, $(\bar{v}_{P/O'})_{\text{rel}}$ is $-3\omega_2\bar{j}$ at this instant, so that

$$(\bar{v}_{P/O'})_{\text{rel}} = -3(90 \text{ rad/min})(1 \text{ min/60 s})\bar{j}$$

$$= -4.5\bar{j} \text{ ft/s}$$

The next item to be considered is the matter of the angular velocity of xyz. Although we have not yet discussed the problem of accelerations, it will be seen later that there are two separate uses for $\bar{\omega}$. One use, which fits the aim of this problem, is to substitute the $\bar{\omega}$ vector into the velocity equation in terms of its components with respect to the xyz axes, in order to perform the necessary computations. A further use is to have an expression for the $\bar{\omega}$ vector that is valid at any arbitrary instant. That expression can then be differentiated with respect to time in order to determine the angular acceleration $\bar{\alpha}$.

Thus, in order to ensure that the procedures we develop for the velocity problem lead directly into the acceleration problem, we shall first express $\bar{\omega}$ in component form according to a description that is always true. Then $\bar{\omega}$ will be evaluated for the desired instant of time for the purpose of substitution into the relative motion formulas. In this example xyz is attached to the arm that is rotating about axis A–A at the rate ω_1. Therefore we have, in general,

$$\bar{\omega} = \omega_1 \bar{J} = \omega_1 \bar{j}$$

At this instant, and also at all instants if ω_1 is constant, this expression becomes

$$\bar{\omega} = 1\bar{j} \text{ rad/s}$$

As a last step we substitute each term into equation (16).

$$\bar{v}_P = \bar{v}_{O'} + (\bar{v}_{P/O'})_{\text{rel}} + \bar{\omega} \times \bar{r}_{P/O'}$$
$$= -40\bar{k} - 4.5\bar{j} + \bar{j} \times (-3\bar{k})$$
$$= -3\bar{i} - 4.5\bar{j} - 40\bar{k} \text{ ft/s}$$

Approach b

In this approach we attach the xyz axes to the cockpit. Again, for ease in describing vector components when the rider is horizontal, we shall orient the xyz axes in the same manner as that shown in Approach a. The relative position vector is therefore

$$\bar{r}_{P/O'} = -3\bar{k} \text{ ft}$$

and the velocity of the origin is again

$$\bar{v}_{O'} = -40\bar{k} \text{ ft/s}$$

With regard to the motion of the rider's eyes with respect to the xyz axes, there is no motion with respect to the cockpit. Therefore, because the xyz axes are fixed to the cockpit, we have

$$(\bar{v}_{P/O'}) = \bar{0}$$

To compute $\bar{\omega}$ of the xyz axes, we note that the cockpit and the axes are rotating with respect to the main arm at ω_1 as the main arm rotates about the fixed axis $A-A$ at ω_2. To determine $\bar{\omega}$ we add the *vector* rates of rotations. Thus, it is generally true that

$$\bar{\omega} = \omega_1 \bar{J} - \omega_2 \bar{I}$$

This expression is always valid because the $A-A$ axis and the Y axis are always parallel. (This form is the one we would use for the determination of angular acceleration.) On the other hand, for the velocity computation we must express $\bar{\omega}$ in terms of $(\bar{i}, \bar{j}, \bar{k})$ components only. At this instant the rider is horizontal, so the y axis is parallel to the Y axis, in which case

$$\bar{\omega} = \omega_1 \bar{j} - \omega_2 \bar{i} = 1\bar{j} - 1.5\bar{i} \text{ rad/s}$$

Finally, the velocity equation (16) gives

$$\begin{aligned}
\bar{v}_P &= \bar{v}_{O'} + (\bar{v}_{P/O'})_{\text{rel}} + \bar{\omega} \times \bar{r}_{P/O'} \\
&= -40\bar{k} + \bar{0} + (\bar{j} - 1.5\bar{i}) \times (-3\bar{k}) \\
&= -3\bar{i} - 4.5\bar{j} - 40\bar{k} \text{ ft/s}
\end{aligned}$$

Because the xyz axes in this approach are parallel at this instant to the xyz axes in Approach a, the components of $\bar{v}_P$ in both approaches should be identical, as they are.

Let us assess the comparative advantages of the two approaches. In Approach a the angular velocity of the reference frame (one component) was easier to describe and understand than the angular velocity of the xyz axes in Approach b, which had two components. However, in Approach b the motion relative to the moving reference system was much simpler to formulate. You should be familiar with both approaches, for when we treat the kinetics of rigid bodies in three-dimensional motion it will be necessary to describe the angular motion of bodies that rotate about more than one axis.

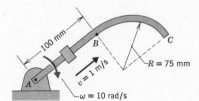

Prob. VI.1

HOMEWORK PROBLEMS

VI.1   Rod ABC rotates clockwise at 10 rad/s as the slider moves outward at 1 m/s. Determine the velocity of the slider when it is at (a) position B, (b) position C.

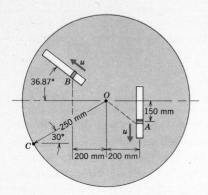

Probs. VI.2 and VI.3

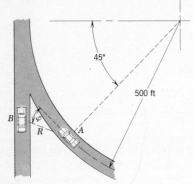

Prob. VI.4

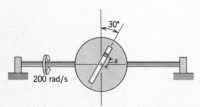

Prob. VI.6

VI.2 Sliders A and B move within their grooves at the same constant speed $u = 100$ mm/s relative to the disk, which is rotating at the constant rate $\omega = 4$ rad/s clockwise about an axis passing through point O. Determine the absolute velocity of each slider in the position shown.

VI.3 Sliders A and B move within their grooves at the same constant speed $u = 100$ mm/s relative to the disk, which is rotating at the constant rate $\omega = 4$ rad/s clockwise about an axis passing through point C. Determine the absolute velocity of each slider in the position shown.

VI.4 A passenger in the front seat of automobile A follows automobile B by observing the separation distance R and the angle ϕ, which is measured with respect to the longitudinal axis of automobile A. In the position shown, automobile A has a speed of 45 miles/hr, $\phi = 45°$, and $\dot\phi = -0.2$ rad/s. Determine the speed of automobile B.

VI.5 Water in the radial flow turbine shown in the sketch enters with a velocity of 30 m/s at an angle of 75° from the radial line. At the inlet A the tangent to the blade forms an angle of 30°. (a) Determine the angular speed ω of the turbine for which the water enters with a relative velocity that is tangent to the blade. (b) Assume that the velocity of the water relative to the blade has constant magnitude and that it is always tangent to the blade. For the conditions of part (a), determine the exit angle θ for which the water leaving the outlet B has only a radial velocity component.

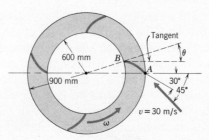

Prob. VI.5

VI.6 The slider oscillates within the groove of the disk, such that the distance $s = 20 \sin 30t$, where s is measured in millimeters and t is in seconds. The disk is fixed to the horizontal axis, which is rotating at 200 rad/s. Determine the velocity of the slider as it passes $s = 10$ mm, moving outward in the groove.

VI.7 The circular hoop is rotating about the vertical axis at 20 rad/s as the block moves relative to the hoop at a constant speed of 4 m/s. Determine the velocity of the slider (a) when $\theta = 0°$, (b) when $\theta = 90°$, (c) when $\theta = 60°$.

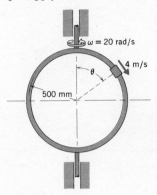

Prob. VI.7

VI.8 Bar BC is pinned at end B to bar AB, which rotates about the vertical axis at 240 rev/min. At the instant when $\theta = 45°$, this angle is decreasing at the rate of 5 rad/s. Determine the velocity of end C at this instant.

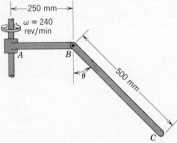

Prob. VI.8

VI.9 The radar antenna rotates about the fixed vertical axis at 2 rev/min as the angle θ oscillates at $\theta = \pi/6 + (\pi/3) \sin \pi t$, where θ is in units of radians and t is in seconds. Determine the velocity of the probe P when $t = 0.25s$, using (a) a moving reference system that is attached to the vertical shaft, (b) a reference system that is attached to the antenna.

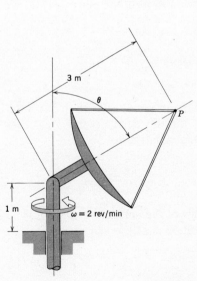

Prob. VI.9

VI.10 A small electric motor pivots about the horizontal pivot A. The disk that is attached to the shaft has an angular speed of 900 rev/min. In the position where $\theta = 30°$, the motor is rotating downward at 5 rad/s.

Determine (a) the angular velocity of the disk, (b) the velocity of point C on the disk. The perpendicular distance from the pivot axis through A to the disk is L.

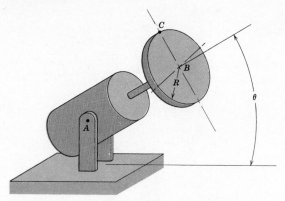

Prob. VI.10

D. ABSOLUTE ACCELERATION

We shall now obtain a formula, comparable to equation (16), for describing the absolute acceleration, $\bar{a}_P$, of a point P. By definition, acceleration is the rate of change of velocity. When we differentiate equation (16) we find that

$$\bar{a}_P = \dot{\bar{v}}_P = \frac{d}{dt}(\bar{v}_{O'}) + \frac{d}{dt}[(\bar{v}_{P/O'})_{\text{rel}}] + \frac{d}{dt}(\bar{\omega} \times \bar{r}_{P/O'})$$

The term $d(\bar{v}_{O'})/dt$ is the acceleration of the origin O', which in our notation is $\bar{a}_{O'}$. Then, by applying the law for the derivative of the cross product of two vectors, we transform this equation to

$$\bar{a}_P = \bar{a}_{O'} + \frac{d}{dt}[(\bar{v}_{P/O'})_{\text{rel}}] + \dot{\bar{\omega}} \times \bar{r}_{P/O'} + \bar{\omega} \times \left[\frac{d}{dt}(\bar{r}_{P/O'})\right] \tag{18}$$

The vector $\dot{\bar{\omega}}$ is the time rate of change of the angular velocity of xyz. In other words, $\dot{\bar{\omega}}$ is the angular acceleration $\bar{\alpha}$ of the xyz system.

$$\boxed{\bar{\alpha} = \dot{\bar{\omega}}} \tag{19}$$

One of the terms in equation (18) is $d(\bar{r}_{P/O'})/dt$. We have encountered this derivative previously; it is given by equation (14). Differentiating $(\bar{v}_{P/O'})_{\text{rel}}$ in equation (18) presents us with a similar problem, because both vectors describe what an observer in the moving reference system, such as the spaceship pilot of Figure 1 is witnessing. This suggests that we again employ the concept of a derivative with respect to the moving reference system, as defined by equation (15). This step gives

$$\frac{d}{dt}[(\bar{v}_{P/O'})_{\text{rel}}] = \frac{\delta}{\delta t}[(\bar{v}_{P/O'})_{\text{rel}}] + \bar{\omega} \times (\bar{v}_{P/O'})_{\text{rel}} \qquad (20)$$

The first term in the right-hand side of equation (20) expresses how an observer in the moving reference system would see the relative velocity change. In other words, it is the relative acceleration,

$$(\bar{a}_{P/O'})_{\text{rel}} \equiv \frac{\delta}{\delta t}[(\bar{v}_{P/O'})_{\text{rel}}] \equiv \frac{\delta^2}{\delta t^2}(\bar{r}_{P/O'})_{\text{rel}} \qquad (21)$$

Substitution of equations (14), (19), (20), and (21) into equation (18) yields the final form of the acceleration equation in relative motion,

$$\boxed{\bar{a}_P = \bar{a}_{O'} + (\bar{a}_{P/O'})_{\text{rel}} + 2\bar{\omega} \times (\bar{v}_{P/O'})_{\text{rel}} + \bar{\alpha} \times \bar{r}_{P/O'} + \bar{\omega} \times (\bar{\omega} \times \bar{r}_{P/O'})} \qquad (22)$$

A study of the terms in this equation will help in understanding the significance of the result. Suppose that the xyz frame is in pure translation, in which case $\bar{\omega}$ and $\bar{\alpha}$ are both zero. Then equation (22) gives the same result as that of Module II for translating reference frames,

$$\bar{a}_P = \bar{a}_{O'} + (\bar{a}_{P/O'})_{\text{rel}} \qquad (\bar{a}_{P/O'})_{\text{rel}} \equiv \bar{a}_{P/O'}$$

Another special case to consider is the situation when point P is a point in a rigid body and xyz is fixed to the body. In this case there is no motion relative to xyz, so equation (22) reduces to

$$\bar{a}_P = \bar{a}_{O'} + \bar{\alpha}_{\text{body}} \times \bar{r}_{P/O'} + \bar{\omega}_{\text{body}} \times (\bar{\omega}_{\text{body}} \times \bar{r}_{P/O'}) \qquad (23)$$

This formula is identical to the one derived in Module IV to describe the kinematics of rigid bodies in planar motion.

When we refer to Figure 7, we can see that the last term in equation (22) is once again the centripetal acceleration directed toward the axis of rotation. The term $\bar{\alpha} \times \bar{r}_{P/O'}$ represents the effect of the angular acceleration. The term $2\bar{\omega} \times (\bar{v}_{P/O'})_{\text{rel}}$ did not occur in our previous investigations of translating reference frame and of rigid body motion. If we had naively tried to construct the general equation (22) on the basis of the foregoing special cases, we surely would have omitted this important effect, which represents the interaction of the motion relative to the xyz frame and the rotation of xyz. We refer to $2\bar{\omega} \times (\bar{v}_{P/O'})_{\text{rel}}$ as the Coriolis acceleration. Note its similarity to the transverse acceleration term $2\dot{R}\dot{\theta}$ in cylindrical coordinates (Module II).

A primary new feature of three-dimensional kinematics is that the axis of rotation need not have a fixed orientation. Hence $\bar{\alpha}$ will generally not be parallel to $\bar{\omega}$, and $\bar{\alpha} \times \bar{r}_{P/O'}$ will not represent solely a transverse acceleration.

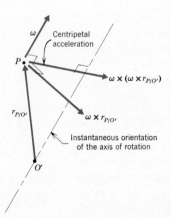

Figure 7

Note that equation (17) for the velocity of a rigid body and equation (23) for the acceleration demonstrate that the superposition principle for the motion of a rigid body is valid for an arbitrary three-dimensional motion. Specifically, this principle states that the general motion of a rigid body can be pictured as the combination of a translatory motion following an arbitrary point O', and a rotation having angular velocity $\bar{\omega}$ and angular acceleration $\bar{\alpha}$ about an instantaneous axis passing through point O'.

The conclusions we arrived at regarding the selection of the moving reference frame for velocity considerations also apply for the determination of acceleration. The only quantity that we have not determined in previous examples is the angular acceleration $\bar{\alpha}$.

Our approach for angular velocity and angular acceleration consists of *first* expressing $\bar{\omega}$ in general terms. This is achieved by first using the unit vectors of either the fixed or moving coordinate system to describe the various axes of rotation, whichever gives a constant description. Thus, if an axis of rotation is fixed in space, we describe it in terms of the fixed $\bar{I}\,\bar{J}\,\bar{K}$, whereas if an axis of rotation has varying orientation, we describe it in terms of the moving $\bar{i}\,\bar{j}\,\bar{k}$. Equally important, we use algebraic variables to denote the rates or rotation. *Second,* we differentiate $\bar{\omega}$ to obtain a general expression for $\bar{\alpha}$, using $\dot{\bar{e}} = \bar{\omega} \times \bar{e}$ to evaluate the derivatives of $\bar{i}, \bar{j}$, and $\bar{k}$. *Finally,* the instantaneous values of $\bar{\omega}$ and $\bar{\alpha}$ are determined by substituting the instantaneous values of the rates of rotation, and expressing the unit vectors of XYZ by their xyz equivalents.

These three steps will be a key feature of the second of the following two examples.

EXAMPLE 3

The turbine rotor of Example 1 is shown again in the accompanying sketch. Find the acceleration of a particle of fluid immediately prior to the instant when it flies off the end of the vane if the rotation rate ω and relative speed u are both constant.

Solution

We shall proceed as we did in Example 1, except that our efforts will be confined to Approaches a and b, in that the other procedures were shown to be inefficient.

Approach a: *xyz* fixed to the disk with origin at A.

Because we have selected the origin of *xyz* to coincide with the fixed point A, we have $\bar{v}_{O'} = \bar{a}_{O'} = \bar{O}$. Also, following our method for describing the angular motion, the general expressions for $\bar{\omega}$ and $\bar{\alpha}$ are

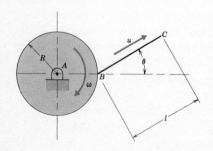

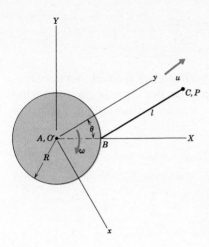

$$\bar{\omega} = \omega(-\bar{K})$$

$$\bar{\alpha} = \dot{\omega}(-\bar{K}) + \omega(-\dot{\bar{K}}) = -\dot{\omega}\bar{K}$$

Because the rate ω is constant and $\bar{K} \equiv \bar{k}$, we have

$$\bar{\omega} = -\omega\bar{k} \qquad \bar{\alpha} = \bar{0}$$

Also, as was determined in Example 1,

$$\bar{r}_{P/O'} = (r \sin \theta)\bar{i} + (R \cos \theta + l)\bar{j}$$

Before we can substitute into the acceleration equation (22), we must determine the relative acceleration. As we saw in Example 1, the vane is fixed to xyz, so the motion of the particle relative to xyz is a simple rectilinear motion tangent to the vane with constant speed. Thus

$$(\bar{v}_{P/O'})_{\text{rel}} = u\,\bar{j}$$

$$(\bar{a}_{P/O'})_{\text{rel}} = \bar{0}$$

Incidentally, you probably realize that the relative acceleration would not have been zero if the speed had not been constant, but do you know what the effect would have been if the vane had a curved shape, such as the arc of a circle? In such a case we could use path variables to describe the relative acceleration. The relative motion would then consist of a velocity $\bar{u}$ tangent to the curved vane, so that

$$(\bar{v}_{P/O'})_{\text{rel}} = u\bar{e}_t$$

$$(\bar{a}_{P/O'})_{\text{rel}} = \dot{u}\bar{e}_t + \frac{u^2}{\rho}\,\bar{e}_n$$

Before these expressions could be used in the computation of acceleration, the unit vector $\bar{e}_t$ (tangent to the vane) and $\bar{e}_n$ (normal to the vane toward the center of curvature of the vane) would have to be expressed in terms of $(\bar{i}, \bar{j}, \bar{k})$ components.

The last step in the solution of the problem at hand is the substitution of the individual terms into equation (22). This gives

$$\bar{a}_P = \bar{0} + \bar{0} + 2(-\omega\bar{k}) \times u\,\bar{j} + (-\omega\bar{k}) \times \{(-\omega\bar{k})$$
$$\times [(R \sin \theta)\bar{i} + (R \cos \theta + l)\bar{j}]\}$$
$$= (2\omega u - R\omega^2 \sin \theta)\bar{i} - \omega^2(R \cos \theta + l)\bar{j}$$

Approach b: xyz fixed to the disk with origin at B
Because xyz is rotating with the disk, we have

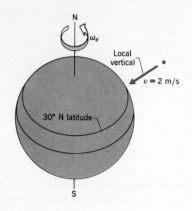

$$\bar{\omega} = -\omega\bar{K} = -\omega\bar{k}$$

$$\bar{\alpha} = \bar{0}$$

In this approach the origin O' is on the perimeter of the disk. Its acceleration can be determined either by describing the circular motion of the point or else by using the rigid body motion formula (23). The latter approach gives

$$\bar{a}_{O'} = \bar{a}_A + \bar{\alpha}_{\text{disk}} \times \bar{r}_{A/O'} + \bar{\omega}_{\text{disk}} \times (\bar{\omega}_{\text{disk}} \times \bar{r}_{A/O'})$$

To compute the $(\bar{\imath}, \bar{\jmath}, \bar{k})$ components of $\bar{a}_{O'}$, we first determine the position vector.

$$\bar{r}_{A/O'} = R\bar{I} = R(\sin\theta\,\bar{\imath} + \cos\theta\,\bar{\jmath})$$

which gives

$$\bar{a}_{O'} = \bar{0} + \bar{0} + (-\omega\bar{k}) \times [(-\omega\bar{k}) \times R(\sin\theta\,\bar{\imath} + \cos\theta\,\bar{\jmath})]$$

$$= -\omega^2 R(\sin\theta\,\bar{\imath} + \cos\theta\,\bar{\jmath})$$

The sketch shows that

$$\bar{r}_{P/O'} = l\,\bar{\jmath}$$

Also, we see that the motion of the fluid particle relative to xyz is rectilinear motion along the vane at constant speed. Therefore

$$(\bar{v}_{P/O'})_{\text{rel}} = u\,\bar{\jmath}$$

$$(\bar{a}_{P/O'})_{\text{rel}} = \bar{0}$$

Substituting the individual terms into equation (22) then gives

$$\bar{a}_P = -\omega^2 R\,(\sin\theta\,\bar{\imath} + \cos\theta\,\bar{\jmath}) + \bar{0} + \bar{0}$$

$$+2(-\omega\bar{k}) \times (u\,\bar{\jmath}) + (-\omega\bar{k}) \times [(-\omega\bar{k}) \times (l\,\bar{\jmath})]$$

$$= (2\omega u - R\omega^2 \sin\theta)\bar{\imath} - \omega^2(R\cos\theta + l)\,\bar{\jmath}$$

The moving reference frame in this approach is parallel to the one in Approach a, so the expressions for $\bar{a}_P$ are identical, as they should be.

EXAMPLE 4

In Example 2 we determined the velocity of the eyes of a test subject in a centrifuge for a given set of rotation rates at a specific instant. In this example we shall determine the acceleration of the rider's eyes at an arbitrary instant when the seat is oriented at an angle θ from the vertical, as shown in the diagram. We shall also allow the angular rates ω_1 and ω_2 to be variables.

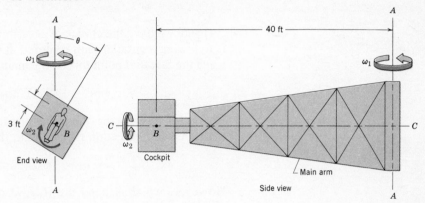

SOLUTION

Without loss of generality, we can select a fixed reference frame by letting the Y axis coincide with the fixed axis A–A, and by letting the X axis coincide with the instantaneous orientation of axis C–C. In both of the methods used to solve Example 2, point B was selected as the origin for the xyz frame. The difference in the methods was that in Approach a, xyz rotated only with the main arm, whereas in Approach b, xyz rotated with the cockpit. We shall again show both of these methods.

We must first decide on the orientation of xyz at the general instant of interest. If we choose the axes of xyz to coincide with the horizontal and vertical directions, the result will be an expression for $\bar{a}_P$ in terms of horizontal and vertical directions. If, on the other hand, we align the axes with the instantaneous orientation of the rider's body, the result will be an expression for $\bar{a}_P$ in terms of what the rider is experiencing. Either choice is correct; we shall employ the latter. Thus for both approaches, the orientations of the reference frames are as shown in the sketch.

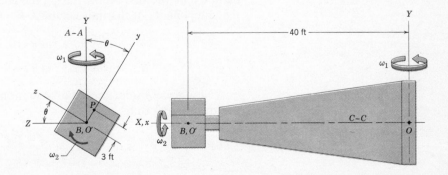

For both methods we shall employ the basic acceleration equation (22), which becomes

$$\bar{a}_P = \bar{a}_B + (\bar{a}_{P/B})_{\text{rel}} + \bar{\alpha} \times \bar{r}_{P/B} + 2\bar{\omega} \times (\bar{v}_{P/B})_{\text{rel}} + \bar{\omega} \times (\bar{\omega} \times \bar{r}_{P/B})$$

As usual we have the choice of describing $\bar{a}_B$ by using the fact that point B is following a circular path centered at point O at angular speed ω_1, or by using the fact that point B is on the main arm. The former approach gives

$$\bar{a}_B = -40(\omega_1)^2(\bar{e}_R)_B + 40\dot{\omega}_1(\bar{e}_\theta)_B$$

where, for this instant, the radial direction coincides with the X axis and the transverse direction is parallel to the negative Z axis. Hence,

$$\bar{a}_B = -40(\omega_1)^2(\bar{e}_R)_B + 40\dot{\omega}_1(\bar{e}_\phi)_B$$

However, we shall need the $(\bar{\imath}, \bar{\jmath}, \bar{k})$ components of $\bar{a}_B$, which we can obtain by using the sketch to transform the components of the unit vectors. The sketch shows that $\bar{I} = \bar{\imath}$, and that $\bar{K}$ is in the yz plane, so

$$\bar{a}_B = -40(\omega_1)^2\bar{\imath} - 40\dot{\omega}_1(\cos\theta\,\bar{k} - \sin\theta\,\bar{\jmath})$$

$$= -40(\omega_1)^2\bar{\imath} + 40\dot{\omega}_1\sin\theta\,\bar{\jmath} - 40\dot{\omega}_1\cos\theta\,\bar{k}$$

We shall now consider how the two approaches for describing the angular motion of xyz affect the determination of $\bar{a}_P$.

Approach a: xyz fixed to the main arm

In this approach the xyz system rotates only about the fixed Y axis at ω_1. Thus, following the method for angular motion detailed on page 344, the general expressions for the rotation of xyz are

$$\bar{\omega} = \omega_1\bar{J} \qquad \text{always}$$

$$\bar{\alpha} = \dot{\omega}_1\bar{J} + \omega_1\dot{\bar{J}} = \dot{\omega}_1\bar{J} \qquad \text{always}$$

The $(\bar{\imath}, \bar{\jmath}, \bar{k})$ components of $\bar{\alpha}$ and $\bar{\omega}$ at this instant are found by transforming the unit vectors. Hence, using the end view in our sketch to express the components of the unit vector $\bar{J}$, we find that

$$\bar{\omega} = \omega_1(\cos\theta\,\bar{\jmath} + \sin\theta\,\bar{k}) \qquad \text{instantaneously}$$

$$\bar{\alpha} = \dot{\omega}_1(\cos\theta\,\bar{\jmath} + \sin\theta\,\bar{k}) \qquad \text{instantaneously}$$

Now we consider the relative motion of the rider's eyes. As we can see from the sketch,

$$\bar{r}_{P/B} = 3\bar{\jmath} \text{ ft}$$

The motion of the cockpit, and thus the rider, relative to xyz (i.e., relative to the main arm) consists of a rotation about the x axis at angular speed ω_2. The rider's eyes are 3 ft from the x axis. Hence, we have

$$(\bar{v}_{P/B})_{\text{rel}} = 3\omega_2(\bar{e}_\theta)_P$$

$$(\bar{a}_{P/B})_{\text{rel}} = -3(\omega_2)^2(\bar{e}_R)_P + 3\dot{\omega}_2(\bar{e}_\theta)_P$$

For this relative motion the radial direction at point P is along the y axis, and the transverse direction in the sense of increasing θ is parallel to the negative z axis. Thus

$$(\bar{v}_{P/B})_{\text{rel}} = 3\omega_2(-\bar{k})$$

$$(\bar{a}_{P/B})_{\text{rel}} = -3(\omega_2)^2\bar{j} + 3\dot{\omega}_2(-\bar{k})$$

To conclude the solution, we substitute the individual terms into equation (22) for $\bar{a}_P$. This gives

$$\begin{aligned}
\bar{a}_P = {} & [-40(\omega_1)^2\bar{i} + 40\dot{\omega}_1 \sin\theta\,\bar{j} - 40\dot{\omega}_1 \cos\theta\,\bar{k}] \\
& + [-3(\omega_2)^2\bar{j} - 3\dot{\omega}_2\bar{k}] + [\dot{\omega}_1(\cos\theta\,\bar{j} \\
& + \sin\theta\,\bar{k}) \times 3\bar{j}] + 2[\omega_1(\cos\theta\,\bar{j} \\
& + \sin\theta\,\bar{k}) \times (-3\omega_2\bar{k})] + \omega_1(\cos\theta\,\bar{j} \\
& + \sin\theta\,\bar{k}) \times [\omega_1(\cos\theta\,\bar{j} + \sin\theta\,\bar{k}) \times 3\bar{j}] \\
= {} & -[40(\omega_1)^2 + 3\dot{\omega}_1 \sin\theta + 6\omega_1\omega_2 \cos\theta]\bar{i} \\
& + [40\dot{\omega}_1 \sin\theta - 3(\omega_2)^2 - 3\omega_1{}^2 \sin^2\theta]\,\bar{j} \\
& + [-40\dot{\omega}_1 \cos\theta - 3\dot{\omega}_2 + 3(\omega_1)^2 \sin\theta \cos\theta]\bar{k}
\end{aligned}$$

The unit of length used in this problem was feet, so if the rotation rates are radians per second, the solution for $\bar{a}_P$ has the units of feet per second squared.

Approach b: xyz fixed to the cockpit

In this approach xyz has two rotational effects: a rotation with the main arm about the fixed Y axis at the rate ω_1 and a rotation relative to the main arm about the axis C–C at the rate ω_2. Note that the axis C–C is always the same as the x axis in this approach, and that the rotation about that axis is negative according to the right hand rule. The general expressions for the rotation of xyz are therefore

$$\bar{\omega} = \omega_1\bar{J} + \omega_2(-\bar{i})$$

$$\bar{\alpha} = \dot{\omega}_1\bar{J} + \omega_1\dot{\bar{J}} - \dot{\omega}_2\bar{i} - \omega_2\dot{\bar{i}}$$

The Y axis is fixed, so $\dot{\bar{J}} = 0$. Because the unit vector $\bar{\imath}$ has the same rotation as the xyz frame, we have $\dot{\bar{\imath}} = \bar{\omega} \times \bar{\imath}$. Thus

$$\bar{\omega} = \omega_1 \bar{J} - \omega_2 \bar{\imath} \qquad \text{always}$$

$$\bar{\alpha} = \dot{\omega}_1 \bar{J} - \dot{\omega}_2 \bar{\imath} - \omega_2 (\bar{\omega} \times \bar{\imath}) \qquad \text{always}$$

We find the instantaneous values of $\bar{\omega}$ and $\bar{\alpha}$ in terms of their $(\bar{\imath}, \bar{\jmath}, \bar{k})$ components by transforming all unit vectors. Thus, using the sketch to express the components of $\bar{J}$, we obtain

$$\bar{\omega} = \omega_1 (\cos \theta\, \bar{\jmath} + \sin \theta\, \bar{k}) - \omega_2 \bar{\imath} \qquad \text{instantaneously}$$

$$\begin{aligned}
\bar{\alpha} &= \dot{\omega}_1 (\cos \theta\, \bar{\jmath} + \sin \theta\, \bar{k}) - \dot{\omega}_2 \bar{\imath} \\
&\quad - \omega_2 [\omega_1 (\cos \theta\, \bar{\jmath} + \sin \theta\, \bar{k}) + \omega_2\, \bar{\imath}] \times \bar{\imath} \\
&= -\dot{\omega}_2 \bar{\imath} + (\dot{\omega}_1 \cos \theta - \omega_1 \omega_2 \sin \theta)\, \bar{\jmath} \\
&\quad + (\dot{\omega}_1 \sin \theta + \omega_1 \omega_2 \cos \theta) \bar{k} \qquad \text{instantaneously}
\end{aligned}$$

Before continuing, let us consider the meaning of the terms containing the product $\omega_1 \omega_2$. The vector representing the rotation of xyz relative to the main arm is $-\omega_2 \bar{\imath}$. This vector rotates with the main arm about the Y axis. Therefore at the instant we are considering, the tip of the unit vector $\bar{\imath}$ is moving out of the plane of the sketch, that is, in the negative z direction. The terms containing $\omega_1 \omega_2$ represent this effect in terms of the $(\bar{\imath}, \bar{\jmath}, \bar{k})$ components.

The relative motion in this approach is very simple. From the diagram we see that

$$\bar{r}_{P/B} = 3\bar{\jmath} \text{ ft}$$

As long as the rider's head is held stationary relative to the cockpit, the person's eyes have no movement relative to xyz. Thus

$$(\bar{v}_{P/B})_{\text{rel}} = \bar{0}$$

$$(\bar{a}_{P/B})_{\text{rel}} = \bar{0}$$

Finally, we substitute into equation (22) for $\bar{a}_P$ to obtain

$$\begin{aligned}
\bar{a}_P &= -40(\omega_1)^2 \bar{\imath} + 40\dot{\omega}_1 \sin \theta\, \bar{\jmath} - 40\dot{\omega}_1 \cos \theta \bar{k} \\
&\quad + \bar{0} + [-\dot{\omega}_2 \bar{\imath} + (\dot{\omega}_1 \cos \theta - \omega_1 \omega_2 \sin \theta) \bar{\jmath} \\
&\quad + (\dot{\omega}_1 \sin \theta + \omega_1 \omega_2 \cos \theta) \bar{k}] \times 3\bar{\jmath} \\
&\quad + \bar{0} + (-\omega_2 \bar{\imath} + \omega_1 \cos \theta\, \bar{\jmath} + \omega_1 \sin \theta\, \bar{k}) \\
&\quad \times [(-\omega_2 \bar{\imath} + \omega_1 \cos \theta\, \bar{\jmath} + \omega_1 \sin \theta\, \bar{k}) \times 3\bar{\jmath}]
\end{aligned}$$

$$= -[40(\omega_1)^2 + 3\dot\omega_1 \sin\theta + 6\omega_1\omega_2 \cos\theta]\bar\imath$$

$$+[40\dot\omega_1 \sin\theta - 3(\omega_2)^2 - 3(\omega_1)^2 \sin^2\theta]\,\bar\jmath$$

$$+[-40\dot\omega_1 \cos\theta - 3\dot\omega_2 + 3(\omega_1)^2 \sin\theta \cos\theta]\bar{k}$$

Note that the answers from Approaches a and b are identical. A comparison of the two approaches leads to the same thinking as that discussed at the conclusion of Example 2. Basically, you should be familiar with both approaches.

Perhaps the most impressive feature of Example 4 is that we were able to solve the problem at all. Such a problem is beyond the scope of the simple kinematical methods of Module II. The two approaches we employed in utilizing the relative motion formulas are not the only ones possible, but they can be regarded as the most direct approaches.

HOMEWORK PROBLEMS

VI.11 In Problem VI.2, determine the absolute acceleration of each slider in the position shown.

VI.12 The pin P slides within the circular groove of the rectangular plate at a constant relative speed of 50 in./s. In the position shown, the plate has an angular velocity of 5 rad/s and an angular acceleration of 1 rad/s, both clockwise. Determine the acceleration of the pin if at this instant it is located at (a) point A, (b) point B.

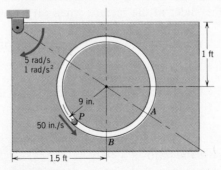

5 rad/s
1 rad/s²

1 ft

9 in.

A

50 in./s

P

B

1.5 ft

Prob. VI.12

VI.13 In Problem VI.3, determine the absolute acceleration of each slider in the position shown.

VI.14 The radar equipment on board aircraft A follows aircraft B by measuring the distance R and angle ϕ with respect to the aircraft's longitudinal axis. Both aircraft are in level flight at the same altitude. Aircraft

A is in a tight horizontal turn having a radius of 1000 m, and its speed is constant at 400 km/hr. At the instant shown, the radar indicates that $R = 500$ m, $\phi = 75°$, $\dot{\phi} = -0.1$ rad/s, $\dot{R} = 80$ m/s, $\ddot{R} = 60$ m/s², $\ddot{\phi} = 0.04$ rad/s². Determine the absolute velocity and absolute acceleration of aircraft B. Show the results in a sketch.

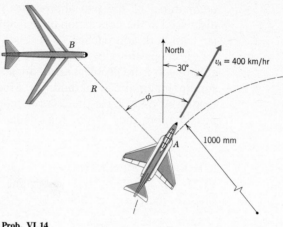

Prob. VI.14

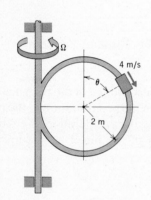

Prob. VI.15

VI.15 At the instant shown, the circular hoop is rotating about the vertical axis at $\Omega = 20$ rad/s, and $\dot{\Omega} = 5$ rad/s². The block is moving at a constant speed of 4 m/s relative to the hoop. Determine the acceleration of the slider at the instant when (a) $\theta = 0°$, (b) $\theta = 90°$, (c) $\theta = 60°$.

VI.16 Determine the acceleration of the slider in Problem VI.6 as a function of time.

VI.17 Determine the acceleration of the probe of the antenna in Problem VI.9 when $t = 0.25$ s using the two methods outlined there.

VI.18 The propeller of an airplane is rotating at 1000 rev/min as the airplane follows a 1000-m vertical circle at 360 km/hr. When the airplane is at the top of the circle, determine the angular acceleration of the propeller.

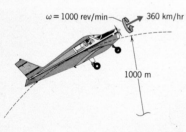

Prob. VI.18

VI.19 In Problem VI.10, the motor is rotating downward at 5 rad/s and this rate is increasing at 10 rad/s². The disk is rotating about the axis of the motor shaft at the constant rate of 900 rev/min. For the instant when $\theta = 30°$ determine (a) the angular acceleration of the disk, (b) the acceleration of point C on the disk.

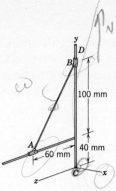

VI.20 The motion of rod AB is constrained by a collar at each end. Collar B moves outward along rod CD at the constant rate of 30 mm/s. Point C is fixed. Determine the velocity and acceleration of each collar for the position shown when $\bar{\omega}$ of the axes is the constant (a) $5\bar{\imath}$ rad/s, (b) $5\bar{\imath} + 5\bar{\jmath}$ rad/s.

Prob. VI.20

E. THE EARTH AS A MOVING REFERENCE FRAME

The principles we have developed in this and previous modules have all depended on the existence of a fixed point that could serve as a point of reference. However, according to Einstein's theory of relativity, no point in the universe is truly fixed. Nevertheless, in most engineering problems concerning phenomena close to the earth's surface, we can neglect the motion of the center of the earth in its orbit about the sun, thereby making the earth's center a fixed point of reference.

In this section we shall study the effect of the fact that the earth is rotating about an axis that contains this fixed point of reference and passes through the north and south poles. Specifically, let us determine the effect of the earth's rotation on the motion of the particle P shown in Figure 8.

We shall consider an elementary model for the earth. In this model the earth is regarded as a homogeneous, perfect sphere of radius $R_e = 6370$ km rotating about its fixed polar axis in a counter-clockwise direction as seen by a fixed observer looking down at the North Pole. This rate of rotation is 1 rev/day, so that

$$\omega_e = \frac{2\pi \text{ rad}}{23.934 \text{ hr}} \times \frac{1 \text{ hr}}{3600 \text{ s}} = 7.292 \times (10^{-5}) \text{ rad/s}$$

Therefore, our model ignores such effects as deviation from a spherical shape, local variation in the mass density of the earth, and the very slight wobble (nutation) of the polar axis.

Let the resultant force on the particle of Figure 8 be $\bar{F}$. From Newton's second law we know that the absolute acceleration of the particle is

$$\bar{a}_P = \frac{\bar{F}}{m}$$

To describe the acceleration we use an xyz reference frame which is attached to the earth with origin O' at the earth's surface. The relative motion equation (22) then gives

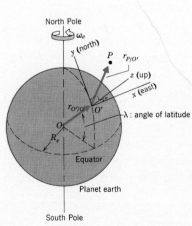

Figure 8

$$\bar{a}_P = \frac{\bar{F}}{m} = \bar{a}_{O'} + (\bar{a}_{P/O'})_{\text{rel}} + \bar{\alpha} \times \bar{r}_{P/O'} + 2\bar{\omega} \times (\bar{v}_{P/O'})_{\text{rel}} + \bar{\omega} \times (\bar{\omega} \times \bar{r}_{P/O'}) \quad (24)$$

Using our knowledge of the earth's motion, let us study the individual terms in this equation.

The angular velocity of the earth is parallel to the polar axis. The xyz frame is attached to the surface of the earth, so the angular velocity vector is

$$\bar{\omega} = \omega_e(\cos \lambda \, \bar{j} + \sin \lambda \, \bar{k}) \qquad (25)$$

Further, because we regard both the polar axis and the rate of rotation to be invariant, it follows that

$$\bar{\alpha} = \bar{0}$$

To determine $\bar{a}_{O'}$ we note that point O' is a point on the earth's surface. It therefore follows a circular path in a plane perpendicular to the polar axis. To obtain an explicit expression for $\bar{a}_{O'}$ we use equation (23) for rigid body motion, which yields

$$\bar{a}_{O'} = \bar{a}_O + \bar{\alpha} \times \bar{r}_{O'/O} + \bar{\omega} \times (\bar{\omega} \times \bar{r}_{O'/O})$$

$$= \bar{\omega} \times (\bar{\omega} \times \bar{r}_{O'/O})$$

Comparing this result to the last term in equation (24), $\bar{\omega} \times (\bar{\omega} \times \bar{r}_{P/O'})$, we can conclude that for a particle near the earth's surface,

$$|\bar{r}_{P/O'}| \ll |\bar{r}_{O'/O}| = R_e = 6370 \text{ km} \qquad (26)$$

Hence the term $\bar{\omega} \times (\bar{\omega} \times \bar{r}_{P/O'})$ is negligible compared to $\bar{a}_{O'}$.

With the foregoing in mind, equation (24) becomes

$$\frac{\bar{F}}{m} = (\bar{a}_{P/O'})_{\text{rel}} + 2\bar{\omega} \times (\bar{v}_{P/O'})_{\text{rel}} + \bar{\omega} \times (\bar{\omega} \times \bar{r}_{O'/O}) \qquad (27)$$

where the kinematical quantities for an observer on the earth at point O' are

$$\bar{r}_{P/O'} = x\bar{i} + y\bar{j} + z\bar{k} \qquad (28)$$

$$(\bar{v}_{P/O'})_{\text{rel}} = \dot{x}\bar{i} + \dot{y}\bar{j} + \dot{z}\bar{k}$$

$$(\bar{a}_{P/O'})_{\text{rel}} = \ddot{x}\bar{i} + \ddot{y}\bar{j} + \ddot{z}\bar{k}$$

In the examples that follow we shall consider two cases where equation (27) has special significance.

EXAMPLE 5

A ball is suspended near the earth's surface from a cord and is brought to rest relative to the earth. Determine the force in the cord and the direction in which the cord is oriented.

Solution

We first draw a free body diagram of the ball that also shows the xyz frame fixed to the surface of the earth.

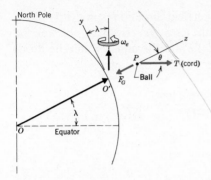

Note that we have given the cord an arbitrary orientation, rather than assuming it to be oriented along the radial line from the center of the earth. Also, we do not assume that the gravitational force $\bar{F}_G$ is the weight. The reasons for doing this will be obvious in the work that follows.

The magnitude of $\bar{F}_G$ is given by the universal law of gravitation, which is given by

$$|\bar{F}_G| = \frac{GM_e m}{R_e^2}$$

Therefore, from the free body diagram we see that the force sum is

$$\bar{F} = \left(T\cos\theta - \frac{GM_e m}{R_e^2}\right)\bar{k} - T\sin\theta\,\bar{j}$$

The ball is constrained to have no motion relative to the earth, so equation (27) reduces to

$$\frac{\bar{F}}{m} = \bar{\omega}\times(\bar{\omega}\times\bar{r}_{O'/O})$$

As we have seen in the general discussion, $\bar{\omega} = \omega_e(\cos\lambda\,\bar{j} + \sin\lambda\,\bar{k})$ and $\bar{r}_{O'/O} = R_e\bar{k}$. The centripetal acceleration is therefore

$$\bar{\omega}\times(\bar{\omega}\times\bar{r}_{O'/O}) = R_e\omega_e^2\cos\lambda(\sin\lambda\,\bar{j} - \cos\lambda\,\bar{k})$$

Notice that the computation shows that the centripetal acceleration is perpendicular to the polar axis, as it should be.

By equating this term to $\bar{F}/m$, we have the two scalar equations

$$\frac{T}{m}\cos\theta - \frac{GM_e}{R_e^2} = -R_e\omega_e^2\cos^2\lambda$$

$$-\frac{T}{m}\sin\theta = R_e\omega_e^2\sin\lambda\cos\lambda$$

Solving for T and θ, we find that

$$T = m\left[\left(\frac{GM_e}{R_e^2}\right)^2 - R_e\omega_e^2\left(2\frac{GM_e}{R_e^2} - R_e\omega_e^2\right)\cos^2\lambda\right]^{1/2} \qquad \triangleleft$$

$$\tan\theta = -\frac{R_e\omega_e^2\sin\lambda\cos\lambda}{(GM_e/R_e^2) - R_e\omega_e^2\cos^2\lambda} \qquad \triangleleft$$

Let us now interpret this solution. In the sense of a statics problem, we would say that the tension T equals the weight mg, where g is the *local* free fall acceleration. We would also say that the cord is oriented vertically. From the results we see that the apparent weight of the body is not solely the force of gravity and that the vertical direction is not truly along the radial line from the center of the earth. However, from the solution for θ we can see that $\theta \equiv 0$ at $\lambda = 0, \pm 90°$, (i.e., at the equator and the north and south poles). Furthermore, because

$$\frac{GM}{R_e^2} = \frac{6.670(10^{-11})(5.976)(10^{24})}{(6.370)^2(10^6)^2} \text{ m/s}^2$$

$$= 9.820 \text{ m/s}^2$$

and

$$R_e\omega_e^2 = (6.371)(10^6)(7.292)^2(10^{-5})^2$$

$$= 3.388(10^{-2}) \text{ m/s}^2$$

we can show that the maximum value $\left|\theta\right| = 0.099°$ occurs at $\lambda = \pm 44.90°$. This value of θ is sufficiently small to consider it negligible in engineering problems. Similarly, if we let $T = mg$, we can see that the value of g increases steadily from $g = 9.786$ m/s² at the equator to $g = 9.820$ m/s² at the poles. This fluctuation is not significant to the usual accuracy required for engineering problems, and hence it has become common practice to use $g = 9.806$ m/s² as an average. However, for completeness, we note that there is an *international gravity formula* that accounts for both the rotation of the earth and the deviation of the earth from a spherical shape. It is

$$g = 9.780(1 + 0.005288 \sin^2 \lambda - 0.000006 \sin^2 2\lambda) \text{ m/s}^2$$

EXAMPLE 6

A low-pressure area forms within a fluid, such as an open drain in a tank of water, or a low-pressure atmospheric weather system. Assuming that the fluid is constrained to move solely in a horizontal plane, describe the resulting fluid motion.

Solution

In the previous example we saw that the term $\bar{a}_{O'} = \bar{\omega} \times (\bar{\omega} \times \bar{r}_{O'/O})$ is essentially a small correction to the weight force. We can therefore neglect this term when we apply equation (27). Hence, we have the following expression for the relative acceleration:

$$(\bar{a}_{P/O'})_{\text{rel}} = \frac{\bar{F}}{m} - 2\bar{\omega} \times (\bar{v}_{P/O'})_{\text{rel}}$$

We know that the relative velocity and acceleration are both parallel to the horizontal (*xy*) plane, because the fluid particles are constrained to move in this plane. We therefore need consider only the horizontal component of the Coriolis acceleration, which depends only on the vertical component of $\bar{\omega}$,

$$(\bar{a}_{\text{cor}})_{\text{hor}} = 2(\omega_z \bar{k}) \times (\bar{v}_{P/O'})_{\text{rel}}$$

Also, from equation (25) we know that

$$\omega_z = \omega_e \sin \lambda$$

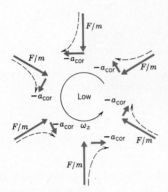

The horizontal component of the force $\bar{F}$ is due to the pressure difference, and is therefore directed toward the center of the low-pressure area.

We now introduce some qualitative reasoning. As a particle of fluid starts from rest relative to the earth, the pressure difference causes the particle to accelerate toward the center of the low pressure area. Therefore, $(\bar{v}_{P/O'})_{\text{rel}}$ initially is also toward the center of the low-pressure area. The initial relative acceleration is the vector sum of the terms $\bar{F}/m$ and $-(\bar{a}_{\text{cor}})_{\text{hor}} = -2(\omega_z \bar{k}) \times (\bar{v}_{P/O'})_{\text{rel}}$; the latter term, of course, is perpendicular to $(\bar{v}_{P/O'})_{\text{rel}}$. This vector sum is illustrated in the sketch, which shows a number of fluid particles around a low-pressure region in the Northern Hemisphere.

The dashed lines are a trace of the path that the particles follow. We see that the Coriolis effect causes a relative acceleration that is to the right of the initial direction of motion, thus causing the particle to assume a counterclockwise motion about the low pressure region. In the Southern Hemisphere the value of ω_z is negative, thus giving a Coriolis term that is opposite in sign from the one shown in the sketch.

We conclude that the fluid motion will consist of an inward movement, plus a counterclockwise rotation in the Northern Hemisphere or a clockwise rotation in the Southern Hemisphere.

Knowledge of this effect is essential to the understanding of meteorological and oceanographic phenomena, such as hurricanes, whirlpools, and ocean currents. However, the preceding discussion does not provide a full description because we have not considered the forces that are generated within the fluid by its motion.

HOMEWORK PROBLEMS

VI.21 A 1500-kg automobile is traveling north at 90 km/hr along a straight road. It is in the Northern Hemisphere, at 45° latitude. Determine the reaction force exerted by the ground on the vehicle in terms of north, east, and vertical components.

VI.22 The automobile in Problem VI.21 is moving (a) south, (b) west. Determine the corresponding reaction forces.

VI.23 A ball bearing rolls on a frictionless horizontal plane located at latitude 60° north. It was given an initial velocity of 20 ft/s east. Determine

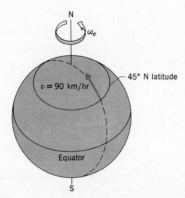

Prob. VI.21

the acceleration of the ball bearing and the radius of curvature of its path, both with respect to the earth, at the instant it was released.

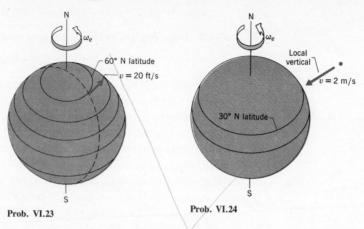

Prob. VI.23   Prob. VI.24

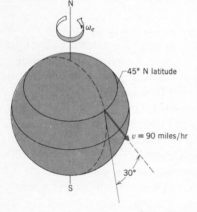

Prob. VI.26

VI.24 An object is falling vertically at 25 m/s in the Northern Hemisphere, at 30° latitude. Neglecting air resistance, determine the acceleration of the object and the radius of curvature of its path, both with respect to the earth.

VI.25 A block of mass m, released from rest, slides down a vertical pole without opposition from friction. Derive expressions for the horizontal components of the contact force exerted by the pole on the block in terms of the time t that the block has been falling and the angle of latitude λ.

VI.26 A baseball is thrown south with a velocity of 90 miles/hr at an angle of elevation of 30° above the horizontal. The location is at latitude 45° north. Neglecting air resistance, determine the initial acceleration of the baseball with respect to the earth.

F. PROBLEM SOLVING

Following the approach of previous modules, we now present a systematic procedure for the determination of expressions for linear and angular velocity and acceleration. These expressions may be the only results required in a problem, or they may represent the kinematical input needed for the study of the kinetics of the system.

1 Sketch the system, showing the axes of rotation and, symbolically, the angular speeds about these axes. This sketch should also show all significant points, including points whose motion is desired and all points that are constrained.

2 Show your choice for the moving reference frame xyz in the sketch. The general criteria for this step are summarized as follows:

a Let the origin O' of xyz be a point in the system that is executing a simple motion, such as a point following a known planar path.

b   Let *xyz* have as many rotational velocities as you can easily comprehend, in order to expedite the description of $(\bar{v}_{P/O'})_{\text{rel}}$ and $(\bar{a}_{P/O'})_{\text{rel}}$. In the situation where it is desired to study the angular motion of a specific body, attach *xyz* to that body.

c   Choose the instantaneous orientation of the *x*, *y*, and *z* axes to give a simple description for the geometry of the system, in particular, the description of the axes of rotation and of the relative position vectors.

When some aspect of the rotation of the system is described with respect to a fixed observer, select a fixed reference frame *XYZ* such that its origin *O* is a convenient fixed point and one of the axes is parallel to the fixed axis of rotation. (For an example of a situation where the fixed *XYZ* system is not required, see Illustrative Problem 3.)

3   Following the method described on page 344, obtain a general description for the angular velocity $\bar{\omega}$ of the *xyz* frame. If the problem involves acceleration, differentiate this expression to obtain a general expression for $\bar{\alpha}\ (= \dot{\bar{\omega}})$. The instantaneous expressions for $\bar{\omega}$ and $\bar{\alpha}$ are then obtained by replacing the components of the $(\bar{I}, \bar{J}, \bar{K})$ unit vectors by the $(\bar{i}, \bar{j}, \bar{k})$ representation of the *xyz* frame. In problems where constraint conditions must be satisfied, it will prove useful to leave all velocity and acceleration rates in algebraic (symbolic) form.

Velocity Problem

4   If any angular speeds were not given, relate the velocities of all constrained points to the rates of rotation using the relative velocity equation (16). The rates of rotation are obtained by satisfying the constraints on the velocities of these points. Then substitute all rotation rates into the expressions for $\bar{\omega}$ and $\bar{\alpha}$.

5   If the problem requires the velocity of any unconstrained points, use equation (16) for these determinations.

Acceleration Problem

6 and 7   Using the relative acceleration equation (22), repeat the procedures of steps 4 and 5 in terms of the acceleration variables and the restrictions on the accelerations of the constrained points.

ILLUSTRATIVE PROBLEM 1

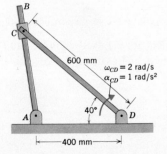

The linkage arm *CD* rotates clockwise about the pivot *D* with an angular velocity of 2 rad/s and an angular acceleration of 1 rad/s². The collar at *C* is pinned to rod *CD* and slides along rod *AB*. For the position shown, determine (a) the angular velocity and acceleration of rod *AB*, (b) the velocity and acceleration of the collar relative to rod *AB*.

Solution

This problem is similar to the planar linkage problems of Module IV where all members were pinned together. The new feature of this problem is that the collar allows for motion of point C *relative* to bar AB. To describe this motion we shall employ a moving reference frame. On the other hand, the collar is pinned to rod CD, so the rigid body motion formulas will be applied to describe the motion of this rod.

Step 1

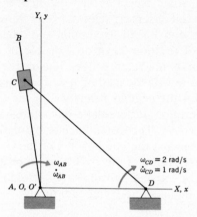

Step 2   The fixed reference frame is selected such that the XY plane is the plane of the problem, and its origin is placed at point A because we shall be treating motion relative to bar AB. Its orientation is not crucial to this problem, but is chosen so that the X axis coincides with the fixed line AD. To describe the motion relative to rod AB, we attach the moving xyz frame to this rod with its origin also at the fixed point A. At the instant of interest the x axis is also selected to coincide with line AD; this expedites the writing of the position vectors. Both the fixed and moving reference frames are shown in the sketch in step 1.

Step 3   Because the xyz frame is attached to rod AB, and because this is a planar motion problem, the rotation of this reference frame is given by

$$\bar{\omega} = \omega_{AB}\,(-\bar{K}) = -\omega_{AB}\bar{k}$$

$$\bar{\alpha} = \dot{\omega}_{AB}(-\bar{K}) = -\alpha_{AB}\bar{k}$$

The magnitudes ω_{AB} and $\dot{\omega}_{AB}$ are two variables that we seek as part of the solution.

Step 4   The constrained points in this problem are the fixed points A and D, and point C on the collar. We can relate points A and C using the relative motion equation (16). Thus

$$\bar{v}_C = \bar{v}_A + (\bar{v}_{C/A})_{\text{rel}} + \bar{\omega} \times \bar{r}_{C/A}$$

However, point C is also a point on rod CD. Its velocity is therefore given by the rigid body equation (17), which becomes

$$\bar{v}_C = \bar{v}_D + \bar{\omega}_{CD} \times \bar{r}_{C/D}$$

Because points A and D are fixed, $\bar{v}_A = \bar{v}_D = \bar{0}$. Also, we know from the given information that

$$\bar{\omega}_{CD} = 2(-\bar{k}) \text{ rad/s}$$

The two expressions for $\bar{v}_C$ will be equated, but we must first describe the individual terms in these expressions. Toward this end we draw a sketch that illustrates the instantaneous geometry of the system.

From this sketch we see that

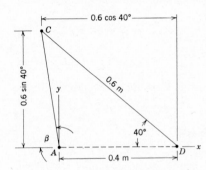

$$\bar{r}_{C/A} = -(0.60 \cos 40° - 0.40)\bar{i} + 0.60 \sin 40° \, \bar{j}$$

$$= -0.0596\bar{i} + 0.3857\bar{j} \text{ m}$$

$$\bar{r}_{C/D} = -0.60 \cos 40° \, \bar{i} + 0.60 \sin 40° \, \bar{j}$$

$$= -0.4596\bar{i} + 0.3857\bar{j} \text{ m}$$

$$\beta = \tan^{-1}\left(\frac{0.3857}{0.0596}\right) = 81.21°$$

The angle β is needed to describe $(\bar{v}_{C/A})_{\text{rel}}$. Noting that the collar is free to slide along rod AB, this relative velocity must be parallel to the rod. The magnitude of this term, denoted by v_{rel}, is one of the unknown quantities that is being sought. Assuming the relative velocity to be outward, we therefore have

$$(\bar{v}_{C/A})_{\text{rel}} = v_{\text{rel}}(-\cos \beta \, \bar{i} + \sin \beta \, \bar{j})$$

$$= v_{\text{rel}}(-0.1528\bar{i} + 0.9883\bar{j})$$

Substituting the appropriate terms into the first expression for $\bar{v}_C$, we find that

$$\bar{v}_C = \bar{0} + v_{\text{rel}}(-0.1528\bar{i} + 0.9883\bar{j}) + (-\omega_{AB}\bar{k}) \times (-0.0596\bar{i} + 0.3857\bar{j})$$

$$= (-0.1528v_{\text{rel}} + 0.3857\omega_{AB})\bar{i} + (0.9883v_{\text{rel}} + 0.0596\omega_{AB})\bar{j} \text{ m/s}$$

Similarly, the second expression for $\bar{v}_C$ gives

$$\bar{v}_C = \bar{0} + (-2\bar{k}) \times (-0.4596\bar{i} + 0.3857\bar{j})$$

$$= 0.7714\bar{i} + 0.9192\bar{j} \text{ m/s}$$

By equating the individual components of $\bar{v}_C$, we obtain the following two simultaneous equations:

$\bar{\imath}$ component: $(v_C)_x = -0.1528v_{rel} + 0.3857\omega_{AB} = 0.7714$

$\bar{\jmath}$ component: $(v_C)_y = 0.9883v_{rel} + 0.0596\omega_{AB} = 0.9192$

The solution of these equations is

$$\omega_{AB} = 2.313 \qquad\qquad \bar{\omega}_{AB} = \bar{\omega} = -2.313\bar{k} \text{ rad/s}$$

$$v_{rel} = 0.7901 \qquad\qquad (\bar{v}_{C/A})_{rel} = -0.1207\bar{\imath} + 0.7809\bar{\jmath} \text{ m/s}$$

Step 5   Not applicable.

Step 6   We repeat the procedures of step 4 using the acceleration variables. Thus,

$$\bar{a}_C = \bar{a}_A + (\bar{a}_{C/A})_{rel} + 2\bar{\omega} \times (\bar{v}_{C/A})_{rel} + \bar{\alpha} \times \bar{r}_{C/A} + \bar{\omega} \times (\bar{\omega} \times \bar{r}_{C/A})$$

$$= \bar{a}_D + \bar{\alpha}_{CD} \times \bar{r}_{C/D} + \bar{\omega} \times (\bar{\omega} \times \bar{r}_{C/D})$$

Noting that points A and D are fixed, $\bar{a}_A = \bar{a}_D = \bar{0}$. From the given information we have

$$\bar{\alpha}_{CD} = 1(-\bar{k}) \text{ rad/s}^2$$

All of the other terms, with the exception of the relative acceleration, are defined in steps 3 and 4. For the relative acceleration we note that because the xyz frame is attached to rod AB, $(\bar{a}_{C/A})_{rel}$ must account for the sliding motion of the collar along the rod. Thus, letting $|(\bar{a}_{C/A})_{rel}| \equiv a_{rel}$, we have

$$(\bar{a}_{C/A})_{rel} = a_{rel}(-\cos\beta\,\bar{\imath} + \sin\beta\,\bar{\jmath}$$

$$= a_{rel}(-0.1528\bar{\imath} + 0.9883\bar{\jmath})$$

When we substitute the individual terms into the first expression for $\bar{a}_C$, we obtain

$$\bar{a}_C = \bar{0} + a_{rel}(-0.1528\bar{\imath} + 0.9883\bar{\jmath})$$
$$+ 2(-2.313\bar{k}) \times (-0.1207\bar{\imath} + 0.7809\bar{\jmath})$$
$$+ (-\alpha_{AB}\bar{k}) \times (-0.0596\bar{\imath} + 0.3857\bar{\jmath})$$
$$+ (-2.313\bar{k}) \times [(-2.313\bar{k}) \times (-0.0596\bar{\imath} + 0.3875\bar{\jmath})]$$
$$= (-0.1528a_{rel} + 3.612 + 0.3857\alpha_{AB} + 0.3189)\bar{\imath}$$
$$+ (0.9883a_{rel} + 0.5584 + 0.0596\alpha_{AB} + 2.064)\bar{\jmath} \text{ m/s}^2$$

The second expression for $\bar{a}_C$ gives

$$\bar{a}_C = \bar{0} + (-\bar{k}) \times (-0.4596\bar{\imath} + 0.3875\bar{\jmath})$$

$$+ (-2\bar{k}) \times [(-2\bar{k}) \times (-0.4596\bar{i} + 0.3875\bar{j})]$$

$$= 2.226\bar{i} - 1.090\bar{j} \text{ m/s}^2$$

The two scalar equations obtained by equating the two expressions for $\bar{a}_C$ are

$\bar{i}$ component: $\quad (a_C)_x = -0.1528a_{\text{rel}} + 3.931 + 0.3875\dot{\omega}_{AB}$

$$= 2.226$$

$\bar{j}$ component: $\quad (a_C)_y = 0.9883a_{\text{rel}} + 2.622 + 0.0596\dot{\omega}_{AB}$

$$= -1.090$$

The solution is

$$\alpha_{AB} = -5.774 \qquad \bar{\alpha}_{AB} = \bar{\alpha} = -5.744\bar{k} \text{ rad/s}$$

$$a_{\text{rel}} = -3.408 \qquad (\bar{a}_{C/A})_{\text{rel}} = 0.5207\bar{i} - 3.368\bar{j} \text{ m/s}^2$$

In closing, let us note the physical significance of the method of the solution. The given rotation values for rod CD define the absolute velocity and acceleration of the slider. The relative motion equations then use the component of these kinematic values perpendicular to rod AB to define the angular velocity and angular acceleration of that rod, whereas the components parallel to rod AB are used to define the velocity and acceleration of the slider relative to rod AB.

ILLUSTRATIVE PROBLEM 2

Cone A is fixed in space and cone B rolls without slipping over the surface of cone A, making five revolutions about the fixed vertical axis each minute. Determine the angular velocity and angular acceleration of cone B.

Solution

The motion of cone B consists of two rotations: the rotation ω_1 about the vertical axis of cone A, which requires 1 minute for 5 rev, plus a rotation of the cone about its own axis at an angular speed ω_2. The value of ω_2 will be determined by satisfying the condition that there is no slipping in the rolling motion.

Step 1 The line of contact CD between the two cones is illustrated by a side view. The angles β and γ defined in the sketch are obtained by solving the appropriate right triangles, specifically,

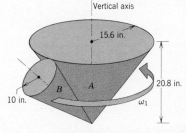

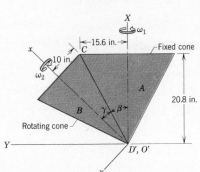

$$\beta = \tan^{-1}\left(\frac{15.6}{20.8}\right) = \tan^{-1}\left(\frac{3}{4}\right) = 36.87°$$

$$\gamma = \sin^{-1}\left(\frac{10}{\sqrt{(15.6)^2 + (20.8)^2}}\right) = \sin^{-1}\left(\frac{5}{13}\right) = 22.62°$$

Step 2 Point D, which is the apex of cone B, is the only point on this cone that is fixed. Therefore, choose this point as the origin of the fixed XYZ system, letting the X axis coincide with the vertical axis and the Y axis be parallel to the plane of the sketch in step 1. Now attach the moving xyz frame to cone B so that the angular motion of the reference frame will be the same as that of the cone. It is convenient to select point D to be the origin of this frame also, aligning the x axis along the axis of cone B and placing the y axis in the plane of the sketch at this instant. This orientation suits the description of the axes of rotation and the position vectors of points on cone B. Note that the sense of the rotation ω_2 shown in the sketch is in accordance with our physical intuition. If our intuition should be wrong, the value we obtain for ω_2 will be negative.

Step 3 The total rotation of cone B is the sum of the rotations about the fixed X axis and moving x axis. The general expressions for angular velocity and acceleration are therefore

$$\bar{\omega} = \omega_1 \bar{I} + \omega_2 \bar{\imath} \qquad \text{always}$$

$$\bar{\alpha} = \dot{\omega}_1 \bar{I} + \dot{\omega}_2 \bar{\imath} + \omega_2 \dot{\bar{\imath}}$$

$$= \dot{\omega}_1 \bar{I} + \dot{\omega}_2 \bar{\imath} + \omega_2 (\bar{\omega} \times \bar{\imath}) \qquad \text{always}$$

The information given in the problem is that

$$\omega_1 = 10\,\pi \text{ rad}/60 \text{ s} = 0.5236 \text{ rad/s}$$

and because ω_1 is constant, $\dot{\omega}_1 = 0$. From the sketch in step 1,

$$\bar{I} = \cos(\beta + \gamma)\,\bar{\imath} - \sin(\beta + \gamma)\,\bar{\jmath}$$

$$= 0.5077\bar{\imath} - 0.8615\bar{\jmath}$$

Hence, with the rotation rates in algebraic form, we have

$$\bar{\omega} = \omega_1(0.5077\bar{\imath} - 0.8615\bar{\jmath}) + \omega_2\bar{\imath}$$

$$= (0.5077\omega_1 + \omega_2)\bar{\imath} - 0.8615\omega_1\,\bar{\jmath} \qquad \text{instantaneously}$$

$$\bar{\alpha} = \dot{\omega}_2\bar{\imath} + \omega_2[(0.5077\omega_1 + \omega_2)\bar{\imath} - 0.8615\omega_1\,\bar{\jmath}] \times \bar{\imath}$$

$$= \dot{\omega}_2\bar{\imath} + 0.8615\omega_1\omega_2\bar{k} \qquad \text{instantaneously}$$

Step 4 We now satisfy the constraint condition that there is rolling without slipping. Physically this requirement means that all points on the line CD of cone B that are in contact with line CD of fixed cone A must have zero velocity. To satisfy this condition, we relate the velocities of points C and D on cone B by using equation (17). Thus

$$\bar{v}_C = \bar{v}_D + \bar{\omega}_{\text{cone}} \times \bar{r}_{C/D} = \bar{0} + \bar{\omega} \times \bar{r}_{C/D} \equiv \bar{0}$$

Interpreting this result, recall that $\bar{\omega}$ is parallel to the instantaneous axis of rotation. Hence, it follows that the condition $\bar{\omega} \times \bar{r}_{C/D} = \bar{0}$ means that the instantaneous axis of rotation must be parallel to line CD.

From the sketch we have

$$r_{C/D} = \sqrt{(15.6)^2 + (20.8)^2} \,(\cos\gamma\,\bar{\imath} - \sin\gamma\,\bar{\jmath}$$

$$= 24\bar{\imath} - 10\bar{\jmath} \text{ in.}$$

so that

$$\bar{\omega} \times \bar{r}_{C/D} = [(0.5077\omega_1 + \omega_2)\bar{\imath} - 0.8615\omega_1\,\bar{\jmath}] \times (24\bar{\imath} - 10\bar{\jmath})$$

$$= (-5.08\omega_1 - 10\omega_2 + 20.68\omega_1)\bar{k} = \bar{0}$$

As a result, we obtain

$$\omega_2 = 1.560\omega_1$$

Recall that the sketch represented an *arbitrary* instant. Therefore, this relationship between ω_1 and ω_2 is valid for all time t, thus enabling us to differentiate it to obtain an expression for $\dot{\omega}_2$.

$$\dot{\omega}_2 = 1.560\dot{\omega}_1$$

Then, for the numerical values appearing in step 3 we have

$$\omega_2 = 1.560(0.5236) = 0.8168 \text{ rad/s}$$

$$\dot{\omega}_2 = 0$$

Placing these results into the general form for $\bar{\omega}$ and $\bar{\alpha}$ yields

$$\bar{\omega} = [0.5077(0.5236) + 0.8168]\bar{\imath} - 0.8615(0.5236)\bar{\jmath}$$

$$= 1.083\bar{\imath} - 0.451\bar{\jmath} \text{ rad/s}$$

$$\bar{\alpha} = 0.8615(0.5236)(0.8168)\bar{k}$$

$$= 0.3684\bar{k} \text{ rad/s}$$

The solution is now complete; the remaining steps are unnecessary.

ILLUSTRATIVE PROBLEM 3

The rotations of an airplane about a set of xyz axes attached to the airplane are termed the roll rate ω_x, the pitching rate ω_y, and the yaw rate ω_z. A passenger on a jumbo jet is 30 m behind and 2 m above the center of mass G of the airplane and at that instant is walking aft at the constant rate of 1 m/s. Determine the difference between the acceleration that this passenger experiences and the acceleration that the center of mass of the airplane undergoes, if the roll rate is 0.5 rad/s, the pitching rate is -0.3 rad/s, and the yaw rate is 0.1 rad/s. All rotation rates are constant.

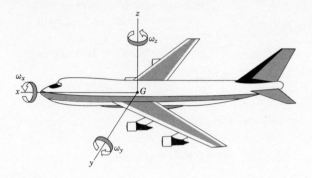

Solution

Step 1 The diagram accompanying the statement of the problem will suffice for the solution.

Step 2 We shall use the moving xyz frame given in the problem, which is attached to the aircraft. Because all rotation rates are referred to the axes of this coordinate system, and because we are not concerned with the absolute acceleration of the origin G, we do not need a fixed reference frame XYZ.

Step 3 The roll, pitch, and yaw rates are with respect to the x, y, and z axes, respectively. Hence

$$\bar{\omega} = \omega_x \bar{\imath} + \omega_y \bar{\jmath} + \omega_z \bar{K} \qquad \text{always and instantaneously}$$

$$\bar{\alpha} = \dot{\omega}_x \bar{\imath} + \omega_x \dot{\bar{\imath}} + \dot{\omega}_y \bar{\jmath} + \omega_y \dot{\bar{\jmath}} + \dot{\omega}_z \bar{k} + \omega_z \dot{\bar{k}}$$

$$= \dot{\omega}_x \bar{\imath} + \dot{\omega}_y \bar{\jmath} \; \dot{\omega}_z \bar{K} + \omega_x(\bar{\omega} \times \bar{\imath}) + \omega_y(\bar{\omega} \times \bar{\jmath}) + \omega_z(\bar{\omega} \times \bar{k})$$

$$= \dot{\omega}_x \bar{\imath} + \dot{\omega}_y \bar{\jmath} + \dot{\omega}_z \bar{k} + \bar{\omega} \times (\omega_x \bar{\imath} + \omega_y \bar{\jmath} + \omega_z \bar{k})$$

$$= \dot{\omega}_x \bar{\imath} + \dot{\omega}_y \bar{\jmath} + \dot{\omega}_z \bar{k} \qquad \text{always and instantaneously}$$

This result for angular acceleration is interesting, because it states that the components of angular acceleration of the airplane are the derivatives of the components of the angular velocity. The fact that this is always true when we fix the moving reference system to the body will have great significance for our treatment of the equations of motion of a rigid body in the next module.

We may now substitute all rotation rates, for all values are given. Thus

$$\bar{\omega} = 0.5\bar{\imath} - 0.3\bar{\jmath} + 0.1\bar{k} \text{ rad/s}$$

The rates of rotation are constant, so $\bar{\alpha} = \bar{0}$.

Steps 4-6 These steps are not applicable, for there are no constraint conditions to satisfy and we do not seek the absolute velocity of any points.

Step 7 We are interested in the value of $(\bar{a}_P - \bar{a}_G)$, where point P is the position of the passenger. This is obtained from equation (22), namely,

$$\bar{a}_P - \bar{a}_G = (\bar{a}_{P/G})_{\text{rel}} + 2\bar{\omega} \times (\bar{v}_{P/G})_{\text{rel}} + \bar{\alpha} \times \bar{r}_{P/G} + \bar{\omega} \times (\bar{\omega} \times \bar{r}_{P/G})$$

From the information given in the problem and the choice of coordinate axes in the sketch, we may write,

$$\bar{r}_{P/G} = -30\bar{i} + 2\bar{k} \text{ m}$$

Also, from the given information we know that the passenger has a constant speed of 1 m/s in the negative x direction relative to the airplane. Thus

$$(\bar{v}_{P/G})_{\text{rel}} = -1.0\bar{i} \text{ m/s}$$

$$(\bar{a}_{P/G})_{\text{rel}} = \bar{0}$$

The solution is then found by substituting the values of $\bar{\omega}$ and $\bar{\alpha}$ in step 4, and the relative motion variables determined above into the acceleration equation. This step gives

$$\bar{a}_P - \bar{a}_G = \bar{0} + 2(0.5\bar{i} - 0.3\bar{j} + 0.1\bar{k}) \times (-1.0\bar{i})$$

$$+ \bar{0} + (0.5\bar{i} - 0.3\bar{j} + 0.1\bar{k}) \times [(0.5\bar{i} - 0.3\bar{j}$$

$$+ 0.1\bar{k}) \times (-30\bar{i} + 2\bar{k})]$$

$$= 3.10\bar{i} + 4.24\bar{j} - 2.78\bar{k} \text{ m/s}^2$$

The magnitude of this result is $0.61g$, which exceeds the value desired in normal operation of passenger aircraft. It should be noted, however, that atmospheric disturbances can cause rotation rates comparable to those considered in this problem.

ILLUSTRATIVE PROBLEM 4

The sketch shows a universal joint of the type commonly employed in the drive train of automobiles. Considering ω_1 as a known value, determine the corresponding value of ω_2 and the angular velocity of the crossbar in the position where arm CD of the crossbar is horizontal and arm AB coincides with the vertical plane, as shown.

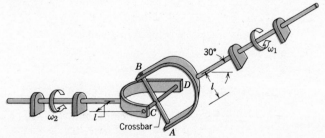

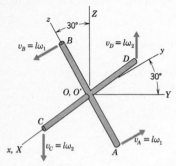

Solution

Step 1   The angular speeds ω_1 and ω_2 are related to each other by the fact that the shafts are connected by a crossbar. For this reason we focus attention on the motion of the crossbar.

Step 2   We place the origin of xyz at the fixed midpoint of the crossbar and attach xyz to that body, so as to eliminate the relative velocity term when we express the velocities of the endpoints. The x and z axes are oriented along the arms in order to expedite writing the position vectors. The fixed XYZ system is oriented horizontally and vertically. This is illustrated in the sketch.

Step 3   Being unsure of the exact nature of the angular motion of the crossbar, we let the angular velocity be an unknown general vector

$$\bar{\omega} = \omega_x \, \bar{i} + \omega_y \, \bar{j} + \omega_z \, \bar{k}$$

Note that we were asked only to relate angular speeds, hence we are not concerned here with accelerations.

Step 4   The constrained points of the crossbar are points B and C (the motion of the other two end points differ only in sign). Point B, being a point that is also on the angled shaft, follows a circular path whose plane is perpendicular to this shaft. Because arm AB is oriented in the vertical plane at this instant, the velocity of point B is $l\omega_1$ horizontally. By similar reasoning the velocity of point C at this instant is $l\omega_2$ vertically. Both velocities are shown in the sketch; in component form they are

$$\bar{v}_B = l\omega_1 \bar{i}$$

$$\bar{v}_C = -l\omega_2 \bar{K} = -l\omega_2(\sin 30° \, \bar{j} + \cos 30° \, \bar{k})$$

These are the (instantaneous) constraint conditions on the velocity. We also know that points B and C are both common to the cross bar. Applying the formula for the velocity of points in a rigid body and using the angular velocity $\bar{\omega}$ from step 3, we then have

$$\bar{v}_C = \bar{v}_B + \bar{\omega} \times \bar{r}_{C/B}$$

$$-l\omega_2(\sin 30° \, \bar{j} + \cos 30° \, \bar{k}) = l\omega_1 \bar{i} + (\omega_x \, \bar{i} + \omega_y \, \bar{j} + \omega_z \, \bar{k}) \times (-l\bar{k} + l\bar{i})$$

$$= (l\omega_1 - l\omega_y)\bar{i} + (l\omega_x + l\omega_z)\bar{j} - l\omega_y \bar{k}$$

Equating components yields

$$\bar{i} \text{ component:} \qquad 0 = l\omega_1 - l\omega_y$$

$$\bar{j} \text{ component:} \qquad -l\omega_2 \sin 30° = l\omega_x + l\omega_z$$

$$\bar{k} \text{ component:} \qquad -l\omega_2 \cos 30° = -l\omega_y$$

Thus

$$\omega_1 = \omega_y = \omega_2 \cos 30°$$

$$\omega_x + \omega_z = -\omega_2 \sin 30°$$

From this we obtain the solution to the problem,

$$\omega_2 = \frac{\omega_1}{\cos 30°} = 1.1547\omega_1$$

Notice that the foregoing equations are not sufficient to define fully the angular motion of the corssbar. A constraint condition we have not employed is that $\bar{v}_0 = \bar{0}$. Relating $\bar{v}_0$ and $\bar{v}_B$, we have

$$\bar{v}_B = \bar{v}_0 + \bar{\omega} \times \bar{r}_{B/0} = l\omega_1\bar{i}$$

$$= \bar{0} + (\omega_x\,\bar{i} + \omega_y\,\bar{j} + \omega_z\,\bar{k}) \times l\bar{k} = l\omega_y\,\bar{i} - l\omega_x\,\bar{j}$$

Matching components then yields

$$\bar{i} \text{ component:} \qquad \omega_1 = \omega_y$$

$$\bar{j} \text{ component:} \qquad 0 = \omega_x$$

We thus find from the earlier equations that

$$\omega_z = -\omega_2 \sin 30° = -\left(\frac{\omega_1}{\cos 30°}\right)\sin 30° = -\omega_1 \tan 30°$$

$$\bar{\omega} = \omega_x\,\bar{i} + \omega_y\,\bar{j} + \omega_z\,\bar{k} = \omega_1(\bar{j} - \tan 30°\,\bar{k})$$

This vector is sketched below. An interesting aspect of this result is that

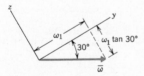

$$\left|\bar{\omega}\right| = \omega_1\sqrt{1 + \tan^2 30°} = \frac{\omega_1}{\cos 30°} \equiv \omega_2$$

Further, note that $\bar{\omega}$ is oriented horizontally.

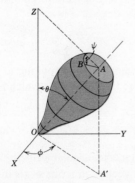

ILLUSTRATIVE PROBLEM 5 (Optional)

In order to define the orientation of a spinning top, a set of *Eulerian angles* is defined. The *precession angle* ϕ describes the angle of rotation about the fixed Z axis of the centerline OA of the top, and the *nutation angle* θ describes the angle between the Z axis and the centerline. Then the *spin angle* ψ defines the rotation of the top about its centerline. Treating these angles as arbitrary functions of time, obtain expressions for the angular velocity and angular acceleration of the top.

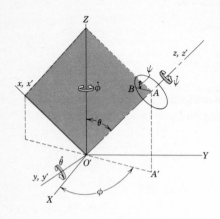

Solution

As we shall see, the new feature of this problem is that the three angles give rise to rotations about three axes, which will cause us to employ two moving reference systems.

Step 1   To simplify the sketch we draw only the centerline OA and the line AB, which are sufficient for illustrating the Eulerian angles.

Steps 2 and 3   The fixed reference frame XYZ is given. The origin O' of the moving reference frame xyz is located at the fixed point O. Further, xyz is attached to the top because we want to describe the angular motion of this body. Let the z axis always coincide with the centerline OA. For convenience we select the instantaneous orientations of the other axes such that the y axis coincides with the XY plane and the x axis coincides with the shaded plane formed by the fixed Z axis and the z axis.

   By this choice of axes we are able to say that the precession rate $\dot{\phi}$ is always about the fixed Z axis and the spin rate $\dot{\psi}$ is always about the moving z axis. What about the nutation rate $\dot{\theta}$? The corresponding axis of rotation is perpendicular to the plane formed by the z and Z axes. At this instant the y axis is perpendicular to this plane. Hence, instantaneously the nutation rate is about this axis. However, when the spin angle ψ changes, the y axis will rotate away from the XY plane and will not longer be perpendicular to the Zz plane. For this reason we introduce another reference frame $x'y'z'$ that does not undergo any spin motion. As shown in the sketch in step 1, the xyz and $x'y'z'$ axes are coincident at the instant under consideration, but only the z and z' axes are coincident at all instants. Then according to the right-hand rule, the nutation rate $\dot{\theta}$ is always about the negative y' axis.

   Let us denote the angular velocity of xyz as $\bar{\omega}$ and the angular velocity of $x'y'z'$ as $\bar{\Omega}$. The general expressions are

$$\bar{\omega} = \dot{\phi}\bar{K} + \dot{\theta}(-\bar{j}') + \dot{\psi}\bar{k} \qquad \text{always}$$

$$\bar{\Omega} = \dot{\phi}\bar{K} + \dot{\theta}(-\bar{j}') \qquad \text{always}$$

As previously mentioned, $\bar{\omega}$ is the angular velocity of the top. To find the angular acceleration we compute $\bar{\alpha} \equiv \dot{\bar{\omega}}$, as follows:

$$\bar{\alpha} = \dot{\bar{\omega}} = \ddot{\phi}\bar{K} - \ddot{\theta}\bar{j}' - \dot{\theta}\,\dot{\bar{j}}' + \ddot{\psi}\bar{k} + \dot{\psi}\dot{\bar{k}} \qquad \text{always}$$

The unit vector $\bar{j}'$ is fixed to the $x'y'z'$ system which is rotating at the angular speed $\bar{\Omega}$, whereas the unit vector $\bar{k}$ is fixed to the xyz system which is rotating at the angular speed $\bar{\omega}$. Hence

$$\bar{\alpha} = \ddot{\phi}\bar{K} - \ddot{\theta}\bar{j}' - \dot{\theta}(\bar{\Omega} \times \bar{j}') + \ddot{\psi}\bar{k} + \dot{\psi}(\bar{\omega} \times \bar{k}) \qquad \text{always}$$

   For the last step we obtain the instantaneous angular velocity and angular acceleration vectors by expressing the unit vectors in terms of their xyz components. Referring to the sketch in step 1, we see that the Z axis coincides with the

xz plane and that θ is the angle between the Z and z axes, so

$$\bar{K} = \sin \theta \, \bar{\imath} + \cos \theta \, \bar{k}$$

Also, the y and y' axes are coincident at this instant, so $\bar{\jmath}' = \bar{\jmath}$. Thus, the general expressions become

$$\bar{\omega} = \dot{\phi}(\sin \theta \, \bar{\imath} + \cos \theta \, \bar{k}) - \dot{\theta}\bar{\jmath} + \dot{\psi}\bar{k}$$

$$= \dot{\phi} \sin \theta \, \bar{\imath} - \dot{\theta} \, \bar{\jmath} + (\dot{\phi} \cos \theta + \dot{\psi})\bar{k}$$

$$\bar{\Omega} = \dot{\phi} \sin \theta \, \bar{\imath} - \dot{\theta}\bar{\jmath} + \dot{\phi} \cos \theta \, \bar{k}$$

$$\bar{\alpha} = \ddot{\phi}(\sin \theta \, \bar{\imath} + \cos \theta \, \bar{k}) - \ddot{\theta}\bar{\jmath} - \dot{\theta}(\dot{\phi} \sin \theta \, \bar{\imath} - \dot{\theta} \, \bar{\jmath} + \dot{\phi} \cos \theta \, \bar{k}) \times \bar{\jmath}$$

$$+ \ddot{\psi}\bar{k} + \dot{\psi}[\dot{\phi} \sin \theta \, \bar{\imath} - \dot{\theta}\bar{\jmath} + (\dot{\phi} \cos \theta + \dot{\psi})\bar{k}] \times \bar{k}$$

$$= (\ddot{\phi} \sin \theta + \dot{\phi} \, \dot{\theta} \cos \theta - \dot{\theta} \, \dot{\psi})\bar{\imath} - (\ddot{\theta} + \dot{\phi}\dot{\psi} \sin \theta) \, \bar{\jmath}$$

$$+ (\ddot{\psi} + \ddot{\phi} \cos \theta - \dot{\phi} \, \dot{\theta} \sin \theta) \, \bar{k}$$

The foregoing expressions are always valid, for the expressions were evaluated at a general instant (the angles are arbitrary). Equations such as these are widely employed for advanced studies of the motion of rigid bodies, especially for computerized studies.

HOMEWORK PROBLEMS

VI.27 Two automobiles are traveling in opposite directions along a curve in a divided highway. Both vehicles are traveling at a constant speed of 55 miles/hr. For the position where $\theta = 0$, determine (a) the absolute velocity and acceleration of car B relative to car A, (b) the velocity and acceleration of car B relative to a passenger in car A who is unaware of the motion of that vehicle.

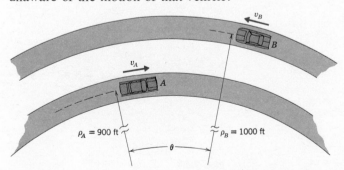

Prob. VI.27

VI.28 Solve Problem VI.27 in the position where $\theta = 20°$.

VI.29 Airplane A is in straight level flight at an altitude of 15 km and airplane B is following a horizontal curve of 1-km radius at an altitude of

12 km. Both aircraft have a constant speed of 750 km/hr. Determine the velocity and acceleration of aircraft A as viewed by the pilot of aircraft B. The aircraft forms the frame of reference for the pilot.

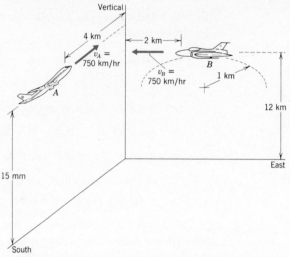

Prob. VI.29

VI.30   A person who is stationed at point B on the perimeter of the rotating platform is observing a baseball in free flight. At the instant shown, the ball is at point C and is moving horizontally with a speed of 40 ft/s. The angular velocity of the platform is 2 rad/s clockwise. For this instant, determine (a) the velocity of the baseball as seen by this person, (b) the acceleration of the baseball as seen by this person, (c) the radius of curvature in the *horizontal plane* of the path of the ball with respect to the person. Draw a sketch of the results.

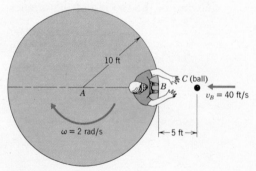

Prob. VI.30

VI.31   Water flows through the arms of the garden sprinkler at the constant rate of 15 m/s as the sprinkler rotates at 20 rad/s. At the nozzle the radius of curvature of the sprinkler arm is 200 m and the tangent to an arm is at a 30° angle from the radial direction. (a) What is the acceleration of the water immediately before it emerges from an arm? (b) What is the actual velocity of the fluid that emerges? Give the magnitude and the angle with respect to the radial line.

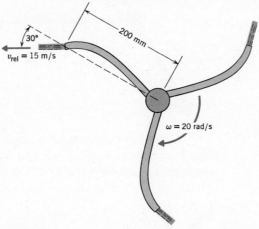

Prob. VI.31

VI.32 The horizontal arm of the amusement park ride rotates about the vertical axis at $\Omega = 6$ rev/min and the cars rotate at the angular speed $\omega_{rel} = 9$ rev/min *relative to that arm*. Determine the velocity and acceleration of the rider's head for the position shown.

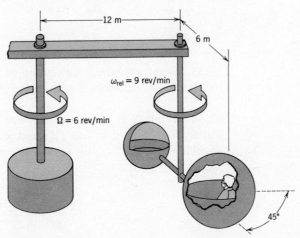

Prob. VI.32

VI.33 Collar B is pinned to bar AB and slides over bar CD. In the position shown, bar CD is rotating clockwise at 50 rad/s and is slowing down at the rate of 20 rad/s². Determine the angular velocity and angular acceleration of bar AB at this instant.

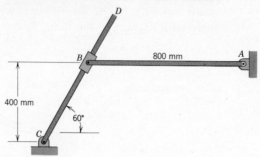

Prob. VI.33

Both bars slide through holes in block E, with the result that the angle between the bars is constant at 45°. Bar AB has a constant counterclockwise angular velocity of 10 rad/s. Determine the corresponding velocity of block E in the position shown.

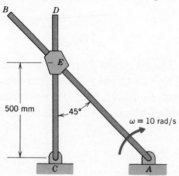

Prob. VI.34

VI.35 Determine the acceleration of block E in Problem VI.34 and the radius of curvature of its path in the position shown.

VI.36 Slider A is moving downward at the constant rate of 600 mm/s. Collar C is pinned to bar CD and slides over bar AB. Determine the angular velocities of both bars in the position shown.

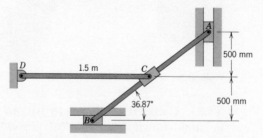

Prob. VI.36

VI.37 Determine the angular acceleration of both bars in Problem VI.36.

VI.38 The bead slides over the bent rod with a constant relative acceleration $\dot{u}$, starting from rest at point A at $t = 0$. The rod is rotating about the vertical axis at a constant angular velocity $\bar{\omega}$. Determine the velocity and acceleration of the bead as a function of t.

VI.39 An insect crawls along the exterior surface of the cylindrical hoop shown at a constant rate of 8 in./s as the hoop rotates about the fixed horizontal axis at the constant rate of 0.25 rad/s. Derive expressions for the components of the acceleration tangent and normal to the *surface* of the hoop in terms of the value of θ. What are the maximum values of each component?

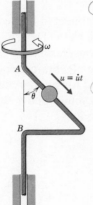

Prob. VI.38

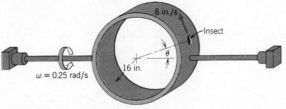

Prob. VI.39

VI.40 Sliders A and B are pinned to bar AB. Slider A has a constant downward velocity of 1.2 m/s as the T-bar assembly rotates about the vertical axis at 3 rad/s. Determine (a) the velocity of slider B, (b) the acceleration of slider B for the position shown.

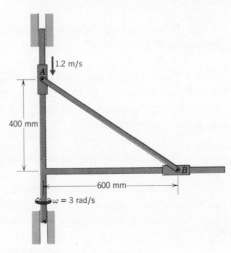

Prob. VI.40

VI.41 For the system in Problem VI.40, determine (a) the angular velocity of bar AB, (b) the angular acceleration of bar AB.

VI.42   Solve Illustrative Problem 4 for the position where arm *AB* is horizontal.

VI.43 and VI.44   Sliders *A* and *B* are connected to bar *AB* by ball joints. Slider *B* has a speed of 300 mm/s to the right. Determine the velocity of slider *A* in the position shown. *Hint:* Considering $\bar{\omega}_{AB}$ as a general vector that is to be determined results in three algebraic equations for four unknowns. Nevertheless, $\bar{v}_A$ may be solved for by a process of elimination. This situation occurs because the ball joints allow the bar to have an arbitrary angular velocity about its own axis.

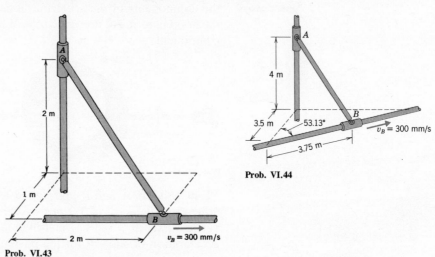

Prob. VI.44

Prob. VI.43

VI.45   The speed of slider *B* in problem VI.43 is constant. What is the corresponding acceleration of slider *A*?

VI.46   Slider *C* and bars *AB* and *BC* are interconnected by ball joints. Bar *AB* is rotating about the fixed point *A* at the constant rate of 5 rad/s. Determine the velocity of slider *C* for the position shown. *Hint:* See the comment regarding Problem VI.43.

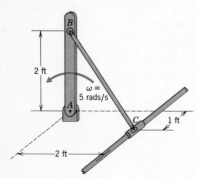

Prob. VI.46

VI.47   Determine the acceleration of slider *C* in Problem VI.46.

VI.48   The disk has a constant angular velocity of 80 rad/s about its axis as the fork that supports the disk rotates about the horizontal axis at 10 rad/s. Determine the velocity and acceleration (a) of point *B*, (b) of point *C*.

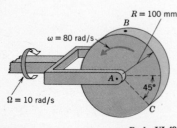

Prob. VI.48

VI.49   The disk is rotating about its axis at a constant rate of 1000 rev/min as arm *AB* pivots about a horizontal axis. In the position shown, arm

AB is inclined 30° from the vertical and it has an angular speed of 10 rad/s which is increasing at the rate of 2 rad/s². Determine the angular velocity and angular acceleration of the disk.

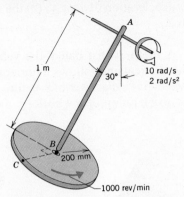

1 m

30°

A

10 rad/s
2 rad/s²

B
200 mm

C

1000 rev/min

Prob. VI.49

VI.50 Determine the velocity and acceleration of point *C* of the disk in Problem VI.49.

VI.51 The three-bladed propeller of an airplane is turning at 1200 rev/min as the airplane follows a horizontal circle of 1000-ft radius at a speed of 180 miles/hr. Determine the angular velocity and angular acceleration of the propeller blades.

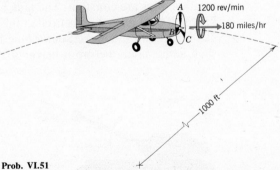

A 1200 rev/min

B
C

180 miles/hr

1000 ft

Prob. VI.51

VI.52 The blades of the airplane in Problem VI.51 are 2 ft long. At the instant when blade *A* is oriented vertically determine the velocity and acceleration (a) of the tip of blade *A* and (b) the tip of blade *C*.

VI.53 Gear *B* rotates freely about its shaft, which has an angular acceleration of 20 rad/s² and an instantaneous angular speed of 10 rad/s about the vertical axis. Gear *A* is stationary. Determine (a) the angular velocity and angular acceleration of gear *B*, (b) the velocity and acceleration of point *C*.

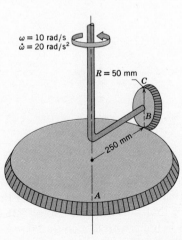

ω = 10 rad/s
ω̇ = 20 rad/s²

R = 50 mm
C

B

250 mm

A

Prob. VI.53

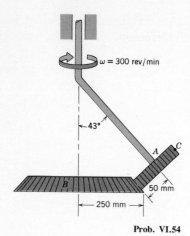

Prob. VI.54

VI.54 The shaft of gear A rotates about the fixed vertical axis at a constant rate of 300 rev/min. This gear rotates freely about its shaft as it meshes with the stationary gear B. Determine (a) the angular velocity and angular acceleration of gear A, (b) the velocity and acceleration of point C on gear A.

VI.55 A motorcyclist banks the vehicle into a turn at an angle of 30° from the vertical and follows a horizontal 30-m-radius curve at 50 km/hr. Each wheel has a diameter of 800 mm. Determine the angular velocity and angular acceleration of a wheel.

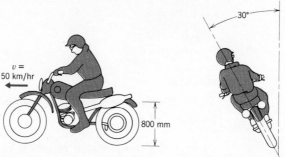

Prob. VI.55

VI.56 In the position shown, the turret on the tank is rotating about the vertical axis of 0.4 rad/s and the barrel is being raised at 0.6 rad/s; both rotation rates are constant. The tank has a constant forward velocity of 60 km/hr. If a cannon shell is fired at this instant with a muzzle speed of 500 m/s relative to the barrel, determine the velocity and acceleration of the shell immediately before it leaves the barrel.

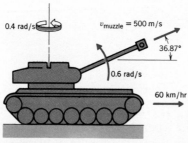

Prob. VI.56

VI.57 The left tread of the crane is moving faster than the right one, with the result that the chassis is following a circular path. The radius of the path of point A is 40 ft and the speed of that point is a constant 5 ft/s. When $\theta = 53.13°$, the 100-ft-long boom is being lowered such that $\dot{\theta} = -4$ rad/min and $\ddot{\theta} = 0.5$ rad/s². Determine the angular velocity and angular acceleration of the boom.

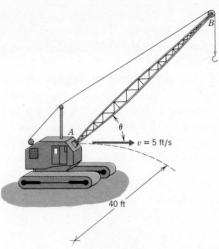

Prob. VI.57

VI.58 For the parameters in Problem VI.57, determine the velocity and acceleration of end B of the boom. The boom is 60 feet long.

VI.59 An oscillating electric fan is aimed at 30° above the horizontal. The rotation of the fan about the vertical axis is defined by $\omega = \sin \pi t/5$, where ω is radians per second and t is in seconds. The fan blades are spinning at 900 rev/min. Determine the angular velocity and angular acceleration of the fan blades as a function of time.

VI.60 The motor is mounted on a bracket that is rotating about the vertical shaft at 10 rev/min. The angular speed of the disk about the axis of the motor shaft is 300 rev/min, and the centerline of the shaft is inclined at a constant 36.87° angle from the horizontal. Determine (a) the angular velocity and the angular acceleration of the disk, (b) the velocity and acceleration of point C on the disk.

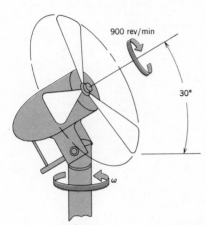

Prob. VI.59

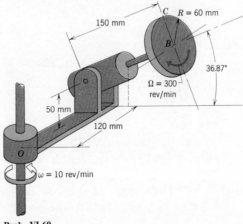

Prob. VI.60

VI.61   Using the moving xyz reference frame shown, derive expressions for the position, velocity, and acceleration in terms of the cylindrical coordinates (R, ϕ, h) and the corresponding unit vectors $(\bar{e}_R, \bar{e}_\phi, \bar{e}_h)$.

VI.62   Spherical coordinates are another useful set of extrinsic coordinates. In this system the location of a point is defined by the distance r from the origin, the angle θ from the Z axis, and the angle ϕ from the X axis. Use the moving xyz reference frame shown to derive expressions for the position, velocity, and acceleration in terms of the spherical coordinates (r, ϕ, θ) and the corresponding orthogonal unit vectors $(\bar{e}_r, \bar{e}_\phi, \bar{e}_\theta)$.

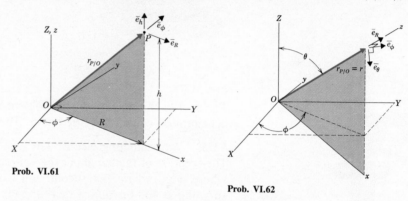

Prob. VI.61

Prob. VI.62

VI.63   In the position shown, the aircraft is executing a diving turn to the right by pitching its nose down at $\omega_P = 0.5$ rad/s and rolling at $\omega_r = 0.3$ rad/s. The roll rate is decreasing at 0.6 rad/s². Determine the angular velocity and angular acceleration of the aircraft.

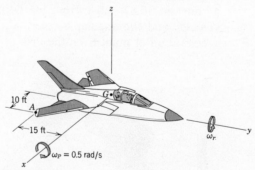

Prob. VI.63

VI.64   In Problem VI.63, determine the velocity and acceleration of point A relative to the center of mass G.

VI.65   The center of mass G of a large ship has a constant forward velocity of 6 m/s. Due to the action of the ocean waves, however, it is pitching, rolling, and yawing about the x, y, and z axes, respectively, which are attached to the ship. At this instant these axes have the horizontal and vertical orientations they would have in a calm sea but $\omega_x = 0.08$ rad/s, $\omega_y = 0.04$ rad/s, $\omega_z = 0.02$ rad/s. Also, all rotation rates have their maximum value at this instant. Determine the absolute velocity and acceleration of the bow of the ship, which is 120 m forward and is 15 m above point G.

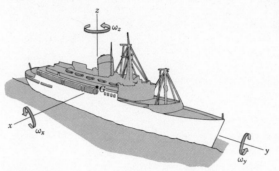

Prob. VI.65

VI.66 In Problem VI.65, an elevator on the ship is 40 m behind, 10 m above, and 5 m to the right (looking forward) of point G. The elevator is moving upward at the constant rate of 4 m/s relative to the ship. Determine the absolute velocity and acceleration of the elevator.

VI.67 The disk is rotating about bar AB at 100 rad/s as the fixed vertical axis rotates at 10 rad/s. In the position shown, $\theta = 90°$, $\dot\theta = 2$ rad/s, $\ddot\theta = 0$. Determine the velocity and acceleration of (a) point C, (b) point D. *Hint:* Follow the development in Illustrative Problem 5, where two moving reference frames are employed.

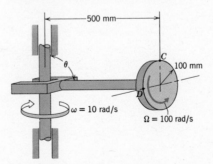

Prob. VI.67

VI.68 The disk of the gyroscope rotates about its own axis at 600 rev/min, and the gimbal support is rotating about the vertical axis at the constant rate of 1 rad/s. In the position where $\theta = 90°$ it is given that $\dot\theta =$

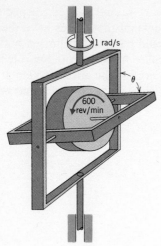

Prob. VI.68

10 rad/s and $\ddot{\theta} = 0$. Determine the angular velocity and angular acceleration of the disk (a) using two moving reference frames as in Illustrative Problem 5, (b) using the formulas for Eulerian angles derived in Illustrative Problem 5.

VI.69 Given that $\dot{\theta} = 5$ rad/s and $\ddot{\theta} = 10$ rad/s² when $\theta = 60°$, solve Problem VI.68.

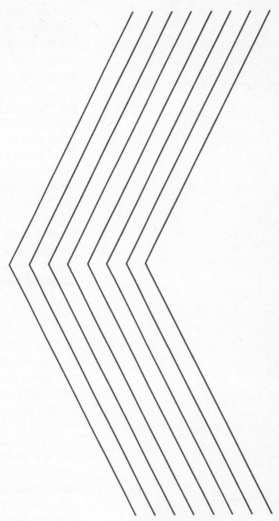

MODULE VII
KINETICS OF A
RIGID BODY IN
THREE-DIMENSIONAL
MOTION

Our purpose in this module is to relate the three-dimensional motion of a rigid body to the forces that cause this motion. In the task confronting us we shall consider a rigid body of known dimensions and mass distribution being acted upon by a general system of forces. Similar to the kinetics studies of particles and rigid bodies in planar motion, this study of the laws of kinetics will consist of investigations of the equations of motion, the work-energy principle, and the linear and angular impulse-momentum principles. The new feature here is that we shall be dealing with the most general case of motion of a rigid system.

When we studied three-dimensional rigid body kinematics in the previous module, we saw that the motion we seek is fully determined once we know both the motion of a point in the body and the angular motion of the body, just as it was in the case of a rigid body in planar motion. However, in contrast to what we observed in the study of planar motion, when a rigid body executes an arbitrary motion in three dimensions its angular velocity can have a variable direction as well as a variable magnitude. We shall see that this variable characteristic produces some fundamental phenomena that are unique to three-dimensional motion.

A.   FUNDAMENTAL PRINCIPLES

Let us pause to recall the principles of statics, where it was shown that there are six scalar equations of equilibrium when a rigid body is subjected to a three-dimensional force system. These equations represent the components of the vector equations obtained by requiring that the resultant of the force system satisfy $\Sigma \bar{F} = \bar{0}$, and that the resultant moment of the force system about an arbitrary point A satisfy $\Sigma \bar{M}_A = \bar{0}$.

We shall find in this module that, correspondingly, there are six scalar equations of motion governing a rigid body undergoing an arbitrary three-dimensional motion. For this study we shall employ the principles in Module III governing a system of particles. Three of the scalar equations we seek are obtained from

$$\boxed{\Sigma \bar{F} = m\bar{a}_G} \tag{1}$$

The mass m of the rigid body is considered to be the collection of an infinite number of infinitesimal masses dm filling the volume $\mathcal{V}$ occupied by the body. Therefore

$$m = \int_{\mathcal{V}} dm$$

Equation (1) serves to describe the motion of one point in the body, the center of mass G. Therefore, the major task confronting us is the determination of the equations of motion governing the rotation of the body. These equations are most conveniently written in terms of the moment equation for a system of particles. Specifically, the vector equation is

$$\Sigma \bar{M}_A = \frac{d}{dt} \bar{H}_A \equiv \dot{\bar{H}}_A \tag{2}$$

where

$$\bar{H}_A = \sum_{i=1}^{N} (\bar{r}_{i/A} \times m_i \bar{v}_{i/A}) \tag{3}$$

Remember that in order for equation (2) to be valid, point A must be either a fixed point, the center of mass of the system, or a point accelerating toward the center of mass.

B. ANGULAR MOMENTUM OF A RIGID BODY

Equation (2) illustrates the importance of the vector $\bar{H}_A$: the angular momentum of a rigid body relative to point A. Hence, our objective in this section is to develop a general formula for the angular momentum of a rigid body undergoing three-dimensional motion.

Figure 1 shows an infinitesimal particle dm within an arbitrary rigid body. In addition to the requirement that point A be one of the three special points mentioned in Section A, we further restrict the choice of this point to be a point that is fixed to the body. Then, because the body is rigid, we know that $\bar{r}_{i/A}$ is a vector with constant magnitude that rotates at the angular velocity $\bar{\omega}$ of the body. Thus, from our study of rigid body kinematics in Module VI, we have

$$\bar{v}_{i/A} = \bar{\omega} \times \bar{r}_{i/A}$$

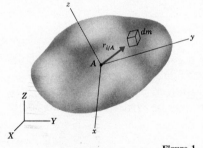

Figure 1

Furthermore, the contribution of each differential element dm to the value of $\bar{H}_A$ must be evaluated by an integral, rather than by a conventional summation process. Equation (3) therefore becomes

$$\bar{H}_A = \int_{\mathscr{V}} [\bar{r}_{i/A} \times (\bar{\omega} \times \bar{r}_{i/A})] dm \tag{4}$$

In order to proceed further, we must choose a coordinate system for describing the vector components and the boundary of the body. For this we could choose the fixed reference system XYZ. Such a choice creates the awkward situation where the boundary of the body, in terms of the (X, Y, Z) coordinates, will change as a function of time. This would cause the computation of the time variation of $\bar{H}_A$ to be difficult. To avoid this problem we attach a moving reference frame xyz to the body, so that all points in the body are fixed with respect to the *xyz* frame. There are some special geometries, such as bodies of revolution, where we could allow the body to have some rotational motion relative to the reference frame. However, attaching the *xyz* frame to the body ensures that our formulation will be valid in all cases.

For simplicity, let the origin of the *xyz* frame coincide with point *A*. Such a coordinate system is shown in Figure 1. (The question of a suitable choice for the orientation of the *xyz* frame will be discussed later.) The position vector of any mass element *dm* at point *i* can then be written as

$$\bar{r}_{i/A} = x\,\bar{\imath} + y\,\bar{\jmath} + z\,\bar{k}$$

To be consistent, we express the angular velocity of the body in terms of its *x*, *y*, and *z* components. Thus

$$\bar{\omega} = \omega_x\,\bar{\imath} + \omega_y\,\bar{\jmath} + \omega_z\,\bar{k}$$

Now, substituting $\bar{r}_{i/A}$ and $\bar{\omega}$ into equation (4) and calculating the cross products, we get

$$\bar{H}_A = \int_{\mathscr{V}} \{[\omega_x(y^2 + z^2) - \omega_y xy - \omega_z xz]\,\bar{\imath}$$

$$+\ [\omega_y(z^2 + x^2) - \omega_z yz - \omega_x xy]\,\bar{\jmath}$$

$$+\ [\omega_z(x^2 + y^2) - \omega_x xz - \omega_z yz]\,\bar{k}\}\,dm$$

Recall that the angular velocity is a property of the motion of the body. Hence, the components of $\bar{\omega}$ do not depend on a particular location within the body, and may therefore be removed from the integrand. As a result, the angular momentum equation may be written more compactly as

$$\bar{H}_A = (I_{xx}\omega_x - I_{xy}\omega_y - I_{xz}\omega_z)\bar{\imath}$$
$$+\ (I_{yy}\omega_y - I_{xy}\omega_x - I_{yz}\omega_z)\bar{\jmath} \qquad (5)$$
$$+\ (I_{zz}\omega_z - I_{xz}\omega_x - I_{yz}\omega_y)\bar{k}$$

where

$$I_{xx} = \int_{\mathscr{V}}(y^2 + z^2)\,dm$$

$$I_{yy} = \int_{\mathscr{V}}(x^2 + z^2)\,dm$$

$$I_{zz} = \int_{\mathscr{V}}(x^2 + y^2)\,dm$$

$$I_{xy} = \int_{\mathcal{V}} xy \, dm$$

$$I_{xz} = \int_{\mathcal{V}} xz \, dm$$

$$I_{yz} = \int_{\mathcal{V}} yz \, dm \qquad (6)$$

The quantities I_{xx}, I_{yy}, and I_{zz} are called the mass *moments of inertia*. The quantities I_{xy}, I_{xz}, and I_{yz} are called the mass *products of inertia*. In the next section we shall treat these inertia properties in detail. For now we shall only make some simple observations regarding them.

Foremost, we can now see the significance of selecting a moving reference frame *xyz* for which the boundaries of the body are constant. Such a selection means that the inertia terms defined by equations (6) will be constant properties of the body. Equally important, we shall find that a proper selection for the orientation of the *xyz* frame with respect to the body can greatly facilitate the calculation of the inertia properties. For instance, we shall see that in some cases the choice for the *xyz* frame will result in $I_{xy} = I_{yz} = I_{xz} = 0$. In such cases we shall refer to *xyz* as *principal axes*. Also, by properly orienting *xyz*, we shall frequently be able to make use of tables that list the inertia properties of basic geometric shapes.

To demonstrate how the angular momentum of a body is determined, we shall now consider two examples of the rotation of a thin disk. For the inertia properties, recall that we defined principal axes as those for which $I_{xy} = I_{yz} = I_{xz} = 0$. It is for such axes that inertia properties of standard geometric shapes are most often tabulated (as they are in Appendix B). Figure 2 shows such information for a thin disk. In the problems that follow, note how we select the axes to make use of this information.

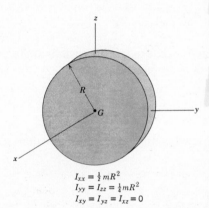

$I_{xx} = \frac{1}{2}mR^2$
$I_{yy} = I_{zz} = \frac{1}{4}mR^2$
$I_{xy} = I_{yz} = I_{xz} = 0$

Figure 2

EXAMPLE 1
A thin disk of radius r is mounted obliquely on a rigid shaft, with the result that the center of mass G is on the axis of the shaft, but the axis of symmetry of the disk is not parallel to the shaft. If the shaft is rotating at the constant rate ω, what is the angular momentum of the disk about its center of mass?

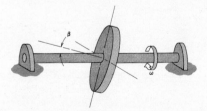

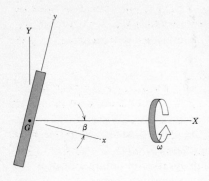

Solution

In order to use the angular momentum equation (5), we must first select an xyz frame. We want the angular momentum about the center of mass. Hence, we place the origin at that point and fix the axes to the disk. Further, orienting the reference frame as shown in the figure allows us to make direct use of the given inertia properties. For the fixed reference frame XYZ, let us choose the X axis to be parallel to the shaft, as shown in the sketch.

Note that because xyz is fixed to the disk, the z axis is not always parallel to the Z axis, but the xy plane always contains the X axis.

The inertia properties are known from Figure 2. Thus, we need only determine the components of angular velocity prior to evaluating the angular momentum. In terms of the chosen coordinate axes, we have

$$\bar{\omega} = \omega \bar{I} \quad \text{always}$$

$$= \omega(\cos \beta\, \bar{i} + \sin \beta\, \bar{j}) \quad \text{instantaneously}$$

The second expression is also always valid, because β is a constant angle between $\bar{i}$ and $\bar{I}$. Thus the components of $\bar{\omega}$ are constant.

$$\omega_x = \omega \cos \beta \qquad \omega_y = \omega \sin \beta \qquad \omega_z = 0$$

Upon substituting the moments of inertia and components of $\bar{\omega}$ into equation (5), we get

$$\bar{H}_G = \tfrac{1}{4}mr^2\omega(2 \cos \beta\, \bar{i} + \sin \beta\, \bar{j})$$

The resulting angular velocity and angular momentum vectors are shown below. As can be seen, the angular momentum vector is not parallel to the angular velocity vector.

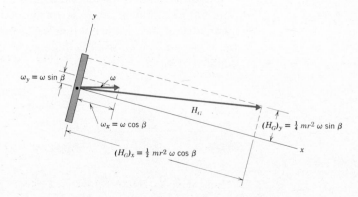

In this problem, the $(\bar{\imath}, \bar{\jmath}, \bar{k})$ components of $\bar{H}_G$ are constant because ω_x, ω_y and ω_z are constant values. Therefore $\bar{H}_G$ is fixed relative to xyz, with the result that $\bar{H}_G$ rotates about the fixed shaft at angular velocity $\bar{\omega}$.

EXAMPLE 2

An electric motor is mounted horizontally on a turntable that rotates about the vertical axis at 300 rev/min as shown. The flywheel of the motor is concentrically mounted on the shaft of the motor, whose rate of rotation is 1500 rev/min. The center of the mass of the flywheel is 0.5 m from the axis of the turntable. Approximate the flywheel as a thin disk of radius R to determine its angular momentum relative to its center of mass.

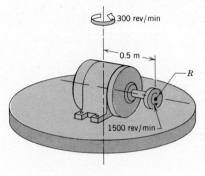

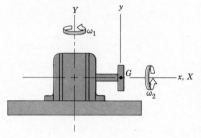

Solution

The relative angular momentum depends only on the angular motion of the body and the location of the origin. As always, we attach the xyz frame to the body with its origin coincident with the point of reference for the angular momentum, in this case the center of mass. We orient the xyz frame relative to the body for greatest convenience in describing the moments and products of inertia. Hence we let the x axis be the axis of the disk, so that we can use the given inertia properties directly. This still leaves the choice of the y and z axes undefined. We make the choice for the instantaneous orientation based on the geometry of the system. Here, we shall let the y axis be vertical at the instant to be considered, as is illustrated in the sketch.

The inertia properties for a thin disk are given in Figure 2. Following the kinematical method developed in the previous module, the general expression for the angular velocity of the disk is

$$\bar{\omega} = \omega_1 \bar{J} + \omega_2 \bar{\imath} \qquad \text{always}$$

Noting that $\bar{J} = \bar{\jmath}$ at the instant illustrated in the sketch, the instantaneous expression for $\bar{\omega}$ is

$$\bar{\omega} = \omega_1 \bar{\jmath} + \omega_2 \bar{\imath} \qquad \text{instantaneously}$$

Converting the given information to the proper units for angular rotation (radians), we have

$$\omega_x = \omega_2 = 1500\left(\frac{2\pi}{60}\right) = 50\pi \text{ rad/s}$$

$$\omega_y = \omega_1 = 300\left(\frac{2\pi}{60}\right) = 10\pi \text{ rad/s}$$

Then, from equation (5) we find that

$$\bar{H}_G = \tfrac{1}{2}mR^2(50\pi)\bar{i} + \tfrac{1}{4}mR^2(10\pi)\bar{j}$$

$$= 2.5\pi mR^2(10\bar{i} + \bar{j})$$

The resulting vectors $\bar{\omega}$ and $\bar{H}_G$ are shown below.

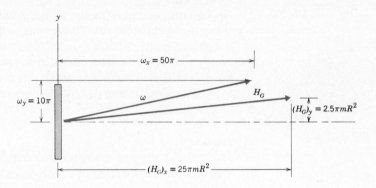

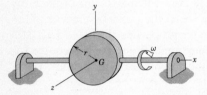

Prob. VII.1

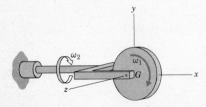

Prob. VII.2

Prob. VII.3

HOMEWORK PROBLEMS

VII.1   A thin disk of mass m and radius r is mounted on a shaft whose axis intersects the center of the disk. The shaft is driven at the constant rate ω. Determine the angular momentum of the disk about its mass center. Sketch the resulting $\bar{H}_G$ and $\bar{\omega}$ vectors.

VII.2   During a maneuver a space capsule is spinning at the angular velocity $\bar{\omega} = 0.2\bar{i} + 0.8\bar{j} + 0.1\bar{k}$ rad/s. The xyz axes are principal axes for which the moments of inertia are $I_{xx} = I_{zz} = 8000$ lb-s²-ft and $I_{yy} = 6000$ lb-s²-ft. Determine the angular momentum $\bar{H}_G$ of the space capsule about its mass center G. Sketch the resulting $\bar{H}_G$ and $\bar{\omega}$ vectors.

VII.3   The thin disk of mass 8 kg and radius $r = 500$ mm rotates at constant angular speed $\omega_1 = 25$ rad/s while the forked rod is driven at constant angular speed $\omega_2 = 10$ rev/min. Determine the angular momentum of the disk about its mass center. Sketch the resulting $\bar{H}_G$ and $\bar{\omega}$ vectors for the disk.

VII.4   Consider a circular disk of mass m and radius r rolling on the ground in a situation where the plane of the disk is oriented vertically. The center of the disk has speed v_0 and there is no slipping. Determine the angular momentum of the disk about its mass center G when this point is

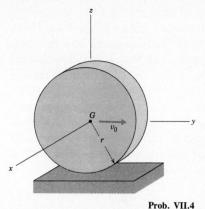

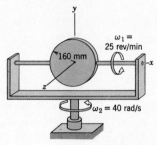

Prob. VII.4

following (a) a straight path, (b) a circular path of radius R. Why are $\bar{H}_G$ and $\bar{\omega}$ parallel in the first case, but not in the second?

VII.5 A 6-kg thin disk of 160 mm radius is welded on a shaft mounted in a rotating frame. The disk rotates at a constant angular speed $\omega_1 = 25$ rev/min about the axis of the shaft, as the frame rotates at a constant angular speed $\omega_2 = 40$ rad/s. Determine the angular momentum of the disk about its mass center. Sketch the resulting $\bar{H}_G$ and $\bar{\omega}$ vectors.

VII.6 A disk of mass m and radius r is mounted on the rigid bar AB. The bar is welded to the vertical shaft, which is rotating at the angular speed ω_1 about the vertical axis. The angular speed of the disk with respect to bar AB is ω_2, as shown. Determine (a) $\bar{H}_G$ for the disk, (b) the angle β between $\bar{H}_G$ and the vertical direction.

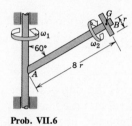

Prob. VII.6

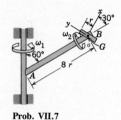

Prob. VII.7

Prob. VII.5

VII.7 The system shown is similar to the one in the preceding problem, except that the disk is mounted obliquely on its shaft. Determine $\bar{H}_G$ for the disk when it is in the position shown.

VII.8 Determine the angular momentum of the disk in Problem VII.7 when it has rotated 180° about bar AB from the position shown.

C. MOMENTS AND PRODUCTS OF INERTIA

From the discussion and examples in the previous section, it is obvious that the evaluation of the inertia properties plays a crucial role in the determination of the angular momentum vector. A question that arises is: What is the physical significance of the inertia properties?

Let us first consider one of the moments of inertia, for instance,

$$I_{zz} = \int_{\mathscr{V}}(x^2 + y^2)\, dm$$

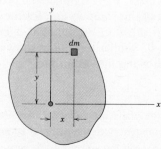

Figure 3

Looking down the z axis at an arbitrary body, such as that in Figure 3, we see that the definition of I_{zz} is the same quantity as the moment of inertia determined for a planar body in Module V. One new feature of the kinetics of rigid bodies in three-dimensional motion is that the moments of inertia about all three coordinate axes are important to the motion. However, the meaning of each term is unchanged. They describe how the mass is distributed about the corresponding axis. For instance, if two bodies have the same mass, the one whose mass is distributed further from the x axis has the greater value for I_{xx}.

One common way of describing a moment of inertia about some axis, such as the z axis, is to give the *radius of gyration* k_z. This quantity is the radius of a fictional circular ring having the same mass and moment of inertia about its centerline as the actual body it stimulates. For such a ring, the mass is concentrated at the perimeter, so that

$$I_{zz} = mk_z{}^2 \qquad (7)$$

where the z axis is normal to the plane of the ring, passing through the center.

Now let us turn our attention to the products of inertia. Considering I_{xy} for the body of Figure 3, by definition,

$$I_{xy} = \int_{\mathcal{V}} xy \; dm$$

In other words, I_{xy} is the sum of the products of the x and y coordinates of a differential element and the mass of that element. Clearly, if the mass of a body is predominantly in one octant, then I_{xy} will have the sign associated with the product of the x and y values in that octant. For example, if a body is entirely situated in an octant where $x < 0$, $y > 0$, then I_{xy} will be negative. Therefore, we can conclude that the products of inertia describe how the mass of the body is distributed within the various octants of the xyz coordinate system.

An important case arises when we deal with bodies having a plane of symmetry. In Figure 4 we have drawn the cross section of a body that is symmetric about the xz plane. For this cross section in the yz plane at some value of x we show a typical element dm_r in the right half of the body and its mirror image mass element dm_l in the left half of the body. The property of symmetry means that there is an element dm_l corresponding to every dm_r. Thus, as is shown in Figure 4, if the coordinates of dm_r are (x, y, z), then the coordinates of dm_l must be $(x, -y, z)$. Hence the product $xy \; dm_l$ for the left element is the negative of the value $xy \; dm_r$ for the right element, and the two terms cancel each other in the integral over the volume $\mathcal{V}$. As a result we find that $I_{xy} = 0$. Similar reasoning shows that $I_{yz} = 0$ also. From this we obtain the following important fact:

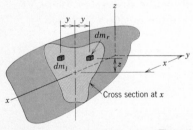

Cross section at x

Figure 4

> If one of the coordinate axes is normal to the plane of symmetry of a body, with the origin of the coordinate system coincident with the plane of symmetry, the two products of inertia having this axis in the subscript are zero.

An important corollary of this statement follows for the case of a body having two perpendicular planes of symmetry, such as a body of revolution.

> A coordinate system whose axes form two perpendicular planes of symmetry constitutes a principal set of axes, for which all products of inertia are zero.

Placing the discussion in this section in perspective, it is sufficient to note that the mass moments and products of inertia are terms that appear in the derivation of the angular momentum formula. Hence, it is necessary to be able to evaluate them. It then becomes expedient to compute the inertia quantities for common geometric shapes and tabulate them for easy reference. A typical set of tables may be found in Appendix B. For uncommon geometric shapes, which do not resemble the tabulated ones, we compute the inertia quantities from the basic definitions. This is the topic of the next section.

HOMEWORK PROBLEMS

VII.9 Select body-fixed axes having an origin at the mass center G that are principal axes of inertia.

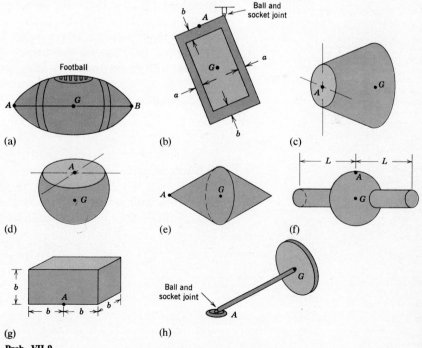

Prob. VII.9

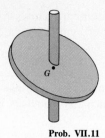

Prob. VII.11

VII.10 For the shapes in Problem VII.9, select body-fixed axes having an origin at point *A* that are principal axes of inertia.

VII.11 The disk is welded obliquely to the vertical shaft. Sketch the principal axes having an origin at the center of mass *G* (a) for the disk, (b) for the shaft. (c) Are the principal axes for the system of the shaft and the disk obvious from symmetry? Explain your answer.

1 Inertia Properties by Integration

Our purpose here is to review the basic calculus necessary to set up and calculate the moments and products of inertia of a rigid body, starting from the definitions. We shall proceed by solving two examples that demonstrate how the inertia properties given in standard tables are obtained, and also how the theorems pertaining to planes of symmetry are utilized.

EXAMPLE 3

Determine the inertia properties of the rectangular parallelepiped relative to the reference frame shown.

Solution

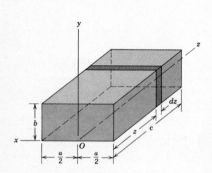

For this body and the selected axes, the *yz* plane is a plane of symmetry. Therefore, because the *x* axis is normal to the plane of symmetry and the origin is coincident with this plane, we find that $I_{xy} = I_{xz} = 0$. To determine the other inertia properties, we shall set up and evaluate the appropriate integrals. Because the surface of the body coincides with planes of constant values for the *x, y,* and *z* coordinates, we shall formulate the integral in terms of Cartesian coordinates.

Let ρ denote the constant mass density of the body. Then

$$dm = \rho \, dx \, dy \, dz$$

To evauate the integrals we shall consider a differential slice of the body at some distance *z* from the origin and integrate first with respect to $x \, (-a/2 \le x \le a/2)$ and then with respect to $y \, (0 \le y \le b)$. Then the integration with respect to $z \, (0 \le z \le c)$ will represent the sum of the contributions of all the cross-sectional slices. Thus

$$I_{xx} = \rho \int_0^c \int_0^b \int_{-a/2}^{a/2} (y^2 + z^2) \, dx \, dy \, dz$$

$$= \rho \int_0^c \int_0^b a(y^2 + z^2) \, dy \, dz$$

$$= \rho a \int_0^c \left(\frac{b^3}{3} + bz^2 \right) dz = \rho ab \left(\frac{b^2 c}{3} + \frac{c^3}{3} \right)$$

$$= \tfrac{1}{3}\rho abc(b^2 + c^2)$$

$$I_{yy} = \rho \int_0^c \int_0^b \int_{-a/2}^{a/2} (x^2 + z^2)\, dx\, dy\, dz$$

$$= \tfrac{1}{12}\rho abc(a^2 + 4c^2)$$

$$I_{zz} = \rho \int_0^c \int_0^b \int_{-a/2}^{a/2} (x^2 + y^2)\, dx\, dy\, dz$$

$$= \tfrac{1}{12}\rho abc(a^2 + 4b^2)$$

$$I_{yz} = \rho \int_0^c \int_0^b \int_{-a/2}^{a/2} yz\, dx\, dy\, dx = \tfrac{1}{4}\rho ab^2c^2$$

These results can be simplified by noting that the volume of the body is $\mathscr{V} = abc$, so that the total mass is

$$m = \rho abc \quad \text{or} \quad \rho = \frac{m}{abc}$$

Thus we can eliminate ρ in the foregoing results to get

$$I_{xx} = \tfrac{1}{3}m(b^2 + c^2) \qquad I_{yy} = \tfrac{1}{12}m(a^2 + 4c^2)$$

$$I_{zz} = \tfrac{1}{12}m(a^2 + 4b^2) \qquad I_{yz} = \tfrac{1}{4}mbc$$

$$I_{xy} = I_{xz} = 0$$

EXAMPLE 4
Determine the inertia properties of a homogeneous right circular cone for the coordinate axes shown.

Solution
The cone is symmetric with respect to both the xy and xz planes, so that xyz are principal axes; that is, $I_{xy} = I_{yz} = I_{xz} = 0$. Because the cone is a body of revolution, we shall formulate the integrals in terms of cylindrical coordinates, with the axial direction coincident with the x axis.

Let us consider a typical slice of the cone at a distance x from the apex. In terms of the cylindrical coordinates (R, θ, x), a differential volume is $R\, dR\, d\theta\, dx$, so that we have

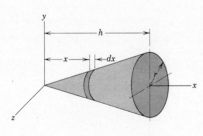

$$y = R \sin \theta \qquad z = R \cos \theta \qquad dm = \rho R\, dR\, d\theta\, dx$$

We can integrate over the cross section by integrating first with respect to θ $(0 \le \theta \le 2\pi)$ and then with respect to R $(0 \le R \le R_0)$. Noting that the radius R_0 of the cross section is proportional to the distance from the apex, we can write

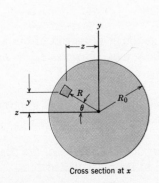

Cross section at x

$$R_0 = \frac{x}{h}r$$

Thus, the moment of inertia about the x axis becomes

$$I_{xx} = \rho \int_0^h \int_0^{R_0} \int_0^{2\pi} (R^2 \sin^2 \theta + R^2 \cos^2 \theta) R \, d\theta \, dR \, dx$$

But $\sin^2 \theta + \cos^2 \theta \equiv 1$, so that

$$I_{xx} = 2\pi\rho \int_0^h \int_0^{xr/h} R^3 \, dR \, dx = 2\pi\rho \int_0^h \tfrac{1}{4}\left(\frac{xr}{h}\right)^4 dx = \tfrac{1}{10}\pi\rho \, r^4 h$$

The volume of a cone is $(\pi/3)r^2 h$, so we know that the mass of the cone is

$$m = \frac{\pi}{3}\rho r^2 h$$

Upon substituting for ρ we get

$$I_{xx} = \tfrac{3}{10} m r^2$$

To complete the solution we need only determine I_{yy}, because $I_{zz} \equiv I_{yy}$ as a result of the symmetry of the body.

$$I_{yy} = \rho \int_0^h \int_0^{R_0} \int_0^{2\pi} (x^2 + R^2 \cos^2 \theta) R \, d\theta \, dR \, dx$$

$$= \rho \int_0^h \int_0^{xr/h} (2\pi x^2 + \pi R^2) R \, dr \, dx$$

$$= \pi\rho r^2 h \left(\frac{h^2}{5} + \frac{r^2}{20}\right)$$

Then substituting for ρ gives

$$I_{yy} = \tfrac{3}{20} m(4h^2 + r^2) = I_{zz}$$

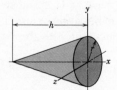

Prob. VII.12

Prob. VII.13

HOMEWORK PROBLEMS

VII.12   Verify by integration that the moment of inertia of the uniform right circular cone of mass m about the y axis shown is $I_{yy} = m(3r^2 + 2h^2)/20$.

VII.13   Determine by integration the moments of inertia of the triangular prism of mass m about the axes shown.

VII.14   Determine by integration the products of inertia of the triangular prism of mass m about the axes shown in Problem VII.13.

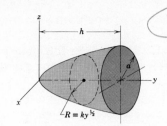

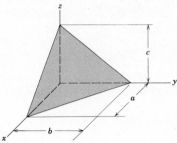

Prob. VII.15

Prob. VII.16.

VII.15 Determine by integration the moments of inertia of the paraboloid of revolution of mass m about the axes shown.

VII.16 Determine by integration the mass moments of inertia of the rectangular tetrahedron of mass m about the given axes. *Hint:* Determine the formula for one moment of inertia and derive the others by induction.

VII.17 Determine by integration the mass products of inertia of the rectangular tetrahedron of mass m about the given axes in Problem VII.16. See the hint in Problem VII.16.

VII.18 Determine by integration the moments of inertia of the circular cylinder of mass m about the axes shown. Then, by taking the limit as the appropriate dimension goes to zero, determine the corresponding formulas for (a) a slender bar, and (b) a thin disk.

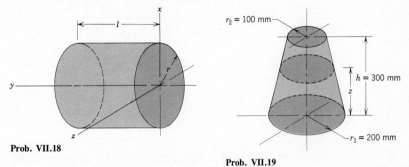

Prob. VII.18

Prob. VII.19

VII.19 The density of the material of the truncated cone is given by $\rho = cz^2$, where z is the distance of the mass element from the base and c is a constant. The mass of this body is 50 kg. Determine (a) the value of c, (b) the location of the mass center G, and (c) the mass moments of inertia about a set of principal axes whose origin is point G.

2 Parallel Axis Theorems

All of the systems we shall study will be composed of bodies whose inertia properties have been compiled in standard tables, such as those in Appendix B. In this way we shall reduce the amount of computations necessary to solve problems, in that it is obviously preferable to use a tabulated value rather than evaluating an integral. This restriction on the types of bodies to be considered will not affect the generality of the presentation of three-dimensional kinetics.

The use of the tabulated information will sometimes require that we transform the inertia properties from the coordinate system given in the tables to a more appropriate set of coordinates for the problem at hand.

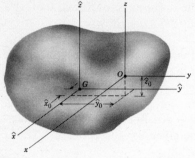

Figure 5

We shall now develop the formulas necessary to perform such transformations.

Figure 5 illustrates an arbitrary body. We consider the moment and products of inertia to be known quantities with respect to the coordinate system $\hat{x}\hat{y}\hat{z}$ whose origin is the center of mass G.

The problem we wish to consider is the determination of the inertia properties of this body with respect to the coordinate system xyz whose axes are parallel to the $\hat{x}$, $\hat{y}$, and $\hat{z}$ axes, respectively. We denote the $(\hat{x}, \hat{y}, \hat{z})$ coordinates of the origin O of the xyz coordinate system by the values $(\hat{x}_0, \hat{y}_0, \hat{z}_0)$. Hence, as can be seen from Figure 5, the (x, y, z) coordinates of any point are related to the $(\hat{x}, \hat{y}, \hat{z})$ coordinates by

$$x = \hat{x} - \hat{x}_0 \qquad y = \hat{y} - \hat{y}_0 \qquad z = \hat{z} - \hat{z}_0 \tag{8}$$

Let us substitute these expressions into the moment and product of inertia equations (6). For I_{xx} we get

$$I_{xx} = \int_{\mathscr{P}} (y^2 + z^2)\, dm$$

$$= \iint_{\mathscr{P}} [(\hat{y} - y_0)^2 + (\hat{z} - \hat{z}_0)^2]\, dm$$

$$= \int_{\mathscr{P}} (\hat{y}^2 + \hat{z}^2)\, dm - 2\hat{y}_0 \int_{\mathscr{P}} \hat{y}\, dm$$

$$-2\hat{z}_0 \int_{\mathscr{P}} \hat{z}\, dm + (\hat{y}_0 + \hat{z}_0{}^2) \int_{\mathscr{P}} dm$$

The first integral is, by definition, the moment of inertia $I_{\hat{x}\hat{x}}$. The second and third integrals are the first moments of mass. These integrals are zero, because the origin of the xyz frame is the center of mass, for which the first moment of mass are

$$\int_{\mathscr{P}} \bar{r}_{i/G}\, dm = \int_{\mathscr{P}} (\hat{x}\bar{i} + \hat{y}\bar{j} + \hat{z}\bar{k})\, dm = m\bar{r}_{G/G} \equiv \bar{0}$$

In addition, the last integral is obviously m, the total mass of the body. Following similar steps for the other moments of inertia, we obtain the final form of the *parallel axis theorems for moments of inertia:*

$$\boxed{\begin{aligned} I_{xx} &= I_{\hat{x}\hat{x}} + m(\hat{y}_0{}^2 + \hat{z}_0{}^2) \\ I_{yy} &= I_{\hat{y}\hat{y}} + m(\hat{x}_0{}^2 + \hat{z}_0{}^2) \\ I_{zz} &= I_{\hat{z}\hat{z}} + m(\hat{x}_0{}^2 + \hat{y}_0{}^2) \end{aligned}} \tag{9}$$

This result takes on a simple meaning once we note that Figure 5 indicates that the distances between the parallel pairs of axes are

$$d_x = (y_0{}^2 + z_0{}^2)^{1/2}$$

$$d_y = (x_0{}^2 + z_0{}^2)^{1/2}$$

$$d_z = (x_0{}^2 + y_0{}^2)^{1/2}$$

Thus equations (9) are the extension to three dimensions of the parallel axis theorem for planar bodies. Specifically, they state that the moment of inertia about any axis equals the moment of inertia about a parallel axis that _passes through the center of mass_ plus the product of the mass and the square of the distance between the axes.

Turning now to the determination of the products of inertia with respect to the xyz coordinate system, we proceed in the same manner as that employed for the moments of inertia. From equations (6) for I_{xy}, we write

$$I_{xy} = \int_{\mathscr{V}} xy \, dm = \int_{\mathscr{V}} (x - \hat{x}_0)(y - \hat{y}_0) \, dm$$

$$= \int_{\mathscr{V}} \hat{x}\hat{y} \, dm - \hat{x}_0 \int_{\mathscr{V}} \hat{y} \, dm - \hat{y}_0 \int_{\mathscr{V}} \hat{x} \, dm + \hat{x}_0\hat{y}_0 \int_{\mathscr{V}} dm$$

Using the fact that the origin of $\hat{x}\hat{y}\hat{z}$ is at the center of mass this, and similarly the other two products of inertia, reduce to the form we call the _parallel axis theorems for products of inertia:_

$$\boxed{\begin{aligned} I_{xy} &= I_{\hat{x}\hat{y}} + m\hat{x}_0\hat{y}_0 \\[4pt] I_{xz} &= I_{\hat{x}\hat{z}} + m\hat{x}_0\hat{z}_0 \\[4pt] I_{yz} &= I_{\hat{y}\hat{z}} + m\hat{y}_0\hat{z}_0 \end{aligned}} \tag{10}$$

Again, it is important to note that in applying both equations (9) and (10), the origin of the $\hat{x}\hat{y}\hat{z}$ reference frame must be the center of mass of the body, and that $(\hat{x}_0, \hat{y}_0, \hat{z}_0)$ represent the coordinates of the origin O of xyz with respect to $\hat{x}\hat{y}\hat{z}$.

3 Composite Bodies

Before we consider an example that illustrates the application of the parallel axis theorems, let us consider one more facet of the determination of the inertia properties, specifically, the matter of composite bodies. By a composite body we mean an arbitrary body whose form is composed of several basic shapes. An example of a composite body is shown in Figure 6, where the cylinder 1 and the rectangular parallelepiped 2 are the components of the composite body.

The basic mass properties of any body are its total mass, the location of the center of mass, and the moments and products of inertia with

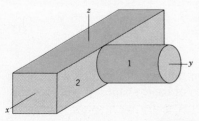

Figure 6

respect to a given coordinate system xyz (whose origin is not necessarily the center of mass). The method for locating the center of mass of an arbitrary composite body is a straightforward extension of the method in Module Five for a planar composite body. Although some of you may have studied this topic as part of statics, we shall review it here for completeness.

The total mass m of the body is obviously the sum of the masses of the components, so that

$$m = \sum_{i=1}^{N} m_i \tag{11}$$

To describe the location of the center of mass G of the composite body, we select any convenient $x'y'z'$ coordinate system; usually the origin of $x'y'z'$ coincides with the center of mass of one of the components. The location of point G is obtained from the first moment of mass. Let (x'_G, y'_G, z'_G) and $(x'_{Gi}, y'_{Gi}, z'_{Gi})$ denote the coordinates relative to the $x'y'z'$ system of point G and of the center of mass of each component G_i, respectively. Then, to obtain the x' coordinate of point G, we have

$$mx'_G = \sum_{i=1}^{N} m_i x'_{Gi} \tag{12}$$

The other coordinates of point G are obtained by permuting the symbol x' in equation (12) to y' or z'.

The next step in the description of the mass characteristics of the composite body is the determination of the moments and products of inertia with respect to the coordinate system xyz whose origin is point A; generally, point A will be the point that was selected for the moment sum in equation (2). To obtain the inertia properties, equations (6) may be decomposed into integrals over the volumes $\mathscr{V}_i$ of the basic components. It follows that, when *evaluated for a given set of axes,* the inertia properties are additive, in which case

$$I_{xx} = \sum_{i=1}^{N} (I_{xx})_i \qquad I_{yy} = \sum_{i=1}^{N} (I_{yy})_i \qquad I_{zz} = \sum_{i=1}^{N} (I_{zz})_i$$

$$I_{xy} = \sum_{i=1}^{N} (I_{xy})_i \qquad I_{xz} = \sum_{i=1}^{N} (I_{xz})_i \qquad I_{yz} = \sum_{i=1}^{N} (I_{yz})_i \tag{13}$$

To use equations (13) properly, it is important to observe that when using tabulated values for the inertia properties of basic shapes, it will probably be necessary to use the parallel axis theorems for each shape in order to transform the basic properties to the desired xyz coordinate system.

All the aforementioned features of the parallel axis theorems and the methods for handling composite bodies are incorporated in the following example.

EXAMPLE 5

Without performing integrations, locate the center of mass of the steel body shown and determine its inertia quantities I_{xx} and I_{xz} with respect to the xyz axes shown in the diagram.

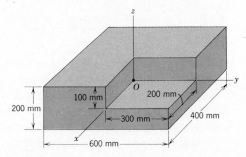

Solution

The composite body can most easily be depicted as a steel block, call it body 1, of dimensions 600 mm × 200 mm × 400 mm, from which a smaller block, call it body 2, of dimensions 300 mm × 100 mm × 200 mm, has been cut away. Hence, for this system we shall subtract the properties of body 2 with respect to the coordinate axes through O from those of body 1. In general, to determine the inertia properties of composite bodies we shall need coordinate axes $\hat{x}_i, \hat{y}_i, \hat{z}_i$ for each component shape i. The origin of $\hat{x}_i, \hat{y}_i, \hat{z}_i$ is the center of mass G_i of that shape, and each $\hat{x}_i, \hat{y}_i, \hat{z}_i$ system should be parallel to the xyz coordinate system of interest. In this problem the center of mass G_1 is coincident with point O, as illustrated in the sketch.

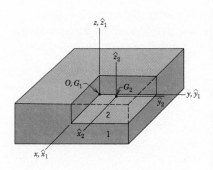

We must first determine the mass of the body. The mass density of steel is $\rho = 7833$ kg/m³, hence in length units of meters,

$$m_1 = \rho \mathcal{V}_1 = 7833(0.6)(0.2)(0.4) = 376.0 \text{ kg}$$

$$m_2 = \rho \mathcal{V}_2 = 7833(0.3)(0.1)(0.2) = 47.0 \text{ kg}$$

$$m = m_1 - m_2 = 329.0 \text{ kg}$$

To obtain the coordinates of the center of mass (x_G, y_G, z_G) with respect to the xyz coordinate system, we calculate the first moment of mass with respect to each coordinate axis. These are

$$mx_G = m_1 x_{G1} - m_2 x_{G2}$$

$$my_G = m_1 y_{G1} - m_2 y_{G2}$$

$$mz_G = m_1 z_{G1} - m_2 z_{G2}$$

where (x_{Gi}, y_{Gi}, z_{Gi}) are the coordinates of point G_i. Substituting the dimensions of the body into these equations yields

$$x_G = \frac{376(0) - 47.0(0.1)}{329} = -0.01429 \text{ meters}$$

$$y_G = \frac{376(0) - 47.0(0.15)}{329} = -0.2143 \text{ meters}$$

$$z_G = \frac{376(0) - 47.0(0.05)}{329} = -0.0714 \text{ meters}$$

Hence, with respect to the xyz system, the coordinates of the center of mass are $(-0.0143, -0.0214, -0.0714)$ meters.

We now turn our attention to the inertia terms. The tables in Appendix B detail the inertia values of a rectangular parallelepiped for a coordinate system with origin at the center of mass, as shown in the sketch.

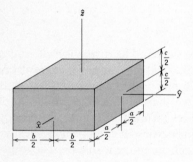

Also, because the $\hat{x}\hat{y}$, $\hat{x}\hat{z}$, and $\hat{y}\hat{z}$ planes are all planes of symmetry for the parallelepiped, we know that $\hat{x}\hat{y}\hat{z}$ are the principal axes for this shape, so that the products of inertia are zero.

From these tabulated values we start by determining the inertia properties of blocks 1 and 2 with respect to the $\hat{x}_1\hat{y}_1\hat{z}_1$ and $\hat{x}_2\hat{y}_2\hat{z}_2$ coordinate systems, respectively. The $\hat{x}_1\hat{y}_1\hat{z}_1$ frame is coincident with the xyz frame. Hence we have

$$(I_{xx})_1 \equiv (I_{\hat{x}_1\hat{x}_1})_1 = \tfrac{1}{12}m_1(0.6^2 + 0.2^2)$$

$$= 12.533 \text{ kg-m}^2$$

$$(I_{xz})_1 \equiv (I_{\hat{x}_1\hat{z}_1})_1 = 0$$

For body 2 the tabulated formulas give

$$(I_{\hat{x}_2\hat{x}_2})_2 = \tfrac{1}{12}m_2(0.3^2 + 0.1^2)$$

$$= 0.392 \text{ kg-m}^2$$

$$(I_{\hat{x}_2\hat{z}_2})_2 = 0$$

To combine the inertia terms of the individual components we must transform the properties of block 2 to the xyz system, using the parallel axis theorems. With respect to $\hat{x}_2\hat{y}_2\hat{z}_2$, the coordinates of the origin O of the xyz system are $(-0.1, -0.15, -0.05)$ meters. Therefore,

$$(I_{xx})_2 = (I_{\hat{x}_2\hat{x}_2})_2 + m_2(\hat{y}_O^2 + \hat{z}_O^2)$$

$$= 0.392 + 47.0[(-0.15)^2 + (-0.05)^2]$$

$$= 1.567 \text{ kg-m}^2$$

$$(I_{xz})_2 = (I_{\hat{x}_2\hat{z}_2})_2 + m_2\hat{x}_O\hat{z}_O$$

$$= 0 + 47.0(-0.1)(-0.05)$$

$$= 0.235 \text{ kg-m}^2$$

As the last step, we obtain the properties of the composite body by combining the results for the two components.

$$I_{xx} = (I_{xx})_1 - (I_{xx})_2 = 12.533 - 1.567$$

$$= 10.966 \text{ kg-m}^2$$

$$I_{xy} = (I_{xz})_1 - (I_{xz})_2 = 0 - 0.235$$

$$= -0.235 \text{ kg-m}^2$$

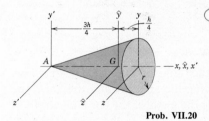

Prob. VII.20

HOMEWORK PROBLEMS

VII.20 Using the formula for I_{yy} given in Problem VII.12, verify the formula in Appendix B for the moment of inertia of a cone about the $\hat{y}$ axis passing through the center of mass G.

VII.21 Using the formula for $I_{\hat{y}\hat{y}}$ given in Appendix B, verify the formula derived in Example 4 for the moment of inertia of the cone in Problem VII.20 about the y' axis passing through the apex A.

VII.22 Use the tables in Appendix B and the parallel axis theorems to determine the moments and products of inertia of the circular cylinder with respect to the xyz coordinate system shown. The cylinder has a mass of 10 kg.

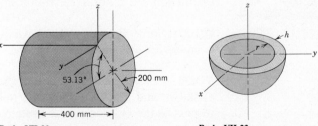

Prob. VII.22   **Prob. VII.23**

VII.23 Use the inertia properties for a hemisphere given in Appendix B to derive formulas for the moments of inertia of the hollow hemisphere with respect to the coordinate system shown, expressing the results in terms of the mass m, radius r, and thickness h. Then, by taking the limit as $h \to 0$, derive the corresponding properties for a thin hemispherical shell.

VII.24 Use the inertia properties for a complete circular cylinder to derive corresponding formulas for a semicircular cylinder. Then use those results to determine the mass moments of inertia of the hollow semicircular cylinder shown with respect to the xyz coordinate system.

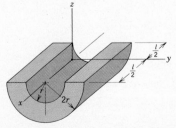

Prob. VII.24

VII.25 and VII.26   The diagrams show aluminum castings. Determine the principal mass moments of inertia for each body for axes whose origin is at the center of mass. How would the results change if the casting was made of some other material?

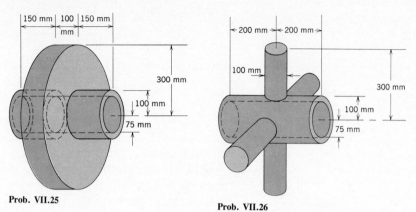

Prob. VII.25

Prob. VII.26

VII.27   Steel spheres are welded to each end of the circular steel bar. Determine the value of I_{zz} (a) considering the spheres and the bar to be three-dimensional bodies, (b) neglecting the mass of the bar and considering the spheres to be particles of the same mass as the actual spheres. What conclusions can you draw from these results?

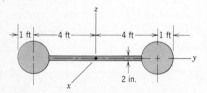

Prob. VII.27

VII.28   The built-up section shown consists of aluminum components. Determine the location of its mass center and the principal moments of inertia about a set of axes whose origin is at the mass center.

VII.29   Solve Problem VII.28 when the cylinder is steel and the other components are aluminum.

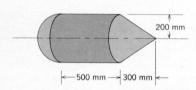

Prob. VII.28

VII.30   Determine the location of the mass center of the shelf and bracket shown and also determine the moments and products of inertia with respect to the *xyz* coordinate system shown. The two pieces are made of wood having a density of 750 kg/m³.

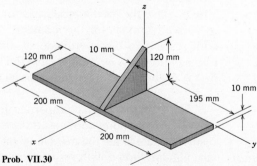

Prob. VII.30

VII.31 Determine the inertia properties with respect to the *xyz* coordinate system of the crankshaft shown, which is formed from a slender wire whose mass per unit length is *m/l*.

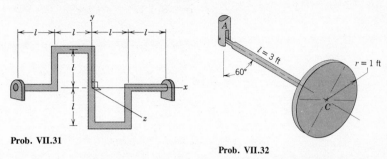

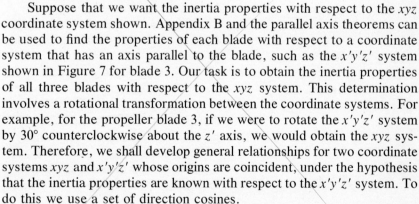

Prob. VII.31

Prob. VII.32

VII.32 The 60-lb thin disk is welded to the 15-lb slender shaft *AC*. Determine the location of the center of mass and the principal moments of inertia of the system for a set of axes having their origin at point *A*.

4 Rotation Transformation of Inertia Properties

The parallel axis theorems detail the transformation of the inertia properties for two coordinate systems having different origins but identical orientations. In some situations it may be necessary to transform the inertia properties between two coordinate systems having the same origin but different orientations. Let us develop such a situation by considering the propeller shown in Figure 7.

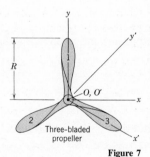

Figure 7

Suppose that we want the inertia properties with respect to the *xyz* coordinate system shown. Appendix B and the parallel axis theorems can be used to find the properties of each blade with respect to a coordinate system that has an axis parallel to the blade, such as the *x'y'z'* system shown in Figure 7 for blade 3. Our task is to obtain the inertia properties of all three blades with respect to the *xyz* system. This determination involves a rotational transformation between the coordinate systems. For example, for the propeller blade 3, if we were to rotate the *x'y'z'* system by 30° counterclockwise about the *z'* axis, we would obtain the *xyz* system. Therefore, we shall develop general relationships for two coordinate systems *xyz* and *x'y'z'* whose origins are coincident, under the hypothesis that the inertia properties are known with respect to the *x'y'z'* system. To do this we use a set of direction cosines.

Let $l_{pq}(\equiv l_{q'p})$ be the cosine of the angle between axis *p* of the *xyz* system and axis *q'* of the *x'y'z'* system. For example, in Figure 8, shown on the following page, we show an arbitrary orientation of *x* with respect to the *x'y'z'* frame, for which we have $l_{xx'} = \cos \alpha$, $l_{xy'} = \cos \beta$, and $l_{xz'} = \cos \gamma$.

As can be seen in Figure 8, the unit vector $\bar{\imath}$ can be written in terms of its components in the *x'y'z'* frame as

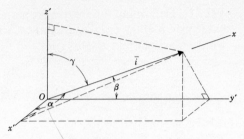

Figure 8

$$\bar{i} = (\cos \alpha)\bar{i}', + (\cos \beta)\bar{j}' + (\cos \gamma)\bar{k}'$$

$$\equiv l_{xx'}\bar{i}' + l_{xy'}\bar{j}' + l_{xz'}\bar{k}' \tag{14a}$$

If we had completed the x, y, z triad in Figure 8, the other unit vectors would be

$$\bar{j} = l_{yx'}\bar{i}' + l_{yy'}\bar{j}' + l_{yz'}\bar{k}' \tag{14b}$$

$$\bar{k} = l_{zx'}\bar{i}' + l_{zy'}\bar{j}' + l_{zz'}\bar{k}' \tag{14c}$$

A rectangular Cartesian coordinate system has the properties that

$$\bar{i} \cdot \bar{i} = \bar{j} \cdot \bar{j} = \bar{k} \cdot \bar{k} = 1$$

$$\bar{i} \cdot \bar{j} = \bar{j} \cdot \bar{k} = \bar{i} \cdot \bar{k} = 0$$

Then in view of equations (14) and the fact that $x'y'z'$ is also a rectangular Cartesian system, the foregoing relations require that

$$\begin{array}{ll} (l_{px'})^2 + (l_{py'})^2 + (l_{pz'})^2 = 1 & p = x, y, \text{ or } z \\ l_{px'}l_{qx'} + l_{py'}l_{qy'} + l_{pz'}l_{qz'} = 0 & p, q = x, y, \text{ or } z, \quad p \neq q \end{array} \tag{15}$$

As an aside, note that equations (15) represent six relationships that the nine direction cosines must satisfy, so that only three of the direction cosines, such as $l_{xx'}$, $l_{yy'}$, and $l_{zz'}$, are independent quantities.

Let us now consider the matter of the computation of the moment of inertia I_{xx}. By definition,

$$I_{xx} = \int_{\mathcal{V}} (y^2 + z^2)\, dm$$

But $(y^2 + z^2)^{1/2} = R$ is the distance from the differential element at point i to the x axis. In other words, it is the component of $\bar{r}_{i/O}$ perpendicular to the x axis, as illustrated in Figure 9 on the next page. From the vector algebra we know that the component of a vector perpendicular to an axis

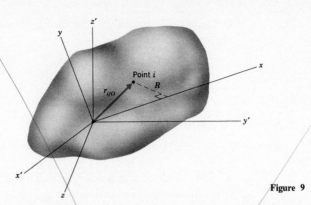

Figure 9

can be obtained from a cross product. Specifically, for our problem, this step gives

$$R = \left| \bar{r}_{i/O} \times \bar{\imath} \right|$$

so that

$$R^2 \equiv (y^2 + z^2) = (\bar{r}_{i/O} \times \bar{\imath}) \cdot (\bar{r}_{i/O} \times \bar{\imath})$$

We shall refer all components to the $x'y'z'$ reference frame, because we are considering the inertia properties to be known with respect to this coordinate system. Hence, using equation (14a),

$$\bar{r}_{i/O} = x'\bar{\imath}' + y'\bar{\jmath}' + z'\bar{k}'$$

$$\bar{\imath} = l_{xx'}\bar{\imath}' + l_{xy'}\bar{\jmath}' + l_{xz'}\bar{k}'$$

it follows that

$$\bar{r}_{i/O} \times \bar{\imath} = (l_{xz'}y' - l_{xy'}z')\bar{\imath}' + (l_{xx'}z' - l_{xz'}x')\bar{\jmath}' + (l_{xy'}x' - l_{xx'}y')\bar{k}'$$

From this we find

$$R^2 \equiv (\bar{r}_{i/O} \times \bar{\imath}) \cdot (\bar{r}_{i/O} \times \bar{\imath})$$

$$= (l_{xz'}y' - l_{xy'}z')^2 + (l_{xx'}z' - l_{xz'}x')^2 + (l_{xy'}x' - l_{xx'}y')^2$$

Upon collecting terms in this equation for R^2, we can form

$$I_{xx} = \int_{\mathscr{V}} R^2 \, dm = \int_{\mathscr{V}} \{ l_{xx'}^2[(y')^2 + (z')^2] + l_{xy'}^2[(x')^2 + (z')^2]$$

$$+ \, l_{xz'}^2[(x')^2 + (y')^2] - 2l_{xx'}l_{xy'}x'y' - 2l_{xy'}l_{xz'}y'z' - 2l_{xx'}l_{xz'}x'z' \} \, dm$$

Comparable formulas for I_{yy} and I_{zz} can be obtained by suitable permutation of the symbols in this equation. Letting p represent either x, y, or z, the general formula for the moments of inertia with respect to the xyz system is then

$$
\begin{aligned}
I_{pp} = {} & l_{px'}^2 I_{x'x'} + l_{py'}^2 I_{y'y'} + l_{pz'}^2 I_{z'z'} \\
& - 2l_{px'}l_{py'}I_{x'y'} - 2l_{py'}l_{pz'}I_{y'z'} \\
& - 2l_{px'}l_{pz'}I_{x'z'} \qquad p = x, y, \text{ or } z
\end{aligned}
\tag{16}
$$

We shall now consider how to obtain a product of inertia, such as I_{xy}, from the inertia properties that are known with respect to the $x'y'z'$ coordinate system. We know that

$$
I_{xy} = \int_{\mathscr{D}} xy \, dm
$$

where x and y are the components of $\bar{r}_{i/O}$ in the $\bar{\imath}$ and $\bar{\jmath}$ directions, respectively. Thus, in view of equations (14a) and (14b),

$$
x = \left| \bar{r}_{i/O} \cdot \bar{\imath} \right| = l_{xx'}x' + l_{xy'}y' + l_{xz'}z'
$$
$$
y = \left| \bar{r}_{i/O} \cdot \bar{\jmath} \right| = l_{yx'}x' + l_{yy'}y' + l_{yz'}z'
$$

By forming the product of this expression and making use of identities (15), we find

$$
\begin{aligned}
x'y' = {} & -(l_{xy'}l_{yy'} + l_{xz'}l_{yz'})(x')^2 - (l_{xx'}l_{yx'} + l_{xz'}l_{yz'})(y')^2 - (l_{xx'}l_{yx'} + l_{xy'}l_{yy'})(z')^2 \\
& + (l_{xx'}l_{yy'} + l_{xy'}l_{yx'})x'y' + (l_{xy'}l_{yz'} + l_{xz'}l_{yy'})y'z' + (l_{xz'}l_{yx'} + l_{xx'}l_{yz'})x'z'
\end{aligned}
$$

We now regroup terms and form the product of inertia.

$$
\begin{aligned}
I_{x'y'} = \int_{\mathscr{D}} \{ & -l_{xy'}l_{yy'}[(x')^2 + (z')^2] - l_{xz'}l_{yz'}[(x')^2 + (y')^2] \\
& - l_{xx'}l_{yx'}[(y')^2 + (z')^2] + (l_{xx'}l_{yy'} + l_{xy'}l_{yx'})x'y' \\
& + (l_{xy'}l_{yz'} + l_{xz'}l_{yy'})y'z' + (l_{xz'}l_{yx'} + l_{xx'}l_{yz'})x'z' \} \, dm
\end{aligned}
$$

Once again, the other products of inertia can be obtained by suitable permutation of the symbols in the foregoing equation. Letting p and q represent x, y, or z, the general formula for the product of inertia with respect to the xyz system is

$$I_{pq} = (l_{px'}l_{qy'} + l_{py'}l_{qx'})I_{x'y'} + (l_{py'}l_{qz'} + l_{pz'}l_{qy'})I_{y'z'} + (l_{px'}l_{qz'} + l_{pz'}l_{qx'})I_{x'z'} \tag{17}$$
$$- l_{px'}l_{qx'}I_{x'x'} - l_{py'}l_{qy'}I_{y'y'} - l_{pz'}l_{qz'}I_{z'z'} \qquad p, q = x, y, \text{ or } z, \text{ but } p \neq q$$

An example of the application of equation (16) follows. The method for equation (17) is essentially the same, so it will not be a feature of the example.

EXAMPLE 6

Consider the propeller of Figure 8. Approximating each propeller blade as a slender bar of mass m, determine the inertia properties of this body with respect to the coordinate system shown in that figure.

Solution

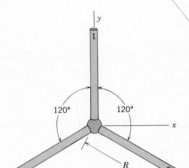

We are told to consider the propeller to be a planar composite body consisting of three uniform bars of mass m. As is illustrated in the sketch, this planar body is symmetric about both the xy and yz planes, so that xyz are principal axes, and

$$I_{xy} = I_{yz} = I_{xz} = 0$$

In order to use the tabulated values for moments of inertia, we treat each blade individually. For a slender uniform bar, Appendix B gives

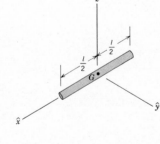

$$I_{\hat{x}\hat{x}} = 0$$
$$I_{\hat{y}\hat{y}} = I_{\hat{z}\hat{z}} = \tfrac{1}{12}ml^2$$

For blade 1, therefore, we need only change the symbols for the axes from those given and apply the parallel axis theorems, equations (9), to find

$$(I_{yy})_1 \equiv (I_{\hat{y}\hat{y}})_1 = 0$$
$$(I_{xx})_1 \equiv (I_{zz})_1 = (I_{\hat{x}\hat{x}})_1 + m\left(\frac{R}{2}\right)^2 = \tfrac{1}{3}mR^2$$

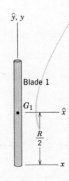

From the symmetry of the system we see that blades 2 and 3 have identical moments of inertia with respect to xyz, so we need only consider blade 2. Our procedure will be first to use the parallel axis theorems to transfer to a coordinate system $x'y'z'$ that has the same origin as that of xyz, and then to use the rotation transformation formulas to obtain the properties with respect to xyz.

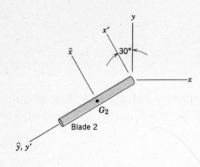

The values for blade 2 with respect to $x'y'z'$ are the same as those of blade 1 with respect to xyz; that is,

$$(I_{x'x'})_2 \equiv (I_{z'z'})_2 = \tfrac{1}{3}mR^2 \qquad (I_{y'y'})_2 = 0$$

By symmetry, $x'y'z'$ are principal axes for blade 2.

In order to transform from $x'y'z'$ to xyz, we need the direction cosines of xyz with respect to $x'y'z'$. These are indicated in the table.

p \\ q'	x'	y'	z'
x	$\cos 120° = -1/2$	$\cos 150° = -\sqrt{3}/2$	$\cos 90° = 0$
y	$\cos 30° = \sqrt{3}/2$	$\cos 120° = -1/2$	$\cos 90° = 0$
z	$\cos 90° = 0$	$\cos 90° = 0$	$\cos 0 = 1$

The transformation equation (16) may now be applied, using the fact that $x'y'z'$ are principal axes. Thus

$$(I_{xx})_2 = l_{xx'}{}^2(I_{x'x'})_2 + l_{xy'}^2(I_{y'y'})_2 + l_{xz'}^2(I_{z'z'})_2$$

$$= \left(-\frac{1}{2}\right)^2\left(\frac{1}{3}\,mR^2\right) + \left(-\frac{\sqrt{3}}{2}\right)^2(0) + (0)^2\left(\frac{1}{3}\,mR^2\right)$$

$$= \tfrac{1}{12}mR^2$$

$$(I_{yy})_2 = l_{yx'}^2(I_{x'x'})_2 + l_{yy'}^2\,(I_{y'y'})_2 + l_{yz'}^2\,(I_{z'z'})_2$$

$$= \left(\frac{\sqrt{3}}{2}\right)^2\left(\frac{1}{3}\,mR^2\right) + 0 + 0 = \tfrac{1}{4}\,mR^2$$

$$(I_{zz})_2 \equiv (I_{z'z'})_2 = \tfrac{1}{3}mR^2$$

Then, because blades 2 and 3 have the same properties, for the system we have

$$I_{xx} = (I_{xx})_1 + (I_{xx})_2 + (I_{xx})_3$$

$$= \tfrac{1}{3}mR^2 + 2(\tfrac{1}{12}mR^2) = \tfrac{1}{2}mR^2$$

$$I_{yy} = (I_{yy})_1 + (I_{yy})_2 + (I_{yy})_3$$

$$= 0 + 2(\tfrac{1}{4}mR^2) = \tfrac{1}{2}mR^2$$

$$I_{zz} = (I_{zz})_1 + (I_{zz})_2 + (I_{zz})_3$$

$$= 3(\tfrac{1}{3}mR^2) = mR^2$$

In the preceding example the tabulated inertia properties were fairly simple because of the symmetry of a slender bar. Also, because of the nature of the problem, it was not necessary to perform a rotation transformation for all of the inertia values. As a result, the calculations for the solution were not unduly complicated. On the other hand, it is not difficult to see that a rotation transformation for a more general problem can be quite awkward. To assist in computations, the transformations defined by equations (16) and (17) can be written in either matrix or tensor form. Matrices and tensors are not presented in this textbook. However, with the aid of either matrices or tensors, we could then determine the principal axes of any body, not just symmetric ones. Principal axes have obvious usefulness, because of the simplifications that arise in the angular momentum equation (5). The student who is interested in these topics is referred to any higher-level textbook in dynamics.

Here, we shall find that elementary problems in kinetics are more easily solved by using the most convenient set of axes, regardless of whether or not they are principal ones. Such an approach avoids the extra step of determining principal axes. In actual engineering practice, the decision to use principal axes should be carefully made, especially when formulating a problem for solution by electronic computer. The computational savings associated with the simplifications in the expression for the angular momentum resulting from principal axes must be weighed against the computations necessary to determine such axes.

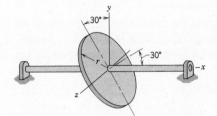

Prob. VII.33

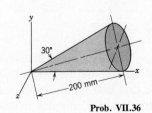

Prob. VII.36

HOMEWORK PROBLEMS

VII.33 Determine I_{xx} for the circular disk of mass m.

VII.34 Determine I_{yy} for the circular disk of mass m in Problem VII.33.

VII.35 Determine I_{xy} for the circular disk of mass m in Problem VII.33.

VII.36 Determine I_{xx} for the 5-kg right circular cone.

VII.37 Determine I_{yy} for the 5 kg right circular cone in Problem VII.36.

VII.38 Determine I_{xy} for the 5-kg right circular cone in Problem VII.36.

D. EQUATIONS OF MOTION

Now that we have studied how the inertia properties of a body are obtained, we can continue with the development of the relationship between angular momentum and the moment exerted by the force system, as given in equation (2).

$$\Sigma \, \bar{M}_A = \dot{H}_A \tag{2}$$

Because they cannot be emphasized enough, we begin by repeating the steps necessary to utilize the derived expression for the angular momentum of a rigid body, equation (5).

a   Restrict the choice of point A to be either a fixed point in space that is also fixed relative to the body, or the center of mass of the body, or else a point in the body that is accelerating directly toward or away from the center of mass, thus assuring the validity of the moment equation (2).
b   Select a moving reference frame xyz that is attached to the body with origin at point A. This choice for xyz ensures that the moments and products of inertia with respect to the reference frame are constants.
c   Orient the reference frame relative to the body in a manner that facilitates the determination of the moments and products of inertia.
d   Substitute these inertia properties and the $(\bar{\imath}, \bar{\jmath}, \bar{k})$ components of the angular velocity $\bar{\omega}$ of the body into equation (5), thus obtaining an expression for the components of $\bar{H}_A$ relative to xyz.

In the case where a body undergoes a translational motion, the only allowable point for summing moments is the center of mass G. Then, because $\bar{\omega} \equiv \bar{0}$ for translation, the moment equation (2) reduces to $\Sigma \, \bar{M}_G = \bar{0}$.

On the other hand, when a body has rotational motion, the procedure described above yields the $(\bar{\imath}, \bar{\jmath}, \bar{k})$ components of $\bar{H}_A$. Therefore, the evaluation of $\dot{H}_A$ requires that we employ the concept of differentiation with respect to a moving reference frame as derived in Module VI. Recall that, for an arbitrary vector $\bar{A}$, the general formula for this procedure is

$$\frac{d\bar{A}}{dt} = \frac{\delta \bar{A}}{\delta t} + \bar{\omega} \times \bar{A} \tag{18}$$

Hence, operating on $\bar{H}_A$, we have

$$\boxed{\Sigma \, \bar{M}_A = \dot{H}_A = \frac{\delta \bar{H}_A}{\delta t} + \bar{\omega} \times \bar{H}_A} \tag{19}$$

where, by definition, the components of $\delta \bar{H}_A / \delta t$ are the derivatives of the components of $\bar{H}_A$. Then, in view of the fact that the moments and products of inertia are constants, differentiating the components of $\bar{H}_A$ in equation (5) gives

$$\begin{aligned}
\frac{\delta \bar{H}_A}{\delta t} = {} & (I_{xx}\dot{\omega}_x - I_{xy}\dot{\omega}_y - I_{xz}\dot{\omega}_z)\bar{\imath} \\[6pt]
& + (I_{yy}\dot{\omega}_y - I_{xy}\dot{\omega}_x - I_{yz}\dot{\omega}_z)\bar{\jmath} \\[6pt]
& + (I_{zz}\dot{\omega}_z - I_{xz}\dot{\omega}_x - I_{yz}\dot{\omega}_y)\bar{k}
\end{aligned} \tag{20}$$

The interpretation of this equation involves an important subtlety. In general, it is a difficult matter to differentiate the components of $\bar{\omega}$ because such an evaluation would require expressions for ω_x, ω_y, and ω_z that are valid for all time instants. However, in the approach we have developed, the xyz reference frame is attached to the body and therefore has the same angular velocity as the body. As a result, equation (18) gives

$$\frac{d\bar{\omega}}{dt} \equiv \bar{\alpha} = \frac{\delta\omega}{\delta t} + \bar{\omega} \times \bar{\omega} \equiv \frac{\delta\bar{\omega}}{\delta t}$$

or, in component form,

$$\alpha_x = \dot{\omega}_x \qquad \alpha_y = \dot{\omega}_y \qquad \alpha_z = \dot{\omega}_z$$

In other words, because the xyz frame is attached to the body, the components of the angular acceleration of the body are identical to the derivatives of the angular velocity components. This allows us to follow the procedures of Module VI to describe angular velocity and angular acceleration, and then simply substitute the results into the equations of kinetics.

As a result, we now write equation (20) in the more explicit form

$$\boxed{\begin{aligned} \frac{\delta\bar{H}_A}{\delta t} &= (I_{xx}\alpha_x - I_{xy}\alpha_y - I_{xz}\alpha_z)\bar{\imath} \\ &+ (I_{yy}\alpha_y - I_{xy}\alpha_x - I_{yz}\alpha_z)\bar{\jmath} \\ &+ (I_{zz}\alpha_z - I_{xz}\alpha_x - I_{yz}\alpha_y)\bar{k} \end{aligned}} \tag{21}$$

Equations (5), (19), and (21) form the basic relationships between the resultant moment applied to a rigid body and the angular motion of the body. Clearly, the moment equation has three components, so that in conjunction with equation (1) governing the effect of the resultant force,

$$\boxed{\Sigma \bar{F} = m\bar{a}_G} \tag{1}$$

we have the six scalar equations of motion alluded to earlier.

For a different perspective of the preceding development, let us see what conditions must exist for equation (19) to reduce to the now familiar $\Sigma \bar{M} = I\bar{\alpha}$ equation for planar motion. For an arbitrary body moving parallel to the fixed XY plane, we select the z axis to be perpendicular to the plane of motion. Then, for planar motion we know that

$$\bar{\omega} = \omega\bar{k} \qquad \bar{\alpha} = \alpha\bar{k}$$

Equations (5) and (20) then give

$$\bar{H}_A = -I_{xz}\omega\bar{\imath} - I_{yz}\omega\bar{\jmath} + I_{zz}\omega\bar{k}$$

$$\frac{\delta\bar{H}_A}{\delta t} = -I_{xz}\alpha\bar{\imath} - I_{yz}\alpha\bar{\jmath} + I_{zz}\alpha\bar{K}$$

By substituting the foregoing expressions, the moment equation (19) is specialized to the case of planar motion, becoming

$$\Sigma \bar{M}_A = \alpha(-I_{xz}\,\bar{\imath} - I_{yz}\,\bar{\jmath} + I_{zz}\,\bar{k}) + \omega\bar{k} \times \omega(-I_{xz}\,\bar{\imath} - I_{yz}\,\bar{\jmath} + I_{zz}\,\bar{k})$$

$$= -(I_{xz}\alpha - I_{xz}\omega^2)\bar{\imath} - (I_{yz}\alpha + I_{xz}\omega^2)\,\bar{\jmath} + I_{zz}\alpha\bar{k} \tag{22}$$

Notice that the $\bar{\imath}$ and $\bar{\jmath}$ components in this equation cause it to be different from the moment equation $\Sigma \bar{M}_A = I_{zz}\alpha\bar{k}$ employed in Module V. The $\bar{\imath}$ and $\bar{\jmath}$ components in equation (22) describe the resultant moments of the force system about the x and y axes necessary to constrain an arbitrary body to planar motion. This is the situation that arises when a body that is rotating about a fixed axis is dynamically unbalanced (as will be illustrated in Example 7 to follow). Only in situations where $I_{xz} = I_{yz} = 0$, that is, where the z axis is a principal axis, will these contraint moments be unnecessary to maintain a body in planar motion. This condition was assured in Module V, because there we considered only bodies that are symmetric about the xy plane.

Let us now return to the study of the equations of motion for a body in three-dimensional motion. In those cases where the chosen xyz reference system does not form principal axes, the most direct way of treating the problem is to evaluate $\bar{H}_A$ and $\delta\bar{H}_A/\delta t$ for the axes chosen and to then substitute these expressions into equation (19) for $\Sigma \bar{M}_A$.

When the xyz frame forms principal axes, the expressions for $\bar{H}_A$ and $\delta\bar{H}_A/\delta t$ simplify considerably, as does equation (19). The result is called *Euler's equations of motion,* named after Leonhard Euler (1707–1783). Thus, *when the xyz coordinate system is a set of principal axes of the body,* the vector equation for the rotational motion of the body is

$$\Sigma \bar{M}_A = [I_{xx}\alpha_x - (I_{yy} - I_{zz})\omega_y\omega_z]\bar{\imath}$$
$$+ [I_{yy}\alpha_y - (I_{zz} - I_{xx})\omega_x\omega_z]\,\bar{\jmath} \tag{23}$$
$$+ [I_{zz}\alpha_z - (I_{xx} - I_{yy})\omega_x\omega_y]\bar{k}$$

It should be evident that for problems where the xyz system forms a principal set of axes, equation (23) involves fewer computations than equations (5), (19), and (21). However, the separate computation of $\bar{H}_A$ and $\delta\bar{H}_A/\delta t$ has the advantage of enabling us to gain insight about the underlying physics of the system being studied. Also, this approach is the one that must be employed if the xyz system is a nonprincipal set of axes. To gain some appreciation of the application of both approaches, let us consider the following two examples.

EXAMPLE 7

The obliquely mounted disk of Example 1 is illustrated again here. The rigid shaft, which is welded to the disk, is rotating at the constant angular speed ω. The bearings are frictionless and can only apply transverse shear forces to restrain the transverse motion of the shaft. Additionally, bearing A can exert a thrust force to restrain axial motion of the shaft. Determine the reaction forces exerted by the bearings due to the motion of the system.

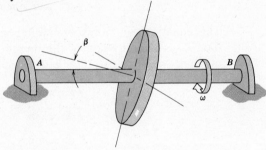

Solution

We begin with a free body diagram of the system that also shows the orientation of xyz relative to the system.

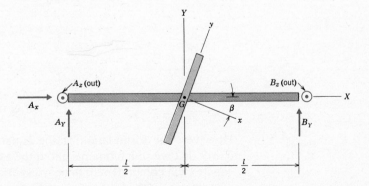

We have not included the gravity force because it is a static effect; if we sought the total reactive forces we would include it. The xyz reference frame is the same as that used in Example 1, so that we may use the results for the vectors $\bar{\omega}$ and $\bar{H}_G$ derived there.

$$\bar{\omega} = \omega \bar{I} \qquad \text{always}$$

$$= \omega(\cos \beta \, \bar{i} + \sin \beta \, \bar{j}) \qquad \text{instantaneously and always}$$

$$\bar{H}_G = \tfrac{1}{4}mr^2\omega(2 \cos \beta \, \bar{i} + \sin \beta \, \bar{j})$$

Both the rate of rotation ω and the unit vector $\bar{I}$ are constants, so we find $\bar{\alpha} \equiv \dot{\bar{\omega}} = \bar{0}$. This means that $\delta\bar{H}_G/\delta t = \bar{0}$, and

$$\dot{\bar{H}}_G \equiv \frac{\delta\bar{H}_G}{\delta t} + \bar{\omega} \times \bar{H}_G = -\tfrac{1}{4}mr^2\omega^2 \sin \beta \cos \beta \, \bar{k}$$

$$= -\tfrac{1}{8}mr^2\omega^2 \sin 2\beta \, \bar{k}$$

We now consider the external force system. Because the center of mass G is on the fixed axis of rotation, it has no acceleration, and

$$\Sigma \, \bar{F} = m\bar{a}_G = \bar{0}$$

This equation can be resolved into components with respect to any coordinate system. In view of the directions of the forces in the free body diagram, we shall use the fixed coordinate system for this step. Thus

$$\bar{I} \text{ direction:} \qquad A_X = 0$$

$$\bar{J} \text{ direction:} \qquad A_Y + B_Y = 0$$

$$\bar{K} \text{ direction:} \qquad A_Z + B_Z = 0$$

We next compute the moment sum, as follows:

$$\Sigma \, \bar{M}_G = \bar{r}_{A/G} \times (A_X\bar{I} + A_Y\bar{J} + A_Z\bar{K})$$

$$+ \, \bar{r}_{B/G} \times (B_Y\bar{J} + B_Z\bar{K})$$

$$= -\frac{l}{2}\bar{I} \times (A_X\bar{I} + A_Y\bar{J} + A_Z\bar{K})$$

$$+\frac{l}{2}\bar{I} \times (B_Y\bar{J} + B_Z\bar{K})$$

$$= \frac{l}{2}(A_Z - B_Z)\bar{J} - \frac{l}{2}(A_Y - B_Y)\bar{K}$$

However, the formulation of the equation for moment equilibrium requires that we sum moments about the axes of the moving system xyz and equate each component to the corresponding component of $\dot{\bar{H}}_G$. To do this we transform the unit vectors, referring to the free body diagram for the orientations of the axes. Thus, using

$$\bar{I} = \cos \beta \, \bar{i} + \sin \beta \, \bar{j}$$

$$\bar{J} = -\sin \beta \, \bar{i} + \cos \beta \, \bar{j}$$

$$\bar{K} = \bar{k}$$

the moment equation becomes

$$\Sigma \bar{M}_G = \frac{l}{2}(A_Z - B_Z)(-\sin \beta\, \bar{i} + \cos \beta\, \bar{j}) - \frac{l}{2}(A_Y - B_Y)\bar{k}$$

$$= \dot{\bar{H}}_G = -\tfrac{1}{8}mr^2\omega^2 \sin 2\beta\, \bar{k}$$

Matching components yields

$\bar{i}$ and $\bar{j}$ components: $\quad A_Z = B_Z = 0$

$\bar{k}$ component: $\quad -\dfrac{l}{2}(A_Y - B_Y) = -\tfrac{1}{8}mr^2\omega^2 \sin 2\beta$

Simultaneous solution of the force and moment equations yields

$$A_X = A_Z = B_Z = 0$$

$$A_Y = -B_Y = \tfrac{1}{16}m\,\frac{r^2}{l}\,\omega^2 \sin 2\beta$$

The force components A_Y and B_Y are a consequence of the fact that $\bar{H}_G$ does not have a fixed direction. Example 1 showed that $\bar{H}_G$ rotates about the shaft at angular velocity $\bar{\omega}$ in the manner illustrated in the sketch.

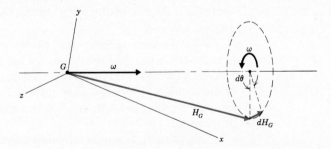

As the system rotates an infinitesimal amount $d\theta$ from its instantaneous position, the angular momentum changes by the amount $d\bar{H}_G$, as shown in the sketch. This demonstrates why $\dot{\bar{H}}_G$ is in the negative $\bar{k}$ direction.

It is common to refer to the component of $\dot{\bar{H}}_G$ that is perpendicular to $\bar{H}_G$ (and therefore a consequence of the variable direction of the angular momentum) as the *gyroscopic moment*. The reaction forces A_Y and B_Y in this example exert the gyroscopic moment on the system.

We selected the orientation of xyz such that the yz plane was the plane of the disk at an arbitrary instant. For this problem the gyroscopic moment is always in the negative $\bar{k}$ direction, so that the reaction forces A_Y and B_Y must always lie in the xy plane. This leads to the conclusion that the reaction forces rotate with the motion of the disk.

It should surprise you that the system is balanced in the static sense (the center of mass is on the axis of rotation), and yet reaction forces are generated by the rotation of the system. Such a system is said to be *dynamically unbalanced*. If we had solved this problem for an arbitrary body, rather than a disk, we could have shown that a system is dynamically balanced only when it is statically balanced *and* the fixed axis of rotation is a principal axis.

EXAMPLE 8

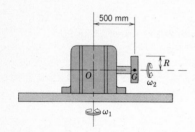

Let us now reconsider the system of Example 2 in which a disk is welded onto the shaft of a motor fixed to a turntable, as shown in the sketch. Initially, the turntable is rotating at the rate $\omega_1 = 10\pi$ rad/s and the motor is rotating at $\omega_2 = 50\pi$ rad/s. If the turntable is suddenly decelerated at the rate of 5π rad/s², what force and couple will the motor shaft exert on the disk?

Solution

The shaft is fixed to the disk, so it restricts the motion of the point of connection between the bodies in three dimensions and also restricts the rotation of the disk about three coordinate axes. Therefore, the shaft exerts an unknown three-dimensional force $\bar{F}_s$ on the disk and also an unknown three-dimensional couple $\bar{M}_s$. This is illustrated in the free body diagram. In this sketch we also show the xyz reference frame.

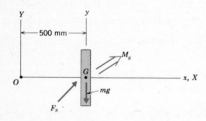

The instantaneous orientation of xyz has been chosen for the convenience it affords in obtaining the inertia properties and in describing the angular velocity.

Before summing forces and moments, we must determine $\bar{a}_G$ and $\dot{\bar{H}}_G$. In order to understand the solution more fully, and also for computational efficiency, we shall retain all results in algebraic form until the completion of the problem. From the given information we know that $\omega_1 = 10\pi$ rad/s, $\omega_2 = 50\pi$ rad/s, $\dot{\omega}_2 = 0$, $\dot{\omega}_1 = -5\pi$ rad/s² ($\dot{\omega}_1$ is negative because ω_1 is decreasing).

The value $\bar{a}_G$ is most easily determined by noting that point G follows a circular path of radius 0.5 m with angular speed ω_1. Furthermore, at the instant shown in the free body diagram, point G is moving into the plane of the diagram, that is, in the negative z direction. Therefore, using cylindrical coordinates with $\bar{e}_R = \bar{\imath}$ and $\bar{e}_\phi = -\bar{k}$, we find that

$$\bar{a}_G = -0.5\omega_1{}^2\bar{e}_R + 0.5\dot{\omega}_1\bar{e}_\phi$$

$$= -0.5\omega_1{}^2\bar{\imath} - 0.5\dot{\omega}_1\bar{k}$$

The angular velocity of the disk was obtained in Example 2. For this problem we shall also require the angular acceleration. The general expressions for these quantities are

$$\bar{\omega} = \omega_1 \bar{J} + \omega_2 \bar{\imath} \qquad \text{always}$$

$$\bar{\alpha} \equiv \dot{\omega}_1 \bar{J} + \dot{\omega}_2 \bar{\imath} + \omega_2 \dot{\bar{\imath}}$$

$$= \dot{\omega}_1 \bar{J} + \bar{0} + \omega_2 (\bar{\omega} \times \bar{\imath}) \qquad \text{always}$$

The instantaneous expressions are obtained by noting that at the instant illustrated in the free body diagram $\bar{J} = \bar{\jmath}$, so that

$$\bar{\omega} = \omega_1 \bar{\jmath} + \omega_2 \bar{\imath} \qquad \text{instantaneously}$$

$$\bar{\alpha} = \dot{\omega}_1 \bar{\jmath} + \omega_2 (\omega_1 \bar{\jmath} + \omega_2 \bar{\imath}) \times \bar{\imath}$$

$$= \dot{\omega}_1 \bar{\jmath} - \omega_1 \omega_2 \bar{k} \qquad \text{instantaneously}$$

The inertia properties of the disk were obtained in Example 2 directly from the tables in Appendix B. By symmetry, xyz are principal axes, and

$$I_{xx} = \tfrac{1}{2} mR^2 \qquad I_{yy} = I_{zz} = \tfrac{1}{4} mR^2$$

We next form the equations of motion. From the free body diagram and the results for $\bar{a}_G$, we have

$$\Sigma F_x = (F_s)_x = m(a_G)_x = -0.5 m \omega_1{}^2$$

$$\Sigma F_y = (F_s)_y - mg = m(a_G)_y = 0$$

$$\Sigma F_z = (F_s)_z = m(a_G)_z = -0.5 m \dot{\omega}_1$$

Thus

$$\bar{F}_s = -0.5 m \omega_1{}^2 \bar{\imath} + mg\,\bar{\jmath} - 0.5 m \dot{\omega}_1 \bar{k}$$

In this example we shall demonstrate the application of Euler's equation (23). Thus, by summing moments about point G and considering the disk to be sufficiently thin to allow us to consider the force $\bar{F}_s$ to be applied at point G, we find that

$$\Sigma M_{Gx} = (M_s)_x = I_{xx} \alpha_x - (I_{yy} - I_{zz}) \omega_y \omega_z$$

$$= \tfrac{1}{2} mR^2 (0) - (0)(\omega_1)(0) = 0$$

$$\Sigma M_{Gy} = (M_s)_y = I_{yy} \alpha_y - (I_{zz} - I_{xx}) \omega_z \omega_x$$

$$= \tfrac{1}{4} mR^2 \dot{\omega}_1 - (\tfrac{1}{4} - \tfrac{1}{2}) mR^2 (0)(\omega_2) = \tfrac{1}{4} mR^2 \dot{\omega}_1$$

$$\Sigma M_{Gz} = (M_s)_z = I_{zz} \alpha_z - (I_{xx} - I_{yy}) \omega_x \omega_y$$

$$= \tfrac{1}{4} mR^2 (-\omega_1 \omega_2) - (\tfrac{1}{2} - \tfrac{1}{4}) mR^2 (\omega_2)(\omega_1)$$

$$= -\tfrac{1}{2} mR^2 \omega_1 \omega_2$$

Thus

$$\bar{M}_s = mR^2(\tfrac{1}{4}\dot{\omega}_1\,\bar{j} - \tfrac{1}{2}\omega_1\omega_2\bar{k})$$

Now, substituting the given values for ω_1, $\dot{\omega}_1$, and ω_2 gives

$$\bar{F}_s = m[-493.5\bar{i} + 9.806\bar{j} - 7.854\bar{k}] \text{ N}$$

$$\bar{M}_s = mR^2[-3.927\bar{j} - 2467\bar{k}] \text{ N-m}$$

These answers, including the large component of $\bar{M}_s$ in the z direction, have simple explanations. The shaft force $\bar{F}_s$ causes the center of mass to follow a circular path. Thus, for this problem $(F_s)_x$ imposes a normal acceleration, $(F_s)_y$ counterbalances gravity to produce planar motion, and $(F_s)_z$ imposes a tangential acceleration. The shaft couple $\bar{M}_s$ causes the disk to assume the proper angular motion. The value of $(M_s)_y$ is independent of the speed of the motor ω_2. It represents the couple that must be applied to the disk if the turntable is to slow down. The large value of $(M_s)_z$ is the gyroscopic moment that must be applied to the disk if its angular motion is to be maintained. This is best illustrated by considering the result for $\bar{H}_G$ from Example 2, which is sketched again here.

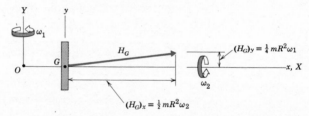

Because of the rotation ω_1 about the Y axis, the tip of $\bar{H}_G$ must move into the plane of the sketch, thus requiring that the force system apply a gyroscopic moment in the negative z direction. Once again, if we had relied on planar kinetics for intuition, we would have completely ignored the gyroscopic moment, thus missing a major effect in the system.

HOMEWORK PROBLEMS

VII.39 Determine $\dot{H}_G$ for the system in Problem VII.1.

VII.40 Determine $\dot{H}_G$ for the system in Problem VII.3.

VII.41 Determine $\dot{H}_G$ for the system in Problem VII.5.

VII.42 The slender bar of mass m and length l is welded to the horizontal shaft, which is rotating at a constant rate ω. In terms of the constant angle θ, determine (a) $\bar{H}_G$ and $\dot{H}_G$, (b) $\bar{H}_A$ and $\dot{H}_A$. Explain why $\dot{H}_G = \dot{H}_A = 0$ for $\theta = 0$ and $\theta = 90°$.

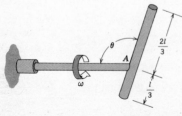

Prob. VII.42

VII.43 For the system in Problem VII.42, determine the total force-couple reaction at point A that the horizontal shaft exerts on the bar when the bar coincides (a) with the horizontal plane, (b) with the vertical plane.

VII.44 The 10-kg thin rectangular plate shown is welded diagonally onto the horizontal shaft AB. This shaft is rotating at ω = 240 rev/min when a couple M is applied, resulting in a deceleration $-\dot{\omega}$. Determine the value of the deceleration rate for which the component of $\dot{H}_G$ associated with $\dot{\omega}$ will have the same magnitude as the gyroscopic component of $\bar{H}_G$.

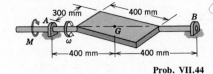

Prob. VII.44

VII.45 For the system in Problem VII.44, the couple M = 20 N-m. Determine the value of $\dot{\omega}$ and the bearing reactions at points A and B (a) when ω = 240 rev/min, (b) when ω = 120 rev/min.

VII.46 Two identical 40-kg disks are welded concentrically to a horizontal shaft that rotates at the angular rate ω. Identical small masses are attached to the perimeter of the disks. Is the system unbalanced, statically balanced only, or dynamically balanced when masses are added (a) nowhere, (b) at point A_1, (c) at points A_1 and A_2, (d) at points A_1 and C_2, (e) at points A_1, B_1, C_2, and D_2, (f) at points A_1, C_1, B_2, and D_2?

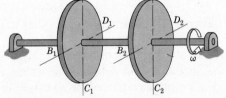

Prob. VII.46

VII.47 A 6-in.-radius sphere weighing 80 lb rotates about bar AB at 3000 rev/min as the bar pivots about pin A. The bar weighs 20 lb. When θ = 45°, $\dot{\theta}$ = 5 rad/s and $\ddot{\theta}$ = −10 rad/s². Determine (a) $\bar{H}_A$ and $\dot{H}_A$ for the sphere, (b) $\bar{H}_A$ and $\dot{H}_A$ for bar AB.

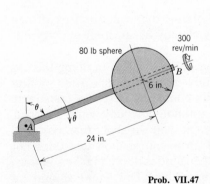

Prob. VII.47

VII.48 For the motion of the system defined in Problem VII.47, determine the couple that the pin at point A applies to the bar (a) if the bar is massless, (b) if the sphere is massless, and (c) if neither body is massless. Compare the results.

VII.49 The system of Problem VII.6 is presented again in the diagram. Determine $\dot{H}_G$ for the disk of mass m, considering ω_1 and ω_2 to be constant.

VII.50 For the system of Problem VII.49, determine $\bar{H}_A$ and $\dot{H}_A$ for the disk of mass m, using a coordinate system parallel to the one shown. Consider ω_1 and ω_2 to be constant. Draw a sketch showing $\bar{H}_A$.

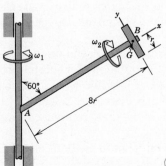

Prob. VII.49

VII.51 For the system of Problem VII.49, at the instant shown $\omega_1 = 5$ rad/s, $\omega_2 = 100$ rad/s, and ω_2 is increasing at 2 rad/s². The mass of the disk is m. Determine $\bar{H}_A$ and $\dot{H}_A$, using a coordinate system parallel to the one shown.

VII.52 For the system of Problem VII.49, the mass of shaft AB is one-half that of the disk. For the case where ω_1 and ω_2 are constant, compare (a) $\dot{H}_A$ for the disk to $\dot{H}_A$ for the shaft, (b) the force-couple reaction exerted upon the shaft at point A when the mass of the shaft is considered to the reaction obtained when the mass of the shaft is neglected.

VII.53 The two identical spheres of mass m and radius r rotate independently at rates ω_1 and ω_2 about shaft AB, as shown. This shaft is supported by a pin joint at its midpoint C and rotates about the vertical axis at the rate Ω. All rates of rotation are constant, and the mass of shaft AB is negligible. Determine (a) $\bar{H}_C$ and $\dot{H}_C$ for each sphere, (b) the couple that must be applied at the pin joint to sustain this motion. Why is this couple zero when $\omega_1 = \omega_2$?

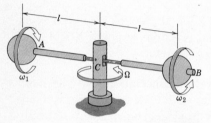

Prob. VII.53

E. PROBLEM SOLVING

The steps for solving problems involving the equations of three-dimensional motion of rigid bodies may already be apparent to you from the examples. They are as follows:

1 Draw a complete free body diagram of the system. In doing this, to be sure that you define the problem fully, remember that if the motion of a point is restricted in a certain direction, then there is a constraint force in that direction. Similarly, if the rotation of a body about a certain axis is restricted, then there is a constraint moment about that axis.

2 Write down, in equation form, all given information that does not appear in the free body diagram. List all quantities desired in the solution.

3 Choose a point A for summing moments. This point should be either the center of mass or a fixed point in space that is also fixed relative to the body. The third type of point is rather special and can generally be ignored.

4 Attach xyz axes to the body with origin at point A and orient xyz relative to the body so as to minimize any computations for inertia properties. If possible, orient xyz so as to make it possible to use the inertia properties in Appendix B and the parallel axis theorems. Select XYZ to fit best the fixed axes of rotation of the body. Show these reference frames in the free body diagram of step 1.

5 Determine the inertia properties. Use integration methods and/or rotation transformations only as a last resort.

6 Determine the angular velocity $\bar{\omega}$, the angular acceleration $\bar{\alpha}$, and the acceleration of the center of mass $\bar{a}_G$. In doing this, be sure to satisfy

any kinematical constraints on the motion of the system. (Here it sometimes proves useful to leave the rotation rates and their derivatives as algebraic symbols.)

7   Compute each component $\Sigma \, \bar{M}_A$ and equate them to the corresponding components of $\dot{H}_A$. Substitute all known quantities into these equations. In general, use equations (5), (19), and (21) to perform this step, but in the special case where the xyz axes are principal axes you can use equation (23) instead.

8   Equate each component of $\Sigma \, \bar{F}$ to the corresponding component of $m\bar{a}_G$.

9   Solve the equations.

As an aside, note that steps 7 and 8 include all the kinematics and kinematical constraints that you determined in step 6. Hence the number of equations and the number of unknowns at this point should balance. If not, the two most common conditions that may have been overlooked are

a   a kinematical constraint condition, or
b   a friction force relationship.

ILLUSTRATIVE PROBLEM 1

The crane shown in the sketch is initially at rest when the vertical support post is given a constant angular acceleration $\dot{\Omega} = 0.5$ rad/s². The boom, which has a mass of 600 kg, is 10 m long, and the cable maintains the angle β at a constant value. Determine the tension in the cable as a function of time.

Solution

Step 1   The cable tension appears as an external force in a free body diagram of the boom. To construct this diagram we note that the pin connection at point A constrains all motions of the boom except for rotations through the angle β. It therefore follows that the force-couple system representing the reaction at this joint consists of a three-dimensional force $\bar{F}_A$ and a moment $\bar{M}_A$ that has no component perpendicular to the plane formed by the post and the boom. The free body diagram of the boom is as shown here.

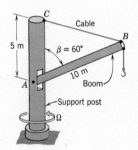

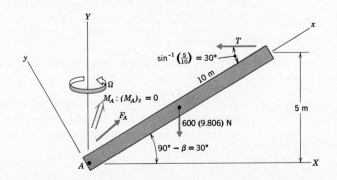

Step 2 The given information tells us that $\Omega = 0$ initially and that $\dot\Omega = 0.5$ rad/s², so that $\Omega = 0.5t$ rad/s. We wish to find the force T as a function of time t.

Step 3 We select the fixed point A for the origin of the moving reference system, thereby eliminating the reaction force $\bar F_A$ from the moment equation.

Step 4 Considering the boom to be a slender rod and selecting the x axis to coincide with the axis of the boom allows us to obtain the inertia properties directly from the tables in Appendix B. For convenience in describing the angular motion, we select the fixed Y axis to coincide with the vertical post and let the y axis lie in the XY plane, as shown in the free body diagram.

Step 5 The inertia properties for a slender rod appearing in Appendix B are repeated here.

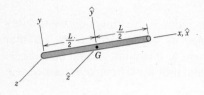

$$I_{\hat x \hat x} = 0 \quad I_{\hat y \hat y} = I_{\hat z \hat z} = \tfrac{1}{12}mL^2$$

To determine the inertia properties with respect to the xyz coordinate system we use the parallel axis theorems, which yields

$$I_{xx} = I_{\hat x \hat x} = 0$$

$$I_{yy} = I_{zz} = I_{\hat y \hat y} + m\left(\frac{L}{2}\right)^4 = \tfrac{1}{3}mL^2$$

$$= \tfrac{1}{3}(600)(10)^2 = 2(10)^4 \text{ kg-m}^2$$

$$I_{xy} = I_{yz} = I_{xz} = 0$$

Note that the xyz system forms principal axes because the mass of a slender rod is considered to be situated entirely along the axis of the rod.

Step 6 The boom is rotating about the vertical axis, so that the general expressions for $\bar\omega$ and $\bar\alpha$ are

$$\bar\omega = \Omega \bar J, \quad \bar\alpha = \dot\Omega \bar J \quad \text{always}$$

Resolving these expressions into $(\bar i, \bar j, \bar k)$ components gives

$$\bar\omega = \Omega(\cos 60° \, \bar i + \sin 60° \, \bar j) = \Omega(0.5\bar i + 0.866\bar j) \quad \text{instantaneously}$$

$$\bar\alpha = \dot\Omega(\cos 60° \, \bar i + \sin 60° \, \bar j) = \dot\Omega(0.5\bar i + 0.866\bar j) \quad \text{instantaneously}$$

We omit calculating $\bar a_G$ because the force equation will only yield the reaction force $\bar F_A$.

Step 7 Because the xyz system is a principal set of axes, we can use Euler's equation (23). Hence, using the inertia values found in step 5 and the values of $\bar\omega$ and $\bar\alpha$ found in step 6, we find that

$$\Sigma \bar{M}_A = [0 - 0]\bar{i} + [2(10^4)(0.5\dot{\Omega}) - 0]\bar{j}$$
$$+ \{0 - [0 - 2(10^4)](0.5\Omega)(0.866\Omega)\}\bar{k}$$
$$= 10^4\dot{\Omega}\bar{j} + 0.866(10^4)\Omega^2\bar{k}$$

We now use the free body diagram to compute the moment of the force system. The reaction couple $\bar{M}_A$ has no component about the z axis, so we find that

$$\Sigma \bar{M}_A = (M_A)_x\,\bar{i} + (M_A)_y\,\bar{j}$$
$$+ [T(5) - 600(9.806)(5 \sin 60°)]\bar{k}$$
$$= (M_A)_x\,\bar{i} + (M_A)_y\,\bar{j} + (5T - 25480)\bar{k}$$

Step 8   We omit this step because we are not interested in the reaction force $\bar{F}_A$.

Step 9   To obtain the value of T we use the $\bar{k}$ component of the moment equation. Thus

$$(\Sigma \bar{M}_A) \cdot \bar{k} = (\dot{H}_A) \cdot \bar{k}$$
$$(5T - 25480) = 0.8660(10^4)\Omega^2$$

Then solving for T and using the expression for Ω in step 2, we finally obtain

$$T = 5095 + \frac{0.8660}{5}\,(10^4)(0.5t)^2$$

$$= 5095 + 433t^2$$

The first term in this result is the tension within the cable caused by gravity, which is a static effect. The second term represents the inertial effect of the rotation.

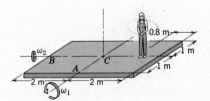

ILLUSTRATIVE PROBLEM 2

A 70-kg person is standing motionless with respect to the 140-kg wooden raft illustrated in the sketch. At the instant when the raft is horizontal, the centroid C of the raft is moving forward with a constant velocity of 2 m/s, while the raft's rider observes the raft to be pitching about the centerline CA at $\omega_1 = 0.2$ rad/s and rolling about the centerline CB at $\omega_2 = 0.5$ rad/s. The derivatives of these rates are zero at this instant. Determine the force and moment about the centroid C that the water must exert on the raft in order to produce this motion. For the purpose of computing inertial properties, the raft may be treated as a thin plate and the rider may be treated as a slender bar 1.80 m tall.

Solution

Step 1   Consider the rider and the raft to form a composite body. Let $\bar{F}$ and $\bar{M}_C$ be the force-couple resultant at point C.

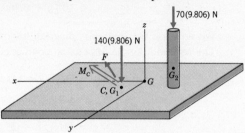

Step 2   Given $\omega_1 = 0.2$ rad/s, $\omega_2 = 0.5$ rad/s, $\dot{\omega}_1 = 0$, $\dot{\omega}_2 = 0$. Find $\bar{F}$ and $\bar{M}_C$.

Step 3   The only allowable point for summing moments is the center of mass G of the *composite* body.

Step 4   Let *xyz* denote the reference frame whose origin is point G. Attach this reference frame to the body and orient the axes to be parallel to the axes of symmetry of the raft, as shown in the free body diagram. We shall need to determine the inertia properties of the composite body, so we also show the *xyz* system and the parallel centroidal axes for each component in a separate sketch.

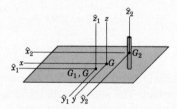

Step 5   Locate point G by computing the first moments of mass about point C. (Note that using point C minimizes the number of terms in the calculations.) With respect to the $\hat{x}_1\hat{y}_1\hat{z}_1$ system, the coordinates of point G_2 are $(-1.2, 0, 0.90)$. Denote the coordinates of point G with respect to the $\hat{x}_1\hat{y}_1\hat{z}_1$ coordinate system as $(\hat{x}, \hat{y}, \hat{z})$. Because the total mass is 210 kg, we find that

$$\hat{x} = \frac{70(-1.2)}{210} = -0.40 \text{ m} \qquad \hat{y} = \frac{70(0)}{210} = 0 \qquad \hat{z} = \frac{70(0.9)}{210} = 0.30 \text{ m}$$

We can now determine the inertia properties of each component with respect to the *xyz* axes. For the raft the tabulated properties of a thin rectangular parallelepiped and the symmetry principles give

$$(I_{\hat{x}_1\hat{x}_1})_1 = \tfrac{1}{12}(140)(2^2 + 0^2) = 46.67 \text{ kg-m}^2$$

$$(I_{\hat{y}_1\hat{y}_1})_1 = \tfrac{1}{12}(140)(4^2 + 0^2) = 186.67 \text{ kg-m}^2$$

$$(I_{\hat{z}_1\hat{z}_1})_1 = \tfrac{1}{12}(140)(2^2 + 4^2) = 233.33 \text{ kg-m}^2$$

$$(I_{\hat{x}_1\hat{y}_1})_1 = (I_{\hat{x}_1\hat{z}_1})_1 = (I_{\hat{y}_1\hat{z}_1})_1 = 0$$

Then, because the $\hat{x}_1\hat{y}_1\hat{z}_1$ coordinates of point G are $(-0.40, 0, 0.30)$, the parallel axis theorems give

$$(I_{xx})_1 = (I_{\hat{x}_1\hat{x}_1})_1 + (140)(0^2 + 0.3^2) = 59.27 \text{ kg-m}^2$$

$$(I_{yy})_1 = (I_{\hat{y}_1\hat{y}_1})_1 + (140)(0.4^2 + 0.3^2) = 221.67 \text{ kg-m}^2$$

$$(I_{zz})_1 = (I_{\hat{z}_1\hat{z}_1})_1 + (140)(0.4^2 + 0^2) = 255.73 \text{ kg-m}^2$$

$$(I_{xy})_1 = (I_{\hat{x}_1\hat{y}_1})_1 + (140)(-0.4)(0) = 0$$

$$(I_{xz})_1 = (I_{\hat{x}_1\hat{z}_1})_1 + (140)(-0.4)(0.3) = -16.80 \text{ kg-m}^2$$

$$(I_{yz})_1 = (I_{\hat{y}_1\hat{z}_1})_1 + (140)(0)(0.3) = 0$$

Treating the rider as a slender bar, the tabulated properties yield

$$(I_{\hat{x}_1\hat{x}_2})_2 = (I_{\hat{y}_2\hat{y}_2})_2 = \tfrac{1}{12}(70)(1.8)^2 = 18.90 \text{ kg-m}^2$$

$$(I_{\hat{z}_2\hat{z}_2})_2 = (I_{\hat{x}_2\hat{y}_2})_2 = (I_{\hat{x}_2\hat{z}_2})_2 = (I_{\hat{y}_2\hat{z}_2})_2 = 0$$

The $\hat{x}_2\hat{y}_2\hat{z}_2$ coordinates of point G are $(0.8, 0, -0.6)$, so the parallel axis theorems give

$$(I_{xx})_2 = (I_{\hat{x}_2\hat{x}_2})_2 + (70)(0^2 + 0.6^2) = 44.10 \text{ kg-m}^2$$

$$(I_{yy})_2 = (I_{\hat{y}_2\hat{y}_2})_2 + (70)(0.8^2 + 0.6^2) = 88.90 \text{ kg-m}^2$$

$$(I_{zz})_2 = (I_{\hat{z}_2\hat{z}_2})_2 + (70)(0.8^2 + 0^2) = 44.80 \text{ kg-m}^2$$

$$(I_{xy})_2 = (I_{yz})_2 = (I_{\hat{y}_2\hat{z}_2})_2 = 0$$

$$(I_{xz})_2 = (I_{\hat{x}_2\hat{z}_2})_2 + 70(0.8)(-0.6) = -33.60 \text{ kg-m}^2$$

We now add the properties of each component to obtain the inertia properties of the system about its mass center G.

$$I_{xx} = (I_{xx})_1 + (I_{xx})_2 = 103.4 \text{ kg-m}^2$$

$$I_{yy} = (I_{yy})_1 + (I_{yy})_2 = 310.6 \text{ kg-m}^2$$

$$I_{zz} = (I_{zz})_1 + (I_{zz})_2 = 300.5 \text{ kg-m}^2$$

$$I_{xy} = I_{yz} = 0$$

$$I_{xz} = (I_{xz})_1 + (I_{xz})_2 = -50.4 \text{ kg-m}^2$$

Step 6 The given angular speeds ω_1 and ω_2 describe the rotation of the body about axes that are fixed to the body. Then, in view of the sense of the rotations shown in the given diagram, it is generally true that

$$\bar{\omega} = \omega_2(-\bar{\imath}) + \omega_1\bar{\jmath} \qquad \text{always}$$

$$\bar{\alpha} = -\dot{\omega}_2 \bar{i} - \omega_2 \dot{\bar{i}} + \dot{\omega}_1 \bar{j} + \omega_1 \dot{\bar{j}}$$

$$= -\dot{\omega}_2 \bar{i} - \omega_2(\bar{\omega} \times \bar{i}) + \dot{\omega}_1 \bar{j} + \omega_1(\bar{\omega} \times \bar{j}) \qquad \text{always}$$

Thus, substituting the given rotation rates, we find that

$$\bar{\omega} = -0.5\bar{i} + 0.2\bar{j} \text{ rad/s}$$

$$\bar{\alpha} = \bar{0} - 0.5(-0.5\bar{i} + 0.2\bar{j}) \times \bar{i} + \bar{0}$$

$$+ 0.2(-0.5\bar{i} + 0.2\bar{j}) \times \bar{j} = \bar{0}$$

(As an alternative procedure, we could also have found $\bar{\alpha} = \bar{0}$ by using the fact that $\bar{\alpha} \equiv \delta\bar{\omega}/\delta t$ when the xyz frame is attached to the body.)

To describe the acceleration of the center of mass G, we use the rigid body motion formula, relating point G to point C, which is moving with constant velocity so that $\bar{a}_C = \bar{0}$. This gives

$$\bar{a}_G = \bar{0} + \bar{\alpha} \times \bar{r}_{G/C} + \bar{\omega} \times (\bar{\omega} \times \bar{r}_{G/C})$$

Substitution of the location of point G and the values of $\bar{\omega}$ and $\bar{\alpha}$ gives

$$\bar{a}_G = \bar{0} + \bar{0} + (-0.5\bar{i} + 0.2\bar{j}) \times [(-0.5\bar{i}$$

$$+ 0.2\bar{j}) \times (-0.4\bar{i} + 0.3\bar{k})]$$

$$= 0.0160\bar{i} + 0.040\bar{j} - 0.0870\bar{k} \text{ m/s}^2$$

Step 7 We now compute $\dot{\bar{H}}_G$. With respect to the nonprincipal axes xyz, equation (5) and the expression for $\bar{\omega}$ found in step 6 give

$$\bar{H}_G = [(103.4)(-0.5) - 0 - 0]\bar{i}$$

$$+ [(310.6)(0.2) - 0 - 0]\bar{j}$$

$$+ [0 - (-50.4)(-0.5) - 0]\bar{k}$$

$$= -51.7\bar{i} + 62.1\bar{j} - 25.2\bar{k} \text{ kg-m}^2/\text{s}$$

Because $\bar{\alpha} = \bar{0}$, equation (21) gives $\delta\bar{H}_G/\delta t = \bar{0}$. Therefore

$$\dot{\bar{H}}_G = \frac{\delta\bar{H}_G}{\delta t} + \bar{\omega} \times \bar{H}_G$$

$$= (-0.5\bar{i} + 0.2\bar{j}) \times (-51.7\bar{i} + 62.1\bar{j} - 25.2\bar{k})$$

$$= -5.04\bar{i} - 12.60\bar{j} - 20.71\bar{k} \text{ N-m}$$

Using the free body diagram of step 1, we compute $\Sigma \bar{M}_G$. The resultant gravitational force acts through the center of mass, so that

$$\Sigma \bar{M}_G = \bar{r}_{C/G} \times \bar{F} + \bar{M}_C$$

$$= (0.4\vec{\imath} - 0.3\vec{k}) \times (F_x\,\vec{\imath} + F_y\,\vec{\jmath} + F_z\,\vec{k}) + \bar{M}_C$$

$$= [0.3F_y + (M_C)_x]\vec{\imath} + [-0.3F_x - 0.4F_z$$

$$+ (M_C)_y]\,\vec{\jmath} + [0.4F_y + (M_C)_z]\,\vec{\jmath}$$

Equating the scalar components of $\Sigma\,\bar{M}_G$ to the scalar components of $\dot{H}_G$ yields the three scalar equations for moment equilibrium

$\vec{\imath}$ component: $0.3F_y + (M_C)_x = -5.04$

$\vec{\jmath}$ component: $-0.3F_x - 0.4F_z + (M_C)_y = -12.60$

$\vec{k}$ component: $0.4F_y + (M_C)_z = -20.71$

Step 8   Using the free body diagram to sum forces, we find that

$$\Sigma\,\bar{F} = F_x\,\vec{\imath} + F_y\,\vec{\jmath} + (F_z - 686.4 - 1373.8)\vec{k}$$

$$= 210(\bar{a}_G) = 3.360\vec{\imath} + 8.400\vec{\jmath} - 18.270\vec{k} \text{ newtons}$$

The scalar equations are therefore

$\vec{\imath}$ component: $F_x = 3.36$

$\vec{\jmath}$ component: $F_y = 8.40$

$\vec{k}$ component: $F_z - 2059 = -18.27$

Step 9   We first solve the equations of step 8 for the force components.

$$F_x = 3.36 \text{ N} \qquad F_y = 8.40 \text{ N} \qquad F_z = 2041 \text{ N} \qquad \triangleleft$$

Now, solving the equations of step 8 for the moment components gives

$$(M_C)_x = -7.6 \text{ N-m} \qquad (M_C)_y = 804.8 \text{ N-m} \qquad (M_C)_z = 24.1 \text{ N-m} \qquad \triangleleft$$

In conclusion, we should note that these computations are similar to those encountered in studying the motion of aircraft and ships when the rotations are described in terms of a set of body-fixed axes.

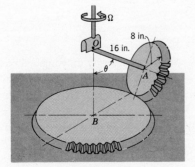

ILLUSTRATIVE PROBLEM 3

Gear A, which meshes with the fixed horizontal gear B, freely rotates about shaft OA. This shaft is attached to the vertical shaft by a clevis joint. The vertical shaft is initially at rest when it is given a constant angular acceleration $\dot{\Omega}$. Obtain an expression for the normal force exerted between the two gears as a function of time. (In this particular design of the gears, the desired force is perpendicular to shaft OA.) The weight of gear A is 15 lb, and the weight of the shaft is negligible.

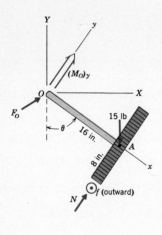

Solution

Step 1 For the free body diagram we isolate gear A and shaft OA in order to utilize the special restraints of a clevis joint. A side view of the system is most suitable.

Acting at the clevis joint is an arbitrary (three-dimensional) force $\bar{F}_O$ that prevents point O from moving. The axis of the pin is perpendicular to the plane of the free body diagram, so there is no couple reaction about the z axis. Also because there is no friction between gear A and shaft OA, there is no tendency for this shaft to twist about the x axis. Hence, the only couple reaction is $(\bar{M}_O)_y$.

Step 2 Given that $\dot{\Omega}$ is constant and $\Omega = 0$ initially, we conclude that $\Omega = \dot{\Omega}t$. The problem is to find the value of N as a function of time.

Step 3 Point O is a fixed point in space that is also fixed relative to gear A. We select this point for summing moments, rather than the center of mass A, in order to eliminate the force $\bar{F}_O$ from the moment equations.

Step 4 In order to utilize the tabulated inertia properties most directly, we attach the xyz frame to gear A, which is approximated as a thin disk, and align one axis to coincide with shaft OA. One of the fixed XYZ axes is chosen to coincide with the fixed vertical axis. The resulting coordinate systems are shown in the free body diagram.

Step 5 For the determination of the inertia properties, we note that the disk is symmetric with respect to both the xy and xz planes, so that the xyz axes are principal axes. To determine the moments of inertia, we use the tabulated properties and the parallel axis theorems. The coordinates of point O with respect to a reference $\hat{x}\hat{y}\hat{z}$ frame (not shown) whose origin is the center of mass A and whose axes are parallel to the xyz system are $(-16$ in., $0, 0)$. Therefore, using inches for the length unit,

$$I_{xx} = I_{\hat{x}\hat{x}} = \frac{1}{2}mR^2 = \frac{1}{2}\left(\frac{15}{386.4}\right)8^2 = 1.242 \text{ in.-lb-s}^2$$

$$I_{yy} = I_{zz} = I_{\hat{y}\hat{y}} + m(16)^2 = m\left(\frac{1}{4}R^2 + 16^2\right)$$

$$= 10.559 \text{ in.-lb-s}^2$$

$$I_{xy} = I_{yz} = I_{xz} = 0$$

Step 6 The study of the kinematics of the system is begun by drawing a sketch that shows the angular speed ω_1 of gear A about shaft OA. Note that the contact point C has no velocity (it is an instant center) because gear B is fixed. Also note that point O is fixed.

The general expressions for the angular velocity and angular acceleration are

$$\bar{\omega} = -\Omega\bar{J} + \omega_1\bar{i} \qquad \text{always}$$

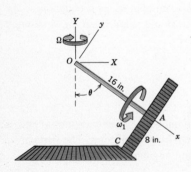

$$\bar{\alpha} = -\dot{\omega}\bar{J} + \dot{\omega}_1\bar{\imath} + \omega_1(\bar{\omega} \times \bar{\imath}) \qquad \text{always}$$

At the instant of the sketch we have

$$-\bar{J} = \cos\theta\,\bar{\imath} - \sin\theta\,\bar{\jmath}$$

Thus, the instantaneous values of $\bar{\omega}$ and $\bar{\alpha}$ are

$$\bar{\omega} = (\omega_1 + \Omega\cos\theta)\bar{\imath} - \Omega\sin\theta\,\bar{\jmath} \qquad \text{instantaneously}$$

$$\bar{\alpha} = (\dot{\omega}_1 + \dot{\Omega}\cos\theta)\bar{\imath} - \dot{\Omega}\sin\theta\,\bar{\jmath} + \omega_1\Omega\sin\theta\,\bar{k} \qquad \text{instantaneously}$$

We shall now determine ω_1 in terms of Ω by satisfying the constraint condition that $\bar{v}_C = \bar{0}$. Point O and point C on gear A are both fixed relative to that gear, so that $(\bar{v}_{C/O})_{\text{rel}} = \bar{0}$, and

$$\bar{v}_C = \bar{v}_O + \bar{\omega} \times \bar{r}_{C/O}, \quad \text{i.e.,} \quad \bar{0} = \bar{\omega} \times \bar{r}_{C/O}$$

This means that the instantaneous axis of rotation of gear A is parallel to line OC. From the sketch we have

$$\bar{r}_{C/O} = 16\bar{\imath} - 8\bar{\jmath}$$

so that using the derived expression for $\bar{\omega}$ gives

$$\bar{\omega} \times \bar{r}_{C/O} = [-8(\omega_1 + \Omega\cos\theta) + 16\,\Omega\sin\theta]\bar{k} = \bar{0}$$

Solving for ω_1, we find that

$$\omega_1 = \Omega(2\sin\theta - \cos\theta)$$

Now, because the sketch describes the system for a general position, this relationship can be differentiated. Thus

$$\dot{\omega}_1 = \dot{\Omega}(2\sin\theta - \cos\theta)$$

Substituting these results into the expressions for $\bar{\omega}$ and $\bar{\alpha}$ gives

$$\bar{\omega} = \Omega\sin\theta(2\bar{\imath} - \bar{\jmath})$$

$$\bar{\alpha} = \dot{\Omega}\sin\theta(2\bar{\imath} - \bar{\jmath})$$

$$+ \Omega^2\sin\theta(2\sin\theta - \cos\theta)\bar{k}$$

The determination of $\bar{a}_G$ is omitted because the equations of motion obtained from summing forces only define the force $\bar{F}_O$. We shall therefore also omit step 8.

Step 7 Substitute the inertia properties from step 5 and the expressions for $\bar{\omega}$ and $\bar{\alpha}$ from step 6 into the equations of kinetics. For principal axes these equations are

$$\bar{H}_O = I_{xx}\omega_x\,\bar{\imath} + I_{yy}\omega_y\,\bar{\jmath} + I_{zz}\omega_z\,\bar{k}$$

$$= (1.242)(2\Omega \sin \theta)\bar{\imath} + (10.559)(-\Omega \sin \theta)\bar{\jmath}$$

$$= 2.484\Omega \sin \theta \,\bar{\imath} - 10.559\Omega \sin \theta \,\bar{\jmath} \text{ in.-lb-s}$$

$$\frac{\delta \bar{H}_O}{\delta t} = I_{xx}\alpha_x\bar{\imath} + I_{yy}\alpha_y\bar{\jmath} + I_{zz}\alpha_z\bar{k}$$

$$= (1.242)(2\dot{\Omega} \sin \theta)\bar{\imath} + (10.559)(-\dot{\Omega} \sin \theta)\bar{\jmath}$$

$$\quad + (10.559)[\Omega^2 \sin \theta(2 \sin \theta - \cos \theta)]\bar{k}$$

$$= \sin \theta[2.484\dot{\Omega}\bar{\imath} - 10.559\dot{\Omega}\bar{\jmath}$$

$$\quad + 10.559\Omega^2(2 \sin \theta - \cos \theta)]\bar{k} \text{ in.-lb}$$

$$\dot{\bar{H}}_O = \frac{\delta \bar{H}_O}{\delta t} + \bar{\omega} \times \bar{H}_O$$

$$= \sin \theta[2.484\dot{\Omega}\bar{\imath} - 10.559\dot{\Omega}\bar{\jmath}$$

$$\quad + \Omega^2(2.484 \sin \theta - 10.559 \cos \theta)\bar{k}]$$

We now use the free body diagram to evaluate the sum of the moments about the origin O.

$$\Sigma \bar{M}_O = (M_O)_y \bar{\jmath} + \bar{r}_{A/O} \times 15(\cos \theta \,\bar{\imath} - \sin \theta \,\bar{\jmath})$$

$$\quad + \bar{r}_{C/O} \times (N\bar{\jmath} + f\bar{k})$$

$$= -8f \,\bar{\imath}$$

$$\quad + [(M_O)_y - 16f] \,\bar{\jmath}$$

$$\quad + (-240 \sin \theta + 16N)\bar{k}$$

Step 8   Omit.

Step 9   The value of N is found from the $\bar{k}$ component of the moment equation. Thus

$$(\Sigma \bar{M}_O) \cdot \bar{k} = (\dot{\bar{H}}_O) \cdot \bar{k}$$

$$-240 \sin \theta + 16N = \Omega^2 \sin \theta(2.484 \sin \theta - 10.559 \cos \theta)$$

We now solve for N and substitute $\Omega = \dot{\Omega}t$ from step 2. Hence

$$N = 15 \sin \theta + \tfrac{1}{16}(\dot{\Omega}t)^2 \sin \theta(2.484 \sin \theta - 10.559 \cos \theta) \text{ lb}$$

The first term on the right side of this result represents the static gravitational effect, whereas the second term represents two opposing inertial effects. If gear A did not mesh with the fixed gear, then the angular speed ω_1 would be zero and the centripetal acceleration of point A would have the effect of decreasing the value of N. However, due to the rotation ω_1, a gyroscopic moment is created that has the effect of forcing the gear downward. If $\theta = \tan^{-1}(10.559/2.484) = 76.76°$, then the two dynamic effects counterbalance each other and N reduces to its static value $N = 15 \sin \theta$.

ILLUSTRATIVE PROBLEM 4

The body shown is in free flight with negligible air resistance. Its properties are arbitrary, with the exception that it is rotationally symmetric about the z axis. From the fact that the gravity force acts at the center of mass, it follows that $\Sigma \bar{M}_G = \dot{\bar{H}}_G = \bar{0}$, so the angular momentum $\bar{H}_G$ is constant. Letting Z be a fixed axis of arbitrary orientation, it is desired to investigate if the motion of the body consists of a constant spinning rotation ω_1 about the z axis, as the z axis itself rotates at the constant angle θ and constant rate ω_2 about the fixed Z axis. Verify that this assumed motion is admissible by showing that, for the foregoing angular motion, the equations of motion require that $\bar{H}_G$ have a constant magnitude, and that it be parallel to the fixed Z axis. Further, determine ω_2 in terms of ω_1 and θ, and derive an expression for the angle β between the angular velocity $\bar{\omega}$ and the z axis in terms of θ.

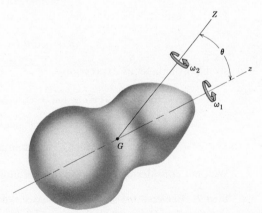

Solution

Steps 1–4 From the given information we know that the only force is gravity. The free body diagram shown here also shows the given information about the rotations. The body is in general motion, so the origin of the xyz axes is located at the center of mass. For convenience, the x axis is chosen such that the Z axis lies in the xz plane.

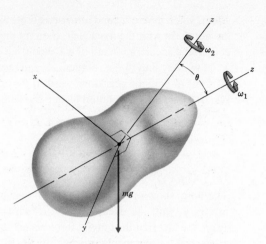

Step 5   For rotational symmetry, $I_{zz} = I_1$, $I_{xx} = I_{yy} = I_2$, where I_1, I_2, and the mass m are arbitrary constants. Also, xyz are principal axes.

Step 6   In the assumed motion, ω_1, ω_2, and θ are constant, so the general expressions for $\bar{\omega}$ and $\bar{\alpha}$ are

$$\bar{\omega} = \omega_2 \bar{K} + \omega_1 \bar{k} \qquad \text{always}$$

$$\bar{\alpha} = \omega_1 (\bar{\omega} \times \bar{k}) \qquad \text{always}$$

For the instant depicted in the sketch in step 1,

$$\bar{K} = \sin \theta \, \bar{\imath} + \cos \theta \, \bar{k}$$

so the instantaneous expressions are

$$\bar{\omega} = \omega_2 \sin \theta \, \bar{\imath} + (\omega_1 + \omega_2 \cos \theta) \bar{k} \qquad \text{instantaneously}$$

$$\bar{\alpha} = \omega_1 [\omega_2 \sin \theta \, \bar{\imath} + (\omega_1 + \omega_2 \cos \theta) \bar{k}] \times \bar{k}$$

$$= -\omega_1 \omega_2 \sin \theta \, \bar{\jmath} \qquad \text{instantaneously}$$

We omit considering $\bar{a}_G$, for we are not interested in it.

Step 7   Because xyz are principal axes, we may use Euler's equation (23). Substituting the inertia properties from step 5 and the values of $\bar{\omega}$ and $\bar{\alpha}$ found in step 6, we find that

$$\Sigma \bar{M}_A \equiv \bar{0} = [I_2(-\omega_1 \omega_2 \sin \theta) - (I_1 - I_2)(\omega_2 \sin \theta)$$

$$(\omega_1 + \omega_2 \cos \theta)] \, \bar{\jmath}$$

$$= \omega_2 \sin \theta [-I_1(\omega_1 + \omega_2 \cos \theta) + I_2 \omega_2 \cos \theta] \, \bar{\jmath}$$

Step 8 The moment equation in step 7 is satisfied when either $\omega_2 = 0$, or $\sin \theta = 0$, or

$$\omega_1 + \omega_2 \cos \theta = \frac{I_2}{I_1} \omega_2 \cos \theta$$

The first two possibilities mean that the only rotation is about the z axis. This is the axis of symmetry, so that $\bar{\omega}$ and $\bar{H}_G$ are then always in the z direction, making $\bar{H}_G$ a constant.

The more interesting possibility is the third one. Substituting the relationship between ω_1, ω_2, and θ for this case into the expression for $\bar{\omega}$ in step 7 gives

$$\bar{\omega} = \omega_2 \sin \theta \, \bar{\iota} + \frac{I_2}{I_1} \omega_2 \cos \theta \, \bar{k}$$

Then substituting these components of $\bar{\omega}$ and the moments of inertia into equation (5) for $\bar{H}_G$ gives

$$\bar{H}_G = I_2(\omega_2 \sin \theta)\bar{\iota} + I_1\left(\frac{I_2}{I_1}\omega_2\cos \theta\right)\bar{k}$$

$$= I_2\omega_2(\sin \theta \, \bar{\iota} + \cos \theta \, \bar{k})$$

$$= I_2\omega_2\bar{K}$$

Thus $\bar{H}_G$ has the constant magnitude $I_2\omega_2$ in the direction of the Z axis, as required.

It is a simple matter to solve the foregoing relationship between ω_1 and ω_2 to find

$$\omega_2 = \frac{I_1\omega_1}{(I_2 - I_1)\cos \theta}$$

To determine the required angle β, we draw a sketch of the components of $\bar{\omega}$. This sketch shows that

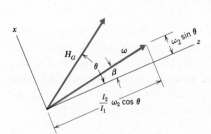

$$\tan \beta = \frac{\omega_2 \sin \theta}{(I_2/I_1)\omega_2 \cos \theta} = \frac{I_1}{I_2} \tan \theta$$

From these results we see that the axis of symmetry, the fixed angular momentum $\bar{H}_G$, and the angular velocity $\bar{\omega}$ are coplanar, with constant angles between them. This type of motion is called *steady free precession;* ω_2 is the rate of precession. There are two kinds of free precession: regular ($\beta < \theta$ and $\omega_2/\omega_1 > 0$) and retrograde ($\beta > \theta$ and $\omega_2/\omega_1 < 0$, depending on whether I_1/I_2 is less than one or not. These topics are treated in depth in more advanced texts on dynamics.

HOMEWORK PROBLEMS

VII.54–VII.57 Each of the systems shown is rotating at $\omega = 200$ rev/min when a constant torque of 50 N-m is applied to slow the system. Determine the initial dynamic bearing reactions and the deceleration rate.

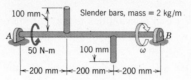

Prob. VII.54

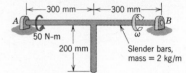

Prob. VII.55

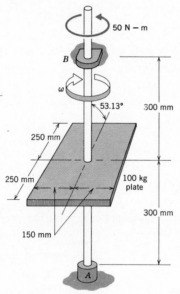

Prob. VII.56

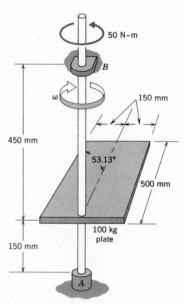

Prob. VII.57

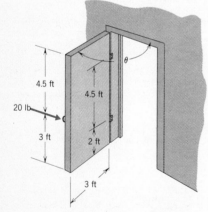

Prob. VII.58

VII.58 A 20-lb force is applied normal to the plane of the 40-lb solid door to shut it. The design of the hinges is such that they only exert reaction forces to support the door. If the door is initially at rest when $\theta = 90°$, determine (a) the angular speed of the door as it closes ($\theta = 0$), (b) the hinge forces as a function of the angle θ as the door swings shut.

VII.59 Two identical 5-kg disks are welded to the rigid bent bar CD. This bar has a uniform cross section and a mass of 4 kg. The entire assembly is rolling without slipping at a constant speed v parallel to the horizontal surface. Determine the friction and normal forces exerted between the ground and each disk in terms of the value of v when bar CD is in the horizontal position shown.

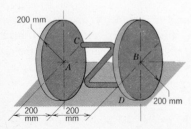

Prob. VII.59

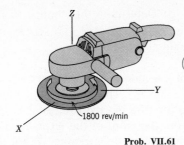

Prob. VII.61

Prob. VII.62

VII.60 Solve Problem VII.59 when the disks have rotated 90° from the position shown.

VII.61 The pad of a portable rotary sander rotates at 1800 rev/min. The pad may be approximated as a 1-lb thin disk of 8 in. diameter. When the plane of the pad is in the horizontal position shown, it is desired to rotate the sander (a) about the fixed X axis at 1 rad/s, (b) about the fixed Y axis at 1 rad/s. For each case of rotation, determine the couple that the person holding the sander must apply to produce the motion.

VII.62 The cutting blade of a rotary lawn mower is rotating at 1200 rev/min, clockwise as viewed from above. The blade is 600 mm long, has a mass of 1 kg, and is dynamically balanced as it rotates about its center. The lawn mower is in the horizontal position when it is rotated backward about its rear wheels at a constant rate of 1 rad/s. Determine the gyroscopic moment that develops in terms of the angle θ.

VII.63 Two bodies in motion are interconnected but can move independently. The force system acting on each body is equivalent to the force-couple system $m_i \bar{a}_{G_i}$ and $\dot{H}_{G_i}$ acting at the center of mass G_i of each body. Derive expressions for the force-couple system acting at an arbitrary point P that is equivalent to the total system of external forces producing the motion of this system.

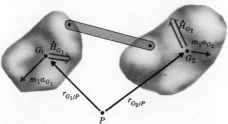

Prob. VII.63

VII.64 An automobile is executing a left turn on a road of radius of curvature $\rho = 60$ m while traveling at 72 km/hr. The 20-kg flywheel is being driven by the engine at 3000 rev/min clockwise as viewed from the front of the vehicle. The radius of gyration of the flywheel about its axis of symmetry is known to be 250 mm, but the other radius of gyration is unknown. The wheelbase of the vehicle is 3 m. Determine (a) the gyroscopic moment $\dot{H}_G$ for the flywheel, and (b) the amount that the normal force exerted by the ground on each of the front tires increases or decreases to provide this gyroscopic moment. (c) Explain why only one radius of gyration need be known to solve this problem.

Prob. VII.64

VII.65 The tires of the automobile in Problem VII.64 have a mass of 10 kg, a diameter of 700 mm, and a radius of gyration about their axle of 250 mm. (a) Determine $\dot{H}_G$ for each tire. (b) Determine the change in the normal reaction on each wheel necessary to produce the gyroscopic moment in part (a). The wheels on each axle are 2 m apart.

VII.66 A jet airplane at the bottom of a 4-mile-radius vertical circle is traveling at a constant speed of 1200 miles/hr. The 200-lb rotor of its single fanjet engine has a radius of gyration of 8 in. and rotates at 15,000 rev/min counterclockwise as viewed from the front of the airplane. The rotor is mounted by bearings in front and back. (a) Determine $\dot{H}_G$ for the rotor. (b) How are the aerodynamic control surfaces adjusted to provide this gyroscopic moment? (c) If the aerodynamic control surfaces were not adjusted, in which direction would the nose of the aircraft tend to shift?

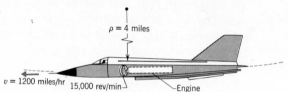

Prob. VII.66

VII.67 The blades of an oscillating electric fan rotate with the motor shaft at the constant angular speed ω_1, as the motor housing rotates about the vertical axis at angular speed ω_2. The angle θ between the motor shaft and the vertical is constant, but it may be adjusted. (a) Obtain an expression for $\dot{H}_G$ of the fan blades in terms of ω_1, ω_2, θ, the mass m of the blade assembly, and the principal radii of gyration. (b) By examining the $\bar{H}_G$ and $\bar{\omega}$ vectors, explain why only the radius of gyration of the fan blades about the motor shaft need be known in the case where $\theta = 90°$.

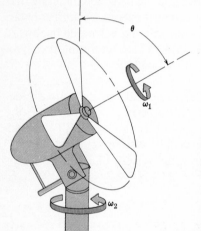

Prob. VII.67

VII.68 A radar antenna rotates about the vertical axis at the constant rate of 1 rev/min as the angle of elevation at which the antenna is aimed varies according to $\theta = (\pi/6)(1 + \cos \pi t)$, where θ is in radians and t is in seconds. This antenna has a mass of 1000 kg, and its radii of gyration with respect to the principal axes shown are $k_x = 600$ mm, $k_y = k_z = 800$ mm. Determine the force-couple reaction at support A for this motion.

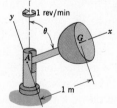

Prob. VII.68

VII.69 The boom AB of a crane may be rotated about the vertical axis at

ω_1 as the angle θ between the boom and the vertical axis is varied. The truck is stationary. The boom may be regarded as a slender rod weighing 5 tons that is supported only at point A. In the position shown, $\theta = 60°$, $\dot{\theta} = 0.5$ rad/s, and $\omega_1 = 0.3$ rad/s. If these rotation rates are constant values, determine the force-couple reaction at point A. Cable BC is slack because no load is being picked up by the hook.

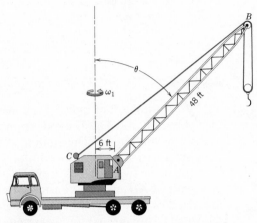

Prob. VII.69

VII.70 Solve Problem VII.69 if the given value of ω_1 is increasing at 0.6 rad/s² and the given value of $\dot{\theta}$ is constant.

VII.71 The 30-kg bar AB is attached at both ends to collars that slide over the smooth fixed horizontal and vertical guides. The system is at rest in the position shown, and it is desired to give collar A an acceleration of 2 m/s² to the left. What force $\bar{F}$ should be applied to collar A parallel to the lower guide in order to obtain this acceleration? The mass of the collars is negligible. *Hint:* Use the body-fixed *xyz* reference frame shown. Refer to Section C.4 for the rotation transformation of inertia properties.

Prob. VII.71

VII.72 Solve Problem VII.71 if collar A is moving to the right at 4 m/s in the position shown.

VII.73 The homogeneous cone is rolling without slipping on a horizontal surface such that it makes 1 rev about the vertical axis every 2 seconds. The cone weighs 2 lb. Determine the force-couple system acting at the apex A that represents the reaction of the ground on the cone.

Prob. VII.73

VII.74 A coin of mass m and radius r rolls without slipping on a horizontal surface. The normal to the plane of the coin is inclined at a constant angle θ and the center of the coin follows a path of radius L. Determine the relationship between the constant speed v, the angle θ, and the radius r.

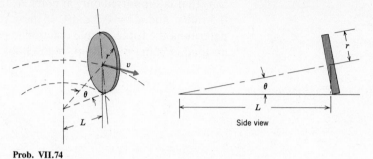

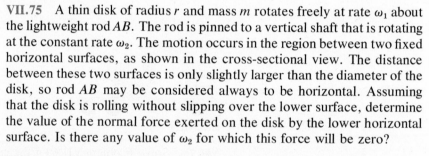

Prob. VII.74

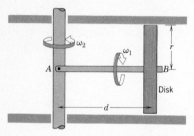

Prob. VII.75

VII.75   A thin disk of radius r and mass m rotates freely at rate ω_1 about the lightweight rod AB. The rod is pinned to a vertical shaft that is rotating at the constant rate ω_2. The motion occurs in the region between two fixed horizontal surfaces, as shown in the cross-sectional view. The distance between these two surfaces is only slightly larger than the diameter of the disk, so rod AB may be considered always to be horizontal. Assuming that the disk is rolling without slipping over the lower surface, determine the value of the normal force exerted on the disk by the lower horizontal surface. Is there any value of ω_2 for which this force will be zero?

VII.76   The disk in Problem VII.75 is rotating without slipping over the upper surface, thereby reversing the direction of the rotation ω_1 of the disk relative to rod AB. Determine the normal force exerted on the disk by the upper horizontal surface. Is there a minimum value of ω_2 for this rotation?

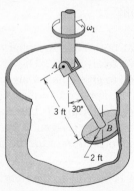

Prob. VII.77

VII.77   The ore grinder consists of a 1000-lb disk mounted on arm AB, as shown. This arm is pinned to the vertical shaft, which is driven at a constant angular speed ω_1. The disk rotates freely on arm AB as it rolls without slipping over the interior surface of the vertical cylindrical container. Determine the minimum value for ω_1 required for contact to be maintained between the disk and the interior wall. Neglect the mass of arm AB.

VII.78   Solve Problem VII.77 if the disk is replaced by a 1000-lb sphere having a radius of 2 ft. The sphere is centered at point B, and the other dimensions given in the diagram are unchanged.

VII.79   A toy gyroscope consists of a spinning disk of radius r that spins about the lightweight arm AB at the constant angular speed ω_1. This arm is supported at point B and rotates about the vertical axis at angular speed ω_2. (a) If θ is constant, derive an equation for θ in terms of ω_1, ω_2, and the

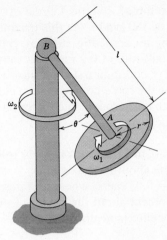

Prob. VII.79

physical properties of the system. (This type of motion is a steady precession about the vertical axis.) (b) Determine the relationship between ω_1 and ω_2 for $\theta = 90°$.

VII.80 For the toy gyroscope in Problem VII.79, $l = 200$ mm, $r = 100$ mm, $\omega_2 = 2$ rad/s, and $\theta = 60°$. What is the value of ω_1?

VII.81 and VII.82 The body shown is pinned at point A to the vertical shaft which is rotating at the constant rate ω_1. (a) Considering the angle θ as a variable, derive a single differential equation of motion for this parameter. (b) Using the result of part (a), determine the relationship between ω_1 and the constant nonzero value that angle θ can be. (c) Letting θ be a very small angle, so that $\sin \theta \approx \theta$ and $\cos \theta \approx 1$, examine the result of part (a) to determine the maximum value of ω_1 for which θ will return to zero if slightly disturbed from the zero value. *Hint:* The equation in part (c) will resemble that for a pendulum.

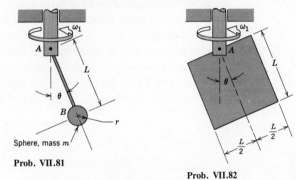

Prob. VII.81

Prob. VII.82

VII.83 The assembly shown is rotating about the vertical axis at a constant rate ω_1. The slender bar CD of mass m is supported by a pin at point B. Derive an expression for the constant angle θ made by the bar.

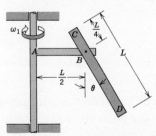

Prob. VII.83

VII.84 In Problem VII.83, the slender bar CD is held at $\theta = 90°$ and then released as the assembly rotates about the vertical axis at the constant rate ω_1. Derive the differential equation of motion for the angle θ.

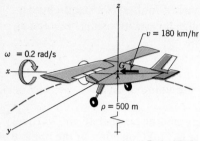

Prob. VII.85

VII.85 A 1200-kg light airplane is at the top of a vertical circle of 500-meter radius, flying at a constant speed of 180 km/hr. At this instant its wings are horizontal and it is performing an aerobatic stunt by rolling about the longitudinal x axis at the constant rate of 0.2 rad/s, as shown. The radii of gyration about the principal xyz coordinate system shown are $k_x = 2$ m, $k_y = 3$ m, $k_z = 3.5$ m. Determine the force-couple system acting at the center of mass G that is equivalent to the aerodynamic forces on the aircraft. The inertia of the propellers is negligible.

VII.86 In another maneuver, the aircraft in Problem VII.85 is rolling about its x axis at $\omega_x = 0.1$ rad/s, pitching about its y axis at $\omega_y = -0.2$ rad/s, and yawing about its z axis at $\omega_z = 0.05$ rad/s, where all rates are constant. Determine the aerodynamic moment about the center of mass for this maneuver. The inertia of the propeller is negligible.

VII.87 Due to the action of ocean waves, a ship is pitching and rolling as its center of mass moves with a constant velocity. At a certain instant these rotation rates about the axes of the body-fixed xyz coordinate system are maximum at $\omega_x = 0.02$ rad/s and $\omega_y = 0.10$ rad/s. This coordinate system forms a set of principal axes with origin at the center of mass G. The mass of the ship is 25,000 metric tons, and the radii of gyration are $k_x = 80$ m, $k_y = 10$ m, and $k_z = 75$ m. Determine the force-couple system at point G that is equivalent to the set of forces causing the motion.

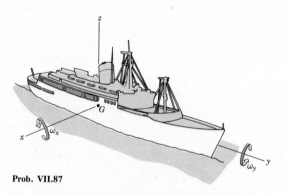

Prob. VII.87

VII.88 In a different motion, the angles of rotation of the ship in Problem VII.87 about the pitch and roll axes are given by $\theta_x = 0.05 \sin(0.5t)$ and $\theta_y = 0.10 \cos(t)$, respectively, where the angles are measured in radians and t is in seconds. The center of mass has a constant velocity. Determine, as a function of time, the force-couple system at point G that is equivalent to the set of forces causing this motion.

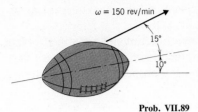

$\omega = 150$ rev/min

15°

10°

Prob. VII.89

VII.89 A football is thrown poorly. When released, the axis of symmetry is 10° above the horizontal and the angular velocity has a magnitude of 150 rev/min at an angle of 15° above the axis of symmetry, as shown. The radius of gyration of the football about its axis of symmetry is two-thirds that about an axis perpendicular to its axis of symmetry passing through the center of mass. Determine (a) the angle between the axis of precession and the horizontal direction, (b) the rate of precession, and (c) the rate of spin. Is this regular or retrograde precession? *Hint:* Refer to Illustrative Problem 4.

VII.90 A coin is thrown into the air. At a certain instant the coin is horizontal and the angular velocity is 450 rev/min in the direction shown. Determine (a) the angle between the precession axis and the centerline of the coin, (b) the rate of precession, and (c) the rate of spin. Is this regular or retrograde precession? *Hint:* Refer to Illustrative Problem 4.

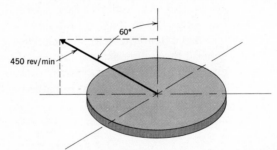

60°

450 rev/min

Prob. VII.90

Note: Before attempting Problems VII.91–94, the kinematical approach presented in Illustrative Problem 5 of Module VI should be fully understood.

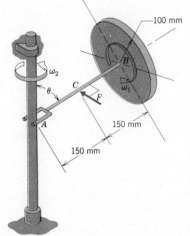

100 mm

ω_2

θ

C

B

ω_1

F

A

150 mm

150 mm

Prob. VII.91

VII.91 The 2-kg disk rotates at a constant rate $\omega_1 = 2100$ rev/min about shaft AB, which is connected to the vertical shaft by a pin at point A. The vertical shaft may rotate freely about its bearings, and the mass of both shafts is negligible. Initially, the angle θ is held at 60°, and the rotation rate ω_2 of the vertical shaft is zero. A horizontal force $F = 50$ N is applied to the midpoint C of shaft AB, and the system is released. This force is always perpendicular to the shaft, as shown. Determine the values of $\ddot{\theta}$ and $\dot{\omega}_2$ at the instant after the system is released.

VII.92 Solve Problem VII.91 if initially the angle θ is 60°, but the vertical axis is rotating at $\omega_2 = 1500$ rev/min when the 50-N force F is applied and the system is released.

VII.93 Solve Problem VII.91 if initially the angle $\theta = 60°$, $\dot{\theta} = 10$ rad/s, and ω_2 is zero when the 50-N force is applied and the system is released.

VII.94 Considering θ and ω_2 as unknown functions of time in Problem VII.91, determine the two differential equations of motion for these variables in terms of the rotation rate ω_1 and the properties of the system.

F. ENERGY AND MOMENTUM PRINCIPLES

1 Kinetic Energy of a Rigid Body in Three-Dimensional Motion

The work-energy principle for the motion of particles was derived in Module III, starting from Newton's second law. Then, in Module V we extended the applicability of this principle to the planar motion of rigid bodies by recognizing that the kinetic energy function for a particle, $T = \frac{1}{2}mv_P{}^2$, had to be rederived to account for the rotational characteristics of the body. Doing that, we obtained $T = \frac{1}{2}mv_G{}^2 + \frac{1}{2}I_{zz}\omega^2$. In this module we shall extend the field of application of the work-energy principle to the general three-dimensional motion of a rigid body by deriving the kinetic energy function for a three-axis rotation. Nevertheless, the basic form of the work-energy equation remains unchanged, being

$$\boxed{T_2 + V_2 = T_1 + V_1 + U_{1\to2}^{(\text{nc})}} \tag{24}$$

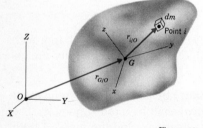

Figure 10

To derive the new expression for the kinetic energy, we have depicted a general body in Figure 10, where xyz is a body-fixed coordinate system, point i is the location of an infinitesimal element of mass, and point G is the center of mass of the rigid body. Summing the kinetic energy of each element of mass dm contained in the volume $\mathscr{V}$, we have

$$T = \tfrac{1}{2} \int_{\mathscr{V}} v_i{}^2 \, dm$$

We know that for a rigid body

$$\bar{v}_i = \bar{v}_G + \bar{\omega} \times \bar{r}_{i/G}$$

where $\bar{\omega}$ is the angular velocity of the rigid body. The kinetic energy expression then becomes

$$T = \tfrac{1}{2} \int_{\mathscr{V}} \bar{v}_i \cdot \bar{v}_i \, dm = \tfrac{1}{2} \int_{\mathscr{V}} (\bar{v}_G + \bar{\omega} \times \bar{r}_{i/G}) \cdot (\bar{v}_G + \bar{\omega} \times \bar{r}_{i/G}) \, dm$$

$$= \tfrac{1}{2}(\bar{v}_G \cdot \bar{v}_G) \int_{\mathscr{V}} dm + \bar{v}_G \cdot \left[\bar{\omega} \times \int_{\mathscr{V}} \bar{r}_{i/G} \, dm \right]$$

$$+ \tfrac{1}{2} \int_{\mathscr{V}} (\bar{\omega} \times \bar{r}_{i/G}) \cdot (\bar{\omega} \times \bar{r}_{i/G}) \, dm$$

In the foregoing we were able to remove $\bar{v}_G$ and $\bar{\omega}$ from the first and second integrals because these quantities are properties of the overall motion of the body, and thus depend only on time. The first integral is readily recognizable as the mass of the body, and the second integral is zero because it is the first moment of mass about the center of mass. Thus, the kinetic energy expression becomes

$$T = \tfrac{1}{2}mv_G{}^2 + \tfrac{1}{2} \int_{\mathscr{V}} (\bar{\omega} \times \bar{r}_{i/G}) \cdot (\bar{\omega} \times \bar{r}_{i/G}) \, dm$$

To proceed further, we express $\bar{\omega}$ and $\bar{r}_{i/G}$ in component form with respect to the body-fixed axes. Hence

$$\bar{\omega} = \omega_x \bar{\imath} + \omega_y \bar{\jmath} + \omega_z \bar{k}$$

$$\bar{r}_{i/G} = x\bar{\imath} + y\bar{\jmath} + z\bar{k}$$

$$\bar{\omega} \times \bar{r}_{i/G} = (\omega_y z - \omega_z y)\bar{\imath} + (\omega_z x - \omega_x z)\bar{\jmath} + (\omega_x y - \omega_y x)\bar{k}$$

$$(\bar{\omega} \times \bar{r}_{i/G}) \cdot (\bar{\omega} \times \bar{r}_{i/G}) = (y^2 + z^2)\omega_x{}^2 + (x^2 + z^2)\omega_y{}^2 + (x^2 + y^2)\omega_z{}^2$$
$$- 2yz\omega_y\omega_z - 2xz\omega_x\omega_z - 2xy\omega_x\omega_y$$

After substituting the last expression into the equation for kinetic energy, we utilize the fact that the components of $\bar{\omega}$ are functions only of time to write

$$T = \tfrac{1}{2}mv_G{}^2 + \tfrac{1}{2}\omega_x{}^2 \int_{\mathscr{V}} (y^2 + z^2) \, dm + \tfrac{1}{2}\omega_y{}^2 \int_{\mathscr{V}} (z^2 + x^2) \, dm$$

$$+ \tfrac{1}{2}\omega_z{}^2 \int_{\mathscr{V}} (x^2 + y^2) \, dm - \omega_x\omega_y \int_{\mathscr{V}} xy \, dm$$

$$- \omega_y\omega_z \int_{\mathscr{V}} yz \, dm - \omega_x\omega_z \int_{\mathscr{V}} xz \, dm$$

We now recognize the integrals as the inertia properties with respect to the xyz axes given in equations (6). Thus

$$T = \tfrac{1}{2}mv_G{}^2 + \tfrac{1}{2}I_{xx}\omega_x{}^2 + \tfrac{1}{2}I_{yy}\omega_y{}^2 + \tfrac{1}{2}I_{zz}\omega_z{}^2$$
$$- I_{xy}\omega_x\omega_y - I_{yz}\omega_y\omega_z - I_{xz}\omega_x\omega_z \qquad (25a)$$

The (alternative) vector form of this equation is

$$T = \tfrac{1}{2}m\bar{v}_G \cdot \bar{v}_G + \tfrac{1}{2}\bar{\omega} \cdot \bar{H}_G \qquad (25b)$$

The vector expression for the kinetic energy of a rigid body emphasizes that the reference system employed for the computation of the inertia properties has its origin at the mass center. Clearly, if the axes are principal axes, the last three terms in equation (25a) become zero.

From equation (25b) it is evident that, as was the case in planar motion, the total kinetic energy is the sum of the translational energy, $mv_G{}^2/2$, associated with the motion of the center of mass and the rotational energy, $\bar{\omega} \cdot \bar{H}_G/2$, associated with the rotation about the center of mass. In the case where the body is in planar motion parallel to the xy plane, the rotational kinetic energy reduces to $I_{zz}\omega^2/2$.

At this juncture let us retrace the steps in deriving equations (25). We began by selecting a body-fixed reference frame whose origin was the center of mass G. As a result, one of the terms in the initial expression for the kinetic energy was zero. Had we located the origin of the reference system at a general point, the derived expression for kinetic energy would be less convenient for computations. However, a special case arises when some point A, having a fixed position with respect to the body, has no velocity. In this case we may locate the origin of the body fixed reference frame at point A, with the result that

$$T = \tfrac{1}{2}I_{xx}\omega_x{}^2 + \tfrac{1}{2}I_{yy}\omega_y{}^2 + \tfrac{1}{2}I_{zz}\omega_z{}^2$$
$$- I_{xy}\omega_x\omega_y - I_{xz}\omega_x\omega_z - I_{yz}\omega_y\omega_z \qquad (26)$$
$$= \tfrac{1}{2}\bar{\omega} \cdot \bar{H}_A$$

The vector form of equation (26) emphasizes that in this case we are considering the kinetic energy function to be the rotational kinetic energy about point A which is at rest, and equally important, that the origin of the xyz reference frame is located at point A.

EXAMPLE 9

A 100-kg model of a space capsule is being tested by mounting it in a ball-and-socket joint at point A. At the instant shown, its angular velocity is $\bar{\omega} = 2\bar{i} - 3\bar{j} + 4\bar{k}$ rad/s, where $\bar{i}, \bar{j}, \bar{k}$ are the unit vectors of the set of

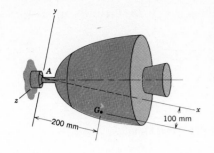

principal axes fixed to the capsule with origin at point A. It is known that, with respect to these axes, $I_{xx} = 30$ kg-m², $I_{yy} = 50$ kg-m², and $I_{zz} = 60$ kg-m². (a) Determine the kinetic energy of the capsule. (b) In a second experiment, point A is towed at a velocity $\bar{v}_A = -5\bar{i} + 2\bar{j} - \bar{k}$ m/s. Determine the kinetic energy of the capsule.

Solution

Part a

In this case the ball joint at point A is fixed, so we use equation (26).

$$T = \tfrac{1}{2}\bar{\omega} \cdot \bar{H}_A$$

The vector $\bar{\omega}$ was given as

$$\bar{\omega} = 2\bar{i} - 3\bar{j} + 4\bar{k} \text{ rad/s}$$

Equation (5) for the angular momentum gives

$$\bar{H}_A = I_{xx}\omega_x\bar{i} + I_{yy}\omega_y\bar{j} + I_{zz}\omega_z\bar{k}$$
$$= 30(2)\bar{i} + 50(-3)\bar{j} + 60(4)\bar{k}$$
$$= 60\bar{i} - 150\bar{j} + 240\bar{k} \text{ kg-m}^2/\text{s}$$

Therefore

$$T = \tfrac{1}{2}\bar{\omega} \cdot \bar{H}_A = \tfrac{1}{2}(120 + 450 + 960) = 765 \text{ J}$$

Part b

In this case the model is in general motion, so we use equation (25).

$$T = \tfrac{1}{2}m\bar{v}_G \cdot \bar{v}_G + \tfrac{1}{2}\bar{\omega} \cdot \bar{H}_G$$

To find $\bar{v}_G$ we write

$$\bar{v}_G = \bar{v}_A + \bar{v}_{G/A}$$
$$= \bar{v}_A + \bar{\omega} \times \bar{r}_{G/A}$$

From the given diagram we see that

$$\bar{r}_{G/A} = 0.20\bar{i} - 0.10\bar{j} \text{ m}$$

so

$$\bar{v}_G = (-5\bar{i} + 2\bar{j} - \bar{k}) + (2\bar{i} - 3\bar{j} + 4\bar{k}) \times (0.20\bar{i} - 0.10\bar{j})$$
$$= -4.60\bar{i} + 2.80\bar{j} - 0.60\bar{k} \text{ m/s}$$

We denote as $\hat{x}\hat{y}\hat{z}$ the coordinate axes with origin at the center of mass whose axes are parallel to the given xyz reference frame. The coordinates of point A with respect to $\hat{x}\hat{y}\hat{z}$ are $(-0.20, 0.10, 0)$ m, so the parallel axis theorems yield

$$I_{\hat{x}\hat{x}} = I_{xx} - m(\hat{y}_A{}^2 + \hat{z}_A{}^2)$$

$$= 30 - 100[(0.10)^2 + 0] = 29.0 \text{ kg-m}^2$$

$$I_{\hat{y}\hat{y}} = I_{yy} - m(\hat{x}_A{}^2 + \hat{z}_A{}^2)$$

$$= 50 - 100[(-0.20)^2 + 0] = 46.0 \text{ kg-m}^2$$

$$I_{\hat{z}\hat{z}} = I_{zz} - m(\hat{x}_A{}^2 + \hat{y}_A{}^2)$$

$$= 60 - 100[(-0.20)^2 + (0.10)^2] = 55.0 \text{ kg-m}^2$$

$$I_{\hat{x}\hat{y}} = I_{xy} - m\hat{x}_A\hat{y}_A$$

$$= 0 - 100(-0.20)(0.10) = 2.0 \text{ kg-m}^2$$

$$I_{\hat{y}\hat{z}} = I_{\hat{x}\hat{z}} = 0$$

With the inertia properties determined, we can evaluate the angular momentum $\bar{H}_G$ using equation (5). This gives

$$\bar{H}_G = (I_{xx}\omega_x - I_{xy}\omega_y)\bar{i} + (I_{yy}\omega_y - I_{xy}\omega_x)\bar{j} + I_{zz}\omega_z\bar{k}$$

$$= [29(2) - 2(-3)]\bar{i} + [46(-3) - 2(2)]\bar{j} + 55(4)\bar{k}$$

$$= 64\bar{i} - 142\bar{j} + 220\bar{k}$$

Thus

$$T = \tfrac{1}{2}(100)[(-4.60)^2 + (2.80)^2 + (0.60)^2]$$

$$+ \tfrac{1}{2}[(2\bar{i} - 3\bar{j} + 4\bar{k}) \cdot (64\bar{i} - 142\bar{j} + 220\bar{k})]$$

$$= 2175 \text{ J}$$

The rotational portion of the kinetic energy is $\bar{\omega} \cdot \bar{H}_G/2 = 717$ J. Note that this is less than that found in part (a), even though the angular velocity is the same.

EXAMPLE 10

The 2-kg homogeneous cone is at rest on the 20° incline in the position shown. It is released and rolls without slipping down the incline. Determine the rate of rotation of the axis of the cone when it passes through the lowest position on the incline.

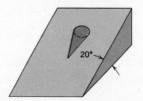

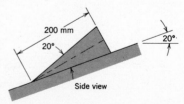

200 mm

20°

20°

Side view

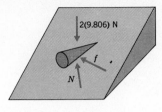

Solution

The solution will be formulated in terms of the work-energy principle because we wish to determine the motion of the cone in the final position for a given set of initial conditions. To evaluate the work term we draw a free body diagram of the cone in a general position.

The friction force $\bar{f}$, parallel to the incline, is the constraint force that prevents slipping. As was shown in planar motion, such a friction force does no work. Also, the contact force $\bar{N}$ is normal to the plane. Therefore, it also does no work, so $U_{1\to2}^{(nc)} = 0$.

The remaining force is gravity, which is conservative. To compute the corresponding potential energy, let us draw sketches of the cone in its initial and final positions. Because the cone is not slipping, its apex remains fixed.

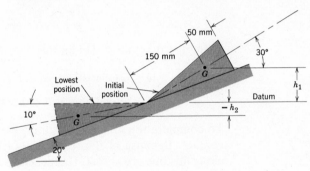

Choosing the elevation of the apex A as the datum, the initial and final elevations of the center of mass are

$$h_1 = 0.150 \sin 30° = 0.0750 \text{ m}$$

$$-h_2 = 0.150 \sin 10° = 0.02605 \text{ m}$$

Thus, the potential energy of gravity is

$$V_1 = mgh_1 = 2(9.806)(0.075) = 1.4709 \text{ J}$$

$$V_2 = mgh_2 = 2(9.806)(-0.02605) = -0.5108 \text{ J}$$

To evaluate the kinetic energy, we note that in the initial position the cone is at rest, so $T_1 = 0$. To find T_2 we locate the origin of the xyz reference frame at the fixed point A in order to employ equation (26). The x, y, and z axes are chosen to be parallel to the $\hat{x}$, $\hat{y}$, and $\hat{z}$ axes shown in the sketch on the next page because the inertia properties with respect to the latter system are known from Appendix B. The parallel axis theorems then give

$$I_{xx} = I_{\hat{x}\hat{x}} = \tfrac{3}{10}mr^2 = \tfrac{3}{10}(2)(0.2 \tan 10°)^2$$

$$= 7.462(10^{-4}) \text{ kg-m}^2$$

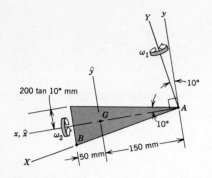

$$I_{yy} = I_{zz} = I_{\hat{y}\hat{y}} + m d_y^2$$

$$= \tfrac{3}{80}(2)[4(0.2 \tan 10°)^2 + (0.2)^2]$$

$$+ (2)(0.15)^2 = 4.837(10^{-2}) \text{ kg-m}^2$$

$$I_{xy} = I_{xz} = I_{yz} = 0$$

To describe the rotation of the cone, we select a fixed reference system XYZ whose Y axis is normal to the incline and whose X axis is coincident with the line of contact, as shown in the foregoing sketch. Then the rotation of the cone consists of a superposition of rotations about the Y axis at angular speed ω_1 and about the x axis at angular speed ω_2. Note that ω_1 is the value we seek. From this we find that the general expression for the angular velocity is

$$\bar{\omega} = \omega_1 \bar{J} - \omega_2 \bar{i} \qquad \text{always}$$

Using the diagram to obtain the instantaneous expression, we have

$$\bar{\omega} = \omega_1(\sin 10° \, \bar{i} + \cos 10° \, \bar{j}) - \omega_2 \bar{i}$$

$$= (0.1736\omega_1 - \omega_2)\bar{i} + 0.9848\omega_1 \, \bar{j} \qquad \text{instantaneously}$$

To complete the kinematics study, we note that the X axis is the line of contact. Because of the restriction of no slipping, all points of the cone along this line are at rest. Hence, choosing point B at the base of the cone, we can write

$$\bar{v}_B = \bar{v}_A + \bar{\omega} \times \bar{r}_{B/A}$$

$$\bar{0} = \bar{0} + [(0.1736\omega_1 - \omega_2)\bar{i} + 0.9848\omega_1 \, \bar{j})$$

$$\times (0.20\bar{i} - 0.20 \tan 10° \, \bar{j})$$

$$\bar{0} = 0.20[-0.1763(0.172\omega_1 - \omega_2)$$

$$- 0.9848\omega_1]\bar{k}$$

We want to find ω_1, hence we solve this equation for ω_2, obtaining

$$\omega_2 = 0.1736\omega_1 + \frac{0.9848}{0.1763}\,\omega_1 = 5.7595\omega_1$$

It then follows that the angular velocity is

$$\bar{\omega} = (0.1736\omega_1 - 5.7595\omega_1)\bar{i} + 0.9848_1 \, \bar{j}$$

$$= \omega_1(-5.586\bar{i} + 0.985\bar{j})$$

We may now determine T_2 by substituting the inertia properties and the angular velocity components into equation (26). This gives

$$T_2 = \tfrac{1}{2}(I_{xx}\omega_x{}^2 + I_{yy}\omega_y{}^2 + I_{zz}\omega_z{}^2)$$

$$= \tfrac{1}{2}(7.462)(10^{-4})(-5.586\omega_1)^2 + \tfrac{1}{2}(4.837)(10^{-2})(0.985\omega_1)^2$$

$$= 3.511(10^{-2})\omega_1{}^2 \text{ J}$$

Finally, the energy principle for a conservative system gives

$$T_2 + V_2 = T_1 + V_1$$

$$3.511(10^{-2})\omega_1{}^2 - 0.5108 = 0 + 1.4709$$

$$\omega_1 = 7.51 \text{ rad/s}$$

HOMEWORK PROBLEMS

VII.95 The homogeneous ellipsoid shown has a mass of 100 kg. Its center of mass is fixed and its instantaneous angular velocity is $\bar{\omega} = 4\bar{\imath} + 3\bar{\jmath} + 7\bar{k}$ rad/s. Determine the kinetic energy of the ellipsoid.

VII.96 The ellipsoid shown in Problem VII.95 has a mass of 100 kg. Point A at the end of the major axis is a fixed point. The instantaneous angular velocity of the ellipsoid is $\bar{\omega} = 4\bar{\imath} + 3\bar{\jmath} + 7\bar{k}$ rad/s. Determine the kinetic energy of the ellipsoid (a) using equation (25), (b) using equation (26).

VII.97 The 20-kg gear A is welded to the 4-kg arm AB. Assuming the gear to be a uniform disk having a radius of 200 mm, determine the maximum angular velocity of arm AB if the system is released from rest at $\theta = 0°$.

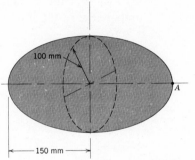

100 mm

150 mm

Prob. VII.95

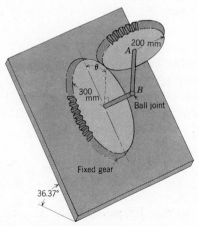

200 mm

A

θ

300 mm

B

Ball joint

Fixed gear

36.37°

Prob. VII.97

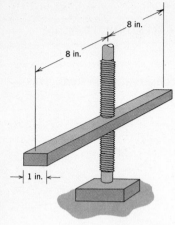

Prob. VII.99

VII.98 The system in Problem VII.97 is started in motion at $\theta = 90°$. What is the minimum initial angular velocity of arm AB for which the system will reach the highest position?

VII.99 The 3-lb bar screws into the single threaded vertical shaft whose pitch is 0.25 in. The bar, which fits loosely onto the shaft, is given an initial angular speed of 20 rad/s to make it descend the shaft. After 32 rev the bar comes to rest. Determine the frictional couple opposing the motion of the bar.

VII.100 The 5-kg sphere rotates about arm AB at the constant rate $\omega_1 = 300$ rad/s. A 20 N-m torque is applied to the vertical shaft, causing it to rotate at the rate ω_2. The masses of the shafts are negligible, and the system is initially at rest. (a) Determine the value of ω_2 after the system has completed 5 rev about the vertical axis. (b) Compare the result of part (a) to that obtained when the angular speed ω_1 is zero.

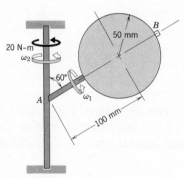

Prob. VII.100

VII.101 The 15-kg bar AB is attached at both ends to lightweight collars that slide over the smooth horizontal and vertical guides. A constant 100-N force, parallel to the horizontal guide, is applied to collar A as shown. If the system is released from rest at $x = 0$, determine the speed of collar A when $x = 1.2$ m.

VII.102 The uniform steel block is moving at 16 ft/s along the smooth angle irons when it encounters the 10° bend in one of the angle irons. Determine the angular velocity of the block after it has moved through a horizontal distance of 4 in. beyond the bend. Assume that edge AB, which rides in the straight angle iron, remains horizontal. *Hint:* If edge AB remains horizontal, then the general motion of the block consists of a

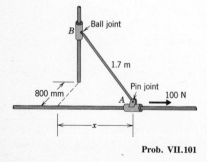

Prob. VII.101

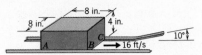

Prob. VII.102

rotation about this edge and a translation parallel to the edge. Also, the component of the velocity of corner C lying in the vertical plane must be parallel to the lower leg of the bent angle iron.

2 Impulse-Momentum Principles

In symbolic form the linear and angular impulse-momentum principles are the same as presented in Module III for a system of independently moving particles and in Module V for a rigid body in planar motion. Specifically, the linear impulse-momentum principle is

$$m(\bar{v}_G)_2 - m(\bar{v}_G)_1 = \int_{t_1}^{t_2} \Sigma \bar{F} \, dt \qquad (27)$$

and the angular impulse-momentum principle is

$$(\bar{H}_A)_2 - (\bar{H}_A)_1 = \int_{t_1}^{t_2} \Sigma \bar{M}_A \, dt \qquad (28)$$

In equation (28) we restrict point A to be one of the allowable points for summing moments (the center of mass or a point in the body that is fixed). Both of the foregoing relations express the fact that the increase in the respective momentum quantities equals the corresponding impulse. Clearly, when we substitute the general expression for $\bar{H}_A$ given by equation (5), the component form of equation (28) is quite lengthy.

Equations (27) and (28) are of particular importance in the case of impulsive forces, that is, very large forces that act over a short interval of time. As was seen when dealing with forces of this type in particle and planar motion problems, the major effect of any impulsive force is described by its impulse and not by its actual variation with time.

EXAMPLE 11

A disk of mass m is struck at point B by a bullet of mass m_b traveling perpendicular to the plane of the disk at a speed v. The disk is free to rotate about the ball joint connection at point A. Determine the instantaneous angular speed and the axis of rotation of the plate immediately after the impact.

Solution

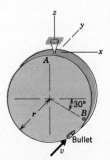

Clearly, this is an impulse-momentum problem. We begin with free body diagrams of the disk and of the bullet, in which we neglect the nonimpul-

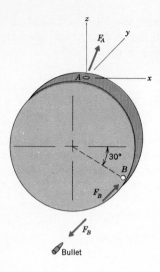

sive weight force in comparison to the impact force $\bar{F}_B$ exerted between the bullet and the disk.

In order to eliminate the impulsive reaction force at point A, we formulate the solution in terms of the angular momentum about point A. Furthermore, to eliminate the interaction force $\bar{F}_B$, we consider the system of particles formed by the disk and the bullet. Thus

$$(\Sigma\,\bar{H}_A)_2 - (\Sigma\,\bar{H}_A)_1 = \int_{t_1}^{t_2} \Sigma\,\bar{M}_A\,dt$$

where, for the principal xyz axes,

$$\Sigma\,\bar{H}_A = (\bar{H}_A)_{\text{disk}} + (\bar{H}_A)_{\text{bullet}}$$

$$= I_{xx}\omega_x\bar{\imath} + I_{yy}\omega_y\,\bar{\jmath} + I_{zz}\omega_z\bar{k} + \bar{r}_{B/A} \times m_b\bar{v}_{\text{bullet}}$$

Also, from the free body diagrams we note that the only impulsive external force acting on this system is $\bar{F}_A$, so $\Sigma\,\bar{M}_A = \bar{0}$. Hence, the angular momentum of the system about point A is conserved,

$$(\Sigma\,\bar{H}_A)_2 = (\Sigma\,\bar{H}_A)_1$$

To evaluate $\Sigma\,\bar{H}_A$ we need the mass moments of inertia. Letting $\hat{x}\hat{y}\hat{z}$ be the axes parallel to xyz with origin at the center of mass of the disk, the mass moments of inertia are found (with the aid of Appendix B and the parallel axis theorems) to be

$$I_{xx} = I_{\hat{x}\hat{x}} + m(\hat{y}_A{}^2 + \hat{z}_A{}^2) = \tfrac{1}{4}mr^2 + mr^2 = \tfrac{5}{4}mr^2$$

$$I_{yy} = I_{\hat{y}\hat{y}} + m(\hat{x}_A{}^2 + \hat{z}_A{}^2) = \tfrac{1}{2}mr^2 + mr^2 = \tfrac{3}{2}mr^2$$

$$I_{zz} \equiv I_{\hat{z}\hat{z}} = \tfrac{1}{4}mr^2$$

We next consider the kinematics of the motion. Initially, the disk is at rest ($\bar{\omega}_1 = \bar{0}$) and the velocity of the bullet is $v\bar{\jmath}$, so

$$(\Sigma\,\bar{H}_A)_1 = \bar{0} + [r\cos 30°\,\bar{\imath} - r(1 + \sin 30°)\bar{k}] \times m_b v\bar{\jmath}$$

$$= rm_b v[1.50\bar{\imath} + 0.8660\bar{k}]$$

The angular velocity of the disk following the impact is a general vector $\bar{\omega}_2$, but because the bullet is embedded in the disk, we know that

$$(\bar{v}_{\text{bullet}})_2 \equiv \bar{v}_B = \bar{v}_A + \bar{\omega}_2 \times \bar{r}_{B/A}$$

$$= 0 + (\omega_x\bar{\imath} + \omega_y\,\bar{\jmath} + \omega_z\bar{k}) \times (0.8660r\bar{\imath} - 1.50r\bar{k})$$

$$= r[-1.500\omega_y\bar{\imath} + (0.8660\omega_z + 1.500\omega_x)\bar{\jmath} - 0.8660\omega_y\bar{k}]$$

Substituting the inertia properties and the foregoing expression for

$(\bar{v}_{\text{bullet}})_2$ into the general equation for $\Sigma \, \bar{H}_A$, we find

$$(\Sigma \, \bar{H}_A)_2 = 1.25mr^2\omega_x\bar{i} + 1.50mr^2\omega_y\,\bar{j} + 0.25mr^2\omega_z\bar{k}$$

$$+ \, r(0.8660\bar{i} - 1.50\bar{k}) \times m_br \, [-1.50\omega_y\bar{i} + (0.8660\omega_z$$

$$+ \, 1.50\omega_x)\bar{j} - 0.8660\omega_y\bar{k}]$$

$$= r^2(1.250m\omega_x + 1.2990m_b\omega_z + 2.250m_b\omega_x)\bar{i}$$

$$+ \, r^2(1.50m\omega_y + 3.00m_b\omega_y)\bar{j}$$

$$+ \, r^2(0.250m\omega_z + 0.750m_b\omega_z + 1.299m_b\omega_x)\bar{k}$$

Equating the corresponding components of $(\bar{H}_A)_2$ and $(\bar{H}_A)_1$ then yields

$\bar{i}$ component: $r^2(1.250m\omega_x + 1.2990m_b\omega_z + 2.250m_b\omega_x)$

$$= 1.50rm_bv$$

$\bar{j}$ component: $r^2(1.50m + 3.00m_b)\omega_y = 0$

$\bar{k}$ component: $r^2(0.250m\omega_z + 0.750m_b\omega_z + 1.299m_b\omega_x)$

$$= 0.8660rm_bv$$

From the second equation, we see that $\omega_y = 0$. The simultaneous solution of the first and third equations is

$$\omega_x = \left(\frac{1.200m_b/m}{1 + 4.800m_b/m}\right)\frac{v}{r}$$

$$\omega_z = \left(\frac{3.464m_b/m}{1 + 4.800m_b/m}\right)\frac{v}{r}$$

Thus

$$\bar{\omega}_2 = \left(\frac{m_b/m}{1 + 4.80m_b/m}\right)\frac{v}{r}(1.20\bar{i} + 3.46\bar{k})$$

This vector defines the instantaneous axis of rotation, because point A is fixed. The sketch below gives the magnitude and orientation of the axis of rotation.

$$\omega_2 = |\bar{\omega}_2| = \left(\frac{3.67m_b/m}{1 + 4.80m_b/m}\right)\frac{v}{r}$$

$$\theta = \tan^{-1}\left(\frac{\omega_x}{\omega_z}\right) = 19.1°$$

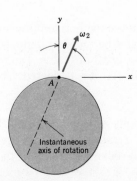

Finally, we should note that an approximation is sometimes made for problems such as this by neglecting the momentum of the bullet following the impact. The result of such a procedure would be to make the denominator in the solution unity, rather than $1 + 4.80m_b/m$. Certainly this is a reasonable approximation for this system if $m_b/m < 0.01$.

HOMEWORK PROBLEMS

VII.103 The angular velocity of the 8-kg disk is initially $\bar{\omega}_1 = -4\bar{k}$ rad/s when it is hit by a 100-gram bullet at point B, as shown. The bullet becomes embedded in the disk, with the result that the disk has no angular velocity component about the z axis after the impact. Determine (a) the initial speed v of the bullet to cause this change, and (b) the angular velocity of the disk after the impact.

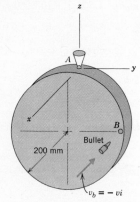

Prob. VII.103

VII.104 The 12,000-kg artificial satellite is initially in pure translation when impacted by a 3-kg meteoroid traveling at a speed of 2 km/s relative to the space vehicle in the direction shown. The radii of gyration of the satellite are $k_x = k_z = 2$ m, $k_y = 1$ m. Determine the angular velocity of the satellite and the change in its velocity immediately after the impact.

VII.105 A square plate of mass m is suspended by a ball-and-socket joint at point A. It is initially at rest when subjected to a large impulsive force F normal to its plane applied at some point P. Determine the relationship between the x and y coordinates of point P if the reaction force at point A is zero.

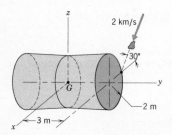

Prob. VII.104

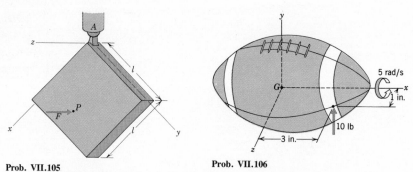

Prob. VII.105

Prob. VII.106

VII.106 A 2-lb football is in free flight, traveling horizontally at 60 ft/s as it spins about its x axis at 5 rad/s. A would-be receiver deflects the ball, applying a 10-lb force as shown for an interval of 0.10 s. Determine the linear and angular velocity of the football after it is deflected. The radii of gyration are $k_x = 2$ in., $k_y = k_z = 3$ in.

VII.107 Steering rockets are fired at points A, B, and C of the 12,000-kg artificial satellite, as shown. The radii of gyration of the satellite are $k_x = k_z = 2$ m, $k_y = 1$ m, and the thrust of each rocket is 1000 N. If the satellite is initially in pure translation, determine the change in the velocity of the center of mass G and the angular velocity after the rockets have been fired for 1 min.

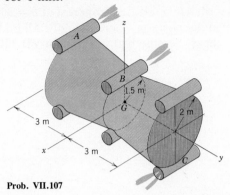

Prob. VII.107

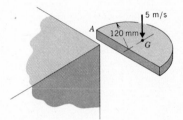

Prob. VII.108

VII.108 The 2-kg semicircular plate is translating downward at 5 m/s when corner A strikes the ledge. Assuming that the collision is perfectly plastic, determine the velocity of the center of mass G and the angular velocity of the plate immediately after the collision.

VII.109 Solve Problem VII.108 if the coefficient of restitution for the collision is 0.50.

VII.110 The 5-lb thin rectangular plate AB is pinned to the vertical shaft, which rotates freely with negligible resistance from friction. Initially, the shaft is rotating at $\omega_1 = 0.5$ rad/s and the angle θ is held constant at 0°. Plate AB is then released. Determine the values of ω_1 and $\dot{\theta}$ when $\theta = 45°$. The mass of the shaft is negligible. *Hint:* The system is conservative. Also, there is no moment exerted on the system about the axis of the shaft.

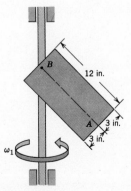

Prob. VII.110

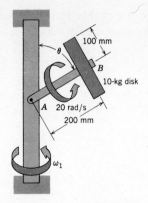

Prob. VII.111

VII.111 The 10-kg disk spins about arm AB at a constant rate of 20 rad/s. Arm AB is pinned to the vertical shaft at point A, the shaft being free to rotate with negligible friction at the (variable) rate ω_1. Initially, $\omega_1 = 0$ and arm AB is released from rest at $\theta = 30°$. Determine the values of $\dot\theta$ and ω_1 in the position where $\theta = 90°$. All bodies except the disk have negligible mass. See the hint in Problem VII.110.

MODULE VIII
SPECIAL TOPICS

A. VIBRATIONS

The general concern in dynamics is with the motion of a system and the forces that cause that motion. This section is devoted to a specialized dynamics investigation entitled *vibrations,* which is the study of the oscillatory motion that may occur in systems having a stable position of static equilibrium. This branch of dynamics is so important that no industrial firm would think of marketing its products, be they household appliances, electronic equipment, airplanes, or automobiles, before thoroughly testing them for vibratory effects.

1 System Model

Consider any of the systems mentioned above, say an automobile. The objective in modeling a system is to create the simplest mathematical description that retains the essential features of the system. The automobile consists of a chassis suspended by springs, energy-dissipative shock absorbers, and pneumatic tires. A primary function of these components is to insulate the chassis from irregularities in the roadway.

As a first attempt to investigate the vertical motion of an automobile, let us model the automobile as a mass suspended by a spring and a viscous damping device, as shown in Figure 1. In this sketch $\bar{F}(t)$ is an excitational force that is causing the motion. The origin of this force could be an unbalanced rotation of the wheels or a rough highway. The damping device is idealized as a linear viscous *dashpot* which produces a force opposing the motion in proportion to its rate of extension. The subject of energy dissipation and material damping is beyond the scope of this book. Thus, suffice it to say that a linear dashpot has been shown by experiments to be a good model for the dissipation occurring in the vibrations of many systems.

A prominent feature of the model appearing in Figure 1 is that only one scalar quantity, the vertical displacement of the mass, need be known to locate the position of the body. Hence, the system of Figure 1 can only be used to treat the motion of the center of mass of the automobile in the vertical direction. A system where one geometric quantity is required to describe its movement is said to have *one degree of freedom.* In our studies here, we shall consider only one-degree-of-freedom systems.

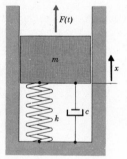

Figure 1

2 General Equation of Motion

Consider the case of the system of Figure 1 resting in static equilibrium, as depicted in Figure 2 on the following page. In the equilibrium position the mass is not moving, so there is no damping force. From the free body diagram shown in Figure 2, we may write

$$mg = k\delta_{st} \tag{1}$$

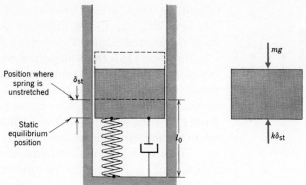

Figure 2

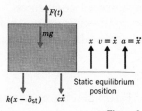

Figure 3

where k is the stiffness of the spring and δ_{st} is the static deflection of the mass.

Now let us draw a free body diagram for the situation where the system is in motion. To measure the displacement of the mass, we need a reference position. For this, let us choose the *static equilibrium position,* as shown in Figure 3. The length of the spring when the mass moves upward from this position is then $(l_0 - \delta_{st}) + x$, where l_0 is the unstretched length of the spring, shown in Figure 2. Thus, the extension of the spring is $[(l_o - \delta_{st}) + x] - l_o \equiv x - \delta_{st}$. Also, because δ_{st} is a constant, the rate of extension of the dashpot is $\dot{x}$.

Note what was done to obtain the direction of each force vector in Figure 3. First, a positive x direction was *arbitrarily* assumed and the kinematical parameters of the motion of the block were expressed in terms of x. Noting that the spring force is opposite the sense of its elongation and the force in the dashpot is opposite the sense of its extensional velocity, Newton's second law gives

$$F(t) - k(x - \delta_{st}) - c\dot{x} - mg = m\ddot{x}$$

so that

$$m\ddot{x} + c\dot{x} + k(x - \delta_{st}) + mg = F(t)$$

Replacing δ_{st} from equation (1), we get

$$\boxed{m\ddot{x} + c\dot{x} + kx = F(t)} \qquad (2)$$

Equation (2) leads to an important observation, which turns out to be true for all systems where the springs are linear and gravity has only a static effect. We see that when we measure the displacement of a system from its static equilibrium position, we may conveniently ignore the weight of the system and the static spring force $k\delta_{st}$ in writing the equation

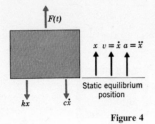

$x \quad v = \dot{x} \quad a = \ddot{x}$

Static equilibrium position

$kx \qquad c\dot{x}$

Figure 4

of motion. Thus, given the system of Figure 1, all that is required is the free body diagram shown in Figure 4. Newton's second law for the free body diagram of Figure 4 also gives equation (2).

Equation (2) is the equation of motion. It is sometimes referred to as the equation of a forced, damped, linear oscillator. It is a second-order, linear ordinary differential equation with constant coefficients. As is true for the motion of any physical system, to determine the response fully we must know the initial conditions, that is, the displacement and velocity when the motion commences.

In its most general form, equation (2) shows the acceleration to be a function of position, velocity, and time. Clearly, we cannot solve this general differential equation by the method of separating variables, although we could use this method for the case where only c, k, or $F(t)$ is nonzero. In the following we shall employ another method, which is valid for all linear differential equations having constant coefficients.

The next three sections focus on three common problems that arise in connection with a system whose motion is defined by equation (2).

3 Free Undamped Motion

Here we shall investigate the motion of a system that is free of an external force, $F(t) = 0$, and where the damping is negligible, $c = 0$. Setting $F(t) = c = 0$ in equation (2) yields the equation of motion for a free undamped system,

$$m\ddot{x} + kx = 0 \tag{3}$$

The theory of linear differential equations states that the solution to the homogeneous equation (3) must have the form

$$x = \mathcal{A}e^{\lambda t} \tag{4}$$

Substituting equation (4) into equation (3) gives

$$(m\lambda^2 + k)\,\mathcal{A}e^{\lambda t} = 0$$

The solution must be valid for all instants t, so we drop the factor $e^{\lambda t}$. Also, we drop the factor $\mathcal{A}$, because $\mathcal{A} = 0$ is the trivial solution. Thus, because k and m are positive numbers, we have

$$\lambda^2 + \frac{k}{m} = 0 \qquad \lambda = \pm\sqrt{-\frac{k}{m}} = \pm i\sqrt{\frac{k}{m}} \tag{5}$$

where $i = \sqrt{-1}$. Hence, there are two possible values of λ. Corresponding to each value of λ, there is a corresponding value of $\mathcal{A}$. As the most general solution of a linear differential equation is a sum of all possible solutions, we have

$$x = \mathcal{A}_1 \exp\left(i\sqrt{\frac{k}{m}}\ t\right) + \mathcal{A}_2 \exp\left(-i\sqrt{\frac{k}{m}}\ t\right) \tag{6}$$

We now recall the identity

$$e^{ip} \equiv \cos p + i \sin p$$

to write

$$x = \mathcal{A}_1\left(\cos\sqrt{\frac{k}{m}}\,t + i\sin\sqrt{\frac{k}{m}}\ t\right) + \mathcal{A}_2\left(\cos\sqrt{\frac{k}{m}}\,t - i\sin\sqrt{\frac{k}{m}}\ t\right) \tag{7}$$

Furthermore, x must be a real quantity. Noting that the coefficients of $\mathcal{A}_1$ and $\mathcal{A}_2$ are complex conjugates of each other, this requirement means that $\mathcal{A}_1$ and $\mathcal{A}_2$ must also be complex conjugates; that is

$$\mathcal{A}_1 = a + bi \qquad \mathcal{A}_2 = a - bi$$

where a and b are real numbers. Equation (7) now becomes

$$x = (a + bi)(\cos\sqrt{\frac{k}{m}}t + i\sin\sqrt{\frac{k}{m}}t)$$

$$+ (a - bi)\left(\cos\sqrt{\frac{k}{m}}\,t - i\sin\sqrt{\frac{k}{m}}\ t\right)$$

$$= 2a\cos\sqrt{\frac{k}{m}}\,t - 2b\sin\sqrt{\frac{k}{m}}\,t$$

Because a and b are arbitrary numbers, the solution may be rewritten as

$$\boxed{x = A\cos\omega_n t + B\sin\omega_n t} \tag{8}$$

where A and B are real numbers, and

$$\boxed{\omega_n = \sqrt{\frac{k}{m}}} \tag{9}$$

For reasons we shall soon discuss, ω_n is called the circular *natural frequency* of the system.

The values of the coefficients A and B are obtained by requiring that equation (8) satisfy the initial conditions for the problem. However, the following alternative form of equation (8), which also has two arbitrary coefficients, has the advantage of being easier to visualize and sketch,

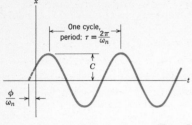

Figure 5

$$x = C \sin(\omega_n t + \phi) \qquad (10)$$

A typical x–t diagram is shown in Figure 5. As can be seen from Figure 5, the parameter C is the *amplitude* of the sinusoidal curve. The parameter ϕ is called the *phase angle;* it defines how far the response is shifted to the left of the origin compared to a simple sine curve.

Let us now consider the meaning of the parameter ω_n. From equation (9) we find that the units of ω_n are radians per unit time. As exhibited in Figure 5, the motion is cyclical, repeating itself when $\omega_n t$ is increased by 2π rad. The parameter $\tau = 2\pi/\omega_n$ is the *period,* which is the time required for the motion to pass through one cycle. Then, because it takes τ time units to complete one cycle, it follows that $1/\tau \equiv \omega_n/2\pi$ cycles will occur in one unit of time. The quantity $\omega_n/2\pi$ is the *cyclical natural frequency.* (The physical unit, cycles per second, has been given the name *hertz,* abbreviated as Hz.)

In contrast, as indicated after equation (9), the parameter ω_n is called the *circular natural frequency,* although the word circular is usually omitted. The reason for the word "natural" is that ω_n defines the frequency at which the system vibrates freely, that is, when it vibrates without any disturbance from external or dissipative forces.

It should be noted that in all computations a frequency ω, be it the natural frequency ω_n or more generally the angular velocity of a rotating body, should carry the units radians per unit time.

Continuing with the introduction of standard terminology, we have already used the term *vibration* and *oscillation,* which are synonomous. Any response $x(t)$, such as that in Figure 6, which represents a fluctuation about some mean value is a vibration (or oscillation).

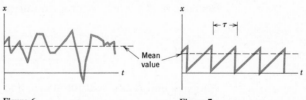

Figure 6   **Figure 7**

When the vibration has the form of a repeated pattern with a constant time interval τ, the oscillation is termed *periodic.* A typical periodic vibration appears in Figure 7.

A periodic vibration that has the shape of a sinusoidal curve is said to be *harmonic.* Thus, Figure 5 represents a harmonic vibration about a zero mean value.

EXAMPLE 1
A pendulum consisting of an 80-g particle suspended by a 2-m cable is known to be returning to the vertical position with a speed of 100 mm/s

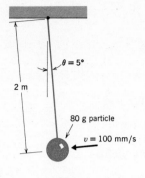

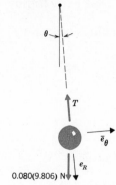

0.080(9.806) N

when $\theta = 5°$. Determine the amplitude and the period of the motion.

Solution

To determine the equation of motion (which is invariably the first step in a vibrations analysis) we draw a free body diagram, using polar coordinates. Remember that $\bar{e}_\theta$ is oriented in the direction of increasing θ. Note that the influence of gravity on the system varies with the angle θ, so that gravity does not have a static effect that may be ignored in formulating the equation of motion.

We are not interested in the value of the cable tension T, so we only sum forces in the transverse direction. The radial distance R has the constant value of 2.0 m, so

$$\Sigma F_\theta = -0.080(9.806) \sin \theta = ma_\theta$$

$$= 0.080(R\ddot{\theta} + 2\dot{R}\dot{\theta}) = 0.080(2.0)\ddot{\theta}$$

Thus

$$\ddot{\theta} + 4.903 \sin \theta = 0$$

This differential equation of motion is nonlinear. We can find a linear approximation to it by limiting the allowable values of θ to be sufficiently small so that $\sin \theta \simeq \theta$. With this restriction, the equation of motion becomes

$$\ddot{\theta} + 4.903 \theta = 0$$

Comparing this equation to equation (3), we see that

$$\omega_n{}^2 = \frac{k}{m} = 4.903 \text{ (rad/s)}^2$$

$$\omega_n = 2.214 \text{ rad/s}$$

The amplitude was expressly asked for, so it is most convenient to use the solution of the equation of motion in the form of equation (10) rather than that of equation (8). Thus

$$\theta = C \sin(\omega_n t + \phi)$$

$$= C \sin(2.214t + \phi)$$

To determine the unknown coefficients we return to the statement of the problem, which tells us that when $t = 0$, $\theta = 5°$, and $v = 100$ mm/s. Recalling that $v = R\dot{\theta}$ for motion in a circular path, we then have

$$\theta\big|_{t=0} = 5° = \frac{5}{180}\pi \text{ rad}$$

$$\dot{\theta}\big|_{t=0} = \frac{v}{R} = 0.050 \text{ rad/s}$$

Now, placing the initial conditions into the general solution gives

$$\theta\big|_{t=0} = C \sin(2.214t + \phi)\big|_{t=0}$$

$$= C \sin \phi = \frac{5}{180}\pi = 8.727(10^{-2})$$

$$\dot{\theta}\big|_{t=0} = 2.214C \cos(2.214t + \phi)\big|_{t=0}$$

$$= 2.214C \cos \phi = 0.050$$

Squaring and adding these two equations yields

$$C^2(\cos^2 \phi + \sin^2 \phi) \equiv C^2 = [(2.258)^2 + (8.727)^2]10^{-4}$$

and upon eliminating C from the two equations we get

$$\frac{\sin \phi}{\cos \phi} = \tan \phi = \frac{8.727}{2.258} = 3.865$$

Therefore the amplitude is

$$C = 9.014(10^{-2}) \text{ rad}$$

Although it was not requested, the phase angle is

$$\phi = \tan^{-1}(3.865) = 1.3176 \text{ rad} = 75.49°$$

We now can write the general solution to the pendulum problem as

$$\theta = 9.014(10^{-2}) \sin(2.214t + 1.3176)$$

The period τ is the time interval required for the motion to repeat itself. This means that the argument of the trigometric function in the general solution must increase by 2π. Thus

$$\omega_n \tau = 2.214\tau = 2\pi$$

$$\tau = 2.838 \text{ s}$$

Note that, in this linear approximation, the period of the motion depends only on the physical properties of the system (k and m) and not on the amplitude of the motion.

Finally, to clarify a matter that can cause some confusion, note that although the natural frequency $\omega_n = 2.214$ has the units of radians per second, it is not the angular velocity of the pendulum. The angular velocity is $\dot{\theta}$, which is found to be

$$\dot{\theta} = (2.214)(9.014)(10^{-2}) \cos(2.214t + 1.3176) \text{ rad/s}$$

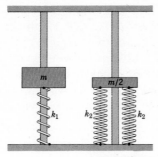

Prob. VIII.1

HOMEWORK PROBLEMS

VIII.1 Determine the ratio of k_2/k_1 required for the two systems shown to have the same natural frequency.

VIII.2 Three different ways of supporting a piston of mass m which rides in a smooth vertical cylinder are shown. Each system has two identical springs of stiffness k. Determine the natural frequency of each system.

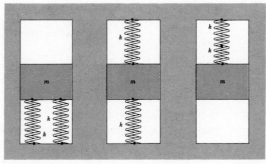

Prob. VIII.2

Prob. VIII.3

VIII.3 The mass of block A and the stiffness of the spring are unknown. It is known that the system vibrates freely at 80 Hz. When a 4-kg block is fastened to block A the frequency drops to 50 Hz. Determine the mass of block A and the stiffness of the spring.

VIII.4 The 200-g collar slides on the smooth horizontal bar. The spring has a stiffness of 2 kN/m, and the system is in the static equilibrium position when $s = 100$ mm. (a) Determine the natural frequency in units of radians per second and hertz. (b) If the collar is released from rest at $s = 80$ mm, determine where it will again come to rest and the elapsed time between these two positions.

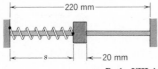

Prob. VIII.4

VIII.5 Two pretensioned springs are attached to the 0.5-kg block. The unstretched length of both springs is 60 mm, and their stiffnesses are $k_1 = k_2 = 1$ kN/m. Determine (a) the value of the distance s corresponding to static equilibrium, (b) the differential equation of motion in terms of the distance s, (c) the differential equation of motion in terms of the displacement x from the static equilibrium position, and (d) the natural frequency of the system. (e) Which, if any, of the foregoing answers were affected by the unstretched length of the springs?

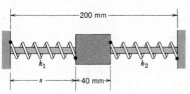

Prob. VIII.5

VIII.6 Solve Problem VIII.5 for $k_1 = 1$ kN/m and $k_2 = 0.5$ kN/m.

VIII.7 A block of mass m is fastened to a lightweight (negligible mass) beam at its midpoint. Unloaded the beam is straight. With the block attached, the static downward displacement of the midpoint of the beam is δ_{st}, as shown. (a) Derive an expression for the period of free vibration of the beam-mass combination. (b) If the system is held in the position where the beam is straight and then released from rest, what will be the amplitude of the subsequent motion? *Hint:* The beam acts, in effect, like a spring.

Prob. VIII.7

VIII.8 Two 1-kg spheres are suspended from the roof of an elevator, as shown. When the elevator is at rest the period of extensional free vibration of the spring-supported sphere is the same as that for the rotation of the pendulum. (a) Determine the stiffness k of the spring. (b) Determine the period of each system when the elevator accelerates upward at $\dot{v} = 0.6g$.

VIII.9 The speed governor shown consists of a block of mass m which may slide within the radial groove in the disk as the disk rotates in the horizontal plane about a vertical axis passing through point O at angular speed Ω. The stiffness of the spring is k. Determine the equation of motion for the radial distance R and also determine the natural frequency of the system. For what value of the angular speed Ω does the block cease to be capable of simple harmonic motion? Explain your answer.

VIII.10 A 2-lb block slides on a smooth horizontal table under the restraint of the two cables. In the static equilibrium position the tensile force within the cables is 40 lb. Because of the (slight) elasticity of the cables, the block can execute small movements x in the direction perpendicular to line AB. (a) Determine the natural frequency of the system in terms of the length l of one of the cables. (b) Determine the value of l that minimizes the natural frequency and also determine this minimum frequency. *Hint:* Formulate the equation of motion in terms of the displacement x, considering the tension to be constant and the angles θ_1 and θ_2 to be very small.

VIII.11 The collar of mass m slides on the smooth horizontal bar. The spring, whose stiffness is k, has a pre-tension force T when the block is in the static equilibrium position. Derive the differential equation of motion that is valid for large values of x and then linearize this result to obtain the equation for small x. Does the value of k affect the latter equation? Explain your answer.

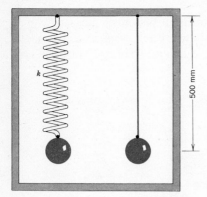

Prob. VIII.8

Prob. VIII.9

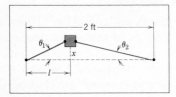

Prob. VIII.10

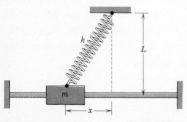

Prob. VIII.11

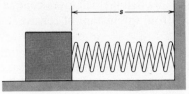

Prob. VIII.12

VIII.12   The 2-lb block slides on the smooth horizontal surface under the restraint of the spring whose stiffness is 40 lb/ft and whose unstretched length is 1 ft. The block is given an initial velocity of 5 ft/s to the right at s = 15 in. Determine (a) the response $x(t)$, where x is the displacement measured from the static equilibrium position, (b) the amplitude, period, and phase angle for the response in part (a). (c) Would the answers to parts (a) and (b) change if the system was inclined upward to the left at 30° above the horizontal?

VIII.13   A 20-gram sphere moves within the smooth curved tube, which is held stationary in the vertical plane. The block is released from rest at θ = −4.5°. (a) Without solving the equation of motion, determine how long it will take for the block to reach θ = 0°. (b) Determine the velocity, acceleration, and elapsed time from release corresponding to the positions where θ = −2.25° and where θ = 2.25°.

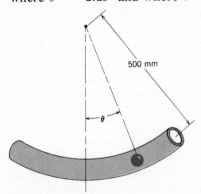

500 mm

θ

Prob. VIII.13

VIII.14   For the system and initial conditions given in Problem VIII.13, locate the following positions: (a) where the acceleration component tangent to the path is maximum, (b) where the acceleration component normal to the path is maximum, (c) where the magnitude of the acceleration is maximum.

Prob. VIII.15

VIII.15   The 4-kg block rests on the spring, but is not attached to it. The block is held at rest 20 mm below the static equilibrium and then released, this distance being the largest initial displacement for which the block never loses contact with the spring. Determine (a) the stiffness of the spring, and (b) the displacement of the block from the static equilibrium position as a function of time.

VIII.16   A 6-oz sphere is to be catapulted into the air vertically by the piston, which rides within the smooth cylinder. The piston weighs 2 oz, and the stiffness of the spring is 80 lb/ft. The piston is held 3 in. below the static equilibrium position and then released. Determine the distance the

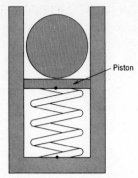

Prob. VIII.16

piston is above the static equilibrium position when the ball loses contact with the piston, and also the elapsed time from the instant of release.

VIII.17 For the system of Problem VIII.16, determine (a) the maximum height attained by the sphere after leaving the piston, (b) the time response of the piston after the sphere flies upward, measuring time t from the instant when the piston was released, and (c) the amplitude, frequency, and phase angle for the motion obtained in part (b).

VIII.18 Two identical 1-kg collars, which may slide on the smooth horizontal bar, are initially at rest in the positions shown. The spring that is attached to block A has a stiffness of 3.6 N/mm and an unstretched length of 500 mm. Collar A is released, causing this body to move to the right and hit collar B. This collision is perfectly elastic. Determine (a) the elapsed time from the instant when the cord is broken until the collars collide, (b) the velocity of collar B immediately after the collision, (c) the amplitude, frequency, and phase angle for the motion of block A after the collision, measuring t from the instant of release.

VIII.19 Solve Problem VIII.18 if the coefficient of restitution for the collision is 0.50.

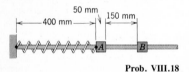

Prob. VIII.18

4 Free Damped Vibrations

Here we wish to learn how damping affects the response of a system that is free of external forces when the system is given an initial disturbance. As the solution is developed, note that the natural frequency $\omega_n = \sqrt{k/m}$ of the undamped system is also an important parameter for the damped system.

The equation of motion for the free, damped system associated with Figure 3 is

$$m\ddot{x} + c\dot{x} + kx = 0 \tag{11}$$

The harmonic solution given by equation (10) does not satisfy this equation because of the presence of the first derivative resulting from damping. To solve equation (11) we use the basic form for the solution of a linear differential equation.

$$x = \mathcal{A}e^{\lambda t} \tag{12}$$

Substituting this expression into equation (11), dividing by m, and replacing k/m by $\omega_n{}^2$ yields

$$\left(\lambda^2 + \frac{c}{m}\lambda + \omega_n{}^2\right)\mathcal{A}e^{\lambda t} = 0$$

As we did in the undamped case, we equate the term multiplied by $\mathcal{A}e^{\lambda t}$ to zero. This yields a quadratic equation for λ, whose solution is

$$\lambda = -\frac{c}{2m} \pm \sqrt{\left(\frac{c}{2m}\right)^4 - \omega_n^2}$$

Clearly the sign of the quantity under the radical will affect the result. For convenience, we define a new parameter,

$$\boxed{\beta = \frac{c}{2m\omega_n} = \frac{c}{2\sqrt{km}}} \tag{13}$$

so the two values for λ appearing above become

$$\lambda = -\omega_n\beta \pm \omega_n\sqrt{\beta^2 - 1} \tag{14}$$

There are now three possibilities to consider, depending on whether β is less than, greater than, or equal to one.

Case a: $\beta < 1$ This is the situation where the coefficient of damping is small. For this case the square root in equation (14) is imaginary, and the two values of λ are complex conjugates of each other; that is,

$$\lambda = -\omega_n\beta \pm i\omega_n\sqrt{1 - \beta^2}$$

We leave it as a homework exercise to follow similar steps to those used in the study of undamped free vibrations to show that the corresponding solution for x can be written in either of the following equivalent forms.

$$\boxed{\begin{aligned} x &= e^{-\omega_n\beta t}(A \cos \omega_n\sqrt{1 - \beta^2}\, t \\ &\quad + B \sin \omega_n\sqrt{1 - \beta^2}\, t) \\ &= Ce^{-\omega_n\beta t} \sin(\omega_n\sqrt{1 - \beta^2}\, t + \phi) \end{aligned}} \tag{15}$$

Once again the unknown coefficients C and ϕ are found by satisfying the initial conditions. Analyzing equation (15), we see that the exponential multiplier means that the response will decay with increasing values of t, whereas the term in parentheses shows that the response is oscillatory. The quantity $\omega_d \equiv \omega_n(1 - \beta^2)^{1/2}$ is sometimes called the *damped natural frequency*. However, referring to the typical $x(t)$ plot in Figure 8 on the next page, note that the response is not periodic. The parameter ω_d is a measure of the number of times x changes sign (or alternatively, reaches an extreme value) in a unit interval of time. The corresponding period is $\tau = 2\pi/\omega_d$.

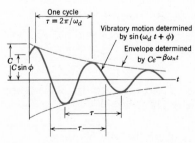

One cycle
$\tau = 2\pi/\omega_d$

Vibratory motion determined by $\sin(\omega_d\, t + \phi)$

Envelope determined by $Ce^{-\beta\omega_n t}$

C
$|C\sin\phi|$

Figure 8

Case b: $\beta > 1$ The coefficient of damping in this situation is large. For this case the square root in equation (14) is real, and hence the two values of λ are real. Corresponding to these two values of λ are two solutions, the sum of which is the general solution

$$x = e^{-\omega_n\beta t}(Ae^{\omega_n\sqrt{\beta^2-1}\,t} + Be^{-\omega_n\sqrt{\beta^2-1}\,t}) \qquad (16)$$

Noting that $(\beta^2 - 1)^{1/2}$ is less than β, the solution given by equation (16) will decay with increasing time. Also, because the solution is a combination of exponential terms, the response is not oscillatory. This is exhibited in Figure 9, where the three curves show the effect of different types of initial conditions on the response.

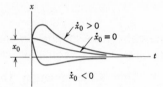

Figure 9

Case c: $\beta = 1$ This represents a transition between the lightly damped situation ($\beta < 1$) and the heavily damped case ($\beta > 1$). Equation (14) now yields only one value for λ, which is

$$\lambda = -\omega_n\beta \equiv -\omega_n$$

However, the theory of linear differential equations tells us that there must be two distinct solutions to a second-order equation. The second solution is obtained by multiplying the basic exponential solution by a factor t. Hence

$$x = e^{-\omega_n t}(A + Bt) \qquad (17)$$

Recalling that

$$\lim_{t\to\infty} (te^{-\omega_n t}) = 0$$

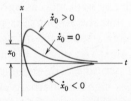

Figure 10

it follows that the response in this case also decays to zero. In fact, Figure 10 shows that the response in this case strongly resembles in all respects the response of Figure 9 for a more heavily damped system, although the case when $\beta = 1$ gives the highest rate of decay for the response.

The terminology used to describe the foregoing three cases is to say that a system for which $\beta < 1$ is *underdamped,* whereas one for which $\beta > 1$ is *overdamped.* As $\beta = 1$ defines the transition between an oscillating response and an exponentially decaying one, we say that $\beta = 1$ is the state of *critical damping.* The corresponding critical damping coefficient is found from equation (13) by setting $\beta = 1$, which gives

$$c_{cr} = 2\sqrt{km} \qquad \text{(18a)}$$

Substituting this expression back into equation (13) yields

$$\beta = \frac{c}{c_{cr}} \qquad \text{(18b)}$$

In other words, the value of β for a system is the ratio of the actual coefficient of damping of the system to the critical coefficient; it is called the *ratio of critical damping.*

A common example is the case of the shock absorbers of an automobile. When new, they do not permit the vehicle to bounce continuously when it encounters an irregularity in the roadway. The automobile is then overdamped. As the shock absorbers wear from age and use, the automobile passes from the overdamped situation, through critical damping, and into the case of underdamping. Disturbances in the roadway now result in the automobile bouncing freely. It is then time to install new shock absorbers.

EXAMPLE 2

An elevator comes into contact with its supports at the bottom of the elevator shaft, as shown, with the result that the elevator vibrates. The period τ is timed electronically, and the displacements measured at two instants τ seconds apart are x_1 and x_2. Determine (a) the ratio of critical damping and the natural frequency in terms of τ, x_1, and x_2, and (b) the percent of critical damping by which the value of c must be increased to prevent the elevator from vibrating when it reaches the bumper.

Solution

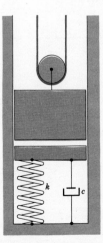

We are measuring x from the static equilibrium position, so we ignore the static gravity force. The equation of motion is

$$m\ddot{x} + c\dot{x} + kx = 0$$

Thus we see that the system gives rise to an equation of motion of the form equation (11). We are told that the system is vibrating, so it must be underdamped. Thus the response is given by equation (15),

$$x = Ce^{-\omega_n \beta t} \sin(\omega_n \sqrt{1 - \beta^2}\, t + \phi)$$

where, from equations (9) and (13), we have

$$\omega_n = \sqrt{\frac{k}{m}} \quad \text{and} \quad \beta = \frac{c}{2\sqrt{km}}$$

The values of m, c, and k are unknown. However, we are told that the period is known. For an underdamped system,

$$\tau = \frac{2\pi}{\omega_d} = \frac{2\pi}{\omega_n \sqrt{1 - \beta^2}}$$

This expression provides one relationship between the unknown ratio of critical damping and the natural frequency. For an additional relationship we use the given information regarding the displacement at the two time instants.

We were not given initial conditions, but we can form a ratio of the two successive displacements x_1 and x_2 in the hope of eliminating the unknown constants C and ϕ. It is given that x_2 is the displacement τ seconds after x_1, hence

$$\frac{x_1}{x_2} = \frac{Ce^{-\omega_n \beta t} \sin(\omega_n \sqrt{1 - \beta^2}\, t + \phi)}{Ce^{-\omega_n \beta(t+\tau)} \sin[\omega_n \sqrt{1 - \beta^2}\, (t + \tau) + \phi]}$$

$$= e^{\omega_n \beta \tau} \frac{\sin(\omega_n \sqrt{1 - \beta^2}\, t + \phi)}{\sin[\omega_n \sqrt{1 - \beta^2}\, (t + \tau) + \phi]}$$

Substituting the expression for τ, we then have

$$\frac{x_1}{x_2} = e^{2\pi\beta/\sqrt{1-\beta^2}} \frac{\sin(\omega_n \sqrt{1 - \beta^2}\, t + \phi)}{\sin(\omega_n \sqrt{1 - \beta^2}\, t + 2\pi + \phi)}$$

$$= e^{2\pi\beta/\sqrt{1-\beta^2}}$$

From this result we see that the ratio x_1/x_2 is a constant (independent of time). Let us denote its natural logarithm by symbol δ, which we call the *logarithmic decrement* of the system. Thus

$$\delta \equiv \ln\left(\frac{x_1}{x_2}\right) = \ln(e^{2\pi\beta/\sqrt{1-\beta^2}})$$

$$= \frac{2\pi\beta}{\sqrt{1 - \beta^2}}$$

Solving for β in terms of the known parameter δ, we find that

$$\beta = \frac{\delta}{\sqrt{4\pi^2 + \delta^2}}$$ ◁

Knowing β, we now can use the expression for the known value τ to determine ω_n. This gives

$$\omega_n = \frac{2\pi}{\tau \sqrt{1 - \beta^2}}$$ ◁

Finally, because the system will not vibrate when critically damped (when $\beta = 1$), the necessary increase in β is

$$\Delta\beta = 1 - \beta = 1 - \frac{\delta}{\sqrt{4\pi^2 + \delta^2}}$$

We also know from equation (18b) that c is proportional to β. Thus, multiplying by 100 to obtain a percentage, we find

$$\%\left(\frac{\Delta c}{c_{cr}}\right) = 100\left(1 - \frac{\delta}{\sqrt{4\pi^2 + \delta^2}}\right)\%$$

HOMEWORK PROBLEMS

VIII.20 An overdamped system for which $\omega_n = 50$ rad/s is released from rest with a positive initial displacement x_0. Determine the time required for the displacement to be $x_0/10$ if the critical damping ratio is (a) $\beta = 1.010$, (b) $\beta = 2.020$.

VIII.21 The result of electronic measurement of the displacement of a system is shown. Determine the damped natural frequency, the ratio of critical damping, and the displacement x_1 indicated.

VIII.22 A system is underdamped. (a) Show that the time interval between two adjacent maximum displacements is exactly equal to the period of damped free vibration. (b) Let x_1 and x_{n+1} denote the maximum displacements occurring at the beginning and end of n cycles of vibration and show that the logarithmic decrement may be obtained from $\delta = (1/n) \ln(x_1/x_{n+1})$.

VIII.23 The block slides on the horizontal surface, which is lubricated by oil, so the friction force is proportional to the velocity of the block. The system is underdamped. When the block is released from rest the spring is compressed a distance x_0. (a) What percentage of the mechanical energy initially given to the system is dissipated in the first cycle of vibration? (b) Does this percentage change for the second and subsequent cycles?

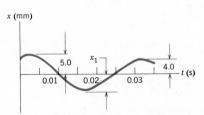

Prob. VIII.21

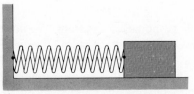

Prob. VIII.23

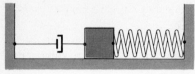

Prob. VIII.24

VIII.24 The 8-oz block is restrained by a spring whose stiffness is 80 lb/ft and a dashpot whose constant is 2 lb-s/ft. Initially, the spring is elongated by 0.2 ft and the block is given a velocity of 10 ft/s to the right. Determine (a) the response as a function of time, (b) the time at which the elongation of the spring first attains the zero value.

VIII.25 Following similar steps to those used to obtain the general undamped free vibration response, equation (10), derive equation (15) for damped free vibration.

VIII.26 A 125-gram block is suspended vertically by a spring whose stiffness is 5 N/mm, and also by a dashpot. The system is underdamped. The block is given an upward velocity of 2 m/s at the static equilibrium position. (a) Determine, in terms of the critical damping ratio β, the time when the upward displacement of the block is maximum and compare this time interval to the period of damped free vibration. (b) Compare the maximum displacement discussed in part (a) to the maximum that would be obtained if damping were not present. (c) Compute the results of parts (a) and (b) for critical damping ratios of 0.90 and 0.10.

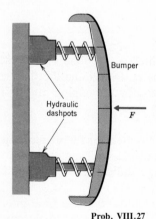

Bumper

Hydraulic
dashpots

F

Prob. VIII.27

VIII.27 In the shock absorbing mechanism for an automobile bumper shown, two springs act concentrically with two hydraulic dashpots. The stiffness of each spring is 3 kN/m, and the bumper has a mass of 40 kg. When subjected to the impulsive force $\bar{F}$, it is desired that the bumper compress in the shortest possible time, without subsequently oscillating when it rebounds. Determine the value of the damping constant μ required for each dashpot to accomplish this.

VIII.28 In Problem VIII.27, the damping constant for each dashpot is μ = 600 N-s/m. A collision force $\bar{F}$, which exerts an impulse of 1.20 kN-s over a very short time interval, is applied to the bumper. Determine (a) the displacement of the bumper as a function of time t, where t is measured from the instant when $\bar{F}$ is applied, (b) the maximum distance the springs are compressed and the corresponding time t, and (c) the maximum compression of the springs if damping were not present.

VIII.29 Solve Problem VIII.28 for μ = 300 N-s/m.

5 Forced Vibration-Harmonic Excitation

Thus far we have concentrated on the free response of a system. We now wish to consider the effect of dynamic excitation of the system. The type of force we shall consider is a harmonic force, that is, one that varies sinusoidally with time. The reasons for studying such a forcing function are many. A prime reason is that if we know the response of the system to harmonic excitation, it is possible to determine the response to any periodic excitation by using a Fourier series.

Letting $F(t) = F \sin \Omega t$, where Ω is the frequency of the excitation (having the units radians per unit time), the equation of motion for the system of Figure 1 is

$$m\ddot{x} + c\dot{x} + kx = F \sin \Omega t \tag{19}$$

There are two types of solutions for equation (19). One, called the *particular solution* and denoted as x_p, corresponds to the nonhomogeneous term in the differential equation. To this solution we add the *complementary solution* x_c, which is the solution corresponding to $F(t) = 0$. For equation (19) the complementary solution has the form of the free response of an undamped, underdamped, overdamped, or critically damped system, depending on the values of m, c, and k. Thus, we shall focus attention here on the determination of the particular solution.

The most direct way to obtain a particular solution is to guess its form. Because x and its first and second derivatives must combine to give a sine term, we guess

$$\boxed{x_p = D \sin(\Omega t + \psi)} \tag{20}$$

In contrast to the homogeneous solution (free vibration), the coefficients D and ψ are not arbitrary values. They are determined by requiring that x_p satisfy equation (19). Upon substituting equation (20) into equation (19), we find that

$$(-m\Omega^2 + k)D \sin(\Omega t + \psi) + c\Omega D \cos(\Omega t + \psi) = F \sin \Omega t$$

The easiest way to proceed at this point is to manipulate the term on the right side using the trigonometric identity for the sine of the difference of two angles, as follows.

$$(-m\Omega^2 + k)D \sin(\Omega t + \psi) + c\Omega D \cos(\Omega t + \psi)$$

$$= F \sin[(\Omega t + \psi) - \psi]$$

$$= F \sin(\Omega t + \psi) \cos \psi - F \cos(\Omega t + \psi) \sin \psi$$

Equating the coefficients of the sine and cosine terms gives

$$(-m\Omega^2 + k)D = F \cos \psi$$

$$c\Omega D = -F \sin \psi$$

Solving these equations for D and ψ in terms of F, Ω, m, c, and k yields

$$D = \frac{F}{\sqrt{(k - \Omega^2 m)^2 + c^2\Omega^2}}$$

$$\tan \psi = \frac{c\Omega}{k - m\Omega^2}$$

This result is now simplified by replacing m, c, and k by the expressions for the natural frequency, $\omega_n = \sqrt{k/m}$, and the critical damping ratio, $\beta = c/2\sqrt{mk}$. Skipping the algebraic details, this yields

$$D = \frac{(F/k)}{\sqrt{(1 - \Omega^2/\omega_n^2)^2 + 4\beta^2\Omega^2/\omega_n^2}}$$

$$\psi = \tan^{-1}\left(\frac{2\beta\Omega/\omega_n}{1 - \Omega^2/\omega_n^2}\right) \tag{21}$$

There are a number of interesting results displayed in equation (21). For instance, notice that F/k is the static displacement that the mass would have if it were loaded by a static force of magnitude F. Thus, the fraction multiplying F/k in the equation for D is a *dynamic magnification factor* for the amplitude D. Furthermore, for a given set of system parameters, the amplitude and phase angle depend only on the ratio of the frequency of excitation to the natural frequency.

To study equations (21) further, let us sketch D and ψ as a function of Ω/ω_n, as is shown in Figure 11. Such plots are called amplitude-frequency (Figure 11a) and phase angle-frequency diagrams (Figure 11b). They are very useful for designing systems to meet dynamic criteria, such as limitations on the amplitude of vibration.

Focusing on Figure 11a, for an undamped system ($\beta = 0$), notice that the value of D becomes infinite as the frequency of excitation approaches the natural frequency ($\Omega/\omega_n \rightarrow 1$). This situation, which is called *resonance,* is obviously quite dangerous, although in reality the response can never be infinite, for all real systems exhibit some damping.

The phase angle for the undamped system is also of interest. When $\Omega/\omega_n < 1$, we see in Figure 11b that, for $\beta = 0$, $\psi = 0$ also. This means that the displacement x_p is positive whenever the force $F(t) = F \sin \Omega t$ is positive. We then say that the response and the excitation are *in phase*. As the resonance condition is passed, so that $\Omega/\omega_n > 1$, ψ suddenly changes to 180°, so the displacement x_p is negative whenever $F(t) = F \sin \Omega t$ is positive. The response and the excitation are then 180° *out of phase*. Another interesting feature of the phase angle-frequency diagram is that when $\Omega/\omega_n = 1$, the response x_p for a damped system is 90° out of phase with the force; that is, the magnitude of the response is a maximum when the force is zero and vice versa.

Let us now return to the problem of determining the total response for the system of equation (19). For this discussion we shall consider the

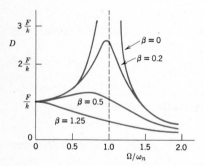

Figure 11a

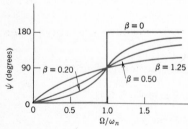

Figure 11b

case of an underdamped system. Thus, the complementary solution is given by equation (15), which in combination with the particular solution of equation (20), gives

$$x = x_c + x_p$$

$$= e^{-\beta\omega_n t}C\,\sin(\omega_n\sqrt{1-\beta^2}\,t + \phi) + D\,\sin(\Omega t + \psi) \qquad (22)$$

A computational point to bear in mind is that the constants C and ϕ are found by using the full expression (22) to satisfy the initial conditions, not just the portion of equation (22) that is the complementary solution.

Examining equation (22), we see that as t becomes large the first term, the complementary solution, goes to zero. Because of this effect, dynamicists call x_c the *transient* solution and x_p the *steady state* solution. It is a steady state vibration in the sense that after the transient decays the system will maintain a steady harmonic oscillation at the amplitude D. The energy put in through the forcing function then balances the energy lost through damping. Note too that the steady state vibration takes place at the frequency of the forcing function. An exception occurs in the case of an undamped system, for then the complementary solution in equation (22) does not decay. In reality, this is of little concern, because all systems have some damping, which causes the transient solution eventually to decay to zero. Thus, we see that it is correct to approximate a system as having no damping only if we are observing it over a comparatively small time interval.

EXAMPLE 3

When $t = 0$, the illustrated system is at rest in its static equilibrium position, where the tension forces in each of the two springs both have the value T. At this instant a harmonic force $F \cos \Omega t$ is applied to the block. Neglecting friction, determine the subsequent motion of the block.

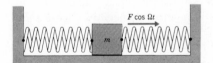

$F \cos \Omega t$

Solution

In the free body diagram let us arbitrarily choose x as the displacement of the block to the right from the static equilibrium position. Notice that a displacement x to the right reduces the tension forces in the right spring by kx, whereas it increases the tension force in the left springs by the same amount.

To obtain the equation of motion, we sum forces to obtain

$$\Sigma F_x = (T - kx) + F \cos \Omega t - (T + kx) = ma_x = m\ddot{x}$$

$$m\ddot{x} + 2kx = F \cos \Omega t$$

As can be seen, the tension force T does not appear in the equation of motion. This is another situation where a static force does not affect a dynamic displacement measured from the static equilibrium position.

The solution of the equation of motion is obtained by adding the complementary and particular solutions for an undamped oscillator. The natural frequency of this system is

$$\omega_n = \sqrt{\frac{2k}{m}}$$

and the complementary solution is found from equation (8) to be

$$x_c = A \cos \omega_n t + B \sin \omega_n t$$

The particular solution for harmonic excitation, given by equations (20) and (21), was derived for a sine force; the force in this problem is a cosine force. However, because this is simply a 90° difference in phase, we may directly employ the derived solution by also shifting the response by 90°, thus replacing the sine function by the cosine function. Hence

$$x_p = D \cos(\Omega t + \psi)$$

The amplitude D in equation (20) is positive, so for $\beta = 0$ we have

$$D = \frac{F/2k}{\left|1 - \Omega^2/\omega_n{}^2\right|}$$

$$\psi = 0 \qquad \text{if } \Omega < \omega_n$$

$$\psi = 180° \qquad \text{if } \Omega > \omega_n$$

It is useful in the case of no damping to remove the phase angle from the problem by setting $\psi \equiv 0$ and letting D be positive or negative. Thus, the equivalent of the foregoing is

$$x_p = D \cos \Omega t$$

$$D = \frac{F/2k}{1 - \Omega^2/\omega_n{}^2}$$

Finally, we satisfy the initial conditions. From the given information we know that $x = 0$ and $\dot{x} = 0$ when $t = 0$. This gives

$$x(0) = 0 = (x_c + x_p)\big|_{t=0}$$

$$= (A \cos \omega_n t + B \sin \omega_n t + D \cos \Omega t)\big|_{t=0}$$

$$= A + D$$

$$\dot{x}(0) = 0 = (\dot{x}_c + \dot{x}_p)\big|_{t=0}$$

$$= [\omega_n(-A \sin \omega_n t + B \cos \omega_n t)$$

$$- \Omega D \sin \Omega t]_{t=0} = \omega_n B$$

Thus, $A = -D$ and $B = 0$, which gives

$$x = D(\cos \Omega t - \cos \omega_n t)$$

$$= \left(\frac{F/2k}{1 - \Omega^2/\omega^2}\right)(\cos \Omega t - \cos \omega_n t)$$

Notice that the transient solution is not zero even though the initial displacement and velocity were zero, thus emphasizing the fact that the initial conditions must be satisfied after first combining the complementary and particular solutions.

EXAMPLE 4

The center of mass G of the 0.5-kg flywheel of an electric motor is at a distance of 20 mm from the motor shaft. The body of the motor is attached to a rigid framework which is supported by four springs, as shown. The framework is capable only of translatory motion in the vertical direction, and the combined mass of the body of the motor and its framework is 4.5 kg. It is desired that the steady state displacement not exceed 2.0 mm when the flywheel is rotating at the natural frequency of the undamped system. Determine the minimum value of the damping coefficient c required to satisfy this criterion.

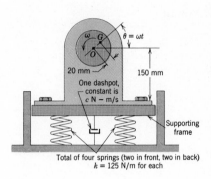

Solution

This problem illustrates a common source of harmonic excitation for a system: an unbalanced rotating mass. The determination of the equation of motion, which is our first task, is not as trivial as in the prevous developments.

The system is composed of two moving bodies, one being the body of the motor and the framework, the other being the flywheel. Therefore, let us draw free body diagrams for both systems. The interaction force exerted between them is the reaction $\bar{F}$ at the shaft.

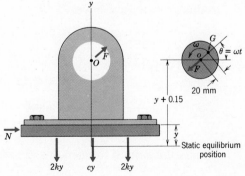

Notice that we are measuring the vertical displacement y from the static equilibrium position, so we ignore the static effect of gravity. The equation for vertical motion of the motor and its framework is

$$\Sigma F_y = -4ky - c\dot{y} + F_y = Ma_y = M\ddot{y}$$

where, from the given information, $M = 4.5$ kg and $k = 125$ N/m.

To determine the reaction component F_y, we sum forces in the vertical direction for the flywheel. This gives

$$\Sigma F_y = -F_y = m(a_G)_y = m\ddot{y}_G$$

where $m = 0.50$ kg and the negative sign arises because of Newton's third law. From the geometry of the free body diagram we can write

$$y_G = y + 0.15 + |\bar{r}_{G/O}| \sin \theta \text{ meters}$$

$$= y + 0.15 + 0.020 \sin \omega t$$

Hence

$$\ddot{y}_G = \ddot{y} - 0.020\omega^2 \sin \omega t$$

and we have

$$F_y = -m\ddot{y} + 0.020m\omega^2 \sin \omega t$$

Now, replacing F_y in the first equation of motion, the basic equation of motion for the system becomes,

$$M\ddot{y} = -4ky - c\dot{y} + (-m\ddot{y} + 0.020m\omega^2 \sin \omega t)$$

$$(M + m)\ddot{y} + c\dot{y} + 4ky = 0.020m\omega^2 \sin \omega t$$

We can determine the response of the system by solving this differential equation, which we recognize to be of the form of equation (19). Because initial conditions are not given, we consider only the steady state response. Let us denote the symbols appearing in the basic differential equation (19) by primes. Then, in comparison with the equation of motion in this problem, we have

$$m' = M + m = 4.5 + 0.5 = 5.0 \text{ kg}$$

$$c' = c \text{ N-s/m}$$

$$k' = 4k = 4(125) = 500 \text{ N/m}$$

$$F' = 0.020m\omega^2 = 0.010\omega^2 \text{ N}$$

$$\Omega = \omega \text{ rad/s}$$

Hence, the natural frequency and ratio of critical damping are

$$\omega_n = \sqrt{\frac{k'}{m'}} = \sqrt{\frac{500}{5.0}} = 10.0 \text{ rad/s}$$

$$\beta = \frac{c'}{2\sqrt{k'm'}} = \frac{c}{2\sqrt{500(5)}} = \frac{c}{100}$$

We may now use equations (20) and (21) to determine the steady state solution. Replacing the coefficients in these equations by their equivalents here, we find that

$$y_p = D \sin (\omega t + \psi)$$

where

$$D = \frac{(0.010\omega^2/500)}{\sqrt{[1 - \omega^2/(10.0)^2]^2 + 4\beta^2\omega^2/(10.0)^2}}$$

$$\psi = \tan^{-1}\left[\frac{2\beta\omega/10.0}{1 - \omega^2/(10.0)^2}\right]$$

To find the value of β, we satisfy the given limitation on the magnitude of the displacement by setting the amplitude D equal to 0.002 meters and $\omega = \omega_n = 10.0$ rad/s. This yields

$$0.002 = \frac{(0.010)(10.0)^2/500}{\sqrt{(1.0 - 1.0)^2 + 4\beta^2(1.0)}} = \frac{0.002}{2\beta}$$

$$\beta = 0.5$$

In other words, the damping in the system must be at least half the critical value. The corresponding minimum value of c is now found from the earlier expression, $\beta = c/100$, to be

$$c = 100\, \beta = 50.0 \text{ N-s/m}$$

HOMEWORK PROBLEMS

Prob. VIII.30

VIII.30 The 4-kg block is restrained by a spring whose stiffness is 4 kN/m. The external force varies harmonically according to $F = 12 \sin 16t$, where F is in units of newtons and t is in seconds. At $t = 0$ the block is at the static equilibrium position and has a velocity of 4 m/s to the left. Determine the response of the block and identify the transient and steady state portions.

VIII.31 Solve Problem VIII.30 if the external force is a combination of two harmonics given by $F = 12(\sin 8t + \sin 16t)$.

VIII.32 The harmonic force $F \cos \Omega t$ is applied to the block at $t = 0$. Initially, the block was at rest in the position where the compressive force

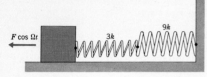

$F \cos \Omega t$

$3k$ $9k$

Prob. VIII.32

in the series-connected spring is F. Determine (a) the response of the block, (b) the amplitude and phase angle of the steady state response for $\Omega = \sqrt{k/m}$ and $\Omega = 2\sqrt{k/m}$.

VIII.33 and VIII.34 A 4-lb collar on a smooth horizontal rod is restrained by a spring whose stiffness is 2 lb/in. The collar is initially at rest in the static equilibrium position when it is subjected to the tangential force F shown in the graph. Determine and plot the displacement $x(t)$ (a) for $0 \le t \le 1$, (b) for $1 \le 1 \le 2$.

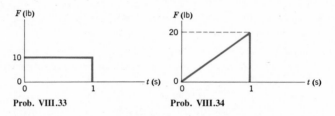

Prob. VIII.33   **Prob. VIII.34**

VIII.35 A 9-kg electric motor is mounted on a lightweight beam. The flywheel of the motor has a mass of 1kg, and its center of mass is 6 mm from the axis of rotation. Knowing that the static deflection of the beam when the motor and flywheel are at rest is 5 mm, determine (a) the angular velocity of the flywheel that will cause the system to resonate, (b) the amplitude of the steady state displacement when the angular speed of the flywheel is eight-tenths of the resonant speed found in part (a).

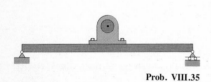

Prob. VIII.35

VIII.36 It is desired to operate the electric motor in Problem VIII.35 at the angular velocity that would cause the system to resonate. To do this, a dashpot is attached to the beam from the ground, directly under the motor. If the steady state displacement of the motor in this condition is to be 10 mm, determine the required constant for the dashpot.

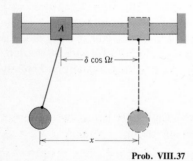

A

$\delta \cos \Omega t$

x

Prob. VIII.37

VIII.37 A simple pendulum formed from a particle of mass m at the end of a cable of length l is attached to collar A. The collar is given a harmonic displacement $\delta \cos \Omega t$. For what range of values of Ω will the steady state amplitude of the horizontal displacement x of the mass be (a) at least three times larger than δ? (b) less than δ?

VIII.38 The runway of an airport can be approximated as a sinusoidal surface whose vertical elevation h is defined by $h = h_0 \sin(\pi x/l)$. An airplane of mass m lands smoothly on the runway with a speed v. Use the model shown to determine the critical landing speed at which the airplane would resonate. *Hint:* Formulate the equation of motion in terms of the absolute vertical displacement y and let $x = vt$ in the expression for the elevation h.

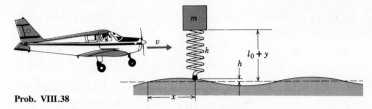

Prob. VIII.38

VIII.39 The automobile gives the 500-kg trailer a constant horizontal velocity component v as the wheels of the trailer follow a wavy road. The elevation of the road surface is defined by $h = 40 \sin \pi x/2$, where h is in millimeters and x is the horizontal distance traveled by the trailer in meters. The trailer does not have shock absorbers, and each of the two springs supporting the trailer has a stiffness of 40 kN/m. Neglecting the mass and the radius of the wheels determine (a) the differential equation of motion for the vertical displacement of the trailer, (b) the speed v at which the trailer would resonate, (c) the amplitude of the vibration of the trailer when $v = 10$ m/s. See the hint given in Problem VIII.38.

Prob. VIII.39

VIII.40 Solve Problem VIII.39 if, in addition to the two springs, two shock absorbers whose constants are 5 kN-s/m each are also connected between the axle and the body of the trailer.

VIII.41 A system whose natural frequency is ω_n and whose critical damping ratio is β is subjected to a harmonic excitation at frequency Ω. (a) Determine the value of Ω at which the amplitude of the steady state displacement is a maximum. (b) What are the amplitude and phase angle at the frequency found in part (a)? (c) For what range of values of β will an increase in the value of Ω always result in a decrease in the steady state amplitude?

VIII.42 In order to isolate an electronic instrument from the vibratory motion of the floor, it is mounted on a platform that is supported by four springs (only two are shown). The mass of the instrument and the platform is 25 kg and the stiffness of each spring is 12.5 kN/m. The upward displacement of the floor is $h = 5 \sin \Omega t$, where h is in millimeters. (a)

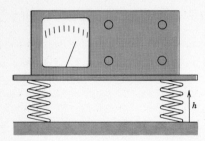

Prob. VIII.42

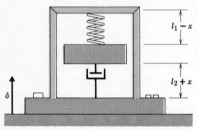

Prob. VIII.44

Letting x denote the upward displacement of the platform from the static equilibrium position when the floor is at rest, determine the steady state expression for x. (b) Determine the maximum acceleration of the instrument when the floor is vibrating at 6.5 Hz and at 7.5 Hz.

VIII.43 Solve Problem VIII.42 if each of the springs is concentric with a dashpot whose constant is 250 N-s/m.

VIII.44 Instrumentation within the accelerometer shown measures the displacement x of the block relative to the surrounding case as an indication of the vertical displacement δ of the floor. In a certain situation the block is initially at rest in the static equilibrium position when the base is given a constant upward acceleration $\ddot{\delta}$. Determine the expression for $x(t)$ in this case and plot the nondimensional displacement parameter $x\omega_n^2/\ddot{\delta}$ in terms of the nondimensional time parameter $\omega_n t$ (a) for no damping, (b) for 50% of critical damping.

VIII.45 In another case of excitation, the base of the accelerometer in Problem VIII.44 is given a harmonic motion $\delta = \delta_0 \cos \Omega t$. Determine the steady state displacement x and sketch diagrams of the amplitude magnification factor x_0/δ_0 and the phase angle ψ in terms of the frequency parameter Ω/ω_n. The system is lightly damped ($\beta \ll 1$).

VIII.46 A system whose natural frequency is ω_n and whose ratio of critical damping is β is subjected to a harmonic force $F_0 \cos \Omega t$. Determine (a) the instantaneous power developed by the force as a function of time in the steady state motion of the system, (b) the total work done by the force in one cycle, (c) the average power developed by the force over a complete cycle. Compare the results of parts (a) and (c).

6 Vibrations of Rigid Bodies

The vibrating systems we have treated so far were successfully modeled as particles. However, as we saw in the basic studies of kinetics, in order to retain the essential features of a physical system it is frequently necessary to model the system as a rigid body. The techniques for investigating the vibration of rigid bodies about their static equilibrium position are essentially the same as for particles, consisting of the formulation and solution of the equation of motion.

There is a small complication that arises in account for the force of a spring that is attached to a specific point in a rotating body. In some respects this complication resembles that encountered in the study of the pendulum in Example 1. We say there that, for large angles of rotation of the pendulum, the equation of motion is nonlinear, and thus difficult to solve. By restricting the angle of rotation to be a small value, we were able to linearize the equation. Similarly, when dealing with rigid bodies that rotate, it will be necessary for us to restrict our attention to the case

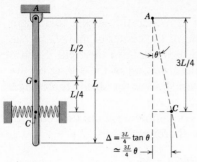

Figure 12a   Figure 12b

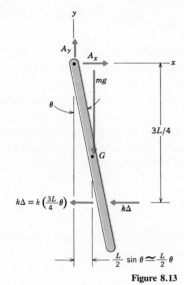

$k\Delta = k\left(\frac{3L}{4}\theta\right)$ $k\Delta$

$\frac{L}{2}\sin\theta \simeq \frac{L}{2}\theta$

Figure 8.13

of small rotations and displacements in order to have a linear equation of motion.

To see how such an approximation is made, consider the system in Figure 12a, where a slender bar may rotate in the vertical plane about the pin A. For small angles of rotation θ of the bar from the vertical position, shown in Figure 12a, we may consider point C to move only horizontally, so, in essence, the springs maintain their horizontal orientations. Furthermore, for small θ, $\tan\theta \simeq \sin\theta \simeq \theta$. Hence, according to Figure 12b, we may approximate the deformation Δ of the springs as

$$\Delta \simeq \frac{3L}{4}\theta$$

For this approximate analysis, the free body diagram of the bar is as shown in Figure 13. Notice that the free body diagram was drawn for a general position of the body, but all linearizing approximations regarding the smallness of the motion are employed.

Continuing, we obtain the equation of motion by summing moments about the z axis passing through the fixed point A, thereby eliminating the unknown reaction forces $\bar{A}_x$ and $\bar{A}_y$. Using the approximation for the length of the lever arm of the gravity force appearing in the free body diagram, this yields

$$\Sigma M_{Az} = -mg\left(\frac{L}{2}\theta\right) - 2k\left(\frac{3L}{4}\theta\right)\left(\frac{3L}{4}\right) = I_{zz}\alpha = \tfrac{1}{3}mL^2(\ddot{\theta})$$

$$\tfrac{1}{3}mL^2\ddot{\theta} + \left(\tfrac{9}{8}kL^2 + \tfrac{1}{2}mgL\right)\theta = 0$$

Thus, in this approximation, the system is equivalent to the model of an undamped freely vibrating oscillator, equation (3).

EXAMPLE 5

The gear shown in the diagram has a mass M and a radius of gyration r. It rolls over a fixed rack and is restrained by a spring. The gear is in static equilibrium in the position shown. Determine the natural frequency of the system.

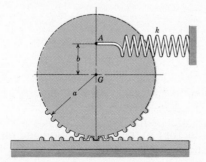

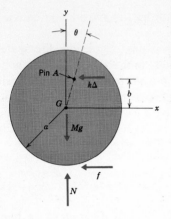

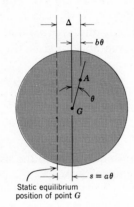

Static equilibrium
position of point G

Solution

For small motions of the gear, we can consider the spring to remain horizontal. Let us denote the displacement of the center of mass G by the variable s, with positive s representing a movement to the right. Such a movement will result in a compressive deformation of the spring by the quantity Δ. Hence in the free body diagram shown, the spring force is $k\Delta$ to the left.

The gear is in general motion, so we sum moments about the center of mass. The basic equations of motion are therefore

$$\Sigma F_x = -k\Delta - f = M(a_G)_x = M\ddot{s}$$

$$\Sigma F_y = N - Mg = M(a_G)_y = 0$$

$$\Sigma M_{Gz} = (k\Delta)b - fa = I_{zz}\alpha = \tfrac{1}{2}Mr^2\alpha$$

In these three equations there are five unknowns: Δ, f, s, N, and α. To find additional relations we use the kinematical constraint that the gear rolls without slipping, which enables us to relate the rotation and the deformation of the springs to the displacement s. We do this by sketching the displaced position of the gear.

Note that in this sketch we employed the restriction of small displacement, allowing us to approximate the horizontal movement of point A with respect to point G as $b\theta$. Then, for rolling without slipping,

$$\theta = \frac{s}{a} \qquad \Delta = s + b\theta = s\left(1 + \frac{b}{a}\right)$$

Furthermore, because a positive θ is clockwise and a positive α is counterclockwise, in accord with the right-hand rule with respect to the z axis we have

$$\alpha = -\ddot{\theta} = -\frac{\ddot{s}}{a}$$

Substituting these expressions into the basic equations of motion yields

$$-ks\left(1 + \frac{b}{a}\right) - f = M\ddot{s}$$

$$N = Mg$$

$$ks\left(1 + \frac{b}{a}\right)b - fa = -\frac{1}{2}Mr^2\frac{\ddot{s}}{a}$$

Solving for f in the first equation and placing it into the third equation, we find that

$$f = -\left[ks\left(1 + \frac{b}{a}\right) + M\ddot{s}\right]$$

$$ks\left(1 + \frac{b}{a}\right)b + ks\left(1 + \frac{b}{a}\right)a = -\frac{1}{2}Mr^2\frac{\ddot{s}}{a}$$

We now simplify this result, and thus obtain the differential equation of motion.

$$\tfrac{1}{2}M\left(\frac{r}{a}\right)^2\ddot{s} + k\left(1 + \frac{b}{a}\right)^2 s = 0$$

This equation has the same form as that for the undamped free vibration of the system model in Figure 1. Hence, the natural frequency is

$$\omega_n = \sqrt{\frac{k(1 + b/a)^2}{M(r/a)^2}} = \left(\frac{1 + b/a}{r/a}\right)\sqrt{\frac{k}{M}}$$

$$= \frac{a + b}{r}\sqrt{\frac{k}{M}}$$

HOMEWORK PROBLEMS

VIII.47–VIII.54 Derive the linear differential equation of motion for the (small) angle θ measuring the rotation of the system shown from its static equilibrium position. Wherever applicable, assume that there is no slipping.

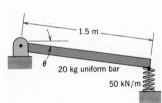

Prob. VIII.49

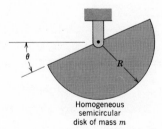

Homogeneous
semicircular
disk of mass m

Prob. VIII.47

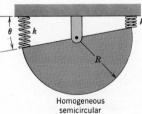

Homogeneous
semicircular
disk of mass m

Prob. VIII.48

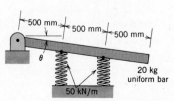

20 kg
uniform bar

50 kN/m

Prob. VIII.50

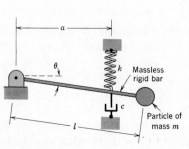

Prob. VIII.51

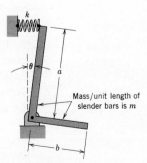

Prob. VIII.52

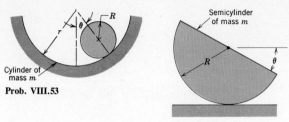

Prob. VIII.53

Prob. VIII.54

VIII.55–VIII.58 The system is shown in its static equilibrium position. Determine its natural frequency.

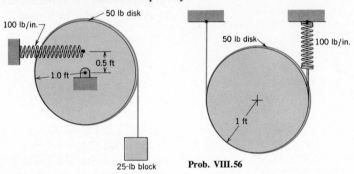

Prob. VIII.55

Prob. VIII.56

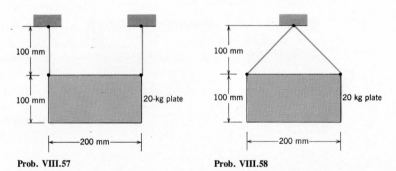

Prob. VIII.57

Prob. VIII.58

VIII.59 A flywheel is suspended from its center by a metal cable. When the flywheel is rotated by an angle θ about the axis of the cable, the cable exerts a restoring torque $k\theta$ opposite the sense of the rotation, due to its elasticity. The free vibration of the system is observed to have a period τ. Derive an expression for the mass moment of inertia of the flywheel about the axis of the cable.

Prob. VIII.59

VIII.60 and VIII.61 Determine the natural frequency of each of the systems shown. Are there any values of the spring stiffness k for which either system will not oscillate freely about the vertical position? Discuss the meaning of these values of k with regard to the static stability of the vertical position.

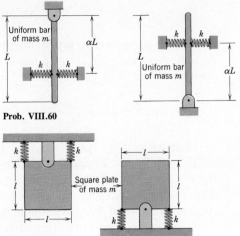

Prob. VIII.60

Prob. VIII.61

B. ORBITAL MOTION

Newton's gravitational law postulates that the force of attraction between two bodies is a function of the distance between them. In our earlier studies we investigated the ballistic motion of a particle for the case where the particle was close to the earth's surface. Thus, the distance from the earth's surface to the particle was small compared to the radius of the earth. This allowed us to approximate gravity as a constant force, thereby simplifying the formulation and solution of the equations of motion.

Our aim here is to account for the varying magnitude and direction of the gravitational force acting on a particle that is located sufficiently far from the attracting body's mass center to make the force a variable quantity.

To place the problem in perspective, consider the gravitational force $\bar{F}_G$ exerted between the two bodies shown in Figure 14.

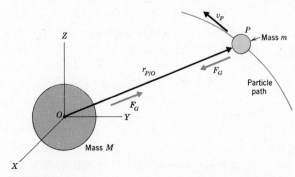

Figure 8.14

According to Newton's law of gravitation,

$$\left|\bar{F}_G\right| = \frac{GMm}{\left|\bar{r}_{P/O}\right|^2}$$

where G is the universal gravitational constant. Also, the attractive force $\bar{F}_G$ acting on the particle at point P is in the opposite direction of the position vector $\bar{r}_{P/O}$. Putting these two facts together, we have

$$\bar{F}_G = \frac{GMm}{\left|\bar{r}_{P/O}\right|^2}\left(-\frac{\bar{r}_{P/O}}{\left|\bar{r}_{P/O}\right|}\right) = -\frac{GMm}{\left|\bar{r}_{P/O}\right|^3}\bar{r}_{P/O} \qquad (23)$$

Let us restrict the discussion to the case where the mass M is sufficiently large compared to the smaller mass m that the gravitational force $-\bar{F}_G$ exerted on M by m has a negligible effect on the body M. This allows us to consider the center point O of body M to be fixed. As a result, the line of action of the gravitational force acting on mass m always passes through the fixed point O. The model we now have, where the force of attraction between two bodies is along the line of their centers, with one center fixed, is called *central force motion*.

EXAMPLE 6
Determine the value of GM in Newton's gravitational law when the attracting body is the earth.

Solution
Newton's gravitational law is

$$\left|\bar{F}_G\right| = \frac{GMm}{\left|r_{P/O}\right|^2}$$

If the attracting body is the earth, then, at the surface of the earth, $\left|\bar{F}_G\right|$ is also mg, and $\left|\bar{r}_{P/O}\right| = R_e$, where R_e is the mean radius of the earth, $R_e = 6370$ km $= 3960$ miles. Thus

$$mg = \frac{GMm}{R_e^2}$$

so

$$GM = gR_e^2 = (9.806)(6370)^2(1000)^2 = 3.979(10^{14})\,\frac{m^3}{s^2}$$

$$= (32.17)(3960)^2(5280)^2 = 1.4064(10^{16})\,\frac{ft^3}{s^2}$$

The value of GM is called the *intensity of the force*.

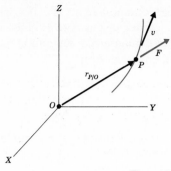

Figure 15

1 Central Force Motion

In this section we shall consider the motion of a particle under the influence of a general central force $\bar{F}$, whose magnitude may be a variable quantity. In Figure 15, point O is the fixed point through which this force always acts, and point P is the location of the particle.

Let the initial position and velocity of the particle be arbitrary and known. Given these initial conditions, we wish to determine the path the particle follows and the relationships that exist among its kinematical variables.

We begin by noting that in central force motion the moment of $\bar{F}$ about the origin O is zero. Hence, the angular impulse-momentum principle for a single particle, formulated about the fixed point O, yields

$$\Sigma \bar{M}_O \equiv \bar{0} \equiv \dot{\bar{H}}_O = \frac{d}{dt}(\bar{r}_{P/O} \times m\bar{v}_P) \tag{24}$$

It immediately follows from equation (24) that the angular momentum $\bar{H}_O$ is constant throughout the motion. For notational convenience let us divide $\bar{H}_O$ by the constant mass m to obtain

$$\bar{h}_O \equiv \frac{\bar{H}_O}{m} = \bar{r}_{P/O} \times \bar{v}_P \equiv \text{constant} \tag{25}$$

where $\bar{h}_O$ is called the angular momentum per unit mass.

In order for the vector $\bar{h}_O$ in equation (25) to be constant, both its magnitude and direction must be constant. Recalling the definition of the direction of the cross product, from equation (25) we see that this constant vector is normal to the plane formed by $\bar{r}_{P/O}$ and $\bar{v}_P$. It follows then that $\bar{h}_O$ can have a constant orientation only if the plane of $\bar{r}_{P/O}$ and $\bar{v}_P$ is constant. Hence, we deduce that

---

A particle moving under the influence of a central force is in planar motion.

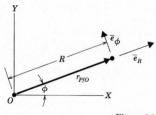

Figure 16

Considering that the central force in this planar motion is always parallel to the radial line from point O to point P, it is logical that, for further study, we employ cylindrical coordinates with $\bar{e}_R$ and $\bar{e}_\phi$ lying in the plane, as illustrated in Figure 16.

Setting $Z \equiv 0$ for planar motion, the kinematical formulas for the position and velocity of particle P in terms of cylindrical coordinates are

$$\bar{r}_{P/O} = R\bar{e}_R \qquad \bar{v}_P = \dot{R}\bar{e}_R + R\dot{\phi}\bar{e}_\phi$$

Substituting these expressions into equation (25) yields

$$\bar{h}_O = R\bar{e}_R \times (\dot{R}\bar{e}_R + R\dot{\phi}\bar{e}_\phi) = R^2\dot{\phi}\bar{K}$$

from which we note that h_O has a single component in the (fixed) $\bar{K}$ direction, which is in agreement with the earlier discussion. Equally important, from the initial conditions we can evaluate the magnitude h_O of the vector $\bar{h}_O$, which is a constant for the entire motion.

$$h_O = |\bar{h}_O| = R_1^2 \dot{\phi}_1 = R^2 \dot{\phi} \tag{26}$$

This result has a geometric interpretation. Consider the particle in two positions that are very close to each other, as illustrated in Figure 17. Let us compute the shaded area $d\mathcal{A}$ in Figure 17, which is the area swept out by the position vector in the time interval from t to $t + dt$. As shown in the figure, the base of the triangle is R and the height is $R\, d\phi$, so that

$$d\mathcal{A} = \tfrac{1}{2}R(R\, d\phi)$$

Dividing by dt and comparing the result to equation (26) gives

$$\frac{d\mathcal{A}}{dt} = \frac{1}{2}R^2 \frac{d\phi}{dt} = \frac{1}{2} R^2 \dot{\phi} = \frac{1}{2} h_O \tag{27}$$

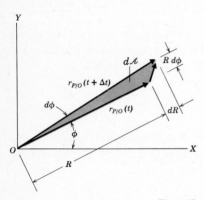

Figure 17

The quantity $d\mathcal{A}/dt$ is called the *areal velocity*. It is the rate at which the position vector sweeps out area. Hence, because h_O is a constant, we see that the areal velocity is constant in central force motion.

We now wish to determine the precise planar path of a particle undergoing central force motion. The central force $\bar{F}$ has only one component (in the radial direction), so from Newton's second law we find the scalar equations of motion to be

$$\Sigma F_R = F = ma_R = m(\ddot{R} - R\dot{\phi}^2)$$

$$\Sigma F_\phi = 0 = ma_\phi = m(R\ddot{\phi} + 2\dot{R}\dot{\phi})$$

$$\Sigma F_Z \equiv 0 \tag{28}$$

Equation (26) may now be used to eliminate $\dot{\phi}$ from the two nontrivial equations of motion. Solving for $\dot{\phi}$ we have

$$\dot{\phi} = \frac{h_O}{R^2}$$

Upon differentiating, this yields

$$\ddot{\phi} = -\frac{2h_O}{R^3} \dot{R}$$

Thus the equations of motion become

$$F = m\left[\ddot{R} - R\left(\frac{h_o}{R^2}\right)^2\right]$$

$$0 = m\left[R\left(-\frac{2h_o}{R^3}\dot{R}\right) + 2\dot{R}\left(\frac{h_o}{R^2}\right)\right] \equiv 0$$

Note that the second equation (the transverse force component) is also trivial. This is a consequence of the fact that the angular impulse-momentum principle for a single particle is equivalent to the transverse component of Newton's second law (see Module III). We are now left with a single scalar equation of motion, the solution to which defines the particle's path.

$$\ddot{R} - \frac{h_o{}^2}{R^3} = \frac{F}{m} \tag{29}$$

Equation (29) is a nonlinear differential equation that is not readily solvable. To facilitate its solution let us transform the equation from its present form, relating R and t, to a simpler one relating R and ϕ.

Considering R to be a function of ϕ, from equation (26) and the chain rule we can form

$$\dot{R} \equiv \frac{dR}{dt} = \frac{d\phi}{dt}\frac{dR}{d\phi} = \frac{h_o}{R^2}\frac{dR}{d\phi}$$

and

$$\ddot{R} \equiv \frac{d}{dt}(\dot{R}) = \frac{d\phi}{dt}\frac{d\dot{R}}{d\phi} = \frac{h_o}{R^2}\frac{d}{d\phi}\left(\frac{h_o}{R^2}\frac{dR}{d\phi}\right)$$

For convenience, we introduce a new variable $q = 1/R$ so that

$$\frac{dR}{d\phi} = \frac{d}{d\phi}\left(\frac{1}{q}\right) = -\frac{1}{q^2}\frac{dq}{d\phi}$$

The expression for $\ddot{R}$ now becomes

$$\ddot{R} = h_o q^2\frac{d}{d\phi}\left[h_o q^2\left(-\frac{1}{q^2}\frac{dq}{d\phi}\right)\right] = -h_o{}^2 q^2\frac{d}{d\phi}\left(\frac{dq}{d\phi}\right)$$

Now substituting these expressions for R and $\ddot{R}$ into equation (29) yields

$$-h_o{}^2 q^2\frac{d^2q}{d\phi^2} - h_o{}^2 q^3 = \frac{F}{m}$$

$$\frac{d^2q}{d\phi^2} + q = -\frac{F}{mh_o{}^2 q^2} \tag{30}$$

Equation (30) is the general form of the equation for the path of a particle undergoing central force motion. To proceed further we must define the functional dependence of the central force. Our concern here is with the motion of a particle under the influence of gravity.

2   Gravitational Central Force Motion — Conic Sections

Restricting our attention to the case where the central force is the gravitational force defined by equation (23), in cylindrical coordinates we have

$$\bar{F} = -\frac{GMm}{R^2}\,\bar{e}_R$$

Upon replacing $1/R$ by q and substituting this force expression into the basic equation (30), we obtain

$$\frac{d^2q}{d\phi^2} + q = -\frac{1}{mh_o{}^2 q^2}(-GMmq^2) = \frac{GM}{h_o{}^2} \tag{31}$$

This is a linear, nonhomogeneous, second-order differential equation with constant coefficients. Its solution is a principal topic in a first course on differential equations. It also is essentially the same as the equation of motion of the vibrating systems considered in Section A of this module, so we shall not repeat the steps in the derivation of the solution. Rather, we call attention to the fact that it is readily verified by direct substitution that the general solution to equation (31) is

$$q = \frac{1}{R} = C\cos(\phi - \beta) + \frac{GM}{h_o{}^2} \tag{32}$$

This is the equation for the *trajectory* of a particle undergoing gravitational central force motion. The constants C and β are constants of integration that, as we shall see, are determined from the initial conditions.

Although it is not obvious, equation (32) is the equation in polar coordinates of the curve formed from a conic section. Hence, it will prove beneficial to review briefly some properties of a conic section. To construct the conic section of Figure 18 on the following page, we first choose a fixed point O, called the *focus,* and a fixed line, called the *directrix*. The line OQ', perpendicular to the directrix and passing through the focus, is the axis of symmetry of the conic section.

A conic section has the property that the radial distance R from the focus to any point P on the curve is in a constant ratio to the perpendicular distance d from the directrix to point P; that is,

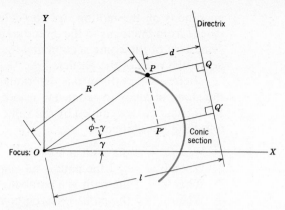

Figure 18

$$e = \frac{R}{d} \tag{33}$$

where the constant e, called the *eccentricity*, is always greater than or equal to zero.

For a specific choice of the focus and directrix, we can determine the constant angle γ and the constant distance l shown in Figure 18 by constructing the line OQ'. Hence, let us express the distance d in terms of these parameters. From the figure we see that

$$l = \left| \bar{r}_{P'/O} \right| + \left| \bar{r}_{Q'/P'} \right| = R \cos(\phi - \gamma) + d$$

Solving this expression for d and substituting the result into equation (33) yields

$$e = \frac{R}{l - R \cos(\phi - \gamma)}$$

or

$$\frac{1}{R} = \frac{1}{el} \left[1 + e \cos(\phi - \gamma) \right] \tag{34}$$

Comparing this expression for R to equation (32) for the gravitational motion gives

$$\boxed{\gamma = \beta \quad , \quad \frac{1}{l} = C \quad , \quad \frac{1}{el} = \frac{GM}{h_0^2}} \tag{35}$$

Thus, once the constants, h_0, β, and C are evaluated from the initial conditions on the motion, we can determine the eccentricity e, the distance l from the focus to the directrix, and the angle γ between the axis of symmetry of the conic section and the X axis. These computations fully define the path of the particle.

Before we perform this determination, let us recall some results of the study of analytic geometry, where it is shown that the type of conic section can be determined by the value of the eccentricity e. We have the following possibilities:

1 When $e = 0$ the path is a circle.
2 When $0 \le e < 1$ the path is an ellipse.
3 When $e = 1$ the path is a parabola.
4 When $e > 1$ the path is a hyperbola.

The ellipse is the only closed conic section (considering the circle as a special ellipse). It is of particular interest to us because many space vehicles, as well as the planets, follow elliptical *orbits*. Basic properties of an elliptical path are the lengths of the semimajor and semiminor axes, which are the values a and b shown in Figure 19. From this figure we see that the minimum and maximum values of R occur when the particle is at point A ($\phi = \gamma$) and when it is at point B ($\phi = 180° + \gamma$), and also that $2a = R_A + R_B$. Solving equation (34) for the values of R_A and R_B yields

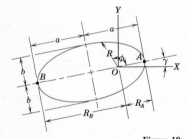

Figure 19

$$\frac{1}{R_A} = \frac{1}{el}(1 + e) \qquad \frac{1}{R_B} = \frac{1}{el}(1 - e)$$

Now, using equation (35) to eliminate the term el gives

$$R_A = \frac{h_o{}^2}{GM}\left(\frac{1}{1 + e}\right) \qquad R_B = \frac{h_o{}^2}{GM}\left(\frac{1}{1 - e}\right)$$

Hence

$$\boxed{\begin{aligned} a &= \frac{1}{2}(R_A + R_B) = \frac{h_o{}^2}{GM}\left(\frac{1}{1 - e^2}\right) \\ R_A &= a(1 - e) \qquad R_B = a(1 + e) \end{aligned}} \tag{36a}$$

The value of b can then be shown to be

$$\boxed{b = a\sqrt{1 - e^2}} \tag{36b}$$

In the case of orbital motion about the earth, point A, where R is a minimum, is called the *perigee*, and point B, where R is a maximum, is called the *apogee*.

3 Initial Conditions

To aid in our understanding of the role of the initial conditions, consider the problem of placing a satellite into orbit, as shown in Figure 20. At the completion of the powered portion of the flight, the space vehicle is at a radial distance R_1 and has a velocity $\bar{v}_1$ directed at an angle θ_1 from the radial line.

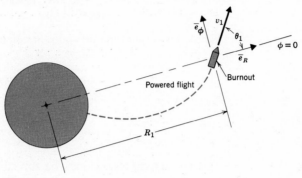

Figure 20

As shown in the figure, ϕ is measured from the initial position, so the initial conditions are stated as

$$R = R_1 \quad \text{and} \quad \bar{v} = v_1(\cos \theta_1 \, \bar{e}_R + \sin \theta_1 \, \bar{e}_\phi) \qquad \text{when } \phi = 0$$

However, when we match the above velocity expression to the cylindrical coordinate expression for velocity, we may rewrite the initial conditions in scalar form as

$$R = R_1 \quad , \quad \dot{R}_1 = v_1 \cos \theta_1 \quad , \quad \dot{\phi}_1 = \frac{v_1 \sin \theta_1}{R_1} \qquad \text{when } \phi = 0 \qquad (37)$$

The task now is to determine the constants in the general solution for the trajectory, equation (32), which match these initial conditions. We do this by first using equation (26) to determine the angular momentum per unit mass. This gives

$$\boxed{h_O = R_1{}^2 \, \dot{\phi}_1 = R_1 v_1 \sin \theta_1} \qquad (38)$$

Next, differentiating equation (32) yields an expression for $\dot{R}$.

$$\frac{d}{dt}\left(\frac{1}{R}\right) \equiv -\frac{1}{R^2}\dot{R} = \frac{d}{dt}\left[C\cos(\phi - \beta) + \frac{GM}{h_O{}^2}\right]$$

$$= -C\dot{\phi}\sin(\phi - \beta) = -C\left(\frac{h_O}{R^2}\right)\sin(\phi - \beta)$$

When we set $\phi = 0$ and substitute equations (37) for R and $\dot{R}$ at the initial position, we obtain

$$\frac{1}{R_1} = C \cos \beta + \frac{GM}{h_0^2}$$

$$-\frac{1}{R_1^2} v_1 \cos \theta_1 = C\left(\frac{h_0}{R_1^2}\right) \sin \beta$$

Employing equation (38) and rearranging terms, the foregoing expressions become

$$C \cos \beta = \frac{1}{R_1} - \frac{GM}{h_0^2}$$

$$C \sin \beta = -\frac{v_1}{h_0} \cos \theta_1 = -\frac{1}{R_1} \cos \theta_1$$

(39)

These expressions may be solved for the values of C and β. In doing so, special care must be taken to place the value of β in the correct quadrant.

Once the value of C is computed, we may determine the properties of the resulting trajectory. In particular, we are interested in the eccentricity. From equations (35) we have

$$e = \frac{h_0^2}{GM} \frac{1}{l} = \frac{h_0^2}{GM} C$$

A formula for e in terms of the initial conditions may now be found by squaring each of the equations (39) and adding them to solve for C^2. This gives

$$C^2(\cos^2 \beta + \sin^2 \beta) = C^2 = \left(\frac{1}{R_1} - \frac{GM}{h_0^2}\right)^2 + \frac{1}{R_1^2} \cot^2 \theta_1$$

$$= \frac{1}{R_1^2} \frac{1}{\sin^2 \theta_1} - \frac{2GM}{h_0^2} \frac{1}{R_1} + \left(\frac{GM}{h_0^2}\right)^2$$

Using equation (38) we then have

$$C = \sqrt{\frac{v_1^2}{h_0^2} - 2\frac{GM}{h_0^2} \frac{1}{R_1} + \left(\frac{GM}{h_0^2}\right)^2}$$

The earlier expressions for the eccentricity then becomes

$$e = \frac{h_0^2}{GM} C = \frac{h_0}{GM} \sqrt{v_1^2 - 2\frac{GM}{R_1} + \left(\frac{GM}{h_0}\right)^2}$$

(40)

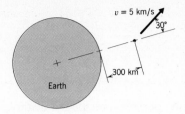

v = 5 km/s
30°
300 km
Earth

EXAMPLE 7

A satellite is placed into its orbit at an altitude of 300 km above the earth's surface at a speed of 5 km/s and an angle $\theta = 30°$, as illustrated. Determine the equation for its trajectory and show that its orbit is elliptical. Also determine the perigee and apogee distances.

Solution

Remember that the radial distance R is the distance from the focus of the central force. Taking the earth's radius to be 6370 km, the given initial conditions are that, when $\phi = 0$,

$$R_1 = 6370 + 300 = 6670 \text{ km}$$

$$v_1 = 5 \text{ km/s} \qquad \theta_1 = 30°$$

The two key parameters for determining the orbit are the angular momentum h_O and the eccentricity e. From equation (38) we have

$$h_O = R_1 v_1 \sin \theta_1 = 6670(5) \sin 30° = 16675 \text{ km}^2/\text{s}$$

Equation (40) then gives the eccentricity and the value of the coefficient C.

$$e = \frac{h_O}{GM} \sqrt{v_1^2 - 2\frac{GM}{R_1} + \left(\frac{GM}{h_O}\right)^2} = \frac{h_O^2}{GM} C$$

Using the value of GM for the earth determined in Example 6, the eccentricity is

$$e = \frac{16675}{3.979(10^5)} \sqrt{(5)^2 - \frac{2(3.979)(10^5)}{6670} + \frac{(3.979)^2(10^{10})}{(16675)^2}} = 0.9134$$

Hence, because $0 < e < 1$, the trajectory is an ellipse.

To describe the equation of the ellipse we need the values of C and β. From the foregoing,

$$C = \frac{GM}{h_O^2} e = \frac{3.979(10^5)}{(16675)^2} (0.9134) = 1.3071(10^{-3}) \text{ 1/km}$$

and from equations (39),

$$\cos \beta = \frac{1}{C}\left(\frac{1}{R_1} - \frac{GM}{h_O^2}\right) = -0.9801$$

$$\sin \beta = -\frac{v_1 \cos \theta_1}{h_O C} = -0.1987$$

Then, because the sine and cosine are both negative, β is in the third quadrant, so

$$\beta = -168.54°$$

Equation (32) for the orbit then gives

$$\frac{1}{R} = C\cos(\phi - \beta) + \frac{GM}{h_0{}^2}$$

$$= 1.3071(10^{-3})\cos(\phi + 168.54°) + 1.4310(10^{-3})\ 1/km$$

Finally, we use equations (36a) to find the perigee and apogee distances. Thus,

$$a = \frac{h_0{}^2}{GM}\frac{1}{(1 - e^2)} = \frac{1}{1.4310(10^{-3})}\frac{1}{[1-(0.9134)^2]} = 4217 \text{ km}$$

$$R_{\text{perigee}} = a(1 - e) = 4217(1 - 0.9134) = 365 \text{ km}$$

$$R_{\text{apogee}} = a(1 + e) = 4217(1 + 0.9134) = 8069 \text{ km}$$

The fact that the perigee distance is smaller than the radius of the earth means that the satellite will crash into the earth before it reaches its perigee; the launch was unsuccessful.

HOMEWORK PROBLEMS

VIII.62 A satellite is launched at an altitude of 500 km above the earth's surface with a velocity of 9 km/s parallel to the earth's surface. Determine the eccentricity of the orbit and the maximum and minimum altitudes for the satellite's trajectory.

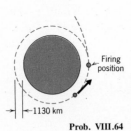

Prob. VIII.62

VIII.63 The altitude of a satellite at apogee is 4040 miles and at perigee it is 540 miles. Determine the equation of the trajectory and the velocity of the satellite at apogee and at perigee.

VIII.64 A satellite in circular orbit at an altitude of 1130 km is to be given a new orbit. To do this the engines are fired tangent to the orbit, imparting an additional velocity of 4 km/s to the satellite. Determine the equation for the new orbit and the eccentricity of the orbit. Is the orbit open or closed?

Prob. VIII.64

VIII.65 Solve Problem VIII.64 if the engines are aligned radially, giving the satellite a radial velocity component of 4 km/s outward.

VIII.66 A satellite in circular orbit at an altitude of 240 miles is to be moved to a higher coplanar circular orbit at an altitude of 6040 miles. To accomplish this the engines are fired tangentially at point P_1, increasing the speed by Δv_1, thus placing the satellite into an elliptical orbit. (This orbit is called the Hohmann transfer ellipse.) At point P_2 the rockets are again fired tangentially, increasing the speed by Δv_2, and thus placing the satellite in the desired circular orbit. Determine Δv_1 and Δv_2.

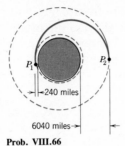

Prob. VIII.67

Prob. VIII.66

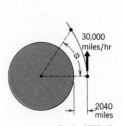

Prob. VIII.68

VIII.67   A satellite is in a circular orbit at an altitude of 630 km when retrorockets are fired tangent to the orbit, slowing the rocket by 5000 km/hr. Determine the angle ϕ locating the point S where the satellite will splash down. Neglect air resistance.

VIII.68   At an altitude of 2040 miles above the earth's surface, a space probe is launched into a hyperbolic trajectory, giving it a velocity of $3(10^4)$ miles/hr tangent to the earth's surface. Determine the largest value of the angle ϕ that is possible for this trajectory.

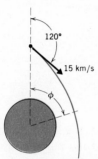

Prob. VIII.69

VIII.69   At an altitude of 30,000 km above the surface of the earth, a meteorite has a velocity of 15 km/s at $\theta = 120°$. Determine (a) the eccentricity of the trajectory of the meteorite, (b) the minimum altitude for the trajectory, (c) the angle ϕ at the position of closest approach. Will the meteorite enter the earth's atmosphere?

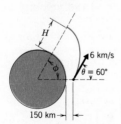

Prob. VIII.70

VIII.70   The engine of a missile is shut off at an altitude of 150 km, at which time the velocity is 6 km/s at an angle $\theta = 60°$. Determine the equation for the trajectory, the maximum altitude H attained by the missile, and the polar angle ϕ at this location.

VIII.71   Determine the angle ϕ defining the location where the missile in Problem VIII.70 returns to the surface of the earth. Air resistance is negligible.

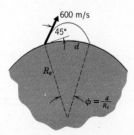

Prob. VIII.71

VIII.72   A bullet is fired from the surface of the earth at 600 m/s at an angle of elevation of 45°. (a) Using the orbit equations, determine the distance d locating where the bullet returns to the earth. (b) Compare the result of part (a) to that obtained when the earth is considered flat and the gravity force is considered constant.

4 Energy

In the preceding section we saw how the trajectory of a particle is determined. The evaluation of the velocity of the particle when it is at a specific position along its trajectory requires further development. There are many ways in which we may proceed. Here we choose the work-energy principles, for aside from being a direct approach, it is a good tool for gaining further insight into the orbital motion problem.

When the gravitational force is defined by the inverse square law, equation (23), the potential energy of this conservative force was shown in Module III to be

$$V_{\text{gr}} = -\frac{GMm}{R}$$

The only force acting on the particle is gravity, and hence mechanical energy is conserved; that is,

$$T + V_{\text{gr}} = \text{constant}$$

Denoting the constant energy per unit mass as E, we then have

$$E \equiv \frac{1}{m}(T + V_{\text{gr}}) = \frac{1}{2}v_P{}^2 - \frac{GM}{R} \qquad (41)$$

The value of E may be obtained from the initial conditions. Then the energy relation, equation (41), will yield the particle speed at any other value of R. The radial component of velocity v_R may then be determined by noting that the speed is the magnitude of the velocity vector. Thus

$$v_P{}^2 = v_R{}^2 + v_\phi{}^2$$

Because $v_\phi = R\dot{\phi}$, we find from equation (26) that

$$v_\phi = R\dot{\phi} = R\left(\frac{h_o}{R^2}\right) = \frac{h_o}{R}$$

$$v_R = \pm \sqrt{v_P{}^2 - \left(\frac{h_o}{R}\right)^2} \qquad (42)$$

Hence, equations (41) and (42) define the velocity as a function of R.

The parameter E assumes greater meaning when we substitute E from equation (41) into equation (40) for the eccentricity of the orbit, for we then find that

$$e = \frac{h_o}{GM} \sqrt{2E + \left(\frac{GM}{h_o}\right)^2}$$

or

$$e = \sqrt{1 + 2E\left(\frac{h_O}{GM}\right)^2}$$

(43)

Recalling the types of trajectories associated with the different ranges of values of the eccentricity, we see that:

1   Circular trajectory: $e = 0$, so $E = -0.5(GM/h_O)^2$
2   Elliptical trajectory: $0 < e \leq 1$, so $-0.5(GM/h_O)^2 \leq E < 0$
3   Parabolic trajectory: $e = 1$, so $E = 0$
4   Hyperbolic trajectory: $e > 1$, so $E > 0$

Thus, the energy level E is a direct indicator of the type of trajectory.

An important consequence of the foregoing is the concept of the *escape speed,* which is the minimum speed that a body must have in order to be in an open trajectory that extends to an infinite value of R. Because the parabola is the open trajectory having the lowest energy level, we set $E = 0$ to find v_{escape}. This yields

$$v_{escape} = \sqrt{\frac{2GM}{R}}$$

(44)

EXAMPLE 8
The system and initial conditions presented in Example 7 are repeated in the accompanying diagram. Once again, determine the perigee and apogee distances, but now use the energy and angular momentum equations derived in this section.

Solution
The given initial conditions allow us to compute the energy parameter E and angular momentum parameter h_O, as follows.

$$E = \frac{1}{2}v_1^2 - \frac{GM}{R_1} = \frac{1}{2}(5)^2 - \frac{3.979(10^5)}{(6370 + 300)}$$

$$= -47.16 \text{ km}^2/\text{s}^2$$

$$h_O = R_1 v_1 \sin\theta_1 = (6370 + 300)5 \sin 30°$$

$$= 16675 \text{ km}^2/\text{s}$$

Equation (41) then gives

$$-47.16 = \frac{1}{2}v_P^2 - \frac{3.979(10^5)}{R}$$

whereas equation (42) gives

$$v_\phi = \frac{16675}{R}$$

Now, imposing the condition that the value of R corresponds to the perigee or apogee (referring to Figure 19), we see that the tangent to the path at these two positions is perpendicular to the radial line. Hence, at these locations $v_R = 0$, so $v_P = v_\phi$. Thus, we have

$$v_P = \frac{16675}{R}$$

Substituting this into the energy equation gives

$$-47.16 = \frac{1}{2}\left(\frac{16675}{R}\right)^2 - \frac{3.979(10^5)}{R}$$

This expression simplifies to a quadratic equation, called the *absidial quadratic,* whose roots are the perigee and apogee distances.

$$R^2 - \frac{3.979(10^5)}{47.16}R + \frac{1}{2}\frac{(16675)^2}{47.16} = 0$$

$$R = 365 \text{ or } 8073 \text{ km} \qquad \triangleleft$$

The smaller value corresponds to perigee. To the accuracy of our calculations these values are identical to those obtained in Example 7.

As an aside, it should be mentioned that the multitude of relations that have been derived often means that there is a choice of methods for solving a problem. No single approach is, in general, optimal.

EXAMPLE 9

A space probe is injected into orbit at position A (altitude of 100 miles) with a velocity that is perpendicular to the radial line and is 10% greater than the local escape velocity. Determine the altitude and velocity of the probe at position B, where the radial line from the center of the earth has swept out a 90° angle.

Solution

First, to avoid any error in units, let us convert the value of GM found in Example 6 to units of miles and hours. Thus,

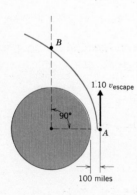

$$GM = 1.4064(10^{16})\ \frac{\text{ft}^3}{\text{s}^2}\left(\frac{3600\text{s}}{1\text{ hr}}\right)^2\left(\frac{1\text{ mile}}{5280\text{ ft}}\right)^3$$

$$= 1.2383(10^{12})\ \text{miles}^3/\text{hr}^2$$

We now list the given initial conditions, using equation (44) for v_{escape}.

Hence when $\phi = 0$,

$$R_1 = (3960 + 100) = 4060 \text{ miles}$$

$$v_1 = (v_\phi)_1 = 1.10 v_{\text{escape}}$$

$$= 1.10 \sqrt{\frac{2GM}{R_1}} = 2.717(10^4) \text{ miles/hr}$$

With these values we can determine the energy and angular momentum parameters, as follows.

$$E = \frac{1}{2} v_1{}^2 - \frac{GM}{R_1} = 6.410(10^7) \text{ miles}^2/\text{hr}^2$$

$$h_O = R_1(v_\phi)_1 = 4060(27170) = 1.1031(10^8) \text{ miles}^2/\text{hr}$$

To determine the distance for a specific value of ϕ we must first determine the coefficients C and β appearing in equation (32) for the trajectory. Because the probe is launched perpendicularly to the radial line ($\theta_1 = 90°$), the launch point A is on the axis of symmetry and $\beta = 0$. This is readily verified from the second of equation (39). Solving equation (40) for e and then for C yields

$$e = \frac{1.1031(10^8)}{1.2383(10^{12})} \left[(2.717)^2(10^8) - 2\frac{1.2383(10^{12})}{4060} + \frac{(1.2383)^2(10^{24})}{(1.1031)^2(10^{16})} \right]^{1/2} = 1.4204$$

$$C = \frac{1.2383(10^{12})}{(1.1031)^2(10^{16})} (1.4204) = 1.4455(10^{-4}) \text{ 1/miles}$$

Equation (32) for the trajectory then becomes, with $\beta = 0$,

$$\frac{1}{R} = C \cos \phi + \frac{GM}{h_O{}^2}$$

$$= 1.4455(10^{-4}) \cos \phi + 1.0176(10^{-4})$$

To find the radial distance of point B we simply substitute $\phi = 90°$. This gives

$$\frac{1}{R_B} = 1.0176(10^{-4}) \qquad R_b = 9827 \text{ miles}$$

so

$$\text{altitude} = 9827 - 3960 = 5867 \text{ miles} \qquad \qquad \triangleleft$$

We can now determine the velocity at this location by employing equations (41) and (42). The former equation yields

$$E = 6.410(10^7) \; \frac{\text{miles}^2}{\text{hr}^2} = \frac{1}{2} \, v_B{}^2 - \frac{GM}{R_B}$$

$$= \frac{1}{2} v_B{}^2 - \frac{1.2383(10^{12})}{9827}$$

Solving this equation, we obtain

$$v_B = 19500 \text{ miles/hr}$$

The first of equations (42) then yields

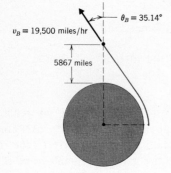

$$(v_\phi)_B = \frac{h_O}{R_B} = \frac{1.1031(10^8)}{9827} = 11,225 \text{ miles/hr}$$

Now, substituting this result into the second of equations (42), we find that

$$(v_R)_B = \sqrt{v_B{}^2 - (v_\phi)_B{}^2} = 15,945 \text{ miles/hr}$$

The resulting velocity is sketched below. The angle between $\bar{v}_B$ and the radial line is $\theta_B = \tan^{-1}[(v_\phi)_B/(v_R)_B] = 35.14°$

HOMEWORK PROBLEMS

VIII.73   An earth satellite has a speed of 5.5 miles/s at its minimum altitude of 400 miles. Determine (a) its angular momentum per unit mass, (b) its energy per unit mass, (c) that its trajectory is an ellipse, (d) its speed at apogee, and (e) its altitude at apogee.

VIII.74   A weather satellite is placed into orbit at an altitude of 330 km above the earth's surface, moving parallel to the earth's surface. Determine (a) the maximum initial velocity that will keep the satellite in the earth's gravitational field. (b) The initial speed that will cause the orbit to be a circle. (c) The minimum initial speed for a successful orbit. (The perigee for this orbit is the radius of the earth, so it is called the *grazing orbit*). Determine the eccentricity and energy per unit mass for such an orbit. (d) The initial speed that will result in a second (different) orbit of the same eccentricity as that of part (c), with the launch position now being the perigee. Find its energy per unit mass and altitude at apogee.

VIII.75   A meteorite at a distance of 60,000 km from the center of the earth has a velocity of 50,000 km/hr at $\theta = 165°$. Determine the velocity of the meteorite at a radial distance of 25,000 km.

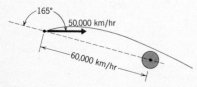

Prob. VIII.75

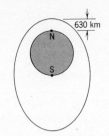

Prob. VIII.77

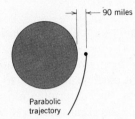

Prob. VIII.78

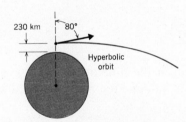

Prob. VIII.79

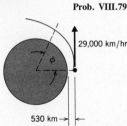

Prob. VIII.80

VIII.76   Determine the velocity of the meteorite in Problem VIII.75 at its closest approach to the earth.

VIII.77   A satellite is in a polar orbit of eccentricity 0.40 for which perigee occurs over the north pole at an altitude of 630 km. Determine the velocity of the satellite as it passes over (a) the south pole, (b) the equator.

VIII.78   A space probe is launched into a parabolic trajectory by shutting down the engines at an altitude of 90 miles, at which point the velocity is tangent to the earth's surface. Determine the velocity of the probe when it reaches an altitude of 10,000 miles.

VIII.79   Determine the rate of change of the speed of the satellite in Problem VIII.78 and the radius of curvature of the trajectory at an altitude of 10,000 miles.

VIII.80   A space probe is following a hyperbolic trajectory of eccentricity 1.50. The engines were shut down at an altitude of 230 km, at which point the velocity was 80° from the outward radial direction. Determine the velocity of the space probe when it is (a) 1000 km above the earth's surface, (b) 10,000 km above the earth's surface.

VIII.81   For the space probe of Problem VIII.80, determine the rate of change of the speed and the radius of curvature of the trajectory at each altitude.

VIII.82   At its closest approach to the earth a satellite is at an altitude of 530 km and has a speed of 36,000 km/hr. Determine the velocity of the satellite when $\phi = 60°$.

VIII.83   Solve Problem VIII.82 for $\phi = 120°$.

5   The Period of an Orbit

Our study of the dynamics of gravitational central force motion is almost completed. We have seen how the position parameters R and ϕ are related through the equations of the trajectory. We have also seen how the velocity at any position along the trajectory may be evaluated. An important parameter we have not yet discussed is time.

In general, the easiest way to bring time into the problem is to use the fact that the angular momentum is constant. From equation (26) we have

$$\dot{\phi} \equiv \frac{d\phi}{dt} = \frac{h_0}{R^2}$$

Eliminating R between this result and equation (32), we obtain a first-order differential equation for $\phi(t)$.

$$\frac{d\phi}{dt} = h_0 \left[C \cos (\phi - \beta) + \frac{GM}{h_0{}^2} \right]^2$$

where the values of C and β are known from the initial conditions. This equation is solved by separating variables and integrating to obtain

$$t = \frac{1}{h_0} \int_0^\phi \frac{d\phi}{[C \cos(\phi - \beta) + GM/h_0{}^2]^2} \qquad (45)$$

The integral may be evaluated with the aid of a table of integrals, but the result is too lengthy to present here.

A very important time parameter is the *period* of an elliptical orbit, that is, the length of time required for a particle to return to its initial position, thus completing one orbit. For this determination we recall equation (27), which shows that h_0 is twice the areal velocity $d\mathcal{A}/dt$. Thus $d\mathcal{A}/dt = \frac{1}{2}h_0$, and upon integrating.

$$\boxed{\mathcal{A} = \tfrac{1}{2}h_0 t} \qquad (46)$$

where $\mathcal{A}$ is the sectorial area swept out in the time interval t. For an elliptical trajectory, the area swept out in a full period τ is the area of an ellipse, which is

$$\mathcal{A} = \pi a b$$

where a and b are the lengths of the semimajor and semiminor axes, respectively. Their values may be found from equations (36). Using the foregoing area and setting $t = \tau$ in equation (46) yields

$$\tau = \frac{2\pi a b}{h_0}$$

At this juncture it is interesting to note the work of the early astronomer Johannes Kepler (1571–1630), who painstakingly studied our solar system. He postulated the following three laws governing planetary motion:

1   The orbit of the planets are ellipses for which the sun is one focus.
2   The areal velocity is constant.
3   The square of the period of the elliptical orbit is in a constant ratio to the cube of its semimajor axis.

We have already proven the first two laws, and the third law may be obtained by using equations (36) to express τ in terms of a. This gives

$$\tau = \frac{2\pi ab}{h_0} = \frac{2\pi a}{h_0} a\sqrt{1-e^2} = \frac{2\pi a^2}{h_0}\left(\frac{h_0}{\sqrt{GMa}}\right)$$

$$\boxed{\tau = \frac{2\pi}{\sqrt{GM}}\,a^{3/2}}$$ (47)

The remarkable aspect of Kepler's laws is that, although we obtained them as mathematical consequences of Newton's laws of gravity and motion, Kepler's work preceded Newton and was the result of careful observation and empiricism.

EXAMPLE 10

The perigee and apogee altitudes for an artificial satellite are 430 and 14,030 km, respectively. Determine the time required for the satellite to travel from point C on the semiminor axis of its elliptical orbit, past the perigee A, and then on to point D, which is the other point on the semiminor axis. Compare this result to the period of the orbit.

Solution

This problem could be solved by integrating equation (45), but a shortcut, based on the concept of the areal velocity, is available. This is so because the sectorial area bounded by the arc CAD and the radial lines OC and OD can be evaluated. In this approach we first determine the eccentricity e and the lengths a and b of the semiaxes of the ellipse.

It is given that

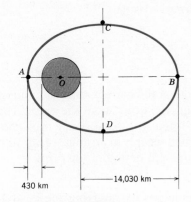

430 km |—— 14,030 km ——|

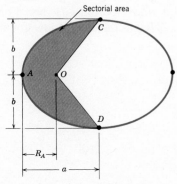

Sectorial area

$$R_A = 6{,}370 + 430 = 6{,}800$$

$$R_B = 6{,}370 + 14{,}030 = 20{,}400 \text{ km}$$

From equations (36) we then find

$$a = \frac{1}{2}(R_A + R_B) = 13{,}600 \text{ km}$$

$$e = \frac{R_B}{a} - 1 = 0.5$$

$$b = a\sqrt{1-e^2} = 11{,}778 \text{ km}$$

The first of equations (36a) provides the value of h_0. Thus

$$h_0 = \sqrt{GMa(1-e^2)} = \sqrt{3.974(10^5)13600(1-0.25)}$$

$$= 6.371(10^4) \text{ km}^2/\text{s}$$

The last step before using equation (46) to compute the time is the evaluation of the sectorial area $\mathcal{A}$, which is shaded in the sketch.

We find $\mathscr{A}$ by deducting the triangular area OCD from the area of the semiellipse. Thus

$$\mathscr{A} = \tfrac{1}{2}\pi ab - \tfrac{1}{2}(2b)(a - R_A) = 1.7152(10^8) \text{ km}^2$$

Denoting the desired time as t^*, equation (46) yields

$$t^* = \frac{2\mathscr{A}}{h_O} = 5384 \text{ s} = 1.496 \text{ hr}$$

In contrast, the orbital period is

$$\tau = \frac{2(\pi ab)}{h_O} = 15,797 \text{ s} = 4.38 \text{ hr}$$

Thus the time t^* is 32% less than $\tau/2$. Why isn't $t^* = \tau/2$?

EXAMPLE 11

The engines of the launch vehicle burn out at the correct altitude to inject a weather satellite into a stationary equatorial orbit about the earth. The satellite also has the proper speed at burnout for a stationary orbit, but due to a malfunction, the velocity at burnout is only at an angle of 80° with the radial line. Determine the eccentricity and period of the orbit.

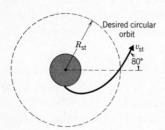

Solution

We must first determine the initial conditions by studying the desired stationary circular orbit, which is an orbit whose period is 1 day. For a circular orbit at radius R_{st} and speed v_{st}, Newton's second law for the radial direction gives

$$\Sigma F_R = -\frac{GMm}{R_{st}^2} = ma_R = -m\frac{v_{st}^2}{R_{st}}$$

so

$$v_{st} = \sqrt{\frac{GM}{R_{st}}}$$

Furthermore, for a circular orbit $a \equiv R_{st}$ and for a stationary orbit $\tau = 23.93$ hr. Hence, setting $GM = 3.979(10^5)$ km³/s², equation (47) yields

$$(R_{st})^{3/2} = \frac{\tau\sqrt{GM}}{2\pi} = \frac{23.93(3600)\sqrt{3.979(10^5)}}{2\pi}$$

$$R_{st} = 4.213(10^4) \text{ km}$$

$$v_{st} = \sqrt{\frac{GM}{R_{st}}} = 3.073 \text{ km/s}$$

These values are also the initial distance and speed for the actual orbit. Due to the 80° launch angle, the initial transverse speed is

$$(v_\phi)_1 = v_{st} \sin 80° = 3.026 \text{ km/s}$$

Thus the value of h_0 for the orbit is

$$h_0 = R_1(v_\phi)_1 = 4.213(10^4)(3.026)$$

$$= 1.2749(10^5) \text{ km}^2/\text{s}$$

and the corresponding value of E is

$$E = \frac{1}{2}(v_1)^2 - \frac{GM}{R_1} = \frac{1}{2}(3.073)^2 - \frac{3.979(10^5)}{4.213(10^4)}$$

$$= -4.723 \text{ km}^2/\text{s}^2$$

We may now compute the eccentricity from equation (43). This yields

$$e = \sqrt{1 + 2(-4.723)\left(\frac{1.2749}{3.979}\right)^2} = 0.1740 \qquad \triangleleft$$

To compute the period, we first evaluate the semimajor axis using equation (36a).

$$a = \frac{h_0{}^2}{GM(1 - e)^2} = \frac{(1.2749)^2(10^{10})}{3.979(10^5)[1 - (0.1740)^2]}$$

$$= 4.213(10^4) \text{ km}$$

The period is then found from equation (47).

$$\tau = \frac{2\pi}{\sqrt{3.979(10^5)}}[4,213(10^4)]^{3/2}$$

$$= 8.615(10^4) \text{ s} = 23.93 \text{ hr} \qquad \triangleleft$$

The fact that a is the same as R_{st}, thus giving the same period for the stationary and elliptical orbits, is not coincidental. By using equation (43) to eliminate the eccentricity from equation (36a), we can show that the value of a depends only on the energy level E. Hence, we can conclude that the period is a function only of the magnitude of the initial velocity, not its direction. Nevertheless, the elliptical orbit in this problem is not stationary. Why?

HOMEWORK PROBLEMS

VIII.84 The altitude of an artificial satellite at apogee is 1230 km and at perigee it is 330 km. Determine the period of the orbit.

VIII.85   A satellite is in an elliptical orbit around the earth whose eccentricity is 0.90. The perigee altitude is 540 miles. Determine the period of the orbit.

VIII.86   The eccentricity of the earth's orbit is 0.0167 and its mean distance from the sun (the average of the perigee and apogee distance) is $1.495(10^8)$ km. Taking the earth's period as 365.3 days, determine (a) the angular momentum per unit mass for the earth, and (b) the maximum and minimum distances from the earth to the sun.

VIII.87   Determine the approximate time required for the earth to traverse from one end of the latus rectum of its orbit (point 2) through the perihelion (point A) to the other end of the latus rectum (point 3) sweeping out the shaded area shown. The eccentricity of the earth's orbit is 0.0167. *Hint:* Approximate the shape of area 1-2-3-4 as a trapezoid. The mass of the sun is $3.328(10^5)$ times that of the earth.

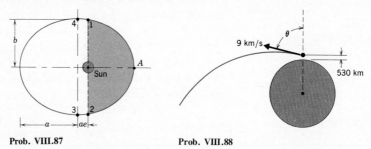

Prob. VIII.87   Prob. VIII.88

VIII.88   When the engines of the launch vehicle of a satellite are shut off, the altitude is 530 km and the velocity is 9 km/s at $\theta = 90°$ from the outward radial line. Determine the period of the resulting orbit.

VIII.89   Solve Problem VIII.88 if $\theta = 85°$ for the launch and all other parameters are unchanged.

VIII.90   A satellite is in a circular orbit at an altitude of 1000 miles. The engines are fired tangentially at point A, placing the satellite into an elliptical orbit whose apogee is at point B, in order to place it in a new circular orbit at an altitude of 10,000 miles. Determine the length of time after the firing required for the satellite to reach (a) point B, (b) point C on the semiminor axis.

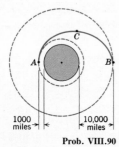

Prob. VIII.90

VIII.91   A satellite is destined for a rendezvous with a space station in a stationary orbit. The engines of the launch vehicle are shut down at point A, where the altitude is 130 km and the velocity is perpendicular to the radial line from the center of the earth. At what angle ϕ should the space station be when the engines of the satellite are turned off if rendezvous is to be accomplished at point B?

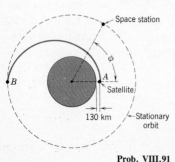

Prob. VIII.91

C. SYSTEMS WITH PARTICLE FLOW

Many engineering problems involve the flow of a system of particles, such as the flow of a fluid through or over a body. Depending on the nature of the information required, and hence the approximations made, a fluid dynamics analysis can be quite complicated. In part, this section is an introduction to the subject of fluid dynamics.

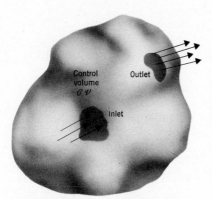

Figure 21

1 Basic Equations

Figure 21 depicts a situation where particles are continually entering and leaving a region of fixed shape, such as the interior of a hollow body. This region is called the control volume, $\mathcal{C}\mathcal{V}$.

We wish to derive the equations of motion for the $\mathcal{C}\mathcal{V}$ system. To accomplish this we shall use the linear and angular impulse-momentum principles for a system of particles. As in the study of all dynamic systems, we must first decide on a model. Here we begin by considering the question of what details of the motion of the particles we must know to write the equations of motion.

To deal with this question, we make some simplifying assumptions in modeling a time interval of flow from t to $t + \Delta t$. In this interval a small group of particles having mass Δm_1 enters the $\mathcal{C}\mathcal{V}$ as another group of particles Δm_2 leaves. Letting m be the mass of the $\mathcal{C}\mathcal{V}$ at time t, it follows that the mass of the $\mathcal{C}\mathcal{V}$ at time $t + \Delta t$ must be $m + \Delta m_1 - \Delta m_2$.

Let us make the following assumptions about the kinematics of the particles as they enter and leave the $\mathcal{C}\mathcal{V}$.

1 Assume that all particles forming Δm_1 have the same velocity $\bar{v}_1$.
2 Assume that all particles forming Δm_2 have the same velocity $\bar{v}_2$.
3 Assume that all particles contained within the $\mathcal{C}\mathcal{V}$ have no motion relative to the $\mathcal{C}\mathcal{V}$. In other words, the $\mathcal{C}\mathcal{V}$ and the particles it contains form a rigid body.

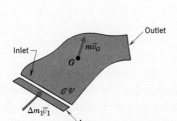

Figure 22a

The first two assumptions follow reasonably from the concept of the motion of a mass center. The third is more difficult to visualize. Experimental data have shown the assumptions to be reasonably accurate in a significant number of engineering problems.

Using these assumptions, in Figure 22 we have the linear momentum for the particles of the $\mathcal{C}\mathcal{V}$ entering at time t and upon leaving the $\mathcal{C}\mathcal{V}$ at time $t + \Delta t$. Note that the velocities depicted in Figure 22b are written as $\bar{v}_G + \Delta\bar{v}_G$ for the center of mass and $\bar{v}_2 + \Delta\bar{v}_2$ for the exiting mass because the velocities need not be constant.

Writing the linear momentum vectors for each of the systems of Figures 22a and 22b, we have

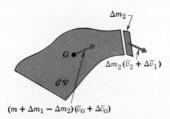

Figure 22b

$$(\Sigma\ m_i\bar{v}_i)_t = m\bar{v}_G + \Delta m_1\bar{v}_1$$

$$(\Sigma\ m_i\bar{v}_i)_{t+\Delta t} = (m + \Delta m_1 - \Delta m_2)(\bar{v}_G + \Delta\bar{v}_G) + (\Delta m_2)(\bar{v}_2 + \Delta\bar{v}_2)$$

Letting $\Sigma\ \bar{F}$ denote the total external force acting on the system, the linear impulse-momentum equation then gives

$$(\Sigma\ m_i v_i)_{t+\Delta t} - (\Sigma\ m_i v_i)_t = \int_t^{t+\Delta t} \Sigma\ \bar{F}\ dt$$

$$m(\Delta\bar{v}_G) + (\Delta m_1)\bar{v}_G + (\Delta m_1)(\Delta\bar{v}_G) - (\Delta m_2)\bar{v}_G - (\Delta m_2)(\Delta\bar{v}_G)$$
$$+ (\Delta m_2)(\bar{v}_2) + (\Delta m_2)(\Delta\bar{v}_2) - (\Delta m_1)\bar{v}_1$$
$$= \int_t^{t+\Delta t} \Sigma\ \bar{F}\ dt$$

Collecting terms and retaining only first-order terms, this reduces to

$$m\ \Delta\bar{v}_G - \Delta m_1(\bar{v}_1 - \bar{v}_G) + \Delta m_2(\bar{v}_2 - \bar{v}_G) = \int_t^{t+\Delta t} \Sigma\ \bar{F}\ dt$$

Dividing this expression by Δt and taking the limit as Δt approaches zero yields

$$\boxed{m\bar{a}_G - \dot{m}_1(\bar{v}_1 - \bar{v}_G) + \dot{m}_2(\bar{v}_2 - \bar{v}_G) = \Sigma\ \bar{F}} \tag{48}$$

where $\dot{m}_1$ is the rate at which mass is entering the inlet and $\dot{m}_2$ is the rate at which it exists in the $\mathscr{C}\mathscr{V}$.

To find the equation for the resultant moment of the force system, we formulate the angular impulse-momentum of the system relative to a point A. For convenience, we restrict point A to be one of the allowable points for summing moments for the rigid body. Also, for simplicity, we shall consider only the case where the $\mathscr{C}\mathscr{V}$ is in planar motion. Thus, the angular momentum for the $\mathscr{C}\mathscr{V}$ is simply $I_{zz}\omega$, where the z axis is perpendicular to the plane of motion passing through point A.

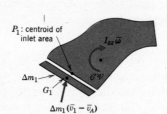

P_1: centroid of inlet area

$I_{zz}\bar{\omega}$

Δm_1

$\mathscr{C}\mathscr{V}$

G_1

$\Delta m_1(\bar{v}_1 - \bar{v}_A)$

Figure 23a

Relative to point A, the velocities of the incoming and outgoing elements of mass are $\bar{v}_1 - \bar{v}_A$ and $\bar{v}_2 - \bar{v}_A$, respectively. Using this notation, the relative and angular momentum vectors for the instant t and $t + \Delta t$ are as shown in Figure 23. The points G_1 and G_2 are the centers of mass of Δm_1 and Δm_2. The thickness of these elements approach zero as $\Delta t \to 0$, so that in the limit, points G_1 and G_2 coincide with the centroids of the inlet and outlet areas, which are points P_1 and P_2. Recalling that, in general,

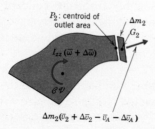

P_2: centroid of outlet area

Δm_2

G_2

$I_{zz}(\bar{\omega} + \Delta\bar{\omega})$

$\mathscr{C}\mathscr{V}$

$\Delta m_2(\bar{v}_2 + \Delta\bar{v}_2 - \bar{v}_A - \Delta\bar{v}_A)$

Figure 23b

$$\bar{H}_A = \Sigma\ \bar{r}_{i/A} \times m_i\bar{v}_{i/A}$$

it follows that the angular momentum vectors are

$$(\bar{H}_A)_t = I_{zz}\bar{\omega} + \bar{r}_{P_1/A} \times \Delta m_1(\bar{v}_1 - \bar{v}_A)$$

$$(\bar{H}_A)_{t+\Delta t} = I_{zz}(\bar{\omega} + \Delta\bar{\omega}) + \bar{r}_{P_2/A} \times \Delta m_2(\bar{v}_2 + \Delta\bar{v}_2 - \bar{v}_A - \Delta\bar{v}_A)$$

Thus, the angular impulse-momentum principle gives

$$(\bar{H}_A)_{t+\Delta t} - (\bar{H}_A)_t = \int_t^{t+\Delta t} \Sigma\,\bar{M}_A\,dt$$

$$I_{zz}\,\Delta\bar{\omega} + \bar{r}_{P_2/A} \times \Delta m_2(\bar{v}_2 + \Delta\bar{v}_2 - \bar{v}_A - \Delta\bar{v}_A)$$

$$- \bar{r}_{P_1/A} \times \Delta m_1(\bar{v}_1 - \bar{v}_A)$$

$$= \int_t^{t+\Delta t} \Sigma\,\bar{M}_A\,dt$$

Again neglecting higher-order terms, dividing by Δt and evaluating the limit as $\Delta t \to 0$ yields

$$\boxed{\begin{aligned}I_{zz}\bar{\alpha} - \bar{r}_{P_1/A} &\times \dot{m}_1(\bar{v}_1 - \bar{v}_A)\\ &+ \bar{r}_{P_2/A} \times \dot{m}_2(\bar{v}_2 - \bar{v}_A) = \Sigma\,\bar{M}_A\end{aligned}} \tag{49}$$

where $\bar{\alpha} \equiv \dot{\bar{\omega}}$ is the angular acceleration of the control volume.

Equations (48) and (49) are the basic equations of motion for the control volume. To understand their meaning further, we can transpose the terms containing $\dot{m}_1$ and $\dot{m}_2$ to the right-hand side of the equations and think of them as forces and moments. Then in equation (48) we see that the terms $\dot{m}_1(\bar{v}_1 - \bar{v}_G)$ and $-\dot{m}_2(\bar{v}_2 - \bar{v}_G)$ add to the sum of the external forces, whereas in equation (49) the terms $\bar{r}_{P_1/A} \times \dot{m}_1(\bar{v}_1 - \bar{v}_A)$ and $-\bar{r}_{P_2/A} \times \dot{m}_2(\bar{v}_2 - \bar{v}_A)$ add to the moment of the external forces about point A. Hence, in the situation where point A is chosen as the center of mass G, the equivalent force system representing the combined effect of the external forces and the particle flow is as depicted in Figure 24.

Figure 24

The term $\dot{m}_1(\bar{v}_1 - \bar{v}_G)$ represents the force the incoming particles exert on the $\mathcal{CV}$. It is the reaction to the force exerted by the $\mathcal{CV}$ on the particles in order to capture them as they enter the inlet. Similarly, the term $-\dot{m}_2(\bar{v}_2 - \bar{v}_G)$ is the force the particles leaving the outlet exert upon the $\mathcal{CV}$ in reaction to the increase of momentum imparted to them by the $\mathcal{CV}$. Notice that these equivalent forces act at the centroids of the inlet and outlet areas.

In summary, equations (48) and (49) describe the general effect of mass flow through a control volume. After studying an example, we shall consider two important cases: the flow of fluids and propulsion by rocket and jet engines.

EXAMPLE 12

When the bottom of a hopper car is opened, coal falls out at the rate of 500 kg/s with an absolute velocity of 4.0 m/s downward, both values being constant. If the hopper car has a velocity of 2.0 m/s to the right on a level track when the bottom is first opened, what is the velocity of the car when it is completely emptied? The car has a mass of 10,000 kg and was loaded with 50,000 kgs of coal.

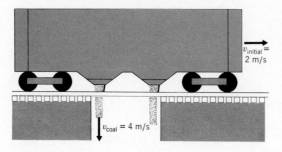

Solution

Clearly, it is best to describe the vectors in equation (48) with respect to an XYZ coordinate system having horizontal and vertical axes, as shown in the free body diagram.

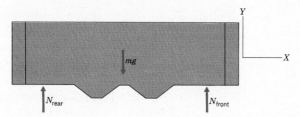

Treating the hopper car as the control volume, the given information is that mass is flowing out at $\dot{m}_2 = 500$ kg/s with a velocity $\bar{v}_2 = -4\bar{J}$ m/s. When the car is fully loaded, the combined mass of the car and the coal is 60,000 kg. Thus, accounting for the loss of mass, the mass within the control volume at an instant t after opening is

$$m = 60,000 - 500t \text{ kg}$$

As the empty mass of the car is 10,000 kg, the time t_f required to empty the car is

$$10,000 = 60,000 - 500t_f \quad \text{or} \quad t_f = 100 \text{ seconds}$$

Also, the control volume is translating to the right, so $\bar{v}_G = v\bar{I}$.

Using the free body diagram to evaluate the force sum, and substituting the foregoing terms, equation (48) becomes

$$m\bar{a}_G + \dot{m}_2(\bar{v}_2 - \bar{v}_G) = \Sigma\,\bar{F}$$

$$(60{,}000 - 500t)\dot{v}\bar{I} + 500(-4\bar{J} - v\bar{I}) = (N_{\text{front}} + N_{\text{rear}} - mg)\bar{J}$$

The Y component of this equation involves the values of the normal forces, which we do not need. The X component gives

$$(60{,}000 - 500t)\dot{v} - 500v = 0$$

To find the value of v at t_f, we separate variables in this differential equation, as follows.

$$(60{,}000 - 500t)\frac{dv}{dt} = 500v$$

$$\frac{dv}{v} = \frac{500\,dt}{(60{,}000 - 500t)} = \frac{dt}{(120 - t)}$$

The initial speed is 2 m/s, so

$$\int_2^v \frac{dv}{v} = \int_0^t \frac{dt}{120 - t}$$

$$\ln v - \ln 2 = -\ln(120 - t) + \ln(120)$$

$$v = 2\left(\frac{120}{120 - t}\right)$$

Hence, for $t_f = 100$ seconds,

$$v_f = 12.0 \text{ m/s} \qquad \triangleleft$$

HOMEWORK PROBLEMS

VIII.92 Solve Example 12 for the case where the given velocity of the coal as it falls out of the hopper car is 4 m/s downward relative to the car.

VIII.93 Sand is dropped from a bin into a 10,000-kg hopper car at a rate of 250 kg/s, attaining a downward velocity of 10 m/s before landing in the car. A horizontal force $\bar{F}$ is applied to the car to give it a constant velocity of 4 m/s to the right. Determine (a) the magnitude of $\bar{F}$, (b) the increase in the total reaction force between the wheels and the tracks due to the action of the sand falling into the car.

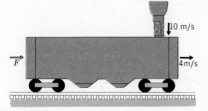

Prob. VIII.93

VIII.94   In Problem VIII.93, the force $\bar{F} = \bar{0}$ and the initial velocity of the hopper car is 4 m/s to the right. Determine the velocity of the hopper car after 20,000 kg of sand have been loaded.

VIII.95   Water from a hose is being collected in a 20-lb cart that rolls freely. The stream of water is discharged from the stationary nozzle at 4 lbs/s with a velocity of 30 ft/s in the direction shown. If the cart is initially at rest, determine (a) the differential equation for the speed v of the cart, (b) the speed of the cart after 0.5 s and 1.0 s.

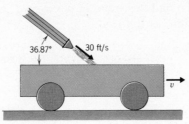

Prob. VIII.95

VIII.96   When $\theta = 20°$ it takes 3 min to dump 6000 kg of sand from the dump truck. The average velocity relative to the truck of the sand as it leaves is 4 m/s tangent to the bed. Determine the traction force that the wheels must supply to move the truck at 10 km/hr to the right.

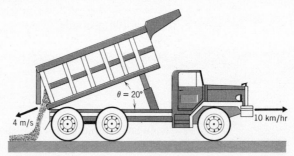

Prob. VIII.96

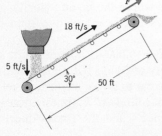

Prob. VIII.97

VIII.97   Portland cement falls from the bin onto the conveyor belt with a velocity of 5 ft/s at a rate of 80 lb/s. Determine the total force F that must be applied to the belt to give it a constant speed of 18 ft/s. Assume that the cement falling off the upper end of the belt is moving parallel to the inclined portion of the belt.

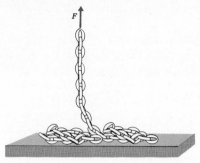

Prob. VIII.98

VIII.98   A chain whose mass per unit length is σ is piled on the ground when a force $\bar{F}$ is applied to its end, with the result that the end is pulled upward with a constant speed v. Determine the magnitude of $\bar{F}$ in terms of the length s of the suspended portion of the chain. *Hint:* For a chain whose links are loose, the links piled on the ground can be considered to change velocity instantaneously when they are pulled. For this reason, these links exert a negligible force on the suspended links.

VIII.99   A chain whose mass per unit length is σ and whose length is l is initially suspended vertically by the force $\bar{F}$. The magnitude of $\bar{F}$ is then adjusted with the result that the end of the chain is lowered with a constant speed v. In terms of the length s of the suspended portion of the chain, determine (a) the magnitude of the normal force exerted on the portion of the chain which is piled on the ground, (b) the magnitude of $\bar{F}$. See the hint for Problem VIII.98.

VIII.100   A chain whose mass per unit length is σ is piled on a table at point A with a section overhanging the edge. The chain is released and slides off the table. In terms of the dimensions d and s and the speed v of end B, derive an expression for the acceleration $\dot{v}$ (a) if the table is smooth, (b) if the coefficient of kinetic friction is μ. See the hint for Problem VIII.98.

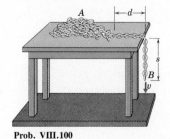

Prob. VIII.100

2   Fluid Flow

The application of the basic force and moment equations (48) and (49) to describe the motion of fluids frequently requires that we first determine the values of $\dot{m}_1$ and $\dot{m}_2$ in terms of the properties of the system. To see how this is done, let us consider the element of fluid Δm_1 which enters the inlet during a small time interval Δt. Clearly, for the fluid to enter the inlet, the relative velocity, $\bar{u}_1 \equiv \bar{v}_1 - \bar{v}_{P_1}$ of the fluid with respect to the centroid of the inlet P_1 must be inward. We then have the

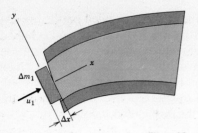

Figure 25

situation depicted in Figure 25, where the x axis was chosen, for convenience, to be parallel to $\bar{u}_1$.

Letting $\mathcal{A}_1$ be the area of the inlet projected onto the yz plane, that is, the inlet area perpendicular to the incoming flow, the volume of the mass element depicted in Figure 25 is $\mathcal{A}_1\Delta x$, which means that

$$\Delta m_1 = \rho_1 \mathcal{A}_1 \Delta x$$

where ρ_1 is the density of the fluid at the inlet. As all of the particles forming Δm_1 must cross the yz plane in the time interval Δt, for small Δt we have

$$\Delta x \simeq u_1 \Delta t$$

Thus

$$\Delta m_1 \simeq \rho_1 \mathcal{A}_1 u_1 \Delta t$$

Dividing by Δt and taking the limit as $\Delta t \to 0$, we find that

$$\boxed{\dot{m}_1 = \rho_1 \mathcal{A}_1 u_1} \tag{50a}$$

By similar reasoning, we can show that the outlet mass flow is given by

$$\boxed{\dot{m}_2 = \rho_2 \mathcal{A}_2 u_2} \tag{50b}$$

where $\bar{u}_2$ is the relative velocity of the fluid leaving the control volume with respect to the outlet, $\bar{u}_2 \equiv \bar{v}_2 - \bar{v}_{P_2}$, and ρ_2 is the density of the fluid at the outlet.

In many cases the fluid merely passes through the control volume, so that the mass of the control volume remains constant and

$$\boxed{\dot{m}_1 = \dot{m}_2} \tag{51}$$

In addition to the simplifications in equations (48) and (49) that result from equation (51), the equation provides a useful formula relating the incoming and outgoing flow velocities. Specifically, placing equations (50) into equation (51) gives

$$\boxed{\rho_1 \mathcal{A}_1 u_1 = \rho_2 \mathcal{A}_2 u_2} \tag{52}$$

For a large class of engineering problems we may consider all liquids, as well as unheated gases flowing at a low speed, to be incompressible, which means that ρ is constant. Hence for this class of problems, equation (52) quantifies the simple fact that, for a fixed mass flow, a fluid must move faster to pass through a smaller opening.

EXAMPLE 13

The blade shown in the diagram is adjusted to scoop water from the surface to a depth of 3 in. by moving to the left at the constant speed of 10 ft/s. The width of the blade (into the diagram) is 18 in. Determine the resultant force exerted by the water on the blade and the power required for this motion. Assume that the layer of water over the blade has constant thickness.

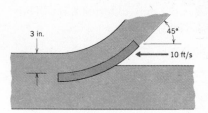

Solution

The desired force is the reaction corresponding to the force the blade applies to the water passing over it. We therefore draw a free body diagram of the layer of water. The sketch does not include the hydrostatic pressure at the inlet A and the weight of the control volume because they are negligible compared to the force required to change the momentum of the fluid. Recall that similar reasoning prevailed when we studied impulsive forces in Module III.

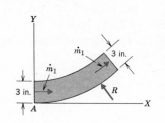

We must first determine the rate at which mass is flowing through the control volume. At the inlet, we know that the fluid is at rest, $\bar{v}_1 = \bar{0}$, so the relative velocity of the fluid with respect to the blade is 10 ft/s to the right. For an 18-in.-wide blade the inlet area is

$$\mathcal{A}_1 = (\tfrac{3}{12})(\tfrac{18}{12}) = 0.375 \text{ ft}^2$$

Hence

$$\dot{m}_1 = \rho_{\text{water}} \mathcal{A}_1 u_1 = \left(1.9397 \, \frac{\text{lb-s}^2}{\text{ft}^4}\right)\left(0.375\text{ft}^2\right)\left(10 \, \frac{\text{ft}}{\text{s}}\right)$$

$$= 7.274 \text{ lb-s/ft}$$

The control volume has constant mass, so $\dot{m}_2 = \dot{m}_1$. The thickness of the layer of water is constant, so the outlet area is the same as the inlet area, which means that $u_2 = u_1$. Thus, the relative velocity of the water with respect to the blade as it leaves the control volume is 10 ft/s tangent to the blade, that is, at a 45° angle above the horizontal.

$$\bar{u}_2 = 10(\cos 45° \, \bar{I} + \sin 45° \, \bar{J}) = 7.071\bar{I} + 7.071\bar{J}$$

The control volume is translating, which means that the outlet is moving with velocity $\bar{v}$. Hence, we may find $\bar{v}_2$ from the relative velocity formula, as follows.

$$\bar{v}_2 - \bar{v} = \bar{u}_2$$

$$\bar{v}_2 = \bar{v} + \bar{u}_2 = -10\bar{I} + 7.071\bar{I} + 7.071\bar{J}$$

$$\bar{v}_2 = -2.929\bar{I} + 7.071\bar{J} \text{ ft/s}$$

The free body diagram shows that $\bar{R}$ is the only external force. Thus, setting $\bar{v}_G = \bar{v}$ and $\bar{a}_G = \bar{0}$, equation (48) yields

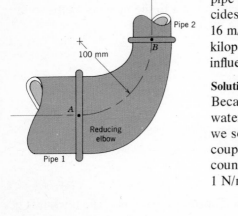

$$-\dot{m}_1(\bar{v}_1 - \bar{v}) + \dot{m}_1(\bar{v}_2 - \bar{v}) \equiv \dot{m}_1(\bar{v}_2 - \bar{v}_1) = \bar{R}$$

$$7.274(-2.929\bar{I} + 7.071\bar{J} - \bar{0}) = \bar{R}$$

$$\bar{R} = -21.31\bar{I} + 51.43\bar{J} \text{ lb}$$

The force applied to the blade is $-\bar{R}$, as shown in the diagram.

The component of fluid force opposite the velocity of a body is called the *drag force*. Thus, a horizontal force of 21.31 lb must be applied in opposition to the drag force to move the blade with a constant velocity of 10 ft/s. The power required to produce this motion is

$$\text{Power} = (21.31)(10) = 213.1 \text{ ft-lb/s} = 0.387 \text{ hp}$$

EXAMPLE 14

The reducing elbow connects pipe 1 whose inside diameter is 100 mm to pipe 2 whose diameter is 50 mm. The centroidal axis of the system coincides with the horizontal plane. The velocity of the inlet water in pipe 1 is 16 m/s to the right. Assuming that the water pressure is constant at 200 kilopascals, determine the force-couple system at point B representing the influence of the water on the piping system.

Solution

Because we want to study the interaction of the water and the pipe, let the water within the elbow be the control volume. The force-couple system we seek is shown in the free body diagram as a force $\bar{R}$ at point B and a couple $\bar{M}_B$. The pressure forces at both ends of the elbow are also accounted for in the free body diagram. Recall that, by definition, a pascal is 1 N/m².

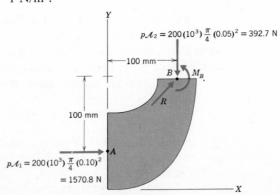

Using the given information that the water in pipe 1 has a speed of 15 m/s, we find that

$$\dot{m}_1 = \rho_{\text{water}}\, \mathcal{A}_1 v_1$$

$$= \left(1000\, \frac{\text{kg}}{\text{m}^3}\right)\left(\frac{\pi}{4}\right)(0.10\text{ m})^2\left(16\, \frac{\text{m}}{\text{s}}\right)$$

$$= 125.66\text{ kg/s}$$

The control volume is always filled with water, so we have

$$\dot{m}_2 = \dot{m}_1$$

Equation (52) then gives

$$\rho_{\text{water}}\, \mathcal{A}_2 v_2 = \rho_{\text{water}}\, \mathcal{A}_1 v_1$$

$$v_2 = \frac{\mathcal{A}_1}{\mathcal{A}_2}\, v_1 = 4(16) = 64\text{ m/s}$$

Thus, the velocity vectors are

$$\bar{v}_1 = 16\bar{I} \qquad \bar{v}_2 = 64\bar{J}\text{ m/s}$$

Noting that the control volume is stationary, we have $\bar{v}_G \equiv \bar{0}$. Thus, using equation (48) to determine the force $\bar{R}$ yields

$$-\dot{m}_1\bar{v}_1 + \dot{m}_1\bar{v}_2 = 125.66(-16\bar{I} + 64\bar{J}) = 1570.8\bar{I} - 392.7\bar{J} + \bar{R}$$

$$\bar{R} = -3581\bar{I} + 8435\bar{J}\text{ N} \qquad \triangleleft$$

To evaluate the couple $\bar{M}_B$ we sum moments. Because the control volume is stationary, $\bar{\alpha} = \bar{0}$ and we may sum moments about any point. In order to eliminate the moment of $\bar{R}$, we choose point B. Equation (49) then gives

$$\dot{m}_1\bar{r}_{A/B} \times (-\bar{v}_1) + \dot{m}_1\bar{r}_{B/B} \times \bar{v}_2 = \Sigma\, \bar{M}_B$$

$$125.66(-0.10\bar{I} - 0.10\bar{J}) \times (-16\bar{I}) + \bar{0} = (-0.10\bar{I} - 0.10\bar{J}) \times 1570.8\bar{I} + M_{BZ}\bar{K}$$

$$-201.1\bar{K} = (157.1 + M_{BZ})\bar{K}$$

$$M_{BZ} = -358.2$$

$$\bar{M}_B = -358.2\bar{K}\text{ N-m} \qquad \triangleleft$$

The force and couple exerted by the water on the pipe are the reactions, $-\bar{R}$ and $-\bar{M}_B$, respectively. These are illustrated in the diagram.

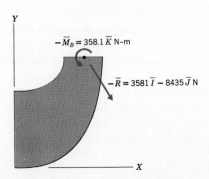

Y

$-\bar{M}_B = 358.1\,\bar{K}$ N-m

$-\bar{R} = 3581\,\bar{I} - 8435\,\bar{J}$ N

X

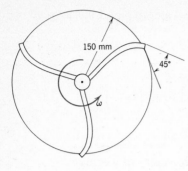

EXAMPLE 15

A rotary lawn sprinkler has three equally spaced curved arms of 5 mm inside diameter. Water flows into the sprinkler at the center from a garden hose, and is divided equally among the arms. Neglecting any resistance from friction, determine the constant rate of rotation ω for a flow rate of 1.20 l/s.

Solution

We consider the entire sprinkler as the control volume in order to study its motion. In the absence of friction, no forces are exerted in the horizontal plane (assuming the sprinkler head is balanced in its rotation). Hence, we do not need a free body diagram; the given sketch will provide the required geometry.

The given flow rate is in terms of volume (1000 liters = 1 m³). The mass rate is then obtained by multiplying by the density of the water.

$$\dot{m}_1 = \left(1000\ \frac{\text{kg}}{\text{m}^3}\right)\left(1.20\ \frac{\text{l}}{\text{s}}\right)\left(\frac{1}{1000}\ \frac{\text{m}^3}{\text{l}}\right)$$

$$= 1.20\ \text{kg/s}$$

This flow is divided equally among the three arms. The outlet has a diameter of 5 mm, so

$$\mathcal{A}_2 = \left(\frac{\pi}{4}\right)(0.005)^2 = 1.9635(10^{-5})\ \text{m}^2$$

Thus, the mass leaving each nozzle is

$$\rho_{\text{water}}\,\mathcal{A}_2 u_2 = \tfrac{1}{3}\dot{m}_1 = 0.40\ \text{kg/s}$$

from which we obtain

$$u_2 = \frac{0.40}{(1000)(1.9635)(10^{-5})} = 20.37\ \text{m/s}$$

The rotation of the system is defined by the moment equation (49), but this equation was derived for the case where fluid is leaving the control volume at only one outlet; the system considered here involves three identical outlets. To handle this we shall consider the mass flow through one arm and multiply the result by three. The control volume we have chosen is the sprinkler. Because it is in fixed-axis rotation about its center O, we shall sum moments about this point. The angular velocity $\bar{\omega}$ is constant, therefore $\bar{\alpha} = \bar{0}$, and we also know that $\bar{v}_G \equiv \bar{0}$. Furthermore, noting that the water enters the sprinkler at its center, the momentum of the incoming fluid has a zero moment arm, and hence drops out of equation (49). Denoting the tip of an arm as point P, equation (49) then reduces to

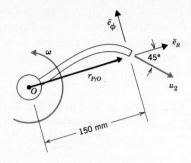

$$3\bar{r}_{P/O} \times (\tfrac{1}{3}\dot{m}_1\bar{v}_2) = \Sigma\ \bar{M}_O = \bar{0}$$

This equation requires that the rotation of the system be such that $\bar{v}_2$ is parallel to $\bar{r}_{P/O}$. In other words, the reaction force from each stream of water must pass through the center point O. To describe the vectors we employ cylindrical coordinates, as sketched here.

From the sketch we see that

$$\bar{r}_{P/O} = 0.150\bar{e}_R\ \text{meters}$$

Earlier, we found that $u_2 = 20.37$ m/s. The vector $\bar{u}_2$ is tangent to the arm, as shown in the sketch. Therefore

$$\bar{u}_2 = 20.37(\cos 45°\ \bar{e}_R - \sin 45°\ \bar{e}_\phi) = 14.404(\bar{e}_R - \bar{e}_\phi)\ \text{m/s}$$

We also know that $\bar{u}_2$ is the velocity of the water relative to the exit area, which means that $\bar{u}_2 = \bar{v}_2 - \bar{v}_P$. Point P is following a circular path of 150 mm radius. Hence we may solve for $\bar{v}_2$.

$$\bar{v}_2 = \bar{v}_P + \bar{u}_2 = 0.15\omega\bar{e}_\phi + 14.404(\bar{e}_R - \bar{e}_\phi)$$

$$= 14.40\bar{e}_R + (0.15\omega - 14.404)\bar{e}_\phi$$

Earlier we showed that to satisfy the moment sum the transverse component of $\bar{v}_2$ must be zero (in order for $\bar{v}_2$ to be parallel to $\bar{r}_{P/O}$). Hence

$$0.15\omega - 14.404 = 0$$

$$\omega = 96.03\ \text{rad/s} = 917\ \text{rev/min}$$

In the following problems a fluid exposed to the atmosphere may be considered to be at atmospheric pressure.

HOMEWORK PROBLEMS

VIII.101 and VIII.102   The nozzle of a 50-mm-diameter hose has an outlet whose opening is 10 mm in diameter. If the water within the hose is flowing at 10 m/s and the internal pressure is 100 kilopascals above atmospheric pressure, determine the force-couple system at point A representing the reaction between the hose and the nozzle.

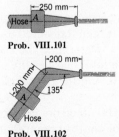

Prob. VIII.101

Prob. VIII.102

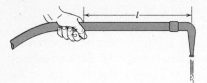

Prob. VIII.103

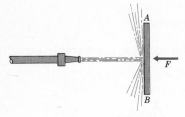

Prob. VIII.104

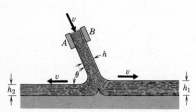

Prob. VIII.106

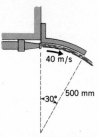

Prob. VIII.107

VIII.103 A 20-mm garden hose is connected to a right-angle nozzle whose opening is 5 mm in diameter. The water within the hose is flowing at 1 l/s. The mass of the nozzle is 100 grams and that of the hose is 0.5 kg/m. Neglecting the rigidity of the hose, what length l of the hose can be supported by a person holding the hose horizontally at point A?

VIII.104 Water flows through a $\frac{5}{8}$-in. inside diameter hose at 20 ft/s, then emerges from the nozzle whose opening is $\frac{1}{4}$ in. in diameter. Determine the magnitude of the force $\bar{F}$ required to hold plate AB stationary.

VIII.105 In Problem VIII.104, determine the magnitude of the force $\bar{F}$ required to translate plate AB at 15 ft/s (a) to the right, (b) to the left.

VIII.106 Water flows between two parallel plates A and B at a speed v. The spacing between these plates is h. The stream is broken into two sheets by a divider at point C on the smooth horizontal plate. Assuming that both sheets also flow at the speed v, determine (a) the force the water exerts on the horizontal plate, (b) the depths h_1 and h_2 of both sheets of water.

VIII.107 Water flowing through a hose at 1000 l/min emerges from the nozzle at 40 m/s. The stream is deflected downward by a vane in the shape of a circular arc of 500 mm radius. Determine the magnitude, direction, and point of application of the force the water exerts on the vane.

VIII.108 Water emerges from a nozzle at the rate of 15 ft³/s with a speed of 100 ft/s and flows at a constant speed over the curved vane. Determine the reactions at supports A and B.

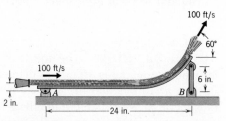

Prob. VIII.108

VIII.109 Grain is carried on the horizontal conveyor belt at 150 kg/s at a speed of 10 m/s. It is then transfered to a curved conveyor belt CD, which is moving at 20 m/s. Determine the components of the reactions at supports A and B. The total mass of the grain on conveyor belt CD and the structure is 1000 kg, and the center of mass is point G.

VIII.110 and VIII.111 Iron ore falls from a hopper at the rate of 300 kg/s, hitting the conveyor chute at point B at a speed of 10 m/s. The ore leaves

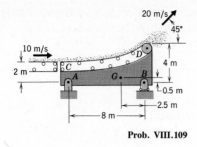

Prob. VIII.109

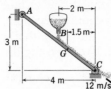

Prob. VIII.110

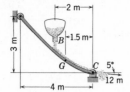

Prob. VIII.111

the chute at point C with a speed of 12 m/s. The total mass of the chute and the ore being transported is 1000 kg, and the center of mass is point G. Determine the reactions at supports A and C.

VIII.112 The slipstream of an electric fan has a diameter of 2 ft and a speed of 40 ft/s. The air weighs 0.07651 lb/ft^3. Determine the thrust force developed by the fan blades. *Hint:* The air approaching the blades, being part of the atmosphere, may be considered to be at rest.

Prob. VIII.112

VIII.113 When the helicoper hovers, the slipstream has a diameter of 10 m and the air within it is flowing at 20 m/s downward. Knowing that the density of the air is 1.0 kg/m^3, determine the thrust being developed by the blades. See the hint for Problem VIII.112.

VIII.114 The slipstream of each engine of a twin-engine propeller airplane has a diameter of 1 m. The airplane has a speed of 80 m/s, and the engines are developing 400 kW of power. If the air has a density of 0.9820 kg/m^3, what is the velocity of the air in the slipstream? See the hint for Problem VIII.112.

Prob. VIII.113

Prob. VIII.114

VIII.115 In Example 13, the blade is moving to the left at the speed v. Determine the required power. Is there a value of v for which the power is maximum?

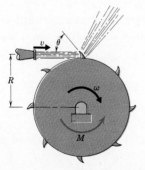

Prob. VIII.116

VIII.116 The Pelton wheel turbine is rotating at angular speed ω due to the impulse of the stream that is flowing at an absolute speed v and volume flow rate q. The tangent to the blade is at angle θ from the incoming stream, as shown. All of the water that strikes a blade is deflected outward tangent to the moving blade. Derive an expression for the couple M resisting the rotation of the turbine in terms of the other parameters.

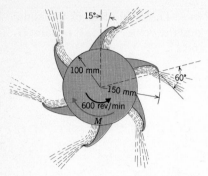

Prob. VIII.118

VIII.117   Solve Example 15 for the case where there is a frictional torque of 0.5 N-m opposing the rotation of the arms of the sprinkler.

VIII.118   The centrifugal-flow turbine is turning counterclockwise at 600 rev/min. Water flows into the center at 20 m³/s and flows outward over each of the six vanes, moving with a speed of 50 m/s relative to the vanes. Determine the torque M opposing the rotation.

3   Jet and Rocket Propulsion

The propulsion of a system by a jet of fluid is accomplished by accelerating the fluid as it passes through the system's engine. In rocket and jet engines this acceleration is achieved by means of the combustion of fuel. One primary difference between a rocket and a jet engine is that the oxidant for the combustion process is carried inside the vehicle for rocket propulsion, whereas it is gathered in from the surrounding atmosphere in the case of a jet engine.

Let us compare the rocket-propelled vehicle in Figure 26a with the jet aircraft in Figure 26b. In normal operation the angular velocities are small, so we shall consider both vehicles to be translating bodies.

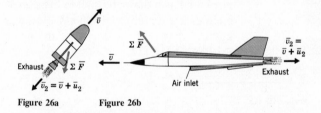

Figure 26a   **Figure 26b**

For both vehicles, $\Sigma \bar{F}$ denotes the combination of the aerodynamic and weight forces. Also, for both vehicles, the gases are exhausted with velocity $\bar{u}_2$ relative to the vehicle, which means that $\bar{v}_2 = \bar{v} + \bar{u}_2$.

Let $\dot{m}_P$ be the rate at which the propellant carried within the vehicle is consumed. Then, for steady operation of the engine, the instantaneous mass of the vehicle is

$$m = m_0 - \dot{m}_P t$$

where m_0 is the initial mass.

In the case of the rocket, the exhaust gases are obtained only from the propellant, so that $\dot{m}_1 = 0$ and $\dot{m}_2 = \dot{m}_P$. Specializing equation (48) for the case of the rocket vehicle then yields

$$(m_0 - \dot{m}_P t)\bar{a} + \dot{m}_P \bar{u}_2 = \Sigma \bar{F} \tag{53}$$

On the other hand, for an air-breathing engine, air enters the inlet at a mass rate $\dot{m}_1 = \dot{m}_{\text{air}}$. The velocity of the air at the inlet is usually negligible, $\bar{v}_1 \simeq \bar{0}$, because it is at atmospheric conditions. The propellant consumed by the engine is the fuel, which is combined in the combustion chamber with the air. Thus, $\dot{m}_2 = \dot{m}_{\text{air}} + \dot{m}_P$, so for an air-breathing engine equation (48) gives

$$(m_0 - \dot{m}_P t)\bar{a} + \dot{m}_{\text{air}}\bar{v} + (\dot{m}_{\text{air}} + \dot{m}_P)\bar{u}_2 = \Sigma \bar{F} \tag{54}$$

In normal aircraft operation the fuel is expended at a slow rate compared to $\dot{m}_{\text{air}}$. For this reason, equation (54) can, at times, be simplified by setting $\dot{m}_P = 0$.

EXAMPLE 16
A single-stage rocket, weighing 1000 lb when unfueled, follows a vertical path. The engine burns propellant at the rate of 300 lb/s, and the gases are exhausted at 5000 ft/s. Determine the maximum speed attained by the rocket if it is launched from rest with 15,000 lb of propellant. Neglect air resistance and the variation in the gravitational attraction resulting from the change of altitude.

Solution
We begin with a free body diagram of the system. It is given that air resistance is negligible, and we shall assume that the pressure at the nozzle is also negligible. Thus, the only force acting on the system is gravity.

From the given information we know that $\dot{m}_P = 300/32.17 = 9.325$ lb-s/ft (be careful to express $\dot{m}_P$ in mass units) and that $u_2 = 5000$ ft/s. In terms of components with respect to the coordinate axes shown, we have

$$\bar{u}_2 = \bar{v}_2 - \bar{v}_G = -5000\bar{J} \text{ ft/s}$$

Also from the given information, we know that the take-off weight is 16,000 lb, which means that $m_0 = 16,000/32.17 = 497.4$ lb-s²/ft.

Substituting these terms into equation (53) and setting $\bar{a} = \dot{v}\bar{J}$ for vertical motion gives

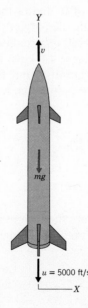

Y

v

mg

$u = 5000$ ft/s

X

$$(m_0 - \dot{m}_P t)\bar{a} + \dot{m}_P(\bar{v}_2 - \bar{v}_G) = -(m_0 - \dot{m}_P t)g\bar{J}$$

$$(497.4 - 9.325t)\dot{v}\bar{J} + 9.325(-5000)\bar{J} = -(497.4 - 9.325t)(32.17)\bar{J}$$

Rearranging terms, this expression yields a differential equation in v and t. The value of v is maximum when all of the propellant has been consumed. Hence the corresponding value of t is given by

$$(\dot{m}_P g)t_{max} = 15,000 \text{ lb} \qquad t_{max} = 50 \text{ seconds}$$

We can solve the differential equation of motion by separating variables.

$$(497.4 - 9.325t)\dot{v} = 46,625 - (497.4 - 9.325t)(32.17)$$

$$dv = \left(\frac{46,625}{497.4 - 9.325t} - 32.17 \right) dt$$

$$\int_0^{v_{max}} dv \equiv v_{max} = \int_0^{50} \left[\frac{5000}{(53.34 - t)} - 32.17 \right] dt$$

$$= 5000[- \ln (53.34 - 50) + \ln (53.34)] - 32.17(50)]$$

$$v_{max} = 12,245 \text{ ft/s}$$

EXAMPLE 17

The two-engine jet aircraft shown in the diagram, whose mass is $4(10^4)$ kg, is in level flight at a constant speed of 600 m/s. Its nose is pitched upward such that the engine exhaust is at an angle of 10° below the horizontal. The inlet area for each engine is 0.40 m², and the density of the atmosphere is 1.10 kg/m³. The exhaust has a speed of 1200 m/s realtive to the aircraft, corresponding to a rate of fuel consumption of 4.0 kg/s. Determine (a) the thrust of the engine, (b) the lift force (the aerodynamic force perpendicular to the velocity), (c) the drag force (the aerodynamic force parallel to the velocity).

Solution

The external forces acting on the aircraft are the weight and the unknown aerodynamic force, as shown in the free body diagram.

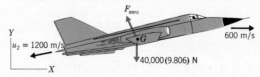

Choosing the XYZ coordinate system shown in the sketch, we may write

$$\bar{v} = 600\bar{I} \text{ m/s}$$

$$\bar{u}_2 = 1200(-\cos 10° \, \bar{I} - \sin 10° \, \bar{J})$$

$$= -1181.8\bar{I} - 208.4\bar{J} \text{ m/s}$$

To account for both engines, we double the given fuel consumption rate to get

$$\dot{m}_P = 8 \text{ kg/s}$$

The rate at which air is taken in is not given, but it can be determined from the formula

$$\dot{m}_{air} \equiv \dot{m}_1 = \rho_1 \mathscr{A}_1 u_1$$

Because $\bar{u}_1 = \bar{v}_1 - \bar{v}$ and the air at the inlet is at rest in the atmosphere, $u_1 = v = 600$ m/s. Doubling the given inlet area (two engines) and using the given value for density, we compute

$$m_{air} = \left(1.10 \frac{\text{kg}}{\text{m}^3}\right)(2)(0.40 \text{ m}^2)\left(600 \frac{\text{m}}{\text{s}}\right) = 528 \text{ kg/s}$$

Hence, equation (54), for the condition $\bar{a}_G = \bar{0}$ gives

$$\dot{m}_{air}\bar{v} + (\dot{m}_{air} + \dot{m}_P)\bar{u}_2 = \Sigma \, \bar{F}$$

$$528(600\bar{I}) + (528+8)(-1181.8\bar{I} - 208.4\bar{J}) = \bar{F}_{aero} - 3.922(10^5)\bar{J}$$

$$-3.166(10^5)\bar{I} - 1.117(10^5)\bar{J} = \bar{F}_{aero} - 3.922(10^5)\bar{J} \text{ N}$$

In our earlier discussion we saw that the mass flow terms on the left side of equation (48) were the negative of the reaction forces corresponding to the flow. Therefore

$$\bar{F}_{engine} = 3.166(10^5)\bar{I} + 1.117(10^5)\bar{J}$$

$$|\bar{F}_{engine}| = 3.357(10^5) \text{ N}$$

Solving the foregoing force equation for $\bar{F}_{aero}$ yields

$$\bar{F}_{aero} = -3.166(10^5)\bar{I} + (3.922 - 1.117)(10^5)\bar{J}$$

$$= -3.166(10^5)\bar{I} + 2.805(10^5)\bar{J} \text{ N}$$

The velocity is in the X direction, so the drag and lift forces are the X and Y components of $\bar{F}_{aero}$. Hence

$$F_{\text{lift}} = 2.805(10^5) \text{ N}$$

$$F_{\text{drag}} = 3.166(10^5) \text{ N}$$

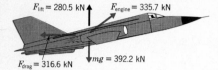

The aerodynamic and thrust forces are shown in the sketch. Notice that because of the pitched-up condition of the nose, the lift force is substantially less than the weight. It is also noteworthy that in this problem the engine thrust is not parallel to the direction of the exhaust velocity.

HOMEWORK PROBLEMS

VIII.119   At take-off a rocket has a weight of 2000 lb, including 1500 lb of fuel. The exhaust speed is a constant 10,000 ft/s relative to the rocket. Compare the thrust of the rocket at launching and at burnout if the fuel is consumed (a) at 30 lb/s, (b) at 60 lb/s.

VIII.120   The rocket in Problem VIII.119 is launched from an airplane moving in level flight at 800 ft/s and follows a horizontal path. Neglecting air resistance, determine the maximum speed attained by the rocket for both rates of fuel consumption.

VIII.121   An atmospheric sounding rocket is fired vertically upward. When launched, the rocket has mass m_0. The exhaust gases from the engine have a constant speed u relative to the vehicle. If the rocket is to have a constant acceleration $\dot{v}$, derive an expression for the required rate of consumption of fuel $\dot{m}_P$ in terms of time. Neglect air resistance.

VIII.122   Two missiles are to be launched at the same time from the ground. The first is a single-stage rocket having a mass of 20,000 kg at take-off, 16,000 of which is fuel. The second is a two-stage rocket, each stage having a mass of 10,000 kg at take-off, including 8000 kg of fuel. The engines of both missiles consume fuel at the rate of 200 kg/s, exhausting them at 3000 m/s relative to the rocket. Compare the maximum speed attained by the two missiles in vertical flight.

VIII.123   Compare the height at burnout for the two missiles in Problem VIII.122.

VIII.124   A 10,000 kg, two-engine jet aircraft whose speed is 250 m/s takes in air at the total rate of 150 kg/s and has a total fuel consumption rate of 1 kg/s. The combustion gases are exhausted at 800 m/s relative to the aircraft in a direction opposite the velocity of the aircraft. (a) If the maximum angle of climb at which the aircraft can maintain its speed is 20°, determine the air resistance. (b) For the air resistance obtained in part (a), determine the rate at which the aircraft gains speed in level flight at an instantaneous speed of 250 m/s.

Prob. VIII.124

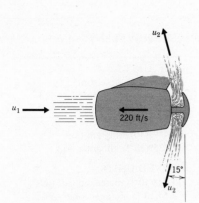

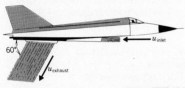

Prob. VIII.125

VIII.125 A jet aircraft may be decelerated by deflecting the exhaust, as shown. Each engine has an intake area of 5 ft², consumes fuel at the rate of 1.5 lb/s, and exhausts the combustion gases at a speed $u = 1500$ ft/s. Determine the thrust of each engine for a ground speed of 220 ft/s when the exhaust is impeded by the deflectors. The air weighs 0.07631 lb/ft³.

VIII.126 A 15,000-kg VTOL aircraft is powered by two jet engines, each of which scoops air in horizontally at 110 kg/s. The inlet area of each engine is 0.5 m². The gases are exhausted at 800 m/s, and each engine consumes fuel at 1 kg/s. At take-off the gases are deflected downward, relative to the aircraft, as shown. Determine the initial acceleration of the aircraft at take-off if the density of the air is 1.10 kg/m. Note: For this type of engine, it is not allowable to consider the air at the inlet to be at rest.

Prob. VIII.126

VIII.127 The aircraft in Example 17 is traveling horizontally at 400 m/s. The aerodynamic drag is 150 kN. If all other parameters are as given in the example, determine the acceleration of the aircraft.

APPENDIX A
SI UNITS

Table 1   Conversion factors from British units

PHYSICAL QUANTITY	U.S.-BRITISH UNIT	= SI EQUIVALENT
		BASIC UNITS
Length	1 foot (ft)	= $3.048(10^{-1})$ meter (m)*
	1 inch (in.)	= $2.54(10^{-2})$ meter (m)*
	1 mile (U.S. statute)	= $1.6093(10^{3})$ meter (m)
Mass	1 slug (lb-s²/ft)	= $1.4594(10)$ kilogram (kg)
	1 pound mass (lbm)	= $4.5359(10^{-1})$ kilogram (kg)
		DERIVED UNITS
Acceleration	1 foot/second² (ft/s²)	= $3.048(10^{-1})$ meter/second² (m/s²)
	1 inch/second² (in./s²)	= $2.54(10^{-2})$ meter/second² (m/s²)
Area	1 foot² (ft²)	= $9.2903(10^{-2})$ meter² (m²)
	1 inch² (in.²)	= $6.4516(10^{-2})$ meter² (m²)*
Density	1 slug/foot³ (lb-s²/ft⁴)	= $5.1537(10^{2})$ kilogram/meter³ (kg/m³)
	1 pound mass/foot³ (lbm/ft³)	= $1.6018(10)$ kilogram/meter³ (kg/m³)
Energy and Work	1 foot-pound (ft-lb)	= 1.3558 joules (J)
	1 kilowatt-hour (kW-hr)	= $3.60(10^{6})$ joules (J)*
	1 British thermal unit (Btu)	= $1.0551(10^{3})$ joules (J)
	(1 joule $\equiv$ 1 meter-newton)	
Force	1 pound (lb)	= 4.4482 newtons (N)
	1 kip (1000 lb)	= $4.4482(10^{3})$ newtons (N)
	(1 newton $\equiv$ 1 kilogram-meter/second²)	
Power	1 foot-pound/second (ft-lb-s)	= 1.3558 watt (W)
	1 horsepower (hp)	= $7.4570(10^{2})$ watt (W)
	(1 watt $\equiv$ 1 joule/second)	
Pressure and Stress	1 pound/foot² (lb/ft²)	= $4.7880(10)$ pascal (Pa)
	1 pound/inch² (lb/in.²)	= $6.8948(10^{3})$ pascal (Pa)
	1 atmosphere (standard, 14.7 lb/in.²)	= $1.0133(10^{5})$ pascal (Pa)
	(1 pascal $\equiv$ 1 newton/m²)	
Speed	1 foot/second (ft/s)	= $3.048(10^{-1})$ meter/second (m/s)*
	1 mile/hr	= $4.4704(10^{-1})$ meter/second (m/s)
	1 mile/hr	= 1.6093 kilometer/hr (km/hr)
Volume	1 foot³ (ft³)	= $2.8317(10^{-2})$ meter³ (m³)
	1 inch³ (in.³)	= $1.6387(10^{-5})$ meter³ (m³)
	1 gallon (U.S. liquid)	= $3.7854(10^{-3})$ meter³ (m³)

*Denotes an exact factor.

Table 2 **Conversion factors from "old" metric units**

PHYSICAL QUANTITY	"OLD" METRIC UNIT	= SI EQUIVALENT
Energy	1 erg	= $1.00(10^{-7})$ joule (J)*
Force	1 kyne	= $1.00(10^{-5})$ newton (N)*
Length	1 angstrom	= $1.00(10^{-10})$ meter (m)*
	1 micron	= $1.00(10^{-6})$ meter (m)*
Pressure	1 bar	= $1.00(10^{5})$ pascal (Pa)*
Volume	1 liter	= $1.00(10^{-3})$ meter³ (m³)*

*Denotes an exact factor

Table 3 **SI prefixes**

PREFIX	SYMBOL	FACTOR BY WHICH UNIT IS MULTIPLED
tera*	T	10^{12}
giga*	G	10^{9}
mega*	M	10^{6}
kilo*	k	10^{3}
hecto	h	10^{2}
deka	da	10
deci	d	10^{-1}
centi	c	10^{-2}
milli*	m	10^{-3}
micro*	μ	10^{-6}
nano*	n	10^{-9}
pico*	p	10^{-12}
femto*	f	10^{-15}
atto*	a	10^{-18}

*Denotes preferred prefixes.

APPENDIX B
INERTIA PROPERTIES

SHAPE	VOLUME	MOMENTS OF INERTIA

Uniform slender bar

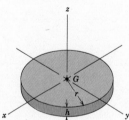

$V = LA$

$I_{xx} = I_{zz} = \frac{1}{12}mL^2$

$I_{yy} = 0$

Thin circular ring

$V = 2\pi rA$

$I_{xx} = I_{yy} = \frac{1}{2}mr^2$

$I_{zz} = mr^2$

Circular plate

$V = \pi r^2 h$

$I_{xx} = I_{yy} = \frac{1}{4}mr^2$

$I_{zz} = \frac{1}{2}mr^2$

Quarter circular plate

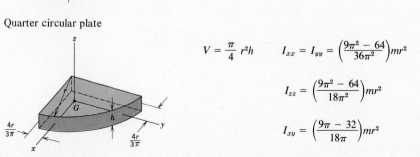

$V = \frac{\pi}{4} r^2 h$

$I_{xx} = I_{yy} = \left(\frac{9\pi^2 - 64}{36\pi^2}\right)mr^2$

$I_{zz} = \left(\frac{9\pi^2 - 64}{18\pi^2}\right)mr^2$

$I_{xy} = \left(\frac{9\pi - 32}{18\pi}\right)mr^2$

Elliptical plate

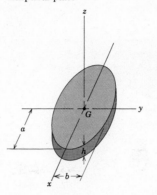

$$V = \pi abh$$

$$I_{xx} = \tfrac{1}{4}mb^2$$
$$I_{yy} = \tfrac{1}{4}ma^2$$
$$I_{zz} = \tfrac{1}{4}m(a^2 + b^2)$$

Triangular plate

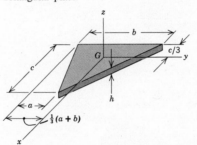

$$V = \tfrac{1}{2}bch$$

$$I_{xx} = \tfrac{1}{18}m(a^2 + b^2 - ab)$$
$$I_{yy} = \tfrac{1}{18}mc^2$$
$$I_{zz} = \tfrac{1}{18}m(a^2 + b^2 + c^2 - ab)$$
$$I_{xy} = \tfrac{1}{36}mc(2a - b)$$

Rectangular plate

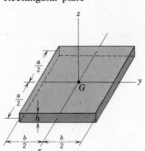

$$V = abh$$

$$I_{xx} = \tfrac{1}{12}mb^2$$
$$I_{yy} = \tfrac{1}{12}ma^2$$
$$I_{zz} = \tfrac{1}{12}m(a^2 + b^2)$$

$= \tfrac{1}{12} m(2a^2)$

Rectangular parallelipiped

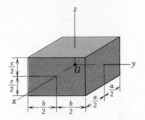

$V = abc$ $I_{xx} = \frac{1}{12}m(b^2 + c^2)$

$I_{yy} = \frac{1}{12}m(a^2 + c^2)$

$I_{zz} = \frac{1}{12}m(a^2 + b^2)$

Sphere

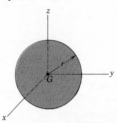

$V = \frac{4}{3}\pi r^3$ $I_{xx} = I_{yy} = I_{zz}$

$= \frac{2}{5}mr^2$

Hemisphere

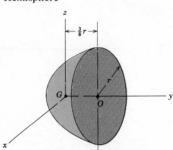

$V = \frac{2}{3}\pi r^3$ $I_{xx} = I_{zz} = \frac{83}{320}mr^2$

$I_{yy} = \frac{2}{5}mr^2$

Cylinder

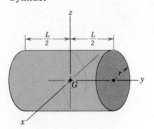

$V = \pi r^2 L$ $I_{xx} = I_{zz} = \frac{1}{12}m(3r^2 + L^2)$

$I_{yy} = \frac{1}{2}mr^2$

Semicylinder

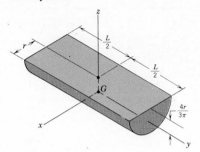

$$V = \tfrac{1}{2}\pi r^2 L \qquad I_{xx} = \left(\frac{9\pi^2 - 64}{36\pi^2}\right)mr^2 + \tfrac{1}{12}mL^2$$

$$I_{yy} = \left(\frac{9\pi^2 - 32}{18\pi^2}\right)mr^2$$

$$I_{zz} = \tfrac{1}{12}m(3r^2 + L^2)$$

Cone

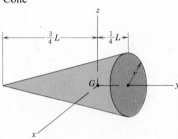

$$V = \tfrac{1}{3}\pi r^2 L \qquad I_{xx} = I_{zz} = \tfrac{3}{80}m(4r^2 + L^2)$$

$$I_{yy} = \tfrac{3}{10}mr^2$$

Semiellipsoid

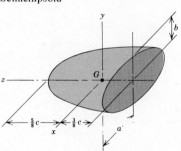

$$V = \tfrac{2}{3}\pi abc \qquad I_{xx} = \frac{m}{320}(64b^2 + 19c^2)$$

$$I_{yy} = \frac{m}{320}(64a^2 + 19c^2)$$

$$I_{zz} = \tfrac{1}{5}m(a^2 + b^2)$$

APPENDIX C
DENSITIES

MATERIAL	SPECIFIC GRAVITY $(\rho/\rho_{water})^*$
Aluminum	2.69
Concrete (av.)	2.40
Copper	8.91
Earth (av. wet)	1.76
(av. dry)	1.28
Glass	2.60
Iron (cast)	7.21
Lead	11.38
Mercury	13.57
Oil (av.)	0.90
Steel	7.84
Water (fresh, liquid)	1.00
(ice)	0.90
Wood (soft pine)	0.48
(hard oak)	0.80

$^*\rho_{water} = 1.00(10^3)$ kg/m^3.

$\gamma_{water} = \rho_{water}g = 62.4$ lb/ft^3.

ANSWERS TO ODD-NUMBERED PROBLEMS

I.1   (a) $1 \text{ kg} - \text{m}^2/\text{s}^2$
(b) $1 \text{ kg} - \text{m}^2/\text{s}^3$
(c) $1 \text{ kg} - \text{m}^2/\text{s}^3$

I.3   (a) $6.45(10^{-4}) \text{ m}^2$
(b) $2.83(10^{-2}) \text{ m}^3$
(c) 0.278 N
(d) $6.90(10^3) \text{ N/m}^2$
(e) 47.9 N/m^2
(f) 0.305 m/s
(g) $7.06(10^{-6}) \text{ m/s}$
(h) $1.242(10^{-4}) \text{ m/s}^2$

I.5   (a) 91.44 m
(b) 9 seconds
(c) 0.450 N

I.7 L^2/T

I.9 $-53.3\bar{I} + 53.3\bar{J}$ N-m

I.11 $\bar{R} = -5576\bar{I} - 4082\bar{J} + 673\bar{K}$ N
$M_D = -217\bar{I} - 313\bar{J} - 1115\bar{K}$ N
$(\bar{I} = \bar{e}_{B/A}, \bar{J} = \bar{e}_{D/A})$

I.13 $\theta_A = \theta_B = 55.55°, \theta_C = 68.90°$
$\bar{e}_n = 0.5145\bar{I} + 0.5145\bar{J} + 0.6860\bar{K}$

I.15 $\bar{R} = 250\bar{I} + 86.6\bar{J}$ N
(a) $M_{OZ} = 45$ N-m
(b) $M_{CZ} = 91.3$ N-m
$(\bar{I} \text{ rt}, \bar{K} \text{ out})$

I.17 $R = 4348 \text{ m}, \theta = 50.04°$

I.19 $\bar{v}_P = -7.9\bar{I} - 19.5\bar{J}$ m/s
$(\bar{I} \text{ rt}, \bar{J} \text{ up})$

I.21 $\bar{a}_P = -0.7\bar{I} - 0.9\bar{J}$ m/s^2
$(\bar{I} \text{ rt}, \bar{J} \text{ dn})$

II.1   139.4 ft

II.3   (a) -14.71 m/s^2
(b) $\bar{e}_n$ is 30° lt of vert, up
(c) 1938 m

II.5   (a) $s = 125$ ft
$v_P = 45$ ft/s
$\dot{v}_P = 6$ ft/s^2
(b) $s = 675$ ft
$v_P = 45$ ft/s
$\dot{v}_P = -6$ ft/s^2

II.7 $2.83 (gl)^{1/2}$

II.9   (a) $2\bar{e}_t + \bar{e}_n$ ft/s^2
(b) $2\bar{e}_t + 6.25\bar{e}_n$ ft/s^2

II.11   (a) $s = 1250$ m
(b) $s = 1229$ m

II.13 $-4.903\bar{e}_t + 0.870\bar{e}_n$ m/s^2

II.15 $\bar{e}_t = -0.930\bar{I} + 0.163\bar{J} + 0.332\bar{K}$
$\bar{e}_n = -0.208\bar{I} - 0.972\bar{J} - 0.105\bar{K}$
$\bar{e}_b = 0.306\bar{I} - 0.167\bar{J} + 0.938\bar{K}$

II.17   (a) $\bar{v}_P = 4\bar{I} + 3.46\bar{J}$ m/s
$\bar{a}_P = -3.46\bar{J}$ m/s^2
(b) $\bar{v}_P = 4\bar{I} + 1.55\bar{J}$ m/s
$\bar{a}_P = -0.31\bar{J}$ m/s^2

II.19 $\bar{v}_P = 6.0\bar{I} + 4.2\bar{J} + .6.36\bar{K}$ m/s
$\bar{a}_P = 180\bar{I} + 126\bar{J} + 159\bar{K}$ m/s^2

II.21 $\bar{e}_t = -\frac{2}{3}\bar{I} - \frac{2}{3}\bar{J} - \frac{1}{3}\bar{K}$
$\bar{e}_n = -0.447\bar{J} + 0.894\bar{K}$
$\dot{v}_P = 0$

II.23 $\bar{v}_P = 0.544\bar{I} - 0.238\bar{J}$ m/s
$\bar{a}_P = -4.93\bar{I} - 5.54\bar{J}$ m/s^2

II.25 $\dot{R} = -2.07$ m/s
$\dot{\phi} = -3.86$ rad/s
$\ddot{R} = 12.48$ m/s^2
$\ddot{\phi} = 12.99$ rad/s^2

II.27 $\dot{\theta} = 0.5$ rad/s
$\ddot{\theta} = -0.573$ rad/s^2

II.29 $\bar{v}_P = 80.8\bar{I} + 9.7\bar{J}$ in./s
$\bar{a}_P = 880\bar{I} - 2017\bar{J}$ in./s^2
$(\bar{I} \text{ lt}, \bar{J} \text{ up})$

II.31 $-5\bar{e}_R + 1.60\bar{e}_\theta + 1.27\bar{e}_h$ m/s^2

II.33 $\bar{v}_P = 4.3\bar{e}_R - 130.9\bar{e}_\theta - 2.5\bar{e}_h$ mm/s
$\bar{a}_P = -1370\bar{e}_R - 91\bar{e}_\theta - \bar{e}_h$ mm/s^2

II.35 $\bar{v}_A = -8.78\bar{e}_t$ m/s
$\bar{a}_A = -24.6\bar{e}_t$ m/s^2

II.37   (a) $\bar{v}_B = 3\bar{e}_R - 3.46\bar{e}_\theta$ in./s
$\bar{a}_B = -0.50\bar{e}_R - 2.98\bar{e}_\theta$ in./s^2
(b) $\dot{\phi} = -0.1455$ rad/s
$\ddot{\phi} = -0.0886$ rad/s^2

II.39 $\bar{v}_A = 5000\bar{e}_t$ in./s
$\bar{a}_A = 10^5[-7.5\bar{e}_t + 10\bar{e}_n]$ in./s^2

II.41 $\dot{\phi} = 0.1483$ rad/s
$\bar{v}_A = 6.69$ m/s

II.43 $\bar{a}_P = 10\bar{e}_t$ in./s^2: t < 0.465 seconds
$= 10\bar{e}_t + 6.67(3 + t)^2\bar{e}_n$: $0.465 < t < 0.789$
$= 10\bar{e}_t + 3.33(3 + t)^2\bar{e}_n$: $t > 0.789$

II.45 $\bar{v}_P = 20\bar{e}_R + 1.57\bar{e}_\phi$ m/s
$\bar{a}_P = -4.93\bar{e}_R + 125.7\bar{e}_\phi$ m/s^2
$\rho = 3.20$ m

II.47 $- (K_1x + K_2x^3)\bar{e}_t$

II.49 $v_P = 20.7$ ft/s
$\left|\bar{a}_P\right| = 138.5$ ft/s²

II.51 $-8.65\bar{e}_R + 4.35\bar{e}_\phi - 0.22\bar{e}_h$ m/s²

II.53 $v_A = 68.6$ m/s
$v_B = 52.3$ m/s
$v_C = 24.3$ m/s

II.55 $\ddot{d} = 12.06$ m/s²
$\ddot{\phi} = 0.0150$ rad/s²

II.57 $\bar{v}_P = 4t[(1 - \cos 4t^2)\bar{I} + \sin 4t^2\bar{J}]$ ft/s
$\bar{a}_P = 4[(1 - \cos 4t^2 + 8t^2 \sin 4t^2)\bar{I}$
$\qquad + (\sin 4t^2 + 8t^2 \cos 4t^2)\bar{J}]$ ft/s²

II.59 $\bar{a}_P = -\dfrac{\pi Ac}{L}\left\{1 - \dfrac{A}{c}\ \sin\left[\dfrac{\pi}{L}(X - ct)\right]\right\}\cos\left[\dfrac{\pi}{L}\ (X - ct)\right]\bar{I}$

II.61 $v_P = 20\pi\ (1 + \sin^2 \pi t/20)^{1/2}$ mm/s
$\dot{v}_P = (\pi^2 \sin \pi t/20)(\cos \pi t/20)(1 + \sin^2 \pi t/20)^{-1/2}$
$\rho = 400(1 + \sin^2 \pi t/20)^{3/2} (5 + 3 \sin^2 \pi t/20)^{-1/2}$

II.63 $\bar{a}_P = -\dfrac{K^2}{a^2 b^2}\ R\ \bar{e}_R$

II.65 $-\dfrac{A^2}{R^3}\ \bar{e}_R$; circle

II.67 $\left|\bar{v}_{P/air}\right| = 273$ km/hr
$\left|\bar{v}_P\right| = 193$ km/hr

II.69 $\dot{R} = -7.24$ m/s
$\ddot{R} = 3.12$ m/s²

II.71 $\bar{v}_{A/B} = -150.5\bar{i} - 40.3\bar{j}$ ft/s
$\bar{a}_{A/B} = -6.51\bar{i} + 0.43\bar{j}$ ft/s²
$[\bar{i} = (\bar{e}_t)_B, j = (\bar{e}_n)_B]$

II.73 $\bar{r}_{B/A} = -107.21\bar{i}$ m
$\bar{v}_{B/A} = -6\bar{i} + 10.39\bar{j}$ m/s
$\bar{a}_{B/A} = -0.373\bar{i} + 0.246\bar{j}$ m/s²
$(\bar{i}$ rt, $\bar{j}$ up)

II.75 (a) $\alpha = \cos^{-1}\left[-\dfrac{v_w}{v} \sin^2 \theta\right.$
$\qquad \left. \pm \cos \theta \left(1 - \dfrac{v_w^2}{v^2} \sin^2\theta\right)^{1/2}\right]$ from $\bar{I}$
(b) $v_0 = v_w \cos \theta \pm (v^2 - v_w^2 \sin^2 \theta)^{1/2}$
(c) $t = R/v_0$
(d) $v_{max} = v/\sin \theta$ for $\theta < 90°$, $v_{max} = v$ for $\theta > 90°$

II.77 $\bar{a}_{B/A} = 21.12\bar{j} - 6.42\bar{k}$ ft/s²

II.79 $s_{car} = 20$ m
$v_{car} = 3.5$ m/s (up)

II.81 0.25 m/s (up)

II.83 (a) 3.2 ft/s (up)
(b) 2.304 ft/s² (up)

III.3 $\dot{v}_{max} = 128.7$ ft/s
$s_{max} = 1222$ ft

III.5 (a) $v_{n+1} = v_n - 0.1\ g$

(b) $t = 9.59$ seconds
$s = 126.5$ ft

III.7 (a) 2.60 m/s (up)
(b) 1.799 m/s (up)

III.9 $v_P = 801.9$ ft/s for variable gravity
$v_P = 802.1$ ft/s for constant gravity

III.11 12.65 m/s (rt)

III.13 (a) 0.862 seconds
(b) 4.80 ft

III.15 $v = v_0 \exp(-kt/m)$
$v = v_0 - \dfrac{k}{m}\ s$
$s = \dfrac{mv_0}{k}\ [1 - \exp(-kt/m)]$

III.17 (a) 30.9 m
(b) 24.2 m/s
(c) 4.92 seconds

III.19 (a) 98.5 m
(b) 44.2 m/s

III.21 $t = 1.355$ seconds
$s_{el} = 4.06$ m

III.23 (a) $\dot{v}_P = 5.51$ ft/s²
$v_P = 141.3$ ft
(b) $\dot{v}_P = -1.126$ ft/s²
$v_P = 8.0$ ft/s

III.25 (a) 133.0 ft/s²
(b) $\rho = 9.95(10^5)$ ft

III.27 $T = \dfrac{mg}{\cos \theta}$, $v_P = \left(\dfrac{gl \sin^2\theta}{\cos \theta}\right)^{1/2}$

III.29 $F_{hor} = 1.207m$ newtons
$\theta = 110.19°$ from $\bar{e}_t$, $20.19°$ from $\bar{e}_n$

III.31 $h = 3.67$ m
$d = 19.26$ m

III.33 $Y = (\tan \alpha)X - \dfrac{g}{2v_0^2 \cos^2\alpha}\ X^2$
$v_P = v_0 \cos \alpha\ \bar{I} + \left(v_0 \sin \alpha - \dfrac{gX}{v_0 \cos \alpha}\right)\bar{J}$

III.35 $N_{cir} = 1956$ newtons
$N_{arm} = 1173$ newtons

III.37 $T = mR\omega^2 \cos \alpha$
$N = mR\omega^2 \sin \alpha$

III.39 0.982 rad/s

III.41 158.1 rad/s

III.43 0.0922 m

III.45 $t = 5.02$ s
$f = 6000$ lb

III.47 $s = 34.1$ m

$t = 6.82$ s

III.49 $F_{tr} = 1.633(10^4)$ N

$C_1 = 8.67(10^3)$ N

$C_2 = 4.33(10^3)$ N

III.51 (a) 2.77 m/s² (up)

(b) 660 N

III.53 $s = 133.8$ ft

III.55 $F = \frac{1}{2} mg\left(1 + \frac{v_0^2 L^2}{16gh^3}\right)\left(1 + \frac{L^2}{4h^2}\right)$

III.57 15.26°

III.59 $v_P = [g(r^2 - h^2)/h]^{1/2}$

III.61 $2.21 < v_P < 2.91$ m/s

III.63 $245\bar{e}_t + 196.1\bar{e}_n + 2\bar{e}_b$ kN

III.65 (a) 126.8 ft/s

(b) −11.68 ft/s²

III.67 $603\bar{e}_t + 627\bar{e}_n$ lb

III.69 (a) & (b) 8.86 m/s

III.71 (a) 7.84 N

(b) 5.79 N

III.73 $-3.01m\,\bar{e}_t + 4.33m\,\bar{e}_n$ N

III.75 $-0.5/X^2$ N

III.77 (a) $\ddot{R} = 100R$

(b) $\dot{R} = 10.01$ m/s

$\bar{v}_P = 10.01\bar{e}_R + 10\bar{e}_\phi$ m/s

III.79 (a) 13.86 m

(b) 3.33 seconds

III.81 (a) $58.7\bar{I} + 44\bar{J}$ ft/s

(b) $\Delta s = 15$ ft

(c) $\Delta h = 1.869$ ft

$(\bar{I}$ rt, $\bar{J}$ up)

III.83 $X = 29.5$ m

$X = 40.8$ m for $k = 0$

III.85 (a) 0.639 seconds

(b) 3.19 m

III.87 $\dfrac{eV_0 l}{mbu^2}\left(\dfrac{l}{2} + h\right)$

III.89 10.52 J

III.91 $-1599\,\dfrac{W}{g}$ ft-lb

III.93 10.36 J

III.95 54.7 J

III.97 (a) 49.02 kJ

(b) 12.26 kJ

III.99 $-mgl\cos\theta$

III.101 0.0922 m

III.103 6.44 m/s

III.105 Cannot reach point B

III.107 (a) 2.52 m/s

(b) 94.44°

III.109 $d = 0.6l$

III.111 $m(999e^{0.2\theta} - 79\cos\theta + 396\sin\theta - 1343)$ J

III.113 2.60 m/s

III.115 $\Delta = 1.773$ ft

$F = 42.6$ lb

III.117 2474 N

III.119 (a) 1.609 in.

(b) 5.44 in.

III.121 14.1 m/s

III.123 (a) 40.5 ft/s

(b) does not reach point B

III.125 3.63 m/s

III.127 ~~179.9 N~~ 238.71N

III.129 4.22 rad/s (cw)

III.131 (a) & (b) 11.34 ft/s

III.133 126 in./s (rt)

III.135 0.976 m/s²

III.137 (a) 122.2 ft/s

(b) 225 seconds

III.139 78.5 J

III.141 $\frac{1}{2} mgv\left(1 + \dfrac{l^2 v^2}{4gd^3}\right)\left(1 + \dfrac{l^2}{d^2}\right)^{1/2}$

III.143 (a) & (b) 2.84(10⁴) lb

III.145 1500 N

III.147 (a) $\bar{v}_{bl} = 13.86\bar{I} - 12.47\bar{J}$ ft/s

$\bar{v}_C = -2.77\bar{I}$ ft/s

(b) $-1.333\bar{I}$ ft

$(\bar{I}$ rt, $\bar{J}$ up)

III.149 (a) 2.86 m/s (uphill)

(b) 38.2 m/s (uphill)

III.151 (a) $-46.7\bar{I} + 50\bar{J} + 30\bar{K}$ mN

(b) 113.0 mN

III.153 (a) 71.4 lb

(b) 14.14 ft/s

(c) 3.11 ft

III.155 1.112 m

III.157 (a) 559 N

(b) 791 N

III.159 $v_{final} = 0.667$ m/s
$F_{AB} = 30$ kN, then 10 kN
$F_{BC} = 0$, then 20 kN

III.161   3 collisions
$\bar{v}_A = \frac{1}{8}v(3 - 3\epsilon + \epsilon^2 - \epsilon^3)\bar{I}$
$\bar{v}_B = \frac{1}{8}v(3 - \epsilon - 3\epsilon^2 + \epsilon^3)\bar{I}$
$\bar{v}_C = \frac{1}{4}v(1 + \epsilon^2)\bar{I}$
$(\bar{I}$ lt)

III.163 $\bar{v}_A = -0.4\bar{i}$ ft/s
$\bar{v}_B = 6.8\bar{i} + 17.32\bar{j}$ ft/s
$(\bar{i}$ rt, $\bar{j}$ up)

III.167 $d = (1 + \epsilon)v_1\left(\dfrac{2h_1}{g}\right)^{1/2}$
$h_2 = \epsilon^2\,h_1$

III.169 $\bar{v}_{ball} = 9.15\bar{i} + 14.14\bar{j}$ ft/s
$\bar{v}_{block} = 1.456\,(-\bar{i} + \bar{j})$ ft/s
$(\bar{i}$ 45° lt of vert, $\bar{k}$ in)

III.171   111 kN

III.173 $\ddot{\theta} + \dfrac{g}{l}\sin\theta = 0$

III.175   (a) 27.8 ft/s
(b) $13.11\bar{e}_R + 18.53\bar{e}_\phi$ ft/s

III.177 $-2.06\bar{e}_R + 1.50\bar{e}_\phi$ m/s

III.179   (a) 55.6 in./s
(b) 111.1 in./s
(c) $v_{circ} = 90.8$ in./s

III.181   15.71 rad/s (ccw)

III.183 $R\,\ddot{R} + 3\,\dot{R}\,\ddot{R} - \dfrac{2M}{mR^{1/2}} = 0$

III.185 $\dot{\omega} = -1.039$ rad/s²
$\dot{v}_C = 0.278$ ft/s² (dn)

III.187 $\ddot{\theta} + \dfrac{5}{9}\dfrac{g}{l}\sin\theta = 0$

IV.1   (a) $18.35\bar{i}$ in./s²
(b) $-4.9\bar{i} - 355.2\bar{j}$ in./s²
(c) $-8.74\bar{i} - 307.5\bar{j}$ in./s²
$(\bar{i}$ rt, $\bar{k}$ out)

IV.3   (a) $-0.707(\bar{i} + \bar{j})$ m/s
(b) $1.414\bar{k}$ rad/s
(c) $-4.83\bar{k}$ rad/s²
$(\bar{i}$ rt, $\bar{k}$ out)

IV.5 $\bar{v}_B = -0.212(\bar{i} + \bar{j})$ m/s
$\bar{a}_B = -0.1016(\bar{i} + \bar{j})$ m/s²
$(\bar{i}$ rt, $\bar{k}$ out)

IV.7 $\omega = 0$, $\alpha = 33.3$ rad/s² (ccw)

IV.9 $-17.38l\bar{i} - 12.57l\bar{j}$
$(\bar{i}$ rt, $\bar{j}$ dn)

IV.11 $\bar{v}_A = \dfrac{vl}{a}\sin^2\theta\cos\theta\,\bar{I} + \left(v - \dfrac{vl}{a}\sin^3\theta\right)\bar{J}$
$\bar{a}_A = \dfrac{v^2 l}{a^2}\sin^3\theta[(3\sin^2\theta - 2)\bar{I} + 3\sin\theta\cos\theta\bar{J}]$
$(\bar{I}$ rt, $\bar{J}$ up)

IV.15   704 ft below the pivot
Independent of θ

IV.17 $\omega = 12$ rad/s (ccw)
$v_0 = 5$ ft/s (lt)

IV.19   (a) $-10.80\bar{k}$ rad/s
(b) $3.60\bar{i} - 1.80\bar{j}$ m/s
$(\bar{i}$ rt, $\bar{k}$ out)

IV.21   25 rad/s (ccw)

IV.23   (a) $-20\bar{k}$ rad/s
(b) $-20\bar{k}$ rad/s²
(c) $40\bar{i}$ m/s
(d) $20\bar{i}$ m/s²
(e) $40\bar{i} - 400\bar{j}$ m/s²
$(\bar{i}$ rt, $\bar{k}$ out)

IV.25 $\bar{v}_0 = \omega r\,\bar{e}_t$
$\bar{a}_0 = \alpha r\,\bar{e}_t + \dfrac{\omega r^2}{R - r}\,\bar{e}_n$

IV.27   (a) $\omega = 16$ rad/s (ccw)
$\alpha = 2$ rad/s² (cw)
(b) $v_C = 8$ m/s (up)
$a_C = 1.0$ m/s² (dn)

IV.29 $(5.40\omega^2 - 12.60\alpha)\bar{i} + (5.40\omega^2 - 5.40\alpha)\bar{j}$
$(\bar{i}$ rt, $\bar{j}$ up)

IV.31 $\omega_{BC} = 1.778$ rad/s (ccw)
$v_C = 1067$ mm/s (up)

IV.33   (a) $\omega_{BC} = \omega\cos\left(\dfrac{l^2}{r^2} - \sin^2\theta\right)^{-1/2}$
(b) $v_C/v_B = \sin\theta + \sin\theta\cos\theta\,(4 - \sin^2\theta)^{-1/2}$

IV.35   54.6 rev/min

IV.37 $\omega_{BC} = \omega$ (cw)
$\omega_{CD} = 2\omega$ (cw)
$\alpha_{BC} = 6\omega^2$ (ccw)
$\alpha_{CD} = 14\omega^2$ (cw)

IV.39 $-128.20\bar{i} - 37.75\bar{j}$ m/s²
$(\bar{i}$ rt, $\bar{j}$ up)

IV.41 $\omega_{BC} = 2.90$ rad/s (cw)
$\alpha_{BC} = 3.43$ rad/s² (cw)

IV.43 $\omega = 1.25/\sin\theta$ (cw)
$\alpha = 1.5625\cos\theta/\sin^3\theta$ (cw)

IV.45   (a) $0.8\bar{i} + 4.8\bar{j}$ m/s
(b) $0.9\bar{i} + 2.4\bar{j}$ m/s²
$(\bar{i}$ lt, $\bar{k}$ out)

IV.47 (a) $48\bar{k}$ rad/s^2
(b) $-32\bar{j}$ ft/s^2
(c) $96\bar{i}$ ft/s^2
($\bar{i}$ rt, $\bar{k}$ in)

IV.49 (a) 0.69 ft/s (rt)
(b) 8.07 ft/s^2 (lt)

IV.51 $-1.93\bar{i} + 3.35\bar{j}$ m/s
($\bar{i}$ rt, $\bar{k}$ out)

IV.55 $\alpha_{AB} = 374$ rad/s^2 (ccw)
$\alpha_{BC} = 872$ rad/s^2 (cw)

IV.57 14.05 m/s^2 (rt)

IV.59 $\bar{v}_E = \omega r \bar{i}$
$\bar{a}_E = -0.75 r \omega^2 \bar{j}$
($\bar{i}$ rt, $\bar{k}$ out)

IV.61 $\bar{v}_C = 10.47\bar{i} + 20.94\bar{j}$
($\bar{i}$ rt, $\bar{k}$ in)

IV.63 $\bar{v}_P = 12.57\bar{i}$ m/s
$\bar{a}_P = -65.8\bar{i} - 79.0\bar{j}$ m/s^2
($\bar{i}$ rt, $\bar{k}$ in)

IV.65 $v_B = 2.78$ m/s (rt)
$a_B = 3.54$ m/s^2 (lt)

IV.67 (a) 0
(b) 1458 rad/s^2 (cw)

IV.69 $\alpha_r = 0$, $\alpha_{fr} = 0.343$ rad/s^2 (ccw)
$\alpha_{body} = 0.0080$ rad/s^2 (cw)

IV.71 2.17 rad/s^2 (cw)

V.1 $x_G = \dfrac{l}{2}$

V.3 $y_G = \dfrac{4r}{3\pi}$

V.5 $y_G = \dfrac{2r \sin \theta}{3\theta}$

V.7 $z_G = \dfrac{l}{2}$

V.9 $\dfrac{ml^2}{12}$

V.11 $0.3199\, mr^2$

V.13 $mr^2 \left(\dfrac{1}{2} - \dfrac{4 \sin^2 \theta}{9 \; \theta^2} \right)$

V.15 $\frac{3}{2}\, mr^2$

V.17 $\bar{r}_G = -3.33\bar{i} - 2.0\bar{j}$ mm
$I_{\hat{z}\hat{z}} = 4.035(10^{-3})$ kg-m^2

V.19 $\bar{r}_G = 129.52\bar{i} + 82.87\bar{j}$ mm
$I_{\hat{z}\hat{z}} = 4.947(10^{-3})$ kg-m^2

V.21 $\bar{r}_G = 78.21\bar{k}$ mm
$I_{\hat{z}\hat{z}} = 4.837(10^{-3})$ kg-m^2

V.23 $\bar{r}_G = 0.038\bar{i} - 7.537\bar{j}$ in.
$I_{\hat{z}\hat{z}} = 119.15$ lb-s^2-ft

V.25 $T = 40.5$ N

V.27 $\mu_k = 0.1$
$T = 0.1\, mg$ (tension)

V.29 (a) $N_{front} = 7.52$ kN
$N_{rear} = 6.96$ kN
(b) 1.70 m/s^2 (downhill)

V.31 $T = 5$ kips
$\bar{v} = 12.35$ ft/s^2 (uphill)

V.33 $0.457 < h < 1.143$ meters

V.35 $0.8 - \dfrac{1}{F}(343 + 784\mu) < h < 0.8$
$+ \dfrac{1}{F}(343 - 784\mu)$ meters
$h = 0.8$ m

V.37 (a) 194.4 ft
(b) 451.4 ft

V.39 $\bar{a}_G = g \cos \theta\, \bar{i}$
$T_{DC} = \dfrac{10g \sin \theta\, (2\sin \theta + \cos \theta)}{2\cos \theta + 3\sin \theta}$
$T_{BA} = \dfrac{10g \sin \theta\, (\cos \theta + \sin \theta)}{2\cos \theta + 3\sin \theta}$
($\bar{j} = \bar{e}_{A/B}$, $\bar{k}$ out)

V.41 1.81 lb-in.

V.43 (a) 7.55 kN
(b) 9.61 kN

V.45 (a) 16 rad/s^2 (cw)
(b) 8.81 rad/s^2 (cw)

V.47 4.99 in.

V.49 3.20 seconds

V.51 $\alpha = 24.1$ rad/s (cw)
$F_A = 3.75$ lb (up)

V.53 (a) $5.14\, \dfrac{P}{mL}$ (ccw)
(b) $0.285\, P\, \bar{i} + mg\, \bar{j}$
($\bar{i}$ rt, $\bar{j}$ up)

V.55 (a) $\ddot{\theta} + 0.947\, \dfrac{g}{b} \sin \theta = 0$
(b) $1.056\, b$

V.57 $f_A = 2.49$ lb (rt)
$f_B = 0$

V.59 (a) 1.810 kN
(b) 16.6 m (wound on)

V.61 (a) $a_A = 0.8915$ m/s^2 (up)
(b) $v_C = 2.67$ m/s

V.63 (a) $a_{axle} = 1.365$ m/s² (up)

$a_{CD} = 4.095$ m/s² (dn)

(b) 8.19 m/s (dn)

V.65 (a) $\frac{2}{3}g \sin \theta$ (dn parallel to incline)

(b) $\frac{1}{3} \tan \theta$

V.67 $a_G = 5.23\bar{i}$ m/s²

$\bar{R} = -6.54\bar{i} + 64.71\bar{j}$ N

$(\bar{j} = \bar{e}_n, \bar{k}$ in)

V.69 $\alpha = 6.84$ rad/s² (ccw)

$a_C = 54.7$ in./s² (lt)

V.71 (a) $8\bar{i}$ m/s

(b) $-66.67\bar{i}$ N

$(\bar{i}$ rt)

V.73 $a_G = 1.657$ m/s² (lt)

$\alpha = 8.29$ rad/s² (ccw)

V.75 (a) $\dfrac{\omega_1 R}{2}$

(b) $\dfrac{\omega_1 R}{2\mu_k g}$

V.77 $\mu = 0.40$

V.79 (a) 0.092

(b) $\dfrac{0.261g}{R}$ (cw)

V.81 0.558 m/s² (lt)

V.83 $T_A = 1245$ lb

$T_B = 1204$ lb

V.85 (a) 3491 kN (up)

(b) 55.3 m/s² (up)

V.87 4.90 m/s² (dn)

V.89 $T_A = 41.34$ lb

$T_B = 38.29$ lb

V.91 (a) $T_A = 0$

$T_B = 28.3$ lb (up)

(b) $-6.10\bar{i} - 5.24\bar{j}$ ft/s²

$(\bar{i}$ lt, $\bar{j}$ dn)

V.93 (a) 0.818 rad/s² (ccw)

(b) $\bar{R}_B = 851\bar{i} + 4255\bar{j}$ N

$\bar{R}_A = 158\bar{i} + 32\bar{j}$ N

$(\bar{i}$ rt, $\bar{j}$ up)

V.95 (a) 0.688 kN-m (ccw)

(b) $\bar{R}_A = 12.17\bar{i} + 9.39\bar{j}$ kN

$\bar{R}_B = -19.77\bar{i} + 13.77\bar{j}$ kN

$(\bar{i}$ lt, $\bar{j}$ up)

V.97 47.72 N

V.99 0

V.101 $a_A = 3.82$ m/s² (lt)

$a_B = 0.98$ m/s² (rt)

V.103 $\ddot{\theta}_1 + \dfrac{9g}{8l} \sin \theta_1 + \dfrac{3}{8} \ddot{\theta}_2 \cos(\theta_1 - \theta_2)$

$+ \dfrac{3}{8} \dot{\theta}_2^2 \sin(\theta_1 - \theta_2) = 0$

$\ddot{\theta}_2 + \dfrac{3g}{2l} \sin \theta_2 + \dfrac{3}{2} \ddot{\theta}_1 \cos(\theta_1 - \theta_2)$

$- \dfrac{3}{2} \dot{\theta}_1^2 \sin(\theta_1 - \theta_2) = 0$

V.105 $T_{disk} = 2m \, r^2\omega^2$

$T_{BC} = \dfrac{m \, \omega^2 r^2}{2}$

$T_{CD} = \dfrac{m \, \omega^2 r^2}{6}$

V.107 525.8 J

V.109 1347 ft-lb

V.111 246.7 N-m

V.113 104.54 rev

V.115 12.81 (dn)

V.117 5.46 m

V.119 (a) 6.69 ft/s (rt)

(b) 4.65 ft/s (rt)

V.121 (a) 8.31 ft/s (tang to circle, dn)

(b) 25.71 lb

V.123 55.25°

V.125 (a) 27.8 ft/s (up)

(b) 23.5 ft/s (up)

V.127 $\omega_{AB} = 3.25$ rad/s (cw)

$\omega_{BC} = 3.25$ rad/s (ccw)

V.129 (a) 3.09 ft/s

(b) 1 ft

V.131 (a) $145.77\bar{i}$ kg-m/s

(b) 163.13 kg-m²/s

$(\bar{i}$ lt, 28° below horiz)

V.133 6.51 seconds

V.135 $v_1 = \frac{2}{3}g \sin \theta t$

$v_{sph} = \frac{5}{7}g \sin \theta t$

V.137 5.11 seconds

V.139 110.95 m/s (uphill)

V.141 (a) 1.732 rad/s

(b) $0.75\bar{e}_R + 1.73\bar{e}_\theta$ ft/s

V.143 $h = 0.450$ m

$\omega = 40$ (cw)

V.145 (a) $8.07\bar{i}$ ft/s

$F_{tt} = 540$ lb, $F_{rt} = 188$ lb

$(\bar{i}$ rt, 30° above horiz)

V.147 (a) 28 mm

(b) 10 m/s (rt)

V.149 $v_{top} = 0$

$v_{bot} = 54.45$ ft/s (rt)

V.151 0.99 ft/s

V.153 $v_A = 7.03$ ft/s (up)

$v_B = 25.49$ ft/s (dn)

V.155 $\bar{v}_G = 5.4\bar{\imath} + 8\bar{\jmath}$

$\bar{\omega} = 100\bar{k}$ rad/s

($\bar{\imath}$ up, $\bar{\jmath}$ rt)

V.157 $\bar{v} = 13.86\bar{\imath} + 10\bar{\jmath}$

($\bar{\imath}$ 60° lt of vert, $\bar{\jmath}$ 30° rt of vert)

VI.1 (a) $1.0\ (\bar{\imath} + \bar{\jmath})$ m/s

(b) $2.75\bar{\imath} - 0.75\bar{\jmath}$ m/s

($\bar{\imath} = \bar{e}_{B/A}$, $\bar{k}$ out)

VI.3 $\bar{v}_A = -0.10\bar{\imath} - 1.766\bar{\jmath}$ m/s

$\bar{v}_B = 1.020\bar{\imath} - 0.006\bar{\jmath}$ m/s

($\bar{\imath}$ rt, $\bar{\jmath}$ up)

VI.5 (a) 17.25 rad/s

(b) 28.12°

VI.7 (a) $4\bar{\imath}$ m/s

(b) $-4\bar{\jmath} - 10\bar{k}$ m/s

(c) $2\bar{\imath} - 3.46\bar{\jmath} - 8.66\bar{k}$ m/s

($\bar{\imath}$ rt, $\bar{\jmath}$ up)

VI.9 (a) & (b) $- 6.98\bar{\jmath} - 0.60\bar{k}$ m/s

($\bar{\imath} = \bar{e}_{P/O}$, $\bar{k}$ out)

VI.11 $\bar{a}_A = -4.0\bar{\imath} + 2.4\bar{\jmath}$ m/s²

$\bar{a}_B = 3.68\bar{\imath} - 1.76\bar{\jmath}$ m/s²

($\bar{\imath}$ rt, $\bar{\jmath}$ up)

VI.13 $\bar{a}_A = -7.46\bar{\imath} + 0.40\bar{\jmath}$ m/s²

$\bar{a}_B = 0.22\bar{\imath} - 3.76\bar{\jmath}$ m/s²

($\bar{\imath}$ rt, $\bar{\jmath}$ up)

VI.15 (a) $-800\bar{\imath} - 8\bar{\jmath} - 170\bar{k}$ m/s²

(b) $-1608\bar{\imath} - 20\bar{k}$ m/s²

(c) $-1500\bar{\imath} - 4\bar{\jmath} - 99\bar{k}$ m/s²

($\bar{\imath}$ rt, $\bar{\jmath}$ up)

VI.17 $-16.35\bar{\imath} - 21.88\bar{\jmath} - 0.88\bar{k}$ m/s²

($\bar{\imath} = \bar{e}_{P/O}$, $\bar{k}$ out)

VI.19 (a) $\bar{\alpha} = 150\pi\bar{\jmath} - 10\bar{k}$ rad/s²

(b) $(10R - 25L)\bar{\imath} - [(900\pi^2 + 25)R + 10L]\bar{\jmath}$

($\bar{\imath} = \bar{e}_{B/A}$, $\bar{k}$ out)

VI.21 $-3.87\bar{\imath} + 1.4709(10^4)\bar{k}$ N

($\bar{\imath}$ east, $\bar{\jmath}$ north)

VI.23 $a = 2.53(10^{-3})$ ft/s² (south)

$\rho = 1.584(10^4)$ ft

VI.25 $-2m\ \omega_e\ gt \cos \lambda\ \bar{\imath}$

($\bar{\imath}$ east)

VI.27 (a) $\bar{v}_{B/A} = -161.3\bar{\imath}$ ft/s

$\bar{a}_{B/A} = -0.723\bar{\jmath}$ ft/s²

(b) $(\bar{v}_{B/A})_{rel} = -152.3\bar{\imath}$ ft/s

$(\bar{a}_{B/A})_{rel} = 27.4\bar{\jmath}$ ft/s²

$[\bar{\imath} = (\bar{e}_t)_A, \bar{\jmath} = (\bar{e}_n)_A]$

VI.29 $(\bar{v}_{A/B})_{rel} = 625\bar{\imath} - 625\bar{\jmath}$ m/s

$(\bar{a}_{A/B})_{rel} = -260\bar{\imath} - 304\bar{\jmath}$ m/s²

($\bar{\imath}$ south, $\bar{\jmath}$ east)

VI.31 (a) $\bar{a}_P = -21\bar{\imath} + 519\bar{\jmath}$ m/s²

(b) $v_P = 13.22$ m/s

$\theta = 10.89°$ (cw from $\bar{\imath}$)

($\bar{\imath} = \bar{e}_R, \bar{\jmath} = \bar{e}_\phi$)

VI.33 $\omega_{AB} = 57.73$ rad/s (ccw)

$\dot{\omega}_{AB} = 15795$ rad/s² (cw)

VI.35 $\bar{a}_E = -100\bar{\imath} - 100\bar{\jmath}$ m/s²

$\rho = 0.354$ ft

($\bar{\imath}$ lt, $\bar{\jmath}$ up)

VI.37 $\alpha_{AB} = 0.1519$ rad/s² (ccw)

$\alpha_{CD} = 0.0969$ rad/s² (cw)

VI.39 $\bar{a}_t = -\sin \theta \cos \theta\ \bar{\jmath} + 4 \cos \theta\ \bar{k}$ in./s²

$\bar{a}_n = -(4 + \sin^2 \theta)\bar{\imath}$ in./s²

$|\bar{a}_t|_{max} = 4$ in./s², $|\bar{a}_n|_{max} = 5$ in./s²

($\bar{\imath} = \bar{e}_n$, $\bar{k}$ out)

VI.41 $\bar{\omega} = 3\bar{\jmath} + 2\bar{k}$ rad/s

$\bar{\alpha} = 6\bar{\imath} - 2.67\bar{k}$ rad/s²

($\bar{\imath}$ rt, $\bar{\jmath}$ up)

VI.43 0.3 m/s (dn)

VI.45 0

VI.47 400 ft/s² (in)

VI.49 $\bar{\omega} = -104.7\bar{\imath} - 10\bar{k}$ rad/s

$\bar{\alpha} = 1047\bar{\jmath} - 2\bar{k}$ rad/s²

($\bar{\imath} = \bar{e}_{B/A}, \bar{\jmath} = \bar{e}_{C/B}$)

VI.51 $\bar{\omega} = -125.66\bar{\imath} - 0.264\bar{k}$ rad/s

$\bar{\alpha} = 33.2\bar{\jmath}$ rad/s²

($\bar{\imath}$ rt, $\bar{k}$ up)

VI.53 (a) $\bar{\omega} = -50\bar{\imath} + 10\bar{\jmath}$ rad/s

$\bar{\alpha} = -100\bar{\imath} + 20\bar{\jmath} + 500\bar{k}$ rad/s²

(b) $\bar{v}_C = -5\bar{k}$ m/s

$\bar{a}_C = -75\bar{\imath} - 125\bar{\jmath} - 10\bar{k}$ m/s²

($\bar{\imath} = \bar{e}_{B/O}, \bar{\jmath}$ up)

VI.55 $\bar{\omega} = -35.2\bar{\jmath} + 0.40\bar{k}$ rad/s

$\bar{\alpha} = 14.01\bar{\imath}$ rad/s²

($\bar{\imath} = -\bar{e}_t$, $\bar{k}$ 60° above $\bar{e}_n$)

VI.57 $\bar{\omega} = -0.0667\bar{\imath} - 0.125\bar{k}$ rad/s

$\bar{\alpha} = 0.5\bar{\imath} + 0.0083\bar{\jmath}$ rad/s²

($\bar{\jmath}$ rt, $\bar{k}$ up)

VI.59 $\bar{\omega} = (0.5 \sin \frac{\pi t}{5} - 30\pi)\bar{\imath} + 0.866 \sin \frac{\pi t}{5} \bar{\jmath}$ rad/s

$\bar{\alpha} = 0.1\pi \cos \frac{\pi t}{5} \bar{\imath} + 0.1732\pi \cos \frac{\pi t}{5} \bar{\jmath} + 25.98\pi \sin \frac{\pi t}{5} \bar{k}$ rad/s²

($\bar{\imath}$ 60° lt of vert, $\bar{\jmath}$ 30° rt of vert)

VI.63 $\bar{\omega} = -0.5\bar{\imath} + 0.3\bar{\jmath}$ rad/s
$\bar{\alpha} = -0.6\bar{\jmath}$ rad/s^2

VI.65 $\bar{v} = -1.80\bar{\imath} + 4.80\bar{\jmath} + 9.60\bar{k}$ m/s
$\bar{a} = 0.408\bar{\imath} - 0.804\bar{\jmath} - 0.024\bar{k}$ m/s^2

VI.67 (a) $\bar{v}_C = 5\bar{\imath} + 0.20\bar{\jmath} - 1.0\bar{k}$ m/s
$\bar{a}_C = -4\bar{\imath} + 148\bar{\jmath} - 1000\bar{k}$ m/s^2
(b) $\bar{v}_D = -5\bar{\imath} + \bar{\jmath} - 11\bar{k}$ m/s
$\bar{a}_D = -1010\bar{\imath} - 92\bar{\jmath}$ m/s^2
($\bar{\jmath}$ rt, $\bar{k}$ up)

VI.69 $\bar{\omega} = -5\bar{\imath} + 63.3\bar{\jmath} + 0.5\bar{k}$ rad/s
$\bar{\alpha} = -64.4\bar{\imath} - 4.3\bar{\jmath} - 311.7\bar{k}$ rad/s^2
($\bar{\jmath}$ 60° rt of vert, $\bar{k}$ 30° lt of vert)

VII.1 $-\frac{1}{4} mr^2\omega\bar{\imath}$

VII.3 $-0.5\omega_2\bar{\imath} - \omega_1\bar{k}$

VII.5 $0.10053\bar{\imath} + 0.16056\bar{\jmath}$ kg-m^2/s

VII.7 $\frac{1}{4}mr^2[1.732(\omega_1 - \omega_2)\bar{\imath} + 0.5(\omega_1 + \omega_2)\bar{\jmath}]$

VII.13 $I_{xx} = \frac{1}{6}m\,(b^2 + c^2)$
$I_{yy} = \frac{1}{6}m\,(2a^2 + c^2)$
$I_{zz} = \frac{1}{6}m\,(2a^2 + b^2)$

VII.15 $I_{xx} = I_{zz} = \frac{1}{6}m\,(3h^2 + a^2)$
$I_{yy} = \frac{1}{3}ma^2$

VII.17 $I_{xy} = \frac{1}{20}mab$, $I_{xz} = \frac{1}{20}mac$
$I_{yz} = \frac{1}{20}mbc$

VII.19 (a) $1.1052(10^5)$ kg/m^5
(b) $z_G = 0.2063$ m
(c) $I_{\hat{x}\hat{x}} = I_{\hat{y}\hat{y}} = 0.4244$ kg-m^2
$I_{\hat{z}\hat{z}} = 0.4419$ kg-m^2

VII.21 $\dfrac{3m}{20}\,(r^2 + 4h^2)$

VII.23 (a) $I_{xx} = I_{yy} = I_{zz}$
$= \frac{2}{5}m\left[\dfrac{5r^4h + 10r^3h^2 + 10r^2h^3 + 5rh^4 + h^5}{3r^2h + 3rh^2 + h^3}\right]$
(b) $I_{xx} = I_{yy} = I_{zz} = \frac{2}{3}mr^2$

VII.25 $I_{\hat{x}\hat{x}} = I_{\hat{y}\hat{y}} = 2.002$ kg-m^2, $I_{\hat{z}\hat{z}} = 3.497$ kg-m^2
($\hat{z}$ is axis of symmetry)

VII.27 (a) 2095 lb-s^2-ft
(b) 2038 lb-s^2-ft

VII.29 $z_G = 0.00641$ (lt of centroid of cylinder)
$I_{xx} = I_{yy} = 24.96$ kg-m^2
$I_{zz} = 10.98$ kg-m^2
(z is axis of symmetry)

VII.31 $I_{xx} = \frac{10}{3}\,ml^2$
$I_{yy} = \frac{22}{3}\,ml^2$
$I_{zz} = \frac{32}{3}\,ml^2$
$I_{xy} = -2ml^2$
$I_{xz} = I_{yz} = 0$

VII.33 $\frac{7}{16}\,mr^2$

VII.35 $\dfrac{-\sqrt{3}}{16}\,mr^2$

VII.37 0.11426 kg-m^2

VII.39 0

VII.41 $-\dfrac{100\pi}{3}\,\bar{k}$ rad/s^2

VII.43 (a) $\bar{F}_s = \frac{1}{8}ml\omega^2\,(\sin\theta\,\bar{\imath} - \cos\theta\,\bar{\jmath}) \pm mg\,\bar{k}$
$\bar{M}_s = \frac{1}{8}ml^2\omega^2\sin\theta\cos\theta\,\bar{k} \pm \frac{1}{8}mgl\,\bar{\jmath}$
(b) $\bar{F}_s = (\frac{1}{8}ml\omega^2 \pm mg)(\sin\theta\,\bar{\imath} - \cos\theta\,\bar{\jmath})$
$\bar{M}_s = (\frac{1}{8}ml^2\omega^2\sin\theta\cos\theta \pm \frac{1}{8}mgl\cos\theta)\bar{k}$
($\bar{\imath}$ along short leg of bar, $\bar{k}$ out)

VII.45 (a) $\bar{A} = -\bar{B} = 22.11\bar{I} + 7.29\bar{K}$ N
(b) $\bar{A} = -\bar{B} = 5.53\bar{I} + 7.29\bar{K}$ N
(a) & (b) $\dot{\omega} = -208$ rad/s^2
($\bar{J} = \bar{e}_{B/A}$, $\bar{K}$ up)

VII.47 (a) $\bar{H}_A = -93.76\bar{\imath} + 611.8\bar{k}$ lb-s-in.
$\dot{\bar{H}}_A = -468.8\bar{\jmath} - 1224\bar{k}$ lb-in.
(b) $\bar{H}_A = 77.72\bar{k}$ lb-s-in.
$\dot{\bar{H}}_A = -155.43\bar{k}$ lb-in.
($\bar{\imath} = \bar{e}_{B/A}$, $\bar{k}$ in)

VII.49 $(0.433\,\omega_1\omega_2 - 0.1083\omega_1{}^2)mr^2\bar{k}$

VII.51 $\bar{H}_A = -48.8\,mr^2\bar{\imath} + 278.2\,mr^2\bar{\jmath}$

VII.53 (a) $(\bar{H}_C)_A = -0.4\,mr^2\omega_1\bar{\imath} + (0.4mr^2 + ml^2)\Omega\bar{\jmath}$
$(\dot{\bar{H}}_C)_A = 0.4mr^2\omega_1\Omega\,\bar{k}$
$(\bar{H}_C)_B = 0.4\,mr^2\omega_2\bar{\imath} + (0.4mr^2 + ml^2)\Omega\,\bar{\jmath}$
$(\dot{\bar{H}}_C)_B = -0.4mr^2\omega_2\Omega\bar{k}$
(b) $0.4r^2(\omega_1 - \omega_2)\Omega\bar{k}$
($\bar{\imath} = \bar{e}_{B/C}$, $\bar{\jmath}$ up)

VII.55 $\dot{\omega} = -9376$ rad/s^2
$\bar{A} = 167\bar{J} - 3563\bar{K}$ N
$\bar{B} = 9\bar{J} - 188\bar{K}$ N
($\bar{I} = \bar{e}_{B/A}$, $\bar{J}$ up)

VII.57 $\dot{\omega} = -8.220$ rad/s^2
$\bar{A} = -3654\bar{I} - 66\bar{K}$ N
$\bar{B} = -5115\bar{I} - 96\bar{K}$ N
($\bar{J} = \bar{e}_{B/A}$, $\bar{K}$ rt)

VII.59 $N_A = N_B = 68.6$ newtons
$f_B = -f_A = -v^2$ newtons

VII.61 (a) $-1.302\bar{\jmath}$ lb-ft
(b) $1.302\bar{\imath}$ lb-ft

VII.63 $\Sigma\bar{F} = m_1\bar{a}_{G_1} + m_2\bar{a}_{G_2}$
$\Sigma\bar{M}_P = \dot{\bar{H}}_{G_1} + \dot{\bar{H}}_{G_2} + \bar{r}_{G_1/P} \times m_1\bar{a}_{G_1} + \bar{r}_{G_2/P} \times m_2\bar{a}_{G_2}$

VII.65 (a) 1.190 N-m (opposite $\bar{v}_{car}$)
(b) $\Delta N_{outer} = -\Delta N_{inner} = 1.190$ N

VII.67 (a) $-m\,[k_1{}^2\omega_1\omega_2\sin\theta + (k_1^2 - k_2^2)\omega_2^2\sin\theta\cos\theta]\bar{k}$
[$\bar{\imath}$ at θ rt of vert, $\bar{\jmath}$ at $(90° - \theta)$ lt of vert]

VII.69 $\bar{F}_A = 2.49(10^3)\bar{\imath} + 9.04(10^3)\,\bar{\jmath} - 1.12(10^3)\bar{k}$ lb
$\bar{M}_A = 3.58(10^4)\,\bar{\jmath} + 2.19(10^5)\bar{k}$ lb-ft
$(\bar{\imath} = \bar{e}_{G/A}, \bar{k}$ out$)$

VII.71 394 N

VII.73 $\bar{M}_A = 18.70\bar{k}$ lb-in.
$\bar{F}_A = 0.526\bar{\imath} + 1.973\bar{\jmath}$ lb
$(\bar{\imath} = \bar{e}_{G/A}, \bar{k}$ out$)$

VII.75 $N = m(g + \frac{1}{2}r\omega_2{}^2) > 0$ always

VII.77 $(\omega_1)_{min} = 3.08$ rad/s

VII.79 (a) $\cos\theta = \dfrac{(3r^2 - 4l^2)\omega_1\omega_2 - 4gl}{(r^2 - 4l^2)\omega_2{}^2}$

(b) $\omega_2 = \dfrac{4gl}{3r^2 - 4l^2}\,\omega_1$

VII.81 (a) $(0.4r^2 + L^2)\ddot{\theta} + gL\sin\theta - L^2\omega_1{}^2\cos\theta\sin\theta = 0$

(b) $\cos\theta = \dfrac{g}{L\omega_1{}^2}$

(c) $\omega_1 < \left(\dfrac{g}{l}\right)^{1/2}$

VII.83 $\omega_1{}^2 L(\frac{1}{2} + \frac{7}{12}\sin\theta)\cos\theta = mg\sin\theta$

VII.85 $\bar{M}_G = -174.0\bar{k}$ N-m
$\bar{F}_G = 5767\bar{k}$ N
$(\bar{\imath} = \bar{e}_t, \bar{k}$ up$)$

VII.87 $\bar{M}_G = -3.15(10^8)\bar{k}$ N-m
$\bar{F}_G = 2.45(10^8)\bar{k}$ N

VII.89 (a) $41.09°$
(b) 7.88 rad/s
(c) 8.43 rad/s

VII.91 $\dot{\omega}_2 = 45.5$ rad/s^2
$\ddot{\theta} = 27.5$ rad/s^2

VII.93 $\dot{\omega}_2 = 1355$ rad/s^2
$\ddot{\theta} = 25.5$ rad/s^2

VII.95 24.34 J

VII.97 7.32 rad/s

VII.99 0.285 lb-ft

VII.101 4.86 m/s

VII.103 (a) 16.16 m/s
(b) $0.80\bar{\jmath}$ rad/s

VII.105 $x_P + y_P = \frac{2}{3}l$

VII.107 $\Delta\bar{v}_G = 5\bar{\imath}$ m/s
$\bar{\omega}_f = 27.5\bar{\jmath} + 7.5\bar{k}$ rad/s

VII.109 $\bar{v}_G = -3.680\bar{k}$ m/s
$\bar{\omega} = 33.0\bar{\imath} + 50.1\bar{\jmath}$ rad/s
$(\bar{k}$ up, $\bar{\jmath}$ lt along flat edge$)$

VII.111 $\dot{\theta} = 8.705$ rad/s
$\omega_1 = 2.038$ rad/s

VIII.1 $k_2/k_1 = 1/4$

VIII.3 $m = 2.56$ kg
$k = 650$ kN/m

VIII.5 (a) 0.10 m
(b) $\ddot{s} + 4(10^3)s = 0.4(10^3)$
(c) $\ddot{x} + 4(10^3)x = 0$
(d) 63.2 rad/s
(e) parts (a) and (b)

VIII.7 (a) $2\pi\left(\dfrac{\delta_{st}}{g}\right)^{1/2}$

(b) δ_{st}

VIII.9 $\ddot{R} + \left(\dfrac{k}{m} - \Omega^2\right)R = \dfrac{k}{m}R_0$

$\omega_n = \left(\dfrac{k}{m} - \Omega^2\right)^{1/2}$

$\Omega < \left(\dfrac{k}{m}\right)^{1/2}$

VIII.11 $m\ddot{x} + k\left[1 + \dfrac{(T - kl)}{k(x^2 + L^2)^{1/2}}\right]x = 0$

$m\ddot{x} + \dfrac{T}{L}x = 0$, for $x << L$

VIII.13 (a) 0.355 seconds
(b) For $\theta = -2.25°$, $t = 0.2364$ seconds
$\bar{v} = 0.151\bar{e}_\theta$ m/s
$\bar{a} = -0.045\bar{e}_R + 0.385\bar{e}_\theta$ m/s^2
For $\theta = 2.25°$, $t = 0.4729$ seconds
$\bar{v} = 0.151\bar{e}_\theta$ m/s
$\bar{a} = -0.045\bar{e}_R + 0.385\bar{e}_\theta$ m/s^2

VIII.15 (a) 1961 N/m
(b) $x = -0.020\cos 22.14t$

VIII.17 (a) 5.0 ft
(b) $x = C\sin(143.5\,t + \phi)$
(c) $C = 0.1251$ ft
$\omega_n = 143.5$ rad/s
$\phi = 0.050$ rad

VIII.19 (a) 0.0349 seconds
(b) 3.90 m/s
(c) $C = 54.9$ mm
$\omega_n = 60$ rad/s
$\phi = 0.932$ rad

VIII.21 $\omega_d = 209.4$ rad/s
$\beta = 0.036$
$x_1 = 4.48$ mm

VIII.23 (a) $100\{1 - \exp[-4\pi\beta/(1 - \beta^2)^{1/2}]\}\%$
(b) T + V decreases by a constant % in each cycle

VIII.27 489 N-s/m

VIII.29 (a) $3.065\sin(9.787t)\exp(-7.363t)$
(b) $t = 0.0946$ seconds, $x = 1.2205$ m
(c) 2.45 m

VIII.31 $x = -129.3 \sin 31.62t + 3.21 \sin 8t$
$+ 4.03 \sin 16t$ mm

VIII.33 (a) $x = 0.417(1 - \cos 13.89t)$ ft
(b) $x = 0.513 \sin (13.89t - 0.663)$ ft

VIII.35 (a) 44.29 rad/s
(b) 1.067 mm

VIII.37 (a) $\Omega > (2g/3R)^{1/2}$
(b) not possible

VII.39 (a) $\ddot{y} + 160\, y = 6.40 \sin \dfrac{\pi vt}{2}$

(b) 8.05 m/s
(c) 73.8 mm

VIII.41 (a) $\Omega = \omega_n(1 - 2\beta^2)^{1/2}$
(b) $D = F/[2k\beta(1 - \beta)]$
$\psi = \tan^{-1}[(1 - 2\beta^2)^{1/2}/\beta]$
$\beta > \sqrt{2}/2$

VIII.43 (a) $x = D \sin (\Omega t + \alpha + \psi)$ where:
$D = 0.005[(1 - \Omega^2/2000)^2 + \Omega^2/2500]^{-1/2}$
$\alpha = \tan^{-1}(0.02\, \Omega)$
$\psi = \tan^{-1}[0.02\, \Omega\, (1 - \Omega^2/2000)^{-1}]$
(b) $\Omega = 6.5$Hz: $a_{max} = 10.0$ m/s^2
$\Omega = 7.5$Hz: $a_{max} = 11.7$ m/s^2

VIII.45 $x = x_0 \cos(\Omega t + \psi)$ where
$\dfrac{x_0}{\delta_0} = \dfrac{\Omega^2}{\omega_n^{\,2}}\left[\left(1 - \dfrac{\Omega^2}{\omega_n^{\,2}}\right)^2 + 4\beta^2\,\dfrac{\Omega^2}{\omega_n^{\,2}}\right]^{-1/2}$
$\psi = \tan^{-1}\left[2\beta\,\dfrac{\Omega}{\omega_n}\left(\dfrac{\Omega^2}{\omega_n^{\,2}} - 1\right)^{-1}\right]$

VIII.47 $\ddot{\theta} + \dfrac{8g}{3\pi R}\,\theta = 0$

VIII.49 $\ddot{\theta} + 7.5(10^3)\theta = 0$

VIII.51 $ml^2\ddot{\theta} + c\, a^2\dot{\theta} + k\, a^2\theta = 0$

VIII.53 $\ddot{\theta} + \dfrac{2g}{3mr}\,\theta = 0$

VIII.55 19.65 rad/s

VIII.57 9.90 rad/s

VIII.59 $I_{zz} = \dfrac{\tau^2}{4\pi^2}\, k$

VIII.61 $(\omega_n)_{el} = \left[\dfrac{6}{5}\left(\dfrac{k}{m} + \dfrac{g}{l}\right)\right]^{1/2}$
□stable for all k
$(\omega_n)_{rt}\left[\dfrac{6}{5}\left(\dfrac{k}{m} - \dfrac{g}{l}\right)\right]^{1/2}$
□unstable when $k < mg/l$

VIII.63 $R^{-1} = 3.29(10^{-8})[0.280 \cos \phi + 1]$
$v_A = 1.548(10)^4$ ft/s
$v_P = 2.75(10)^4$ ft/s

VIII.65 $R^{-1} = 7.32(10^{-5}) \cos (\phi - 89.04°) + 1.333(10^{-4})$
1/km
$e = 0.549$, closed orbit

VIII.67 18.75°

VIII.69 (a) 1.695
(b) 2.49(10^4) km
(c) 31.69°

VIII.71 51.46°

VIII.73 (a) 2.398(10^4) mile^2/s
(b) -6.790 mile^2/s
(c) 0.380
(d) 2.469 mile/s
(e) 9712 miles

VIII.75 $\bar{v} = -11.34\bar{e}_R + 8.63\bar{e}_\phi$ km/s

VIII.77 (a) $3.82\bar{e}_\phi$ km/s
(b) $\pm 2.55\bar{e}_R + 6.37\bar{e}_\phi$ km/s

VIII.79 $\dot{v} = -5.35(10^3)$ mile/hr^2
$\rho = 5.18(10^4)$ miles

VIII.81 (a) $\dot{v} = -2.85(10^{-3})$ km/s^2
$\rho = 2.06(10^4)$ km
(b) $\dot{v} = -1.273(10^{-3})$ km/s^2
$\rho = 9.90(10^4)$ km

VIII.83 $\bar{v} = 3.67\bar{e}_R + 3.65\bar{e}_\phi$ km/s

VIII.85 53.90 hrs

VIII.87 178.75 days

VIII.89 3.452 hours

VIII.91 85.08°

VIII.93 (a) 1000 N
(b) 2500 N

VIII.95 (a) $(20 + 0.12434\, t)\dot{v} + 0.12434\, v = 2.984$
(b) at $t = 0.5$ s, $v = 0.0744$ ft/s
at $t = 1.0$ s, $v = 0.1483$ ft/s

VIII.97 50.98 lb more than tang comp of weight

VIII.99 (a) $\sigma[g(l - s) + v^2]$
(b) $\sigma g s$

VIII.101 $R = 4516$ N (rt)
$M = 0$

VIII.103 12.51 m

VIII.105 (a) 9.09 lb
(b) 11.57 lb

VIII.107 $|\bar{R}| = 345.1$ N, midpoint of vane along outward normal

VIII.109 $\bar{A} = 621.3\bar{I} + 3750\bar{J}$ N
$\bar{B} = 8177\bar{J}$ N
($\bar{I}$ rt, $\bar{J}$ up)

VIII.111 $\bar{A} = 6903\bar{I}$ N
$\bar{C} = -3170\bar{I} + 12{,}492\bar{J}$ N
($\bar{I}$ rt, $\bar{J}$ up)

VIII.113 $3.14(10^4)$ N

VIII.115 $0.2131v^3$ ft-lb/s
increases monotonically

VIII.117 40.47 rad/s

VIII.119 (a) for $\dot{m} = 30/g$, $T = 9325$ lb
(b) for $\dot{m} = 60/g$, $T = 18{,}650$ lb

VIII.121 $\dot{m}_P = \dfrac{m_0\,(\dot{v} + g)}{(\dot{v} + g)t + u}$

VIII.123 single stage: $h_{burnout} = 1.1205(10^5)$ m
two stage: $h_{burnout} = 2.497(10^5)$ m

VIII.125 $T_{engine} = 4555$ lb (opp $\bar{v}$)

VIII.127 $\bar{a} = 3.37\bar{I} - 0.92\bar{J}$ m/s²
($\bar{I}$ rt, $\bar{J}$ up)

COMBINED INDEX

Numbers in *italics* denote pages in *Dynamics*.

555

CONVERSION FACTORS

ALPHABETICAL LIST OF UNITS
U.S.–BRITISH UNITS TO SI UNITS

To convert from	to	Multiply by
atmosphere	pascal (Pa)	$1.013\ 25\ (10^5)$
bar	pascal (Pa)	$1.000\ 000\ (10^5)$*
British thermal unit	joule (J)	$1.055\ 056\ (10^3)$
Btu/hour	watt (W)	$2.930\ 711\ (10^{-1})$
centimeter of mercury (0°C)	pascal (Pa)	$1.333\ 22\ (10^3)$
dyne	newton (N)	$1.000\ 000\ (10^{-5})$*
dyne-centimeter	newton-meter (N-m)	$1.000\ 000\ (10^{-7})$*
dyne/centimeter^2	pascal (Pa)	$1.000\ 000\ (10^{-1})$*
erg	joule (J)	$1.000\ 000\ (10^{-7})$*
erg/second	watt (W)	$1.000\ 000\ (10^{-7})$*
foot	meter (m)	$3.048\ 000\ (10^{-1})$*
foot^2	meter^2 (m^2)	$9.290\ 304\ (10^{-2})$*
foot^3	meter^3 (m^3)	$2.831\ 685\ (10^{-2})$
foot/second	meter/second (m/s)	$3.048\ 000\ (10^{-1})$*
foot-pound	joule (J)	$1.355\ 818$
foot-pound/second	watt (W)	$1.355\ 818$
foot/second^2	meter/second^2 (m/s^2)	$3.048\ 000\ (10^{-1})$*
gallon (U.S. liquid)	meter^3 (m^3)	$3.785\ 412\ (10^{-3})$
gram/centimeter^3	kilogram/meter^3 (kg/m^3)	$1.000\ 000\ (10^3)$*
horsepower (550 ft-lb/s)	watt (W)	$7.456\ 999\ (10^2)$
inch	meter (m)	$2.540\ 000\ (10^{-2})$*
inch^2	meter^2 (m^2)	$6.451\ 600\ (10^{-4})$*

*Denotes an exact quantity.